HANDBOOK OF RECURSIVE MATHEMATICS

Volume 1: Recursive Model Theory

STUDIES IN LOGIC

AND

THE FOUNDATIONS OF MATHEMATICS

VOLUME 138

ELSEVIER
AMSTERDAM • LAUSANNE • NEW YORK • OXFORD • SHANNON • SINGAPORE • TOKYO

HANDBOOK OF RECURSIVE MATHEMATICS

Volume 1: Recursive Model Theory

Edited by

Yu. L. ERSHOV
Sobolev Institute of Mathematics
pr. Akademika Koptuga, 4
Novosibirsk, 630090, Russia

A. NERODE
Department of Mathematics
Cornell University
Ithaca, NY 14853-7901, U.S.A.

S.S. GONCHAROV
Sobolev Institute of Mathematics
pr. Akademika Koptuga, 4
Novosibirsk, 630090, Russia

J.B. REMMEL
Department of Mathematics
University of California at San Diego
La Jolla, CA 92093, U.S.A.

Associate Editor

V.W. MAREK
Department of Computer Science
University of Kentucky
Lexington, KY 40506, U.S.A.

N·H

1998

ELSEVIER
AMSTERDAM • LAUSANNE • NEW YORK • OXFORD • SHANNON • SINGAPORE • TOKYO

ELSEVIER SCIENCE B.V.
Sara Burgerhartstraat 25
P.O. Box 211, 1000 AE Amsterdam, The Netherlands

Library of Congress Cataloging in Publication Data
A catalog record from the Library of Congress has been applied for.

ISBN: 0 444 50003 0 volume 1 (SiL vol. 138)

ISBN: 0 444 50106 1 volume 2 (SiL vol. 139)
ISBN: 0 444 50107 X (set: vols. 1 & 2)

⊗The paper used in this publication meets the requirements of ANSI/NISO Z39.48-1992 (Permanence of Paper).

Printed in The Netherlands.

Contents

Part Two Recursive Algebra, Analysis, and Combinatorics

Introduction to the Handbook of Recursive Mathematics

Y. L. Ershov, S. S. Goncharov,
A. Nerode, and J. B. Remmel

Preface

Recursive or Computable Mathematics is the study of the effective or computable content of the techniques and theorems of Mathematics. Recursive Mathematics has been an active area of research for the last 25 years. Its tools are the techniques of modern Computability Theory. These tools have been applied to analyze the effective content of results in a wide variety of mathematical fields including Algebra, Analysis, Topology, Combinatorics, Logic, Model Theory, Algebraic Topology, and Mathematical Physics.

This volume consists of surveys and new results from many of the leading researchers in the field. Almost all of the major subfields of Recursive Mathematics are represented in the volume. There are contributions which analyze the effective content of results from Algebra, Analysis, Combinatorics, Orderings, Logic, Non-monotonic Logics, Topology and Model Theory.

In this introduction, we will describe the historical roots of Recursive Mathematics as well as summarize some of the major themes that have arisen in the study of Recursive Mathematics and that are covered by the articles in this volume. We also provide a brief description of the articles of this volume. Finally we will end with what the editors believe are promising areas for further research.

HANDBOOK OF RECURSIVE MATHEMATICS
Edited by Yu. L. Ershov, S. S. Goncharov,
A. Nerode, and J. B. Remmel

Historical roots of recursive and constructive mathematics

Mathematics consists of constructions of new mathematical objects from old and of proofs that these objects have certain properties and lack other properties. Cantor's work on set theory in the late nineteenth century presented to the mathematical world what Hilbert called the paradise of set theory. Although Cantor's work was controversial at that time, the set theoretic point of view is at present widely accepted. Thus in the twentieth century, most working mathematicians have accepted Zermelo-Fraenkel set theory as the foundation of mathematics within which their work takes place. That is, almost all working mathematicians generally accept that Zermelo-Fraenkel set theory provides a foundation for mathematics with the caveat that occasionally one might have to extend ZF or ZFC by new axioms to accommodate certain proofs and constructions.

Within Zermelo-Fraenkel set theory, sets are built up by operations which construct sets from sets such as union, intersection, and power set. In addition, one employs comprehension-like schemes which assert that a set exists consisting of all elements having a property expressed in the language of set theory. The expression of the property in the language of set theory may itself involve sets as is the case with the axiom of replacement. Moreover, one assumes that the basic objects, sets, are extensional. That is, if two different properties are satisfied by the same elements, and one property defines a set, then the other property defines the same set.

However there are much narrower conceptions of mathematical objects, constructions, and proofs proposed by finitists and constructivists. In one way or another, they can be more or less adequately described within the classical Zermelo-Fraenkel set theory, but some of them are better viewed as alternate foundations of mathematics.

The earliest instance of such constructivist tendencies is the classical Greek fascination with what can be constructed by limited means. For instance consider the concept of a ruler and compass construction. Here one is naturally led to ask questions like: "What are the ruler and compass constructible regular polygons?" or "Is a segment the length of a circumference of circle ruler and compass constructible from a segment the length of a diameter?".

Algebra as developed by the Greek Diophantus and by the Persian al-Khwârismî and his successors was wholeheartedly constructive. They wanted to find formulas for the solution of algebraic equations. In fact, it is thought by some that the word "algorithm" comes from the name al-Khwârismî. Descartes' seventeenth century coordinatization of Geometry was intended as a reduction of geometric construction problems to the problem of solving algebraic equations. This endeavor was highly constructive. In fact, the explicit formula-based approach to Algebra continued into the nineteenth century to the time of Galois and Abel. By then negative results could be obtained since, through Lagrange, the notion of a solution by radicals of the fifth degree equation became precise and Galois' theory showed there were no such solutions for the general fifth degree equation.

Of the two strongest algebraists of the middle of the nineteenth century, Kronecker and Dedekind, Kronecker was the one to develop explicit formula and algorithms for all of his complex and Diophantine Algebraic Geometry. Kronecker's philosophy was that the true subject matter of mathematics is concretely representable structures whose operations are combinatorial. Dedekind preferred a set theoretic approach, which was not algorithmic. One need merely contrast Dedekind's set-theoretic definition of prime ideals and primary ideal factorizations with Kronecker's wholly constructive description via polynomials found by factoring algorithms. Dedekind explicitly said that he wanted set-theoretic definitions which are independent of the algorithm used. Thus Dedekind distinguishes himself from Kronecker and started to lead us away from algorithmic mathematics and toward Cantor's absolutist set theory.

In Analysis, Newton and Leibniz put great emphasis on solving geometric and physical problems by translating them to purely symbolic forms. Newton especially emphasized power series. For physicists ever since, publishing proofs without algorithms is just not satisfactory. Almost all of Analysis was highly constructive till the Weierstrassian revolution in the 1850's when exact definitions and proofs were developed for real n–dimensional Analysis and pure existence proofs using compactness arguments started to become popular. After the time of Dedekind in Algebra and Weierstrass in Analysis, general non-algorithmic set-theoretic methods took hold.

However, very early on, mathematicians, including Cantor himself, realized that there were paradoxes within naive set theory. The response to these paradoxes led to an extensive interest in the foundation of mathematics by many of the leading mathematicians including the likes of Hilbert and Poincaré. Of course, one reaction to the paradoxes was to develop a set of axioms for set theory which was free of the paradoxes and yet was strong enough to capture almost all mathematical constructions. This work led to the axiomatization of set theory by Zermelo and Fraenkel, and by Gödel and Bernays. Quite another reaction was to abandon naive set theory and return a completely constructive foundation for mathematics as reflected in the work of Brouwer and Heyting. In 1918, Brouwer [11, 12] announced his constructivism, and played a role in Analysis similar to that of Kronecker in Algebra. He developed n–dimensional calculus with proofs and constructions that can actually be carried out by (mental) algorithms. Since he denounced classical logic and the resulting classical mathematics as being wrong-headed, rather than simply saying that it was a less constructive species of thought, he encountered a great deal of hostility. One of the reasons for this hostility is that he used all of the words of the conventional mathematician, both for mathematics and for proofs, but with a different meaning, and asked everyone else to accept his meanings. He was understandably hostile to formalizing constructive logic, because as a humble human being, he thought that the human mind might extend his or anyone else's notion of construction by finding new ones at any time.

Hilbert's Program and the development of computable functions

The conflict between the philosophy of constructive mathematics which was being pursued by Brouwer [11, 12] and by his student Heyting [65, 66] and the overwhelming success of abstract methods, in part developed by Hilbert himself, lead, in the 1920's, to Hilbert's program to provide a finitistic proof of the consistency of mathematics.

Of course, Hilbert's program of proving mathematics consistent by finitary means was generally accepted as coming to an end with Gödel's famous incompleteness results [51]. Indeed, Gödel observed that his theorems applied to any "sufficiently rich" theory and that the notion of "sufficiently rich" could be rephrased as systems which could compute a wide class of functions by employing the proof procedures of the system. Within a few years, the idea of a computable function on the integers had emerged through the efforts of Church [19], Kleene [79], and Turing [154]. Later significant versions of the definition of computability were published by Post [120], Markov [97] and Mal'tsev [92]. By the end of the 1940's, all the proposed definitions up to that time had been proven equivalent and the formal definition of computable function was widely accepted as a plausible formalization of the intuitive notion of a function of integers whose values can be computed by a fixed program or algorithm. Moreover great effort was spent to show that every definition of program that one could think of yielded a function which was computable in the formal sense. Indeed, it remains the case today that no one has produced a function on the natural numbers which can be computed by an algorithm or program that is not included within the formal definition of computable function. However, just as Brouwer was humble enough to understand that there may be people with new constructions which we will all accept, so too we have to admit that it is possible that some computation scheme, not hitherto thought of, might extend the class of computable functions. But till then, we accept Church's thesis that the formal definition of recursive function captures the intuitive notion of computable function.

Accepting Church's Thesis has profound implications as well as limitations. To understand these implications and limitations, consider Hilbert's 10–th problem which was to find a procedure which, in a finite number of steps, determines whether or not a Diophantine equation has an integral solution. We regard Matijasevič [100] as having solved this problem because we have accepted that such a procedure must be represented by a program computing a computable function with the currently accepted definition. If someone in the future extends the notion of computable function beyond our current formal definition of computable or recursive function, we would have to revisit Hilbert's 10–th problem. The proof of the unsolvability of the word problem for groups, Church's theorem on the undecidability of formalized arithmetic itself, etc., all have this limitation, as does most of the work presented in this volume.

We should note that Hilbert never gave up on his program. Gödel said that perhaps the only way to close the issue is to find universally acceptable axioms for finitary methods and then to show that consistency cannot be proved by any method obeying the axioms. This has not been done yet. Heyting's formalization of constructive proofs by intuitionistic predicate calculus was not accepted as the last word by Brouwer for the same reasons as expressed above. That is, one cannot rule out the possibility that new constructive rules of reasoning might be discovered tomorrow and accepted universally. Nevertheless, Heyting's formalization remains important since it has been seen to have very deep connections with Church's lambda calculi, and the Church-Curry theories, in their typed reincarnations, have had wide impact in Computer Science. In the end, the significance of both the formal definitions of computable functions and of intuitionistic logic will have to rest on the insights they give us about the nature of constructions and their usefulness in mathematics, computer science, linguistics, and other areas of science, and not on their problematical philosophical significance as final descriptions of intuitive computability and intuitive constructive reasoning.

While the historical roots and motivation of Recursive Mathematics can be found in constructivist philosophies of mathematics, Recursive Mathematics does not lie within Constructive Mathematics. Researchers in Recursive Mathematics use classical reasoning unacceptable to constructivists to prove theorems about the non-existence of computable procedures. They accept Church's Thesis that the modern definition of a recursive function captures all functions which are intuitively computable. Constructivists simply do not accept the classical logic behind Recursive Mathematics. Bishop was the first to capture the constructive content of modern Functional Analysis, (Brouwer's work was finite dimensional). Bishop did not view Recursive Mathematics as within the constructive philosophy of mathematics espoused in his book on Constructive Analysis [6]. He was, however, interested in Recursive Analysis and was helpful to researchers in Recursive Mathematics, including Metakides, Nerode, and Remmel. Despite the fact that Recursive Mathematics is not part of Constructive Mathematics, the work on Recursive Mathematics does have interesting implications for Constructive Mathematics. For example, many of the counterexamples in the literature of Recursive Mathematics can easily be translated into counterexamples accepted by a constructivist as espoused say in the work of Troelstra and van Dalen [152, 153]. The converse is also true that constructivist counterexamples can be used as the basis for counterexamples in Recursive Mathematics.

Development of recursive mathematics

With the development of a coherent notion of computable function, it was quite natural that researchers would begin to apply it to other areas of mathematics.

In the 1930's, Church and Kleene raised the question of effective content of the theory of ordinals, namely, which ordinals and operations on ordinals are constructive from the point of view of recursive functions? This they called the theory of constructive ordinals. The Church-Kleene constructive ordinals were later shown by Markwald [99] to be the ordinals represented by recursive well-orderings. What is noteworthy is the non-extensional character of the Church-Kleene theory. Their theory is a theory of recursive operations on notations (indices) for ordinals. The questions of whether a number is a notation for an ordinal, or whether two notations for ordinals are notations for the same ordinal, are highly non-constructive. This means that all algorithms should act on notations (indices) for ordinals, not on the ordinals themselves. Thus Church-Kleene theory of recursive ordinals contains a basic insight which is shared with constructivists, namely, constructive operations should cannot be restrained to be extensional.

Brouwer-Heyting intuitionistic mathematics continued on in the 1930's. Brouwer influenced the algebraist van der Waerden. In his 1937 book, Modern Algebra [157], supposedly based on lectures of the highly abstract algebraist Emil Artin, he said in a most un-Artinian manner "*a field* Δ *is given explicitly* if its elements are uniquely represented by distinguishable symbols with which addition, subtraction, multiplication, and divisions can be performed by a finite number of operations", and proved that if a field Δ is given explicitly, then every simple algebraic extension $\Delta(t)$ and every simple algebraic extension $\Delta(\nu)$ defined by an irreducible polynomial φ is given explicitly. On the other hand, van der Waerden states that the construction of a splitting field requires that the field have a factorization algorithm for polynomials. He states explicitly that there is no known universal method for factorization and "there are reasons for the assumption that such a general method is impossible" and refers to his 1930 paper [155] as justification. With the development of recursion theory it was natural to apply its definitions and methods, and analogous results were proven in the recursive mathematics context by Frölich and Shepherdson [48] in the 1950's.

As for determining the effective content of Analysis, in the 1940's Goodstein [58] asked what it means to be a primitive recursively computable function of reals. Pre-World War II work of Banach-Mazur [101] was published in the same area in 1963.

The development of Recursion Theory in the 1940's and early 50's by Kleene [80, 82], Post [121], Peter [119], Myhill [109], Rice [132], and many others, provided the basic tools for the first results in Recursive Mathematics which appeared in the 1950's and early 60's. In particular, Frölich and Shepherdson [48] and Rabin [124, 125] provided explicit counterexamples of recursive fields which failed to have factorization algorithms, which justified van der Waerden's claim. Spector analyzed the effective content of recursive well orderings [144]. Kreisel analyzed the Cantor-Bendixson theorem [85]. Dekker and Myhill began to develop Isol Theory which can be viewed as an effective version of the theory of cardinals [33]. Lacombe [88], Goodstein [58] and Markov [98] started their work on Recursive Analysis. Similarly Julia Robinson [133] and Davis, Putnam and Robinson [29] started their work on Hilbert's 10-th problem which was finally culminated in Matijasevič's result [100, 28] that there is no recursive algorithm to decide whether an arbitrary Diophantine equation has an integer solution.

Modern history of Recursive Mathematics

While there was not a formally recognized subject of Recursive Mathematics, a number of results in Recursive Mathematics appeared in 1960's and early 70's. For example, in Isol Theory, the work of Ellentuck [34] and Nerode [110] answered many of the fundamental questions about the theory of effective cardinality on the natural numbers so that researchers begin to apply effective cardinality theory to algebraic structures. In particular, Crossley [23], and independently Manaster, developed a theory of constructive order types. Ellentuck [36], Applebaum [2] and Hasset [62] developed a theory of isolic groups, Dekker [30, 31] developed isolic vector spaces. Finally Crossley and Nerode gave a general theory for isolic structure in their book Combinatorial Functors [25]. Remmel [126] extended Hay's theory of co-r.e. isols to the general setting of Crossley and Nerode. Similarly there was considerable work on the effective content of theorems of combinatorics. Jockusch [72] studied the effective content of Ramsey's theorem. Jockusch and Soare produced their fundamental papers on Π_1^0-classes [74, 73] which can be viewed as the study of the effective content of König's lemma on trees. Manaster and Rosenstein studied the effective content [95, 96] of Hall's matching theorem and graph colorings. The study of the effective content of Analysis continued to flourish throughout this period. In particular, there was the work of Šanin [137] and others in his Leningrad school, including Orevkov [118] and Ceĭtin [14]. Other contributions include the work of Kušner [87] and others in Moscow who were influenced by Markov. In the West, key contributions were made by Aberth [1], Hauck [63], and, much later in the 1980's, by Pour-El and Richards [122].

The foundations for much of the work represented in this volume can be traced to fundamental work by Ershov and Nerode in the early 1970's. First in Russia, Ershov building on Mal'tsev's work on constructive algebra [92] published his theory of enumerations [38, 39] in which he defined a *constructivization* of a model $\mathcal{A}$ with universe A to be a surjective function $\alpha : \omega \to A$ such that for any atomic formulae φ in the underlying language $\mathcal{L}$, the relations $\{\overline{a} : \mathcal{A} \models \varphi(\alpha(\overline{a}))\}$ are recursive uniformly in φ. In fact, Ershov's theory of constructive models was first written in 1967 at Novosibirsk but was not published until after 1973. Ershov's work led to a vast amount of research on constructive Model Theory and Algebra in the former Soviet Union. This effort was lead by Ershov, Goncharov and their students Peretyat'kin, Nurtazin, Dobritsa, Khisamiev, Kudaibergenov, Ventsov, Fedoryaev, Morozov, and many others.

Independently, at Cornell University in the United States in 1972, Nerode began his program to determine the effective content of mathematical constructions and started to develop a systematic theory of recursive structures. Let $\varphi_{e,n}$ denote the partial recursive function of n variables computed by the e–th Turing machine. Here we say that a structure

$$\mathcal{A} = (A, \{R_i^{\mathcal{A}}\}_{i \in S}, \{f_i^{\mathcal{A}}\}_{i \in T}, \{c_i^{\mathcal{A}}\}_{i \in U}),$$

(where the universe A of $\mathcal{A}$ is a subset of the natural numbers ω), is *recursive* if A is a recursive subset of ω, S, T, and U are initial segments of ω, the set of relations $\{R_i^{\mathcal{A}}\}_{i \in S}$ is uniformly recursive in the sense that there is a recursive function G such that for all $i \in S$, $G(i) = [n_i, e_i]$ where $R_i^{\mathcal{A}}$ is an n_i–ary relation and φ_{e_i,n_i} computes the characteristic function of $R_i^{\mathcal{A}}$, the set of functions $\{f_i^{\mathcal{A}}\}_{i \in T}$ is uniformly recursive in the sense that there is a recursive function F such that for all $i \in T$, $F(i) = [n_i, e_i]$ where $f_i^{\mathcal{A}}$ is an n_i–ary function and φ_{e_i,n_i} restricted to A^{n_i} computes $f_i^{\mathcal{A}}$, and there is a recursive function interpreting the constant symbols in the sense that there is a recursive function H such that for all $i \in U$, $H(i) = c_i^{\mathcal{A}}$. Note that if $\mathcal{A}$ is a recursive structure, then the atomic diagram of $\mathcal{A}$ is recursive. Thus a recursive structure which is isomorphic to a structure $\mathcal{B}$ can be viewed a constructivization of $\mathcal{B}$. It follows that recursive structures and constructivizations are essentially interchangeable.

In particular, in a series of papers with his student Metakides [102, 103, 104, 105], Nerode introduced the systematic use of the finite injury priority argument applied to algebraic requirements to construct recursive models which could be used to give effective counterexamples to classical mathematical theorems. This allowed the whole technology of modern Recursion Theory as developed by Friedberg, Rogers, Sacks, Lachlan, Soare, Jockusch, Lerman, and others, to be applied to these problems. Many applications of the priority argument to Recursive Mathematics were produced by Nerode's students at that time: Remmel, Metakides, Millar, Kalantari, Retzlaff, Lin, and subsequently by many others in North America, and also by Crossley, Ash, Downey, Moses, Hird and others in Australia. Another theme that emerged from this work was a systematic theory of the lattice of recursively enumerable substructures of a recursive structure (see the 1982 survey paper by Nerode and Remmel [112]).

During the late 1970's and 80's, Cold War politics allowed almost no communications between the schools in Novosibirsk of Ershov and the school of Nerode in the West, so that often results in Recursive Model Theory and Algebra were duplicated. Nevertheless, 25 years of work by researchers associated in one way or another with these two schools have generated a large body of results which is the main subject of most of the papers in this volume. Unfortunately, it is impossible within a single volume to cover all the areas of Recursive Mathematics and its connections with other subjects such as Reverse Mathematics, Constructive Set Theory, and Decidability Theory. However, we feel that the articles in this volume will provide the reader with a good overview of a significant portion of the work that has been produced and continues to be produced in Recursive Mathematics. In addition, the bibliographies on Recursive Algebra and Model Theory compiled by Kalantari and on Recursive Analysis compiled by Brattka and Kalantari are valuable resources for current and future researchers in Recursive Mathematics. Recursive Mathematics continues to be a vital field and our hope is that this volume will inspire others to deepen our understanding and broaden the fields of application of Recursive Mathematics.

Some general themes of Recursive Mathematics

A number of important common mathematical themes emerged from the past work on Recursive Mathematics which the reader will find reflected in this volume. These include the following.

1. Fix a given recursive structure, ideally one which is universal for a large class of recursive structures, and study the complexity of various model-theoretical and algebraic constructions on that structure.

2. Find necessary and sufficient conditions for the existence of recursive or constructive models of a theory with given properties.

3. Classify the class of recursive models within a classical isomorphism type that are unique up to recursive isomorphisms. More generally, find necessary and sufficient conditions for existence of recursive or constructive representations of a given structure with nonequivalent algorithmic properties.

4. Study the lattice of recursively enumerable substructures of a given recursive structure. In particular, investigate the similarities and differences with the lattice of r.e. substructures of an algebraic structure and the so-called Post's zoo of maximal, simple, h–simple, hh–simple sets in the lattice of r.e. sets.

5. Determine whether the class of all recursive models of a given structure is computable.

6. Study the complexity of the set of solutions of a recursive instance of a classical combinatorial, algebraic or analytic problem.

7. Study the possible degrees of models of a given theory or the set of degrees of all structures which are isomorphic to a given structure.

8. Study the differences between r.e. presented structures and recursive structures. Here, an r.e. presented structure consists of a recursive structure modulo an r.e. congruence relation. For example, an r.e. Boolean algebra can be represented as a recursive Boolean algebra modulo an r.e. ideal. More generally, study the differences between recursive structures and structures recursive in some oracle set A.

In the first area, many algebraic structures have been studied. In the West, the computability of ordered sets was studied by Ash, Case, Chen, Crossley, Downey, Feiner, Feldman, Fellner, Hay, Hingston, Hird, Jockusch, Kierstead, Knight, Lerman, Manaster, Metakides, McNulty, Moses, Remmel, Richter, Rosenstein, Roy, Schmerl, Schwarz, Soare, Tennenbaum, Trotter and Watnick; the computability of vector spaces by Ash, Dekker, Downey, Guhl, Guichard, Hamilton, Kalantari, Remmel, Retzlaff, Shore, Smith and Welch; the computability of rings and fields by Ash, Hodges, Jockusch, MacIntyre, Madison, Marker, Metakides, Nerode, Mines, Remmel, Rosenthal, Seidenberg, Shlapentokh, Smith, Staples, Tucker and van den Dries; the computability of the structures with a dependence relation by Baldwin, Downey, Metakides, Nerode, and Remmel. Other mathematical structures that were also extensively studied include groups by Ash, Barker, Ge, Kent, Knight, Lin, Oates, Richards, Richman and Smith; graphs by Aharoni, Bean, Beigel, Burr, Carstens, Gasarch, Golze, Kierstead, Lockwood, Manaster, Magidor, Päppinghaus, Remmel, Rosenstein, Schmerl and Shore; Boolean algebras by Carroll, Downey, Feiner, La Roche, Remmel, Soare and Thurber; topological spaces by Kalantari, Leggett, Remmel, Retzlaff and Weitkamp. Computable Ramsey's theory has been studied by Clote, Hummel, Jockusch, Seetapun, Simpson, Solovay and Specker. In the East, Goncharov and his students extensively studied recursive Boolean algebras and orderings, Morozov studied groups of recursive automorphisms, Khisamiev studied Abelian p-groups establishing connections between constructivizability and Ulm's quotients, and Odintsov studied semi-lattices of recursively enumerable subalgebras.

For the second type of problem, the following principal results were obtained: the Ershov theorems about kernels and the theorem about the existence of constructive models for a theory with finite obstacles, Baur's theorem about the existence of constructive model for $\forall\exists$-theories, the Goncharov and Harrington theorem about the decidability of prime models, the Morley theorem about the decidability of homogeneous models, the Goncharov-Peretyat'kin criterion for decidability of homogeneous models, the Khisamiev-Millar theorems about omitting recursive types, and theorems due to Lachlan, Peretyat'kin, Millar, and Ash-Read about decidability of Ehrenfeucht theories.

For the third problem various equivalence types on constructivizations of models were studied. In particular: recursive equivalence, autoequivalence, algebraic equivalence, program equivalence, uniform equivalence were investigated by Goncharov, Peretyat'kin, Nurtazin, Ash, Nerode, Ventsov, Uspenskiĭ, Remmel, Kudinov and others. In this direction, powerful criteria for the existence of nonautoequivalent recursive representations of models and criteria for autostability were found by Goncharov, Nurtazin, and Kudinov. Goncharov constructed a series of examples of nonautostable models of finite algebraic dimension. Goncharov, Cholak, Shore, Khoussainov found examples of nonautostable models with autostable enrichment of constants. Ash and Nerode established close connections between autostability and definability of hereditarily enumerable relations. Ash developed the method of labeling system that provided an essential advantage in the description of hereditarily arithmetic relations. This in turn, led to series of applications by Ash and Knight. Harizanov began important investigation of Turing spectra of relations.

Extensive studies of the lattice of r.e. substructures for a variety of recursive structures have been carried out. For example, there is a large body of work on the lattice of r.e. subspaces $\mathcal{L}(V_\infty)$ of an infinite dimensional recursive vector space by Ash, Bäuerle, Downey, Hird, Kalantari, Kurtz, Metakides, Nerode, Remmel, Retzlaff, Smith, Shore, Welch, and others. The lattices of r.e. subalgebras and r.e. ideals of a recursive Boolean algebra have been studied by Downey, Carroll, Goncharov, Morozov, Remmel, Yang and others. The lattice of r.e. open sets of a recursive topological space by Kalantari, Leggett, Retzlaff, Remmel, and others. Lattices of r.e. suborderings have been studied by Metakides, Remmel, Roy, and others. The lattices of r.e. affine spaces of a recursive vector space over an ordered field have been studied by Downey, Kalantari, Remmel and others. In addition, there has been significant work on providing a unified setting for the study of that lattice of r.e. substructures. For example, the study of the lattice of r.e. substructures of effective Steinitz systems, which cover both $L(V_\infty)$ and the lattice of r.e. algebraically closed subfields of an algebraically closed recursive field with an infinite transcendence basis, has been carried out by Baldwin, Downey, Metakides, Nerode, Remmel, and others. Remmel provided an even more general setting which covers all the lattices described above, see the article by Downey and Remmel in this volume. For a survey of results up to 1982, see the survey article by Nerode and Remmel [112].

For the fifth problem, Nurtazin proved the computability of the class of all constructive models of a fixed signature without function symbols and demonstrated that such a class is not computable if the signature contains functional symbols. Furthermore, Goncharov proved that the class of constructivizations for a decidable nonautostable model is not computable. A number of results concerning the complexity of index sets and the infinity of the Rogers semi-lattices of computable enumerations of classes of constructive models were obtained by Dobritsa. On the basis of the Ash method, the complexity of index sets in the arithmetic hierarchy was studied by Ash and Knight.

For the sixth type of problem see the survey articles by Gasarch for combinatorics, Downey for linear orderings, and especially the survey article by Cenzer and Remmel on Π_1^0 classes and Recursive Mathematics in this volume.

For results on the seventh type of problem, we refer the reader to the survey article on degrees of models by Knight in this volume.

Finally, prototypical results for the eighth type of problem include Feiner's result [43] that there is an r.e. Boolean algebra which is not isomorphic to any recursive Boolean algebra, and a recent result of Thurber [150] that every low_2 Boolean algebra is isomorphic to a recursive Boolean algebra. Also see the survey article by Downey on linear orderings in this volume for various results of this type for linear orders. There has been a significant number of papers on r.e. presented structures. For example, Ash, Downey, Goncharov, Jockusch, Knight, Metakides, Nerode, Remmel, Stob, Thurber, and others, have studied r.e. presented Boolean algebras; Ash, Downey, Goncharov, Jockusch, Knight, Metakides, Moses, Nerode, Remmel, Roy, Rosenstein, Soare, and others, have studied r.e. presented linear orderings; and Downey, Metakides, Nerode, and Remmel have studied r.e. presented vector spaces.

The list above is by no means complete nor is the list of contributors listed in each area complete. Nevertheless the reader will find that keeping these themes in mind will explain the motivation for many of the articles in this volume.

The questions above arose mainly in the study of Recursive Model Theory and Algebra. There is a large body of work in Recursive Analysis and Topology, as can be seen from the bibliography on Recursive Analysis and Topology at the end of this volume. The emphasis in the research on Recursive Analysis and Topology continues to be on analyzing the effective content of constructions and theorems of Analysis and Topology based on the definitions of computable real numbers, computable functions of the real numbers, and effectively closed sets.

Just as the formalization of computable functions in the 1930's naturally lead to Recursive Mathematics, the development of Complexity Theory by Blum, Cobham, Cook, Hartmanis, Karp, Ladner, Lewis, Sterns, and others in the 1960's and 1970's in Computer Science has led to development of a polynomial-time, and more generally, a feasible version of Recursive Mathematics which is called Polynomial-time or Feasible Mathematics. Of course, the definition of the polynomial-time hierarchy, the polynomial-space hierarchy, etc., were defined by analogy with the arithmetic hierarchy from

Computability Theory. The work of Karp on NP–complete problems has lead the ever expanding list of NP–complete problems, see Garey and Johnson [49], and to the increasing importance of the fundamental question of whether P equals NP. The paper by Baker, Gill, and Solovay [4] introduced oracle arguments into Complexity Theory. All these ideas, plus ideas from Recursive Mathematics, have played an important part in the development of Feasible Mathematics.

The work on Feasible Mathematics started with the work of Friedman and Ko [84, 46], where they developed a coherent notion of a polynomial-time computable function of the reals, and related a number of classical complexity-theoretic questions to questions about complexity of operations on polynomial computable functions. This subject has continued to develop (see the survey article by Ko in this volume). Remmel and Nerode developed a theory of the lattice of NP–substructures of a polynomial-time presented structure, and showed that priority arguments on oracles could play a fundamental role in the analysis of such lattices. The driving analogy in the Nerode-Remmel work is that recursive is to recursively enumerable as polynomial-time is to nondeterministic polynomial-time. Thus, for example, the polynomial-time analogue of Dekker's [30] result, that every r.e. subspace of a recursively presented infinite dimensional vector space over a recursive field has a recursive basis, is that every NP–subspace of a polynomial-time presented infinite dimensional vector space over a polynomial-time field has a basis in P. This analogy is true over certain infinite polynomial-time fields, but is oracle dependent over finite fields (see [115] and the article by Cenzer and Remmel on Complexity Theoretic Model Theory and Algebra in this volume).

Cenzer and Remmel have developed a rich theory of Polynomial-time and Feasible Model Theory which is also outlined in their paper in this volume. A number of interesting refinements of questions in Recursive Model Theory have arisen in Feasible Model Theory. For example, the questions of when a recursive model is isomorphic or recursively isomorphic to a polynomial-time model, or when a recursive model is isomorphic or recursively isomorphic to a polynomial-time model with a standard universe, such as the binary representation of the natural numbers or the unary representation of the natural numbers, leads to a surprisingly rich theory. Other developments in Feasible Mathematics include the work of Crossley, Nerode and Remmel [27, 116] on a polynomial-time analogue of Isol Theory which is far from a mere imitation of standard Isol Theory. Also other models of computation have been studied. For example, Blum, Smale and Shub [8] developed a theory of complexity for computing with real numbers and Khoussainov and Nerode [77] have developed a theory of automata representable structures.

Overview of the Handbook

The Handbook of Recursive Mathematics contains over 1350 pages and hence we were forced to split the Handbook into two volumes. The editors decided that it was best to split the papers between the two volumes according to subject matter. The first volume covers Recursive Model Theory and the second volume covers Recursive Algebra, Analysis, Combinatorics and a variety of other topics. Since we did not originally plan for two volumes, the partition of the papers between the two volumes could not be accomplished without some overlap. That is, there are a number of papers in the second volume which are relevant to Recursive Model Theory and use recursive model theoretic techniques, and there are a number of papers in the first volume which are relevant to Recursive Algebra and use recursive algebraic techniques. We shall provide brief summaries of the papers in the two volumes below.

Volume 1: Recursive Model Theory

The first two papers of volume 1 were chosen because they provide good general introductions to Recursive or Computable Model Theory as developed in both the West, mainly the United States and Australia, and the East, mainly in the former Soviet Union. The rest of volume 1 is devoted to more specialized topics in Recursive Model Theory. The last paper of the volume is a survey paper by Cenzer and Remmel on Polynomial Time Model Theory and Algebra. The Cenzer and Remmel paper shows how seriously taking into account the resource bounds of computations greatly affects the types of questions and results that one obtains when considering effective content of model theoretic questions. It also surveys results on resource bounded versions of algebraic constructions, and thus provides a nice segue into the second volume.

In the first paper of volume 1, Harizanov provides a very valuable survey of results in Recursive or Computable Model Theory mainly from the Western perspective in her article "Pure Computable Model Theory". In particular, she supplies the model theoretic background as well as the proofs of a large number of basic results in computable model theory, including the basic effective completeness theorem and effective omitting types theorem, various results on conditions which ensure the existence of various types of decidable models including prime, homogeneous, and saturated models or models with effective sets of indiscernibles, results on decidable theories with only finitely many models or finitely many recursive models (the so-called Ehrenfeucht theories, or effective Ehrenfeucht theories), a theory of the degrees of models, and classification results on the number of computable models. Thus her article is an excellent place for a student who is interested in Recursive Model Theory as well as a valuable reference for established researchers in the field.

The article by Ershov and Goncharov, "Elementary Theories and Their Constructive Models", is a good introductory article for those who are unfamiliar with the approach of the Ershov school to Recursive Model Theory. They provide a survey of some of the basic existence theorems for constructive and strongly constructive models. In particular they provide a proof of the Goncharov-Peretyat'kin criterion for the existence of a decidable homogeneous model. They also apply their theory to Boolean algebras, which is a rich source of examples and for which a well developed theory of constructive models exists.

The paper, "Isomorphic Recursive Structures" by Ash, surveys several results on when a recursive structure is unique up to recursive isomorphisms, Δ_2^0 isomorphisms, etc.. In a similar spirit, Ash looks at conditions which ensure that a given relation is always recursive, r.e., Σ_2^0, etc., in any recursive structure which is isomorphic to the original recursive structure. Such relations are called intrinsically recursive, r.e., etc.. A number of characterizations of intrinsically recursive and r.e. relations can be found in the Ash article.

The article "Computable Classes and Constructive Models" by Dobritsa presents a large body of results on when the set of recursive models, which are isomorphic to a given recursive model or are extensions of a given recursive model, are computable; in the sense that one can effectively list all such models. In addition, he surveys a number of results on conditions which ensure that the class of such recursive models lies in the arithmetic hierarchy. A theory of effective reductions of one class of models to another class of models is also presented.

In the paper "Σ–Definability of Algebraic Systems", Ershov proves a number of interesting results on the extension of his theory of numerations and constructivizations where recursive and computable functions are replaced by definability notions from the theory of admissible sets. He thus defines the notion of a Σ–definable algebraic structure of an admissible set, and proves a number of analogues of results on constructivizations of algebraic structures in this setting. This work opens up another extension of Recursive Mathematics using notions from α–recursion theory and admissible sets.

In the article "Autostable Models and Algorithmic Dimensions", Goncharov defines several reducibilities which can be defined on the class of constructivizations, $Con(\mathfrak{M})$, of a given model $\mathfrak{M}$. That is, given two constructivizations $\nu, \mu : \omega \to \mathfrak{M}$, we say that

1. $\nu \leqslant_K \mu$, if there exists a recursive function f such that $\nu = \mu f$ (Kolmogorov reducibility),

2. $\nu \leqslant \mu$, if there is an automorphism φ of $\mathfrak{M}$ such that $\varphi\nu \leqslant_K \mu$ (autoreducibility),

3. $\nu \leqslant_U \mu$, if there exists a computable operator F such that $F(\chi_{\mu^{-1}(S)}) = \chi_{\nu^{-1}(S)}$ for all relations S which are stable in the sense that S is invariant under automorphisms of $\mathfrak{M}$ (uniform reducibility),

4. $\nu \leqslant_P \mu$, if there is a partial recursive function f such that if φ_n is the characteristic function of $\mu^{-1}(S)$ for some stable relation S, then $\varphi_{f(n)}$ is the characteristic function of $\nu^{-1}(S)$ (program reducibility), and

5. $\nu \leqslant_{Alg} \mu$, if every stable relation of $\mathfrak{M}$ which is decidable under the constructivization μ is also decidable under the constructivization ν (algebraic reducibility).

Goncharov then discusses the relations between these reducibilities, and surveys a number of results on conditions which ensure that the set of reducibility classes of a model has cardinality 1, ω, or is finite.

In her article "Degrees of Models", Knight presents another significant theme in recursive model theory, namely, she surveys a number of results on the sets of degrees of structures which are isomorphic to a given structure or the set of degrees of models of a particular theory. For example, let $DI(\mathcal{A}) = \{\deg(\mathcal{B}) : \mathcal{B} \text{ is isomorphic to } \mathcal{A}\}$. Generally $DI(\mathcal{A})$ is closed upwards so that it is natural to ask whether there is a least element in $DI(\mathcal{A})$, i.e., when is there a degree $\mathbf{c}$ such that $DI(\mathcal{A}) = \{\mathbf{d} : \mathbf{d} \geqslant_T \mathbf{c}\}$? Knight surveys a number of results on structures for which such a degree $\mathbf{c}$ exists, and on structures for which no such degree $\mathbf{c}$ exists. More generally, one can define a structure to have α–th jump degree $\mathbf{d}$ for a given recursive ordinal α if $\{\mathbf{b}^{(\alpha)} : \mathbf{b} \in DI(\mathcal{A})\} = \{\mathbf{c} : \mathbf{c} \geqslant_T \mathbf{d}\}$, where $\mathbf{b}^{(\alpha)}$ denotes the α–th jump of $\mathbf{b}$. Knight also surveys a number of interesting results on structures which have an α–th jump degree. Finally Knight presents results on the possible degrees of non-standard models of arithmetic and on results which guarantee the existence of recursive models of nonrecursive theories.

In his article "Groups of Computable Automorphisms", Morozov provides a survey of results on the groups of all automorphisms, of all recursive automorphisms, of all arithmetic automorphisms, etc., of a recursive or constructible model over an effective language. He starts his survey with a number of general results on the set of recursive automorphisms of a recursive model. For example, there exist many examples of recursive models which have $2^{\aleph_0}$ automorphisms but only a single recursive automorphism. In fact, one can construct a homogeneous strongly constructive model which has $2^{\aleph_0}$ automorphisms, but every constructivization of that model has only one recursive automorphism. Similarly, one can construct a hyperarithmetic model with $2^{\aleph_0}$ automorphisms, but which has a unique hyperarithmetical automorphism. He also surveys results about when a recursive model $\mathfrak{M}$ can have a computable set of recursive automorphisms $\mathrm{Aut}_R(\mathfrak{M})$. He ends his survey with a number of interesting results on the set of recursive automorphisms of recursive Boolean algebras and vector spaces. For example, it is known that if B is a decidable atomic Boolean algebra and B' is any recursive Boolean algebra, then the fact that $\mathrm{Aut}_r(B)$ is isomorphic to $\mathrm{Aut}_r(B')$ implies that B is isomorphic to B'. On the other hand, a result of Remmel shows that for any recursive Boolean algebra with infinitely many atoms, there exists a recursive Boolean algebra C isomorphic to B such that every recursive automorphism of C moves only finitely many atoms. Morozov's article shows that there is a surprisingly rich theory of the various automorphism groups of recursive models.

In his article "Constructive Models and Finitely Axiomatizable Theories", Peretyat'kin surveys a number of results on finitely axiomatizable theories. Finitely axiomatizable theories have been extensively studied in model theory, and Peretyat'kin provides a nice summary of such results in his article.

He then surveys results on the complexity of the the prime, countably homogeneous, and countable saturated models of such theories as well as a large number of index set results for finitely axiomatizable theories. He also surveys a number of interesting results on the complexity of the Lindenbaum algebra of finitely axiomatizable theories. Finally he presents a long list of open problems in the area.

As mentioned above, the last article of volume 1 looks at how resource bounds affect questions in Recursive Model Theory. That is, one restricts one's attention to models where the underlying universe, functions, and relations are limited to be in some natural complexity class such as polynomial-time or polynomial-space, and studies how such restrictions affect the types of results developed in Recursive Model Theory. As the reader can see from the article by Cenzer and Remmel in this volume, and the article by Ko on Polynomial Time Analysis in the second volume, restricting one's attention to feasible functions does not produce a theory which is a mere imitation of results in Recursive Mathematics. Cenzer and Remmel provide an introduction to the types of questions which arise in Feasible Mathematics in their article "Complexity Theoretic Model Theory and Algebra". Cenzer and Remmel give a survey of their theory of polynomial-time and feasible models in this paper. They also survey the theory of Polynomial-time Algebra and the theory of the lattices of NP–substructures of a polynomial-time structure that has been developed by Nerode and Remmel. This paper provides a good introduction to an area that is really in its infancy. Nevertheless, the results achieved so far show that this is an interesting area of research in which a number of new phenomena arise which do not appear in Recursive Algebra. Thus Polynomial-time Model Theory and Algebra offer a rich opportunity for further research.

Finally, volume 1 ends with two extremely valuable bibliographies. The first is a bibliography on Recursive Algebra which was compiled by Kalantari. The second is a bibliography on Recursive Analysis and Topology that was compiled by Kalantari and Brattka. Both of these bibliographies were put together in response to a rather late request by the editors. Because of the lack of time, the bibliographies could not be as exhaustive as desired and we apologize to those whose work has been inadvertently omitted. Nevertheless, these bibliographies are valuable additions to this volume and provide those researchers and students who want a deeper treatment than is presented in this volume a valuable guide to the literature.

Volume 2: Recursive Algebra, Analysis, and Combinatorics

In the first paper of the second volume, "Π_1^0 Classes in Mathematics", Cenzer and Remmel provide a fairly comprehensive survey of the uses of Π_1^0 classes in Recursive Mathematics. It is well known to recursion theorists that Π_1^0 classes are ubiquitous in many areas of mathematics. The reason why Π_1^0 classes arise so naturally in Recursive Mathematics is because it is often the case that the set of solutions to a recursive problem can be viewed as the set of paths through a recursive tree. For example, the set of proper k–colorings of a recursive graph G can be viewed as the set of paths through a recursive tree T_G, and hence is a Π_1^0 class. The question then becomes whether for every recursive tree T, the set of paths through T is in one-to-one degree preserving correspondence with the set of k–colorings of some recursive graph. If so, then one can transfer a vast number of results on the possible degrees of elements of Π_1^0 classes and index set results for Π_1^0 classes to results about the set of k–colorings of a recursive graph. Cenzer and Remmel give a large number of problems for which this type of correspondence, and weaker correspondences, exist. They also give an overview of basic results on Π_1^0 classes, and explain how such results can be used to give a complexity analysis of a large number of problems considered in Recursive Mathematics.

Downey, in his paper "Recursion Theory and Linear Orderings", provides an extensive survey of results on recursive orderings. A number of very interesting questions have arisen in the theory of recursive orderings, including the question of classifying when certain recursive orderings are unique up to recursive isomorphisms, the question of finding the effective dimension of a recursive partial ordering, the question of when an r.e. presented linear ordering or low linear ordering is isomorphic to a recursive linear ordering, the question of when a linear ordering has an effective ω or ω^* sequence, and many others. Downey's paper is an excellent starting point for those who are not conversant with Recursive Mathematics, as his paper starts out with a good introduction to many basic theorems and techniques of modern computability theory.

In the paper, "Effective Algebras and Closure Systems: Coding Properties", Downey and Remmel show that a large number of results on the lattice of r.e. substructures of various recursive structures can be given uniform proofs which often involve simple coding arguments based on results from the lattice of r.e. sets. They work in a general setting due to Remmel [129] called effective closure systems, which cover a large number of lattices, including the lattice of r.e. sets, the lattice of r.e. subspaces of an infinite dimensional recursive vector space over a recursive field, the lattice of r.e. algebraically closed subfields of a recursive algebraically closed field with an infinite transcendence basis, and the lattice of r.e. ideals and the lattice of r.e. subalgebras of a recursive Boolean algebra. They also give some examples of lattices which cannot be covered by their general setting.

Gasarch provides an extensive survey of results on the effective content of theorems in combinatorics in his article "Recursive Combinatorics". Many results in infinite combinatorics are not effective, and there is a large body of work on classifying which of the various theorems from graph theory, orderings, Ramsey theory, matching theory, etc., are or are not effective. Moreover, there are many beautiful recursive variations of combinatorial problems that provide many avenues of interesting research. For example, Dilworth's theorem that every partial ordering of width n can be covered by n chains is not effective. Indeed, it is relatively easy to construct recursive partial orders of width 2 which cannot be covered by 2 recursive chains. However Kierstead [78] showed that every recursive partial order of width n could be covered by $\frac{1}{4}(5^n - 1)$ recursive chains. The exact bound on how many recursive chains are required to cover a recursive partial order of width n is not known. Gasarch's article has extensive connections with the articles of Downey on recursive orderings, of Cenzer and Remmel on Π^0_1 classes and Recursive Mathematics, and of Kierstead on recursive and on-line colorings.

In his article "Constructive Abelian Groups", Khisamiev looks at the effective content of the theory of Abelian groups. First he surveys a large number of results on various classes of groups which have a constructivization, i.e., groups which have a recursive presentation, and groups which have a strong constructivization, i.e., groups which have a decidable presentation. This provides a number of nice examples of the differences between constructivizations and strong constructivizations. He then goes on to classify which groups have a unique recursive presentation up to recursive isomorphisms. He also presents a number of results which show that Ulm's theorem for p-groups is not effective. Finally he surveys results on the constructibility of torsion-free abelian groups and results on the constructibility of subgroups and factor groups in constructible groups. The Khisamiev article thus provides a fascinating chapter of the interactions of Recursive Mathematics with a classical Algebra.

The article by Kierstead, "Recursive and On-Line Graph Coloring", studies another very interesting connection of recursive mathematics with computer science, namely the theory of on-line algorithms. In the on-line coloring problem for a graph G, the set of vertices of the graph are presented one at a time and the *on-line* algorithm is required to decide the color of that vertex based only on the knowledge of the colors assigned to, and the edges between, the points which have previously appeared plus the current vertex. The connection with finding recursive colorings is due to the fact that to give a procedure to recursively color a graph, one can only use local information, and hence an on-line coloring algorithm can be translated into results about recursive colorings. It turns out that many of the techniques developed in the study of recursive colorings of recursive graphs have applications to the on-line coloring of graphs. In particular, certain recursive counterexamples can be translated to give counterexamples for on-line colorings. It should be noted that the theory of on-line colorings contains many subtle questions which have no analogues in recursive graph theory, and Kierstead's article presents a number of such questions and surveys a number of interesting results in this area.

Ko presents a survey of results in Feasible Analysis in his article "Polynomial Time Analysis". After a brief survey of results on recursive analysis and the basic definitions of the complexity theoretic hierarchies, Ko presents the Friedman-Ko model of complexity of computable functions $f : [0,1] \to \mathbb{R}$. He then studies the complexity of various operations on computable real functions including maximization, root-finding, integration, solving ordinary differential equations, and solving integral equations. In each of these cases, there are very strong connections with standard Complexity Theory. For example, one of the first results in this area is due to Friedman, and states that the following are equivalent:

1. $\mathrm{P} = \mathrm{NP}$.
2. For each polynomial-time computable function $f : [0,1]^2 \to \mathbb{R}$, the function $g(x) = \max\{f(x,y) : 0 \leqslant y \leqslant 1\}$ is polynomial-time computable.
3. For each polynomial-time computable function $f : [0,1] \to \mathbb{R}$, the function $g(x) = \max\{f(y) : 0 \leqslant y \leqslant x\}$ is polynomial-time computable.
4. For each polynomial-time computable function $f : [0,1] \to \mathbb{R}$ that is infinitely differentiable, the function $g(x) = \max\{f(x,y) : 0 \leqslant y \leqslant 1\}$ is polynomial-time computable.

For each of the operations listed above, Ko surveys results on conditions which guarantee polynomial-time or polynomial-space solutions, as well as results connecting the problems of finding polynomial-time or polynomial-space solutions in general with well known separation problems from Complexity Theory. Thus, Ko presents another interesting area of Feasible Mathematics which is a fruitful area for further research.

The theory of recursive or constructive Boolean algebras provides an immensely rich area for Recursive Mathematics, see for example Remmel's survey article [130] and Goncharov's book [57]. In his article, "Generally Constructive Boolean Algebras", Odintsov surveys a number of results on extensions of constructible Boolean algebras. A strongly constructive Boolean algebra is a Boolean algebra B such that there exists a map $\nu : \omega \to B$ such that the set of all $\langle n, b_0, \dots, b_{n-1} \rangle$, where n is an index of a first order formula φ_n such that $B \vdash \varphi_n(\nu(b_0), \dots, \nu(b_{n-1}))$, is recursive. One can generalize this notion in several ways. For example, one could restrict the set of formulas φ_n to consists only of Σ_n formulas or Π_n formulas. Similarly, one could insist that the set of all $\langle n, b_0, \dots, b_{n-1} \rangle$, where n is an index of a first order formula φ_n such that $B \vdash \varphi_n(\nu(b_0), \dots, \nu(b_{n-1}))$, be arithmetic rather than recursive. One could also consider formulas in some fragment of second order logic like $L(Q)$, where Q is the Mostowski quantifier, "there exist infinitely many". Finally one can add extra predicates such as predicates for various types of filters. Odintsov surveys results on all of these possible extensions and shows that there are many natural extensions of the basic notions of constructivizability and strong constructivizability which yield fruitful areas of study.

The article "Reverse Algebra", by Simpson and Rao, provides an introduction and survey of results on Reverse Mathematics and Algebra. The goal of Reverse Mathematics is to answer the question about which set existence axioms are needed to prove theorems in ordinary mathematics. To this end, five subsystems of second order arithmetic have been introduced: RCA_0 which consists of the usual axioms for the ordered semi-ring $(\omega, +, \cdot, 0, 1, <)$ plus Δ^0_1–comprehension and Σ^0_1–induction, WKL_0 which is RCA_0 plus the statement of weak König's Lemma (every infinite n–ary branching tree has an infinite path), ACA_0 which is RCA_0 plus comprehension axioms for all arithmetical formulas, ATR_0 which includes ACA_0 plus the axiom that arithmetical comprehension can iterated along any countable well ordering, and Π^1_1–CA which consists of RCA_0 plus comprehension axioms for all Π^1_1 formulas. At this point, there are a large number of theorems of classical mathematics that have been classified according to these five subtheories. For example, consider the theorem that every commutative ring has a maximal ideal. Friedman, Simpson, and Smith [47] have shown that this theorem can be proved in ACA_0, and moreover that any theorem which can be proved in ACA_0 can be proved in the theory RCA_0 plus the axiom that every commutative ring has a maximal ideal. Thus one could say that this theorem has the same power as arithmetic comprehension. Simpson and Rao survey a number of theorems in Algebra that have been classified with this framework and provide an interesting section on the connections between Reverse Mathematics and Recursive Mathematics.

Relations with other subjects and future developments

We have attempted to show in this introduction that there is a natural relation between Recursive Mathematics and other areas of Mathematics. For example, there is a natural relation between Recursive Mathematics and Constructive Mathematics. There is also an intimate relation between Recursive Mathematics and Reverse Mathematics which was started by Friedman to classify theorems of mathematics by their proof theoretic strength over weak theories of arithmetics (see Simpson's article in this volume). Similarly, some of the work on Recursive Combinatorics is naturally connected with the theory of on-line algorithms (see Kierstead's article in this volume).

There are many important unsolved problems in Recursive Mathematics and many areas which require further development. For example, the effectiveness of model-theoretic constructions continues to supply a rich source of problems. Pour-El and Richards [122] looked at the effective content of results from mathematical physics, but their work has just scratched the surface of a large area of interesting problems concerning the effective content of other applied areas, such as physics, control theory (see the article by Ge and Nerode [50]), and statistics. The work on Feasible Model Theory still lacks the sort of general sufficient conditions for the construction of polynomial-time models that abound in Recursive Model Theory. Similarly a theory of Polynomial-space Model Theory and Algebra is yet to be developed. In general, the work on Feasible Algebra has not yet yielded the equivalences with classical complexity theoretic questions that can be found in the theory of Polynomial-time Analysis. In contrast, the use of the priority method which has many applications in Recursive Model Theory and Algebra has not found any uses in Recursive Analysis. This raises the question of whether there is an essential difference between Recursive Model Theory and Algebra versus Recursive Analysis, or whether we have just not studied sufficiently deep questions in Recursive Analysis to require the priority method at this time. Many open questions remain on the effective content of fields and topological vector spaces.

The exact relation between Recursive Mathematics and other areas is not well understood. For example, are there relations between Recursive Mathematics and typed lambda calculus interpretations of higher order intuitionistic logic, as in the work of Girard, or between Recursive Mathematics and untyped lambda calculus models of intuitionistic Zermelo-Fraenkel set theory, as in the work of McCarty [91]? In the same vein, the relation between Friedman-Simpson's weak systems which have only limited comprehension, the systems of Reverse Mathematics, and Recursive Mathematics is still not completely understood. For example, does the finer analysis of the effective content of König's lemma, i.e., the full theory of Π_1^0 classes, give a finer analysis of proof theoretic strengths? For example, can Reverse Mathematics distinguish between problems which can represent an arbitrary recursively bounded Π_1^0 class as opposed to the class of separating sets of a pair of r.e. sets (see the article by Cenzer and Remmel on Π_1^0 classes in this volume)? We hope that the methods surveyed in this book will help to clarify these matters.

Acknowledgments

There are number of people who deserve special thanks for this volume. First we must thank all the authors who contributed articles. Our hope was to have a volume which would first exhibit the beauty and depth of the subject and second inspire a new generation of researchers to go into the field of Recursive Mathematics. Only time will tell if the latter goal will be met but we are confident that reader will agree that our first goal has been achieved splendidly. Many of the authors also refereed articles for the volume. A very special note of thanks must go to Victor Marek who took over the job of coordinating the volume when the last editor was on leave in industry for the last two years. Without Victor's tireless efforts on behalf of the volume, it would not have been possible to produce it this year.

We are grateful to numerous colleagues who helped us edit this volume. Our LaTeX wizard, Professor Graeme Bailey of Cornell University, helped us transform the variety of styles supplied by authors into a coherent whole. Our thanks go to a number of young researchers at Cornell University, Jennifer Davoren, Dennis Hirschfeldt, Robert Milnikel, David Solomon, and Anke Walz, who assisted the authors in Kazakhstan and in Russia to improve the readability and linguistic correctness of their contributions. We are grateful to Diana Drake who provided administrative support at Cornell. When the electronic connections between the United States and Novosibirsk failed, we were assisted by Professor Sergiey Artemov (of Moscow State University and Cornell University) and Professor Lev Beklemishev (of Moscow State University) who helped to provide the required connections often via the telephone. Finally a special thanks is due to Julia Knight who edited the contribution by the late Christopher Ash.

We gratefully acknowledge support for our work on the volume. In particular, J. Remmel has been partially supported by US Department of Commerce agreement 70-NANB5H1164 and NSF grant DMS-9306427. V. Marek has been partially supported by NSF grant IRI-9619233 and USARO contract DAAH-04-96-1-0398. A. Nerode and his collaborators at Cornell University have been partially supported by USARO MURI DAAH-04-96-10341, Center for Foundations of Intelligent Systems at Cornell University.

References

[1] O. Aberth, Analysis in the computable number field, J. Assoc. Comput. Mach., **15** (1968) 275–299.

[2] C. H. Applebaum, Isomorphisms of ω–groups, Notre Dame J. Formal Logic, **12** (1971) 238–248.

[3] E. Artin and O. Schreier, Algebraische Konstruktion reeller Körper, Abh. Math. Sem. Univ. Hamburg, **5** (1927) 85–99.

[4] T. Baker, J. Gill and R. Solovay, Relativizations of the P = ? NP question, SIAM J. Comput., **4** (1975) 431–442.

[5] P. Bernays and D. Hilbert, Grundlagen der Mathematik, Vol. I, II, (1934, 1939), Grundlehren Math. Wiss.; [2nd. edn. (1968, 1970)].

[6] E. Bishop, Foundations of Constructive Analysis, (McGraw-Hill, New York, Toronto, London, 1967).

[7] E. Bishop and D. S. Bridges, Constructive Analysis, Grundlehren Math. Wiss., **279** (1985).

[8] L. Blum, M. Shub and S. Smale, On a theory of computation and complexity over the real numbers: NP–completeness, recursive functions and universal machines, Bull. Amer. Math. Soc. (N.S.), **21** (1989) 1–46.

[9] M. Blum, A machine independent theory of the complexity of recursive functions, J. Assoc. Comput. Mach., **14** (1967) 322–336.

[10] W. W. Boone, Certain simple unsolvable problems in group theory, I–VI, Nederl. Akad. Wentensch. Proc. Ser. A, **57** (1954) 231–237, 492–497; **58** (1955) 252–256, 571–577; **60** (1957) 22–27, 227–232.

[11] L. E. J. Brouwer, Collected Works, Vol. 1, Philosophy and Foundations of Mathematics, A. Heyting, (ed.), (North-Holland, Amsterdam, Oxford, 1975).

[12] L. E. J. Brouwer, Collected Works, Vol. 2, Geometry, Analysis, Topology and Mechanics, H. Freudenthal, (ed.), (North-Holland, Amsterdam, Oxford, 1976).

[13] L. E. J. Brouwer and B. de Loor, Intuitionistischer Beweis des Fundamentalsatzes der Algebra, Koninklijke Nederl. Akad. Wentensch. Proc., **27** (1924) 186–188.

[14] G. S. Ceĭtin, Algorithmic operators in constructive complete separable metric spaces (Russian), Dokl. Akad. Nauk SSSR, **128** (1959) 49–52.

[15] D. Cenzer and J. B. Remmel, Polynomial-time versus recursive models, Ann. Pure Appl. Logic, **54** (1991) 17–58.

[16] D. Cenzer and J. B. Remmel, Feasible graphs and colorings, Math. Logic Quart., **41** (1992) 327–352.

[17] D. Cenzer and J. B. Remmel, Polynomial-time abelian groups, Ann. Pure Appl. Logic, **56** (1992) 313–363.

[18] C. C. Chang and H. J. Keisler, Model Theory, 3rd. edn., Stud. Logic Found. Math., **73** (1990); [1st. edn. 1973, 2nd. edn. 1977].

[19] A. Church, An unsolvable problem of elementary number theory, Amer. J. Math., **58** (1936) 345–363.

[20] A. Church and S. C. Kleene, Formal definitions in the theory of ordinal numbers, Fund. Math., **28** (1937) 11–21.

[21] A. Cobham, The intrinsic computational difficulty of functions, in: Proc. Inter. Cong. for Logic, Methodology, and Philos. Sci., (Jerusalem, 1984), North Holland, (1965) 24–30.

[22] S. A. Cook, The complexity of theorem proving procedures, ACM Sympos. Theory of Computation, (1971) 151–158.

[23] J. N. Crossley, Constructive Order Types, (North-Holland, Amsterdam, 1969).

[24] J. N. Crossley, Recursive equivalence, Bull. London Math. Soc., **2** (1970) 129–151.

[25] J. N. Crossley and A. Nerode, Combinatorial Functors, Ergeb. Math. Grenzgeb. Ser. 3, **81** (1974).

[26] J. N. Crossley and A. Nerode, Effective dimension, J. Algebra, **41** (1976) 398–412.

[27] J. N. Crossley and A. Nerode, Cancellation laws for polynomial-time p–isolated sets, Ann. Pure Appl. Logic, 56 (1992) 147-172.

[28] M. D. Davis, Yu. Matijasevič and J. Robinson, Hilbert's tenth problem. Diophantine equations: positive aspects of a negative solution, in: Mathematical Developments Arising from Hilbert Problems, (N. Illinois Univ., De Kalb, Ill, 1974), Proc. Sympos. Pure Math., **28** (1976) 323–378.

[29] M. D. Davis, H. Putnam and J. Robinson, The decision problem for exponential diophantine equations, Ann. of Math., Ser. 2, **74** (1961) 425–436.

[30] J. C. E. Dekker, Countable vector spaces with recursive operations. Part I, J. Symbolic Logic, **34** (1969) 363–387.

[31] J. C. E. Dekker, Countable vector spaces with recursive operations. Part II, J. Symbolic Logic, **36** (1971) 477–493.

[32] J. C. E. Dekker, Recursive equivalence types and cubes, in: Aspects of Effective Algebra, (Proc. Conf. Monash Univ., Clayton, Australia, Aug. 1–4, 1979), J. N. Crossley (ed.), (Upside Down A Book Co., Yarra Glen, Victoria, Australia, 1981) 87–121.

[33] J. C. E. Dekker and J. Myhill, Recursive equivalence types, Publ. Math. Univ. California, **3** (1960) 67–214.

[34] E. Ellentuck, The first order properties of Dedekind finite integers, Fund. Math., **63** (1968) 7–25.

[35] E. Ellentuck, Nonrecursive combinatorial functions, J. Symbolic Logic, **37** (1972) 90–95.

[36] E. Ellentuck, Sylow subgroups of a regressive group, Houston J. Math., **5** (1979) 46–67.

[37] Yu. L. Ershov, Constructive models (Russian), in: Izbr. Vopr. Alg. i Log., [Selected Questions in Algebra and Logic], Sbornik posvjascen. pamjati A. I. Mal'ceva, [A collection dedicated to the memory of A. I. Mal'tsev], Izdat. Nauk Sibirsk. Otdel., Novosibirsk, (1973) 111–130.

[38] Yu. L. Ershov, Skolem functions and constructive models (Russian), Algebra i Logika, **12** (1973) 644–654, 735; [translated in: Algebra and Logic, **12** (1973) 368–373].

[39] Yu. L. Ershov, Theory of Numerations III (Constructive Models) (Russian), Lib. Dept. Algebra and Math. Logic Novosibirsk Univ., **13**, (Novosibirsk Gosudarstr. Univ., Novosibirsk, 1974).

[40] Yu. L. Ershov, Hereditarily effective operations (Russian), Algebra i Logika, **15** (1976) 642–654. [translated in: Algebra and Logic, **15** (1976) 400–409.

[41] Yu. L. Ershov, Theory of Numerations (Russian), Monographs in Math. Logic and Foundations of Math., (Nauka, Moskva, 1977).

[42] Yu. L. Ershov, Theorie der Numerierungen, III, (translated from Russian and edited by G. Asser and H.-D. Hecker), Z. Math. Logik Grundlag. Math., **23** (1977) 289–371.

[43] L. J. Feiner, Hierarchies of Boolean algebras, J. Symbolic Logic, **35** (1970) 365–374.

[44] R. M. Friedberg, Two recursively enumerable sets of incomparable degrees of unsolvability, Proc. Nat. Acad. Sci. U.S.A., **43** (1957) 236–238.

[45] H. M. Friedman, Set theoretic foundations for constructive analysis, Ann. of Math., Ser. 2, **105** (1977) 1–28.

[46] H. M. Friedman, The computational complexity of maximization and integration, Adv. Math., **53** (1984) 80–98.

[47] H. M. Friedman, S. G. Simpson and R. L. Smith, Countable algebra and set existence axioms, Ann. Pure Appl. Logic, **25** (1983) 141–181; addendum, ibid. **27** (1985) 319–320.

[48] A. Fröhlich and J. C. Shepherdson, Effective procedures in field theory, Philos. Trans. Roy. Soc. London, Ser. A, **248** (1956) 407–432.

[49] M. R. Garey and D. S. Johnson, Computers and Intractability, a guide to the theory of NP–completeness, (W. H. Freeman, San Francisco, 1979).

[50] X. Ge and A. Nerode, Effective content of the calculus of variations I: semicontinuity and the chattering lemma, Ann. Pure Appl. Logic, **78** (1996) 127–146.

[51] K. Gödel, Über formal unentscheidbare Sätze der *Principia mathematica* und verwandter Systeme I, Monatsh. Math. Phys., **38** (1931) 173–198; [translated as: On formally undecidable propositions of *Principia Mathematica* and related systems I, (translated by B. Meltzer, with an introduction by R. B. Braithwaite), (Basic Books, NY, 1963, reprinted Dover, NY, 1992); also in: From Frege to Gödel: A Source Book in Logic, 1879–1931, J. van Heijenroot, (ed.), (Harvard Univ. Press, Cambridge, Mass. and Oxford Univ. Press, London, 1967) 592–617.]

[52] S. S. Goncharov, The constructivizability of superatomic Boolean algebras (Russian), Algebra i Logika, **12** (1973) 31–40, 120; [translated in: Algebra and Logic, **12** (1973) 17–22].

[53] S. S. Goncharov, Autostability and computable families of constructivizations (Russian), Algebra i Logika, **14** (1975) 647–680; [translated in: Algebra and Logic, **14** (1975) 392–409].

[54] S. S. Goncharov, Certain properties of the constructivization of Boolean algebras (Russian), Sibirsk. Math. Zh., **16** (1975) 264–278; [translated in: Siberian Math. J., **16** (1975) 203–214].

[55] S. S. Goncharov, Constructive models of $\aleph_1$-categorical theories (Russian), Mat. Zametki, **23** (1978) 885–888; [translated in: Math. Notes, **23** (1978) 486–487].

[56] S. S. Goncharov, Strong constructivizability of homogeneous models (Russian), Algebra i Logika, **17** (1978) 363–388, 490; [translated in: Algebra and Logic, **17** (1978) 247–263].

[57] S. S. Goncharov, Countable Boolean Algebras (Russian), (Nauka. Sibirsk. Otdel., Novosibirsk, 1988).

[58] R. L. Goodstein, Recursive analysis, in: Constructivity in Mathematics, (Proc. Colloq., Amsterdam, Aug. 26–31, 1957), A. Heyting, (ed.), Stud. Logic Found. Math., (1959) 37–42.

[59] L. Harrington, Recursively presentable prime models, J. Symbolic Logic, **39** (1974) 305–309.

[60] J. Hartmanis, On sparse sets in NP − P, Inform. Process. Lett., **16** (1983) 55–60.

[61] J. Hartmanis and R. E. Stearns, On the computational complexity of algorithms, Trans. Amer. Math. Soc., **117** (1965) 285–306.

[62] M. J. Hassett, Recursive equivalence types and groups, J. Symbolic Logic, **34** (1969) 13–20.

[63] J. Hauck, Funktional-Rekursion, Z. Math. Logik Grundlag. Math., **18** (1972) 31–36.

[64] L. S. Hay, The co-simple isols, Ann. of Math., Ser. 2, **83** (1966) 231–256.

[65] A. Heyting, Die formalen Regeln der intuitionistischen Logik, Sitz. Preuss. Akad. Wiss. Phys.-Math. Kl., (1930) 42–56; Die formalen Regeln der intuitionistischen Mathematik, ibid., (1930) 57–71; Die formalen Regeln der intuitionistischen Mathematik III, ibid., (1930) 143, 158–169.

[66] A. Heyting, Intuitionism, An Introduction, (North-Holland, Amsterdam, 1956); [2nd. rev. edn., 1966].

[67] A. G. Higman, Subgroups of finitely presented groups, Proc. Roy. Soc. London Ser. A, **262** (1961) 455–475.+

[68] D. Hilbert, Mathematische Probleme, Archiv. f. Math. und Phys., Ser. 3, **1** (1901) 44–63, 213–237.

[69] D. Hilbert and W. Ackermann, Principles of Mathematical Logic, (translated from German by L. M. Hammond, G. G. Lechie and F. Steinhardt, edited with notes by R. E. Luce), (Chelsea, 1950).

[70] S. Homer and W. Maass, Oracle-dependent properties of the lattice of NP sets, Theoret. Comput. Sci., **24** (1983) 279–289.

[71] J. E. Hopcroft and J. D. Ullman, Formal Languages and their Relations to Automata, (Addison-Wesley, Reading, Mass, 1969).

[72] C. G. Jockusch Jr., Ramsey's theorem and recursion theory, J. Symbolic Logic, **37** (1972) 268–280.

[73] C. G. Jockusch Jr. and R. I. Soare, Degrees of members of Π_1^0 classes, Pacific J. Math., **40** (1972) 605–616.

[74] C. G. Jockusch Jr. and R. I. Soare, Π_1^0 classes and degrees of theories, Trans. Amer. Math. Soc., **173** (1972) 33–56.

[75] I. Kalantari and A. Retzlaff, Maximal vector spaces under automorphisms of the lattice of recursively enumerable vector spaces, J. Symbolic Logic, **42** (1977) 481–491.

[76] R. M. Karp, Reducibilities among combinatorial problems, in: Complexity of Computer Computations, (Proc. Sympos. IBM Watson Res. Ctr., Yorktown Heights, NY, Mar. 20–22, 1972), R. E. Miller, J. W. Thatcher and J. D. Bohlinger, (eds.), (Plenum Press, New York, 1972) 85–103.

[77] B. M. Khoussainov and A. Nerode, Automatic presentations of structures, in: Logic and Computational Complexity, (Papers from Interntl. Workshop, LCC '94, Oct. 13–16, 1994, Indianapolis, IN), D. Leivant (ed.), Lecture Notes in Comput. Sci., **960** (1995) 367–392.

[78] H. A. Kierstead, An effective version of Dilworth's theorem, Trans. Amer. Math. Soc., **268** (1981) 63–77.

[79] S. C. Kleene, λ-definability and recursiveness, Duke Math. J., **2** (1936) 340–353.

[80] S. C. Kleene, Recursive predicates and quantifiers, Trans. Amer. Math. Soc., **53** (1943) 41–73.

[81] S. C. Kleene, Introduction to Metamathematics, (D. Van Nostrand, Princeton, 1952, and Elsevier, Amsterdam, 1964, 1971).

[82] S. C. Kleene, Hierarchies of number-theoretic predicates, Bull. Amer. Math. Soc., **61** (1955) 193–213.

[83] K. Ko, The maximum value problem and NP real numbers, J. Comput. System Sci., **24** (1982) 15–35.

[84] K. Ko and H. M. Friedman, Computational complexity of real functions, Theoret. Comput. Sci., **20** (1982) 323–352.

[85] G. Kreisel, Analysis of the Cantor-Bendixson theorem by means of the analytic hierarchy, Bull. Acad. Polon. Sci. Sér. Sci. Math. Astronom. Phys., **7** (1959) 621–626.

[86] L. Kronecker, Grundzüge einer arithmetischen Theorie der algebraischen Grössen, J. Reine Angew. Math., **92** (1882) 1–122.

[87] B. Kušner, Riemann integration in constructive analysis (Russian), Dokl. Akad. Nauk SSSR, **156** (1964) 255–257; [translated in: Soviet Math. – Dokl., **5** (1964) 628–630].

[88] D. Lacombe, Classes récursivement fermées et fonctions majorantes, C. R. Acad. Sci. Paris, **240** (1955) 716–718.

[89] R. Ladner, On the structure of polynomial time reducibility, J. Assoc. Comput. Mach., **22** (1975) 155–171.

[90] C. R. Lin, The effective content of Ulm's Theorem, in: Aspects of Effective Algebra, (Proc. Conf. Monash Univ., Clayton, Australia, Aug. 1–4, 1979), J. N. Crossley (ed.), (Upside Down A Book Co., Yarra Glen, Victoria, Australia, 1981) 147–160.

[91] C. McCarty, Realizability and recursive set theory, Ann. Pure Appl. Logic, **32** (1986) 153–183.

[92] A. I. Mal'tsev, Constructive algebras I (Russian), Uspekhi Mat. Nauk, 16 (1961) 3–60; [translated in: Constructive algebras I, Russian Math. Surveys, **16**:3 (1961) 77–129; also in: The Metamathematics of Algebraic Systems, Collected Papers: 1936–1967, translated and edited by B. F. Wells III, Stud. Logic Found. Math., **66** (1971), Ch. 18, 148–200].

[93] A. I. Mal'tsev, Algorithms and Recursive Functions (Russian), (Izdat. Nauka, Moscow, 1965); [translated: by L. F. Boron, L. E. Sanchis, J. Stilwell and K. Iseki, (Wolters-Noordhoff Publishing, Groningen, 1970)].

[94] A. I. Mal'tsev, Algebraic Systems (Russian), (posth. edn.), V. D. Smirnov and M. Taĭclin, (eds.), (Izdat. Nauka, Moscow, 1970); [translated: by B. D. Seckler and A. P. Doohovskoy, Grundlehren Math. Wiss., **192** (1973).]

[95] A. B. Manaster and J. G. Rosenstein, Effective matchmaking (recursion theoretical aspects of a theorem of Philip Hall, Proc. London Math. Soc., Ser. 3, **25** (1972) 615–654.

[96] A. B. Manaster and J. G. Rosenstein, Effective matchmaking and k–chromatic graphs, Proc. Amer. Math. Soc., 39 (1973) 371–378.

[97] A. A. Markov, On the representation of recursive functions (Russian), Dokl. Akad. Nauk SSSR (N.S.), **58** (1947) 1891–1892.

[98] A. A. Markov, The Theory of Algorithms (Russian), Izdat. Akad. Nauk SSSR, Moscow, Trudy Mat. Inst. Steklov., **42** (1954); [translated: by J. J. Schorr-Kon and PST staff, publ. for NSF, Washington, DC, and the Dept. Commerce by (Israel Prog. for Scientific Translations, Jerusalem, 1961).]

[99] W. Markwald, Zur Theorie der konstruktiven Wohlornungen, Math. Ann., **127** (1954) 135–149.

[100] Yu. V. Matijasevič, Enumerable sets are diophantine (Russian), Dokl. Akad. Nauk SSSR, **191** (1970) 279–282; [translated in: Soviet Math. – Dokl., **11** (1970) 354–357].

[101] S. Mazur, Computable Analysis, (based on lecture notes academic year 1949/50, Inst. Math., Polish Acad. Sci., Warsaw), A. Grzegorczyk and H. Rasiowa, (eds.), Rozprawy Matematyczne, Vol. 33, (Państwowe Wydawn. Naukowe, Warszawa, 1963).

[102] G. Metakides and A. Nerode, Recursion theory and algebra, in: Algebra and Logic, (Proc. 14th. Summer Res. Inst. Austral. Math. Soc., Monash Univ., Clayton, Vic., Australia, Jan. 6 – Feb. 16, 1974), J. N. Crossley, (ed.), Lecture Notes in Math., **450** (1975) 209–219.

[103] G. Metakides and A. Nerode, Recursively enumerable vector spaces, Ann. Math. Logic, **11** (1977) 147–171.

[104] G. Metakides and A. Nerode, Effective content of field theory, Ann. Math. Logic, **17** (1979) 289–320.

[105] G. Metakides and A. Nerode, Recursion theory on fields and abstract dependence, J. Algebra, **65** (1980) 36–59.

[106] T. S. Millar, The Theory of Recursively Presented Models, Ph.D. Thesis, Cornell Univ., Ithaca, NY, (1976).

[107] T. S. Millar, Foundations of recursive model theory, Ann. Math. Logic, **13** (1978) 305–320.

[108] A. A. Mučnik, On the unsolvability of the problem of reducibility in the theory of algorithms (Russian), Dokl. Akad. Nauk SSSR, **108** (1956) 194–197.

[109] J. Myhill, Creative sets, Z. Math. Logik Grundlag. Math., **1** (1955) 97–108.

[110] A. Nerode, Extensions to isols, Ann. of Math., **73** (1961) 362–403.

[111] A. Nerode and J. B. Remmel, Recursion theory on matroids. in: Patras Logic Symposion, (Proc. Logic Sympos. Patras, Greece, Aug. 18–22, 1980) G. Metakides (ed.), Stud. Logic Found. Math., **109** (1982) 41–65.

[112] A. Nerode and J. B. Remmel, A survey of lattices of r.e. substructures, in: Recursion Theory, (Proc. AMS-ASL Summer Inst., Ithaca, NY, June 28 – July 16, 1982), A. Nerode and R. A. Shore, (eds.), Proc. Sympos. Pure Math., **42** (1985) 323–375.

[113] A. Nerode and J. B. Remmel, Complexity theoretic algebra I: Vector spaces over finite fields, in: Structure in Complexity Theory, (Proc. 2nd. Annual IEEE Conf. Struct. Complexity Theory, Cornell Univ., Ithaca NY, June 16–19, 1987), (IEEE Comput. Soc. Press, Washington, DC, 1987), 218–239.

[114] A. Nerode and J. B. Remmel, Complexity-theoretic algebra II: Boolean algebras, (Special issue: 3rd. Asian Conf. Math. Logic, Beijing, Oct. 26–30, 1987), D. P. Yang, (ed.), Ann. Pure Appl. Logic, **44** (1989) 71–99.

[115] A. Nerode and J. B. Remmel, Complexity-theoretic algebra: vector space bases, in: Feasible Mathematics, (Proc. Workshop, June 16–18, 1989, Cornell Univ. Ithaca, NY), S. Buss and P. J. Scott (eds.), Progr. Comput. Sci. Appl. Logic, **9** (1990) 293–319.

[116] A. Nerode and J. B. Remmel, Polynomial time equivalence types, in: Logic and Computation, (Proc. Workshop, CMU, Pittsburg, June 30 – July 2, 1987), W. Sieg, (ed.), Contemp. Math., **106** (1990) 221–249.

[117] P. S. Novikov, On the algorithmic unsolvability of the word problem in group theory (Russian), Trudy Mat. Inst. Steklov., **44** (1955).

[118] V. P. Orevkov, Certain types of continuity of constructive operators (Russian), Trudy Mat. Inst. Steklov., **93** (1967) 164–186.

[119] R. Peter, Rekursive Funktionen, (Akademischer Verlag, Budapest, 1951).

[120] E. L. Post, Formal reductions of the general combinatorial decision problem, Amer. J. Math., **65** (1943) 197–215.

[121] E. L. Post, Recursively enumerable sets of positive integers and their decision problems, Bull. Amer. Math. Soc. (N.S.), **50** (1944) 284–316.

[122] M. B. Pour-El and J. I. Richards, Computability in Analysis and Physics, Perspect. in Math. Logic, (1989).

[123] M. Presburger, Über die Vollständigkeit eines gewissen Systems der Arithmetik ganzer Zahlen, in welchem die Addition als einzige Operation hervortritt, Comptes-Rendus du I Congres des Mathématiciens des Pays Slaves, Warszawa, (1929) 92–101, 393.

[124] M. O. Rabin, Recursive unsolvability of group theoretic problems, Ann. of Math., **67** (1958) 172–194.

[125] M. O. Rabin, Computable algebra, general theory and theory of computable fields, Trans. Amer. Math. Soc., **95** (1960) 341–360.

[126] J. B. Remmel, Combinatorial functors on co-r.e. structures, Ann. Math. Logic, **10** (1976) 261–287.

[127] J. B. Remmel, Co-hypersimple structures, J. Symbolic Logic, **41** (1976) 611–625.

[128] J. B. Remmel, Recursively enumerable Boolean algebras, Ann. Math. Logic, **15** (1978) 75–107.

[129] J. B. Remmel, Recursion theory on algebraic structures with independent sets, Ann. Math. Logic, **18** (1980) 153–191.

[130] J. B. Remmel, Recursive Boolean algebras, in: Handbook of Boolean Algebras, Vol. 3, J. D. Monk, R. Bonnet and S. Koppelberg, (eds.), (North-Holland, Amsterdam, NY, 1989), Ch. 25, 1097–1165.

[131] A. Retzlaff, Direct summands of recursively enumerable vector spaces, Z. Math. Logik Grundlag. Math., **25** (1979) 363–372.

[132] H. G. Rice, Recursive and recursively enumerable orders, Trans. Amer. Math. Soc., **83** (1956) 277–300.

[133] J. Robinson, Definability and decision problems in arithmetic, J. Symbolic Logic, **14** (1949) 98–114.

[134] H. Rogers Jr., Theory of Recursive Functions and Effective Computability, (1st. edn., McGraw-Hill, New York-Toronto, Ont.-London, 1967; 2nd. edn., MIT Press, Cambridge, Mass., London, 1987).

[135] J. G. Rosenstein, Recursive linear orderings, in: Orders: Description and Roles, Ann. Discrete Math., **23** (1984) 465–475.

[136] G. E. Sacks, The recursively enumerable degrees are dense, Ann. of Math., **80** (1964) 300–312.

[137] N. A. Šanin, Some problems of mathematical analysis in the light of constructive logic, Z. Math. Logik Grundlag. Math., **2** (1956) 27–36.

[138] W. Savitch, Relationships between nondeterministic and deterministic tape complexities, J. Comput. System Sci., **4** (1970) 177–192.

[139] J. R. Shoenfield, Degrees of formal systems, J. Symbolic Logic, **23** (1958) 389–392.

[140] J. R. Shoenfield, Degrees of models, J. Symbolic Logic, **25** (1960) 233–237.

[141] R. A. Shore, Controlling the dependence degree of a recursively enumerable vector space, J. Symbolic Logic, **43** (1978) 13–22.

[142] R. I. Soare, Recursion theory and Dedekind cuts, Trans. Amer. Math. Soc., **140** (1969) 271–294.

[143] E. Specker, Ramsey's Theorem does not hold in recursive set theory, in: Logic Colloq. '69, (1971) 439–442.

[144] C. Spector, Recursive well-orderings, J. Symbolic Logic, **20** (1955) 151–163.

[145] L. J. Stockmeyer, The polynomial-time hierarchy, Theoret. Comput. Sci., **3** (1976) 1–22 (1977).

[146] A. Tarski, Undecidability of group theory, J. Symbolic Logic, **14** (1949) 76–77.

[147] A. Tarski, A Decision method for elementary algebra and geometry, 2nd. edn., revised, (Univ. California Press, Berkeley and Los Angeles, CA, 1951).

[148] A. Tarski, Logic, Semantics, Metamathematics, Papers from 1923 to 1938, (Oxford Univ. Press, 1956).

[149] S. Tennenbaum, Non-archimedean models for arithmetic, Notices Amer. Math. Soc., **6** (1959) 270.

[150] J. Thurber, Degrees of Boolean Algebras, Ph.D. Thesis, Univ. Notre Dame, Notre Dame, IN, (1994).

[151] J. J. Thurber, Recursive and r.e. quotient Boolean algebras, Arch. Math. Logic, **33** (1994) 121–129.

[152] A. S. Troelstra and D. van Dalen, Constructivism in Mathematics, Vol. 1, Stud. Logic Found. Math., **121** (1988).

[153] A. S. Troelstra and D. van Dalen, Constructivism in Mathematics, Vol. 2, Stud. Logic Found. Math., **123** (1988).

[154] A. M. Turing, On computable numbers, with an application to the Entscheidungsproblem, Proc. London Math. Soc., Ser. 2, **42** (1936) 230–265; corr. ibid., **43** (1937) 544–546.

[155] B. L. van der Waerden, Eine Bemerkung über die Unzerlegbarkeit von Polynomen, Math. Ann., **102** (1930) 738–739.

[156] R. L. Vaught, Sentences true in all constructive models, J. Symbolic Logic, **25** (1960) 39–58.

[157] B. L. van der Waerden, Moderne Algebra, Parts I and II, (G. E. Stechert and Co., New York, 1943); [translated as: Modern Algebra, Vols. I and II, (translated from the 2nd. rev. German edition by F. Blum, with revisions and additions by the author), Frederick Ungar Publishing Co., NY, (1949)].

Chapter 1

Pure Computable Model Theory

Valentina S. Harizanov*

Contents

*This work was partially supported by the NSF RP grant DMS-9210443.

HANDBOOK OF RECURSIVE MATHEMATICS
Edited by Yu. L. Ershov, S. S. Goncharov,
A. Nerode, and J. B. Remmel

Introduction

Exploiting the fundamental concepts of computability theory, computable model theory introduces effective analogues of model-theoretic notions. By combining methods from both fields, it has enabled the development of machinery for investigating the *effective* content of model-theoretic constructions. While some model-theoretic constructions can be replaced by effective ones, for others such replacement is impossible. Thus, another important objective for computable model theory is the discovery of effective counterexamples to model-theoretic results. For instance, Vaught's theorem (no complete theory has exactly two non-isomorphic countable models) cannot be effectivized.

The article begins with the foundations of computable model theory: the definitions and examples of decidable theories, and computable and decidable models. It then presents the effective completeness theorem and the effective omitting types theorem; and characterizations of decidable theories with decidable prime models, and then with decidable saturated models. The next sections characterize decidable homogeneous models, and give examples of decidable theories with exactly two non-isomorphic decidable models. The following sections present the results on decidable theories with only finitely many, and on decidable theories with only countably many, non-isomorphic countable models, and investigate the model-theoretic nature and the computability-theoretic complexity of models of such theories. Later sections study indiscernibles from the computability-theoretic point of view, and the degrees of models. Finally, we consider the isomorphisms of effective models and related subtopics, such as intrinsically c.e. relations, computably stable models, and computably categorical models.

Computable model theory was developed simultaneously and for the most part independently in the West, mainly in the United States and Australia, and in Russia. Because of poor communication between the two groups, many results were independently discovered by both groups. This article looks at computable model theory from the Western perspective. (There are articles in this volume on the Russian approach.) However, the article also presents some results of the Russian group, and often emphasizes the connections with and gives references to their results.

Almost every section contains a detailed proof with a survey of the computability-theoretic and model-theoretic background needed. The bibliography contains both Western and Russian papers in pure computable model theory, but not papers in computable algebra nor in computable combinatorics. Another survey article on this subject has been recently and independently written by Millar [148].

1 History

The goal of computable mathematics is to find the extent to which certain classical results of mathematics are effectively true. Although many consider the modern study of computability of algebraic constructions to have started with Fröhlich and Shepherdson in 1955–56 and Rabin in 1958–60, even van der Waerden in his book [210] from 1930, see also [211], discussed the problem of carrying out certain field-theoretic procedures effectively. He also defined an *explicitly* given field as one whose elements are uniquely represented by distinguishable symbols with which one can perform the field operations effectively. In a pioneering paper from 1930, van der Waerden [209] proved that there does not exist a splitting algorithm applicable to every explicit field. Fröhlich and Shepherdson [62, 63] used the precise notion of a *computable* function to obtain a collection of results and examples about explicit fields.

Rabin [176, 177] did a systematic study of *computable* groups and *computable* fields. In Russia, a systematic study of *constructive* algebraic systems and their *enumerations* was initiated by Mal'cev [124] in the 1960's, and continued by Ershov and his collaborators, see [56, 57].

In the 1970's, Nerode and his collaborators revived the study of computability of algebraic constructions. At the 1974 *Recursive Model Theory Symposium* at Monash University (Melbourne, Australia), Metakides and Nerode announced that, in addition to other computability-theoretic tools, they started using the priority method as an important tool in the algorithmic part of computable mathematics, see [128]. Thus, they founded in the West the field of the post-Friedberg-Muchnik computable mathematics. Metakides and Nerode used the priority method in their systematic study of the effective content of specific structures, such as vector spaces [129], fields [130], and structures with a dependence relation [131]. For more information on the development of computable mathematics see [38, 132, 183]. In Russia, the post-Friedberg-Muchnik constructive mathematics was founded by Nurtazin and Goncharov in the 1970's [79, 159].

In the West, the computability of ordered sets has also been studied by Ash, Case, Chen, Crossley, Downey, Feiner, Feldman, Fellner, Hay, Hingston,

Hird, Jockusch, Kierstead, Knight, Lerman, Manaster, Metakides, McNulty, Moses, Remmel, Richter, Rosenstein, Roy, Schmerl, Schwarz, Soare, Tennenbaum, Trotter and Watnick; the computability of vector spaces by Ash, Guhl, Guichard, Dekker, Downey, Hamilton, Kalantari, Remmel, Retzlaff, Shore, Smith and Welch; the computability of rings and fields by Ash, Hodges, Jockusch, MacIntyre, Madison, Marker, Mines, Rosenthal, Seidenberg, Shlapentokh, Smith, Staples, Tucker and van den Dries; the computability of the structures with a dependence relation by Baldwin, Downey and Remmel. The computability in other mathematical structures is also extensively studied: in groups by Ash, Barker, Ge, Kent, Knight, Lin, Oates, Richards, Richman and Smith; in graphs by Aharoni, Bean, Beigel, Burr, Carstens, Gasarch, Golze, Kierstead, Lockwood, Manaster, Magidor, Päppinghaus, Remmel, Rosenstein, Schmerl and Shore; in Boolean algebras by Carroll, Feiner, LaRoche, Remmel and Thurber; in topological spaces by Kalantari, Legett, Remmel, Retzlaff and Weitkamp. Computable Ramsey's theory has been studied by Clote, Hummel, Jockusch, Seetapun, Simpson, Solovay and Specker. Computability in analysis and physics has also been studied, see [175].

The generalization of the definition of a particular computable algebraic structure to an arbitrary model yields one of the basic concepts of pure computable model theory, an area of logic developed in the last twenty-five years. That is, the notion of a *computable* model, and a stronger notion of a *decidable* model. Lerman and Schmerl have given examples of theories with computable models. The first general results in computable model theory have been obtained by following the fundamental notions and results of classical model theory. For example, Millar has obtained the effective version of the omitting types theorem, and Harrington, Goncharov and Nurtazin have found when a complete decidable theory with a prime model has a decidable prime model. Millar and Morley have characterized decidable theories with decidable saturated models, and Goncharov and Peretyat'kin have characterized decidable homogeneous models. Barwise, Schlipf and Ressayre have introduced the notion of a computably saturated model. Although developed in the context of admissible sets and admissible fragments of infinitary logic, computably saturated models have also provided a useful tool for research and exposition in classical model theory.

In the West, Millar has further produced an extensive body of work on topics including effective Vaught's theorem, the structure of types in decidable models, decidability and prime, saturated and homogeneous models, decidable theories with finitely many and decidable theories with countably many non-isomorphic countable models. Reed has also studied decidable theories with finitely many non-isomorphic countable models. Kierstead and Remmel have investigated the degrees of sets of indiscernibles in decidable models. Ash, Knight, Macintyre, Marker, Nadel, Nies, Richter, Jockusch, Lachlan, Scott, Shoenfield, Shore, Soare and Tennenbaum have studied the degrees of models of various theories, including the theory of linear orders, Peano arithmetic, true arithmetic, and the theory of Boolean algebras.

The whole spectrum of questions involving the isomorphisms of abstract computable models has been investigated by Ash, Barker, Chisholm, Cholak, Crossley, Downey, Eisenberg, Guichard, Harizanov, Hird, Khoussainov, Knight, Manasse, Manaster, Millar, Moses, Nerode, Remmel, Shore, Slaman and Wehner. The lattices of computably enumerable submodels have been studied by Ash, Guichard, Carroll, Downey, Metakides, Nerode, Remmel and Smith. More recently, Nerode, Remmel and Cenzer [31, 157] have been developing feasible model theory (as a part of feasible mathematics), the theory of models with bounded space and time resources. They have investigated how feasible models differ from computable models. The feasible models studied include Boolean algebras, abelian groups, linear orders, models of arithmetic, and graphs.

2 Notation and Basic Definitions

The set $\{0, 1, 2, \dots\}$ of all natural numbers is denoted by ω. Unless explicitly stated otherwise, it is assumed that all languages considered are first-order and computable (hence at most countable), and that the domains of the considered models are subsets of ω. For a set of sentences T, by $L(T)$ we denote its language. A set of sentences T is deductively closed if T contains every sentence σ of $L(T)$ such that $T \vdash \sigma$. A consistent deductively closed set of sentences is called a *theory*.

Models are denoted by script letters, and their domains by the corresponding capital Latin letters. By $\mathcal{A} \subseteq \mathcal{B}$ we denote that $\mathcal{A}$ is a submodel of $\mathcal{B}$, and by $\mathcal{A} \preceq \mathcal{B}$ that $\mathcal{A}$ is an elementary submodel of $\mathcal{B}$. By $\mathcal{A} \equiv \mathcal{B}$ we denote that $\mathcal{A}$ and $\mathcal{B}$ are elementarily equivalent, and by $\mathcal{A} \cong \mathcal{B}$ that $\mathcal{A}$ and $\mathcal{B}$ are isomorphic. A model is *prime* if it can be elementarily embedded in every model of its theory. Hence a prime model for a countable language must be countable. Two prime models of the same complete theory are isomorphic.

Let $\mathcal{A}$ be a model (with domain A) for L. By $\mathrm{Th}(\mathcal{A})$ we denote the theory of $\mathcal{A}$. For $X \subseteq A$, let L_X be the language $L \cup \{\mathbf{a} : a \in X\}$, L expanded by adding a constant $\mathbf{a}$ for every $a \in X$. Let $\mathcal{A}_X = (\mathcal{A}, a)_{a \in X}$ be the expansion of $\mathcal{A}$ to the language L_X such that for every $a \in X$, $\mathbf{a}$ is interpreted by a. The *atomic diagram* of $\mathcal{A}$ is the set of all atomic and negated atomic sentences of L_A which are true in $\mathcal{A}_A$. It is denoted by $\Delta_{\mathcal{A}}$. The *complete diagram* of $\mathcal{A}$ is the set of all sentences of L_A which are true in $\mathcal{A}_A$. The complete diagram of $\mathcal{A}$ is often called an elementary diagram of $\mathcal{A}$.

A sequence of variables displayed after a formula or after a set of formulae includes all the free variables occurring in any of the formulae. For two sequences $\overline{x}$ and $\overline{y}$ of the same length k, by writing $\overline{x}/\overline{y}$ after a formula or a set of formulae, we denote the result of replacing every occurrence of $\overline{y}(i)$ by $\overline{x}(i)$ for $i < k$. To simplify the notation, instead of $\theta(y)(x/y)$ we often write only $\theta(x)$. For a set of formulae Θ, $\wedge\Theta$ is the conjunction of all formulae in Θ. For a formula θ, let $\theta^1 =_{\text{def}} \theta$ and $\theta^0 =_{\text{def}} \neg\theta$.

A formula is in a $\Sigma^0_0 = \Pi^0_0$ form if it contains no quantifiers. For $n > 0$, a formula is in a Σ^0_n (Π^0_n, respectively) form if it is logically equivalent to a formula in a prenex normal form which begins with an existential (universal) quantifier and has $n-1$ alternations of quantifiers. Σ^0_1 (Π^0_1, respectively) sentences are also called existential (universal, respectively). $T_\exists$ ($T_\forall$, respectively) denotes the set of all existential (universal, respectively) sentences in T. For infinite cardinals κ and λ, $L_{\kappa\lambda}$ denotes the infinitary logic which has κ individual variables, allows conjunction and disjunction of a set of $< \kappa$ formulae, and allows universal and existential quantification over a set of $< \lambda$ individual variables. In particular, $L_{\omega\omega}$ is classical first-order logic, and $L_{\omega_1\omega}$ allows countable conjunctions and disjunctions but only finite quantification. For more information on infinitary logic see [98].

A *type* of a theory T in variables $x_0, \dots, x_{n-1}$ is a maximal consistent set of formulae containing T, with free variables among $x_0, \dots, x_{n-1}$. To emphasize its maximality, it is often called a *complete* type in the literature. An n–type is a type in n variables, and a (finite) type is an n–type for some $n \in \omega$. If $\Gamma(x_0, \dots, x_{n-1})$ is a type and $x_{i_0}, \dots, x_{i_{k-1}} \in \{x_0, \dots, x_{n-1}\}$, then $\Gamma \restriction \{x_{i_0}, \dots, x_{i_{k-1}}\}$ is the subtype of Γ in variables $x_{i_0}, \dots, x_{i_{k-1}}$. A formula $\theta(x_0, \dots, x_{n-1})$ is *complete* in T if for every formula $\psi(x_0, \dots, x_{n-1})$, exactly one of

$$T \vdash (\forall x_0) \dots (\forall x_{n-1})[\theta(x_0, \dots, x_{n-1}) \Rightarrow \psi(x_0, \dots, x_{n-1})],$$

$$T \vdash (\forall x_0) \dots (\forall x_{n-1})[\theta(x_0, \dots, x_{n-1}) \Rightarrow \neg\psi(x_0, \dots, x_{n-1})]$$

holds. That is, there is exactly one (complete) type of T in $x_0, \ldots, x_{n-1}$ which contains θ. A type which contains a complete formula is called *principal.* The set of all (complete finite) types realized in a model $\mathcal{A}$ is called the *type spectrum* of $\mathcal{A}$. A type spectrum of a theory is the type spectrum of one of its models.

Let X be a set. $|X|$ denotes the cardinality of X. X is *countable* if $|X| = \omega$. X is *at most countable* if $|X| \leqslant \omega$. Let κ be an infinite cardinal, and let T be a complete theory in a countable language. T is called *stable in power* κ, or κ*-stable*, if for an arbitrary model $\mathcal{U}$ of T, for every subset X of U with $|X| = \kappa$, the model $\mathcal{U}_X$ realizes exactly κ many 1-types. T is called *stable* if it is stable in some power. If T is $\aleph_0$-stable, then T is stable in every infinite power (see Chapter VII of [32]). T is called *superstable* if it is κ-stable for every $\kappa \geqslant 2^{\aleph_0}$. For more information on stability theory see [22, 120, 174].

The quantifier $\exists! x$ abbreviates "there exists a unique x". The empty set is denoted by $\emptyset$. For a set X, $\mathcal{P}(X)$ is its power set. If f is a partial function, then $\mathrm{dom}(f)$ is the domain of f, $\mathrm{rng}(f)$ is the range of f, and $f(a) \downarrow$ denotes that $a \in \mathrm{dom}(f)$. The length of a sequence $\overline{x}$ is denoted by $\mathrm{lh}(\overline{x})$. If $\overline{x} = (x_0, \ldots, x_{n-1})$ and f is a unary function, then $f(\overline{x}) =_{\mathrm{def}} (f(x_0), \ldots, f(x_{n-1}))$. The concatenation of sequences is denoted by $\frown$. A set $\mathcal{T}$ of sequences of numbers is a *tree* if it is closed under subsequences. The empty sequence is the *root* of $\mathcal{T}$. Elements of $\mathcal{T}$ are also called *nodes.* A *branch* of $\mathcal{T}$ is a maximal linearly ordered subset of $\mathcal{T}$. The terminal node of a finite branch of $\mathcal{T}$ is a *leaf.*

Let $\varphi_0^{(n)}, \varphi_1^{(n)}, \varphi_2^{(n)}, \ldots$ be a fixed effective enumeration of all n-ary partial computable functions. If $X \subseteq \omega$, let $\varphi_0^{(n),X}, \varphi_1^{(n),X}, \varphi_2^{(n),X}, \ldots$ be a fixed effective enumeration of all n-ary X-computable functions. The superscripts are usually omitted for $n = 1$ or when it is clear from the context. φ_e (φ_e^X) is also denoted by $\{e\}$ ($\{e\}^X$), and e is called the Gödel number or index of φ_e. We write $\varphi_{e,s}(n) = m$ if $e, n, m < s$ and m is the output of $\varphi_e(n)$ after $< s$ steps in the corresponding computation. Let $W_e =_{\mathrm{def}} \mathrm{dom}(\varphi_e)$ and $W_{e,s} =_{\mathrm{def}} \mathrm{dom}(\varphi_{e,s})$. Thus, $W_0, W_1, W_2, \ldots$ is a computable enumeration of all c.e. sets. We fix $\langle \cdot , \cdot \rangle$ to be a computable bijection from ω^2 onto ω, which is strictly increasing with respect to both arguments. For $X \subseteq \omega$ and $i \in \omega$, we define $X^{[i]} = \{k : \langle k, i \rangle \in X\}$.

Let $X \subseteq \omega$ and $Y \subseteq \omega$. The join $X \oplus Y$ is

$$\{2n : n \in X\} \cup \{2n + 1 : n \in Y\}.$$

By $X \leqslant_T Y$ we denote that X is Turing reducible to Y. $X <_T Y$ denotes $X \leqslant_T Y$ but $Y \nleqslant_T X$. $X \equiv_T Y$ if $X \leqslant_T Y$ and $Y \leqslant_T X$. $\deg(X)$ denotes the Turing degree of X. Let $\mathbf{0} =_{\text{def}} \deg(\emptyset)$. If $\mathbf{x} = \deg(X)$ and $n \geqslant 1$, then $\mathbf{x}^{(n)} =_{\text{def}} \deg(X^{(n)})$, where $X^{(n)}$ is the n-th jump of X. Define $X^{(\omega)} = \{\langle k,n\rangle : k \in X^{(n)} \wedge k,n \in \omega\}$ and $\mathbf{x}^{(\omega)} = \deg(X^{(\omega)})$. A degree $\mathbf{x}$ is *low* if $\mathbf{x}' = \mathbf{0}'$. Turing degrees $\mathbf{x}$ and $\mathbf{y}$ form a *minimal pair* if they are nonzero and for every Turing degree $\mathbf{z}$,

$$(\mathbf{z} \leqslant \mathbf{x} \wedge \mathbf{z} \leqslant \mathbf{y}) \Longrightarrow \mathbf{z} = \mathbf{0}.$$

The set of all Turing degrees is denoted by $\mathcal{D}$. For more information on classical computability theory see [108, 160, 187, 196]. An ordinal is *computable* if it is finite or is the order type of a computable well-ordering on ω. The computable ordinals form a countable initial segment of the ordinals. Kleene's $\mathcal{O}$ is the set of notations for computable ordinals, with the corresponding partial ordering $<_{\mathcal{O}}$, see [187, 189]. The least non-computable ordinal is denoted by ω_1^{CK}, where CK stands for Church-Kleene. To obtain hyperarithmetic sets, we define the representative sets in the hyperarithmetic hierarchy, H_a for $a \in \mathcal{O}$. The definition is recursive, and is based on iterating the Turing jump:

$$H_1 = \emptyset,$$

$$H_{2^a} = (H_a)',$$

$$H_{3\cdot 5^e} = \{2^x \cdot 3^n : x \in H_{\{e\}(n)}\}.$$

A set of natural numbers X is *hyperarithmetic* if $(\exists a \in \mathcal{O})[X \leqslant_T H_a]$. The hyperarithmetic sets coincide with the Δ_1^1 sets.

3 Decidable Theories, and Computable and Decidable Models

Computable model theory explores the effectiveness of constructions and theorems in model theory, see [32, 44, 92, 188], and in universal algebra, see [37, 80, 125]. It begins by defining effective analogues of classical concepts of algebra and model theory. Three of its fundamental concepts are: decidable theories, computable models and decidable models. One of the basic problems is determining whether computable or decidable models satisfying certain conditions exist.

Definition 3.1

(i) A theory T is *decidable* if T is a computable set of sentences.

(ii) A model $\mathcal{A}$ is *computable* if its domain A is computable and its relations and functions are uniformly computable. That is, $\mathcal{A}$ is *computable* if A is computable and there is a computable enumeration $(a_i)_{i\in\omega}$ of A such that the atomic diagram of $\mathcal{A}$ is decidable.

(iii) A model $\mathcal{A}$ is *decidable* if A is computable and there is a computable enumeration $(a_i)_{i\in\omega}$ of A such that the complete diagram of $\mathcal{A}$ (that is, $\mathrm{Th}((\mathcal{A}, \mathbf{a}_i)_{i\in\omega})$) is decidable.

We assume that a formula is identified with its Gödel number, so a set of formulae is thought of as a subset of ω. Thus, a theory is decidable (resp. belongs to $\mathcal{P}$, where $\mathcal{P}$ is a complexity class) if the set of Gödel numbers of its sentences is computable (resp. belongs to $\mathcal{P}$). Hence, if Ax is a set of axioms of a theory T, then T is decidable if there is an algorithm which determines for every sentence σ of L, whether $Ax \vdash \sigma$. Clearly, a computably axiomatizable theory is computably enumerable. Hence a complete computably axiomatizable theory is decidable. In particular, a complete finitely axiomatizable theory is decidable. An example of such a theory is the theory of dense linear order. Peretyat'kin [167, 168, 170, 169, 171, 172, 173] has developed intricate methods for constructing finitely axiomatizable theories satisfying various additional properties. In [167], he constructed a complete, finitely axiomatizable, $\aleph_1$-categorical theory which is not $\aleph_0$-categorical. Well-known and important examples of decidable theories in mathematics include the theory of equality, the theory of unary predicates, the additive number theory, the theory of the field of real numbers, the theory of the field of complex numbers, the theory of algebraically closed fields, the theory of real-closed fields, the theory of p-adic fields, the theory of Boolean algebras, the theory of linear order, the theory of abelian groups, and the theory of free commutative algebras. Well-known and important examples of undecidable theories in mathematics include number theory, the theory of simple groups, the theory of semigroups, the theory of rings, the theory of fields, the theory of distributive lattices, and the theory of partial order. For more information on decidable and undecidable theories see [58] and Part III in [149]. For computability-theoretic complexity of various sets of sentences satisfied in certain classes of models see [201].

A model $\mathcal{A}$ is *computable* if A is computable, and if there is a computable enumeration $(a_i)_{i\in\omega}$ of A and an algorithm which determines, for every quantifier-free formula $\theta(x_0, \dots, x_{n-1})$ and every sequence $(a_{i_0}, \dots, a_{i_{n-1}}) \in A^n$, whether $\mathcal{A}_A \models \theta(\mathbf{a}_{i_0}, \dots, \mathbf{a}_{i_{n-1}})$. A model $\mathcal{A}$ is *decidable* if A is computable and there is a computable enumeration $(a_i)_{i\in\omega}$ of A and an algorithm which determines for every formula $\theta(x_0, \dots, x_{n-1})$ and every sequence $(a_{i_0}, \dots, a_{i_{n-1}}) \in A^n$, whether $\mathcal{A}_A \models \theta(\mathbf{a}_{i_0}, \dots, \mathbf{a}_{i_{n-1}})$. Clearly, every decidable model is computable. The converse is not true. For example, $(\omega, +, \times)$ is a computable model which is not decidable (by Gödel's incompleteness theorem [64]). Peretyat'kin [162] has constructed a decidable linear order without a computable proper elementary extension. In [159], Nurtazin characterized decidable models which are isomorphic to computable non-decidable models. Peretyat'kin [165] has shown that there is a complete decidable theory T which is neither $\aleph_0$-categorical nor $\aleph_1$-categorical, and which has, up to isomorphism, a unique decidable model. Moreover, all computable models of T are decidable.

A model is *computably presentable* if it is isomorphic to a computable model. Goncharov [67] has constructed an $\aleph_1$–categorical theory which is not $\aleph_0$–categorical and whose only computably presentable model is the prime model. On the other hand, Khoussainov, Nies and Shore [104] have shown that there is an $\aleph_1$–categorical theory which is not $\aleph_0$–categorical and whose only countable non-computably presentable model is the prime model. It is sometimes convenient to call a model computable (decidable, etc.) even if it is only computably (decidably, etc.) presentable.

Morozov [151, 152, 153] has extensively studied the automorphisms of computable models. He constructed a decidable model [151] whose theory is $\aleph_0$–categorical and which does not have non-trivial computable automorphisms. He also constructed a computable model [152] with $2^{\aleph_0}$ many automorphisms and without a non-trivial hyperarithmetic automorphism.

The notion of a computable (resp. decidable) model corresponds to the notion of a *constructive* (resp. *strongly constructive*) model used by the group in Novosibirsk. A constructive (resp. strongly constructive) model is a pair $(\mathcal{A}, \nu)$, where $\mathcal{A}$ is a countable model, and ν is a function from ω onto the domain of $\mathcal{A}$, such that the model "induced on ω by $\mathcal{A}$ via ν^{-1}" is computable (resp. decidable). ν is called a *constructivization* (resp. *strong constructivization*) of $\mathcal{A}$. For example, the field of rational numbers has a constructivization, while the group of all computable permutations of ω does not.

In general, the *Turing degree* of a model $\mathcal{A}$ with finite language is the least upper bound of the Turing degrees of its universe, and its relations and functions. Hence a model is computable if its Turing degree is zero. Isomorphic models may have different Turing degrees. Tennenbaum [199] has proved that there is no computable nonstandard model of Peano arithmetic. Scott and Tennenbaum [193] have established that every degree $\mathbf{d}$ such that $\mathbf{d} > \mathbf{0}'$ is a degree of a complete extension of Peano arithmetic, and that no computably enumerable degree $\mathbf{d}$ such that $\mathbf{d} < \mathbf{0}'$ can be a degree of a complete extension of Peano arithmetic. Jockusch and Soare [95] have shown that there is a nonstandard model of Peano arithmetic of low degree. Jockusch and Soare [96] have proved that for every non-zero c.e. degree $\mathbf{d}$, there is a linear order of degree $\mathbf{d}$ which is not isomorphic to any computable linear order. Lerman and Schmerl [121] have given a number of examples of important theories with computable models.

By a *theory of linear order* we mean a theory whose language consists of a binary relation symbol, and which contains the axioms of linear order. Lerman and Schmerl [121] have extended Peretyat'kin's [164] result that every c.e. (Σ_1^0) theory of linear order has a computable model, by showing that every Σ_2^0 theory of linear order has a computable model. They have also constructed a Δ_3^0 theory of linear order without a computable model. Lerman and Schmerl have further shown that if $\mathbf{x}$ is a Turing degree such that $\mathbf{x} \not\leq \mathbf{0}''$, then there is a theory of linear order of degree $\mathbf{x}$ without a computable model.

Definition 3.2 (Millar [143]) Let $\mathcal{P}$ be a class of theories. A theory T is *persistently* $\mathcal{P}$ if for every $n \in \omega$, for every complete n–type $\Gamma(x_0, \ldots, x_{n-1})$ of T and a sequence $c_0, \ldots, c_{n-1}$ of new constants, the theory $\Gamma(c_0, \ldots, c_{n-1})$ belongs to $\mathcal{P}$.

In [55], Ershov has studied persistently $\forall$*–finitely axiomatizable* theories. A theory T is $\forall$–finitely axiomatizable if for every theory S extending T, $S_\forall$ is finitely axiomatizable. For examples of persistently $\forall$–finitely axiomatizable theories see [55, 100]. Ershov [55] has established that every c.e. theory extending a persistently $\forall$–finitely axiomatizable theory has a computable model. This result implies the previously mentioned result that every c.e. theory of linear order has a computable model. It also implies that every c.e. (Σ_1^0) theory of trees has a computable model. By a *theory of trees* we mean a theory whose language consists of a binary relation symbol, and which contains the axioms of a partially ordered set such that the set of all predecessors of any element is linearly ordered. Lerman and Schmerl [121] have constructed a Δ_2^0 theory of trees without a computable model. Lerman and Schmerl have further shown that for every Turing degree $\mathbf{x}$ such that $\mathbf{x} \not< \mathbf{0}'$, there is a complete theory of trees of degree $\mathbf{x}$ without a computable model.

Lerman and Schmerl [121] have proved that if T is an arithmetic $\aleph_0$-categorical theory such that for every $n \in \omega$, the set of all Σ_{n+2}^0 sentences in T is a Σ_{n+1}^0 set, then T has a computable model. They have also shown that for every $n \in \omega$, and a Turing degree $\mathbf{x}$ such that $\mathbf{x} \not\leqslant \mathbf{0}^{(n)}$, there is an $\aleph_0$-categorical theory T of degree $\mathbf{x}$ such that the set of all Σ_{n+1}^0 sentences in T is computable and T does not have a computable model. In particular, for every Turing degree $\mathbf{x}$, there is an $\aleph_0$-categorical theory of degree $\mathbf{x}$ such that the set of all existential sentences in T is computable and every model of T has the degree $\geqslant \mathbf{x}$.

Feldman [60, 61] has constructed a complete decidable $\aleph_0$-categorical theory T of a partial order with the greatest lower bound operator. T has a decidable model in which every countable lower semilattice can be embedded. Knight [109] has constructed a complete, decidable, superstable theory T with $2^{\aleph_0}$ many types, such that no independent sequence of formulae (with respect to T) is computable in a type of T. A sequence $(\sigma_n(\overline{x}))_{n\in\omega}$ of formulae in $L(T)$ is independent with respect to T if for every $\alpha \in 2^{<\omega}$,

$$T \vdash (\exists \overline{x})[\bigwedge_{\alpha(n)=1} \sigma_n(\overline{x}) \wedge \bigwedge_{\alpha(n)=0} \neg\sigma_n(\overline{x})].$$

Hurlburt [93] has given some general conditions which are sufficient to construct computable models for highly non-decidable theories.

According to the Ryll-Nardzewski theorem, a complete theory T is $\aleph_0$-categorical if and only if for every $n \in \omega$, the set of all n-types of T is finite. For such a theory T, the function which assigns to every n the number of all n-types of T is called *Ryll-Nardzewski function.* Schmerl [191], Herrmann [90] and Venning [202] have proved independently that a complete decidable $\aleph_0$-categorical theory does not necessarily have a computable Ryll-Nardzewski function. More generally, the following relativized result holds.

Theorem 3.1 (Schmerl [191]) *For every Turing degree* $\mathbf{x}$, *there is a function* $f : \omega \to \omega$ *of degree* $\mathbf{x}$ *such that for every Turing degree* $\mathbf{y}$ *with the property that* $\mathbf{x}$ *is c.e. in* $\mathbf{y}$, *there is a complete* $\aleph_0$*-categorical theory of degree* $\mathbf{y}$ *(in a language consisting of one binary relation symbol) whose Ryll-Nardzewski function is* f.

We can assume that the characteristic function of a consistent set $\Gamma(\overline{x})$ of formulae in L is a function $\chi : \omega \to \{0, 1\}$, defined by:

$$\chi_{\Gamma(\overline{x})}(k) = \begin{cases} 1 & \text{if } \theta_k(\overline{x}) \in \Gamma(\overline{x}), \\ 0 & \text{if } \theta_k(\overline{x}) \notin \Gamma(\overline{x}), \end{cases}$$

where $\theta_0(\overline{x}), \theta_1(\overline{x}), \theta_2(\overline{x}), \ldots$ is an effective enumeration of all formulae in L whose free variables are among those in $\overline{x}$. The set $\Gamma(\overline{x})$ is computable if its characteristic function is computable. Equivalently, $\Gamma(\overline{x})$ is computable if the set $\{n : \theta_n(\overline{x}) \in \Gamma(\overline{x})\}$ is computable.

Proposition 3.2 *Every type realized in a decidable model is computable.*

Proof. Let $\mathcal{A}$ be a decidable model such that a type $\Gamma(x_0, \ldots, x_{n-1})$ of $\mathrm{Th}(\mathcal{A})$ is realized in $\mathcal{A}$ by some $a_0, \ldots, a_{n-1} \in A$. Since $\mathcal{A}$ is decidable and

$$\gamma(x_0, \ldots, x_{n-1}) \in \Gamma \iff \mathcal{A}_A \models \gamma(\mathbf{a}_0, \ldots, \mathbf{a}_{n-1}),$$

Γ must be computable. □

A set of codes of a set of computable (complete) types of a theory T is a set of Gödel numbers of characteristic functions (which are computable) of these types, containing at least one index for each type. We say that a set of computable types belongs to $\mathcal{P}$, where $\mathcal{P}$ is a complexity class, if it has a set of codes which belongs to $\mathcal{P}$. The following proposition follows from a more general proposition in the theory of enumerations (see Chapter VI of [57]).

Proposition 3.3 *Every Σ^0_{n+1} set of codes of a set of computable types of a theory T is a Π^0_n set of codes.*

Hence, every c.e. set of codes of a set of computable types is a computable set of codes. To determine the complexity of the set of types realized in a decidable model, we need from *computability theory* the *s-m-n* theorem.

Theorem 3.4

(i) (*s-m-n* theorem) *For every $m, n \geqslant 1$, there is an $(m+1)$-ary computable function, denoted by s^m_n, such that*

$$\varphi_e^{(m+n)}(l_1, \ldots, l_m, k_1, \ldots, k_n) = \varphi^{(n)}_{s^m_n(e, l_1, \ldots, l_m)}(k_1, \ldots, k_n),$$

where $e, l_1, \ldots, l_m, k_1, \ldots, k_n \in \omega$.

(ii) (Relativized s-m-n theorem) *For every $m, n \geqslant 1$ and every oracle $X \subseteq \omega$, there is an $(m+1)$-ary computable function, denoted by s_n^m, such that*

$$\varphi_e^{(m+n),X}(l_1, \dots, l_m, k_1, \dots, k_n) = \varphi_{s_n^m(e,l_1,\dots,l_m)}^{(n),X}(k_1, \dots, k_n),$$

where $e, l_1, \dots, l_m, k_1, \dots, k_n \in \omega$.

Proposition 3.5 *The set of all types of T realized in a decidable model of T is computable.*

Proof. Let $\mathcal{A}$ be a decidable model of T and let $a_0, a_1, a_2, \dots$ be an effective enumeration of A. Choose $g : A^{<\omega} \to \omega$ to be a computable bijection. Define a computable function $h : \omega^2 \to \{0, 1\}$ by:

$$h(n, k) = \begin{cases} 1 & \text{if } \mathcal{A} \models \theta_k[\overline{a}], \\ 0 & \text{if } \mathcal{A} \not\models \theta_k[\overline{a}], \end{cases}$$

where $g(\overline{a}) = n$, and $\theta_0, \theta_1, \theta_2, \dots$ is an effective enumeration of all formulae of $L(T)$ whose free variables are among $\overline{x} = (x_{i_0}, \dots, x_{i_{l-1}})$, corresponding to $\overline{a} = (a_{i_0}, \dots, a_{i_{l-1}})$. By the s-m-n theorem, $h(n, k) = \varphi_{f(n)}(k)$ for some computable function f. Clearly, $\{f(n) : n \in \omega\}$ is a c.e. set which is a set of codes of all (computable) types of T realized in $\mathcal{A}$. □

Proposition 3.6 *Let T be a decidable theory.*

(i) *The set of all Gödel numbers of all computable types of T is a Π_2^0 set.*

(ii) *Every principal type of T is a computable type, and the set of all principal types of T is a Π_1^0 set.*

Proof. For a given sequence $\overline{x}$ of variables, let $\theta_0(\overline{x}), \theta_1(\overline{x}), \theta_2(\overline{x}), \dots$ be a computable enumeration of all formulae of $L(T)$ with all free variables contained in $\operatorname{ran}(\overline{x})$.

(i) For $e \in \omega$, φ_e is the characteristic function of a computable type of T in variables $\overline{x}$ if and only if

$$\forall n \exists s \forall j \leqslant n \exists k_j \in \{0, 1\}[\varphi_{e,s}(j) \downarrow = k_j \wedge T \vdash \exists \overline{x}(\bigwedge\{\theta_j^{k_j}(\overline{x}) : j \leqslant n\})].$$

(ii) Every principal type of T is computable because it is generated by a complete formula. For every $i \in \omega$, use $\emptyset'$ to determine whether $\theta_i(\overline{x})$ is a complete formula. That is, $\theta_i(\overline{x})$ is a complete formula if and only if

$$\forall j \exists k \in \{0,1\}[T \vdash \forall \overline{x}(\theta_i(\overline{x}) \Rightarrow \theta_j^k(\overline{x}))].$$

Hence by the relativized *s-m-n* theorem, we can enumerate with oracle $\emptyset'$:

The principal type that $\theta_i(\overline{x})$ generates, if $\theta_i(\overline{x})$ is a complete formula;

Any fixed principal type of T, if $\theta_i(\overline{x})$ is not a complete formula.

Thus, since the sets which are computably enumerable in $\emptyset'$ are Σ_2^0, it follows that the set of all principal types is Σ_2^0. □

Proposition 3.7 (Millar [134]) *Every Σ_2^0 set of computable types of a decidable theory T is contained in a computable set of computable types of T.*

Nerode and his collaborators have also initiated the study of the lattice of all computably enumerable submodels of a computable model. Models whose computably enumerable submodels have been investigated include vector spaces, fields, Boolean algebras, and linear orders. For more information see [6, 30, 45, 47, 48, 81, 82, 156, 158].

Moses [154] has generalized the concepts of computable and decidable models to "Γ–computably enumerable models", where Γ is a computably enumerable set of formulae. For such a set Γ, a model $\mathcal{A}$ for $L(\Gamma)$ is Γ–computably enumerable if the universe of $\mathcal{A}$ is computable, and its satisfaction predicate restricted to Γ is computably enumerable. For other notions of an "effective model" and of an "effective isomorphism", see [180] and [50].

4 Effective Completeness Theorem

One of the major tasks of computable model theory is to obtain effective versions of or effective counterexamples to various classical model-theoretic results. To obtain an effective version of the completeness theorem, we use from *model theory*, Henkin's method of constructing models; and from *computability theory*, the notion of a computable set and Church's thesis.

Theorem 4.1 (Effective Completeness Theorem) *A decidable theory has a decidable model.*

Proof. Let T be a decidable theory. A corresponding model of T will be obtained in an effective way by Henkin's method. Let $c_0, c_1, c_2, \dots$ be an effective one-to-one enumeration of an infinite set C of new constants. Let $\sigma_0, \sigma_1, \sigma_2, \dots$ be an effective enumeration of all sentences in $L(T) \cup C$. We will construct effectively, by induction, a complete theory Ψ in $L(T) \cup C$ such that $\Psi \supseteq T$. Ψ will be the complete diagram of a model $\mathcal{A}_A$, where $\mathcal{A}$ is a desired model for T. As usual, the domain A consists of the equivalence classes of the constants in C, where two constants $c, d \in C$ are equivalent if and only if $(c = d) \in \Psi$. We will arrange that $\Psi = \{\delta_0, \delta_1, \delta_2, \dots\}$, where δ_s is defined at stage s. For $s > 0$, let $\psi^s =_{\text{def}} \delta_0 \wedge \delta_1 \wedge \dots \wedge \delta_{s-1}$.

Construction

STAGE 0:

Let $\delta_0 =_{\text{def}} (c_0 = c_0)$.

STAGE $s = 2e + 1$ for $e \in \omega$ (Henkin's witnesses requirement):

If δ_e is of the form $\delta_e = \exists x \theta(x)$, we effectively find the least i such that c_i does not occur in ψ^s and let $\delta_s =_{\text{def}} \theta(c_i)$. Otherwise, let $\delta_s =_{\text{def}} (c_0 = c_0)$.

STAGE $s = 2e + 2$ for $e \in \omega$ (Completeness of the diagram requirement):

Let $\bar{c}$ be a sequence of all constants in C which occur in $(\psi^s \Rightarrow \sigma_e)$. Let $\bar{x}$ be the first sequence of variables of the same length as $\bar{c}$ (in some fixed effective enumeration of the finite sequences of all variables) which do not occur in $(\psi^s \Rightarrow \sigma_e)$. We effectively check whether

$$T \vdash \forall \bar{x}[(\psi^s \Rightarrow \sigma_e)(\bar{x}/\bar{c})]. \qquad (*)$$

If this is true, let $\delta_s =_{\text{def}} \sigma_e$. Otherwise, let $\delta_s =_{\text{def}} \neg\sigma_e$. End of the construction.

Condition $(*)$ can be verified effectively because T is a decidable theory. We describe the action at stage $2e + 1$ as effectively providing a Henkin's witness for δ_e, and the action at stage $2e+2$ as effectively satisfying the e–th completeness requirement. □

Proposition 4.2 (Millar [134]) *Every computable type of a theory T is realized in some decidable model of T.*

Proof. Assume that $\Gamma = \Gamma(x_0, \dots, x_{n-1})$ is a computable type of a theory T. Let $c_0, \dots, c_{n-1}$ be constants which do not occur in Γ. $T \cup \Gamma(c_0, \dots, c_{n-1})$ is a complete decidable theory in $L(T) \cup \{c_0, \dots, c_{n-1}\}$, so it has a decidable model $\mathcal{A}$. The reduct of $\mathcal{A}$ to $L(T)$ is a decidable model of T realizing Γ. □

5 Model Completeness and Decidability

Many examples of decidable theories constructed to illustrate certain model-theoretic or computability-theoretic properties are obtained as model completions of universal theories, which allow the elimination of quantifiers.

Definition 5.1 A theory T is *model complete* if for any two models $\mathcal{A}$ and $\mathcal{B}$ of T,

$$\mathcal{A} \subseteq \mathcal{B} \Rightarrow \mathcal{A} \preceq \mathcal{B}.$$

Neither one of completeness and model completeness implies the other.

Theorem 5.1 *A theory T in a language L is model complete*

$\iff$ *For every $\mathcal{A} \models T$, the theory $T \cup \Delta_{\mathcal{A}}$ is complete in L_A.*

$\iff$ *If $\mathcal{A}$ and $\mathcal{B}$ are models of T and $\mathcal{A} \subseteq \mathcal{B}$, then every existential sentence of L_A true in $\mathcal{B}_A$ is also true in $\mathcal{A}_A$.*

$\iff$ *For every formula $\theta(\overline{x})$, there is a universal formula $\psi(\overline{x})$ such that*

$$T \vdash \forall \overline{x}[\theta(\overline{x}) \Leftrightarrow \psi(\overline{x})].$$

Definition 5.2 T is a *model completion* of a theory T' if

$(\forall \mathcal{A} \models T)(\mathcal{A} \models T')$,

$(\forall \mathcal{A} \models T')(\exists \mathcal{B} \models T)[\mathcal{A} \subseteq \mathcal{B}]$, and

$(\forall \mathcal{D} \models T')(\forall \mathcal{A}, \mathcal{B} \models T)[(\mathcal{D} \subseteq \mathcal{A} \wedge \mathcal{D} \subseteq \mathcal{B}) \Rightarrow \mathcal{A}_D \equiv \mathcal{B}_D]$.

A model completion of a theory is a model complete theory.

Theorem 5.2 (Robinson) *If T_1 and T_2 are model completions of T', then $T_1 = T_2$.*

A theory T^* is a *model companion* of a theory T if T^* is model complete and $T_\forall = T^*_\forall$. For example, the theory of atomless Boolean algebras is a model companion of the theory of Boolean algebras, and the theory of algebraically closed fields is a model companion of the theory of fields. Both the theory of atomless Boolean algebras and the theory of algebraically closed fields are decidable. Burris [29] has established some general criteria for a model companion to be decidable.

Definition 5.3 T is *submodel complete* if for every model $\mathcal{B}$ of T and every $\mathcal{A} \subseteq \mathcal{B}$, the theory $T \cup \Delta_\mathcal{A}$ is complete in L_A.

Hence a submodel complete theory is both complete and model complete.

Theorem 5.3 (Robinson) *A model completion of a universal theory is submodel complete.*

We say that T *admits the elimination of quantifiers* if for every formula $\theta(x_0, \ldots, x_{n-1})$, there is a quantifier-free formula $\psi(x_0, \ldots, x_{n-1})$ such that

$$T \vdash \forall x_0, \ldots, x_{n-1}[\theta(x_0, \ldots, x_{n-1}) \Leftrightarrow \psi(x_0, \ldots, x_{n-1})].$$

If there is an algorithm which for every formula $\theta(\overline{x})$ finds the corresponding quantifier-free formula $\psi(x)$, then we say that T *effectively admits the elimination of quantifiers.*

Proposition 5.4

(i) *Let T be a theory which effectively admits the elimination of quantifiers. Then every computable model of T is a decidable model of T.*

(ii) *Let T be a computably enumerable theory which admits the elimination of quantifiers. Then every computable model of T is a decidable model of T.*

Proof.

(i) The statement follows immediately from the definitions of a computable and of a decidable model.

(ii) The statement follows from (i) because if T is a computably enumerable theory which admits the elimination of quantifiers, then T effectively admits the elimination of quantifiers. □

Theorem 5.5 *A theory T is submodel complete*

$\iff$ $(\forall \mathcal{A}, \mathcal{B} \models T)(\forall \mathcal{D} \subseteq \mathcal{A}, \mathcal{B})$ [$\mathcal{A}$ *and* $\mathcal{B}$ *satisfy the same existential sentences in* $L(T)$ *with parameters from* D]

$\iff$ T *admits the elimination of quantifiers.*

Millar has characterized universal theories which have decidable model completions, thus providing a uniform approach for producing specific examples of decidable theories.

To state this characterization, we fix a language L and let $\theta_0, \theta_1, \theta_2, \ldots$ be an effective enumeration of all quantifier-free formulae of L in variables $x_0, x_1, x_2, \ldots; y_0, y_1, y_2, \ldots$. The convention will be that if the free variables of a formula are displayed, then the free x–variables (if any) are displayed before the free y–variables (if any).

Theorem 5.6 (Millar [137]) *Assume that T' is a universal theory in L. T' has a (complete) decidable model completion if and only if there is a unary computable function f such that for every $i \in \omega$, $\theta_{f(i)}$ does not contain any y–variable and for all $i, j \in \omega$:*

(i) (θ_i *is inconsistent with* T') $\Leftrightarrow \theta_{f(i)} = \neg(x_0 = x_0)$,

(ii) $T' \vdash \forall \overline{x}[\exists \overline{y} \theta_i(\overline{x}, \overline{y}) \Rightarrow \theta_{f(i)}(\overline{x})]$,

(iii) *If θ_i does not contain any x–variable and is consistent with T', then* $\theta_{f(i)} = (x_0 = x_0)$,

(iv) ($T' \cup \{\theta_{f(i)}(\overline{x}), \theta_j(\overline{x}, \overline{y}*)\}$ *is consistent*)
$\Longrightarrow$ ($T' \cup \{\theta_i(\overline{x}, \overline{y}), \theta_j(\overline{x}, \overline{y}*)\}$ *is consistent*),

where $\mathrm{rng}(\overline{y}) \cap \mathrm{rng}(\overline{y}*) = \emptyset$.

Notice that, by (ii), the implication in (iv) can be replaced by the equivalence. Property (iv) is often called the *amalgamation property.*

Proof. Assume that T is a decidable model completion of T'. By Theorem 5.3, T admits the elimination of quantifiers. Thus, there is a unary computable function f which has the following properties:

(a) (θ_i is inconsistent with T') $\Leftrightarrow \theta_{f(i)} = \neg(x_0 = x_0)$;

(b) If θ_i is consistent with T and does not contain an x–variable, then $\theta_{f(i)} = (x_0 = x_0)$;

(c) If θ_i is consistent with T, then $T \vdash \forall\overline{x}[\exists\overline{y}\theta_i(\overline{x},\overline{y}) \Leftrightarrow \theta_{f(i)}(\overline{x})]$.

Clearly, (i) and (iii) are satisfied.

Let us prove (ii). Assume otherwise. If follows that

$$T' \cup \{\exists\overline{x}\exists\overline{y}[\theta_i(\overline{x},\overline{y}) \wedge \neg\theta_{f(i)}(\overline{x})]\}$$

is consistent. Hence it has a model $\mathcal{A}$. Since T is a model completion of T', there is a model $\mathcal{B}$ of T such that $\mathcal{A} \subseteq \mathcal{B}$. Hence $\mathcal{B}$ is a model of

$$T \cup \{\exists\overline{x}\exists\overline{y}[\theta_i(\overline{x},\overline{y}) \wedge \neg\theta_{f(i)}(\overline{x})]\}.$$

Since T is complete, we have

$$T \vdash \exists\overline{x}\exists\overline{y}[\theta_i(\overline{x},\overline{y}) \wedge \neg\theta_{f(i)}(\overline{x})].$$

That is,

$$T \vdash \neg\forall\overline{x}[\forall\overline{y}\neg\theta_i(\overline{x},\overline{y}) \vee \theta_{f(i)}(\overline{x})]$$

or, equivalently,

$$T \vdash \neg\forall\overline{x}[\exists\overline{y}\theta_i(\overline{x},\overline{y}) \Rightarrow \theta_{f(i)}(\overline{x})],$$

thus contradicting (c). Hence (ii) holds.

Finally, let us prove (iv). Assume that $\mathrm{rng}(\overline{y}) \cap \mathrm{rng}(\overline{y}\star) = \emptyset$. Let

$$T' \cup \{\exists\overline{x}\exists\overline{y}\star\,[\theta_{f(i)}(\overline{x}) \wedge \theta_j(\overline{x},\overline{y}\star)]\}$$

be consistent. By the same argument as in the proof of (ii), we conclude that $T \vdash \exists\overline{x}\exists\overline{y}\star\,[\theta_{f(i)}(\overline{x}) \wedge \theta_j(\overline{x},\overline{y}\star)]$. Hence, by (c),

$$T \vdash \exists\overline{x}\exists\overline{y}\star\,[\exists\overline{y}\theta_i(\overline{x},\overline{y}) \wedge \theta_j(\overline{x},\overline{y}\star)].$$

That is, $T \cup \{\theta_i(\overline{x},\overline{y}), \theta_j(\overline{x},\overline{y}\star)\}$ is consistent. Thus, since T is a model completion of T', $T' \cup \{\theta_i(\overline{x},\overline{y}), \theta_j(\overline{x},\overline{y}\star)\}$ is consistent.

To prove the converse, we assume that a universal theory T' and a unary computable function f satisfy (i)-(iv). Let T be obtained by adding to T' the following two sets of axioms:

Ax I $\forall\overline{x}\forall\overline{y}\neg\theta_i(\overline{x},\overline{y})$ for all $i \in \omega$ such that $\theta_{f(i)} = \neg(x_0 = x_0)$;

Ax II $\forall \overline{x}[\theta_{f(i)}(\overline{x}) \Rightarrow \exists \overline{y}\theta_i(\overline{x}, \overline{y})]$ for all $i \in \omega$ such that $\theta_{f(i)} \neq \neg(x_0 = x_0)$.

Clearly $T \supseteq T'$. We will show that T is a decidable model completion of T'.

Lemma 5.7 *T is consistent.*

Proof. We will prove that the union of T' and the two sets of axioms is consistent. If σ is an axiom in Ax I, then $T' \vdash \sigma$ by (i). Therefore, by the compactness argument, it is enough to prove that for every finite set S of axioms in Ax II, $T' \cup S$ is consistent. Let

$$S = \{\forall \overline{x} \exists \overline{y}[\theta_{f(i_k)}(\overline{x}) \Rightarrow \theta_{i_k}(\overline{x}, \overline{y})] : 0 \leqslant k \leqslant n-1\}$$

for some $n \geqslant 1$. We will construct a model $\mathcal{A}$ for $T' \cup S$ by Henkin's method.

We choose an infinite set C of new constants. For each $k \in \{0, \ldots, n-1\}$, let C_k be an enumeration of all sequences of elements in C of the same length as the length of $\overline{x}$ in $\theta_{i_k}(\overline{x}, \overline{y})$, such that every such sequence appears in C_k infinitely often. Let $\sigma_0, \sigma_1, \sigma_2, \ldots$ be an enumeration of all sentences in $L(T') \cup C$. We will construct the complete diagram Ψ of $\mathcal{A}_A$, where A will consist of the equivalence classes of the constants in C. We will arrange that $\Psi = \{\delta_0, \delta_1, \delta_2, \ldots\}$, where δ_s is defined at stage s. For $s > 0$, let $\Psi^s = \{\delta_0, \delta_1, \ldots, \delta_{s-1}\}$ and let ψ^s be $\wedge\Psi^s$.

Construction

STAGE 0:

Let $\delta_0 =_{\text{def}} (c_0 = c_0)$.

STAGE $s = (n+2)e$ for $e \geqslant 1$:

Satisfy the $(e-1)$–st completeness of the diagram requirement.

STAGE $s = (n+2)e + 1$ for $e \in \omega$:

Provide a Henkin's witness for δ_e.

STAGE $s = (n+2)e + k$ for $e \in \omega$ and $k \in \{2, \ldots, n+1\}$:

Let C_{k-2} be $\overline{c}_0, \overline{c}_1, \overline{c}_2, \ldots$. If $\theta_{f(i_{k-2})}(\overline{c}_e) \notin \Psi^s$, then $\delta_s =_{\text{def}} (c_0 = c_0)$. If $\theta_{f(i_{k-2})}(\overline{c}_e) \in \Psi^s$, then $\delta_s =_{\text{def}} \theta_{i_{k-2}}(\overline{c}_e, \overline{c})$, where $\overline{c}$ is a sequence of constants in C which do not occur in ψ^s such that $\overline{c}$ is of the same length as $\overline{y}$ in $\theta_{i_{k-2}}(\overline{x}, \overline{y})$. End of the construction.

We can prove inductively that for every $s \in \omega$, $T' \cup \Psi^s$ is consistent. In the proof, we use property (iv) at stages of the form $(n+2)\,e + k$ for $k \in \{2, \ldots, n+1\}$. Hence Ψ is consistent. The corresponding model $\mathcal{A}$ satisfies $T' \cup S$. □

Lemma 5.8 *Every model of T' can be isomorphically embedded in a model of T.*

Proof. Assume that $\mathcal{B}$ is a model of T'. Let Ω be the atomic diagram of $\mathcal{B}$. To prove that there is a model for $\Omega \cup T$, we use the same argument as in the proof of Lemma 5.7 to construct a model for $\Omega \cup S$, where S is a finite set of axioms in Ax II. □

Lemma 5.9 $(\forall \mathcal{A}, \mathcal{B} \models T)(\forall \mathcal{D} \subseteq \mathcal{A}, \mathcal{B})$ [*$\mathcal{A}$ and $\mathcal{B}$ satisfy the same existential sentences in $L(T)$ with parameters from D*]

Proof. If follows from (ii) and Ax II that T admits the elimination of quantifiers, which is equivalent to this statement. □

Although the following lemma follows from Theorem 5.3, we also give an easy direct proof.

Lemma 5.10 *T is complete.*

Proof. Let σ be a sentence in L such that $T \cup \{\sigma\}$ is consistent. Since T admits the elimination of quantifiers, the formula $\sigma \wedge (y_0 = y_0)$ is T–equivalent to $\theta_i = \theta_i(y_0)$ for some $i \in \omega$. By (iii), $\theta_{f(i)} = (x_0 = x_0)$. By the definition of axioms in Ax II, $T \vdash \forall x_0[x_0 = x_0 \Rightarrow \exists y_0(\sigma \wedge (y_0 = y_0))]$. Hence $T \vdash \sigma$. □

T is decidable because it is complete and the given sets of axioms are computable. T is a model completion of T' by Lemma 5.7, Lemma 5.8 and Lemma 5.9. □

6 Omitting Types and Decidability

Let Γ be a nonprincipal type of a complete theory T. Then there is a countable model $\mathcal{A}$ of T which omits Γ. However, $\mathcal{A}$ does not have to be computable even if Γ is. The following theorem shows that if T is decidable and Γ is computable, then Γ is omitted in some decidable model of T.

Theorem 6.1 *Let Γ be a computable nonprincipal type of a complete decidable theory T. There is a decidable model of T which omits Γ.*

Proof. Without loss of generality, we assume that Γ is a 1–type, $\Gamma(x)$. Let C, $(\sigma_i)_{i\in\omega}$, $\Psi = \{\delta_0, \delta_1, \delta_2, \ldots\}$, ψ^s and $\mathcal{A}$ be as in the proof of Theorem 4.1.

Construction

STAGE 0:
Let $\delta_0 =_{\text{def}} (c_0 = c_0)$.

STAGE $s = 3e + 1$ for $e \in \omega$:
We effectively provide a Henkin's witness for δ_e.

STAGE $s = 3e + 2$ for $e \in \omega$ (Omitting the types requirement):
Let ψ^s be of the form $\psi^s(c_e, \bar{c})$, where c_e does not occur in $\bar{c}$. We effectively find the first formula $\gamma(x) \in \Gamma$ such that

$$(\circ) \quad T \nvdash \forall z[\exists \bar{y}\psi^s(z, \bar{y}) \Rightarrow \gamma(z)],$$

where $(z, \bar{y})$ is an appropriate effectively chosen sequence of new variables. Let $\delta_s =_{\text{def}} \neg\gamma(c_e)$.

STAGE $s = 3e + 3$ for $e \in \omega$:
We effectively satisfy the e–th completeness of the diagram requirement. End of the construction.

At stage $3e + 2$, the corresponding formula γ exists because Γ is a nonprincipal type and, by the construction, $T \cup \{\exists z \exists \bar{y}\psi^s(z, \bar{y})\}$ is a consistent set. Condition $(\circ)$ can be verified effectively because T is a decidable theory.

Stage $3e + 2$ guarantees that the interpretation of c_e in $\mathcal{A}$ does not realize Γ. Since every element in the domain of $\mathcal{A}$ is the interpretation of some constant in C, $\mathcal{A}$ omits Γ.

Clearly, for an arbitrary n–type Γ, stage $3e + 2$ should be modified so that instead of $(c_i)_{i\in\omega}$ some effective enumeration of all n–tuples of elements of C is considered. □

A *partial type* of T is a subset of a (complete) type of T. Millar has established the following general result.

Theorem 6.2 (Effective Omitting Types Theorem, Millar [140]) *Let T be a complete decidable theory. If Φ_1 is a Σ^0_2 set of computable nonprincipal partial types of T, and Φ_2 is a Σ^0_2 set of computable types of T, then there is a decidable model of T which omits all types in Φ_1 and all nonprincipal types in Φ_2.*

The completeness of types in Φ_2 plays an important role in Theorem 6.2, as demonstrated by the next theorem.

Theorem 6.3 (Millar [140]) *There is a complete decidable theory T and a computable set Φ of computable partial types of T such that no decidable model of T omits all nonprincipal types in Φ.*

The following two theorems can be obtained using the Effective Omitting Types Theorem.

Theorem 6.4 (Millar [140]) *Let T be a complete decidable theory without a decidable prime model. There are infinitely many distinct decidable models of T such that the set of all types realized in any two of these models simultaneously is exactly the set of all principal types of T.*

Theorem 6.5 (Millar [140]) *Let T be a complete decidable theory and let Φ be a Σ_2^0 set of computable nonprincipal types of T. Assume that for every decidable model $\mathcal{A}$ of T which omits Φ, for every finite $X \subseteq A$, $\mathcal{A}_X$ is not a prime model. Then there are $2^{\aleph_0}$ distinct type spectra of decidable models of T which omit Φ.*

7 Decidable Prime Models

Definition 7.1 Let $\mathcal{U}$ be an arbitrary (possibly uncountable) model. $\mathcal{U}$ is *atomic* if every n–tuple of elements of U satisfies a complete formula in the theory of $\mathcal{U}$.

Proposition 7.1 *Let T be a complete theory in at most countable language.*

(i) *A countable model $\mathcal{A}$ of T is prime if and only if $\mathcal{A}$ is atomic.*

(ii) *T has a prime model if and only if every formula consistent with T is a member of a principal type of T.*

Definition 7.2 An arbitrary model $\mathcal{U}$ is $\aleph_0$*–homogeneous* if for every two sequences of elements of U of the same length,

$$(a_0, \ldots, a_{n-1}) \text{ and } (b_0, \ldots, b_{n-1}),$$

with the property

$$(\mathcal{U}, a_0, \dots, a_{n-1}) \equiv (\mathcal{U}, b_0, \dots, b_{n-1}),$$

for every $a \in U$, there is $b \in U$ such that

$$(\mathcal{U}, a_0, \dots, a_{n-1}, a) \equiv (\mathcal{U}, b_0, \dots, b_{n-1}, b).$$

A countable model $\mathcal{A}$ which is $\aleph_0$–homogeneous is also called a *homogeneous* model.

Proposition 7.2

(i) *Every atomic model is $\aleph_0$–homogeneous.*

(ii) *Two countable homogeneous models which realize exactly the same types are isomorphic.*

Proposition 7.3 *Let T be a complete theory in at most countable language.*

(i) *If T has $> \aleph_0$ types, then T has $2^{\aleph_0}$ non-isomorphic countable homogeneous models.*

(ii) *If all countable models of T are homogeneous, then the number of non-isomorphic countable models of T is either 1, or $\aleph_0$, or $2^{\aleph_0}$.*

The following theorem, obtained independently by Harrington, and Goncharov and Nurtazin, is an effective version of Proposition 7.1 (ii). It establishes that a complete decidable theory T has a decidable prime model if there is an algorithm which for a given formula $\theta(\overline{x})$ consistent with T, outputs Gődel number of the characteristic function of a computable principal type $\Gamma(\overline{x})$ containing $\theta(\overline{x})$. In the proof of this result we use from *model theory*, Henkin's method of constructing models; and from *computability theory*, the finite injury priority method.

Theorem 7.4 (Goncharov-Nurtazin [79], Harrington [88]) *Let T be a complete decidable theory. The following are equivalent.*

(i) *T has a decidable prime model.*

(ii) *T has a prime model and the set of all principal types of T is computable.*

Proof. ($\Longrightarrow$): The conclusion follows from Proposition 3.5, since the set of all types realized in a decidable prime model of T is the set of all principal types of T.

($\Longleftarrow$): Let f be a computable function such that $\{f(n) : n \in \omega\}$ is a set of codes of the set of all principal types $\{\Gamma_n : n \in \omega\}$ of T, where $\varphi_{f(n)} = \chi_{\Gamma_n}$. We will use Henkin's method to construct a decidable prime model of T. Let $C = \{c_0, c_1, c_2, \dots\}$ be a set of new constants, and let $\sigma_0, \sigma_1, \sigma_2, \dots$ be an effective enumeration of all sentences in $L(T) \cup C$. As usual, the domain of the resulting model $\mathcal{A}$ will be $\{[c_0], [c_1], \dots\}$, where $[c]$ is the corresponding equivalence class of c. We will ensure that in $\mathcal{A}$, for every $e \geqslant 0$, $([c_0], \dots, [c_e])$ realizes a principal type of T, that is, a type from $\{\Gamma_n : n \in \omega\}$. This is sufficient since, for example, if a (complete) formula $\xi(x_0, x_1)$ generates a principal 2–type, then $\exists x_0 \xi(x_0, x_1)$ generates a principal 1–type. That is because $T \vdash \xi(x_0, x_1) \Rightarrow \zeta(x_1)$ implies $T \vdash \exists x_0 \xi(x_0, x_1) \Rightarrow \zeta(x_1)$. Hence every finite sequence of elements in the domain of $\mathcal{A}$ will satisfy a principal type.

We will construct the complete diagram Ψ of $\mathcal{A}$. At every stage s ($s \geqslant 0$) of the construction, we will have a finite set Ψ^s of sentences such that

$$\Psi^0 \subseteq \Psi^1 \subseteq \Psi^2 \subseteq \cdots \qquad \text{and} \qquad \Psi = \bigcup_{s \geqslant 0} \Psi^s.$$

Let $\psi^s = \bigwedge \Psi^s$. If $\psi^s = \psi^s(c_0, \dots, c_{n_s})$, then for every $e \in \{0, \dots, n_s\}$, we set

$$\psi_e^s =_{\text{def}} \exists y_{e+1} \dots \exists y_{n_s} \psi^s(c_0, \dots, c_e, y_{e+1}, \dots, y_{n_s}).$$

For every $e \geqslant 0$, at almost every stage s of the construction, we have a type $\Omega_e^s \in \{\Gamma_n : n \in \omega\}$ which is a candidate for a principal type realized by $([c_0], \dots, [c_e])$. We will allow Ω_e^s to be undefined for finitely many s. Because of the consistency property, if Ω_e^s is defined then $\psi_e^s(\overline{x}/\overline{c}) \in \Omega_e^s$. The construction will satisfy the following requirements for every $e \geqslant 0$.

P_e^1: $\sigma_e \in \Psi$ or $\neg\sigma_e \in \Psi$;

P_e^2: If $\sigma_e \in \Psi$ and $\sigma_e = \exists x \theta(x)$, then $\theta(c) \in \Psi$ for some $c \in C$;

Q_e: $([c_0], \dots, [c_e])$ realizes a principal type of T.

The priority ranking of the requirements in the decreasing order is:

$$P_0^1, P_0^2, Q_0, \dots, P_e^1, P_e^2, Q_e, \dots$$

We attempt to satisfy the requirements in the order of their priority. We say that at stage $s > 0$:

P_e^1 *requires attention* if $\sigma_e \notin \Psi^{s-1}$ and $\neg\sigma_e \notin \Psi^{s-1}$;

P_e^2 *requires attention* if $\sigma_e \in \Psi^{s-1}$ and $\sigma_e = \exists x\theta(x)$ for some θ such that $\theta(c) \notin \Psi^{s-1}$ for every $c \in C$;

Q_e *requires attention* if Ω_e^{s-1} is undefined.

Once satisfied at some stage, requirements P_e^1 and P_e^2 are never injured again. However, we say that

Q_e *is injured* at stage $s > 0$ if Ω_e^{s-1} is defined, but $\psi_e^s(\overline{x}/\overline{c}) \notin \Omega_e^{s-1}$.

Construction

STAGE 0:

Let $\Psi^0 = \emptyset$ and let Ω_e^0 be undefined for every $e \in \omega$.

STAGE $s > 0$:

Let *Req* be the highest priority requirement which requires attention at stage s. We now attack *Req* as follows.

Let $Req = P_e^1$

(a) If $T \vdash \forall\overline{x}[(\psi^{s-1} \Rightarrow \sigma_e)(\overline{x}/\overline{c})]$, then $\Psi^s = \Psi^{s-1} \cup \{\sigma_e\}$.

(b) If $T \vdash \forall\overline{x}[(\psi^{s-1} \Rightarrow \neg\sigma_e)(\overline{x}/\overline{c})]$, then $\Psi^s = \Psi^{s-1} \cup \{\neg\sigma_e\}$.

The properties on the left-hand side of (a) and (b) can be checked effectively because T is decidable.

(c) If neither (a) nor (b) is satisfied, we add either σ_e or $\neg\sigma_e$ to Ψ^{s-1} such that if some Q–requirement must be injured, then the first such injured requirement is of the highest priority. (Since the types in $\{\Gamma_n : n \in \omega\}$ are computable, we can effectively check whether a given Q–requirement is injured.)

We effectively check whether some Q_n is injured at stage s. Let n_0 be the least such n, if it exists. For every $n \geqslant n_0$, Ω_n^s will be undefined.

Let $Req = P_e^2$

Thus, $\sigma_e \in \Psi^{s-1}$ and $\sigma_e = \exists x\theta(x)$ for some θ. Let c be the first constant in C which has not been used in the construction before stage s. We define $\Psi^s = \Psi^{s-1} \cup \{\theta(c)\}$.

Let $Req = Q_e$

Thus, Ω_e^{s-1} is undefined. We find the first $(e+1)$–type $\Gamma(x_0, \dots, x_e) \in \{\Gamma_n : n \in \omega\}$ such that $\psi_e^{s-1}(x_0/c_0, \dots, x_e/c_e) \in \Gamma(x_0, \dots, x_e)$, and set $\Omega_e^s =_{\text{def}} \Gamma(x_0, \dots, x_e)$. This can be done effectively because of the following two facts.

(1) Such Γ exists because T has a prime model, so it is an atomic theory, hence every formula consistent with T belongs to some principal type.

(2) For every computable complete type, we can effectively decide whether a given formula or its negation belongs to that type. End of the construction.

Lemma 7.5 *For every e, $\Omega_e =_{\text{def}} \lim_s \Omega_e^s$ exists. Hence every Q_e is satisfied.*

Proof. Assume that $e = 0$. Let t_0 be the least stage such that $\Omega_0^{t_0}$ is defined. Let $\Omega_0^{t_0} = \Gamma_{n_0}$. Then $\psi_0^{t_0-1}(x_0/c_0) \in \Gamma_{n_0}$. Hence, by construction, Q_0 will never be injured, so $\Omega_0 = \Gamma_{n_0}$.

Assume that $e = 1$. Let ξ_0 be a complete formula such that $\xi_0 \in \Omega_0 (= \Gamma_{n_0})$. Clearly, $\neg\xi_0(x_0)$ is inconsistent with Ω_0, and Q_0 is never injured. Choose the least stage s_0 such that $\xi_0(c_0) \in \Psi^{s_0}$. Let t_1 be the least stage $> s_0$ such that $\Omega_1^{t_1}$ is defined. If $\Omega_1^{t_1} = \Gamma_{n_1}$, then $\psi_1^{t_1-1}(x_0/c_0, x_1/c_1) \in \Gamma_{n_1}$. Since every formula consistent with $\Psi^{s_0}(\overline{x}/\overline{c})$ is also consistent with Ω_0, it follows that Q_1 is not injured after t_1.

The general proof is by induction on e. If $e > 0$, choose the least s such that

$$\forall t > s(\Omega_{e-1}^t = \Omega_{e-1}^s = \Omega_{e-1}),$$

$$\psi_{e-1}^s(x_0/c_0, \dots, x_{e-1}/c_{e-1}) \text{ is a complete formula for } \Omega_{e-1},$$

$$\Omega_e^s \text{ is defined.}$$

Let $t > s$. It follows that $\Omega_e^t = \Omega_e^s$ since $\psi_e^s(x_0/c_0, \dots, x_{e-1}/c_{e-1}, x_e/c_e) \in \Omega_e^s$, so $\psi_{e-1}^s(x_0/c_0, \dots, x_{e-1}/c_{e-1}) \in \Omega_e^s$, and hence $\Omega_{e-1}^s \subseteq \Omega_e^t$. □

Theorem 7.6 (Millar [134]) *There is a complete decidable theory T with a prime model which does not have a computable prime model. In addition, all types of T are computable.*

Proof. The language of T is $L = \{P_n(\cdot) : n \in \omega\}$, where every P_n is a unary relation symbol. Let $\theta_0(x), \theta_1(x), \theta_2(x), \ldots$ be a computable enumeration of all quantifier-free formulae in L whose only free variable is x. For a quantifier-free formula $\theta(x)$ in L, let

$$\lceil \theta(x) \rceil =_{\text{def}} \mu k(\theta(x) = \theta_k(x)).$$

For a finite sequence $\alpha \in 2^m$, let

$$\theta_\alpha(x) =_{\text{def}} \bigwedge \{P_k(x)^{\alpha(k)} : 0 \leqslant k \leqslant m-1\}.$$

The set of sentences T is defined using a computable tree $\mathcal{T} \subseteq 2^{<\omega}$ which will be constructed later. The idea is to use the nodes in $\mathcal{T}$ to define certain formulae which are consistent with T and to use the nodes in $(2^{<\omega} - \mathcal{T})$ to define certain formulae which are inconsistent with T. Namely, the axioms of T fall into the following two groups:

Ax I $\forall x \neg \theta_\beta(x)$ for every $\beta \in 2^{<\omega} - \mathcal{T}$,

Ax II $\exists x_0 \ldots \exists x_{n-1} [\bigwedge_{0 \leqslant i < j \leqslant n-1} x_i \neq x_j \wedge \bigwedge_{0 \leqslant i \leqslant n-1} \theta_\alpha(x_i)]$ for every $\alpha \in \mathcal{T}$ and every $n \geqslant 1$.

In addition to being a computable tree, $\mathcal{T}$ will satisfy the following condition:

$$\forall \beta [\beta \in \mathcal{T} \Rightarrow \beta\hat{\ }1 \in \mathcal{T}]. \qquad (*)$$

This allows us to conclude that T has the properties stated in the following four lemmas.

Lemma 7.7 *T is consistent. Hence T is a theory.*

Proof. Consistency of T will follow easily from the construction of $\mathcal{T}$. We can also use the compactness theorem to prove that the set of all axioms of T has a model. Assume that S is a finite set of axioms. Let $\sigma_0, \ldots, \sigma_{k-1}$

For any a, let a^0 denote the empty sequence, and for $m \geqslant 1$, let a^m abbreviate the sequence of m consecutive a's.

Construction

STAGE 0:

$\mathcal{T}_0$ is the tree consisting of the nodes $1^e\hat{\ }0$ for all $e \in \omega$, and of their initial segments. For every $e \in \omega$, the node $1^e\hat{\ }0$ is e–marked.

STAGE $s+1$:

STEP 1: For every $\beta \in \mathcal{T}_s$:

Enumerate $\beta\hat{\ }1$ into $\mathcal{T}_{s+1}$; and

Declare that $\beta\hat{\ }0$ (and hence every $\gamma \in \mathcal{T}_s$ such that $\beta\hat{\ }0$ is an initial segment of γ) is in the complement of $\mathcal{T}$.

STEP 2: Consider each $e \leqslant s$. Let α be the e–marked node at s.

Case (a): No e–marked element has been defined at any previous stage. *Action*: Search for the least $j \leqslant s$ (if it exists) such that

$$\varphi^{(2)}_{e,s}(\ulcorner \theta_\alpha(x) \urcorner, j) \downarrow = 1.$$

If such j does not exist, then α is the e–marked node at $s+1$. If such j is found, define the e–marked element to be j. Let β be the node, enumerated in $\mathcal{T}_{s+1}$ by Step 1, of the maximal length such that α is the initial segment of β. The construction guarantees the uniqueness of β. Define β to be the e–marked node at $s+1$, and enumerate both $\beta\hat{\ }0$ and $\beta\hat{\ }1$ into $\mathcal{T}_{s+1}$.

Case (b): Assume that j is the e–marked element. *Action*: Let $lh(\alpha) = n$. Find the least $b \in \{0,1\}$ (if it exists) such that

$$\varphi^{(2)}_{e,s}(\ulcorner \theta_\alpha(x) \wedge P_n(x)^b \urcorner, j) \downarrow = 1.$$

If such b does not exist, then α is the e–marked node at $s+1$. Now assume that b exists. Let β be the node, enumerated in $\mathcal{T}_{s+1}$ by Step 1, of the maximal length such that $\alpha\hat{\ }b$ is an initial segment of β. The construction guarantees the uniqueness of β. Define β to be the e–marked node at $s+1$, and enumerate both $\beta\hat{\ }0$ and $\beta\hat{\ }1$ into $\mathcal{T}_{s+1}$. End of the construction.

Let $\mathcal{T} =_{\text{def}} \bigcup_{s\in\omega} \mathcal{T}_s$. $\mathcal{T}$ is a computable tree by construction.

Lemma 7.11

(i) *Every* 1*–type of* T *is computable.*

(ii) *Every type of* T *is computable.*

Proof.

(i) It is easy to see that the principal types of T are computable. Assume that $\Gamma(x)$ is a nonprincipal type of T. Since T admits the elimination of quantifiers, $\Gamma(x)$ is uniquely determined by a function $f_\Gamma \in 2^\omega$. That is, for every $i \in \omega$,

$$P_i(x) \in \Gamma \Longleftrightarrow f_\Gamma(i) = 1.$$

To prove that $\Gamma(x)$ is computable, it is sufficient to prove that f is computable. Let $e \geqslant 0$ be such that $1^{e}\hat{\ }0$ is an initial segment of f. If there were only finitely many e–marked nodes on the infinite branch of $\mathcal{T}$ determined by f, then, since every e-marked node "branches", f would determine a principal type. Therefore, there is an infinite set E of such e–marked nodes. Since E is computable by the construction of $\mathcal{T}$, f is computable.

(ii) Let $\Omega(x_0, \ldots, x_{n-1})$ be an arbitrary type of T. Since T admits the elimination of quantifiers, Ω is uniquely determined by its 1–subtypes and by the set Ψ of all inequalities among $x_0, \ldots, x_{n-1}$, which are in Ω. Since, by (i), all 1–types are computable, and Ψ is finite, it follows that Ω is computable. □

Lemma 7.12 *Every decidable model of* T *realizes a nonprincipal type of* T.

Proof. Assume that $\mathcal{A}$ is a decidable model of T. Then there is an effective enumeration $(a_i)_{i\in\omega}$ of A, and $e \in \omega$ such that for every formula $\theta(x)$ of L and every $i \in \omega$:

$$\mathcal{A} \models \theta(\mathbf{a}_i) \Longleftrightarrow \varphi_e^{(2)}(\ulcorner\theta(x)\urcorner, i) \downarrow = 1.$$

Let α be the e–marked node of the least length. Since $T \vdash \exists x \theta_\alpha(x)$, there is $i \in \omega$ such that $\mathcal{A}_A \models \theta_\alpha(\mathbf{a}_i)$. Hence there is $j \in \omega$ such that j is the e–marked element and for some stage $s \in \omega$, $\varphi^{(2)}_{e,s}(\ulcorner\theta_\alpha(x)\urcorner, j) \downarrow= 1$. It follows from the construction of $\mathcal{T}$ that there are infinitely many e–marked nodes. For every such node β, both $\beta\hat{\ }0$ and $\beta\hat{\ }1$ belong to $\mathcal{T}$. Hence $\mathcal{T}$ has an infinite branch determining a nonprincipal type which is realized in $\mathcal{A}$ by a_j. □

Now, let us show that the Σ^0_2 complexity assumption in the Effective Omitting Types Theorem (Theorem 6.2) cannot be replaced by a Π^0_2 one. Assume that a theory T is as in the previous theorem. Then the set of all nonprincipal types of T is not omitted in any decidable model of T. However, by Proposition 3.6, the set of all nonprincipal types of T is a Π^0_2 set.

Goncharov and Nurtazin [79] have also given an example of a decidable theory without a computable prime model. The language of the theory is infinite, and the theory is $\aleph_0$–stable. In [73], Goncharov has established a criterion for the computability of a prime model of a complete decidable theory. Let us first state a model-theoretic result about $\aleph_0$–stable theories.

Theorem 7.13 *Let T be an $\aleph_0$–stable theory in at most countable language, and let $\mathcal{U}$ be an arbitrary model of T. For every set $X \subseteq U$, the complete theory of $\mathcal{U}_X$ has an atomic prime model.*

Theorem 7.14 (Goncharov [73]) *There is a complete decidable $\aleph_0$–stable theory in a finite language (consisting of four unary relation symbols and one binary relation symbol), which does not have a computable prime model.*

Theories obtained in Theorem 7.6 and in Theorem 7.14 have infinite sets of axioms. However, Peretyat'kin [168] has found a finitely axiomatizable complete (hence decidable) theory T with a prime model, which does not have a decidable prime model. In Peretyat'kin's example, T is associated with a computably enumerable binary tree $\mathcal{T}$ which has the following properties. For every node α of $\mathcal{T}$, either both or none of $\alpha\frown 0$ and $\alpha\frown 1$ belong to $\mathcal{T}$. Every node of $\mathcal{T}$ is an initial segment of a leaf of $\mathcal{T}$, and the set of all finite branches of $\mathcal{T}$ is non-computable. A tree with these properties was first used by Goncharov and Nurtazin [79]. To prove that the described tree suffices for the result, Peretyat'kin has invented a general method for constructing finitely axiomatizable theories whose properties are determined by Turing machine computations.

Definition 7.3 Let $X \subseteq \omega$. A model $\mathcal{A}$ is *decidable in* X if $A \leqslant_T X$ and there is an enumeration $(a_i)_{i\in\omega}$ of A such that the theory of $(\mathcal{A}, a_i)_{i\in\omega}$ is $\leqslant_T X$.

Theorem 7.15 (Denisov [43], Millar [134], Drobotun [49]) *Let T be a complete decidable theory with a prime model. Then T has a prime model which is decidable in $\emptyset'$.*

Theorem 7.16 (Drobotun [49]) *For every set X such that $X <_T \emptyset'$, there is a complete decidable theory with a prime model which is not decidable in X.*

Millar [145] has introduced a different concept of the effectiveness of a model, which is weaker than the concept of decidability.

Definition 7.4 A countable model $\mathcal{A}$ for L is *almost decidable* if there is a computable function F which assigns to every finite binary sequence α a finite set $F(\alpha)$ of formulae in $L \cup \{c_0, c_1, c_2, \dots\}$, where $c_0, c_1, c_2, \dots$ are new constants, such that the following conditions are satisfied.

(1) For $\alpha \in 2^\omega$, if β is an initial segment of α , then $F(\beta) \subseteq F(\alpha)$.

(2) We can assign to every $f \in 2^\omega$ a model $\mathcal{A}_f$ such that

$$\bigcup \{F(\alpha) : \alpha \text{ is an initial segment of } f\}$$

determines the complete diagram of $\mathcal{A}_f$, and for all but countably many $f \in 2^\omega$, $\mathcal{A}_f$ is isomorphic to $\mathcal{A}$.

Every decidable model is almost decidable, and there is an almost decidable model which is not decidable. In fact, the concept of almost decidability is introduced to capture a class of models which fail to be decidable because, although there are computable strategies for their construction, the strategies are not uniformly computable.

Theorem 7.17 (Millar [145])

(i) *If a complete decidable theory T has fewer than continuum many complete types, then T has an almost decidable prime model.*

(ii) *There is a complete decidable theory which has a prime model but does not have an almost decidable prime model.*

8 Computable Saturated Models and Computably Saturated Models

In 1961, Vaught introduced the notion of a countably saturated model. In 1970's, Barwise, Schlipf and Ressayre [28, 27, 184, 190] introduced the notion of a computably saturated model. Barwise and Schlipf have extensively used computably saturated models to study computability over admissible sets.

Definition 8.1

(i) Let $\mathcal{U}$ be an arbitrary model. $\mathcal{U}$ is $\aleph_0$*–saturated* if for every finite subset X of its domain, $\mathcal{U}_X$ realizes every type $\Phi(x)$ of the theory $\mathrm{Th}(\mathcal{U}_X)$.

(ii) Let $\mathcal{U}$ be a model for a computable language L. $\mathcal{U}$ is *computably saturated* if for every finite subset X of its domain, every computable set of formulae $\Phi(x)$ in L_X consistent with $\mathrm{Th}(\mathcal{U}_X)$ is realized in $\mathcal{U}_X$.

Hence, every $\aleph_0$–saturated model for a computable language is computably saturated. A countable $\aleph_0$–saturated model is simply called *saturated.*

Theorem 8.1

(i) *A complete theory in a computable language whose models are infinite has a countable computably saturated model.*

(ii) *Every computably saturated model is* $\aleph_0$*–homogeneous.*

(iii) *Elementarily equivalent countable saturated models are isomorphic.*

(iv) *A complete theory with a countable saturated model has a prime model.*

Hence it follows from (ii) of the previous theorem that every countable saturated model is homogeneous. It follows from (i) of the previous theorem that every countable model for a computable language has a countable computably saturated elementary extension.

Theorem 8.2 (Engeler, Ryll-Nardzewski, Svenonius) *The following statements are equivalent for a complete theory* T.

(i) T *is* $\aleph_0$*–categorical.*

(ii) *There is a countable model of T which is both prime and saturated.*

(iii) *All types of T are principal.*

(iv) *For every finite sequence $\overline{x}$ of variables, there are only finitely many types of T in $\overline{x}$.*

(v) *For every finite sequence $\overline{x}$ of variables, there are only finitely many formulae with free variables among the elements of the sequence $\overline{x}$, which are not pairwise equivalent with respect to T.*

(vi) *All models of T are atomic.*

Theorem 8.3 *A complete theory T has a countable saturated model if and only if for every $n \in \omega$, T has only countably many n–types.*

Hence, every complete theory with only countably many non-isomorphic countable models has a countable saturated model. While countable saturated models do not exist for complete theories with uncountably many types, countable computably saturated models always exist. Thus, the proofs of many early results in model theory are simplified if countable computably saturated models are used to replace certain large models which exist only under specific assumptions of set theory.

Theorem 8.4 (Millar [137]) *Let T be a theory in a computable language L. Suppose that T has a complete extension T' in the language $L \cup \{c_0, \ldots, c_{n-1}\}$, where $c_0, \ldots, c_{n-1}$ are new constants, such that T' does not have an atomic model. Then T has a model which is not computably saturated.*

Proof. Since T does not have an atomic model, there is a formula

$$\psi(c_0, \ldots, c_{n-1}; x_0, \ldots, x_{m-1}) \text{ in } L(T')$$

which is consistent with T' and not contained in any principal type of T'. Let $\theta_0, \theta_1, \theta_2, \ldots$ be a computable enumeration of all formulae in L in free variables $(\overline{y}, \overline{x}) = (y_0, \ldots, y_{n-1}, x_0, \ldots, x_{m-1})$. We define a computable set of formulae $\Phi(\overline{y}, \overline{x}) = \{\psi_0, \psi_1, \psi_2, \ldots\}$ by:

$$\psi_0 = \psi(\overline{y}, \overline{x}),$$

$$\psi_{k+1} = [\theta_k(\overline{y}, \overline{x}) \Leftrightarrow \exists \overline{z}(\theta_k(\overline{y}, \overline{z}) \wedge \psi_0(\overline{y}, \overline{z}) \wedge \ldots \wedge \psi_k(\overline{y}, \overline{z}))] \text{ for } k \geqslant 0.$$

$\Phi(c_0, \dots, c_{n-1}, \overline{x})$ generates an m–type of T'. It must be a nonprincipal type since no principal type of T' contains $\psi(c_0, \dots, c_{n-1}, \overline{x})$. So there is a model $(\mathcal{A}, a_0, \dots, a_{n-1})$ of T' omitting $\Phi(c_0, \dots, c_{n-1}, \overline{x})$, such that $\mathcal{A}$ is a model of T. Since $\Phi(\overline{y}, \overline{x})$ is a computable set of formulae, $\mathcal{A}$ is not computably saturated. □

Proposition 8.5 (Millar [134]) *Let T be a complete decidable theory with a countable saturated model. Every consistent computably enumerable set $\Phi(x_0, \dots, x_{n-1})$ of formulae, $n \in \omega$, is contained in a computable type of T.*

Proof. Assume that $\Phi(x_0, \dots, x_{n-1})$ is a consistent computably enumerable set of formulae which is not contained in any computable type of T. Then there is no formula $\varphi = \varphi(x_0, \dots, x_{n-1})$ of $L(T)$ such that $\Phi \cup \{\varphi\}$ is contained in exactly one n–type of T in variables $x_0, \dots, x_{n-1}$. We can use the splitting along the nodes of a binary tree to show that T has $2^{\aleph_0}$ many n-types. Hence T does not have a countable saturated model, contradicting the assumption. □

Theorem 8.6 (Morley [150], Millar [133, 134]) *Let T be a complete decidable theory such that all types of T are computable. If the set of all types of T is computably enumerable, then T has a decidable saturated model.*

Proof. Let $\Gamma_0, \Gamma_1, \Gamma_2, \dots$ be an effective enumeration of all types of T such that every type appears infinitely often. Also, consider an effective enumeration of all finite sequences of constants from an infinite set C of new constants. Modify the construction in the proof of Theorem 7.4 so that the constructed decidable model is saturated. □

Theorem 8.7 (Millar [134]) *There is a complete decidable theory T with a countable saturated model which does not have a computable saturated model. In addition, all types of T are computable.*

Proof. The example in the proof of Theorem 7.6 can be modified to guarantee that every decidable model of T omits a (nonprincipal) 1–type of T. □

As Millar [134] has pointed out, there is no connection between the decidability of a prime model and the decidability of a countable saturated model (if they exist) of a complete decidable theory.

Theorem 8.8 (Millar [134]) *Let T be a complete decidable theory.*

(i) *If all types of T are computable, then T has a countable saturated model which is decidable in $\emptyset'$.*

(ii) *If T has a countable saturated model, then T has a saturated model which is decidable in some hyperarithmetic set.*

Theorem 8.9 (Denisov [43]) *For every set $X \subseteq \omega$ such that $X < \emptyset'$, there is a complete decidable theory T with a countable saturated model, such that T has neither a countable saturated model decidable in X, nor a prime model decidable in X.*

Morley introduced a very important classification of formulae according to their complexity.

Definition 8.2 Let $\mathcal{U}$ be an $\aleph_1$–saturated model for a countable language L.

(i) Let $\theta(x)$ be a formula in L_U. We say that an ordinal α is the *Morley rank* or the *transcendence rank* of $\theta(x)$ if the set of formulae

$$\{\theta(x)\} \cup \{\psi(x) : \neg\psi(x) \text{ has the Morley rank } < \alpha\}$$

in L_U is consistent and has finitely many maximal consistent extensions in the theory of $\mathcal{U}_U$.

We assign ∞ to $\theta(x)$ as its Morley rank if $\theta(x)$ is consistent with the theory of $\mathcal{U}_U$, but no ordinal is assigned to it as its Morley rank.

(ii) The *Morley rank* of $\mathcal{U}$ is the Morley rank of the formula $x = x$.

It is convenient to work with $\aleph_1$–saturated models because a formula has the same Morley rank in two elementary equivalent $\aleph_1$-saturated models. The valid formula $x = x$ is chosen to "represent the model" because it has the largest Morley rank. Clearly, a formula $\theta(x)$ in L_U has the Morley rank 0 if it is satisfied in $\mathcal{U}$ by at least one but at most finitely many distinct elements. Such a formula is also called *algebraic.*

It follows that all $\aleph_1$–saturated models of a complete theory T have the same Morley rank, called the *Morley rank of T*. It can be shown that if Morley rank of T is not ∞, then it is a countable ordinal. A theory T whose Morley rank is not ∞ is called *totally transcendental.* A complete theory is totally transcendental if and only if it is $\aleph_0$–stable.

Theorem 8.10 (Peretyat'kin [169]) *There is a complete finitely axiomatizable $\aleph_0$-stable theory of finite Morley rank, which has neither a computable prime model nor a computable saturated model.*

Schlipf [190] has established that if $\mathcal{A}$ is a countable, computably saturated model and S is a computably axiomatizable theory consistent with $\mathrm{Th}(\mathcal{A})$, then $\mathcal{A}$ can be expanded to a computably saturated model of S. For example, a countable nonstandard model of additive number theory can be expanded to a model of Peano arithmetic if and only if it is computably saturated (see [122]). For applications of computably saturated models see [112, 117, 123].

9 Decidable Homogeneous Models

While countable homogeneous models are relatively simple objects in model theory, they can be very complex from a computability-theoretic point of view. Classical model theory has established that every countable model has a countable homogeneous elementary extension. Two countable homogeneous models are isomorphic if and only if they realize the same finite types. Thus, a countable homogeneous model is uniquely determined, up to isomorphism, by a set of types it realizes. Therefore, the following question, first posed by Morley, is a very natural one.

> *Let T be a complete decidable theory. Assume that the type spectrum of a countable homogeneous model $\mathcal{A}$ of T consists only of computable types and is computable. Is $\mathcal{A}$ necessarily decidable?*

(The converse is obviously true.)

Goncharov, Peretyat'kin and Millar have independently answered Morley's question negatively by providing examples of a non-computable countable homogeneous model of a complete decidable theory such that the type spectrum of the model consists only of computable types and is computable. Millar [136] has used the infinite injury priority method to construct his counterexample. In addition, Goncharov [68] and Peretyat'kin [166] have characterized a decidable countable homogeneous model of a complete decidable theory. While Peretyat'kin's counterexample has not used this characterization, Goncharov has used the characterization to find his counterexample.

Goncharov [72] has later given an example of a complete $\aleph_0$–stable decidable theory which does not have a computable homogeneous model. Notice that such theory has neither a computable prime nor a computable saturated model. Another consequence of Goncharov's example is the existence of a decidable model without a homogeneous computable elementary extension.

To present Peretyat'kin's counterexample to Morley's question, we use from *model theory*, the Łoś-Vaught Test and a characterization of submodel complete theories from Theorem 5.5; and from *computability theory*, the notion of an *approximable* set and the existence of a non-approximable c.e. set, both of which are due to Peretyat'kin.

Theorem 9.1 (Łoś-Vaught Test) *If a theory S of an arbitrary language has only infinite models and for some infinite cardinal $\kappa \geqslant |L(S)|$, S is κ–categorical, then S is complete.*

Definition 9.1 Let $X \subseteq \omega$. X is *approximable* if there is a computable function f such that for every $m \in \omega$,

$$|\{0, \dots, m-1\} \cap X| \geqslant f(m)$$

and for infinitely many m,

$$|\{0, \dots, m-1\} \cap X| = f(m).$$

(If $m = 0$, then $\{0, \dots, m-1\} =_{def} \emptyset$.)

Hence, a set X is not approximable if and only if for every computable function f,

$$[\forall m \, |\{0, \dots, m-1\} \cap X| \geqslant f(m)] \Longrightarrow$$
$$[\exists m_0 \forall m \geqslant m_0 \, |\{0, \dots, m-1\} \cap X| > f(m)].$$

Theorem 9.2 (Peretyat'kin [166]) *There is a computably enumerable set X which is not approximable.*

Proof. We will algorithmically enumerate X at stages.

Construction

STAGE 0: Let $X_0 =_{\text{def}} \emptyset$.

STAGE $s+1$: Let X_s be the part of X enumerated by stage s. For every $e \leqslant s$, consider all k such that

for every $n \in \{\langle e,k\rangle+1,\dots,\langle e,k+1\rangle\}$,

$$\varphi_{e,s}(n)\downarrow \wedge |\{0,\dots,n-1\}\cap X_s| \geqslant \varphi_{e,s}(n).$$

Enumerate all such $\langle e,k\rangle$ in X_{s+1}. End of the construction.

Let $X = \bigcup_{s\in\omega} X_s$.

Lemma 9.3 *X is not approximable.*

Proof. Let $e \in \omega$ be such that φ_e is total. Assume that

$$\forall n\, |\{0,\dots,n-1\}\cap X| \geqslant \varphi_e(n).$$

Let $m_0 =_{\text{def}} \langle e,0\rangle + 1$. We will prove that

$$\forall m \geqslant m_0[|\{0,\dots,m-1\}\cap X| > \varphi_e(m)].$$

Let $m \geqslant m_0$. There is a unique k such that $m \in I$, where

$$I = \{\langle e,k\rangle+1,\dots,\langle e,k+1\rangle\}.$$

Consider the least $s \geqslant e$ such that

$$(\forall n \in I)[\varphi_{e,s}(n)\downarrow \wedge |\{0,\dots,n-1\}\cap X_s| \geqslant \varphi_{e,s}(n)].$$

By construction, $\langle e,k\rangle \in X_{s+1} - X_s$. Hence

$$|\{0,\dots,m-1\}\cap X| > |\{0,\dots,m-1\}\cap X_s| \geqslant \varphi_{e,s}(m) = \varphi_e(m)]. \quad \square$$

Theorem 9.4 (Goncharov [68], Millar [136], Peretyat'kin [166]) *There is a complete decidable theory T and a countable homogeneous model $\mathcal{M}$ of T such that $\mathcal{M}$ is not computable, and the type spectrum of $\mathcal{M}$ consists only of computable types and is computable.*

Proof. We will present the example from [166]. First, we will define a complete and decidable theory T in L which admits the elimination of quantifiers. Then for an arbitrary c.e. set X, we will define a c.e. set $\mathcal{S}_X$ of types of T such that there is a homogeneous model $\mathcal{M}$ with the following properties. $\mathcal{M}$ realizes precisely the types in $\mathcal{S}_X$; and if $\mathcal{M}$ is computable, then X must be approximable. However, by Theorem 9.2, X can be chosen to be non-approximable, thus forcing $\mathcal{M}$ to be non-computable.

Theory T

Definition of T. The language of T is $L = \{=, R, P_0, P_1, P_2, \dots\}$, where R is a binary relation symbol and for $i \in \omega$, P_i is a unary predicate symbol. We will also consider the finite sublanguages $L_0 = \{=, R\}$, and $L_s = \{=, R, P_0, \dots, P_{s-1}\}$ for $s > 0$. Let $T =_{\text{def}} \bigcup_{s \geqslant 0} T_s$, where T_s is a set of sentences in L_s defined as follows.

T_0 has the following two axioms.

Ax 1 $\forall x \neg R(x, x)$;

Ax 2 $\forall x \forall y [R(x, y) \Rightarrow R(y, x)]$.

For $s > 0$, T_s has, in addition to the above two axioms, the following axiom schema:

Ax$_s$ $\forall x_0 \dots \forall x_{n-1} [\delta'(x_0, x_1, \dots, x_{n-1}) \Longrightarrow \exists x_n \delta(x_0, x_1, \dots, x_{n-1}, x_n)]$, where δ is a conjunction of atomic formulae and negations of atomic formulae in L_s which is consistent with $\{Ax\ 1, Ax\ 2\}$, and δ' is a subformula of δ. We will call δ a finite diagram in L_s and δ' a subdiagram of δ.

Lemma 9.5 *For $s \geqslant 0$, T_s is consistent.*

Proof. Clearly, T_0 is consistent. Assume that $s > 0$. Let $\mathcal{A}_0$ be a model in L_s of axioms Ax 1 and Ax 2. We will construct a model $\mathcal{A}$ of T_s. Let $\theta_1, \theta_2, \dots$ be an enumeration of all axioms Ax$_s$ in which each axiom appears infinitely often. Let θ_1 be of the form

$$\forall x_0 \dots \forall x_{n-1} [\delta'(x_0, x_1, \dots, x_{n-1}) \Longrightarrow \exists x_n \delta(x_0, x_1, \dots, x_{n-1}, x_n)].$$

We will extend $\mathcal{A}_0$ to $\mathcal{A}_1$ in such a way that $\mathcal{A}_1$ satisfies the matrix of θ_1 on all n–tuples from A_0. Let $A_1 = A_0 \cup \{a\}$, where $a \notin A_0$. Let

$$\mathcal{A}_0 \models \delta'(\mathbf{a}_0, \mathbf{a}_1, \ldots, \mathbf{a}_{n-1})$$

for $a_0, a_1, \ldots, a_{n-1} \in A_0$. Extend the definitions of the predicates in L_s to the set $\{a_0, a_1, \ldots, a_{n-1}, a\}$ so that

$$\mathcal{A}_1 \models \delta(\mathbf{a}_0, \mathbf{a}_1, \ldots, \mathbf{a}_{n-1}, \mathbf{a}).$$

Continuing in a similar fashion, we construct a chain of models

$$\mathcal{A}_0 \subseteq \mathcal{A}_1 \subseteq \mathcal{A}_2 \subseteq \cdots .$$

Let $\mathcal{A} =_{\text{def}} \bigcup_{s \geqslant 0} \mathcal{A}_s$. □

Lemma 9.6 *For $s \geqslant 0$, T_s is $\aleph_0$–categorical.*

Proof. Let $\mathcal{A}$ and $\mathcal{B}$ be countable models of T_s. We will prove that they are isomorphic. Assume that f is a finite (partial) isomorphism from $\mathcal{A}$ to $\mathcal{B}$ and $\text{dom}(f) = \{a_0, a_1, \ldots, a_{n-1}\}$. Let $\delta'(x_0, x_1, \ldots, x_{n-1})$ be the finite diagram of $\mathcal{A}$ determined by $\text{dom}(f)$, let $a \in A - \text{dom}(f)$, and $\delta(x_0, x_1, \ldots, x_{n-1}, x_n)$ be the finite diagram of $\mathcal{A}$ determined by $\text{dom}(f) \cup \{a\}$. We have that $\mathcal{B} \models \delta'[f(a_0), f(a_1), \ldots, f(a_{n-1})]$. Thus, there is $b \in B$ such that

$$\mathcal{B} \models \delta[f(a_0), f(a_1), \ldots, f(a_{n-1}), b].$$

Then $f_1 = f \cup \{(a, b)\}$ is a finite isomorphism from $\mathcal{A}$ to $\mathcal{B}$. Similarly, if $b_1 \in B - ran(f)$, there is $a_1 \in A$ such that $f_2 = f_1 \cup \{(a_1, b_1)\}$ is a finite isomorphism from $\mathcal{A}$ to $\mathcal{B}$. □

Lemma 9.7 *For $s \geqslant 0$, T_s is complete.*

Proof. T_s has no finite models since $\{x_i \neq x_j : i \neq j \wedge i, j \leqslant n\}$ belongs to a finite diagram of T_s. Since T_s is $\aleph_0$–categorical, by the Łoś-Vaught Test, it is complete. □

Lemma 9.8 *For $s \geqslant 0$, T_s is decidable.*

Proof. T_s is complete and computably axiomatizable. Hence, it is decidable. □

Lemma 9.9 *For $s \geqslant 0$, T_s admits the elimination of quantifiers.*

Proof. We will prove that T_s is submodel complete by showing that

$$(\forall \mathcal{A}, \mathcal{B} \models T_s)(\forall \mathcal{D} \subseteq \mathcal{A}, \mathcal{B})\ [\mathcal{A} \text{ and } \mathcal{B} \text{ satisfy the same existential sentences in } L(T_s) \text{ with parameters from } D].$$

Let $\mathcal{A} \models \theta(\mathbf{d}_0, \ldots, \mathbf{d}_{n-1}, \mathbf{a}_0, \ldots, \mathbf{a}_{m-1})$, where

$$d_0, \ldots, d_{n-1} \in D \qquad \text{and} \qquad a_0, \ldots, a_{m-1} \in A - D.$$

Extend the identity function on $\{d_0, \ldots, d_{n-1}\}$ to a finite isomorphism f from $\mathcal{A}$ to $\mathcal{B}$ such that $a_0, \ldots, a_{m-1} \in \mathrm{dom}(f)$. Then

$$\mathcal{B} \models \exists x_0 \ldots \exists x_{m-1} \theta(\mathbf{d}_0, \ldots, \mathbf{d}_{n-1}, x_0, \ldots, x_{m-1}).$$

□

Types of T

Description of the types of T. The fact that T admits the elimination of quantifiers allows us to easily describe all finite types of T.

A 1–type $\Gamma(x)$ of T is uniquely determined by the sequence $f \in \{0,1\}^\omega$ such that $\{P_0^{f(0)}(x), P_1^{f(1)}(x), P_2^{f(2)}(x), \ldots\} \subseteq \Gamma$.

For $n \geqslant 2$, an n–type $\Gamma(x_0, \ldots, x_{n-1})$ of T is uniquely determined by the 1–types $\Gamma \restriction \{x_0\}, \ldots, \Gamma \restriction \{x_{n-1}\}$ and some finite L_0–diagram $\delta(x_0, \ldots, x_{n-1})$.

Description of a set $\mathcal{S}_X$ of types of T. Let X be a c.e. set of natural numbers.

Definition 9.2

(i) A sequence $f \in \{0,1\}^\omega$ is compatible with X if there is $l \in \omega$ such that for $i \in \{0, \ldots, l-1\}$,

$$f(l+i) = 1 \text{ if } i \in X \text{ , and } f(l+i) = 0 \text{ if } i \notin X,$$

$$f(2l) = 0 \text{ and for } i > 2l, \ f(i) = 1.$$

(ii) A 1–type is compatible with X if the infinite binary sequence which determines it is compatible with X.

That is, f is compatible with X if f is defined arbitrarily on some initial segment of length l, then "follows" X on length l, after that has value 0, and then its value becomes and remains 1 forever.

A 1–type belongs to $\mathcal{S}_X$ if and only if it is determined by an almost constant 1–sequence (that is, $\exists n_0 \forall n \geqslant n_0 f(n) = 1$).

A 2–type $\Gamma = \Gamma(x, y)$ of T with $\Gamma_1(x) = \Gamma \restriction \{x\}$ and $\Gamma_2(y) = \Gamma \restriction \{y\}$ belongs to $\mathcal{S}_X$ if and only if $\Gamma_1, \Gamma_2 \in \mathcal{S}_X$ and the following condition is satisfied: $\neg R(x, y) \in \Gamma(x, y)$, or neither $\Gamma_1(x)$ nor $\Gamma_2(y)$ is determined by the constant 1–sequence, or if one of $\Gamma_1(x)$ and $\Gamma_2(y)$ is determined by the constant 1–sequence, then the other one is compatible with X.

An n–type belongs to $\mathcal{S}_X$ if and only if each of its 2-subtypes belongs to $\mathcal{S}_X$.

Lemma 9.10 *There is a homogeneous model which realizes precisely the types in* $\mathcal{S}_X$.

Proof. The existence of such a model follows from the next two properties. Property (1) will guarantee Henkin's witnesses, and property (2) will guarantee the homogeneity of the model which can be constructed by Henkin's method.

(1) If $\Gamma(x_0, \ldots, x_{n-1}) \in \mathcal{S}_X$ and $\theta(x_0, \ldots, x_{n-1}, x)$ is a formula consistent with Γ (that is, $\exists x \theta(x_0, \ldots, x_{n-1}, x) \in \Gamma$)), then there is an $(n+1)$–type $\Omega(x_0, \ldots, x_{n-1}, x) \in \mathcal{S}_X$ containing Γ and θ.

Let us prove (1). Let the language of $\theta(x_0, \ldots, x_{n-1}, x)$ be L_s. Since T_s eliminates the quantifiers, there is a finite diagram $\delta(x_0, \ldots, x_{n-1}, x)$ of T_s such that

$$\vdash \delta(x_0, \ldots, x_{n-1}, x) \Rightarrow \theta(x_0, \ldots, x_{n-1}, x).$$

We can extend $\Gamma(x_0, \ldots, x_{n-1}) \cup \{\delta(x_0, \ldots, x_{n-1}, x)\}$ to a type

$$\Omega(x_0, \ldots, x_{n-1}, x)$$

in $\mathcal{S}_X$. If $x = x_i$ for some $i < n$, then the required extension Ω is unique. Otherwise, choose Ω in such a way that $\Omega(x_0, \dots, x_{n-1}, x) \upharpoonright \{x\}$ is compatible with X.

(2) If $\Gamma_1(x_0, \dots, x_{n-1}, x_n) \in \mathcal{S}_X$ and $\Gamma_2(x_0, \dots, x_{n-1}, x_n) \in \mathcal{S}_X$ and if

$$\Gamma_1 \upharpoonright \{x_0, \dots, x_{n-1}\} = \Gamma_2 \upharpoonright \{x_0, \dots, x_{n-1}\},$$

then there is an $(n+2)$–type $\Omega(x_0, \dots, x_{n-1}, x_n, x) \in \mathcal{S}_X$ such that Ω contains $\Gamma_1(x_0, \dots, x_{n-1}, x_n)$ and $\Gamma_2(x_0, \dots, x_{n-1}, x)$.

Let us prove (2). Let

$$\Gamma(x_0, \dots, x_{n-1}, x_n, x) =_{\text{def}} \Gamma_1(x_0, \dots, x_{n-1}, x_n) \cup \Gamma_2(x_0, \dots, x_{n-1}, x).$$

If for some $i < n$,

$$(x_i = x_n) \in \Gamma_1(x_0, \dots, x_{n-1}, x_n) \cup \Gamma_2(x_0, \dots, x_{n-1}, x_n),$$

then the required extension Ω of Γ is uniquely determined. Otherwise, Ω will be determined by Γ, $x_n \neq x$, and $\neg R(x_n, x)$. □

Lemma 9.11 *The set $\mathcal{S}_X$ of types is computably enumerable.*

Proof. It is sufficient to prove that the set of all 2–types in $\mathcal{S}_X$ is computably enumerable. To prove this fact, it is enough to prove that a family $\mathcal{T}$ of 2–types in $\mathcal{S}_X$ is computably enumerable, where

$$\begin{aligned}\mathcal{T} \supseteq \{\Gamma(x,y) : R(x,y) \in \Gamma \\ &\wedge (\Gamma \upharpoonright \{x\} \text{ is determined by the constant 1–sequence}) \\ &\wedge (\Gamma \upharpoonright \{y\} \text{ is determined by a sequence compatible with } X)\}.\end{aligned}$$

Let $\{X_t\}_{t\in\omega}$ be a computable enumeration of X. For every pair $(\overline{p}, t)$ of a finite sequence $\overline{p} = (p_0, \dots, p_{l-1})$ and a number t, we define $f, g \in 2^\omega$ as follows:

$$\begin{aligned}&f(0) = p_0, \dots, f(l-1) = p_{l-1},\\ &f(l) = 1 \text{ if } 0 \in X_t, \text{ and } \quad f(l) = 0 \text{ if } 0 \notin X_t,\\ &\dots\\ &f(2l-1) = 1 \text{ if } l-1 \in X_t, \text{ and } \quad f(2l-1) = 0 \text{ if } l-1 \notin X_t,\\ &f(2l) = 0, \text{ and } f(i) = 1 \quad \text{if } i > 2l.\end{aligned}$$

Clearly, f is compatible with X if no new elements among $\{0,\dots,l-1\}$ are enumerated in X after stage t. The sequence g keeps track of that part of the enumeration. Namely,

$$g(s) = 0 \text{ if } (X_s - X_{s-1}) \cap \{0,\dots,l-1\} \neq \emptyset \text{ for } s > t\text{ , and}$$

$$g(s) = 1 \text{ otherwise.}$$

Notice that g is determined by an almost constant 1–sequence. Also, by the above remark, if g is determined by the constant 1–sequence, then f is compatible with X.

Let $\Gamma_{(\overline{p},t)}(x,y)$ be the 2–type such that

$$\Gamma_{(\overline{p},t)}(x,y) \supseteq \{R(x,y), P_k^{f(k)}, P_k^{g(k)} : k \geqslant 0\}.$$

Then $\mathcal{T} = \{\Gamma_{(\overline{p},t)}(x,y) : \overline{p} \in 2^{<\omega} \wedge t \in \omega\}$. □

Lemma 9.12 *Assume that a homogeneous model of T realizing precisely the types in $\mathcal{S}_X$ is computable. Then the set of types*

$$\{\Gamma : \Gamma \text{ is a 1-type of } T \text{ compatible with } X\}$$

is computably enumerable.

Proof. Let $\mathcal{M}$ be a homogeneous model of T realizing precisely the types in $\mathcal{S}_X$. Let a' be an element of M which realizes in $\mathcal{M}$ the 1–type Θ of T determined by the constant 1–sequence. For every $a \in M$, let Γ_a be the 1–type realized in $\mathcal{M}$ by a. Since $\mathcal{M}$ is computable by assumption, it is enough to prove that

$$\{\Gamma : \Gamma \text{ is a 1-type of } T \text{ compatible with } X\} =$$

$$\{\Gamma_a : a \in M \wedge \mathcal{M} \models R(\mathbf{a}',\mathbf{a})\}.$$

We first assume that $\mathcal{M} \models R(\mathbf{a}',\mathbf{a})$ for some $a \in M$. Since the 2–type realized in $\mathcal{M}$ by (a',a) belongs to $\mathcal{S}_X$, Γ_a is compatible with X (by the choice of a').

We now assume that Γ is a 1–type of T compatible with X. Let a 2–type $\Omega(x,y)$ be such that $R(x,y) \in \Omega$, $\Omega \restriction \{x\} = \Theta(x)$ and $\Omega \restriction \{y\} = \Gamma(y)$. Since Ω belongs to $\mathcal{S}_X$, it is realized in $\mathcal{M}$ by some (b',b). Thus, a' and b' realize the same 1–type in the homogeneous model $\mathcal{M}$. Let f be an automorphism of $\mathcal{M}$ such that $f(b') = a'$. Let $f(b) = a$ for some $a \in M$. Since (a', a) and (b', b) realize in $\mathcal{M}$ the same 2–types, it follows that $\Gamma = \Gamma_a$. Also, since $R^{\mathcal{M}}(\mathbf{b}', \mathbf{b})$, we have that $R^{\mathcal{M}}(\mathbf{a}', \mathbf{a})$. □

Lemma 9.13 *Let $\mathcal{M}$ be a homogeneous model realizing precisely the types in $\mathcal{S}_X$. If $\mathcal{M}$ is computable, then X is approximable.*

Proof. By Lemma 9.12, the set

$$\{\Gamma : \Gamma \text{ is a 1–type of } T \text{ compatible with } X\}$$

is computably enumerable. Therefore, we can algorithmically enumerate the infinite binary sequences which determine the 1–types of T compatible with X. We choose such a computable enumeration $\alpha_0, \alpha_1, \alpha_2, \ldots$ in such a way that for every $e \geqslant 0$, the length of agreement of α_e with X is at least e. Hence $(\exists i \geqslant 2e)[\alpha_e(i) = 0]$.

We will define, by recursion, a unary computable function g as follows.

$$g(0) = 0$$

For $n > 0$, $g(n)$ is the least number such that

$$g(n) > g(n-1), \qquad g(n) > 2n + 2,$$

and for every $e < n$, there is $l = l_{e,n}$ which satisfies the following conditions:

$$\begin{aligned} & g(n) > 2l, \\ & \alpha_e(2l) = 0, \alpha_e(2l+1) = \alpha_e(2l+2) = \ldots = \alpha_e(g(n)) = 1, \qquad (*) \\ & \{i : 0 \leqslant i \leqslant l - 1 \wedge \alpha_e(l+i) = 1\} = \{0, 1, \ldots, l-1\} \cap X_{g(n)}. \end{aligned}$$

Thus, at stage n, we look at the initial segment of length $g(n)$ of each of the sequences $\alpha_0, \alpha_1, \ldots, \alpha_{n-1}$ and, within this segment, obtain the compatibility of the sequences with $X_{g(n)}$. However, past this initial segment, it is still possible to have value 0 in certain α_e's for $e < n$.

We define a function f by $f(n) = |\{0, 1, \ldots, n-1\} \cap X_{g(n)}|$. Clearly, f is computable and for every $n \in \omega$, $f(n) \leqslant |\{0, 1, \ldots, n-1\} \cap X|$. Let $e \in \omega$. We define n_e to be the least number such that $n_e > n$ and α_e does not have a value 0 past the initial segment of length $g(n_e)$. We will prove that $f(n_e) = |\{0, 1, \ldots, n_e - 1\} \cap X|$.

Since α_e is compatible with X, we have the following equality for $l = l_{e,n_e}$:

$$\{i : 0 \leqslant i \leqslant l - 1 \wedge \alpha_e(l + i) = 1\} = \{0, 1, \ldots, l - 1\} \cap X.$$

On the other hand, by the definition of g, we have

$$\{i : 0 \leqslant i \leqslant l - 1 \wedge \alpha_e(l + i) = 1\} = \{0, 1, \ldots, l - 1\} \cap X_{g(n_e)}.$$

Therefore, to prove that $f(n_e) =| \{0, 1, \ldots, n_e - 1\} \cap X |$, it is enough to prove that $n_e \leqslant l_{e,n_e}$.

Assume that $e = n_e - 1$. The required inequality follows from the length of the compatibility of α_e.

Now assume that $e < n_e - 1$. By the definition of g, $g(n_e - 1) > 2n_e$. By the definition of n_e, α_e must have value 0 past the initial segment of length $g(n_e - 1)$. Hence, the desired inequality follows from the condition (*) in the definition of g. □

In [142], Millar has given an example of a complete decidable theory with only computable complete types and with only countably many non-isomorphic countable models, which has an undecidable countable homogeneous model.

To state Goncharov's and Peretyat'kin's characterization of a decidable countable homogeneous model of a complete decidable theory, we introduce the following definition.

Definition 9.3 A computable set $\mathcal{T}$ of computable types of a theory T in L has the *effective extension property* if the following condition is satisfied for an effective enumeration $\Gamma_0, \Gamma_1, \Gamma_2, \ldots$ of all types in $\mathcal{T}$, and an effective enumeration $\theta_0, \theta_1, \theta_2, \ldots$ of all formulae in L.

There is a partial computable binary function f such that for every $n, i \in \omega$, if $\Gamma_n = \Gamma_n(x_0, \ldots, x_{k-1})$ for some $k \in \omega$, and Γ_n is consistent with $\theta_i = \theta_i(x_0, \ldots, x_{k-1}, x_k)$, then $f(n, i)$ is defined, $\Gamma_{f(n,i)}$ is a $(k+1)$–type and

$$(\Gamma_n \cup \{\theta_i\}) \subseteq \Gamma_{f(n,i)}.$$

Theorem 9.14 (Goncharov [68], Peretyat'kin [166]) *Let $\mathcal{A}$ be a countable homogeneous model with the type spectrum $\mathcal{T}$. Then $\mathcal{A}$ is decidable if and only if $\mathcal{T}$ is a computable set of computable types and $\mathcal{T}$ has the effective extension property.*

As a consequence of this characterization, Goncharov-Nurtazin's, and Harrington's characterization of a decidable prime model, as well as Morley's and Millar's characterization of a decidable countably saturated model can be obtained. Another consequence of this characterization is the next theorem, also obtained by Millar [139] as a consequence of a more general result.

Theorem 9.15 (Goncharov [68], Millar [139]) *Let the set of all computable types of a complete theory T be computable. If the set of all complete types realized in a countable homogeneous model $\mathcal{A}$ of T is a Σ_2^0 set of computable types, then $\mathcal{A}$ is decidable.*

Theorem 9.16 (Millar [147]) *Assume that T is a complete decidable theory all of whose types are computable and which has only countably many type spectra. Let $\mathcal{A}$ be a countable homogeneous model of T. If the type spectrum of $\mathcal{A}$ is Σ_2^0, then $\mathcal{A}$ is almost decidable.*

Algorithmic complexity of countable homogeneous models has also been studied by Denisov [41, 42, 43]. The following result is a computable analogue of the classical model-theoretic result that every theory in a countable language has a countable homogeneous model.

Theorem 9.17 (Denisov [43])

(i) *Every complete decidable theory has a countable homogeneous model which is decidable in $\emptyset'$.*

(ii) *For every $X \subseteq \omega$ such that $X <_T \emptyset'$, there is a complete decidable theory which does not have a countable homogeneous model decidable in X.*

Theorem 9.18 (Tusupov [200]) *Let $\mathcal{A}$ be a countable homogeneous model of a decidable theory, such that the type spectrum of $\mathcal{A}$ is a computable family of computable types. Then $\mathcal{A}$ is decidable in $\emptyset'$.*

While every countable model has a countable homogeneous elementary extension, Goncharov and Drobotun [76] have constructed a computable linear order which does not have a computable homogeneous elementary extension. They have also constructed a decidable model which does not have a computable homogeneous elementary extension. (Also see [162].)

10 Vaught's Theorem Computably Visited

Theorem 10.1 (Vaught) *There is no complete theory which has exactly two non-isomorphic countable models.*

Proof. By contradiction. Assume that T has exactly two non-isomorphic countable models. Then T must have a countable saturated model $\mathcal{A}$ and a prime model $\mathcal{C}$. Clearly, $\mathcal{A}$ and $\mathcal{C}$ are not isomorphic. Since $\mathcal{A}$ is not prime, there is an n–tuple of elements of A which realizes a nonprincipal type of T. Without loss of generality, assume that $n = 1$. Thus, there is $a \in A$ which realizes a nonprincipal type $\Gamma(x)$. Let c be a new constant. Since $(\mathcal{A}, a)$ is a countable saturated model of $\Gamma(c)$, $\Gamma(c)$ also has a prime model $(\mathcal{B}, b)$. However, $\mathcal{B}$ is not prime because it realizes a nonprincipal type $\Gamma(x)$. Hence $\mathcal{B}$ and $\mathcal{C}$ are not isomorphic. Finally, $(\mathcal{B}, b)$ is not saturated because T is not $\aleph_0$–categorical, so T and, hence, $\Gamma(c)$ satisfy (v) of Theorem 8.2, so $\Gamma(c)$ is not $\aleph_0$-categorical. Hence $\mathcal{B}$ and $\mathcal{A}$ are not isomorphic. The existence of $\mathcal{A}$, $\mathcal{B}$ and $\mathcal{C}$ contradicts the assumption at the beginning of the proof. □

Theorem 10.2 (Ehrenfeucht) *For every $n \geqslant 3$, there is a complete theory with exactly n non-isomorphic models.*

On the other hand, Millar and Kudaibergenov have constructed a complete decidable theory with exactly two non-isomorphic decidable models. However, the effective version of Ehrenfeucht's result remains true [118]. To present Millar's and Kudaibergenov's result, we use from *model theory,* a characterization of submodel complete theories from Theorem 5.5; and from *computability theory,* the existence of two computably inseparable c.e. sets. That is, there are c.e. sets X and Y such that

$$X \cap Y = \emptyset \quad \text{and} \quad \neg(\exists R)[R \text{ is computable} \wedge X \subseteq R \wedge R \cap Y = \emptyset].$$

Theorem 10.3 (Millar [135], Kudaibergenov [118]) *There is a complete decidable theory T with exactly two non-isomorphic decidable models.*

Proof. We present the example from [135]. We will define a theory T such that the following conditions are satisfied.

(1) T has only one nonprincipal 1–type, $\Gamma(x)$. $\Gamma(x)$ is a computable type. (Notice that $T \subseteq \Gamma(x)$.)

(2) There is no computable 2–type $\Omega(x, y)$ of T such that

$$\Gamma(x) \cup \Gamma(y) \cup \{x \neq y\} \subseteq \Omega(x, y).$$

(3) $\Gamma(c)$ has a decidable prime model, where c is a new constant.

(4) If a model of T realizes a computable nonprincipal type of T, then it realizes all computable nonprincipal types of T.

Lemma 10.4 *Conditions* (1)–(4) *imply the theorem.*

Proof. Let $(\mathcal{B}, b)$ be a decidable prime model of $\Gamma(c)$, $\Gamma(c) = T \cup \Gamma(c)$, which exists by (3). Then $\mathcal{B}$ is a decidable prime model of T which realizes $\Gamma(x)$. $\Gamma(x)$ is a nonprincipal type, hence there is a decidable model $\mathcal{A}$ which omits $\Gamma(x)$. Since $\mathcal{A}$ is decidable, all types realized in $\mathcal{A}$ are computable. Since every type realized in $\mathcal{A}$ is principal, $\mathcal{A}$ is a prime model of T. We will prove that every decidable model of T is either isomorphic to $\mathcal{A}$ or to $\mathcal{B}$. Let $\mathcal{D}$ be a decidable model of T.

Case (*a*): $\mathcal{D}$ omits $\Gamma(x)$. Since $\mathcal{D}$ is decidable, all types realized in $\mathcal{D}$ are computable. By (4), $\mathcal{D}$ omits all nonprincipal types of T. Since every type realized in $\mathcal{D}$ is principal, $\mathcal{D}$ is a prime model of T, hence $\mathcal{D} \cong \mathcal{A}$.

Case (*b*): $\mathcal{D}$ realizes $\Gamma(x)$. Let $d \in D$ be such that $\mathcal{D} \models \Gamma(x)[d]$. We claim that $(\mathcal{D}, d) \cong (\mathcal{B}, b)$ and, hence, that $\mathcal{D} \cong \mathcal{B}$. Assume otherwise, that is, $(\mathcal{D}, d) \not\cong (\mathcal{B}, b)$. Then $(\mathcal{D}, d)$ is not a prime model of $\Gamma(c)$. Thus, $(\mathcal{D}, d)$ must realize a nonprincipal type $\Omega(c, \overline{x}) = \Omega(c, x_1, \ldots, x_n)$ of $\Gamma(c)$. Hence $\Gamma(x) \subseteq \Omega(x, \overline{x})$, and $\Omega(x, \overline{x})$ is a computable type. Also, $\Omega(x, \overline{x})$ is a nonprincipal type of T, hence it is realized in $\mathcal{B}$. Let $b', b'_1, \ldots, b'_n \in B$ be such that $\mathcal{B} \models \Omega(x, \overline{x})[b', b'_1, \ldots, b'_n]$. It follows by (2) that $\Gamma(x)$ cannot be realized in a decidable model $\mathcal{B}$ by two different elements, b and b', since the 2-type determined by (b, b') in $\mathcal{B}$ would be computable. Thus, $(\mathcal{B}, b) \models \Omega(c, \overline{x})[b'_1, \ldots, b'_n]$. This is a contradiction, since $(\mathcal{B}, b)$ is a decidable prime model of $\Gamma(c)$, and $\Omega(c, \overline{x})$ is a nonprincipal type of $\Gamma(c)$. □

The language of T is $L = \{P_n(\cdot), S_n(\cdot,\cdot) : n \in \omega\}$, where each $P_n(\cdot)$ is a unary relation symbol and each $S_n(\cdot,\cdot)$ is a binary relation symbol.

Let $X \subseteq \omega$ and $Y \subseteq \omega$ be computably inseparable c.e. sets. We will encode X and Y into 2–types of T. Let $(X_t)_{t\in\omega}$ and $(Y_t)_{t\in\omega}$ be computable enumerations of X and Y, respectively, such that if $n \in X_t$ or $n \in Y_t$, then $n < t$. (We have $X_0 \subseteq X_1 \subseteq X_2 \subseteq \dots$ and $\bigcup_{t\in\omega} X_t = X$, and similar relations for Y.)

We first define T' such that $T' \subseteq T$. The axioms of T' are the universal closures with respect to x and y of the following formulae. Let $n, t \in \omega$.

Ax 1 $P_t(x) \Rightarrow P_{t+1}(x)$;

Ax 2 $\neg S_n(x,x)$;

Ax 3 $S_n(x,y) \Rightarrow S_n(y,x)$;

Ax 4 $P_t(x) \Rightarrow \neg S_t(x,y)$;

Ax 5 $(\neg P_t(x) \wedge \neg P_t(y) \wedge x \neq y) \Rightarrow S_n(x,y)$ if $n \in X_t$;

Ax 6 $(\neg P_t(x) \wedge \neg P_t(y) \wedge x \neq y) \Rightarrow \neg S_n(x,y)$ if $n \in Y_t$.

Clearly, T' is a universal set of sentences. T' is obviously consistent, since a nonempty set A with $P_t^{\mathcal{A}} = A$ for $t \in \omega$, and $S_n^{\mathcal{A}} = \emptyset$ for $n \in \omega$ is a model of T'.

We will now extend T' to T in such a way that T is submodel complete, and, therefore, admits the elimination of quantifiers. We will add a new set of axioms. First we introduce some notation.

Let $\mathcal{M}$ be a finite model of T'. Let $\Delta_{\mathcal{M}^n}(\overline{a})$ be the conjunction of all atomic and negated atomic sentences true in $\mathcal{M}_M$, in which only relation symbols $P_0, \dots, P_n$ and $S_0, \dots, S_n$ may occur. There are only finitely many such sentences.

T' is extended to T by adding a new group of axioms for every $n \in \omega$:

Ax 7 $(\forall \overline{x})(\exists \overline{y})[\Delta_{\mathcal{M}^n}(\overline{x}) \Rightarrow \Delta_{\mathcal{N}^n}(\overline{x},\overline{y})]$,

where $\mathcal{M}$ and $\mathcal{N}$ are finite models of T' such that $\mathcal{M} \subseteq \mathcal{N}$, allowing $M = \emptyset$.

Lemma 10.5 *T is consistent.*

Proof. By compactness. □

Lemma 10.6 *T is computably axiomatizable.*

Lemma 10.7 *T is submodel complete.*

Proof. We will prove

$$(\forall \mathcal{A}, \mathcal{B} \models T)(\forall \mathcal{D} \subseteq \mathcal{A}, \mathcal{B})[\mathcal{A} \text{ and } \mathcal{B} \text{ satisfy the same existential sentences with parameters from } D].$$

Let $\mathcal{A}, \mathcal{B} \models T$ and $\mathcal{D} \subseteq \mathcal{A}, \mathcal{B}$. Let $\delta(\overline{x}, \overline{y})$ be a conjunction of atomic and negated atomic formulae, and $\overline{d} \in D^{\mathrm{lh}(\overline{x})}$ such that $\mathcal{A} \models (\exists \overline{y})\delta(\overline{x}, \overline{y})[\overline{d}]$. Let $\overline{a} \in A^{lh(\overline{y})}$ be such that $\mathcal{A} \models \delta(\overline{x}, \overline{y})[\overline{d}, \overline{a}]$. Let $\mathcal{M}$ be the submodel of $\mathcal{A}$ whose domain consists of the elements in $\overline{d}$. Let $\mathcal{N}$ be the submodel of $\mathcal{A}$ whose domain consists of the elements in $\overline{d}$ and $\overline{a}$. Since T' is a universal theory, we have $\mathcal{M}, \mathcal{N} \models T'$. Let n be the largest subscript of a P–predicate symbol or of an S–predicate symbol occurring in $\delta(\overline{x}, \overline{y})$. (If no P–predicate symbol and no S–predicate symbol occurs in $\delta(\overline{x}, \overline{y})$, let $n = 0$.) Clearly,

$$\models [\Delta_{\mathcal{N}^n}(\overline{x}, \overline{y}) \Rightarrow \delta(\overline{x}, \overline{y})].$$

Since $(\forall \overline{x})(\exists \overline{y})[\Delta_{\mathcal{M}^n}(\overline{x}) \Rightarrow \Delta_{\mathcal{N}^n}(\overline{x}, \overline{y})]$ is an axiom of T, we have that

$$T \models (\forall \overline{x})(\exists \overline{y})[\Delta_{\mathcal{M}^n}(\overline{x}) \Rightarrow \delta(\overline{x}, \overline{y})].$$

Since $\mathcal{B} \models \Delta_{\mathcal{M}^n}(\overline{x})[\overline{d}]$ and $\mathcal{B} \models T$, we have $\mathcal{B} \models (\exists \overline{y})\delta(\overline{x}, \overline{y})[\overline{d}]$. □

Let us now prove that T satisfies the conditions (1)–(4).

(1) *T has only one nonprincipal 1–type, $\Gamma(x)$. Furthermore, $\Gamma(x)$ is a computable type.*

Let $\Gamma^*(x)$ be a 1–type of T. Since T admits the elimination of quantifiers and $\neg S_n(x, x)$ is an axiom, every formula in $L(T)$ with one free variable is equivalent to a quantifier-free formula whose only relation symbols are from $\{P_n : n \in \omega\}$. Therefore, $\Gamma^*(x)$ is uniquely determined by the set $\{t : P_t(x) \in \Gamma^*(x)\}$. Assume that this set is nonempty, and let t_0 be its smallest element. Since

$$P_{t_0}(x) \Rightarrow P_{t_0+1}(x) \Rightarrow P_{t_0+2}(x) \Rightarrow \cdots,$$

we have that $\neg P_0(x) \wedge \ldots \wedge \neg P_{t_0-1}(x) \wedge P_{t_0}(x)$ is a complete formula of $\Gamma^*(x)$. Hence, $\Gamma^*(x)$ is a principal type. Thus, there is exactly one nonprincipal 1–type $\Gamma(x)$, where $\Gamma(x)$ contains $\{\neg P_t(x) : t \in \omega\}$. Since T is decidable and admits the elimination of quantifiers, we can effectively find for each formula a corresponding quantifier-free formula. Thus, Γ is computable.

(2) *There is no computable 2–type $\Omega(x, y)$ of T such that*

$$\Gamma(x) \cup \Gamma(y) \cup \{x \neq y\} \subseteq \Omega(x, y).$$

Assume otherwise for some $\Omega(x, y)$. Define

$$R = \{n \in \omega : S_n(x, y) \in \Omega(x, y)\}.$$

R is a computable set since $\Omega(x, y)$ is a computable type. By Ax 5 and Ax 6, $X \subseteq R$ and $Y \cap R = \emptyset$, contradicting the computable inseparability of X and Y.

(3) *Let c be a new constant. Then $\Gamma(c)$ has a decidable prime model.*

To prove (3), it is enough to prove the following lemma.

Lemma 10.8 *Every computable type of $\Gamma(c)$ is principal.*

Let us first prove that Lemma 10.8 implies (3). Since $\Gamma(x)$ is a computable type, $\Gamma(c)$ has a decidable model $(\mathcal{B}, b)$. Since every type realized in a decidable model is computable, by Lemma 10.8, every type realized in $(\mathcal{B}, b)$ is principal. Hence $(\mathcal{B}, b)$ is a prime model of $\Gamma(c)$.

Let us next prove Lemma 10.8.

Proof. Assume otherwise. Let $\Xi = \Xi(c, x_2, \ldots, x_n)$ be a computable nonprincipal type of $\Gamma(c)$. It is an $(n-1)$–type for $n \geqslant 2$. By (2), for each $i \in \{2, \ldots, n\}$, there is the least $k(i)$ with $P_{k(i)}(x_i) \in \Xi$. Thus,

$$[\bigwedge_{2 \leqslant i \leqslant n} P_{k(i)}(x_i)] \in \Xi.$$

Since Ξ is nonprincipal, $\bigwedge_{2 \leqslant i \leqslant n} P_{k(i)}(x_i)$ is not a complete formula. Hence there are infinitely many distinct $(n-1)$–types of $\Gamma(c)$ which contain

$$\{P_{k(i)}(x_i) : 2 \leqslant i \leqslant n\}.$$

Thus, there are infinitely many distinct n–types of T which contain

$$\{P_{k(i)}(x_i) : 2 \leqslant i \leqslant n\} \cup \Gamma(x_1).$$

Since T eliminates quantifiers, every formula of $L(T)$ in n free variables for $n \geqslant 2$ is uniquely determined by the $\frac{1}{2}n(n-1)$ many 2–types it determines. Hence infinitely many 2–types of T contain

$$\{P_{k(i)}(x_i), P_{k(j)}(x_j), x_i \neq x_j\}$$

for some $i, j \in \{2, \dots, n\}$, or infinitely many 2–types of T contain $\{P_{k(i)}(x_i)\} \cup \Gamma(x_1)$ for some $i \in \{2, \dots, n\}$. Again, by the elimination of quantifiers, each of these implies that for some $k(i)$ and for infinitely many n, both $\{P_{k(i)}(x_i), S_n(x, y)\}$ and $\{P_{k(i)}(x_i), \neg S_n(x, y)\}$ are consistent. This contradicts the axioms of T'. □

(4) *If a model of T realizes a computable nonprincipal type of T, then it realizes all computable nonprincipal types of T.*

For every nonprincipal computable type $\Theta(x_1, \dots, x_n)$ of T, there is $i \in \{1, \dots, n\}$ such that $\Gamma(x_i) \subseteq \Theta(x_1, \dots, x_n)$. Therefore, any model of T realizing a nonprincipal computable type must realize Γ. As before, we can conclude that all decidable models of T realizing Γ are isomorphic. Hence the statement follows. □

While the theory T constructed in the previous theorem has only two non-isomorphic decidable models, it has $2^{\aleph_0}$ non-isomorphic countable models. Millar [138] has further shown that there is a complete decidable theory with only countably many non-isomorphic countable models, which has exactly two non-isomorphic decidable models.

11 Decidable Ehrenfeucht Theories

Definition 11.1 An *Ehrenfeucht* theory is a complete theory with only finitely many non-isomorphic countable models.

Clearly, if a complete decidable theory T is $\aleph_0$–categorical, then T has, up to isomorphism, only one countable model which can be chosen to be decidable. In 1971, Baldwin and Lachlan [23] established Vaught's conjecture that every complete $\aleph_1$–categorical theory has either exactly one or exactly ω many non-isomorphic countable models. The following result is an effective version of the Baldwin-Lachlan's result.

Theorem 11.1 (Harrington [88], Khisamiev [99]) *If a complete decidable theory T is $\aleph_1$–categorical, then every countable model of T is isomorphic to a decidable model.*

Proof. Every countable model of T can be viewed as a prime model of some other $\aleph_1$–categorical decidable theory. □

Nerode posed the following problem:

> *Let T be a complete decidable theory which has only finitely many non-isomorphic countable models. Can all of these models be chosen to be decidable?*

By Vaught's theorem, T cannot have exactly two non-isomorphic countable models. We will prove that T must have a decidable prime model. Assume otherwise. Then T has a decidable model realizing a nonprincipal type, which is omitted in another decidable model of T realizing another nonprincipal type, etc. Here we use the fact that every finite set of nonprincipal types of T is omitted in some decidable model of T. Thus, contrary to the assumption, T has infinitely many non-isomorphic decidable models.

Morley gave an example of a theory T with exactly six non-isomorphic countable models, of which only the prime one can be chosen to be computable (even decidable). Moreover, Lachlan has found a simple example of such a theory, using the fact that there is a computable linear ordering of order type $\omega + \omega*$ whose ω–part is not computable. Peretyat'kin [163] has generalized this result. He has obtained for every $n \geqslant 3$, a theory T in a finite language, with exactly n non-isomorphic countable models, of which only the prime one can be chosen to be computable (decidable). To construct such theories, Peretyat'kin has used a least upper bound operator to obtain an underlying $\aleph_0$–categorical theory which admits the elimination of quantifiers, and in which a binary tree can be distinguished by constants.

The countable non-isomorphic models of decidable Ehrenfeucht theories in all mentioned examples can be chosen to be decidable in $\mathbf{0}'$. The question then arises whether all countable models of an arbitrary complete decidable

Ehrenfeucht theory can be chosen to be, for example, arithmetic. Millar has answered this question negatively by showing that there is a complete decidable theory with only finitely many non-isomorphic countable models, some of which must be chosen to be of arbitrarily high hyperarithmetic degree. Moreover, the theory in Millar's example is persistently Ehrenfeucht (see Definition 3.2). Persistently Ehrenfeucht theories are also called *persistently finite* theories and have been introduced and first studied by Benda. It can be shown that every persistently decidable Ehrenfeucht theory has a decidable saturated model.

Definition 11.2 Let $X \subseteq \omega$. We say that a model $\mathcal{A}$ is *decidable exactly in* X if $\mathcal{A}$ is decidable in X and for every $Y \subseteq \omega$, if $\mathcal{A}$ is decidable in Y then $X \leqslant_T Y$.

Theorem 11.2 (Millar [141]) *Let H_n be a hyperarithmetic set, where $n \in \omega$. Then there is a complete decidable persistently Ehrenfeucht theory T with an undecidable countable model, such that for every undecidable countable model $\mathcal{A}$ of T, $\mathcal{A}$ is decidable exactly in H_n.*

For every H_n, the corresponding theory in the previous theorem has eighteen countable non-isomorphic models, exactly three of which are decidable. To define such a theory, Millar has used the existence of a computable subtree of $\omega^{<\omega}$ having exactly one infinite branch f, where $f \equiv_T H_n$ (see [187], page 456).

In Morley's, Lachlan's and Peretyat'kin's counterexamples to Nerode's question, the theories have non-computable types. Therefore Morley raised the following question:

> *Let T be a complete decidable theory with all types computable, which has only finitely many non-isomorphic countable models. Can all of these models be chosen to be decidable?*

Assume that one of the finitely many non-isomorphic models of T must be undecidable. Clearly, T is not $\aleph_0$–categorical. The fact that T has a finite number of non-isomorphic decidable models has several implications. As shown before, T has a decidable prime model. The set of all types of T is computably enumerable, because every computable type of T is realized in some decidable model of T. Since the set of all types of T is computably enumerable, T has a decidable saturated model. We can mimic Vaught's construction to obtain a third non-isomorphic decidable model. Therefore, if the answer to Morley's question is negative, then a counterexample must have at least four non-isomorphic countable models. Indeed, Goncharov has recently announced a negative answer to Morley's question.

Theorem 11.3 (Goncharov [75]) *There is a decidable Ehrenfeucht theory with all types computable, whose non-isomorphic countable models cannot be chosen to be all decidable.*

Millar asked the following question:

> *If T is an arithmetic Ehrenfeucht theory whose types are all arithmetic, are all countable models of T arithmetic?*

Ash and Millar have proven that if the answer to this question is negative, then a counterexample must have at least five non-isomorphic countable models. Ash and Millar have also proven that the answer to this question is positive when every type of T is realized in only finitely many non-isomorphic countable models.

Theorem 11.4 (Ash-Millar [20]) *If T is a complete, arithmetic, persistently Ehrenfeucht theory with a countable non-arithmetic model, then T has at least five non-isomorphic countable models.*

Theorem 11.5 (Ash-Millar [20]) *If T is a complete, persistently Ehrenfeucht theory with only arithmetic complete types, then all countable models of T are arithmetic.*

Theorem 11.6 (Millar [143]) *If T is a decidable Ehrenfeucht theory with a countable model which is not decidable in $\emptyset''$, then T has at least five non-isomorphic countable models.*

Theorem 11.7 (Reed [178, 179]) *Let H_n be a hyperarithmetic set, where $n \in \omega$. Then there is a decidable persistently Ehrenfeucht theory T with exactly five non-isomorphic countable models:*

(i) *A decidable prime model;*

(ii) *A decidable non-homogeneous model which is the reduct of the prime model of $\Gamma(c)$, where c is a new constant and $\Gamma(x)$ is a computable nonprincipal type of T;*

(iii) *A decidable homogeneous model which realizes all computable types of T;*

(iv) *A non-homogeneous model decidable exactly in H_n, which is the reduct of the prime model of $\Omega(d)$, where d is a new constant and $\Omega(x)$ is the only non-computable 1–type of T;*

(v) *A saturated model decidable exactly in H_n.*

Thus, the theory in the previous theorem has, up to isomorphism, three decidable models and two models which are decidable exactly in H_n. It follows from Theorem 11.6 that this is an example of a decidable Ehrenfeucht theory with the least possible number of non-isomorphic countable models which are not all decidable in $\emptyset''$. It is not known whether a decidable Ehrenfeucht theory whose undecidable countable models are decidable exactly in $\emptyset''$ can have fewer than five countable models.

Closely related to the notion of an $\aleph_0$–homogeneous model is the notion of an almost homogeneous model.

Definition 11.3 A model is *almost homogeneous* if some finite expansion of the model by constants is $\aleph_0$–homogeneous.

It is not known whether there is an Ehrenfeucht theory with a model which is not almost homogeneous. Millar [143] has shown that if T is a persistently Ehrenfeucht, persistently decidable theory whose every model is almost homogeneous, then every countable model of T is isomorphic to a decidable model.

12 Decidable Theories with Countably Many Countable Models

Millar has constructed a complete decidable theory with exactly two non-isomorphic decidable models and only countably many non-isomorphic countable models. To present Millar's construction, we use from: *computable model theory*, a characterization of a universal theory with a complete decidable model completion, as stated in Theorem 5.6; and from *computability theory*, the existence of a certain computable binary tree, as will be established by Theorem 12.1.

Definition 12.1 For a tree $\mathcal{T}$, an infinite branch f of $\mathcal{T}$ is called a limit branch if for every initial segment α of f, there is a node $\beta \in \mathcal{T}$ such that α is an initial segment of β, and β is not an initial segment of f.

For finite or infinite binary sequences f and g, we write $f <_L g$ if there is a (finite) binary sequence α such that $\alpha\hat{}0$ is an initial segment of f and $\alpha\hat{}1$ is an initial segment of g.

Theorem 12.1 (Millar [138]) *There is a computable binary tree $\mathcal{T}$ whose leaves form a computable set $\mathcal{L}$, and a unary computable function h such that the following conditions are satisfied.*

(i) $\forall \alpha \in \mathcal{T}[\alpha \notin \mathcal{L} \Rightarrow \alpha\hat{}1 \in \mathcal{T}]$

(ii) *There is exactly one limit branch of $\mathcal{T}$, which we denote by f. Moreover, f is not computable.*

(iii) *If g is an infinite branch of $\mathcal{T}$ different from f, then all but finitely many values of g are 1, and $f <_L g$.*

(iv) *If $\beta \in \mathcal{L}$ and $\gamma \in \mathcal{T}$ are such that $\gamma <_L \beta$, then* $\mathrm{lh}(\gamma) < \mathrm{lh}(\beta)$.

(v) *There is at most one element of a given length in $\mathcal{L}$. If $\alpha_0, \alpha_1, \alpha_2, \ldots$ is an enumeration of $\mathcal{L}$ in the order of the increasing length of nodes, then for all $i, j \in \omega$, $h(i) = lh(\alpha_i)$ and $(i < j \Rightarrow \alpha_i <_L \alpha_j)$.*

Theorem 12.2 (Millar [138]) *There is a complete decidable theory T with exactly two non-isomorphic decidable models, which has only countably many non-isomorphic countable models.*

Proof. We will define a complete decidable theory T such that the following conditions are satisfied.

(1) T has a nonprincipal computable 1–type, $\Gamma(x)$.

(2) Every countable model of T is homogeneous.

(3) There is a sequence $(\Gamma_n)_{n\in\omega}$ of types of T such that:

(3.1) $\Gamma_0 = \Gamma$;

(3.2) Γ_1 is non-computable;

(3.3) If $i < j$, then every model which realizes Γ_j also realizes Γ_i;

(3.4) For every $n \in \omega$, there is a countable model which realizes Γ_n and omits Γ_{n+1};

(3.5) The type spectrum of a countable model $\mathcal{A}$ of T is exactly the set of all types in $\{\Gamma_n : n \in \omega\}$ which $\mathcal{A}$ realizes.

Lemma 12.3 *Conditions* (1)–(3) *imply the Theorem.*

Proof. Since Γ is a computable type, T has a decidable model $\mathcal{A}$ which realizes Γ. Since the computable type Γ is nonprincipal, T has a decidable model $\mathcal{B}$ which omits Γ. Clearly, $\mathcal{A}$ and $\mathcal{B}$ are non-isomorphic. Let $\mathcal{D}$ be a decidable model of T. $\mathcal{D}$ must omit Γ_1, because Γ_1 is not computable. Hence, by (3.3), $\mathcal{D}$ omits every Γ_k for $k \geqslant 1$. Thus, since all countable models of T are homogeneous, if $\mathcal{D}$ realizes Γ, $\mathcal{D}$ is isomorphic to $\mathcal{A}$, and if $\mathcal{D}$ omits Γ, $\mathcal{D}$ is isomorphic to $\mathcal{B}$. Hence, T has exactly two decidable non-isomorphic models.

For every $n \in \omega$, let $\mathcal{A}_n$ be a countable model of T which realizes Γ_n and omits Γ_{n+1}. Hence, by (3.3), $\mathcal{A}_n$ realizes every Γ_k for $k \leqslant n$, and omits every Γ_k for $k > n$. Hence, by (3.5), $\{\Gamma_k : k \leqslant n\}$ is the type spectrum of $\mathcal{A}_n$. Thus, since all countable models are homogeneous, T has exactly countably many countable models.

The language of T is $L = \{P_n(\cdot), S_n(\cdot,\cdot) : n \in \omega\}$, where for $n \in \omega$, $P_n(\cdot)$ is a unary relation symbol and $S_n(\cdot,\cdot)$ is a binary relation symbol.

Let a computable binary tree $\mathcal{T}$ whose leaves form a computable set $\mathcal{L}$, and a unary function h be as in Theorem 12.1. We first define T' such that $T' \subseteq T$. The axioms of T' are the universal closures of the following formulae:

Ax 1 $P_t(x) \Rightarrow P_{t+1}(x)$ for $t \in \omega$;

Ax 2 $\neg S_n(x,x)$ for $n \in \omega$;

Ax 3 $S_n(x,y) \Rightarrow S_n(y,x)$ for $n \in \omega$;

Ax 4 $P_t(x) \Rightarrow \neg S_n(x,y)$ for $n \geqslant h(t)$;

Ax 5 $[\neg\, P_t(x) \wedge \neg\, P_t(y) \wedge x \neq y] \Longrightarrow \neg \bigwedge_{i<lh(\alpha)} S_i(x,y)^{\alpha(i)}$
for $\alpha \notin \mathcal{T}$ such that $\mathrm{lh}(\alpha) = h(t+1)$;

Ax 6 $\bigwedge_{i<lh(\alpha)} S_i(x,y)^{\alpha(i)} \Longrightarrow \neg S_n(x,y)$
for $\alpha \notin \mathcal{T} - \mathcal{L}$ such that $\mathrm{lh}(\alpha) \leqslant n$;

Ax 7 $\bigwedge_{i<lh(\alpha)} S_i(x,y)^{\alpha(i)} \Longrightarrow [P_t(x) \Leftrightarrow P_t(y)]$
for $\alpha \in \mathcal{T} - \mathcal{L}$ such that $\mathrm{lh}(\alpha) \leqslant h(t)$;

Ax 8 $\bigwedge_{i<lh(\alpha)} S_i(x,y)^{\alpha(i)} \Longrightarrow [\ P_t(x) \vee P_t(y)]$
for $\alpha \in \mathcal{L}$ such that $\mathrm{lh}(\alpha) = h(t)$;

Ax 9 $[\bigwedge_{i<\mathrm{lh}(\alpha)} S_i(x,y)^{\alpha(i)} \wedge \bigwedge_{i<\mathrm{lh}(\beta)} S_i(y,z)^{\beta(i)} \wedge x \neq y]$
$\Longrightarrow \bigwedge_{i<\mathrm{lh}(\alpha)} S_i(x,z)^{\alpha(i)}$ for $\alpha, \beta \in 2^{<\omega}$ such that $\alpha <_L \beta$.

Now it can be shown that Theorem 5.6 applies to T'. T will be a complete decidable model completion of T'. □

Let T be a complete decidable theory with all complete types computable. It is known that there is such a theory which has, up to isomorphism, $2^{\aleph_0}$ countable models. Hence it has undecidable models. Millar (see Theorem 6.5) has shown that if T does not have a decidable model whose finite expansion by constants is prime, then T must have, up to isomorphism, $2^{\aleph_0}$ countable models. The question then arises whether there is a T with only countably many non-isomorphic countable models and with an undecidable countable model. First we introduce the following

Definition 12.2 Let Γ and Ω be types of T. The *type ordering* is defined by

$$\Gamma \leqslant \Omega \Longleftrightarrow (\forall \mathcal{A} \models T)[\mathcal{A} \text{ realizes } \Gamma \Rightarrow \mathcal{A} \text{ realizes } \Omega].$$

Theorem 12.4 (Millar [142]) *There is a complete decidable theory T with all complete types computable, and with only countably many non-isomorphic countable models such that its countable saturated model is undecidable.*

Proof. The theory T is the model completion of a universal theory T', obtained using Theorem 5.6. The ordering of all nonprincipal 1-types of T is linear with order type ω^*. The set of all complete types of T is not Σ^0_2. Every decidable model of T omits a type of T, and, therefore, every countable saturated model of T is undecidable. □

Theorem 12.5 (Millar [144]) *There is a complete decidable theory T with all complete types computable, and with only countably many non-isomorphic countable models, such that T has a decidable saturated model and a countable undecidable homogeneous model.*

Proof. The theory T is the model completion of a universal theory T', obtained using Theorem 5.6. The set of all complete types of T is computably enumerable. This guarantees the existence of a decidable saturated model. The set of all types realized by a countable undecidable homogeneous model is not Σ_2^0. However, both the set of all 1-types realized and the set of all 1–types omitted by a countable undecidable homogeneous model are linearly ordered by the type ordering relation, with order type ω^*. □

13 Indiscernibles and Decidability

The notion of order indiscernibles, introduced by Ehrenfeucht and Mostowski, plays an important role in generating models with certain properties.

Let T be a fixed complete theory in L and let $\mathcal{U}$ be an $\aleph_1$–saturated model of T. Since all countable models of T are elementarily embeddable in $\mathcal{U}$, we can assume that all countable models considered in this section are elementary submodels of $\mathcal{U}$.

Definition 13.1 Let $D \subseteq U$.

(i) A set of formulae $\Gamma = \Gamma(x_0, \ldots, x_{n-1})$ is a *type over* D if there are $a_0, \ldots, a_{n-1} \in U$ such that for every formula $\theta(x_0, \ldots, x_{n-1})$ in L_D we have

$$\theta(x_0, \ldots, x_{n-1}) \in \Gamma \Longleftrightarrow \mathcal{U}_D \models \theta(x_0, \ldots, x_{n-1})[a_0, \ldots, a_{n-1}].$$

(ii) Let $B \subseteq U$, where $B = \{b_0, b_1, b_2, \ldots\}$ is a fixed enumeration of B. A set Γ of formulae with free variables among $x_0, x_1, x_2, \ldots$ is the *ω–type of B over D* (with respect to the enumeration of B) if for every $n \in \omega$, for every finite sequence $(k_0, \ldots, k_{n-1})$ of natural numbers and every formula θ in L_D in n free variables, we have

$$\theta(x_{k_0}, \ldots, x_{k_{n-1}}) \in \Gamma \Longleftrightarrow \mathcal{U}_D \models \theta(x_{k_0}, \ldots, x_{k_{n-1}})[b_{k_0}, \ldots, b_{k_{n-1}}].$$

Definition 13.2 Let $D \subseteq U$ and $I \subseteq U$, where $I = \{i_0, i_1, i_2, \ldots\}$ is a fixed enumeration of I.

(i) I is a set of (*order*) *indiscernibles over* D if for every $n \in \omega$ and every increasing n–tuple $k_0 < \ldots < k_{n-1}$ of natural numbers:

$$(i_0, \ldots, i_{n-1}) \text{ and } (i_{k_0}, \ldots, i_{k_{n-1}}) \text{ satisfy the same formulae in } L_D.$$

(ii) I is a set of *total indiscernibles over* D if for every $n \in \omega$ and every n–tuple $(k_0, \ldots, k_{n-1})$ of distinct natural numbers:

$$(i_0, \ldots, i_{n-1}) \text{ and } (i_{k_0}, \ldots, i_{k_{n-1}}) \text{ satisfy the same formulae in } L_D.$$

(iii) The indiscernibles over $\emptyset$ are simply called the *indiscernibles.*

Proposition 13.1 *Every theory with an infinite model has a model* $\mathcal{A}$ *with an infinite set* I *of indiscernibles such that* $I \subseteq A$.

Kierstead and Remmel [107] have studied computable analogues of the previous proposition. They have shown that the problem of determining whether a decidable model of T has an infinite set of indiscernibles is a Σ_1^1 question. They have investigated decidable theories which have decidable models with infinite computable sets of indiscernibles, as well as the possible Turing degrees of the sets of indiscernibles in decidable models.

Let us recall that an ω–branching tree is a tree whose nodes belong to $\omega^{<\omega}$. Kierstead and Remmel [107] have shown that the problem of finding an infinite set of indiscernibles in an infinite decidable model of T is, in some sense, equivalent to the problem of finding an infinite branch in a computable ω–branching tree. More precisely, a decidable model $\mathcal{A}$ of T is equivalent to an ω–branching tree $\mathcal{T}$ if there are oracle algorithms $\varphi_{e_1}^{(\ )}$ and $\varphi_{e_2}^{(\ )}$, such that the following is true:

(i) For every infinite set I of indiscernibles in $\mathcal{A}$, $\varphi_{e_1}^{(I)}$ outputs an infinite branch f_I of $\mathcal{T}$;

(ii) For every infinite branch f of $\mathcal{T}$, $\varphi_{e_2}^{(f)}$ outputs an infinite set I_f of indiscernibles in $\mathcal{A}$;

(iii) For every infinite branch f of $\mathcal{T}$, $f_{I_f} = f$.

Kierstead and Remmel have proven that for every decidable model of a complete theory, there exists an equivalent computable ω–branching tree; and for every computable ω–branching tree $\mathcal{T}$, there exists a complete decidable theory whose every decidable model is equivalent to $\mathcal{T}$.

Definition 13.3 Let $\mathcal{A}$ be a countable model of T, and let $D \subseteq U$ be such that $A \subseteq D$. Let $\Gamma = \Gamma(x_0, \ldots, x_{n-1})$ be a type of T over D.

(i) Γ is *definable over* $\mathcal{A}$ if for every $k \in \omega$, for every formula

$$\theta(x_0, \ldots, x_{n-1}, y_0, \ldots, y_{k-1})$$

in L_A there exists a formula $\delta_\theta = \delta_\theta(y_0, \ldots, y_{k-1})$ in L_A such that for every $d_0, \ldots, d_{k-1} \in D$

$$\theta(x_0, \ldots, x_{n-1}, \mathbf{d}_0, \ldots, \mathbf{d}_{k-1}) \in \Gamma \iff \mathcal{U} \models \delta_\theta(y_0, \ldots, y_{k-1})[d_0, \ldots, d_{k-1}].$$

We call the set

$$\{\delta_\theta(y_0, \ldots, y_{k-1}) : \theta(x_0, \ldots, x_{n-1}, y_0, \ldots, y_{k-1}) \text{ is a formula in } L_A\}$$

a definition of Γ over $\mathcal{A}$.

(ii) Γ is *computably definable over* $\mathcal{A}$ if there is an algorithm which assigns to every formula $\theta(x_0, \ldots, x_{n-1}, \overline{y})$ in L_A a formula $\delta_\theta(\overline{y})$ such that $\{\delta_\theta(\overline{y}) : \theta(x_0, \ldots, x_{n-1}, \overline{y}) \text{ is a formula in } L_A\}$ is a definition of Γ over $\mathcal{A}$.

To prove that certain theories have decidable models with infinite computable sets of indiscernibles, we use from *model theory*, a result in stability theory which establishes that the range of a sequence whose every member realizes a certain type, forms an infinite set of indiscernibles.

This result is stated in part (ii) of the following theorem, and it uses the basic fact about the unique definable extensions, stated in part (i) of the same theorem.

Theorem 13.2

(i) *Let $\mathcal{A}$ be a countable model of T, and let B and D be subsets of U such that $A \subseteq B \subseteq D$. Let $\Gamma(\overline{x})$ be a type of T over B which is definable over $\mathcal{A}$. There is a unique type over D, denoted by $\Gamma_D(\overline{x})$, which is definable over $\mathcal{A}$, such that $\Gamma(\overline{x}) \subseteq \Gamma_D(\overline{x})$.*

(ii) *Let $\mathcal{A}$ be a countable model of T such that there is a type $\Gamma(x)$ which is definable over $\mathcal{A}$ and not realized in $\mathcal{A}$. Let $(b_0, b_1, b_2, \ldots)$ be a sequence of elements in U such that for every $n \in \omega$, b_n realizes $\Gamma_{A_n}(x)$, where $A_n = A \cup \{b_k : k < n\}$. Then $\{b_0, b_1, b_2, \ldots\}$ is an infinite set of indiscernibles over $\mathcal{A}$.*

Proof.

(i) Let $\{\delta_\theta(\overline{y}) : \theta(\overline{x}, \overline{y}) \text{ is a formula in } L_A\}$ be a definition of $\Gamma(\overline{x})$ over $\mathcal{A}$. Define $\Gamma_D(\overline{x})$ to be the following set of formulae in L_A.

$$\{\theta(\overline{x}, \mathbf{d}_0, \dots, \mathbf{d}_{k-1}) : (k \in \omega) \wedge (d_0, \dots, d_{k-1} \in D) \\ \wedge (\mathcal{U} \models \delta_\theta(y_0, \dots, y_{k-1})[d_0, \dots, d_{k-1}])\}.$$

Since $B \subseteq D$, we have that $\Gamma(\overline{x}) \subseteq \Gamma_D(\overline{x})$.

$\Gamma_D(\overline{x})$ is a consistent set of formulae by compactness. $\Gamma_D(\overline{x})$ is complete because $\mathcal{U} \models (\neg\delta_\theta \Leftrightarrow \delta_{\neg\theta})$. The uniqueness of $\Gamma_D(\overline{x})$ follows from the definition of a type over a model.

(ii) Notice that, by (i), if $i, j \in \omega$ are such that $i < j$, then the restriction of the type $\Gamma_{A_j}(x)$ to A_i is the type $\Gamma_{A_i}(x)$.

Let $\{\delta_\theta(\overline{y}) : \theta(x, \overline{y}) \text{ is a formula in } L_A\}$ be a definition of $\Gamma(x)$ over $\mathcal{A}$. To show that $\{b_0, b_1, b_2, \dots\}$ is a set of indiscernibles over $\mathcal{A}$, it is enough to show that for every two increasing sequences $f, g \in 2^\omega$, for every $n \geqslant 1$, and every formula θ in L_A in n free variables

$$\mathcal{U} \models \theta[b_{f(0)}, \dots, b_{f(n-1)}] \iff \mathcal{U} \models \theta[b_{g(0)}, \dots, b_{g(n-1)}]. \qquad (*)$$

Let such f and g be given, and fix n. Assume that $f(n) < g(n)$. Let $\{\gamma_\theta(\overline{y}) : \theta(x, \overline{y}) \text{ is a formula in } L_A\}$ be a definition of $\Gamma_{A_{g(n)}}$ over $\mathcal{A}$. Then for a formula θ in L_A in $(n+1)$ free variables, we have

$$\begin{aligned}
\mathcal{U} \models \theta[b_{f(0)}, \dots, b_{f(n-1)}, b_{g(n)}] &\iff \mathcal{U} \models \gamma_\theta[b_{f(0)}, \dots, b_{f(n-1)}] \\
&\iff \theta(\mathbf{b}_{f(0)}, \dots, \mathbf{b}_{f(n-1)}, x) \in \Gamma_{A_{g(n)}}(x) \\
&\iff \theta(\mathbf{b}_{f(0)}, \dots, \mathbf{b}_{f(n-1)}, x) \in \Gamma_{A_{f(n)}}(x) \\
&\iff \mathcal{U} \models \theta[b_{f(0)}, \dots, b_{f(n-1)}, b_{f(n)}].
\end{aligned}$$

Now the equivalence in $(*)$ follows inductively.

To prove that $\{b_0, b_1, b_2, \dots\}$ is an infinite set, consider the formula $\theta(x, y) = \neg(x = y)$. To prove that $\mathcal{A} \models \forall y \delta_\theta(y)$, we assume otherwise.

Hence

$$\mathcal{A} \models \neg\delta_\theta(y)[a] \text{ for some } a \in A \implies \mathcal{U} \models \neg\delta_\theta(y)[a]$$

$$\implies (x = \mathbf{a}) \in \Gamma(x)$$

$$\implies a \text{ realizes } \Gamma(x) \text{ in } \mathcal{U}$$

$$\implies a \text{ realizes } \Gamma(x) \text{ in } \mathcal{A}.$$

However, the last statement contradicts the assumption of the theorem. Hence $\mathcal{A} \models \forall y \delta_\theta(y)$ and, thus, $\mathcal{U} \models \forall y \delta_\theta(y)$. Let $i, j \in \omega$ be such that $i < j$. Then $\mathcal{A} \models \delta_\theta(y)[b_i]$ and $\neg(x = \mathbf{b}_i) \in \Gamma_{A_j}(x)$. Since b_j realizes $\Gamma_{A_j}(x)$, we have that $b_i \neq b_j$. □

Theorem 13.3 (Kierstead-Remmel [106]) *Let $\mathcal{A}$ be a decidable model of T such that there is a computably definable type $\Gamma(x)$ over $\mathcal{A}$, which is not realized in $\mathcal{A}$. Then T has a decidable model with an infinite computable set of indiscernibles.*

Proof. Let $\{\delta_\theta(\overline{y}) : \theta(x, \overline{y}) \text{ is a formula in } L_A\}$ be a definition of $\Gamma(x)$ over $\mathcal{A}$ such that there is an algorithm which to every formula $\theta(x, \overline{y})$ in L_A assigns $\delta_\theta(\overline{y})$. Let L' be $L_A \cup \{c_0, c_1, c_2, \ldots\}$, where $c_0, c_1, c_2, \ldots$ is a computable enumeration of new constants. We inductively define the following sets of sentences in L'

$$T_0 = \text{the theory of } \mathcal{A}_A \text{ in } L_A$$

$$T_{n+1} = T_n \cup \{\theta(c_n, c_{n-1}, \ldots, c_0) : (\theta(x_n, x_{n-1}, \ldots, x_0) \text{ is in } L_A) \\ \wedge \delta_\theta(c_{n-1}, \ldots, c_0) \in T_n\} \text{ for } n \geqslant 0.$$

Let $T' =_{\text{def}} \bigcup_{n \in \omega} T_n$. T' is a consistent complete theory in L'. T' is decidable because $\mathcal{A}$ is decidable and the considered definition of $\Gamma(x)$ is algorithmic. By the Effective Completeness Theorem, there is a decidable model $\mathcal{B}$ of T'. As mentioned before, we assume that $\mathcal{B} \preceq \mathcal{U}$. Let $I =_{\text{def}} \{b_0, b_1, b_2, \ldots\}$, where for every $i \in \omega$, b_i is the interpretation in $\mathcal{B}$ of the constant c_i. Since $\mathcal{B}$ satisfies T_0, we can assume that $\mathcal{A} \preceq \mathcal{B}$. Clearly, I is a computable set.

We use Theorem 13.2 (ii) to show that I is an infinite set of indiscernibles. Let $A_0 = A$, and $A_{n+1} = A \cup \{b_0, \ldots, b_n\}$. We show that

$$\Gamma_{A_n}(x) = \{\theta(x, c_{n-1}, \ldots, c_0) : (\theta(x_n, x_{n-1}, \ldots, x_0) \text{ is in } L_A) \\ \wedge \delta_\theta(c_{n-1}, \ldots, c_0) \in T_n\}.$$

Thus, b_n realizes $\Gamma_{A_n}(x)$. □

Examples of theories to which Theorem 13.3 applies are the theory of dense linear order without endpoints, and the theory of real closed fields.

Theorem 13.4 (Kierstead-Remmel [106]) *Let Q be a generalized quantifier whose interpretation is "there are infinitely many". Assume that T is a stable theory which also satisfies the following decidability condition* (D).

> *There is an effective procedure which decides for every formula in L of the form $\varphi(x, y_0, \ldots, y_{k-1})$, whether*
>
> $$T \cup \{(\exists y_0) \ldots (\exists y_{k-1})(Qx)\varphi(x, y_0, \ldots, y_{k-1})\}$$
>
> *has a model.*

Then T has a decidable model with an infinite computable set of total indiscernibles.

Proof. Such a theory T has a decidable model $\mathcal{A}$ and a type Γ over A, such that Γ is computably definable over $\mathcal{A}$, and not realized in $\mathcal{A}$. Also, since T is stable, every set of order indiscernibles is a set of total indiscernibles. □

The strong decidability condition (D) in the previous theorem cannot be replaced by the usual decidability condition. Also, the stability condition cannot be omitted from the assumption of the theorem, as shown by the following counterexample.

Proposition 13.5 (Kierstead-Remmel [106]) *There is a complete theory T satisfying the decidability condition (D) such that T has infinitely many decidable models, none of which has an infinite computable set of indiscernibles, although each of them has an infinite set of indiscernibles.*

It is well known from the classical model theory that every $\aleph_0$–stable theory is stable in all infinite powers. It is easy to show that there is an $\aleph_0$–stable decidable theory which does not satisfy the decidability condition in Theorem 13.4.

Theorem 13.6 (Kierstead-Remmel [106]) *If T is an $\aleph_0$–stable and decidable theory, then T has a decidable model with an infinite computable set of total indiscernibles.*

Kierstead and Remmel have shown that $\aleph_0$–stability in the previous theorem can be replaced neither by stability nor even by superstability.

Proposition 13.7 (Kierstead-Remmel [106]) *There is a complete decidable superstable theory which has an infinite decidable model, but it does not have a decidable model with an infinite computable set of indiscernibles.*

The following result illustrates an application of Theorem 13.6.

Theorem 13.8 (Kierstead-Remmel [106]) *If T is $\aleph_0$–stable and decidable, then T has models of arbitrarily large cardinality, which realize only computable types.*

Proof. By Theorem 13.6, T has a decidable model $\mathcal{A}$ with an infinite computable set of indiscernibles I. Let κ be an arbitrary infinite cardinal. There are a model $\mathcal{B}$ of T, and a subset J of B of cardinality κ such that J is the set of indiscernibles satisfying the same ω–type of T as I. Since T is $\aleph_0$–stable, by a result from model theory, there is a prime model $\mathcal{C}$ over J. Clearly, $\mathcal{A}$ and $\mathcal{C}$ realize the same types. Since $\mathcal{A}$ is decidable, the types that they realize are computable. □

Theorem 13.9 (Kierstead-Remmel [107]) *If $\mathcal{A}$ is a decidable model with an infinite set of indiscernibles, then $\mathcal{A}$ has an infinite set I of indiscernibles such that the hyperdegree of I is strictly less than the hyperdegree of Kleene's $\mathcal{O}$.*

Kierstead and Remmel have also investigated the degrees of sets of indiscernibles in decidable models of $\aleph_0$–categorical theories.

Definition 13.4 A decidable theory T has *decidable atoms* if there is an effective procedure which decides whether a given formula is an atom in the Lindenbaum algebra of formulae with the corresponding free variables.

Kierstead and Remmel [107] have shown that the problem of finding an infinite set of indiscernibles in an infinite decidable model of an $\aleph_0$–categorical theory with decidable atoms is, in some sense, equivalent to the problem of finding an infinite branch in an infinite computable tree. In particular, for every infinite computable binary tree $\mathcal{T}$, there is a decidable model $\mathcal{A}$ of an $\aleph_0$–categorical decidable theory with decidable atoms, such that there is an effective one-to-one correspondence between the infinite branches of $\mathcal{T}$ and the ω–types of infinite sets of indiscernibles in $\mathcal{A}$.

Thus, the set of Turing degrees realized by the sets of ω–types of infinite sets of order indiscernibles in a decidable model of an $\aleph_0$–categorical theory coincides with the set of degrees realized by recursively bounded Π_1^0 classes. Thus, the following result follows from Jockusch-Soare's work [94] on Turing degrees of Π_1^0 classes.

Theorem 13.10 (Kierstead-Remmel [107]) *Let $\mathcal{A}$ be a decidable model of an $\aleph_0$–categorical theory with decidable atoms. $\mathcal{A}$ has an infinite set of indiscernibles of low Turing degree, and $\mathcal{A}$ has an infinite set of indiscernibles of a c.e. degree. If $\mathcal{A}$ does not have an infinite computable set of indiscernibles, then there are continuum many ω–types of infinite sets of indiscernibles, which have mutually incomparable Turing degrees.*

There are decidable models of T with infinite sets of indiscernibles which have no hyperarithmetic infinite sets of indiscernibles. However, it is not true if T is $\aleph_0$–categorical.

Theorem 13.11 (Kierstead-Remmel [107]) *If $\mathcal{A}$ is a decidable model of an $\aleph_0$–categorical complete theory, then $\mathcal{A}$ has an infinite set I of indiscernibles such that* $\deg(I) \leqslant \mathbf{0}'$.

14 Degrees of Models

Clearly, a computably axiomatizable complete theory is computably enumerable. Kleene [108] and Hasenjaeger [89] have independently shown that if T is a computably axiomatizable theory, then T has a countable model whose domain is a set of natural numbers, such that every relation and function of the model is Δ_2^0. On the other hand, there is a computably axiomatizable theory which does not have a model in which every relation and function is c.e. or co-c.e.

Unless otherwise stated, we consider only models whose domain is ω. (For such a model $\mathcal{A}$, a set of formulae in L_A can be thought of as a set of natural numbers.) This allows us to define the (*Turing*) *degree* of $\mathcal{A}$, denoted by $\deg(\mathcal{A})$, as the Turing degree of the atomic diagram $\Delta_{\mathcal{A}}$ of $\mathcal{A}$. Thus, $\mathcal{A}$ is computable if and only if $\deg(\mathcal{A}) = \mathbf{0}$.

It is easy to see that the theory of a model $\mathcal{A}$ is computable in the complete diagram of $\mathcal{A}$, and that the complete diagram of $\mathcal{A}$ is computable in $(\Delta_{\mathcal{A}})^{(\omega)}$. Henkin's construction of a model of a complete theory T produces a model $\mathcal{B}$ whose atomic diagram and complete diagram are both computable in T (see Theorem 4.1). Hence T and the complete diagram of $\mathcal{B}$ have the same Turing degree. The atomic diagram of a model of T may be of much lower Turing degree than T. For example, *true arithmetic* is the theory of the standard model of natural numbers, and its Turing degree is $\mathbf{0}^{(\omega)}$.

Shoenfield has used the following lemma from computability theory to improve Hasenjaeger's and Kleene's result.

Lemma 14.1 (Kreisel's Basis Lemma) *An infinite computable binary tree has a Δ_2^0 infinite branch.*

Shoenfield has first strengthened Kreisel's Basis Lemma by proving that an infinite computable binary tree has an infinite branch of Turing degree $< \mathbf{0}'$.

Theorem 14.2 (Shoenfield [194]) *If T is a computably axiomatizable theory, then T has a countable model whose degree is $< \mathbf{0}'$.*

Proof. Extend T to a complete theory S in the same language such that the Turing degree of S is $< \mathbf{0}'$. This can be done using Shoenfield's strengthening of Kreisel's Basis Lemma. □

Jockusch and Soare [95] have generalized Kreisel-Shoenfield Basis Theorem.

Theorem 14.3 (Low Basis Theorem) *An infinite computable binary tree has an infinite branch of low Turing degree.*

The Low Basis Theorem implies that every computably axiomatizable theory has a model of low Turing degree.

Knight [110] has shown that for a model $\mathcal{A}$, either there is a finite set $S \subseteq A$ such that all bijections of A that fix S are automorphisms of $\mathcal{A}$; or for every Turing degree $\mathbf{d} \geqslant \deg(\mathcal{A})$, there is a model $\mathcal{B}$ isomorphic to $\mathcal{A}$ such that $\deg(\mathcal{A}) = \mathbf{d}$. Wehner [212] and Slaman [195] have independently found a countable model $\mathcal{A}$ such that the Turing degrees of models isomorphic to $\mathcal{A}$ are exactly the non-computable degrees.

Since the degree of a model is not invariant under isomorphisms, Jockusch has introduced the following complexity measure of the isomorphism type of a model. The isomorphism type of a model $\mathcal{A}$ is the set of all models isomorphic to $\mathcal{A}$.

Definition 14.1 (Richter [185]) The *degree of the isomorphism class* of $\mathcal{A}$, if it exists, is the least Turing degree in $\{\deg(\mathcal{B}) : \mathcal{B} \cong \mathcal{A}\}$.

The following theorem establishes that the degree of the isomorphism class of a model satisfying certain general computable condition cannot be different from $\mathbf{0}$.

Theorem 14.4 (Richter [186]) *Assume that a model $\mathcal{A}$ satisfies the following computable embeddability condition.*

> *For every finite model $\mathcal{C}$ isomorphic to a submodel of $\mathcal{A}$ and every embedding f of $\mathcal{C}$ into $\mathcal{A}$, there is an algorithm which determines whether a given finite model $\mathcal{D}$ extending $\mathcal{C}$ can be embedded into $\mathcal{A}$ by an embedding extending f.*

Then if the degree of the isomorphism class of $\mathcal{A}$ exists, it must be $\mathbf{0}$.

Proof. If $\mathcal{A}$ is a computable model, then the statement follows immediately. Assume that $\mathcal{A}$ is not computable. We will prove that there is model $\mathcal{B}$ isomorphic to $\mathcal{A}$ such that $\deg(\mathcal{A})$ and $\deg(\mathcal{B})$ form a minimal pair. Hence $\mathbf{0}$ will be the only possible degree of the isomorphism class of $\mathcal{A}$. A model $\mathcal{B}$ and an isomorphism h from $\mathcal{B}$ onto $\mathcal{A}$ will be constructed in stages by finite extension. Let L be the language of $\mathcal{A}$.

Construction

STAGE 0: Let $\mathcal{B}_0 = \emptyset$ and $h_0 = \emptyset$.

STAGE $s = 2e + 1$: First assume that there is a finite model $\mathcal{C}$ for L extending $\mathcal{B}_s$ and an embedding g of $\mathcal{C}$ into $\mathcal{A}$ extending h_s, such that for some $n \in \omega$, both $\{e\}^{\mathcal{C}}(n)$ and $\{e\}^{\mathcal{A}}(n)$ are defined and

$$\{e\}^{\mathcal{C}}(n) \neq \{e\}^{\mathcal{A}}(n).$$

In this case, for some such $\mathcal{C}$ and g, let $\mathcal{B}_{s+1} =_{\mathrm{def}} \mathcal{C}$ and $h_{s+1} =_{\mathrm{def}} g$. Otherwise, let $\mathcal{B}_{s+1} =_{\mathrm{def}} \mathcal{B}_s$ and $h_{s+1} =_{\mathrm{def}} h_s$.

STAGE $s = 2e + 2$: If $e \in A - \mathrm{rng}(h_s)$, let $h_{s+1} =_{\mathrm{def}} h_s \cup \{(u, e)\}$, where $u \in \omega$ is the least number such that $u \notin \mathrm{dom}(h_s)$. Otherwise, let $h_{s+1} =_{\mathrm{def}} h_s$. In both cases, extend $\mathcal{B}_s$ to $\mathcal{B}_{s+1}$ such that h_{s+1} is an embedding of $\mathcal{B}_{s+1}$ into $\mathcal{A}$. End of the construction.

Let $\mathcal{B} =_{\mathrm{def}} \bigcup_{s\in\omega} \mathcal{B}_s$ and $h =_{\mathrm{def}} \bigcup_{s\in\omega} h_s$. Clearly, h is an isomorphism from $\mathcal{B}$ to $\mathcal{A}$. Now, let us prove that $\deg(\mathcal{A})$ and $\deg(\mathcal{B})$ form a minimal pair. Since $\mathcal{A}$ is not computable, by Posner's Lemma, it is enough to prove that for every $e \in \omega$:

$$\{e\}^{\mathcal{A}} = \{e\}^{\mathcal{B}} = f \text{ total} \implies f \text{ is computable.}$$

Thus, assume $\{e\}^{\mathcal{A}} = \{e\}^{\mathcal{B}} = f$, where f is total. By construction, there is a stage s such that for every finite extension $\mathcal{C}$ of $\mathcal{B}_s$ which can be embedded into $\mathcal{A}$, and every $n \in \omega$ such that $\{e\}^{\mathcal{C}}(n)$ is defined, we have $\{e\}^{\mathcal{C}}(n) = \{e\}^{\mathcal{A}}(n)$. Hence $f(n) = \{e\}^{\mathcal{A}}(n)$. By the computable embeddability condition, f must be computable. □

The previous theorem can be applied to show that the isomorphism class of a countable tree which is not isomorphic to a computable tree, does not have a degree. Hence, the isomorphism class of a countable linear ordering which is not isomorphic to a computable linear ordering does not have a degree.

Theorem 14.5 (Richter [186]) *Let S be a theory in a finite language L such that there is a computable sequence $\mathcal{A}_0, \mathcal{A}_1, \mathcal{A}_2, \ldots$ of finite models for L which are pairwise non-embeddable. Assume that for every $X \subseteq \omega$, there is a countable model $\mathcal{A}_X$ of S which is computable in X and*

$$(\forall i)[\mathcal{A}_i \text{ is embeddable in } \mathcal{A}_X \Leftrightarrow i \in X].$$

Then for every Turing degree $\mathbf{d}$, *there is a countable model of S whose isomorphism class has degree* $\mathbf{d}$.

Proof. Let $\mathbf{d}$ be a Turing degree and $D \subseteq \omega$ be such that $\deg(D) = \mathbf{d}$. Let $X =_{\mathrm{def}} D \oplus \overline{D}$. We will show that $\mathcal{A}_X$ is a countable model of S whose isomorphism class has degree $\mathbf{d}$. Clearly,

$$\mathcal{A}_{D\oplus\overline{D}} \leqslant_T D \oplus \overline{D} \leqslant_T D,$$

so $\deg(\mathcal{A}_X) \leqslant \mathbf{d}$. Let $\mathcal{B}$ be a model isomorphic to $\mathcal{A}_X$. It is enough to prove that $\deg(\mathcal{B}) \geqslant \mathbf{d}$. This follows from the fact that

$$(i \in D \Leftrightarrow \mathcal{A}_{2i} \text{ is embeddable in } \mathcal{B}) \wedge$$
$$(i \notin D \Leftrightarrow \mathcal{A}_{2i+1} \text{ is embeddable in } \mathcal{B}).$$

□

The previous theorem can be used to show that for every Turing degree **d**, there is a countable abelian group whose isomorphism class has degree **d**. A corresponding sequence of finite models consists of cyclic groups of every prime order, and the abelian group assigned to an arbitrary set of natural numbers is obtained by forming countable direct sums.

Theorem 14.3 implies that there is a nonstandard model of Peano arithmetic of low degree. McAloon has asked whether there is a nonstandard model $\mathcal{A}$ of Peano arithmetic such that the theory of $\mathcal{A}$ is not arithmetic and the degree of $\mathcal{A}$ is arithmetic. Harrington has given the answer by establishing the following result.

Theorem 14.6 (Harrington) *There is a nonstandard model $\mathcal{A}$ of Peano arithmetic such that the theory of $\mathcal{A}$ has degree $\mathbf{0}^{(\omega)}$ and the degree of $\mathcal{A}$ is $\leqslant \mathbf{0}'$.*

The construction uses Harrington's *worker method* with infinitely many workers. The n–th worker produces the Σ_n–part of the complete diagram of the model, using $\emptyset^{(n)}$ as an oracle. To assure coherence, every n–th worker constantly guesses what the $(n+1)$–st worker has done. In [111], Knight has improved Harrington's result by showing that there is a nonstandard model $\mathcal{A}$ of Peano arithmetic such that the theory of $\mathcal{A}$ has degree $\mathbf{0}^{(\omega)}$, and the degree of $\mathcal{A}$ is low. This result follows from a general theorem of Knight [111] for which she has used Harrington's worker method with infinitely many workers to produce a model of a theory T, which realizes a certain set of types of bounded complexity.

Feferman [59] has stated that every arithmetic set is computable in the degree of every nonstandard model of true arithmetic. In fact, his proof yields a stronger result. First we need the following definition.

Definition 14.2 A Turing degree **d** is a *subuniform upper bound* for the arithmetic sets if there is $X \subseteq \omega$ such that $\deg(X) \leqslant \mathbf{d}$ and

$$(\forall n)(\exists i)[X^{[i]} = \emptyset^{(n)}].$$

Theorem 14.7 (Feferman [59]) *If $\mathcal{A}$ is a nonstandard model of true arithmetic of degree* $\mathbf{d}$*, then* $\mathbf{d}$ *is a subuniform upper bound for the arithmetic sets.*

As Marker has pointed out, certain results on the degrees of nonstandard models of true arithmetic are analogous to the results on the degrees of nonstandard models of Peano arithmetic. From the fact that the degree of true arithmetic is $\mathbf{0}^{(\omega)}$, it follows that there is a nonstandard model of true arithmetic of degree $\leqslant \mathbf{0}^{(\omega)}$. Knight has shown that there is such a model of degree $< \mathbf{0}^{(\omega)}$. Marker has used a modification of Harrington's worker method with three workers to obtain the following result.

Theorem 14.8 (Marker [127]) *Let* $\mathbf{d}$ *be a Turing degree such that for every* $n \geqslant 0$, $\mathbf{d} > \mathbf{0}^{(n)}$*. Then there is a nonstandard model $\mathcal{A}$ of true arithmetic such that* $\deg(\mathcal{A}) \leqslant \mathbf{d}'$.

It follows from the previous theorem that there is a nonstandard model of true arithmetic whose degree $\mathbf{d}$ is such that $\mathbf{d}' = \mathbf{0}^{(\omega)}$. Marker has also shown that for a nonstandard model $\mathcal{A}$ of Peano arithmetic, the set of degrees of all models isomorphic to $\mathcal{A}$ is closed upward. In particular, the set of degrees of all nonstandard models of true arithmetic is closed upward.

Knight, Lachlan and Soare [116] have strengthened Theorem 14.8 by showing that, given $\mathbf{d}$ as in Theorem 14.8, there is a nonstandard model $\mathcal{A}$ of true arithmetic such that $(\deg(\mathcal{A}))' \leqslant \mathbf{d}'$. As a consequence of their result, they have obtained

Corollary 14.9 (Knight-Lachlan-Soare [116]) *There is a nonstandard model of true arithmetic of degree* $\mathbf{d}$ *such that* $\mathbf{d}'' = \mathbf{0}^{(\omega)}$.

Proof. By a result of Sacks, there is a Turing degree $\mathbf{d}$ such that $\mathbf{d}'' = \mathbf{0}^{(\omega)}$, and for every $n \in \omega$, $\mathbf{d} > \mathbf{0}^{(n)}$. Fix such $\mathbf{d}$. Let $\mathcal{A}$ be a nonstandard model of true arithmetic such that $(\deg(\mathcal{A}))' \leqslant \mathbf{d}'$. Hence $(\deg(\mathcal{A}))'' \leqslant \mathbf{0}^{(\omega)}$. □

Knight attempted to answer Jockusch's question about a characterization of the degrees of nonstandard models of true arithmetic, by conjecturing that if a degree $\mathbf{d}$ is such that $(\forall n \geqslant 1)[\mathbf{d} > \mathbf{0}^{(n)}]$, then $\mathbf{d}$ is the degree of a model of true arithmetic. This conjecture is refuted by the following theorem.

Theorem 14.10 (Knight-Lachlan-Soare [116]) *There is a Turing degree* $\mathbf{d}$ *which is not a subuniform upper bound for the arithmetic sets, such that* $(\forall n \geqslant 1)[\mathbf{d} > \mathbf{0}^{(n)}]$. *In addition,* $\mathbf{d}'' = \mathbf{0}^{(\omega)}$.

If $\mathbf{d}$ is as in Theorem 14.10, then, by Theorem 14.7, $\mathbf{d}$ is not the degree of a model of true arithmetic.

In the 1984 *Logic Colloquium* material, Solovay gave a characterization of the degrees of nonstandard model of true arithmetic. Solovay's characterization is in terms of the effective enumerations of families of the so-called Scott sets.

Let $\alpha_0, \alpha_1, \alpha_2, \dots$ be a computable enumeration without repetition of all nodes in $2^{<\omega}$.

Definition 14.3 A set $\mathcal{S} \subseteq \mathcal{P}(\omega)$ is called a *Scott set* if it satisfies the following conditions for all $X, Y \subseteq \omega$:

(1) $(X \in \mathcal{S} \wedge Y \leqslant_T X) \Rightarrow Y \in \mathcal{S}$;

(2) $(X \in \mathcal{S} \wedge Y \in \mathcal{S}) \Rightarrow X \oplus Y \in \mathcal{S}$;

(3) $[X \in \mathcal{S} \wedge (\mathcal{T} = \{\alpha_n : n \in X\}$ is an infinite tree$)]$
$\Longrightarrow (\exists Z \in \mathcal{S})[\{\alpha_n : n \in Z\}$ is an infinite branch of $\mathcal{T}]$.

For $n \in \omega$, let $\theta_n(x)$ be a formula in the language of Peano arithmetic which expresses that "x is divisible by the n–th prime number". If $\mathcal{A}$ is a nonstandard model of Peano arithmetic, then

$$\{\{n : \mathcal{A} \models \theta_n(x)[a]\} \mid a \in A\}$$

is a Scott set. It is called *the Scott set of* $\mathcal{A}$ and is denoted by $Scott(\mathcal{A})$.

Definition 14.4 Let T be a complete extension of Peano arithmetic and let $X \subseteq \omega$. X is *representable* with respect to T if for some formula $\theta(x)$ of $L(T)$ and every $n \in \omega$:

$$[T \vdash \theta(\mathbf{n})] \Leftrightarrow n \in X.$$

Scott [192] has proven that the family of all Scott sets coincides with the family of sets which are representable with respect to some complete extension of Peano arithmetic.

An *enumeration* of a countable family $\mathcal{S} \subseteq \mathcal{P}(\omega)$ is a binary relation ν such that $\mathcal{S} = \{\nu_0, \nu_1, \nu_2, \dots\}$, where for every $i \in \omega$, $\nu_i =_{\text{def}} \{n : (i, n) \in \nu\}$. By "effectivizing" conditions $(1)-(3)$ in Definition 14.3, we obtain the notion of an effective enumeration of a Scott set.

Definition 14.5 Let $\mathcal{S}$ be a countable Scott set. An enumeration ν of $\mathcal{S}$ is an effective enumeration if there are computable functions $f(\cdot,\cdot)$, $g(\cdot,\cdot)$ and $h(\cdot,\cdot)$ such that the following conditions are satisfied for all $i,j \in \omega$:

(1) $(\nu_i = X \wedge Y = \{e\}^X) \Rightarrow Y = \nu_{f(i,e)}$;

(2) $\nu_i \oplus \nu_j = \nu_{g(i,j)}$;

(3) $[\nu_i = X \wedge (\mathcal{T} = \{\alpha_n : n \in X\}$ is an infinite tree$) \wedge \nu_{h(i)} = Y]$
$\Longrightarrow [\{\alpha_n : n \in Y\}$ is an infinite branch of $\mathcal{T}]$.

Theorem 14.11 (Solovay) *Let* **d** *be a Turing degree.*

(i) **d** *is the degree of a nonstandard model of true arithmetic*

$\Leftrightarrow$ **d** *is the degree of an effective enumeration of a countable Scott set which contains all arithmetic sets.*

(ii) *Let* $\mathcal{S}$ *be a countable Scott set.*

d *is the degree of a nonstandard model of true arithmetic with the Scott set* $\mathcal{S}$

$\Leftrightarrow$ **d** *is the degree of an effective enumeration of* $\mathcal{S}$.

In the following theorem, Knight has established a general sufficient condition for a Turing degree to be the degree of a model representing a given Scott set. We will use the following notation in the theorem. For a theory T and $n \in \omega$, we define T_n to be the set of Gődel numbers of all Σ_n sentences in T.

Theorem 14.12 (Knight [113]) *Let* ν *be an effective enumeration of a (countable) Scott set* $\mathcal{S}$, *and let* T *be a complete theory such that for every* $n \in \omega$, $T_n \in \mathcal{S}$. *Assume that there is an algorithm which on every input* $n \in \omega$, *using the* n*–th jump of* ν, *outputs* $i \in \omega$ *such that* $\nu_i = T_{n+1}$. *Then there is a model* $\mathcal{A}$ *of* T *which represents* $\mathcal{S}$, *such that the atomic diagram of* $\mathcal{A}$ *is Turing reducible to* ν.

This theorem gives Theorem 14.11 as a corollary. Another corollary is the following strengthening of Theorem 14.6.

Theorem 14.13 (Knight [113]) *Let ν be an effective enumeration of a (countable) Scott set $\mathcal{S}$. Let* $\mathbf{d}$ *be the Turing degree of ν. There is a nonstandard model $\mathcal{A}$ of Peano arithmetic with the Scott set $\mathcal{S}$ such that the theory of $\mathcal{A}$ has degree $\geqslant \mathbf{d}^{(\omega)}$ and the degree of $\mathcal{A}$ is $\leqslant \mathbf{d}$.*

Since many models do not have the degree of their isomorphism class, Jockusch has introduced another measure of model complexity which is invariant under isomorphisms. This measure uses jumps of the degrees of models.

Definition 14.6 Let α be a computable ordinal. The *α–th jump degree* of a model $\mathcal{A}$ is, if it exists, the least Turing degree among $\{\deg(\mathcal{B})^{(\alpha)} : \mathcal{B} \cong \mathcal{A}\}$.

Obviously, the notion of the 0–th jump degree of $\mathcal{A}$ coincides with the notion of the degree of the isomorphism class of $\mathcal{A}$. While Richter [186] has shown that the only possible 0–th jump degree of a linear ordering is $\mathbf{0}$, Knight [110] has shown that the only possible first jump degree of a linear ordering is $\mathbf{0}'$. No nonstandard model of Peano arithmetic has 0–th jump degree. There is a nonstandard model of Peano arithmetic with a 1–st jump degree. We have the following general results for jump degrees of linear orderings and Boolean algebras.

Theorem 14.14 (Knight [110], Ash-Knight [9], Jockusch-Soare [96], Ash-Jockusch-Knight [8], Downey-Knight [46]) *Let $\alpha \geqslant 1$ be a computable ordinal and let* $\mathbf{d}$ *be a Turing degree such that $\mathbf{d} \geqslant \mathbf{0}^{(\alpha)}$. Then there is a linear ordering $\mathcal{A}$ whose α–th jump degree is* $\mathbf{d}$ *and such that $\mathcal{A}$ does not have β–th jump degree for any $\beta < \alpha$.*

Theorem 14.15 (Jockusch-Soare [97])

(i) *Let* $\mathbf{d}$ *be a Turing degree such that $\mathbf{d} \geqslant \mathbf{0}^{(\omega)}$. Then there is a Boolean algebra $\mathcal{A}$ whose ω–th jump degree is* $\mathbf{d}$.

(ii) *Let $n \in \omega$, and* $\mathbf{d}$ *be a Turing degree such that $\mathbf{d} > \mathbf{0}^{(n)}$. Then there is no Boolean algebra $\mathcal{A}$ whose n–th jump degree is* $\mathbf{d}$.

Result (i) of Theorem 14.15 is a straightforward application of a method by Feiner, see [97].

15 Automorphisms and Computable Models

One of the important and interesting questions in computable model theory is how a specific aspect of a computable model may change if the model is isomorphically transformed so that it remains computable. A model $\mathcal{B}$ isomorphic to a computable model $\mathcal{A}$ is not necessarily computable. However, even if $\mathcal{B}$ is computable, it can still lose many of the computable properties of $\mathcal{A}$.

A computable property of a computable model $\mathcal{A}$ which Ash and Nerode have considered is an additional computable relation R on the domain of $\mathcal{A}$ (that is, R is not named in the language of $\mathcal{A}$). For example, Ash and Nerode have studied conditions under which the image of R under any isomorphism from $\mathcal{A}$ to another computable model is necessarily a computable or a c.e. relation.

Definition 15.1 Let R be an additional relation on the domain of a computable model $\mathcal{A}$.

(i) (Ash-Nerode [21]) R is *intrinsically c.e.* on $\mathcal{A}$ if the image of R under every isomorphism from $\mathcal{A}$ to a computable model is c.e.

(ii) Let $\mathcal{P}$ be a certain class of relations. R is called *intrinsically* $\mathcal{P}$ on $\mathcal{A}$ if the image of R under every isomorphism from $\mathcal{A}$ to a computable model belongs to $\mathcal{P}$.

For example, Moses [155] has established that relations which are intrinsically computable on a computable linear order $\mathcal{A}$ are precisely those that are equivalent in $\mathcal{A}$ to quantifier-free formulae with finitely many parameters. Let $\mathcal{A}$ be a computable Boolean algebra and let R be a computable subalgebra of $\mathcal{A}$. Odintsov [161] has established that R is intrinsically c.e. if and only if R is generated by a finite set of elements and a finite set of principal ideals of $\mathcal{A}$. This characterization implies that if R is intrinsically c.e. then R is intrinsically computable. However, it is easy to see that there are intrinsically c.e. relations which are not intrinsically computable.

Ash and Nerode have introduced a computable syntactic condition for a new relation on the domain of a computable model, to be called a formally c.e. relation.

Definition 15.2

(i) An $L_{\omega_1\omega}$ formula with free variables among $\overline{x}$ is a (computable) Σ_1 formula if it is equivalent to a formula of the form

$$\bigvee_{n\in\omega} \exists\overline{y}_n\theta_n(\overline{x},\overline{y}_n),$$

where $(\theta_n(\overline{x},\overline{y}_n))_{n\in\omega}$ is a (computable) sequence of quantifier-free formulae.

(ii) (Ash-Nerode [21]) Let R be an additional m–ary computable relation on the domain of a computable model $\mathcal{A}$. R is *formally c.e.* on $\mathcal{A}$ if and only if there is a finite sequence $(b_0,\dots,b_{k-1})$ of elements in A and a computable Σ_1 formula $\mathcal{F}(x_0,\dots,x_{m-1},\mathbf{b}_0,\dots,\mathbf{b}_{k-1})$ such that the following equivalence holds for every $a_0,\dots,a_{m-1}\in A$:

$$R(a_0,\dots,a_{m-1}) \Leftrightarrow \mathcal{A}_A \models \mathcal{F}(\mathbf{a}_0,\dots,\mathbf{a}_{m-1},\mathbf{b}_0,\dots,\mathbf{b}_{k-1}).$$

R is *formally computable* on $\mathcal{A}$ if both R and its complement are formally c.e. on $\mathcal{A}$.

That is, R is formally c.e. on $\mathcal{A}$ if and only if R is equivalent to an infinite disjunction of a computable sequence of existential formulae with finitely many fixed parameters from A. A formally c.e. relation is also called a *formally* Σ_1 relation.

Clearly, every formally c.e. relation on a computable model is intrinsically c.e. Ash and Nerode have proven, under a certain decidability condition (D), the converse, thus establishing the equivalence of a syntactic and a semantic condition. For an m–ary relation R on a model $\mathcal{A}$, the condition (D) is:

There is an algorithm which determines for $k\in\omega$, for an existential formula $\psi(x_0,\dots,x_{m-1},y_0,\dots,y_{k-1})$ and a sequence $(b_0,\dots,b_{k-1})$ of elements of A, whether the following implication holds for every $a_0,\dots,a_{m-1}\in A$:

$$[\mathcal{A}_A \models \psi(\mathbf{a}_0,\dots,\mathbf{a}_{m-1},\mathbf{b}_0,\dots,\mathbf{b}_{k-1})] \Longrightarrow R(a_0,\dots,a_{m-1}).$$

Condition (D) implies that R is a computable relation. It also implies that $\mathcal{A}$ is 1–computable, which is a property of a model defined as follows.

Definition 15.3 A model $\mathcal{A}$ is 1–*computable* if there is an algorithm which determines for every existential formula $\psi(x_0,\dots,x_{n-1})$ and every sequence $(a_0,\dots,a_{n-1})$ of elements of A, whether $\psi(\mathbf{a}_0,\dots,\mathbf{a}_{n-1})$ is true in $\mathcal{A}_A$.

Let $\mathcal{P}$ be a class of formulae. Define the $\mathcal{P}$*–diagram* of a model $\mathcal{A}$ for language L to be the set of all $\mathcal{P}$–sentences in L_A which are true in $\mathcal{A}_A$. Thus, a model is 1–computable if its existential diagram is computable or, equivalently, its universal diagram is computable.

Theorem 15.1 (Ash-Nerode [21]) *Let R be an additional m–ary relation on the domain of a model $\mathcal{A}$, satisfying the decidability condition (D). Then*

$$R \text{ is intrinsically c.e. on } \mathcal{A} \Leftrightarrow R \text{ is formally c.e. on } \mathcal{A}.$$

Proof. ($\Leftarrow$) Always true for a relation R on the domain of any computable model.

($\Rightarrow$) Without loss of generality, let R be a unary relation. Assume that R is not formally c.e. We assume that ω is the domain of all considered computable models. We will construct a computable model $\mathcal{B}$ and an isomorphism $f : B \to A$ such that $f^{-1}(R)$ is not c.e. Let s be an arbitrary stage of the construction. We will define a finite set Ψ^s of formulae of the open diagram of $\mathcal{B}$, and a finite partial isomorphism f_s from $\mathcal{B}$ to $\mathcal{A}$.

By a finite partial isomorphism from $\mathcal{B}$ to $\mathcal{A}$ at stage s, we understand an injective function g with a finite domain such that for every $\theta \in \Psi^s$, if $\theta = \theta(\mathbf{b}_0, \dots, \mathbf{b}_{n-1})$ for some $b_0, \dots, b_{n-1} \in \omega$, then $g(b_0)\downarrow, \dots, g(b_{n-1})\downarrow$ and $\mathcal{A} \models \theta[g(b_0), \dots, g(b_{n-1})]$.

Define $\Psi^{-1} = \emptyset$ and $f_{-1} = \emptyset$. Let $X_s = f_s^{-1}(R)$ for $s \in \omega$. At the end of the construction, we will have that $f = \lim_s f_s$ exists. Let $X =_{\text{def}} f^{-1}(R)$ and $\Psi = \bigcup_{s \geqslant -1} \Psi^s$. The construction will ensure that X is not c.e.

Let $(\theta_e)_{e \in \omega}$ be an effective list of all atomic and negated atomic formulae in the language of $\mathcal{A}$, augmented with the constants for the elements of ω. The construction will meet the following requirements for every $e \geqslant 0$,

$$P_e^0 : \quad \theta_e \in \Psi \text{ or } \neg\theta_e \in \Psi;$$

$$P_e^1 : \quad e \in \text{dom}(f);$$

$$P_e^2 : \quad e \in \text{rng}(f);$$

$$Q_e : \quad X \neq W_e.$$

The strategy for meeting a single requirement Q_e is to wait for a stage s such that for some $b \in \omega$, $b \in W_{e,s}$. Define $f_s(b)$ such that $f_s(b) \notin R$. Hence $b \notin X_s$. Now, let $n_e^s =_{\text{def}} b$. Let n_e^{-1} be undefined for every $e \in \omega$.

We say that at stage s,

P^0_e *requires attention* if $\theta_e \notin \Psi^{s-1}$, $\neg\theta_e \notin \Psi^{s-1}$ and all elements of ω occurring in θ_e are in the domain of f_{s-1};

P^1_e (P^2_e) *requires attention* if $e \notin \mathrm{dom}(f_{s-1})$ ($e \notin \mathrm{rng}(f_{s-1})$);

Q_e *requires attention* if n^{s-1}_e is undefined;

P^1_e (P^2_e) *is injured* if $f_s(e) \neq f_{s-1}(e)$ ($f^{-1}_s(e) \neq f^{-1}_{s-1}(e)$);

Q_e *is injured* if n^{s-1}_e is defined and $f_s(n^{s-1}_e) \neq f_{s-1}(n^{s-1}_e)$.

Construction

STAGE s: For a requirement Req, we have the following clauses in the definition of Req *is attacked* at stage s.

Req= P^0_e Let $\theta_e = \theta_e(\mathbf{b}_0, \dots, \mathbf{b}_{n-1})$ for some $b_0, \dots, b_{n-1} \in \omega$. Define $\Psi^s = \Psi^{s-1} \cup \{\theta^k_e\}$, where $k \in \{0,1\}$ is such that

$$\mathcal{A} \models \theta^k_e[f_{s-1}(b_0), \dots, f_{s-1}(b_{n-1})].$$

Let $f_s =_{\mathrm{def}} f_{s-1}$.

Req= P^1_e Define $\Psi^s = \Psi^{s-1}$, and $f_s = f_{s-1} \cup \{(e,a)\}$, where $a \in \omega$ is the least new element at stage s.

Req= P^2_e Define $\Psi^s = \Psi^{s-1}$, and $f_s = f_{s-1} \cup \{(b,e)\}$, where $b \in \omega$ is the least new element at stage s.

Req= Q_e Let $\Psi^s =_{\mathrm{def}} \Psi^{s-1}$. There exists $b \in W_{e,s}$ and a partial isomorphism from $\mathcal{B}$ to $\mathcal{A}$ at stage s which maps b into an element from $(\omega - R)$. Choose the least such b, and then define f_s to be the least corresponding partial isomorphism (in some effective ordering of all finite functions on ω). Hence $b \notin X_s$. Define $n^s_e = b$.

Attack the highest priority requirement Req which requires attention at stage s, and which can be attacked without injuring any requirement of a higher priority. Whether this can be done for a Q–requirement can be checked effectively because the decidability condition (D) holds. If some lower priority requirement Q_i is injured at s, then n^s_i becomes undefined. End of the construction.

It is not difficult to show that each requirement is attacked and injured only finitely often, and that all P-requirements are met. Thus, we have a computable model $\mathcal{B}$ and an isomorphism f.

Lemma 15.2 *Every Q-requirement is satisfied.*

Proof. Assume otherwise. For example, let Q_e be the requirement of the highest priority which is not satisfied. Then $X = W_e$. Let s_0 be a stage by which all requirements of higher priority than Q_e have been attacked for the last time, and at which the sequences of numbers coming from the higher priority requirements have reached their final values, $\overline{d}$ and $f(\overline{d})$. Let $b_0, b_1, b_2, \ldots$ be a computable enumeration of W_e. Consider an arbitrary b_k. Find the least corresponding stage s. Let $\psi_k(x, \overline{\mathbf{d}})$ be the corresponding existential formula. That is, $\psi_k(x, \overline{\mathbf{d}}) = (\exists \overline{y})\delta(x, \overline{\mathbf{d}}, \overline{y})$, where $\delta(b_k, \overline{\mathbf{d}}, \overline{\mathbf{d}}')$ is the conjunction of all formulae of Ψ^{s-1}, and $\mathrm{lh}(\overline{y}) = \mathrm{lh}(\overline{d}')$. Clearly, $\mathcal{B} \models \psi_k[b_k, \overline{d}]$, so $\mathcal{A} \models \psi_k[f(b_k), f(\overline{d})]$. Let $f(\overline{d}) = (a_0, \ldots, a_{n-1})$. Since Q_e is not attacked at s, we have for every $a \in A$

$$[\mathcal{A}_A \models \psi_k(\mathbf{a}, \mathbf{a}_0, \ldots, \mathbf{a}_{n-1})] \implies R(a).$$

Conversely, for every $a \in R$, there is $k \in \omega$ such that $a = f(b_k)$. Thus, the following equivalence holds for every $a \in A$

$$[\mathcal{A}_A \models \bigvee_{k \in \omega} \psi_k(\mathbf{a}, \mathbf{a}_0, \ldots, \mathbf{a}_{n-1})] \iff R(a).$$

This is a contradiction since R is not formally c.e. on $\mathcal{A}$. □

As an immediate consequence, we have that if both R and its complement satisfy the decidability condition (D), then

R is intrinsically computable on $\mathcal{A}$ $\iff$ R is formally computable on $\mathcal{A}$.

The decidability condition (D) cannot be omitted from the previous theorem. Goncharov [70] and Manasse [126] have shown that there are computable models with intrinsically c.e. relations which are not formally c.e. Chisholm [34] has established the best possible result on the definability of intrinsically c.e. relations on 1-computable models.

Definition 15.4 Let $\mathcal{F}$ be an $L_{\omega_1\omega}$ formula with free variables among $\overline{x}$. $\mathcal{F}$ is a (computable) Σ_2 formula if it is equivalent to a formula of the form

$$\bigvee_{n\in\omega} \exists \overline{y}_n \bigwedge_{m\in\omega} \forall \overline{z}_{mn} \theta_{mn}(\overline{x}, \overline{y}_n, \overline{z}_{mn})$$

and $\mathcal{F}$ is a (computable) Π_2 formula if it is equivalent to a formula of the form

$$\bigwedge_{n\in\omega} \forall \overline{y}_n \bigvee_{m\in\omega} \exists \overline{z}_{mn} \theta_{mn}(\overline{x}, \overline{y}_n, \overline{z}_{mn}),$$

where $(\theta_{mn}(\overline{x}, \overline{y}_n, \overline{z}_{mn}))_{n,m\in\omega}$ is a (computable) sequence of quantifier-free formulae.

This definition has been extended by Ash [1] to all (computable) Σ_α and Π_α formulae, where α is a computable ordinal.

Definition 15.5 Let R be an additional m–ary computable relation on the domain of a computable model $\mathcal{A}$. R is *formally* Σ^0_2 (Π^0_2, respectively) on $\mathcal{A}$ if and only if there is a finite sequence $(b_0, \dots, b_{k-1})$ of elements in A and a computable Σ_2 (Π_2, respectively) formula $\mathcal{F}(x_0, \dots, x_{m-1}, \mathbf{b}_0, \dots, \mathbf{b}_{k-1})$ such that the following equivalence holds for every $a_0, \dots, a_{m-1} \in A$.

$$[\mathcal{A}_A \models \mathcal{F}(\mathbf{a}_0, \dots, \mathbf{a}_{m-1}, \mathbf{b}_0, \dots, \mathbf{b}_{k-1})] \iff R(a_0, \dots, a_{m-1}).$$

R is *formally* Δ_2 on $\mathcal{A}$ if R is both formally Σ^0_2 and formally Π^0_2 on $\mathcal{A}$.

Theorem 15.3 (Chisholm [34])

(i) *Let R be an additional relation on the domain of the 1–computable model $\mathcal{A}$. Then*

$$R \text{ is intrinsically c.e. on } \mathcal{A} \implies R \text{ is formally } \Pi^0_2 \text{ on } \mathcal{A}.$$

(ii) *There is a decidable model $\mathcal{A}$ and an additional relation R on its domain, such that R is intrinsically c.e. and not formally Σ^0_2 on $\mathcal{A}$. Moreover, R is not definable by any Σ_2 formula.*

Barker [24] has extended Theorem 15.1 to Σ^0_2 relations. He has proved that if certain extra decidability conditions are satisfied, then R is intrinsically Σ^0_2 if and only if R is formally Σ^0_2. Barker [25] has further proved an analogous result for all Σ^0_α relations, where α is a computable ordinal.

Let $\mathcal{A}$ be a computable model. Davey [40] has considered two additional, disjoint, computable relations, R_1 and R_2, on the domain A. He has studied conditions under which there is a computable model $\mathcal{B}$ isomorphic to $\mathcal{A}$ such that the corresponding isomorphic images of R_1 and R_2 are Δ^0_α–inseparable. For example, let R_1 and R_2 be infinite, disjoint, computable subsets of ω such that $R_1 \cup R_2$ is coinfinite. Then, there is a computable model isomorphic to $(\omega, <)$ such that the images of R_1 and R_2 are computably inseparable.

While all the previous results address only levels of the arithmetic or hyperarithmetic hierarchy, Harizanov has also considered Turing degrees of the images of a computable relation on the domain of a computable model $\mathcal{A}$, under all isomorphisms from $\mathcal{A}$ to computable models.

Definition 15.6 (Harizanov [83]) Let R be an additional relation on the domain of a computable model $\mathcal{A}$. The *(Turing) degree spectrum* of R on $\mathcal{A}$, in symbols $Dg_{\mathcal{A}}(R)$, is the set of Turing degrees of the images of R under all isomorphisms from $\mathcal{A}$ to computable models.

For a computable model $\mathcal{B}$ isomorphic to $\mathcal{A}$, the (Turing) degree spectrum of R on $\mathcal{A}$ with respect to $\mathcal{B}$, in symbols $Dg_{\mathcal{A},\mathcal{B}}(R)$, is the set of Turing degrees of the images of R under all isomorphisms from $\mathcal{A}$ to $\mathcal{B}$.

Harizanov has studied various aspects of degree spectra, such as: the structure of uncountable degree spectra, the effect of decidability condition (D) on the cardinality of a degree spectrum, realizing c.e. degrees in a degree spectrum via c.e. and, in general, via Δ^0_2 isomorphic images of R, and finite degree spectra.

To state results about uncountable degree spectra we assume, without loss of generality, that R is unary. Let $\mathcal{B}$ be a computable model isomorphic to $\mathcal{A}$. By $\mathcal{I}(\mathcal{A},\mathcal{B})$ we denote the set of all isomorphisms from $\mathcal{A}$ to $\mathcal{B}$. We say that a partial function p from A to B is a *finite isomorphism* from $\mathcal{A}$ to $\mathcal{B}$ if p is one-to-one, dom(p) is finite and for every atomic formula $\alpha = \alpha(x_0, \dots, x_{n-1})$ in $L(\mathcal{A})$, and every $a_0, \dots, a_{n-1} \in \text{dom}(p)$, we have

$$\mathcal{A} \models \alpha[a_0, \dots, a_{n-1}] \iff \mathcal{B} \models \alpha[p(a_0), \dots, p(a_{n-1})].$$

By $\mathcal{I}_{\text{fin}}(\mathcal{A},\mathcal{B})$ we denote the set of all finite isomorphisms from $\mathcal{A}$ to $\mathcal{B}$. We define the R–equivalence relation $\sim_R$ on $\mathcal{I}_{\text{fin}}(\mathcal{A},\mathcal{B})$ as follows:

$$q \sim_R r \iff (\forall b \in \text{ran}(q) \cap \text{ran}(r))[q^{-1}(b) \in R \Leftrightarrow r^{-1}(b) \in R].$$

Theorem 15.4 (Harizanov [85])

(i) *The following are equivalent:*

(0) $Dg_{\mathcal{A}}(R)$ *is uncountable.*

(1) $Dg_{\mathcal{A},\mathcal{B}}(R)$ *is uncountable.*

(2) $Dg_{\mathcal{A},\mathcal{B}}(R)$ *has cardinality* 2^ω.

(3) *There is a nonempty set* $\mathbb{S} \subseteq \mathcal{I}_{\text{fin}}(\mathcal{A},\mathcal{B})$ *such that the following two conditions are satisfied:*

$$\text{(A)}\quad (\forall p \in \mathbb{S})(\forall a \in A)(\forall b \in B)(\exists q \in \mathbb{S})$$
$$[(q \supseteq p) \wedge (a \in \text{dom}(q)) \wedge (b \in \text{ran}(q))];$$

$$\text{(B)}\quad (\forall p \in \mathbb{S})(\exists q, r \in \mathbb{S})[(q \supseteq p) \wedge (r \supseteq p) \wedge \neg(q \sim_R r)].$$

(ii) *Let* $\mathbb{S}$ *be as in* (i)(3). *Then for every set* $C \geqslant_T \mathbb{S}$, *there is an isomorphism* f *from* $\mathcal{A}$ *to* $\mathcal{B}$ *such that*

$$C \equiv_T f(R) \oplus \mathbb{S} \equiv_T f \oplus \mathbb{S}.$$

In particular, if $\mathbb{S}$ *is computable, then* $Dg_{\mathcal{A},\mathcal{B}}(R) = \mathcal{D}$ *and, moreover, for every set* $C \subseteq \omega$, *there is an isomorphism* f *from* $\mathcal{A}$ *to* $\mathcal{B}$ *such that*

$$C \equiv_T f(R) \equiv_T f.$$

Theorem 15.5 (Harizanov [87]; Ash, Cholak and Knight [5]) *The following are equivalent:*

(1) $Dg_{\mathcal{A},\mathcal{B}}(R) = \mathcal{D}$ *and, moreover, for every set* $C \subseteq \omega$, *there is an isomorphism* f *from* $\mathcal{A}$ *to* $\mathcal{B}$ *such that* $C \equiv_T f(R) \equiv_T f$.

(2) *There is* $e \in \omega$ *and* $p \in 2^{<\omega}$ *such that the set*

$$\mathbb{S}_{e,p} =_{\text{def}} \{\varphi_e^q : q \in 2^{<\omega} \wedge q \supseteq p\}$$

has the following properties:

$\mathbb{S}_{e,p} \subseteq \mathcal{I}_{\text{fin}}(\mathcal{A},\mathcal{B})$,

Condition (3)(A) *from Theorem* 15.4 *is satisfied for* $\mathbb{S} = \mathbb{S}_{e,p}$, *and*

$(\exists i \in \omega)(\forall q \supseteq p)(\forall a \in \text{dom}(q))[\varphi_i^{\varphi_e^q(R)}(a) \downarrow = q(a)]$.

(3) *There is a nonempty computable set* $\mathbb{S} \subseteq \mathcal{I}_{\text{fin}}(\mathcal{A}, \mathcal{B})$ *such that the conditions* (A) *and* (B) *from Theorem* 15.4 *are satisfied.*

In the proof of $\neg(2) \Rightarrow \neg(1)$ for Theorem 15.5 in [87], the construction of C can be done computably in $\emptyset''$. Hence $C \in \Delta^0_3$. Thus, if not every Turing degree is obtained in a degree spectrum $Dg_{\mathcal{A},\mathcal{B}}(R)$ via an isomorphism of the same Turing degree, then there is such a Δ^0_3 degree. This conclusion also follows from the proof in [5] since there is a generic Δ^0_3 set.

In [84], the priority method has been used to establish how the Ash-Nerode decidability condition affects the cardinality of the degree spectrum.

Theorem 15.6 (Harizanov [84])

(i) *If the Ash-Nerode decidability condition* (D) *holds for a non-intrinsically c.e. relation* R *on a model* $\mathcal{A}$*, then the degree spectrum of* R *on* $\mathcal{A}$ *is infinite.*

(ii) *There is a computable non-intrinsically c.e. relation* R *on a computable model* $\mathcal{A}$ *such that the degree spectrum of* R *on* $\mathcal{A}$ *has exactly two degrees.*

Also, in [84] some new computable syntactic conditions have been introduced, which have allowed the use of the permitting method to obtain every c.e. degree in the degree spectrum. Ash, Cholak and Knight [5] have generalized this result to include in the degree spectrum all α–c.e. degrees in Ershov's hierarchy of Δ^0_2 degrees, see [52, 53, 54]. For a computable ordinal α, a Turing degree is α–c.e. if it contains an α–c.e. set. A set $C \subseteq \omega$ is α–c.e. if there exists a computable function $f : \omega^2 \to \{0, 1\}$ and a computable function $o : \omega^2 \to \alpha + 1$ with the following properties:

$$(\forall x)[\lim_{s\to\infty} f(x, s) = C(x) \wedge f(x, 0) = 0],$$

$$(\forall x)(\forall s)[o(x, s+1) \leqslant o(x, s) \wedge o(x, 0) = \alpha], \text{ and}$$

$$(\forall x)(\forall s)[f(x, s+1) \neq f(x, s) \Rightarrow o(x, s+1) < o(x, s)].$$

In particular, 1–c.e. sets are c.e. sets, and 2–c.e. sets are d–c.e. sets. For other characterizations of α–c.e. sets, also see [51, 15]. In [15], Ash and Knight have studied intrinsically α–c.e. relations. For other generalizations of a syntactic condition in [84], see [14, 16].

In [86], Goncharov's infinite injury method has been modified to construct a computable non-intrinsically c.e. relation with a two-element degree spectrum whose nonzero degree is $\leqslant \mathbf{0}'$. First, a family $\mathcal{S}$ of c.e. sets and a computable set P, which have certain required properties, have been constructed. A function ν from ω onto $\mathcal{S}$ is called a *computable enumeration* of $\mathcal{S}$ if there is a uniformly computable sequence $\{\nu_t\}_{t\in\omega}$ of functions from ω to the set of finite subsets of ω such that for every $n \in \omega$, $\nu(n) = \cup\{\nu_t(n) : t \in \omega\}$. The family $\mathcal{S}$ constructed has two injective computable enumerations, ν and μ, such that every other injective computable enumeration λ of $\mathcal{S}$ is computably equivalent to ν or μ. Here, λ is *computably equivalent* to ν if the function $f : \omega \to \omega$ such that $\nu = \lambda f$ is recursive. The set Y defined by $Y = \{n \in \omega : (\exists m \in P)[\nu(m) = \mu(n)]\}$ is a non-c.e. Δ^0_2 set. The enumeration ν has then been encoded into a rigid computable model $\mathcal{A}$. The category of injective computable enumerations of $\mathcal{S}$, whose morphisms are equivalences (computable equivalencies, respectively) of the enumerations, is equivalent to the category of computable models isomorphic to $\mathcal{A}$ whose morphisms are isomorphisms (computable isomorphisms, respectively) of the models. The set R which encodes P in $\mathcal{A}$ is computable and its degree spectrum on $\mathcal{A}$ has the required property.

The ideas described in the previous paragraph have originated in Goncharov's work [69, 70] on the dimension of a computable model (see Theorem 15.7). Similar ideas have also been used by Ventsov [203, 204, 206], as well as by Cholak, Goncharov, Khoussainov and Shore [36].

Definition 15.7 Let $\mathcal{P}$ be a certain class of functions. A computable model $\mathcal{A}$ is *$\mathcal{P}$–categorical* if for every computable model $\mathcal{B}$ isomorphic to $\mathcal{A}$, there exists an isomorphism from $\mathcal{A}$ to $\mathcal{B}$, which belongs to $\mathcal{P}$.

An example of a computably categorical model is the ordered set of rationals. In general, a computable linear ordering is computably categorical if and only it has only finitely many elements with an immediate successor [77, 181]. A computable Boolean algebra is computably categorical if and only if it has finitely many atoms ([182], also see Theorem 1 in [77]). For more examples of computably categorical models see [39].

Ash [3] has established for every ordinal $\alpha < \omega_1^{\mathrm{CK}}$, under certain extra decidability assumptions, a necessary and sufficient condition for a computable model $\mathcal{A}$ to be Δ^0_α–categorical, termed $\mathcal{A}$ has a Σ^0_α *Scott family*. (The extra decidability assumptions are needed only for establishing the necessary condition.) For $\alpha = 1$, this result has been first obtained by Goncharov [65].

16 Acknowledgments

I thank Terry Millar for teaching me computable model theory. I thank Richard Shore for a careful proofreading and for many helpful comments and suggestions. I thank Tim McNicholl for proofreading parts of an early draft. I thank Graeme Bailey for technical assistance with word-processing. Finally, I thank Victor Marek for his heroic efforts in getting this volume to print.

References

[1] C. J. Ash, Recursive labelling systems and stability of recursive structures in hyperarithmetical degrees, Trans. Amer. Math. Soc., **298** (1986) 497–514; errata ibid., **310** (1988) 851.

[2] C. J. Ash, Stability of recursive structures in the arithmetical degrees, Ann. Pure Appl. Logic, **32** (1986) 113–135.

[3] C. J. Ash, Categoricity in hyperarithmetical degrees, Ann. Pure Appl. Logic, **34** (1987) 1–14.

[4] C. J. Ash, Labelling systems and r.e. structures, Ann. Pure Appl. Logic, **47** (1990) 99–119.

[5] C. J. Ash, P. Cholak, and J. F. Knight. Permitting, forcing, and copying of a given recursive relation. Ann. Pure Appl. Logic, **86** (1997) 219–236.

[6] C. J. Ash and R. G. Downey, Decidable subspaces and recursively enumerable subspaces, J. Symbolic Logic, **49** (1984) 1137–1145.

[7] C. J. Ash and S. S. Goncharov, Strong Δ_2^0-categoricity (English), Algebra i Logika, **24** (1985) 718–727; [also in: Algebra and Logic, **24** (1985) 471–476].

[8] C. J. Ash, C. G. Jockusch Jr. and J. F. Knight, Jumps of orderings, Trans. Amer. Math. Soc., **319** (1990) 573–599.

[9] C. J. Ash and J. F. Knight, Pairs of recursive structures, Ann. Pure Appl. Logic, **46** (1990) 211–234.

[10] C. J. Ash and J. F. Knight, Relatively recursive expansions, Fund. Math., **140** (1992) 137–155.

[11] C. J. Ash and J. F. Knight, A completeness theorem for certain classes of recursive infinitary formulas, Math. Logic Quart., **40** (1994) 173–181.

[12] C. J. Ash and J. F. Knight, Mixed systems, J. Symbolic Logic, **59** (1994) 1383–1399.

[13] C. J. Ash and J. F. Knight, Ramified systems, Ann. Pure Appl. Logic, **70** (1994) 205–221.

[14] C. J. Ash and J. F. Knight, Possible degrees in recursive copies, Ann. Pure Appl. Logic, **75** (1995) 215–221.

[15] C. J. Ash and J. F. Knight, Recursive structures and Ershov's hierarchy, Math. Logic Quart., **42** (1996) 461–468.

[16] C. J. Ash and J. F. Knight, Possible degrees in recursive copies II. Ann. Pure Appl. Logic, **87** (1997) 151–165.

[17] C. J. Ash, J. F. Knight, M. Manasse and T. A. Slaman, Generic copies of countable structures, Ann. Pure Appl. Logic, **42** (1989) 195–205.

[18] C. J. Ash, J. F. Knight and J. B. Remmel, Quasi-simple relations in copies of a given recursive structure, Ann. Pure Appl. Logic, **86** (1997) 203–218.

[19] C. J. Ash, J. F. Knight and T. A. Slaman, Relatively recursive expansions II, Fund. Math., **142** (1993) 147–161.

[20] C. J. Ash and T. S. Millar, Persistently finite, persistently arithmetic theories, Proc. Amer. Math. Soc., **89** (1983) 487–492.

[21] C. J. Ash and A. Nerode, Intrinsically recursive relations, in: Aspects of Effective Algebra, (Proc. Conf. Monash Univ., Clayton, Australia, Aug. 1–4, 1979), J. N. Crossley (ed.), (Upside Down A Book Co., Yarra Glen, Victoria, Australia, 1981), 26–41.

[22] J. T. Baldwin, Fundamentals of Stability Theory, Perspect. in Math. Logic, (1988).

[23] J. T. Baldwin and A. H. Lachlan, On strongly minimal sets, J. Symbolic Logic, **36** (1971) 79–96.

[24] E. J. Barker. Intrinsically Σ_2^0–relations. M.Sc. Thesis, Monash Univ., Clayton, Victoria, Australia, 1985.

[25] E. J. Barker, Intrinsically Σ_α^0 relations, Ann. Pure Appl. Logic, **39** (1988) 105–130.

[26] E. J. Barker, Back and forth relations for reduced abelian p–groups, Ann. Pure Appl. Logic, **75** (1995) 223–249.

[27] J. Barwise and J. Schlipf. On recursively saturated models of arithmetic, in: Model Theory and Algebra, D. H. Saracino and V. B. Weispfenning, (eds.), Lecture Notes in Math., **498** (1975) 42–55.

[28] J. Barwise and J. Schlipf, An introduction to recursively saturated and resplendent models, J. Symbolic Logic, **41** (1976) 531–536.

[29] S. Burris, Decidable model companions, Z. Math. Logik Grundlag. Math., **35** (1989) 225–227.

[30] J. S. Carroll, Some undecidability results for lattices in recursion theory, Pacific J. Math., **122** (1986) 319–331.

[31] D. Cenzer and J. B. Remmel, Polynomial-time versus recursive models, Ann. Pure Appl. Logic, **54** (1991) 17–58.

[32] C. C. Chang and H. J. Keisler, Model Theory, 3rd. edn., Stud. Logic Found. Math., **73** (1990); [1st. edn. 1973, 2nd. edn. 1977].

[33] J. Chisholm, Effective Model Theory vs. Recursive Model Theory, Ph.D. Thesis, Univ. Wisconsin, Madison, WI, (1988).

[34] J. Chisholm, The complexity of intrinsically r.e. subsets of existentially decidable models, J. Symbolic Logic, **55** (1990) 1213–1232.

[35] J. Chisholm, Effective model theory vs. recursive model theory, J. Symbolic Logic, **55** (1990) 1168–1191.

[36] P. A. Cholak, S. S. Goncharov, B. M. Khoussainov and R. A. Shore, Computably categorical structures and expansions by constants, (to appear).

[37] P. M. Cohn. Universal Algebra. (Harper and Row, Publishers, NY, 1965).

[38] J. N. Crossley, Fifty years of computability, Southeast Asian Bull. Math., **11** (1988) 81–99.

[39] J. N. Crossley, A. B. Manaster and M. Moses, Recursive categoricity and recursive stability, (Special issue: 2nd. Southeast Asian Logic Conf., Bangkok, 1984), Ann. Pure Appl. Logic, **31** (1986) 191–204.

[40] K. J. Davey, Inseparability in recursive copies, Ann. Pure Appl. Logic, **68** (1994) 1–52.

[41] A. S. Denisov, Constructive homogeneous extensions, Sibirsk. Math. Zh., **25** (1984) 60–69 (Russian); [translated in: Siberian Math. J., **25** (1984) 879–888].

[42] A. S. Denisov, Every decidable theory has a $\mathbf{0}'$–strongly constructive homogeneous model (Russian), in: Computable Invariants in the Theory of Algebraic Systems, V. N. Remeslennikov, (ed.), Akad. Nauk SSSR Sibirsk. Otdel., Vyčisl. Centr, Novosibirsk, (1987) 13–21.

[43] A. S. Denisov, Homogeneous $\mathbf{0}'$–elements in structural pre-orders (Russian). Algebra i Logika, **28** (1989) 619–639, 743; [translated in: Algebra and Logic, **28** (1989) 405–418].

[44] K. Doets, Basic Model Theory. (Center for the Study of Language and Information, Stanford, CA, and FOLLI: European Assoc. Logic, Language and Informatics, Amsterdam, 1996).

[45] R. G. Downey, Undecidability of $L(F_\infty)$ and other lattices of r.e. substructures, Ann. Pure Appl. Logic, **32** (1986) 17–26; corr. ibid., **48** (1990) 299–301.

[46] R. G. Downey and J. F. Knight, Orderings with α–th jump degree $\mathbf{0}^\alpha$, Proc. Amer. Math. Soc., **114** (1992) 545–552.

[47] R. G. Downey and J. B. Remmel, Automorphisms and recursive structures, Z. Math. Logik Grundlag. Math., **33** (1987) 339–345.

[48] R. G. Downey and J. B. Remmel, Classification of degree classes associated with r.e. subspaces, Ann. Pure Appl. Logic, **42** (1989) 105–124.

[49] B. N. Drobutun, Enumerations of simple models (Russian), Sibirsk. Math. Zh., **18** (1977) 1002–1014, 1205; [translated in: Siberian Math. J., **18** (1977) 707–716].

[50] E. F. Eisenberg and J. B. Remmel, Effective isomorphisms of algebraic structures, in: Patras Logic Symposion, (Proc. Logic Sympos. Patras, Greece, Aug. 18–22, 1980) G. Metakides (ed.), Stud. Logic Found. Math., **109** (1982) 95–122.

[51] R. L. Epstein, R. Haas and R. L. Kramer, Hierarchies of sets and degrees below $\mathbf{0}'$, in: Logic Year 1979-1980, (Proc. Seminars and Conf. Math. Logic, Univ. Connecticut, Storrs, CT, 1979/80), M. Lerman, J. H. Schmerl and R. I. Soare, (eds.), Lecture Notes in Math., **859** (1981) 32–48.

[52] Yu. L. Ershov, A hierarchy of sets I (Russian), Algebra i Logika, **7** (1968) 47–74 [translated in: Algebra and Logic, **7** (1968) 25–43].

[53] Yu. L. Ershov, A hierarchy of sets II (Russian), Algebra i Logika, **7** (1968) 15–47 [translated in: Algebra and Logic, **7** (1968) 212–232].

[54] Yu. L. Ershov, A hierarchy of sets III (Russian), Algebra i Logika, **9** (1970) 34–51 [translated in: Algebra and Logic, **9** (1970) 20–31].

[55] Yu. L. Ershov, Skolem functions and constructive models (Russian), Algebra i Logika, **12** (1973) 644–654, 735; [translated in: Algebra and Logic, **12** (1973) 368–373].

[56] Yu. L. Ershov, Theory of Numerations (Russian), Monographs in Math. Logic and Foundations of Math., (Nauka, Moskva, 1977).

[57] Yu. L. Ershov, Decision Problems and Constructivizable Models (Russian), (Mathematical Logic and Foundations of Mathematics, Nauka, Moskva, 1980).

[58] Yu. L. Ershov, On elementary theories of regularly closed fields (Russian), Dokl. Akad. Nauk SSSR, **257** (1981) 271–274; [translated in: Soviet Math. - Dokl., **23** (1981) 259–262].

[59] S. Feferman, Arithmetically definable models of formalized arithmetic, Notices Amer. Math. Soc., **5** (1958) 679–680 (abstract).

[60] A. Feldman. Recursion Theory in a Partial Order with Greatest Lower Bound. Ph.D. thesis, Univ. Wisconsin, Madison, 1985.

[61] A. Feldman, Recursion theory in a lower semilattice, J. Symbolic Logic, **57** (1992) 892–911.

[62] A. Fröhlich and J. C. Shepherdson, On the factorisation of polynomials in a finite number of steps, Math. Z., **62** (1955) 331–334.

[63] A. Fröhlich and J. C. Shepherdson, Effective procedures in field theory, Philos. Trans. Roy. Soc. London, Ser. A, **248** (1956) 407–432.

[64] K. Gödel, Über formal unentscheidbare Sätze der *Principia mathematica* und verwandter Systeme I, Monatsh. Math. Phys., **38** (1931) 173–198; [translated as: On formally undecidable propositions of *Principia Mathematica* and related systems I, (translated by B. Meltzer, with an introduction by R. B. Braithwaite), (Basic Books, NY, 1963, reprinted Dover, NY, 1992); also in: From Frege to Gödel: A Source Book in Logic, 1879–1931, J. van Heijenroot, (ed.), (Harvard Univ. Press, Cambridge, Mass. and Oxford Univ. Press, London, 1967) 592–617.]

[65] S. S. Goncharov, Autostability and computable families of constructivizations (Russian), Algebra i Logika, **14** (1975) 647–680; [translated in: Algebra and Logic, **14** (1975) 392–409].

[66] S. S. Goncharov, The quantity of nonautoequivalent constructivizations (Russian), Algebra i Logika, **16** (1977) 257–282, 377; [translated in: Algebra and Logic, **16** (1977) 169–185].

[67] S. S. Goncharov, Constructive models of $\aleph_1$–categorical theories (Russian), Mat. Zametki, **23** (1978) 885–888; [translated in: Math. Notes, **23** (1978) 486–487].

[68] S. S. Goncharov, Strong constructivizability of homogeneous models (Russian), Algebra i Logika, **17** (1978) 363–388, 490; [translated in: Algebra and Logic, **17** (1978) 247–263].

[69] S. S. Goncharov, Computable numerations (Russian), Algebra i Logika, **19** (1980) 507–551, 617; [translated in: Algebra and Logic, **19** (1980) 325–356].

[70] S. S. Goncharov, On the problem of the number of nonautoequivalent constructivizations (Russian), Algebra i Logika, **19** (1980) 621–639, 745; [translated in: Algebra and Logic, **19** (1980) 401–414].

[71] S. S. Goncharov, The problem of the number of nonautoequivalent constructivizations (Russian), Dokl. Akad. Nauk SSSR, **251** (1980) 271–274; [translated in: Soviet Math. - Dokl., **21** (1980) 411–414].

[72] S. S. Goncharov, A totally transcendental decidable theory without constructivizable homogeneous models (Russian), Algebra i Logika, **19** (1980) 137–149, 250; [translated in: Algebra and Logic, **19** (1980) 85–93].

[73] S. S. Goncharov, Totally transcendental theory with a nonconstructivizable prime model (Russian), Sibirsk. Math. Zh., **21** (1980) 44–51; [translated in: Siberian Math. J., **21** (1980) 32–37].

[74] S. S. Goncharov, Limit equivalent constructivizations (Russian), Trudy Inst. Mat. Nauka Sibirsk. Otdel., Novosibirsk, **2** (1982) 4–12.

[75] S. S. Goncharov, Morley's problem, Bulletin of Symbolic Logic, **3** (1997) 99 (abstract).

[76] S. S. Goncharov and B. N. Drobotun, Numerations of saturated and homogeneous models (Russian), Sibirsk. Math. Zh., **21** (1980) 25–41, 236; [translated in: Siberian Math. J., **21** (1980) 164–176].

[77] S. S. Goncharov and V. D. Dzgoev, Autostability of models (Russian), Algebra i Logika, **19** (1980) 45–58, 132; [translated in: Algebra and Logic, **19** (1980) 28–37].

[78] S. S. Goncharov and A. A. Novikov, Examples of nonautostable systems (Russian), Sibirsk. Math. Zh., 25 (1984) 37–45; [translated in: Siberian Math. J., 25 (1984) 538–545].

[79] S. S. Goncharov and A. T. Nurtazin, Constructive models of complete decidable theories (Russian), Algebra i Logika, **12** (1973) 125–142, 243; [translated in: Algebra and Logic, **12** (1973) 67–77].

[80] G. Grätzer. Universal Algebra. (Springer-Verlag, New York and Heidelberg, 2nd. edn., 1979); [1st. edn. D. Van Nostrand Reinhold, Princeton, Toronto and London, 1968].

[81] D. Guichard, Automorphisms and Large Submodels in Effective Algebra, Ph.D. Thesis, Univ. Wisconsin, Madison, WI, (1982).

[82] D. Guichard, Automorphisms of substructure lattices in recursive algebra, Ann. Pure Appl. Logic, **25** (1983) 47–58.

[83] V. S. Harizanov, Degree Spectrum of a Recursive Relation on a Recursive Structure, Ph.D. Thesis, Univ. Wisconsin, Madison, WI, (1987).

[84] V. S. Harizanov, Some effects of Ash-Nerode and other decidability conditions on degree spectra, Ann. Pure Appl. Logic, **55** (1991) 51–65.

[85] V. S. Harizanov, Uncountable degree spectra, Ann. Pure Appl. Logic, **54** (1991) 255–263.

[86] V. S. Harizanov, The possible Turing degree of the nonzero member in a two element degree spectrum, Ann. Pure Appl. Logic, **60** (1993) 1–30.

[87] V. S. Harizanov, Turing degrees of certain isomorphic images of computable relations, Ann. Pure Appl. Logic, (to appear).

[88] L. Harrington, Recursively presentable prime models, J. Symbolic Logic, **39** (1974) 305–309.

[89] G. Hasenjaeger, Eine Bemerkung zu Henkin's Beweis für die Vollständigkeit des Prädikatenkalküls der ersten Stufe, J. Symbolic Logic, **18** (1953) 42–48.

[90] E. Herrmann, On Lindenbaum functions of $\aleph_0$–categorical theories of finite similarity type, Bull. Acad. Polon. Sci. Sér. Sci. Math. Astronom. Phys., **24** (1976) 17–21.

[91] G. Hird, Recursive properties of intervals of recursive linear orders, in: Logical Methods, (Papers from Conf. in honor of Anil Nerode's Sixtieth Birthday, June 1–3, 1992, Cornell Univ., Ithaca, NY), J. N. Crossley, J. B. Remmel, R. A. Shore and M. E. Sweedler, (eds.), Progr. Comput. Sci. Appl. Logic, **12** (1993) 422–437.

[92] W. Hodges, Model Theory, Encyclopedia of MAthematics and its Applications, **42**, (Cambridge Univ. Press, 1993).

[93] K. R. Hurlburt, Sufficiency conditions for theories with recursive models, Ann. Pure Appl. Logic, **55** (1992) 305–320.

[94] C. G. Jockusch Jr. and R. I. Soare, Degrees of members of Π_1^0 classes, Pacific J. Math., **40** (1972) 605–616.

[95] C. G. Jockusch Jr. and R. I. Soare, Π_1^0 classes and degrees of theories, Trans. Amer. Math. Soc., **173** (1972) 33–56.

[96] C. G. Jockusch Jr. and R. I. Soare, Degrees of orderings not isomorphic to recursive linear orderings, (Special issue: Internatl. Sympos. Math. Logic and its Applications, Nagoya, 1988), Ann. Pure Appl. Logic, **52** (1991) 39–64.

[97] C. G. Jockusch Jr. and R. I. Soare, Boolean algebras, Stone spaces, and the iterated Turing jump, J. Symbolic Logic, **59** (1994) 1121–1138.

[98] H. J. Keisler. Model Theory for Infinitary Logic. Logic with countable conjuctions and finite quantifiers. Stud. Logic Found. Math., **62** (1971).

[99] N. G. Khisamiev, Strongly constructive models of a decidable theory (Russian), Izv. Akad. Nauk Kazakh. SSR, Ser. Fiz.-Mat., **1** (1974) 83–84.

[100] B. M. Khoussainov, Strongly ∀–finite theories of unars (Russian), Mat. Zametki, 41 (1987) 265–271, 288; [translated in: Math. Notes, **41** (1987) 151–154].

[101] B. M. Khoussainov, Algorithmic degree of unars (Russian), Algebra i Logika, **27** (1988) 479–494, 499; [translated in: Algebra and Logic, **27** (1988) 301–312].

[102] B. M. Khoussainov, Algorithmic dimensions of homomorphic images of models (Russian), Algebra i Logika, **31** (1992) 317–337; [translated in: Algebra and Logic, **31** (1992) 195–203].

[103] B. M. Khoussainov and R. Dadajanov, Algorithmic stability of models, in: Logical Methods, (Papers from Conf. in honor of Anil Nerode's Sixtieth Birthday, June 1–3, 1992, Cornell Univ., Ithaca, NY), J. N. Crossley, J. B. Remmel, R. A. Shore and M. E. Sweedler, (eds.), Progr. Comput. Sci. Appl. Logic, **12** (1993) 438–466.

[104] B. M. Khoussainov, A. Nies and R. A. Shore, Computable models of theories with few models, Notre Dame J. Formal Logic, **38** (1997) 165–178.

[105] B. M. Khoussainov and R. A. Shore, Categoricity and Scott families, (to appear).

[106] H. A. Kierstead and J. B. Remmel, Indiscernibles and decidable models, J. Symbolic Logic, **48** (1983) 21–32.

[107] H. Kierstead and J. B. Remmel, Degrees of indiscernibles in decidable models, Trans. Amer. Math. Soc., **289** (1985) 41–57.

[108] S. C. Kleene, Introduction to Metamathematics, (D. Van Nostrand, Princeton, 1952, and Elsevier, Amsterdam, 1964, 1971).

[109] J. F. Knight, Degrees of types and independent sequences, J. Symbolic Logic, **48** (1983) 1074–1081.

[110] J. F. Knight, Degrees coded into jumps of orderings, J. Symbolic Logic, **51** (1986) 1034–1042.

[111] J. F. Knight, Effective construction of models, in: Logic Colloquium '84, (Manchester, July 15–24, 1984), J. B. Paris, A. J. Wilkie and G. M. Wilmers, (eds.), Stud. Logic Found. Math., **120**, (1986) 105–119.

[112] J. F. Knight, Saturation of homogeneous resplendent models, J. Symbolic Logic, **51** (1986) 222–224.

[113] J. F. Knight, Degrees of models with prescribed Scott set, in: Classification Theory, (Proc. Joint U.S.–Israel Workshop on Model Theory in Math. Logic, Chicago, Dec. 15–19, 1985), J. T. Baldwin, (ed.), Lecture Notes in Math., **1292** (1987) 182–191.

[114] J. F. Knight, Constructions by transfinitely many workers, Ann. Pure Appl. Logic, **48** (1990) 237–259.

[115] J. F. Knight, A metatheorem for constructions by finitely many workers, J. Symbolic Logic, **55** (1990) 787–804.

[116] J. F. Knight, A. H. Lachlan and R. I. Soare, Two theorems on degrees of models of true arithmetic, J. Symbolic Logic, **49** (1984) 425–436.

[117] J. F. Knight and M. Nadel, Expansions of models and Turing degrees, J. Symbolic Logic, **47** (1982) 587–604.

[118] K. Zh. Kudaibergenov, A theory with two strongly constructivizable models (Russian), Algebra i Logika, **18** (1979) 176–185, 253; [translated in: Algebra and Logic, **18** (1979) 111–117].

[119] O. V. Kudinov, An autostable 1–decidable model without a computable Scott family of ∃–formulas (Russian), Algebra i Logika, **35** (1996) 458–467, 498; [translated in: Algebra and Logic, **35** (1996) 255–260].

[120] D. Lascar, Stabilité en théorie des modèles, Monographies de Mathématique, Institut de Mathématique Pure et Appliquée, Université Catholique de Louvain, (Cabay Libraire-Éditeur S.A., Louvain-la-Neuve, Belgium, 1986); [translated as: Stability in Model Theory, translated by J. E. Wallington, Pitman Monogr. Surveys Pure Appl. Math., **36** (1987)].

[121] M. Lerman and J. H. Schmerl, Theories with recursive models, J. Symbolic Logic, **44** (1979) 59–76.

[122] L. Lipshitz and M. Nadel, The additive structure of models of arithmetic, Proc. Amer. Math. Soc., **68** (1978) 331–336.

[123] A. Macintyre and D. Marker, Degrees of recursively saturated models, Trans. Amer. Math. Soc., **282** (1984) 539–554.

[124] A. I. Mal'tsev, Constructive algebras I (Russian), Uspekhi Mat. Nauk, 16 (1961) 3–60; [translated in: Constructive algebras I, Russian Math. Surveys, **16**:3 (1961) 77–129; also in: The Metamathematics of Algebraic Systems, Collected Papers: 1936–1967, translated and edited by B. F. Wells III, Stud. Logic Found. Math., **66** (1971), Ch. 18, 148–200].

[125] A. I. Mal'tsev, Algebraic Systems (Russian), (posth. edn.), V. D. Smirnov and M. Taĭclin, (eds.), (Izdat. Nauka, Moscow, 1970); [translated: by B. D. Seckler and A. P. Doohovskoy, Grundlehren Math. Wiss., **192** (1973).]

[126] M. S. Manasse, Techniques and Counterexamples in Almost Categorical Recursive Model Theory, Ph.D. Thesis, Univ. Wisconsin, Madison, WI, (1982).

[127] D. Marker, Degrees of models of true arithmetic, in: Logic Colloq. '81. (Proc. Herbrand Sympos., Marseilles, July, 16–24, 1981), J. Stern (ed.), Stud. Logic Found. Math., **107** (1982) 233–242.

[128] G. Metakides and A. Nerode, Recursion theory and algebra, in: Algebra and Logic, (Proc. 14th. Summer Res. Inst. Austral. Math. Soc., Monash Univ., Clayton, Vic., Australia, Jan. 6 – Feb. 16, 1974), J. N. Crossley, (ed.), Lecture Notes in Math., **450** (1975) 209–219.

[129] G. Metakides and A. Nerode, Recursively enumerable vector spaces, Ann. Math. Logic, **11** (1977) 147–171.

[130] G. Metakides and A. Nerode, Effective content of field theory, Ann. Math. Logic, **17** (1979) 289–320.

[131] G. Metakides and A. Nerode, Recursion theory on fields and abstract dependence, J. Algebra, **65** (1980) 36–59.

[132] G. Metakides and A. Nerode, The introduction of nonrecursive methods into mathematics, in: The L. E. J. Brouwer Centenary Sympos., (Noordwijkerhout, June 8–13, 1981), A. S. Troelstra and D. van Dalen, (eds.), Stud. Logic Found. Math., **110** (1982) 319–335.

[133] T. S. Millar, The Theory of Recursively Presented Models, Ph.D. Thesis, Cornell Univ., Ithaca, NY, (1976).

[134] T. S. Millar, Foundations of recursive model theory, Ann. Math. Logic, **13** (1978) 305–320.

[135] T. S. Millar, A complete, decidable theory with two decidable models, J. Symbolic Logic, **44** (1979) 307–312.

[136] T. S. Millar, Homogeneous models and decidability, Pacific J. Math., **91** (1980) 407–418.

[137] T. S. Millar, Counterexamples via model completions, in: Logic Year 1979-1980, (Proc. Seminars and Conf. Math. Logic, Univ. Connecticut, Storrs, CT, 1979/80), M. Lerman, J. H. Schmerl and R. I. Soare, (eds.), Lecture Notes in Math., **859** (1981) 215–229.

[138] T. S. Millar, Vaught's theorem recursively revisited, J. Symbolic Logic, **46** (1981) 397–411.

[139] T. S. Millar, Type structure complexity and decidability, Trans. Amer. Math. Soc., **271** (1982) 73–81.

[140] T. S. Millar, Omitting types, type spectrums, and decidability, J. Symbolic Logic, **48** (1983) 171–181.

[141] T. S. Millar, Persistently finite theories with hyperarithmetic models, Trans. Amer. Math. Soc., **278** (1983) 91–99.

[142] T. S. Millar, Decidability and the number of countable models, Ann. Pure Appl. Logic, **27** (1984) 137–153.

[143] T. S. Millar, Decidable Ehrenfeucht theories, in: Recursion Theory, (Proc. AMS-ASL Summer Inst., Ithaca, NY, June 28 – July 16, 1982), A. Nerode and R. A. Shore, (eds.), Proc. Sympos. Pure Math., **42** (1985) 311–321.

[144] T. S. Millar, Bad models in nice neighborhoods, J. Symbolic Logic, **51** (1986) 1043-1055.

[145] T. S. Millar, Prime models and almost decidability, J. Symbolic Logic, **51** (1986) 412–420.

[146] T. S. Millar, Recursive categoricity and persistence, J. Symbolic Logic, **51** (1986) 430–434.

[147] T. S. Millar, Homogeneous models and almost decidability, J. Austral. Math. Soc., Ser. A, **46** (1989) 343–355.

[148] T. S. Millar, Abstract recursive model theory, in: Handbook of Recursion Theory, E. R. Griffor, (ed.), North-Holland, (to appear).

[149] J. D. Monk. Mathematical Logic. Grad. Texts in Math., **37** (1976).

[150] M. Morley, Decidable models, Israel J. Math., **25** (1976) 233–240.

[151] A. S. Morozov, A countably categorical decidable model without nontrivial recursive automorphisms (Russian), **30** (1989) 221–224; [translated in: Siberian Math. J., 30 (1989) 346–348].

[152] A. S. Morozov, Functional trees and automorphisms of models (Russian), Algebra i Logika, 32 (1993) 54–72; [translated in: Algebra and Logic, 32 (1993) 28–38].

[153] A. S. Morozov, Automorphism groups of decidable models (Russian), Algebra i Logika, **34** (1995) 437–447; [translated in: Algebra and Logic, **34** (1995) 242–248].

[154] M. Moses, Recursive properties of isomorphism types, J. Austral. Math. Soc., Ser. A, **34** (1983) 269–286.

[155] M. Moses, Relations intrinsically recursive in linear orders, Z. Math. Logik Grundlag. Math., **32** (1986) 467–472.

[156] A. Nerode and J. B. Remmel, A survey of lattices of r.e. substructures, in: Recursion Theory, (Proc. AMS-ASL Summer Inst., Ithaca, NY, June 28 – July 16, 1982), A. Nerode and R. A. Shore, (eds.), Proc. Sympos. Pure Math., **42** (1985) 323–375.

[157] A. Nerode and J. B. Remmel, Complexity-theoretic algebra II: Boolean algebras, (Special issue: 3rd. Asian Conf. Math. Logic, Beijing, Oct. 26–30, 1987), D. P. Yang, (ed.), Ann. Pure Appl. Logic, **44** (1989) 71–99.

[158] A. Nerode and R. L. Smith, The undecidability of the lattice of recursively enumerable subspaces, in: Proc. 3rd. Brazilian Conf. on Math. Logic, (Inst. Math., Fed. Univ. Pernambuco, Recife, 1979), A. I. Arruda, N. C. A. da Costa and A. M. Sette, (eds.), Soc. Brasil Logica, Sao Paulo, (1980) 245–252.

[159] A. T. Nurtazin, Strong and weak constructivizations and computable families (Russian), Algebra i Logika, **13** (1974) 311–323, 364; [translated in: Algebra and Logic, **13** (1974) 177–184].

[160] P. Odifreddi, Classical Recursion Theory. The Theory of Functions and Sets of Natural Numbers, Stud. Logic Found. Math., **125** (1989).

[161] S. P. Odintsov, Hereditarily recursively enumerable subalgebras of recursive Boolean algebra (Russian), Algebra i Logika, **31** (1992) 38–46, 96; [translated in: Algebra and Logic, **31** (1992) 24–29].

[162] M. G. Peretyat'kin, Strongly constructive models and enumerations of the Boolean algebra of recursive sets (Russian), Algebra i Logika, **10** (1971) 535–557; [translated in: Algebra and Logic, **10** (1971) 332–345].

[163] M. G. Peretyat'kin, Complete theories with a finite number of countable models (Russian), Algebra i Logika, **12** (1973) 550–576, 618; [translated as: On complete theories with a finite number of denumerable models, Algebra and Logic, **12** (1973) 310–326].

[164] M. G. Peretyat'kin, Every recursively enumerable extension of a theory of linear order has a constructive model (Russian), Algebra i Logika, **12** (1973) 211–219, 244; [translated in: Algebra and Logic, **12** (1973) 120–124].

[165] M. G. Peretyat'kin, Strongly constructive model without elementary submodels and extensions (Russian), Algebra i Logika, **12** (1973) 312–322, 364; [translated in: Algebra and Logic, **12** (1973) 178–183].

[166] M. G. Peretyat'kin, A criterion for strong constructivizability of a homogeneous model (Russian), Algebra i Logika, **17** (1978) 436–454, 491; [translated in: Algebra and Logic, **17** (1978) 290–301.]

[167] M. G. Peretyat'kin, Example of an ω_1–categorical complete finitely axiomatizable theory (Russian), Algebra i Logika, **19** (1980) 314–347, 382–383; [translated in: Algebra and Logic, **19** (1980) 202–229].

[168] M. G. Peretyat'kin, Calculations on Turing machines in finitely axiomatizable theories (Russian), Algebra i Logika, **21** (1982) 410–441; [translated as: Turing machine computations in finitely axiomatizable theories, Algebra and Logic, **21** (1982) 272–295.]

[169] M. G. Peretyat'kin, Finitely axiomatizable theories (Russian), in: Proc. Int. Congr. Math., Berkeley, CA, USA, 1986, **1** (1987) 322–330; [translated in: Nine Papers from the International Congress of Mathematicians, Aug. 3–11, 1986, Berkeley, CA, USA, B. Silver, (ed.), Amer. Math. Soc. Transl. Ser. 2, **147** (1990) 11–19.]

[170] M. G. Peretyat'kin, The similarity of properties of recursively enumerable and finitely axiomatizable theories (Russian), Dokl. Akad. NaukSSSR, **308** (1989) 788–791; [translated in: Soviet Math. – Dokl., **40** (1990) 372–375].

[171] M. G. Peretyat'kin, Analogues of Rice's theorem for semantic classes of propositions (Russian), Algebra i Logika, **30** (1991) 517–539, 626; [translated in: Algebra and Logic, **30** (1991) 332–348].

[172] M. G. Peretyat'kin, Semantically universal classes of models (Russian), Algebra i Logika, **30** (1991) 414–431, 507; [translated in: Algebra and Logic, **30** (1991) 271–282.]

[173] M. G. Peretyat'kin, Semantic universality of theories over a superlist (Russian), Algebra i Logika, **31** (1992) 47–73, 96; [translated in: Algebra and Logic, **31** (1992) 30–47].

[174] A. Pillay. An Introduction to Stability Theory. (Oxford Univ. Press, NY, 1983).

[175] M. B. Pour-El and J. I. Richards, Computability in Analysis and Physics, Perspect. in Math. Logic, (1989).

[176] M. O. Rabin, Recursive unsolvability of group theoretic problems, Ann. of Math., **67** (1958) 172–194.

[177] M. O. Rabin, Computable algebra, general theory and theory of computable fields, Trans. Amer. Math. Soc., **95** (1960) 341–360.

[178] R. C. Reed. A Decidable Ehrenfeucht Theory with Exactly Two Hyperarithmetic Models, Ph.D. Thesis, Univ. Wisconsin, Madison, WI, (1986).

[179] R. C. Reed, A decidable Ehrenfeucht theory with exactly two hyperarithmetic models, Ann. Pure Appl. Logic, **53** (1991) 135–168.

[180] J. B. Remmel, Effective structures not contained in recursively enumerable structures, in: Aspects of Effective Algebra, (Proc. Conf. Monash Univ., Clayton, Australia, Aug. 1–4, 1979), J. N. Crossley (ed.), (Upside Down A Book Co., Yarra Glen, Victoria, Australia, 1981) 206–225.

[181] J. B. Remmel, Recursively categorical linear orderings, Proc. Amer. Math. Soc., **83** (1981) 387–391.

[182] J. B. Remmel, Recursive isomorphism types of recursive Boolean algebras, J. Symbolic Logic, **46** (1981) 572–594.

[183] J. B. Remmel and J. N. Crossley, The work of Anil Nerode: a retrospective, in: Logical Methods, (Papers from Conf. in honor of Anil Nerode's Sixtieth Birthday, June 1–3, 1992, Cornell Univ., Ithaca, NY), J. N. Crossley, J. B. Remmel, R. A. Shore and M. E. Sweedler, (eds.), Progr. Comput. Sci. Appl. Logic, **12** (1993) 1–85.

[184] J. P. Ressayre, Boolean models and infinitary first order languages, Ann. Math. Logic, **6** (1973) 41–92.

[185] L. J. Richter, Degrees of Unsolvability of Models, Ph.D. Thesis, Univ. Illinois at Urbana-Champaign, Urbana, IL, (1977).

[186] L. J. Richter, Degrees of structures, J. Symbolic Logic, **46** (1981) 723–731.

[187] H. Rogers Jr., Theory of Recursive Functions and Effective Computability, (1st. edn., McGraw-Hill, New York-Toronto, Ont.-London, 1967; 2nd. edn., MIT Press, Cambridge, Mass., London, 1987).

[188] G. E. Sacks, Saturated Model Theory, (Mathematics Lecture Note Series, W. A. Benjamin, Inc., Reading, Mass., 1972).

[189] G. E. Sacks, Higher Recursion Theory, Perspect. in Math. Logic, (1990).

[190] J. Schlipf, Toward model theory through recursive saturation, J. Symbolic Logic, **43** (1978) 183–206.

[191] J. H. Schmerl, A decidable $\aleph_0$–categorical theory with a nonrecursive Ryll-Nardzewski function, Fund. Math., **98** (1978) 121–125.

[192] D. Scott, Algebra of sets binumerable in complete extensions of arithmetic, Proc. Sympos. Pure Math., **5** (1962) 117–121.

[193] D. Scott and S. Tennenbaum, On the degrees of complete extensions of arithmetic, Notices Amer. Math. Soc., **7** (1960) 242–243.

[194] J. R. Shoenfield, Degrees of models, J. Symbolic Logic, **25** (1960) 233–237.

[195] T. A. Slaman, Relative to any non-recursive set, Proc. Amer. Math. Soc., (to appear).

[196] R. I. Soare, Recursively Enumerable Sets and Degrees. A Study of Computable Functions and Computably Generated Sets, Perspect. in Math. Logic, (1987).

[197] I. N. Soskov, Intrinsically hyperarithmetical sets, Math. Logic Quart., **42** (1996) 469–480.

[198] I. N. Soskov, Intrinsically Π_1^1 relations, Math. Logic Quart., **42** (1996) 109–126.

[199] S. Tennenbaum, Non-archimedean models for arithmetic, Notices Amer. Math. Soc., **6** (1959) 270.

[200] D. A. Tusupov, Numerations of homogeneous models of decidable complete theories with a computable family of types (Russian), Vyčisl. Systemy, **129**, (Akad. Nauk SSSR Sibirsk. Otdel., Inst. Mat., Novosibirsk) (1989) 152–171, 197–198.

[201] R. L. Vaught, Sentences true in all constructive models, J. Symbolic Logic, **25** (1960) 39–58.

[202] M. C. Venning, Type Structures of $\aleph_0$–Categorical Theories, Ph.D. Thesis, Cornell Univ., Ithaca, NY, (1977).

[203] Yu. G. Ventsov, A family of recursively enumerable sets with finite classes of nonequivalent univalent computable numerations (Russian), Vyčisl. Systemy, **120**, Logich. Metody v Program., (Novosibirsk, 1987) 105–142.

[204] Yu. G. Ventsov, Nonuniform autostability of models (Russian), Algebra i Logika, **26** (1987) 684–714; [translated in: Algebra and Logic, **26** (1987) 422–440].

[205] Yu. G. Ventsov, Algorithmic dimension of models (Russian), Dokl. Akad. Nauk, **305** (1989) 21–24; [translated in: Soviet Math. - Dokl., **39** (1989) 237–239].

[206] Yu. G. Ventsov, Computable classes of constructivizations of models of infinite algorithmic dimension (Russian), Algebra i Logika, **33** (1994) 37–75; [translated in: Algebra and Logic, **33** (1994) 22–45].

[207] Yu. G. Ventsov, Constructive models of regularly infinite algorithmic dimension (Russian), Algebra i Logika, 33 (1994) 135–146, 228; [translated in: Algebra and Logic, **33** (1994) 79–84].

[208] A. D. Vlach, Hyperarithmetical relations in expansions of recursive structures, Ann. Pure Appl. Logic, **66** (1994) 163–196.

[209] B. L. van der Waerden, Eine Bemerkung über die Unzerlegbarkeit von Polynomen, Math. Ann., **102** (1930) 738–739.

[210] B. L. van der Waerden, Moderne Algebra, Parts I and II, (G. E. Stechert and Co., New York, 1943); [translated as: Modern Algebra, Vols. I and II, (translated from the 2nd. rev. German edition by F. Blum, with revisions and additions by the author), Frederick Ungar Publishing Co., NY, (1949)].

[211] B. L. van der Waerden, Algebra, Vols. I and II, (7th. edn. (1966) Vol. I, 5th. edn. (1967) Vol II), Heidelberger Taschenbücher **12** and **23**, Springer-Verlag, Berlin and NY; [Vol. I translated by F. Blum and J. R. Schulenberger, Vol. II translated by J. R. Schulenberger, (Frederick Ungar Publishing Co., NY, 1970), reprinted: Springer-Verlag, NY, 1991].

[212] S. Wehner. Enumerations, countable structures and Turing degrees, to appear.

Valentina S. Harizanov
Department of Mathematics
The George Washington University
Washington, D. C. 20052, USA
email: val@math.gwu.edu

Chapter 2

Elementary Theories and their Constructive Models

Yu. L. Ershov and S. S. Goncharov*

Contents

Introduction

The theory of constructive models is a branch of mathematical logic concerned with the study of algorithmic properties of algebraic structures. In this paper, we consider the algorithmic complexity of relations defined on algebraic structures. We investigate the algorithmic properties of various structures by studying their numberings and considering the algorithmic complexity of the set of numbers of elements appearing in the relations under consideration.

In Section 1, on the basic notions of constructive model theory, we discuss the decidability of basic relations and elementary properties of structures. Models equipped with numberings in which these relations (and elementary properties) are decidable are called constructive (strongly constructive), while the models themselves are referred to as constructivizable (strongly constructivizable). Section 2 is concerned with the existence problem for

*This work was supported by the Russian Foundation for Basic Research, Grant No. 96-01-01525

HANDBOOK OF RECURSIVE MATHEMATICS
Edited by Yu. L. Ershov, S. S. Goncharov,
A. Nerode, and J. B. Remmel

decidable models with given model-theoretic properties. The main result of this section is the Goncharov–Peretyat'kin criterion for the decidability of homogeneous models. Section 3 is a study of decidable Boolean algebras, with particular focus on the existence of strongly constructive elementary extensions.

1 Basic notions of constructive model theory

We use the terminology of the theory of constructive models and computability theory, following [17, 15, 14, 29, 41, 43, 76, 78].

We study the notion of constructibility of arbitrary algebraic structures. The notion is considered with respect to B, where B is some subset of $\mathbb{N}$, or a class of sets or functions. This means that, in computations below, we can use information about the membership of natural numbers appearing in algorithms to the set B (or a set of the class B) which may have any algorithmic complexity. As we show below, such a generalization allows us to obtain interesting results in the case of ordinary constructivization.

There are two equivalent approaches to the study of algorithmic properties of algebraic structures. The first approach deals with abstract structures of arbitrary elements. Various numberings of universes by natural numbers are chosen and their properties are investigated. The second approach admits only those structures whose universes consist of natural numbers. The first approach leads to constructive and strongly constructive structures and the second one leads to (equivalent) recursive and decidable structures.

Let $\mathfrak{A} = \langle A\,; P_0\,, \ldots\,, P_r\,; F_0\,, \ldots\,, F_s\,; c_0\,, \ldots\,, c_t\rangle$ be an algebraic structure of a signature

$$\sigma \leftrightharpoons \langle P_0^{n_0}\,, \ldots\,, P_r^{n_r}\,; F_0^{m_0}\,, \ldots\,, F_s^{m_s}\,; c_0\,, \ldots\,, c_t\rangle$$

where each P_i, $i \leqslant r$, is an n_i–ary predicate symbol; each F_i, $i \leqslant s$, is an m_i–ary function symbol; and each c_i, $i \leqslant t$, is a constant symbol. If the signature is infinite, we require that the functions $i \to n_i$ and $i \to m_i$ be recursive.

A pair $\langle \mathfrak{A}, \nu\rangle$, where ν is a mapping of $\mathbb{N}$ (or of an initial segment of $\mathbb{N}$) onto the universe A of the structure $\mathfrak{A}$ is called an *enumerated structure* and ν is called a *numbering* of $\mathfrak{A}$.

Let B be either a subset of $\mathbb{N}$ or else a family of subsets of $\mathbb{N}$, and let $\mathfrak{A}$ be an algebraic structure of signature σ. An enumerated structure $\langle \mathfrak{A}, \nu \rangle$ is called *B–positive* if the numeration equivalence under ν in $\mathfrak{A}$,

$$\eta_\nu \leftrightharpoons \{\langle n, m \rangle \mid \nu n = \nu m \text{ in } \mathfrak{A}\}$$

together with the inverse images under ν of the relations $P_i^{\mathfrak{A}}$ defined on $\mathfrak{A}$ by predicates P_i,

$$\nu^{-1}(P_i) = \{\langle l_1, \dots, l_{m_i} \rangle \mid (\nu l_1, \dots, \nu l_{m_i}) \in P_i^{\mathfrak{A}}\}$$

for all $i \leqslant r$, are all recursively enumerable with respect to B, and there exist B–recursive functions f_i, $i \leqslant s$, such that

$$\nu f_i(\langle l_1, \dots, l_{m_i} \rangle) = F_i^{\mathfrak{A}}(\nu l_1, \dots, \nu l_{m_i}), \quad \text{for any } l_1, \dots, l_{m_i} \in \mathbb{N}$$

A B–positive structure $\langle \mathfrak{A}, \nu \rangle$ is called *B–constructive* if the sets η_ν and $\nu^{-1}(P_i)$ for $i \leqslant r$ are B–recursive (instead of just r.e.–in–B).

If σ is infinite, then in the definitions of B–positive and B–constructive, we impose the extra conditions of uniform recursive enumerability and uniform recursivity in the indices i.

In particular, an enumerated structure $\langle \mathfrak{A}, \nu \rangle$ is called *constructive* if the set η_ν, the numeration equivalence under ν in $\mathfrak{A}$, together with the sets $\nu^{-1}(P_i)$, the inverse images under ν of all the relations $P_i^{\mathfrak{A}}$, are all recursive, and there are recursive functions f_i representing the functions $F_i^{\mathfrak{A}}$, in the sense that $\nu \circ f_i = F_i^{\mathfrak{A}} \circ \nu$. Thus an enumerated structure $\langle \mathfrak{A}, \nu \rangle$ is constructive if the set of all atomic formulas satisfied in $\mathfrak{A}$ is decidable.

A numbering ν of a structure $\mathfrak{A}$ such that $\langle \mathfrak{A}, \nu \rangle$ is constructive, is called a *constructivization* of the structure $\mathfrak{A}$.

We define a numbering γ of all formulas of the signature σ in variables of the set $\{v_0, v_1, \dots\}$ so that all questions concerning the structure of formulas can be decided by means of the numbers of these formulas. Such a numbering is called a *Gödel numbering* [19].

Let σ be a signature of the form

$$\langle P_0^{n_0}, \dots, P_r^{n_r}, \dots; F_0^{m_0}, \dots, F_s^{m_s}, \dots; c_0, \dots, c_t, \dots \rangle$$

such that there exist partial recursive functions $[n]$ and $[m]$ defined as follows:

$$[n](i) = n_i, \quad \text{where } n_i \text{ is the arity of the predicate symbol } P_i,$$

$$[m](i) = m_i, \quad \text{where } m_i \text{ is the arity of the function symbol } F_i.$$

If there is only a finite number of predicate and function symbols, such functions $[n]$ and $[m]$ exist. If the language has a countably infinite number of symbols, then we must be able to effectively determine the arities n_i and m_i from the index i in the signature. This is possible due to the requirement of recursivity imposed on the functions $i \to n_i$ and $i \to m_i$.

Let V be the set of variables $v_0, v_1, \ldots, v_n, \ldots$; let $\mathrm{Term}_\sigma(\mathrm{V})$ denote the set of terms of the signature σ in variables of V, and let $\mathrm{Form}_\sigma(\mathrm{V})$ denote the set of formulas of the signature σ in variables of V. A *Gödel numbering* is a mapping

$$\gamma_\sigma : \mathrm{Term}_\sigma(\mathrm{V}) \cup \mathrm{Form}_\sigma(\mathrm{V}) \xrightarrow{1-1} \mathbb{N}$$

such that we can effectively recognize if a given number is the number of a formula or that of a term, and from a number, we can recover the structure of the corresponding formula or term.

A Gödel numbering γ is constructed by induction on the complexity of formulas. Once we have a Gödel numbering, we can discuss the notion of decidability from the mathematical point of view. A subset

$$X \subseteq \mathrm{Term}_\sigma(\mathrm{V}) \cup \mathrm{Form}_\sigma(\mathrm{V})$$

is called *decidable* if the set of Gödel numbers $\gamma(X) = \{\gamma(q) \mid q \in X\}$ is recursive, and it is called *enumerable* if $\gamma(X)$ is recursively enumerable. With some complexity hierarchy of subsets of $\mathbb{N}$ (e.g., the arithmetic hierarchy, the analytic hierarchy, the Ershov hierarchy, etc.) in mind, we say that X *belongs to some complexity class* Δ if the set $\gamma(X)$ belongs to Δ.

Given the Gödel number of a formula, we can recognize whether the formula is an axiom of the predicate calculus PC^σ. Given a sequence of Gödel numbers of formulas, we also can recognize whether one formula is obtained from others in the sequence by a deductive rule of PC^σ. Hence we can recognize if a sequence of formulas with given Gödel numbers is a proof in PC^σ.

Let σ be a signature and let σ' be an extension of σ. If, from the indices of predicate symbols, function symbols and constants of the signature σ, we can compute the indices of these symbols regarded as symbols of the signature σ', then, starting from Gödel numbers of terms and formulas of the signature σ, we can compute the Gödel numbers of the corresponding terms and formulas with respect to the signature σ'. Moreover, if, starting from indices of predicate symbols, function symbols, and constants of the signature σ', we can recognize whether they belong to σ, then, starting from any term or formula of the signature σ', we can recognize whether this term or formula is of the signature σ and find its number with respect to σ. We only consider extensions of signatures which satisfy these conditions. In this case, we say that the Gödel numbering of formulas and terms of the signature σ' *extends* the Gödel numbering of formulas and terms of the signature σ.

For a decidable theory T in a signature σ, we define the principal computable numbering $p_0(\overline{x}_0), \ldots, p_n(\overline{x}_n), \ldots$ of the set of all partial enumerable types consistent with the theory T.

A *partial type* $p(\overline{x})$ of a theory T is the set of formulas in variables of $\overline{x}$ such that the set $p(\overline{x}) \cup T$ is consistent. A numbering $d_0(\overline{x}_0), \ldots, d_n(\overline{x}_n), \ldots$ of partial types of the theory T is called *computable* if $\{\overline{d}_i\}_i$ is a computable numbering of recursively enumerable sets, where

$$\overline{d}_i = \{k \mid k \text{ is the Gödel number of a formula in } d_i(\overline{x}_i)\}$$

and there exists a recursive function v such that for any n, the value $v(n)$ is equal to the code of the sequence $\langle i_1, \ldots, i_{m_n}\rangle$ of indices such that

$$\overline{x}_n = (v_{i_1}, \ldots, v_{i_{m_n}}).$$

A numbering $p_0(\overline{x}_0), \ldots, p_n(\overline{x}_n), \ldots$ of partial types of the theory T is called *principal* if for any computable numbering $d_0(\overline{x}'_0), \ldots, d_n(\overline{x}'_n), \ldots$ of partial types of the theory T, there is a recursive function $f(n)$ such that $d_n(\overline{x}'_n) = p_{f(n)}(\overline{x}_{f(n)})$ for any n.

To define the $p_n(\overline{x}_n)$ of a principal numbering, for each n we construct an expanding sequence of finite sets

$$\emptyset = p_n^0(\overline{x}_n) \subseteq p_n^1(\overline{x}_n) \subseteq \cdots \subseteq p_n^t(\overline{x}_n) \subseteq \cdots$$

such that

$$p_n(\overline{x}_n) \leftrightharpoons \bigcup_t p_n^t(\overline{x}_n).$$

For any $n \in \mathbb{N}$, we take i and k such that $c(i,k) = n$, where c is some fixed pairing function on the natural numbers. We use i as the number of the ith recursively enumerable set W_i and k as the code of the sequence $\langle i_1, \ldots, i_s\rangle$, with respect to a fixed coding of all sequences of finite length. For any $i, k, t \in \mathbb{N}$, we set

$$W_i^t \restriction k \leftrightharpoons \{m \in W_i^t \mid m \text{ is the Gödel number of a formula in free variables with indices in } \langle i_1, \ldots, i_s\rangle \text{ where the sequence of indices is coded by } k\}$$

and

$$p_n^t(\overline{x}_n) \leftrightharpoons \{\Phi \mid \text{the Gödel number of } \Phi \text{ is in } W_{l(n)}^m \upharpoonright r(n)\}$$

where m is the greatest number less than $t+1$ such that the set

$$T \cup \{\Phi \mid \text{ the Gödel number of } \Phi \text{ is in } W_{l(n)}^m \upharpoonright r(n)\}$$

is consistent. By the decidability of T, this consistency condition is decidable. Consequently, given n and t, we can recognize whether a formula belongs to $p_n^t(\overline{x}_n)$ and compute the Gödel number of such formulas; i.e., write out the list of all formulas in $p_n^t(\overline{x}_n)$. It is clear that the Gödel number of any such formula is less than $t+1$, in view of the condition imposed on W_n^t. The definition of p_n in terms of $W_{l(n)}$, the possibility of computing exactly a sequence of free variables with respect to a computable numbering of a family of finite types, and the fact that $\{W_n\}_{n\in\mathbb{N}}$ is principal, together imply that the numbering $\{p_n\}_{n\in\mathbb{N}}$ of partial types is principal.

Proposition 1.1 *A family S of types of a theory T is computable, i.e., there exists a computable numbering $d_0(\overline{x}'_0), \ldots, d_n(\overline{x}'_n), \ldots$ such that*

$$S = \{d_n(\overline{x}'_n) \mid n \in \mathbb{N}\}$$

if and only if there exists a recursively enumerable set W such that $S = \{p_n(\overline{x}_n) \mid n \in W\}$.

Proof. The claim follows from the definition since the numbering $\{p_n\}_{n\in\mathbb{N}}$ is principal. □

In the study of algorithmic properties of models, it is necessary to solve the problem of the existence of an algorithm that verifies the truth in a model of *all elementary formulas*, not just atomic formulas. This leads us to a notion that is stronger than that of constructibility.

We adopt the convention that a quantifier-free formula of signature σ *has* 0 *alternating quantifiers.* We say that a formula Φ of σ *has n alternating quantifiers* if the prenex normal form of Φ has n blocks of alternating quantifiers. Let $\mathfrak{F}_n$ denote the set of all formulas of σ with n alternating quantifiers and let $\mathfrak{F}_\omega$ denote the union over n of the sets $\mathfrak{F}_n$. The sets $\mathfrak{F}_n$ are said to be *restricted fragments* (of the language of σ).

Let $\mathfrak{F}$ be a set of formulas of signature σ. Let $\mathcal{B}$ be a family of subsets of $\mathbb{N}$. An enumerated structure $\langle\mathfrak{A},\nu\rangle$ (in signature σ) is called $\mathcal{B}$–$\mathfrak{F}$*–constructive* or

$\mathfrak{F}$*-constructive with respect to* $\mathcal{B}$ if the set of Gödel numbers of the $\mathfrak{F}$*-diagram of* $\langle\mathfrak{A},\nu\rangle$

$$\mathfrak{F}\mathcal{D}\langle\mathfrak{A},\nu\rangle = \{\langle s, l_1, \dots, l_k\rangle \mid s \text{ is the Gödel number of a formula } \Phi(x_1, \dots, x_k) \text{ of } \mathfrak{F} \text{ with } k \text{ free variables and } \mathfrak{A} \models \Phi(\nu l_1, \dots, \nu l_k)\}$$

belongs to $\mathcal{B}$. Since $\mathfrak{F}_0\mathcal{D}\langle\mathfrak{A},\nu\rangle$ is the set of Gödel numbers of the quantifier-free formulas satisfied in $\mathfrak{A}$, and this set can be $\mathcal{B}$–effectively recovered from the corresponding set for atomic formulas (positive diagram of $\langle\mathfrak{A},\nu\rangle$), it follows that structures $\mathfrak{F}_0$–constructive with respect to $\mathcal{B}$ are exactly $\mathcal{B}$–constructive structures.

A $\mathcal{B}$–$\mathfrak{F}$–constructive structure is called $\mathfrak{F}$*–constructive* if $\mathcal{B}$ is the class of all recursive subsets of $\mathbb{N}$, and is called $\mathcal{B}$*–constructive* if $\mathfrak{F} = \mathfrak{F}_0$. If we do not specify $\mathfrak{F}$ or $\mathcal{B}$, we mean that $\mathfrak{F} = \mathfrak{F}_0$ or $\mathcal{B}$ is the class of all recursive sets, respectively. $\mathcal{B}$–$\mathfrak{F}_\omega$–constructive structures are referred to as *strongly* $\mathcal{B}$*–constructive* or $\mathcal{B}$*–*ω*–constructive* and $\mathcal{B}$–$\mathfrak{F}_n$–constructive structures are called $\mathcal{B}$*–*n*–constructive.*

In particular, an enumerated structure $\langle\mathfrak{A},\nu\rangle$ is *strongly constructive* if the elementary diagram of $\langle\mathfrak{A},\nu\rangle$ is decidable; i.e. the set $\mathfrak{F}\mathcal{D}\langle\mathfrak{A},\nu\rangle$ is recursive, where $\mathfrak{F} = \mathfrak{F}_\omega$ is set of all first-order formulas in signature σ.

A structure $\mathfrak{A}$ is called *decidable* if there exists a numbering ν of $\mathfrak{A}$ such that $\langle\mathfrak{A},\nu\rangle$ is *strongly constructive*; in this case, such a numbering *nu* is called a *strong constructivization* of $\mathfrak{A}$.

We now move on to the second approach to the study of algorithmic properties of algebraic structures, in which we consider only those structures whose universes consist of natural numbers. Let $\mathfrak{A}$ be a structure of a signature σ whose universe A is a subset of $\mathbb{N}$. In this setting, we can discuss the effectiveness of various relations without handling numbers.

A structure $\mathfrak{A}$ is called $\mathcal{B}$*–recursive* if its basic predicates and functions are of class $\mathcal{B}$. For many classes of computability $\mathcal{B}$, an abstract structure is $\mathcal{B}$–constructivizable if and only if it is isomorphic to a $\mathcal{B}$–recursive structure. Given a $\mathcal{B}$–constructive structure, we can construct a $\mathcal{B}$–recursive structure effectively with respect to $\mathcal{B}$ provided that we can choose a single number from each set in the family $\mathcal{B}$. The passage from a $\mathcal{B}$–recursive structure to a $\mathcal{B}$–constructive one can usually be realized with the help of a $\mathcal{B}$–recursive function that enumerates the universe of a $\mathcal{B}$–recursive model and defines the $\mathcal{B}$–constructivization of this structure.

For structures that are $\mathfrak{F}$–constructive with respect to $\mathcal{B}$, we can define $\mathcal{B}$–$\mathfrak{F}$–recursive models in a similar way. If a model is isomorphic to a $\mathcal{B}$–$\mathfrak{F}_\omega$–recursive model, then it is $\mathcal{B}$–decidable. When $B \subseteq \mathbb{N}$, we use the notation $\mathcal{B}(B)$ for the class of B–recursive sets. $\mathcal{B}(B)$–$\mathfrak{F}$–recursive (respectively $\mathcal{B}(B)$–$\mathfrak{F}$–constructive) structures are called $\mathfrak{F}$–recursive (resp. $\mathfrak{F}$–constructive) with respect to B.

Proposition 1.2 *If the structures $\langle \mathfrak{M}, \nu \rangle$ and $\langle \mathfrak{N}, \mu \rangle$ are n–constructive (strongly constructive) with respect to B, then the direct product $\mathfrak{M} \times \mathfrak{N}$ with the numbering $(\nu \otimes \mu)(n) = (\nu(l(n)), \mu(r(n)))$ is a model which is n–constructive (strongly constructive) with respect to B.*

Proof. The assertion follows from the algorithm which reduces the question of the truth of formulas with respect to the direct product to that of the truth of formulas, with the same number of alternations of quantifiers, with respect to the components of the direct product [15]. □

Let $\langle \mathfrak{A}, \nu \rangle$ and $\langle \mathfrak{B}, \mu \rangle$ be enumerated structures, and let $\varphi \colon \mathfrak{A} \to \mathfrak{B}$ be a homomorphism. A homomorphism φ from $\langle \mathfrak{A}, \nu \rangle$ into $\langle \mathfrak{B}, \mu \rangle$ is called *C–recursive* if there exists a C–recursive function f such that $\varphi\nu = \mu f$, i.e., the diagram

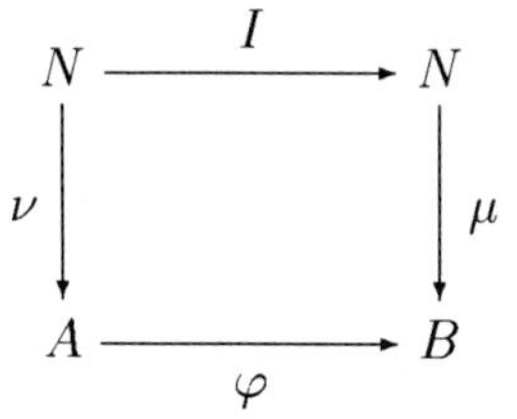

commutes. In this case, the function f represents φ and the latter is called a *C–homomorphism*. If there exists a C–recursive embedding (injective homomorphism) φ of $\langle \mathfrak{A}, \nu \rangle$ into $\langle \mathfrak{B}, \mu \rangle$, then $\langle \mathfrak{B}, \mu \rangle$ is called a *C–extension* of $\langle \mathfrak{A}, \nu \rangle$ *with respect to* φ. If $\mathfrak{A} \subseteq \mathfrak{B}$ and the identity embedding of $\mathfrak{A}$ into $\mathfrak{B}$ is C–recursive, then $\langle \mathfrak{B}, \mu \rangle$ is a C–extension of $\langle \mathfrak{A}, \nu \rangle$.

Proposition 1.3 *Let $\langle \mathfrak{B}, \mu \rangle$ be an enumerated structure, D be a C–recursively enumerable subset of $\mathbb{N}$, and the set $\mu(D)$ be closed under the functions of $\mathfrak{B}$ and contain the values of constants of $\mathfrak{B}$. Then the substructure $\mathfrak{A} \preccurlyeq \mathfrak{B}$ with universe $\mu(D)$ has a numbering ν such that $\langle \mathfrak{B}, \mu \rangle$ is a C–extension of $\langle \mathfrak{A}, \nu \rangle$.*

Proof. We obtain the required numbering if we put $\nu(n) \leftrightharpoons \mu f(n)$, where f is a C–recursive function enumerating the set D. □

Proposition 1.4 *If $\langle \mathfrak{B}, \mu \rangle$ is a C–extension of $\langle \mathfrak{A}, \nu \rangle$ with respect to φ, and $\langle \mathfrak{B}, \mu \rangle$ is constructive with respect to C, then $\langle \mathfrak{A}, \nu \rangle$ is also constructive with respect to C.*

Proof. The assertion follows from the fact that the verification of the truth of the predicates and equalities of $\langle \mathfrak{A}, \nu \rangle$ can be reduced to the verification of the truth of the corresponding predicates and equalities of $\langle \mathfrak{B}, \mu \rangle$ by means of a reducing function. □

Proposition 1.5 *If $\langle \mathfrak{A}, \nu \rangle$ is constructive with respect to B, and the theory* $\mathrm{Th}(\mathfrak{A}, a_1, \dots, a_n)$ *of the structure $\mathfrak{A}$, enriched by a finite number of constants $a_1, \dots, a_n$, is model-complete and B–decidable, then $\langle \mathfrak{A}, \nu \rangle$ is strongly constructive with respect to B.*

Proof. It suffices to note that, in view of the B–decidability of the theory $\mathrm{Th}(\mathfrak{A}, a_1, \dots, a_n)$, we can verify effectively the equivalence of formulas. However, any formula is equivalent (with respect to a model-complete theory) to an $\exists$–formula. Therefore, from any formula Φ, we can effectively find two $\exists$–formulas $\exists \overline{y} \Psi_1$ and $\exists \overline{y} \Psi_2$ that are equivalent to Φ and $\neg\Phi$ respectively. To verify the truth of Φ on elements with numbers $\overline{l} = (l_1, \dots, l_n)$, it suffices to search through all possible finite sequences $\overline{c}$ and $\overline{b}$ of numbers of elements, to find the first sequence such that $\Psi_1(\nu\overline{l}, \nu\overline{c})$ or $\Psi_2(\nu\overline{l}, \nu\overline{b})$ is satisfied in $\mathfrak{A}$. If $\Psi_1(\nu\overline{l}, \nu\overline{c})$ holds for some $\overline{c}$, then, by the above equivalences, Φ is true in $\mathfrak{A}$ on the sequences with numbers $\overline{l}$, and is false if $\Psi_2(\nu\overline{l}, \nu\overline{b})$ holds for some $\overline{b}$. □

Proposition 1.5 and the model completeness of the theory of Boolean algebras with respect to a signature extended by predicate symbols A_n, B_n, C_n, and I_n, for $n \in \mathbb{N}$, lead to the following proposition.

Proposition 1.6 *If $\langle \mathfrak{B}, \nu \rangle$ is a B–constructive Boolean algebra in an extended signature σ^*, then $\langle \mathfrak{B}, \nu \rangle$ is B–strongly constructive.*

The two approaches to effective representations of structures allow us to investigate the same properties, choosing a language as a function of the situation. We intend to establish that these approaches are equivalent. To this end, we show the equivalence of the corresponding categories. In our use of notions from category theory, we follow [1].

Let **Num** be the category of all enumerated structures in a fixed signature σ, with homomorphisms as morphisms, and let **Nat** be the category of all structures over σ whose universes are subsets of $\mathbb{N}$, with homomorphisms as morphisms. We define a functor $\mathbf{R} : \mathbf{Num} \to \mathbf{Nat}$ as follows. For each enumerated structure $\langle \mathfrak{M}, \nu \rangle$, with universe $M = |\mathfrak{M}|$, let

$$\mathbf{R}\,\langle \mathfrak{M}, \nu \rangle \;\leftrightharpoons\; \mathfrak{N}_{\mathfrak{M}}$$

with universe $N_{\mathfrak{M}}$, where

$$N_{\mathfrak{M}} \leftrightharpoons \{n \mid n \text{ is the least element of } \nu^{-1}(x), \text{ for some } x \in |\mathfrak{M}|\}$$

$$P_{\mathfrak{N}} \leftrightharpoons \{\langle n_1, \dots, n_k \rangle \in N_{\mathfrak{M}} \mid \mathfrak{M} \models P(\nu n_1, \dots, \nu n_k)\}$$

$$F_{\mathfrak{N}}(n_1, \dots, n_k) \leftrightharpoons \min\{n \in N_{\mathfrak{M}} \mid F_{\mathfrak{M}}(\nu n_1, \dots, \nu n_k) = \nu n\}$$

$$c_{\mathfrak{N}} \leftrightharpoons \min\{n \in N_{\mathfrak{M}} \mid \nu n = c_{\mathfrak{M}}\}$$

for each predicate symbol $P \in \sigma$, function symbol $F \in \sigma$ and constant $c \in \sigma$. It is clear from the definition of $\mathbf{R}$ that $\mathbf{R}\,\langle \mathfrak{M}, \nu \rangle$ and $\mathfrak{M}$ are isomorphic structures.

If φ is a homomorphism from $\langle \mathfrak{M}, \nu \rangle$ into $\langle \mathfrak{N}, \mu \rangle$, then define

$$\text{graph}\,\mathbf{R}\,(\varphi) \leftrightharpoons \{\langle n, m \rangle \mid n \in N_{\mathfrak{M}},\ m \in N_{\mathfrak{N}} \text{ and } \varphi(\nu n) = \mu m\}$$

in which case $\mathbf{R}\,(\varphi)$ is clearly a homomorphism from $\mathbf{R}\,\langle \mathfrak{M}, \nu \rangle$ into $\mathbf{R}\,\langle \mathfrak{N}, \mu \rangle$. If φ is an isomorphism, then $\mathbf{R}\,(\varphi)$ is also an isomorphism.

Proposition 1.7 *The functor* **R** *determines an equivalence between the categories* **Num** *and* **Nat**.

Next, we define a functor $\mathbf{C} : \mathbf{Nat} \to \mathbf{Num}$ as follows. For each structure $\mathfrak{N}$ in **Nat**, let $\mathbf{C}\,(\mathfrak{N}) \leftrightharpoons \langle \mathfrak{N}, \nu \rangle$, where ν is an enumeration of elements of the universe $|\mathfrak{N}|$ in ascending order. If $|\mathfrak{N}|$ is finite, then all the numbers that do not appear in this enumeration are mapped onto the last element of the enumeration. The numbering ν obtained in this way will be denoted $\nu_{\mathbf{C}}$.

For each subset $B \subseteq \mathbb{N}$, we define a subcategory $\mathbf{Con}^B$ of the category **Num** consisting of B–constructive enumerated structures with B–recursive homomorphisms as morphisms, and a subcategory $\mathbf{Rec}^B$ of the category **Nat** consisting of B–recursive structures with B–recursive homomorphisms as morphisms.

Theorem 1.8 *The restriction* $\mathbf{R}^B$ *of the functor* $\mathbf{R}$ *to the subcategory* $\mathbf{Con}^B$ *determines an equivalence between the categories* $\mathbf{Con}^B$ *and* $\mathbf{Rec}^B$.

Proof. It is easy to check that the model $\mathbf{R}\langle\mathfrak{M},\nu\rangle$ is B–recursive, provided that $\langle\mathfrak{M},\nu\rangle$ is a B–constructive structure. Now if φ is a B–recursive homomorphism from $\langle\mathfrak{M},\nu\rangle$ into $\langle\mathfrak{N},\mu\rangle$, where both of these enumerated structures are B–constructive, then the universes of $\mathbf{R}\langle\mathfrak{M},\nu\rangle$ and $\mathbf{R}\langle\mathfrak{N},\mu\rangle$ are B–recursive and there exists a B–recursive function f such that $\varphi\nu = \mu f$; together, these conditions imply that $\mathbf{Rec}\,\varphi$ is a partial B–recursive function whose graph is B–recursive. Thus $\mathbf{R}^B$ is a functor from $\mathbf{Con}^B$ into $\mathbf{Rec}^B$. Moreover, it is readily verified that the restriction $\mathbf{C}^B$ of the functor $\mathbf{C}$ is itself a functor from $\mathbf{Rec}^B$ into $\mathbf{Con}^B$, and there exist isomorphic natural transformations $\alpha : \mathbf{R}^B \circ \mathbf{C}^B \to 1_{\mathbf{Con}^B}$ and $\beta : \mathbf{C}^B \circ \mathbf{R}^B \to 1_{\mathbf{Rec}^B}$ such that $\mathbf{R}^B \circ \alpha = \beta \circ \mathbf{C}^B$. Hence $\mathbf{R}^B$ determines an equivalence of categories, as required. □

Corollary 1.9 *Two enumerated structures* $\langle\mathfrak{M},\nu\rangle$, $\langle\mathfrak{N},\mu\rangle$ *are isomorphic if and only if* $\mathbf{R}\langle\mathfrak{M},\nu\rangle$ *and* $\mathbf{R}\langle\mathfrak{N},\mu\rangle$ *are isomorphic.*

Corollary 1.10 *Two B–constructive enumerated structures* $\langle\mathfrak{M},\nu\rangle$, $\langle\mathfrak{N},\mu\rangle$ *are B–isomorphic if and only if* $\mathbf{R}\langle\mathfrak{M},\nu\rangle$ *and* $\mathbf{R}\langle\mathfrak{N},\mu\rangle$ *are B–isomorphic.*

Thus, the study of constructivizations of a model $\mathfrak{M}$, up to autoequivalence, is equivalent to the study of recursive models, up to a recursive isomorphism, that are isomorphic to $\mathfrak{M}$.

Proposition 1.11 *If ν and μ are constructivizations of a model $\mathfrak{M}$, then* $\mathbf{R}\langle\mathfrak{M},\nu\rangle$ *and* $\mathbf{R}\langle\mathfrak{M},\mu\rangle$ *are recursive models isomorphic to $\mathfrak{M}$. Moreover, ν and μ are autoequivalent if and only if* $\mathbf{R}\langle\mathfrak{M},\nu\rangle$ *and* $\mathbf{R}\langle\mathfrak{M},\mu\rangle$ *are recursively isomorphic and, for any recursive model $\mathfrak{N}$ isomorphic to $\mathfrak{M}$, there exists a constructivization ξ such that* $\mathbf{R}\langle\mathfrak{M},\xi\rangle$ *and $\mathfrak{N}$ are recursively isomorphic.*

Corollary 1.12 *Any infinite constructive algebra is recursively isomorphic to a constructive algebra with a single-valued numbering.*

We now consider numberings of quotient structures by congruences. If η is a congruence on an algebraic structure $\mathfrak{A}$ and ν is a numbering of $\mathfrak{A}$, then we can define a quotient numbering ν/η of the quotient structure $\mathfrak{A}/\eta$ by setting: $(\nu/\eta)(n) \leftrightharpoons \nu(n)/\eta$. If it is clear which congruence we mean, then we write simply ν instead of ν/η.

Proposition 1.13

(a) *If η is a B–decidable strict congruence and $\langle\mathfrak{A},\nu\rangle$ is a B–constructive algebraic structure, then $\langle\mathfrak{A}/\eta\,,\,\nu/\eta\rangle$ is a B–constructive algebraic structure.*

(b) *If $\langle\mathfrak{A},\nu\rangle$ is a B–positive structure and η is a B–enumerable congruence, then $\langle\mathfrak{A}/\eta\,,\,\nu/\eta\rangle$ is a B–positive structure.*

Let $\mathfrak{M} = \langle M;\sigma\rangle$ be a finite model in a finite signature

$$\sigma \;=\; \langle P_0^{n_0}\,,\,\dots\,,\,P_r^{n_r}\,,\,a_0\,,\,\dots\,,\,a_s\rangle$$

without function symbols. Let ν be a mapping from the finite set of numbers $[0,n] = \{i \mid 0 \leqslant i \leqslant n\}$ onto M, so the universe M has at most $n+1$ elements. Such a pair $\langle\mathfrak{M},\nu\rangle$ is called a *finitely enumerated* or n*–enumerated* model.

To each finite signature $\sigma = \langle P_0^{n_0}\,,\,\dots\,,\,P_r^{n_r}\,,\,a_0\,,\,\dots\,,\,a_s\rangle$, we assign a number (code of the sequence)

$$\langle\,\langle\langle 0,n_0\rangle\,,\,\dots\,,\,\langle r,n_r\rangle\rangle\,,\,s\,\rangle.$$

Given an n–enumerated model $\langle\mathfrak{M},\nu\rangle$ in σ, we extend the signature by new constant symbols $c_0\,,\,\dots\,,\,c_n$ to $\sigma^n = \sigma\cup\{c_0\,,\,\dots\,,\,c_n\}$ so that each element of the model is named by one of the c_i's. Let $\mathfrak{M}^n$ denote the enrichment of $\mathfrak{M}$ in σ^n, in which the new constant c_i denotes the element νi of the universe M, for $i \leqslant n$.

We define the *diagram* $\mathcal{D}\langle\mathfrak{M},\nu\rangle$ of an n–enumerated model $\langle\mathfrak{M},\nu\rangle$ by reference to $\mathfrak{M}^n$ as follows:

$$\mathcal{D}\langle\mathfrak{M},\nu\rangle \;=\; \big\{\Phi \;\big|\; \Phi \text{ is an atomic sentence of } \sigma^n \text{ or the negation of an atomic sentence of } \sigma^n, \text{ and } \mathfrak{M}^n \models \Phi\big\}$$

Let $G\mathcal{D}\langle\mathfrak{M},\nu\rangle$ be the set of Gödel numbers of the formulas of $\mathcal{D}\langle\mathfrak{M},\nu\rangle$. If u is the canonical number of the finite set $D_u = G\mathcal{D}\langle\mathfrak{M},\nu\rangle$, then the sequence

$$\langle\,n\,,\,\langle\langle\langle 0,n_0\rangle\,,\,\langle 1,n_1\rangle\,,\,\dots\,,\,\langle r,n_r\rangle\rangle\,,\,s\rangle\,,\,u\,\rangle,$$

is called *the Gödel number* of the enumerated finite model $\mathfrak{M}$ in the finite signature σ.

We note the following obvious properties of numberings of finitely enumerated models.

(1) *The set of the Gödel numbers of all finitely enumerated models is recursive.*

(2) *From the Gödel number of a finitely enumerated model* $\langle \mathfrak{M}, \nu \rangle$*, we can compute the number of elements of* $|\mathfrak{M}|$.

(3) *From the Gödel number of a finitely enumerated model, we can compute how many predicate symbols and constants are contained in the signature* σ *and compute the arities of the predicate symbols as well.*

(4) *From the Gödel number of a finitely enumerated model* $\langle \mathfrak{M}, \nu \rangle$*, we can compute the set* $G\mathcal{D}\langle \mathfrak{M}, \nu \rangle$ *and hence compute the diagram* $\mathcal{D}\langle \mathfrak{M}, \nu \rangle$.

(5) *Given two Gödel numbers of finitely enumerated models, we can recognize whether these models are of the same signature.*

(6) *The set of numbers of finite signatures is recursive.*

(7) *If* n *is the Gödel number of a finitely enumerated model and* m *is the number of a finite signature, then we can recognize whether a finitely enumerated model with Gödel number* n *is a model in the signature with number* m.

(8) *If* n *is the Gödel number of a finitely enumerated model and* m *is the number of a finite signature, then we can recognize whether a finitely enumerated model with Gödel number* n*, in a signature*

$$\sigma = \langle P_0^{n_0}, \ldots, P_k^{n_k}, a_0, \ldots, a_s \rangle$$

has some enrichment to a finitely enumerated model in a signature σ' *with the number* m.

A finite n–enumerated model $\langle \mathfrak{M}, \nu \rangle$ is called an *extension* of an m–enumerated model $\langle \mathfrak{N}, \mu \rangle$ if $m < n$, the models $\mathfrak{M}$ and $\mathfrak{N}$ are of the same signature, and the set $\{\langle \nu(i), \mu(i) \rangle \mid i \leqslant m\}$ is an isomorphic embedding of $\mathfrak{M}$ into $\mathfrak{N}$.

(9) *Given two Gödel numbers* n *and* m *of finitely enumerated models, we can recognize whether a finitely enumerated model with Gödel number* m *is an extension of a finitely enumerated model with Gödel number* n.

If $\langle \mathfrak{M}, \nu \rangle$ is an n–enumerated finite model in the signature

$$\sigma = \langle P_0^{n_0}, \ldots, P_r^{n_r}, a_0, \ldots, a_s \rangle$$

and $\langle \mathfrak{N}, \mu \rangle$ is a m–enumerated finite model in the signature

$$\sigma' = (P_0^{m_0}, \ldots, P_{r'}^{m_{r'}}, a_0, \ldots, a_{s'}),$$

we say that $\langle \mathfrak{M}, \nu \rangle$ is an *extension* of $\langle \mathfrak{N}, \mu \rangle$ if $m < n$; $r' \leqslant r$; $s' \leqslant s$; for any $0 \leqslant i \leqslant r'$, the arities of predicates P_i are such that $n_i = m_i$; and $\{\langle \mu(i), \nu(i) \rangle \mid i \leqslant m\}$ is an isomorphism of $\mathfrak{N}$ into $\mathfrak{M} \restriction \sigma'$.

(10) *Given two Gödel numbers n and m of finitely enumerated models, we can recognize whether a model with Gödel number n is an extension of a finitely enumerated model with Gödel number m.*

Now let $\langle \mathfrak{M}, \nu \rangle$ be any enumerated model in the signature

$$\sigma = \langle P_0^{n_0}, \ldots, P_r^{n_r}, a_0, \ldots, a_s, \ldots \rangle$$

with r predicate symbols and possibly infinitely many constants. Given $n \in \operatorname{dom} \nu$, we define $M_n \leftrightharpoons \{\nu i \mid i \leqslant n\}$ and consider a signature σ_n obtained from the signature σ by setting:

$$\sigma_n \leftrightharpoons \langle P_0^{n_0}, \ldots, P_q^{n_q}, a_{j_1}, \ldots, a_{j_p} \rangle$$

where $q = n$ if $n \leqslant r$, i.e., σ contains at least n predicates, and $q = r$ otherwise, and the set $\{j_1, \ldots, j_p\}$ consists of all indices j of constants of σ such that $j \leqslant n$ and the value of the constant a_j belongs to M_n; i.e., $(a_j)_{\mathfrak{M}} = \nu i$ for some $i \leqslant n$. Note that σ_n has at most $n+1$ predicate symbols and at most $n+1$ constants.

Now let $\mathfrak{M}_n$ be any submodel of the restriction $\mathfrak{M} \restriction \sigma_n$ of the model $\mathfrak{M}$ to the signature σ_n such that the universe

$$|\mathfrak{M}_n| = M_n = \{\nu i \mid i \leqslant n\}.$$

Finitely enumerated models $\langle \mathfrak{M}_n, \nu_n \rangle$, where $\nu_n(i) \leftrightharpoons \nu(i)$ for $i \leqslant n$, will be called *finitely enumerated submodels* or *n–enumerated submodels* of $\langle \mathfrak{M}, \nu \rangle$. By a *representation* of an enumerated model $\langle \mathfrak{M}, \nu \rangle$ we mean the set $W\langle \mathfrak{M}, \nu \rangle$ of the Gödel numbers of all finitely enumerated submodels of $\langle \mathfrak{M}, \nu \rangle$.

Note that in considering the Gödel number of a finitely enumerated submodel $\langle\mathfrak{M}_n, \nu_n\rangle$, we actually refer to the enrichment $(\mathfrak{M}_n)^n$ of $\mathfrak{M}_n$ in the signature $(\sigma_n)^n = \sigma_n \cup \{c_0, \dots, c_n\}$, in which each new constant c_i denotes the element νi of the universe M_n, for $i \leqslant n$, and the diagram of atomic sentences or their negations true in $(\mathfrak{M}_n)^n$.

Proposition 1.14 *An enumerated model $\langle\mathfrak{M}, \nu\rangle$ is constructive if and only if its representation $W\langle\mathfrak{M}, \nu\rangle$ is recursively enumerable.*

Proof. To prove the proposition, it suffices to compare the decidability problem for a model with that of its representation. First, assume $\langle\mathfrak{M}, \nu\rangle$ in signature σ is constructive. Then $\mathfrak{F}_0\mathcal{D}\langle\mathfrak{M}, \nu\rangle$, the set of Gödel numbers of the quantifier-free formulas satisfied in $\mathfrak{M}$, is recursive. It is then straightforward to show that $W\langle\mathfrak{M}, \nu\rangle$ is recursively enumerable.

Conversely, let $\langle\mathfrak{M}, \nu\rangle$ be an enumerated model in signature σ, and assume its representation $W = W\langle\mathfrak{M}, \nu\rangle$ is recursively enumerable. We define a model $\mathfrak{M}_W$ in the same signature σ together with a numbering ν_W as follows. First, consider the set

$$M_W^0 = \{c_i \mid \text{some finitely enumerated submodel of } \langle\mathfrak{M}, \nu\rangle \text{ has Gödel number } n+1 \text{ in } W, \text{ and } i \leqslant n\}$$

On M_W^0, we introduce the following equivalence relation: $c_i \sim_W c_j$ if the atomic sentence $(c_i = c_j)$ is included in the diagram of some n–enumerated submodel with Gödel number in $W\langle\mathfrak{M}, \nu\rangle$. Let M_W be the quotient set $M_W^0/{\sim_W}$. We define the numbering ν_W by setting $\nu_W(i) = c_i/{\sim_W}$, where $c_i \in M_W^0$. We set $\nu_W(i) = a_j$ for each constant a_j of the signature σ for which the equality $(c_i = a_j)$ is included in the diagram of some n–enumerated submodel with Gödel number in W. For the model $\mathfrak{M}_W$, each $i \leqslant r$, define

$$\mathfrak{M}_W \models P_i(\nu_W n_0, \dots, \nu_W n_k) \Leftrightarrow \text{the formula } P_i(c_{n_0}, \dots, c_{n_k}) \text{ is in the diagram of some } n\text{–enumerated model with Gödel number in } W.$$

This concludes our definition of $\langle\mathfrak{M}_W, \nu_W\rangle$. We also define $\varphi : \mathfrak{M}_W \to \mathfrak{M}$ by $\varphi(c_i/{\sim_W}) \leftrightharpoons \nu i$. It is readily established that φ is a recursive isomorphism of the enumerated model $\langle\mathfrak{M}_W, \nu_W\rangle$ onto $\langle\mathfrak{M}, \nu\rangle$. Thus if $W\langle\mathfrak{M}, \nu\rangle$ is recursively enumerated, then the model $\langle\mathfrak{M}_W, \nu_W\rangle$ is constructive, and hence $\langle\mathfrak{M}, \nu\rangle$ is constructive. □

We now fix a finite signature σ without function symbols and define a numbering κ^σ of all constructive models in the signature σ as follows. Let $\{W_n\}_{n\in\mathbb{N}}$ be the principal numbering of all recursively enumerable subsets of $\mathbb{N}$. As usual, W_n^t is that part of the set W_n that is already enumerated by step t in numbering. Recall that we enumerate only numbers $x < t$ into W_n^t. From W_n, we construct a new set $V_n = \bigcup V_n^t$, where the V_n^t is defined as follows. At stage 0, we set $V_n^0 = \emptyset$. At stage $t+1$, we verify the following (recursive) conditions:

(i) every element of W_n^{t+1} is the Gödel number of some finitely enumerated model of the signature σ,

(ii) for every $x, y \in W_n^{t+1}$, either the finitely enumerated model with Gödel number x is an extension of the finitely enumerated model with Gödel number y, or vice-versa.

We set $V_n^{t+1} \leftrightharpoons V_n^t \cup W_n^{t+1}$ if both these conditions are satisfied, and $V_n^{t+1} \leftrightharpoons V_n^t$ otherwise. To complete the construction, we set $V_n \leftrightharpoons \bigcup\limits_t V_n^t$.

It is clear that this sequence of sets $\{V_n\}_{n\in\mathbb{N}}$ is computable. Hence there is a recursive function ρ such that $V_n = W_{\rho(n)}$ for any $n \in \mathbb{N}$. Furthermore, $W_{\rho(\rho(n))} = W_{\rho(n)}$ for any $n \in \mathbb{N}$.

It is easy to see that for any $n \in \mathbb{N}$, the set V_n is recursively enumerable and is the representation of some constructive model in the signature σ. Using the construction in the proof of the last proposition, we recover the constructive model $\mathfrak{M}_{V_n}$ and its constructivization ν_{V_n} by V_n. We set $\kappa^\sigma(n) \leftrightharpoons \langle \mathfrak{M}_{V_n}, \nu_{V_n} \rangle$. Subsequently, the model $\mathfrak{M}_{V_n}$ will be denoted $\mathfrak{M}_n^\kappa$ and the numbering ν_{V_n} will be written ν_n^κ.

From the characterization of recursive models via their representations, we see that $\kappa^\sigma(n)$ enumerates all the constructive models of the signature σ, including finitely enumerated models and the empty one.

If σ is an infinite signature and the function $i \to n_i$ is recursive, where n_i denotes the arity of the ith predicate symbol, then by a similar construction, we obtain representations $W\langle \mathfrak{M}, \nu \rangle$ and the numbering κ^σ enumerating all the constructive models of the signature σ, including finite constructive models in finite initial segments σ_n of the signature σ. In constructing the sets V_n, we must require that finitely enumerated models are models of finite initial segments of the signature σ and take this into account in the notion of an extension.

The following notion of a computable sequence of constructive models is often used in the literature on constructive models. A sequence of constructive models $\langle \mathfrak{M}_0, \nu_0 \rangle$, ..., $\langle \mathfrak{M}_n, \nu_n \rangle$, ... (in a signature σ) is called *computable* if these models are uniformly constructive, i.e., there is a recursive function which uniformly computes from n the indices of the (necessarily recursive) domain, predicates and functions of the model $\langle \mathfrak{M}_n, \nu_n \rangle$.

Proposition 1.15 *If a sequence of constructive models*

$$\langle \mathfrak{M}_0, \nu_0 \rangle, \ldots, \langle \mathfrak{M}_n, \nu_n \rangle, \ldots$$

is computable, then there exist recursive functions f and g such that

(i) *the models $\langle \mathfrak{M}_n, \nu_n \rangle$ and $\kappa(f(n))$ are recursively isomorphic for any n, and*

(ii) *the index $g(n)$ of the recursive function $\kappa_{g(n)}$ that defines this recursive isomorphism is computed from n by means of the function g,*

i.e., defining $\varphi_n(\nu_n(m)) \leftrightharpoons \nu^{\kappa}_{f(n)}(\kappa_{g(n)}(m))$, we conclude that φ_n is an isomorphism of $\mathfrak{M}_n$ onto $\mathfrak{M}^{\kappa}_{f(n)}$, where κ_n is the principal numbering of the set of all partially recursive functions.

Proof. The existence of functions f and g follows from the construction of the recursively enumerable representations $W\langle \mathfrak{M}_n, \nu_n \rangle$ of the constructive models $\langle \mathfrak{M}_n, \nu_n \rangle$. The computability of the sequence $\{\langle \mathfrak{M}_n, \nu_n \rangle\}_{n \in \mathrm{N}}$ implies the computability of the sequence $\{W\langle \mathfrak{M}_n, \nu_n \rangle\}_{n \in \mathrm{N}}$. Since W is principal, we obtain the reducing function f. The function g can be constructed from the diagrams since it is the same for $\langle \mathfrak{M}_n, \nu_n \rangle$ and $\kappa(n)$. □

The converse assertion is a consequence of the following simple propositions.

Proposition 1.16 *The sequence $\langle \mathfrak{M}^{\kappa}_n, \nu^{\kappa}_n \rangle$ of constructive models is computable.*

Proof. The assertion follows from the construction of $\langle \mathfrak{M}^{\kappa}_n, \nu^{\kappa}_n \rangle$ and the definition of a computable sequence, since from the index n, we can recover the diagram of $\langle \mathfrak{M}^{\kappa}_n, \nu^{\kappa}_n \rangle$ that is defined from $V_n = W\langle \mathfrak{M}^{\kappa}_n, \nu^{\kappa}_n \rangle$. □

Let $\alpha = \{\langle \mathfrak{M}_n, \nu_n \rangle\}_{n \in \mathrm{N}}$ and $\beta = \{\langle \mathfrak{N}_n, \mu_n \rangle\}_{n \in \mathrm{N}}$ be two sequences of enumerated models. The sequence α is a *reduction* of the sequence β, written $\alpha \leqslant \beta$, if there exists a recursive function f such that the constructive models $\langle \mathfrak{M}_n, \nu_n \rangle$ and $\langle \mathfrak{N}_{f(n)}, \mu_{f(n)} \rangle$ are recursively isomorphic (in the sense of constructive models). We say that the sequence α is an *effective reduction* of the sequence β if there exist recursive functions f and g such that for any n, the function $\kappa_{g(n)}$ defines an isomorphism of the constructive model $\langle \mathfrak{M}_n, \nu_n \rangle$ onto the model $\langle \mathfrak{N}_{f(n)}, \mu_{f(n)} \rangle$, i.e., the mapping $\varphi_n(\nu_n(m)) \leftrightharpoons \mu_{f(n)}(\kappa_{g(n)}(m))$ is well defined and is an isomorphism between $\mathfrak{M}_n$ and $\mathfrak{N}_{f(n)}$.

From the definitions, we immediately obtain the following claim.

Proposition 1.17 *If a sequence $\{\langle \mathfrak{M}_n, \nu_n \rangle\}_{n \in \mathrm{N}}$ of enumerated models is an effective reduction of a computable sequence $\{\langle \mathfrak{N}_n, \mu_n \rangle\}_{n \in \mathrm{N}}$ of constructive models, then $\{\langle \mathfrak{M}_n, \nu_n \rangle\}_{n \in \mathrm{N}}$ is a computable sequence of constructive models.*

Let $\mathcal{S}$ be the class of constructive models of a signature σ and let K be a construction defining some finite sequence of constructive models and homomorphisms of them into constructive models of $\mathcal{S}$. We say that a construction K is *uniformly effective* on the class $\mathcal{S}$ if there exists a partially recursive function g_K such that if the constructive model $\kappa^\sigma(n)$ belongs to $\mathcal{S}$ then $g_K(n)$ is defined and $g_K(n)$ is equal to the code of a pair of two finite sequences, where the first sequence consists of the numbers κ^σ of constructive models that are constructed by K and the second sequence consists of the indices of recursive functions that define the required recursive homomorphisms with respect to the numbering κ.

Proposition 1.18 *Let K be a uniformly effective construction on $\mathcal{S}$ that, given a constructive model $\langle \mathfrak{N}, \nu \rangle$ of $\mathcal{S}$, constructs a constructive model $\langle \mathfrak{M}, \mu \rangle$ of $\mathcal{S}$ and a homomorphism (φ, f) from $\langle \mathfrak{N}, \nu \rangle$ into $\langle \mathfrak{M}, \mu \rangle$ such that $K\langle \mathfrak{N}, \nu \rangle = \langle \langle \mathfrak{M}, \mu \rangle, (\varphi, f) \rangle$. Then, for any constructive model $\langle \mathfrak{N}, \nu \rangle$ of $\mathcal{S}$, there is a computable sequence $\langle \mathfrak{N}_0, \nu_0 \rangle, \ldots, \langle \mathfrak{N}_n, \nu_n \rangle, \ldots$ of constructive models of $\mathcal{S}$ and a computable sequence $g_0, g_1, \ldots, g_n, \ldots$ of recursive functions such that*

(1) $\langle \mathfrak{N}_0, \nu_0 \rangle = \langle \mathfrak{N}, \nu \rangle$,

(2) *for any $n \in \mathbb{N}$, the function g_n defines a homomorphism φ_n from $\mathfrak{N}_n$ into $\mathfrak{N}_{n+1}$, where $\varphi_n(\nu_n(m)) = \nu_{n+1} g_n(m)$,*

(3). *for any $n \in \mathbb{N}$, the construction K defines the model $\langle \mathfrak{N}_{n+1}, \nu_{n+1} \rangle$ and the homomorphism (φ_n, g_n) from $\langle \mathfrak{N}_n, \nu_n \rangle$.*

Proof. The assertion follows from the recursive definition of the required indices by means of the function g_K representing the construction K on the class $\mathcal{S}$. □

2 Decidable models

The notion of a strongly constructive model was introduced by Ershov [17, 15, 14]. Later, a similar notion of decidable models was studied by Harrington [36] and Morley [50]. The notions are equivalent, but they have some methodological differences. These notions are more natural for an effectively prescribed model in studies of model-theoretic properties of decidable theories. A model $\mathfrak{M}$ is called *decidable* if there exists a numbering ν of $\mathfrak{M}$ such that the model $\langle \mathfrak{M}, \nu \rangle$ is *strongly constructive.* In considering a decidable model, we generally have some fixed strong constructivization ν in mind. We begin with some general facts about strongly constructive models and then proceed to the decidability of algorithmic properties of strong constructivizations.

In this section, we consider the existence problem for decidable models with given model-theoretic properties. It is easy to see that the theory $\mathrm{Th}(\mathfrak{M})$ of a decidable model $\mathfrak{M}$ is decidable. The converse assertion, which can be regarded as a "constructive" analog of an existence theorem for models of consistent theories, also holds.

Theorem 2.1 (Existence theorem for decidable models, [14]) *Any decidable theory has a decidable model.*

If a type $p(x_1, \dots, x_n)$ is realized in a decidable model, then it is decidable, i.e., there exists an algorithm determining whether a formula $\varphi(x_1, \dots, x_n)$ belongs to $p(x_1 \dots, x_n)$. Hence any undecidable type is omitted in any decidable model. By Theorem 2.1, any decidable type is realized in some decidable model.

Theorem 2.2 (Omitting-decidable-types theorem, [38, 37, 49]) *If T is a decidable theory and $\{p_i(\overline{x}_i) \mid i \in \mathbb{N}\}$ is a computable family of decidable nonprincipal types, then there exists a decidable model of T omitting all the types $p_i(\overline{x}_i)$, $i \in \mathbb{N}$.*

Theorems 2.1 and 2.2 can be proved by a step-by-step construction of a Henkin theory with the required properties. We extend the signature by new constants $c_0, c_1, \ldots, c_k, \ldots$ and construct step-by-step finite parts $\mathcal{D}^t$ of the Henkin theory by adding φ or $\neg\varphi$ at even steps in order to keep consistency with T. The consistency condition is effectively verified since T is decidable. To add a formula φ of the form $(\exists y)\,\psi(y)$, we take a constant c_k that has not been used and add the formula $\psi(c_k)$, which does not violate the consistency. At odd steps, for every type $p_i(\overline{x})$ and any sequence of constants $\overline{c}$ we find $\varphi(\overline{x}) \in p_i(\overline{x})$ such that $\neg\varphi(\overline{c})$ is consistent with the part $\mathcal{D}^t$ constructed above, and add the result to $\mathcal{D}^t$. We obtain $\mathcal{D} = \cup\mathcal{D}^t$ which is consistent with T. Thus, we obtain the Henkin theory T and the corresponding Henkin model $\mathfrak{M}_T$, which has the required properties. The model is decidable, since for every sentence φ, we can find a step on which φ or $\neg\varphi$ is added, and determine what is true on $\mathfrak{M}_T$.

From the above construction, we obtain the following claim.

Corollary 2.3 *If T_i is a computable family of complete decidable theories, then there exists a computable sequence of strongly constructive models $\langle\mathfrak{M}_n, \nu_n\rangle$ such that $\mathfrak{M}_n \models T_n$.*

As shown in [2], the family of types $S_{\mathfrak{M}}$ realized in a homogeneous model and the homogeneity condition define such a countable homogeneous model uniquely up to an isomorphism. Therefore, it is natural to try to find the decidability condition in terms of types realized in a homogeneous model. First we present the following simple condition.

Proposition 2.4 *If $\mathfrak{M}$ is a decidable model, then the family of types $S_{\mathfrak{M}}$ realized in $\mathfrak{M}$ is computable.*

Proof. Taking any strong constructivization of $\mathfrak{M}$ and enumerating all the finite sequences of numbers of elements $\overline{n} = (n_0, \ldots, n_k)$, we define the type $p_{\overline{n}}(x_0, \ldots, x_k)$ consisting of those formulas that are true on elements with given numbers. Thereby, we define a computable family $S_{\mathfrak{M}}$ of types. □

In the general case, the above condition is necessary but not sufficient for decidability. Let us consider a computable principal numbering of all partial enumerable types $p_0(\overline{x}_0), p_1(\overline{x}_1), \ldots$ that are consistent with a decidable theory T and are such that, given a number n, we can compute a sequence of variables $\overline{x}_n$ of a partial type with number n, as constructed in Section 1.

A family of types S is called a *computable family with an effective extension* if it is computable and there exists a partially recursive function $R(n, m)$ such that, if a type $p_n(\overline{x}_n) \in S$ and a formula $\varphi_m(\overline{x}_n, \overline{x})$ with the Gödel number m are consistent, then $R(n, m)$ is defined, $p_{R(n,m)} \supseteq p_n$, $p_{R(n,m)} \in S$, and $\varphi_m(\overline{x}_n, \overline{x}) \in p_{R(n,m)}$.

Without loss of generality, we can assume that for a computable family S with an effective extension, there exists a recursively enumerable set W such that $S = \{p_n \mid n \in W\}$, and $R(n, m) \in W$ for any $n \in W$ and m such that φ_m is consistent with p_n.

Theorem 2.5 (The Goncharov–Peretyat'kin criterion for the decidability of homogeneous models, [25, 64]) *A countable homogeneous model $\mathfrak{M}$ with the family $S_{\mathfrak{M}}$ of types realized in $\mathfrak{M}$ is decidable if and only if $S_{\mathfrak{M}}$ is a computable family with effective extension.*

Proof. If $\mathfrak{M}$ is a decidable homogeneous model, then $S_{\mathfrak{M}}$ is a computable family of types, in view of Proposition 2.4. To prove the effectiveness of extensions, we define step-by-step the type

$$d_{c(n,m)}(\overline{x}_n, \overline{x}) = \bigcup d^t(\overline{x}_n, \overline{x}).$$

At every step $t+1$, we find the least sequence $\overline{m}_n$ such that p_n^t is true, where p_n^t is that part of p_n that is already computed by step t. We consider the conjunction $\bigwedge d^{t+1}(\overline{x}_n, \overline{x}) \wedge \varphi_m(\overline{x}_n, \overline{x})$. If a formula

$$(\exists \overline{x}) \left(\bigwedge d^{t+1}(\overline{x}_n, \overline{x}) \wedge \varphi_m(\overline{x}_n, \overline{x})\right)$$

belongs to p_n^t, then we find the least sequence $\overline{m}$ such that this formula holds on a sequence of elements with numbers $(\overline{m}_n, \overline{m})$. We consider a formula $\varphi(\overline{x}_n, \overline{x})$ with the least number among the formulas that do not belong to d^t, but hold on the sequence of elements with numbers $(\overline{m}_n, \overline{m})$. If

$$(\exists \overline{x}) \left(\bigwedge d^t(\overline{x}_n, \overline{x}) \wedge \varphi_m(\overline{x}_n, \overline{x}) \wedge \varphi(\overline{x}_n, \overline{x})\right)$$

belongs to $p_n(\overline{x}_n)$, then we add $\varphi(\overline{x}_n, \overline{x})$ to d^t and pass to the following step; otherwise, we set $d^{t+1} = d^t$. Proceeding in this way, we arrive at a computable sequence of partial types. Since p_n is universal, there is a recursive function $\alpha(n, m)$ such that $d_{c(n,m)} = p_{\alpha(n,m)}$, $n, m \in \mathbb{N}$. It is easy to see that α possesses the required properties.

The proof of the sufficiency is more complicated. We construct a Henkin theory $\mathcal{D}$ in the enrichment of the signature σ of T by new constant symbols $c_0, c_1, \ldots, c_n, \ldots$ such that the following conditions on types hold:

(1) For any type $p(\overline{x})$ of S there exists a sequence of constants $\overline{c}$ such that $p(\overline{c}) \subseteq \mathcal{D}$,

(2) for any sequence of constants $\overline{c}$ there exists a type $p(\overline{x})$ of S such that $p(\overline{c}) \subseteq \mathcal{D}$,

(3) for any types $p(\overline{x})$ and $q(\overline{x}, y)$ of S and a sequence of constants $\overline{c}$ there is a constant c such that $q(\overline{c}, c) \subseteq \mathcal{D}$ provided that $p(\overline{x}) \subseteq q(\overline{x}, y)$ and $p(\overline{c}) \subseteq \mathcal{D}$.

By (1) and (2), all the types of S (and no other types) are realized in the Henkin model $\mathfrak{M}_{\mathcal{D}}$ of the theory $\mathcal{D}$. By (3), $\mathfrak{M}_{\mathcal{D}}$ is homogeneous in the signature σ. Furthermore, (3) implies (1). Thus, to complete the construction it suffices to require the validity of (2) and (3).

We arrange the set of all 1–types

$$\{\Gamma_n(A) \mid n \in \omega \text{ and } A \text{ is a g–finite subset of the set } \{c_0, c_1, \ldots\}\}$$

as an effective sequence $\Xi_0, \Xi_1, \ldots$ in such a way that, given the number of Ξ, we can find the corresponding $n \in \omega$ and A (i.e., $\Xi = \Gamma_n(A)$). Let $\Xi_n = \Gamma_{\sigma(n)}(A_n)$, and let σ be a general recursive function. If c is a constant symbol, then

$$\Xi_n(c) \leftrightharpoons \Gamma_{\sigma(n)}(A_n)(c) = \{[\varphi]_c^{x_0} \mid \varphi \in \Xi_n\}.$$

We will construct sets $\Delta_j(A)$ for $j \in \omega$ and finite $A \subseteq \{c_0, c_1, \ldots\}$. The sets Ξ_i and $\Delta_j(A)$ will be effectively enumerated according to a step-by-step procedure. Let Ξ_i^n and $\Delta_j^n(A)$ denote those finite subsets of Ξ_i and $\Delta_j(A)$ that are already enumerated to the end of the step n. So we have

$$\Xi_i^n \subseteq \Xi_i^{n+1}, \qquad n \in \omega, \qquad \bigcup_{n\in\omega} \Xi_i^n = \Xi_i,$$

as well as

$$\Delta_j^n(A) \subseteq \Delta_j^{n+1}(A), \qquad n \in \omega, \qquad \bigcup_{n\in\omega} \Delta_j^n(A) = \Delta_j(A).$$

We will effectively construct step-by-step a sequence of formulas $\theta_0, \theta_1, \ldots$ such that $\theta_{n+1} = \theta_n \wedge \varphi_n$ or $\theta_{n+1} = \theta_n \wedge \neg\varphi_n$, where $\varphi_0, \varphi_1, \ldots$ (see (2)) is an effective sequence of all sentences in the signature $\sigma \cup \langle c_0, c_1, \ldots\rangle$. For any n, the theory $T \cup \{\theta_n\}$ is consistent. Consequently, the theory T_ω in the signature $\sigma \cup \langle c_0, c_1, \ldots\rangle$, that is defined by the set of axioms $\{\theta_n \mid n \in \omega\}$, is a complete extension of the theory T. We will prove that

$$\varphi_n = \exists x\, \varphi'_n(x) \in T_\omega \quad \text{implies} \quad \varphi'_n(c_s) \in T_\omega$$

for some $s \in \omega$. Over the constants $\{c_0, c_1, \ldots\}$ we will construct a strongly constructive model $\mathfrak{M}'$ of the theory T_ω in a standard way. By construction, $\mathfrak{M}' \restriction \sigma \simeq \mathfrak{M}_S$.

We now proceed with the construction. At the end of each step n, the following objects are defined: the natural number m_n, the finite set

$$I_n \subseteq [m_n]\ (\leftrightharpoons \{i \mid i \leqslant m_n\}),$$

and the functions

$$g^n : [m_n] \longrightarrow \omega, \qquad f^n : I_n \longrightarrow \omega, \qquad h^n : I_n \longrightarrow \omega.$$

We introduce the following notation, for $i \leqslant m_n$:

$$B_i^n \leftrightharpoons \big(\bigcup_{j \leqslant i} A_j\big) \cup \{f^n(j) \mid j \in I_n,\ j < i\},$$

and for $i \in I_n$:

$$C_i^n \leftrightharpoons B_i^n \cup \{f^n(i)\}.$$

We note that $B_i^n \subseteq B_j^n$ for $i < j \leqslant m_n$; moreover, if $i \in I_n$, then

$$B_i^n \subseteq C_i^n \subseteq B_j^n.$$

We require that the following conditions hold for any $i \leqslant m_n$:

(1_i^n) $\theta_n(B_i^n) \in \Delta_{g^n(i)}(B_i^n)$.

(2_i^n) If $i \in I_n$, then

$$\Xi_i^n \cup \Delta_{g^n(i)}^n(B_i^n) \subseteq \Gamma_{h^n(i)}(B_i^n)$$

and

$$\theta_n(C_i^n) \in \Gamma_{h^n(i)}(B_i^n)(c_{f^n(i)}).$$

(3_i^n) If $i \notin I_n$, then the set

$$\{\theta_n(B_i^n)\} \cup \Xi_i^n \cup \Delta_{g^n(i)}^n(B_i^n)$$

is inconsistent with T.

STEP 0:

$\theta_0 \leftrightharpoons (c_0 = c_0)$, $m_0 \leftrightharpoons 0$, $I_0 \leftrightharpoons \{0\}$, $g^0(0)$ is the least $s \in \omega$ such that $\Delta_s(A_0) \subseteq \Xi_0$, $f^0(0)$ is the least $t \in \omega$ such that $c_t \notin A_0 \cup \{c_0\}$, $h^0(0) \leftrightharpoons \sigma(0)$ (hence $\Gamma_{h^0(0)}(A_0) = \Xi_0$).

STEP $n+1$:

We find the greatest $m \leqslant m_n$ such that for either $\theta_{n+1} \leftrightharpoons \theta_n \wedge \varphi_n$, or $\theta_{n+1} \leftrightharpoons \theta_n \wedge \neg\varphi_n$, the following conditions hold for any $i \leqslant m$:

(1) $\theta_{n+1}(B_i^n) \in \Delta_{g^n(i)}(B_i^n)$,

(2) if $i \in I_n$, then

$$\Xi_i^{n+1} \cup \Delta_{g^n(i)}^{n+1}(B_i^n) \subseteq \Gamma_{h^n(i)}(B_i^n)$$

and

$$\theta_{n+1}(C_i^n) \in \Gamma_{h^n(i)}(B_i^n)(c_{f^n(i)}).$$

Furthermore, if $m < m_n$, then we choose (if possible) θ_{n+1} such that it satisfies the condition $\theta_{n+1}(B_{m+1}^n) \in \Delta_{g^n(m+1)}(B_{m+1}^n)$.

It is easy to see that for $m = 0$, conditions (1) and (2) can be satisfied, it is necessary to choose θ_{n+1} so that $\theta_{n+1}(A_0 \cup \{c_{f^n(0)}\}) \in \Gamma_{h^n(0)}(A_0)(c_{f^n(0)})$. Hence such an m exists. The choice of θ_{n+1} is determined by the conditions mentioned above (the maximality of m and the additional requirement for $m < m_n$); if both $\theta_n \wedge \varphi_n$ and $\theta_n \wedge \neg\varphi_n$ satisfy all the conditions, then we put $\theta_{n+1} \leftrightharpoons \theta_n \wedge \varphi_n$.

We define $m_{n+1} \leftrightharpoons m + 1$,

$$I'_{n+1} \leftrightharpoons I_n \cap [m], \qquad I_{n+1} \cap [m] \leftrightharpoons I'_{n+1},$$

and

$$g^{n+1} \restriction [m] \leftrightharpoons g^n \restriction [m], \quad f^{n+1} \restriction I'_{n+1} \leftrightharpoons f^n \restriction I'_{n+1}, \quad h^{n+1} \restriction I'_{n+1} \leftrightharpoons h^n \restriction I'_{n+1}.$$

We note that B^{n+1}_{m+1} has been defined and $B^{n+1}_{m+1} = B^n_{m+1}$ if $m < m_n$. If $m < m_n$ and $\theta_{n+1}(B^{n+1}_{m+1}) \in \Delta_{g^n(m+1)}(B^{n+1}_{m+1})$, then we set

$$g^{n+1}(m_{n+1}) \leftrightharpoons g^n(m_{n+1}) = g^n(m+1).$$

If $m = m_n$ or if $m < m_n$ and $\theta_{n+1}(B^{n+1}_{m+1}) \notin \Delta_{g^n(m+1)}(B^{n+1}_{m+1})$, then we find $t \in \omega$ such that $\theta_{n+1} = \varphi_t$. If $m \notin I_n$, then we set

$$g^{n+1}(m+1) \leftrightharpoons G(B^n_m\,,\, B^{n+1}_{m+1}\,,\, g^n(m)\,,\, t).$$

We note that the assertion (2) and the condition $\theta_{n+1}(B^n_m) = \varphi_t(B^n_m) \in \Delta_{g^n(m)}(B^n_m)$ imply

$$\theta_{n+1}(B^{n+1}_{m+1}) = \varphi_t(B^{n+1}_{m+1}) \in \Delta_{g^{n+1}(m+1)}(B^{n+1}_{m+1})$$

and

$$\Delta_{g^n(m)}(B^n_m) \subseteq \Delta_{g^{n+1}(m+1)}(B^{n+1}_{m+1}).$$

If $m \in I_n$, then $s \leftrightharpoons H(B^n_m\,,\, f^n(m)h^n(m))$ and

$$g^{n+1}(m+1) \leftrightharpoons G(C^n_m\,,\, B^{n+1}_{m+1}\,,\, s\,,\, t).$$

We note that $\theta_{n+1}(B^{n+1}_{m+1}) \in \Delta_{g^{n+1}(m+1)}(B^{n+1}_{m+1})$, and

$$\Gamma_{h^n(m)}(B^n_m)(c_{f^n(m)}) = \Delta_s(C^n_m) \subseteq \Delta_{g^{n+1}(m+1)}(B^{n+1}_{m+1}).$$

We consider the following finite set of formulas:

$$\{\theta_{n+1}(B^{n+1}_{m+1})\} \cup \Xi^{n+1}_{m+1} \cup \Delta^{n+1}_{g^{n+1}(m+1)}(B^{n+1}_{m+1}).$$

If this set is inconsistent with T, then we put $I_{n+1} \leftrightharpoons I'_{n+1}$, and everything is defined. Otherwise, we set $I_{n+1} \leftrightharpoons I'_{n+1} \cup \{m+1\}$. We find the least $r \in \omega$ such that $c_r \notin B^{n+1}_{m+1} \cup c(\theta_{n+1})$ and put $f^{n+1}(m+1) \leftrightharpoons r$. We now find the least t such that

$$\{\theta_{n+1}(B^{n+1}_{m+1})\} \cup \Xi^{n+1}_{m_n} \cup \Delta^{n+1}_{g^{n+1}(m+1)}(B^{n+1}_{m+1}) \subseteq \Gamma_t(B^{n+1}_{m+1}),$$

and set $h^{n+1}(m+1) \leftrightharpoons t$. The construction is completed.

We now show that the construction is stable. We show that for any $m \in \omega$, there exists $s \in \omega$ such that for any $n \geqslant s$:

(1) $m_n > m$,

(2) $I_n \cap [m] = I_s \cap [m]$,

(3) $g^n \restriction [m] = g^s \restriction [m]$,

(4) $f^n \restriction I_s \cap [m] = f^s \restriction I_s \cap [m]$, and

(5) $h^n \restriction I_s \cap [m] = h^s \restriction I_s \cap [m]$.

For $m = 0$, the stabilization takes place at step zero. We assume that for some $m \in \omega$, after step s, the stabilization already takes place on $[m]$. In the following, for $i \leqslant m$ we write $i \in I$ instead of $i \in I_s$, $g(i)$ instead of $g^s(i)$, and B_i instead of B_i^s. If, in addition, $i \in I$, then we write $f(i)$ and $h(i)$ instead of $f^s(i)$ and $h^s(i)$ respectively, and C_i instead of C_i^s.

We note that $B_0 \subseteq B_1 \subseteq \cdots \subseteq B_m$. Let us show that

$$\Delta_{g(0)}(B_0) \subseteq \Delta_{g(1)}(B_1) \subseteq \cdots \subseteq \Delta_{g(m)}(B_m).$$

Indeed, if at least one of the inclusions fails, then there exist $i < m$ and $t > s$ such that $\varphi_t \in \Delta_{g(i)}(B_i) \setminus \Delta_{g(i+1)}(B_{i+1})$ and at the step $t+1$, the conditions (1_i^{t+1}) and (1_j^{t+1}) cannot hold simultaneously. Consequently, $m_t \leqslant m$, which is impossible. Similarly, we can prove that if $i \in I$, $i < m$, then

$$\Xi_i(c_{f(i)}) \subseteq \Gamma_{h(i)}(B_i)(c_{f(i)}) \subseteq \Delta_{g(m)}(B_m).$$

If in addition, $m \in I$, then

$$\Xi_m(c_{f(m)}) \subseteq \Gamma_{h(m)}(B_m)(c_{f(m)}), \qquad \text{and}$$

$$\Delta_{g(m)}(B_m) \subseteq \Gamma_{h(m)}(B_m)(c_{f(m)}) = \Delta_{s_m}(c_m),$$

where $s_m \leftrightharpoons H(B_m, f(m), h(m))$. We suppose that there is a step $n \geqslant s$ such that $m_n = m+1$. (Otherwise, for $m+1$ we have the stabilization at step s.) We note that $B_{m+1}^n = B_m \cup A_{m+1}$ (or $B_{m+1} = B_m \cup A_m \cup \{c_{f(m)}\}$ if $m \in I$) has been stabilized ($B_{m+1}^n = B_{m+1}^s$).

We show that $g^n(m+1)$ can change its value at most once for $n \geqslant s$. Indeed, we assume that $g^n(m+1) \neq g^{n+1}(m+1)$ for $n \geqslant s$. It is clear that $m_{n+1} = m+1$ and, by construction, $g^{n+1}(m+1) = G(B_m, B_{m+1}, g(m), t)$ (or $g^{n+1}(m+1) = G(C_m, B_{m+1}, s_m, t)$ if $m \in I$), where t denotes the number of the formula θ_{n+1}. By definition of G,

$$\Delta_{g(m)}(B_m) \subseteq \Delta_{g^{n+1}(m+1)}(B_{m+1})$$

(or

$$\Delta_{g(m)}(B_m) \subseteq Delta_{s_m}(C_m) \subseteq \Delta_{g^{n+1}(m+1)}(B_{m+1})$$

if $m \in I$). Therefore, at any step $n' > n$ we can choose $\theta_{n'}$ such that all the conditions $(1_i^{n'})$–$(3_i^{n'})$ for $i \leqslant m$, and the condition $(1_{m+1}^{n'})$ are satisfied. Indeed, choosing $\theta_{n'}$ such that $\theta_{n'}(B_{m+1}) \in \Delta_{g^{n+1}(m+1)}(B_{m+1})$, we see that all the conditions $(1_i^{n'})$–$(3_i^{n'})$ for $i \leqslant m$, and $(1_{m+1}^{n'})$ hold. Consequently, there is a step $n_0 \geqslant s$ such that the value of $g^{n_0}(m+1)$ is stabilized. We write $g(m+1)$ instead of $g^{n_0}(m+1)$, and consider two cases.

CASE 1: Ξ_{m+1} is inconsistent with $\Delta_{g(m+1)}(B_{m+1})$.

Then $\Xi_{m+1}^{n_1} \cup \Delta_{g(m+1)}^{n_1}(B_{m+1})$ is inconsistent with T for some $n_1 \geqslant n_0$ and $m+1 \notin I_n$ for all $n \geqslant n_1$. Hence the stabilization can be started at step n_1.

CASE 2: Ξ_{m+1} is consistent with $\Delta_{g(m+1)}(B_{m+1})$.

Since $\Xi_{m+1} = \Gamma_{\sigma(m+1)}(A_{m+1})$ is consistent with $\Delta_{g(m+1)}(B_{m+1})$ and $A_{m+1} \subseteq B_{m+1}$, the property (3) implies that there exists $t \in \omega$ such that

$$\Delta_{g(m+1)}(B_{m+1}) \cup \Xi_{m+1} \subseteq \Gamma_t(B_{m+1}).$$

Let t_0 be the least t with the above properties. Let $n_2 \geqslant n_0$ be such that $\Delta_{g(m+1)}^{n_2}(B_{m+1}) \cup \Xi_{m+1}^{n_2} \nsubseteq \Gamma_t(B_{m+1})$ for all $t < t_0$. If we have no stabilization at step n_2 on the set $[m+1]$, then there exists $n_3 > n_2$ such that $m_{n_3} = m+1$. Since $\theta_{n_3}(B_{m+1}) \in \Delta_{g(m+1)}(B_{m+1})$, we have that

$$\{\theta_{n_3}(B_{m+1})\} \cup \Xi_{m+1}^{n_3} \cup \Delta_{g(m+1)}^{n_3}(B_{m+1}) \subseteq \Gamma_{t_0}(B_{m+1})$$

is consistent with T. Consequently, $m+1 \in I_{n_3}$. By choice of n_2, $h^{n_3}(m+1)$ is equal to t_0, $f^{n_3}(m+1) = r$, and $c_r \notin B_{m+1} \cup c(\theta_n)$. Since $\Delta_{g(m+1)}(B_{m+1}) \subseteq \Gamma_{t_0}(B_{m+1})$ and $\Delta_{g(m+1)}(B_{m+1}) \subseteq \Gamma_{t_0}(B_{m+1})(c_r)$, after step n_3 we can choose θ such that the conditions (1_i)–(3_i) hold for $i \leqslant m+1$. More precisely, for $n > n_3$, θ_n must be chosen such that $\theta_n(B_{m+1} \cup \{c_r\}) \in \Gamma_{t_0}(B_{m+1})(c_r)$. Hence $m+1$ becomes a stable element of I_n and the values of the functions f and h are unchanged at $m+1$.

We omit the subscripts and write the stable values as I^n, g^n, f^n, h^n and B^n. We have $B_0 \subseteq B_1 \subseteq \cdots$, and

$$\Delta_{g(0)}(B_0) \subseteq \Delta_{g(1)}(B_1) \subseteq \cdots.$$

Since $A_i \subseteq B_i$ for all $i \in \omega$, we have $\bigcup_{n\in\omega} B_n = \{c_0, c_1, \ldots\}$. Condition (1_i^n) shows that $T_\omega \leftrightharpoons \bigcup_{i\in\omega} \Delta_{g(i)}(B_i)$ is a complete consistent extension of the theory $T(A)$. Let Ξ_m be consistent with the theory T_ω for $m \in \omega$. Then $\Xi_m \cup \Delta_{g(m)}(B_m)$ is consistent with T. The proof of stability shows that in this case $m \in I$ and

$$\Xi_m(c_{f(m)}) \subseteq \Delta_{g(m+1)}(B_{m+1}) \subseteq T_\omega.$$

Let $\varphi = \exists x_0 \varphi'(x_0) \in T_\omega$, $A \leftrightharpoons c(\varphi)$, $n \in \omega$ be such that $\Delta_n(A) = T_\omega \restriction A$. Since $\varphi'(x_0)$ and $\Delta_n(A)$ are consistent, there exist $t, m \in \omega$ such that

$$\{\varphi'(x_0)\} \cup \Delta_n(A) \subseteq \Gamma_t(A) = \Xi_m.$$

Since $\Xi_m \supseteq \Delta_n(A) = T_\omega \restriction A$, we see that Ξ_m is consistent with T_ω. Consequently, $\Xi_m(c_{f(m)}) \subseteq T_\omega$, but $\varphi'(x_0) \in \Xi_m$, whence $\varphi'(c_{f(m)}) \in T_\omega$. Thus, on the constants, $\{c_0, c_1, \ldots\}$, the strongly constructivizable model $\mathfrak{M}'$ of the theory T_ω can be constructed in a standard way. The model $\mathfrak{M} \leftrightharpoons \mathfrak{M}' \restriction \sigma$ is the strongly constructivizable model of the theory T. For any finite $A \subseteq \{c_0, c_1, \ldots\}$, there exists $m \in \omega$ such that $A \subseteq B_m$, and consequently $T_\omega \restriction A \subseteq \Delta_{g(m)}(B_m)$. Therefore, any (finite) type that is realized in $\mathfrak{M}$ belongs to S. The property that every type of Ξ_m, that is consistent with T_ω $(= FD(\mathfrak{M}'))$, is realized in $\mathfrak{M}'$, shows that $\mathfrak{M}$ is an S–homogeneous model, i.e., $\mathfrak{M} \simeq \mathfrak{M}_S$. Thus Theorem 2.5 is proved. □

Corollary 2.6 (The Morley criterion for decidability of countably saturated models, [50]) *A countably saturated model of a decidable theory T is decidable if and only if the family of all types of the theory T is computable.*

Proof. Since a countably saturated model is homogeneous, it suffices to note that the effective extension condition also holds in the case in which the family of all types is computable. □

Corollary 2.7 [36, 34] *A prime model of a decidable theory T is decidable if and only if the family of principal types of the theory T is computable.*

Proof. One merely verifies the effective extension condition for the family of principal types. □

Theorem 2.8 *If the family S_T of all decidable types consistent with a theory T is computable, then T has a decidable prime model.*

Proof. We need to show that the theory T is atomic and that the family of all principal types is computable. To this end, for any formula $\varphi(\overline{x})$ consistent with T we construct a computable sequence of formulas $\varphi_n(\overline{x})$ such that $\varphi_0(\overline{x}) = \varphi(\overline{x})$, and for any n the set $T \cup \{\varphi_n(\overline{x})\}$ is consistent and

$$T \models (\forall \overline{x})\,(\varphi_{n+1}(\overline{x}) \to \varphi_n(\overline{x})).$$

Furthermore, for any formula $\varphi(\overline{x})$ there exists n such that $\varphi_n(\overline{x})$ is complete. Consequently,

$$d_n(\overline{x}) \;\leftrightharpoons\; \{\psi(\overline{x}) \mid \text{there exists } n \text{ such that } T \models \varphi_n(\overline{x}) \to \psi(\overline{x})\}$$

is a principal type containing a formula $\varphi(\overline{x})$ with the Gödel number n.

Let $p_{s_0}(\overline{x}), \ldots, p_{s_n}(\overline{x}), \ldots$ be a computable sequence of all decidable types consistent with the theory T.

STEP 0:

We set $\varphi_0(\overline{x}) = \varphi(\overline{x})$ if $\varphi(\overline{x})$ is consistent with T, and $\varphi_0(\overline{x}) \leftrightharpoons x_0 = x_0$ otherwise.

STEP $t+1$:

We consider a formula ψ with the Gödel number t. If it has free variables different from the variables occurring in the formula $\varphi(\overline{x})$, then we set $\varphi_{t+1} = \varphi_t$. Otherwise, we consider the following two cases:

CASE 1: Either $\psi(\overline{x}) \wedge \varphi_t$ or $\neg\psi(\overline{x}) \wedge \varphi_t$ is not consistent with T. Then we set $\varphi_{t+1} = \varphi_t$.

CASE 2: Both $\psi(\overline{x}) \wedge \varphi_t$ as well as $\neg\psi(\overline{x}) \wedge \varphi_t$ are consistent with T. We find the least numbers i, j such that $\psi(\overline{x}) \wedge \varphi_t \in p_{s_i}$ and $\neg\psi(\overline{x}) \wedge \varphi_t \in p_{s_j}$. We set $\varphi_{t+1} = \varphi_t \wedge \neg\psi(\overline{x})$ if $i < j$, and $\varphi_{t+1} = \varphi_t \wedge \psi(\overline{x})$ if $j < i$.

If Case 2 occurs infinitely many times, then the sequence $\{\varphi_t \mid t \in \mathbb{N}\}$ defines a decidable type which is consistent with the theory T, and is different from each of the types in the enumeration $p_{s_0}, p_{s_1}, \ldots, p_{s_n}, \ldots$. This contradicts the assumption, which provides the validity of the conditions imposed on φ_t. □

Corollary 2.9 *If the family of all decidable types consistent with a theory T is computable, then T has a prime model $\mathfrak{N}$, and the family $S_{\mathfrak{N}}$ of types realized in $\mathfrak{N}$ is computable.*

A theory T is called an *Ehrenfeucht theory* if it has only a finite number of countable models.

Corollary 2.10 *If T is a decidable Ehrenfeucht theory, then the family of all decidable types of T is computable, T has a decidable prime model $\mathfrak{N}$, and the family $S_{\mathfrak{N}}$ of types realized in this model is computable.*

Theorem 2.11 *If the family of all decidable types of a theory T is computable, and $\mathfrak{N}$ is a homogeneous countable model of the theory T with the computable family of types $S_{\mathfrak{N}}$, then the model $\mathfrak{N}$ is decidable.*

Proof. By the criterion for the decidability of homogeneous models, Theorem 2.5, it suffices to show that the family $S_{\mathfrak{N}}$ has an effectively extending function. From a type $p_n(\overline{x})$ and a formula $\varphi(\overline{x},\overline{y})$ we construct the required type step by step. Let $p_{s_0}(\overline{x}_0), \ldots, p_{s_n}(\overline{x}_n), \ldots$ denote the computable family of all decidable types consistent with the theory T. We construct $d_{k(n,m)} = \bigcup_t d^t_{c(n,m)}$ from a partial type $p_n(\overline{x})$ and a formula $\varphi(\overline{x},\overline{y})$ with the Gödel number m.

Step 0:
We set $d^0_{c(n,m)} = \{\varphi(\overline{x},\overline{y})\}$.

Step $t+1$:
If the formula $(\exists\overline{y})(\bigwedge d^t \wedge \varphi(\overline{x},\overline{y}))$ is not in p^t_n, we set $d^{t+1}_{c(n,m)} = d^t_{c(n,m)}$. Otherwise, we consider a formula $\psi(\overline{x},\overline{y})$ with the Gödel number $l(t)$.

Case 1: T satisfies the conditions

$$T \models \bigwedge p^t_n(\overline{x}) \to (\exists\overline{y})(\psi(\overline{x},\overline{y}) \wedge \bigwedge d^t_{c(n,m)})$$
$$T \models \bigwedge p^t_n(\overline{x}) \to (\exists\overline{y})(\neg\psi(\overline{x},\overline{y}) \wedge \bigwedge d^t_{c(n,m)}),$$

and the formula ψ contains only those free variables that occur in $\overline{x},\overline{y}$. We find the least i and j such that

$$\psi(\overline{x},\overline{y}) \wedge \bigwedge d^t_{c(n,m)} \in p_{s_i} \quad \text{and} \quad \neg\psi(\overline{x},\overline{y}) \wedge \bigwedge d^t_{c(n,m)} \in p_{s_j}.$$

We set $d^{t+1}_{c(n,m)} = d^t_{c(n,m)} \cup \{\neg\psi(\overline{x},\overline{y})\}$ if $i < j$, and $d^{t+1}_{c(n,m)} = d^t_{c(n,m)} \cup \{\psi(\overline{x},\overline{y})\}$ if $j < i$.

Case 2: T satisfies the condition

$$T \models \bigwedge p^t_n(\overline{x}) \to (\exists\overline{y})\,\psi(\overline{x},\overline{y}) \;\wedge\; \neg(\exists\overline{y})\,\neg\psi(\overline{x},\overline{y}).$$

We set $d^{t+1}_{c(n,m)} = d^t_{c(n,m)} \cup \{\psi\}$.

CASE 3: T satisfies the condition

$$T \models \bigwedge p_n^t(\overline{x}) \to (\exists \overline{y}) \neg\psi(\overline{x}, \overline{y}) \ \wedge \ \neg(\exists \overline{y})\, \psi(\overline{x}, \overline{y}).$$

We set $d^{t+1}_{c(n,m)} = d^t_{c(n,m)} \cup \{\neg\psi(\overline{x}, \overline{y})\}$.

CASE 4: Otherwise, we set $d^{t+1}_{c(n,m)} = d^t_{c(n,m)}$.

The type $p_n(\overline{x}) \in S_{\mathfrak{N}}$ is complete. Therefore, if $\varphi(\overline{x}, \overline{y})$ is consistent with $p_n(\overline{x})$, then the partial type $d_{c(n,m)} = \bigcup d^t_{c(n,m)}$ is also complete. If Case 1 occurs infinitely many times, then the type just constructed is decidable and different from each of the types $p_{s_0}, p_{s_1}, \ldots, p_{s_n}, \ldots$. However, the last sequence contains all decidable types, and consequently, Case 4 can occur only a finite number of times. Then the type $d_{c(n,m)}(\overline{x}, \overline{y})$ is principal over $p(\overline{x})$. Consequently, this type is realized in every model in which $p_n(x)$ is realized. Therefore, $d_{c(n,m)}(\overline{x}, \overline{y}) \in S_{\mathfrak{N}}$. The function reducing the sequence of partial types to $\{p_n \mid n \in \mathbb{N}\}$ gives an effectively extending function. □

Corollary 2.12 *If the family of all decidable types of a decidable theory T is computable, then a prime model $\mathfrak{N}$ of T exists and is decidable. If the family $S_{\mathfrak{N}}$ of types realized in $\mathfrak{N}$ is computable, then T has a decidable homogeneous countable model.*

Corollary 2.13 *If some universal model of a theory T is decidable, then a prime model of T exists and is decidable. Moreover, the family of all decidable types consistent with T is computable.*

Corollary 2.14 *If T is a decidable Ehrenfeucht theory, and the family of all types of a countable homogeneous model $\mathfrak{N}$ contains only decidable types, then the family $S_{\mathfrak{N}}$ is computable and the model $\mathfrak{N}$ is decidable.*

It was proved by Reed [66] that for any hyperarithmetic A, there exists a decidable Ehrenfeucht theory with five countable models and with a non-A-decidable countable model. The first examples of decidable Ehrenfeucht theories with non-decidable countable models were constructed by Lachlan and Peretyatkin [63]. But these theories have non-decidable types. Goncharov has constructed a decidable Ehrenfeucht theory with a decidable saturated model and non-decidable countable models.

3 Decidable Boolean algebras

The study of strongly constructive Boolean algebras was instigated by Ershov [14, 15]. The existence theorem for decidable models and the decidability of complete elementary theories of all elementary characteristics imply the following claim.

Proposition 3.1 *For each elementary characteristic* $\chi = (\alpha, \beta, \gamma)$ *of Boolean algebras, there exists a strongly constructive Boolean algebra* $\langle \mathcal{L}_\chi, \nu_\chi \rangle$ *such that* $\mathrm{ch}(\mathcal{L}_\chi) = \chi$*, which can be constructed uniformly in* χ.

Theorem 3.2 [52] *A countably saturated model of any elementary characteristic is decidable.*

Proof. In view of the Morley criterion (Corollary 2.6) for the decidability of countably saturated models, it suffices to show the computability of the family of all types of the theory T_χ of a given elementary characteristic χ. □

Corollary 3.3 [46] *Any prime model of the theory of Boolean algebras of any elementary characteristic is decidable.*

Corollary 3.4 [51] *Any countable homogeneous Boolean algebra* $\mathfrak{A}$ *such that* $\mathrm{ch}_1(\mathfrak{A}) < \infty$ *is decidable.*

Proof. By the decidability of a countably saturated model of the theory $\mathrm{Th}(\mathfrak{A})$ and Theorem 2.11, it suffices to show that the family of types realized in $\mathfrak{A}$ is computable. □

In the case of Boolean algebras of the first infinite characteristic, there is a continuum of homogeneous models, so only some of these homogeneous models can be decidable. As is known, the prime models and the countably saturated model are decidable. However, there are other decidable homogeneous models.

Theorem 3.5 [51] *If* $\mathfrak{A}$ *is a countable homogeneous Boolean algebra of elementary characteristic* $\langle \infty, 0, 0 \rangle$ *and of homogeneity type* $\langle \alpha\,, \langle p_n \mid n \in \mathbb{N} \rangle \rangle$*, then* $\mathfrak{A}$ *is a decidable model if and only if* $\{n \mid p_n = 0\} \in \Pi^0_2$.

The following simple theorem provides an effective method for establishing the decidability and the strong constructibility of models.

Theorem 3.6 *If a complete theory T is model-complete and decidable, then all of its constructive models are strongly constructive.*

We now examine the existence problems for strongly constructive elementary extensions and elementary saturated extensions of strongly constructive boolean algebras [15, 14]. This problem was solved by Ershov [15] for elementary extensions and by Goncharov for saturated extensions. We will describe the proof here, and as a corollary, construct a proper elementary constructive extension.

Proposition 3.7 *For any constructive atomless Boolean algebra there exists a proper constructive elementary extension to a strongly constructive Boolean algebra.*

Proof. In view of Thorem 3.6 and the model completeness of the theory of atomless Boolean algebras in the signature of Boolean algebras, it suffices to find a constructive isomorphic embedding of a constructive atomless Boolean algebra $\langle \mathfrak{A}, \nu \rangle$ into a constructive atomless Boolean algebra $\langle \mathfrak{B}, \mu \rangle$ such that $\varphi : \mathfrak{A} \to \mathfrak{B}$ is an isomorphic embedding but not an epimorphism and there exists a recursive function f such that $\varphi\nu(n) = \mu f(n)$ for any $n \in \mathbb{N}$.

For η, we take the set of rational numbers $\mathbb{Q}$ with the natural order. We define an embedding φ of $\mathfrak{B}_{\mathbf{L}}$ into $\mathfrak{B}_{\eta\times\mathbf{L}}$ as follows:

$$\varphi(\cup\,[a_i\,, b_i[\,) \;\leftrightharpoons\; \bigcup_i\,[\alpha_i\,, \beta_i[$$

where

$$\begin{array}{lll} \alpha_i & = (0, a_i) & \text{if } a_i \in \mathbf{L} \\ \alpha_i & = -\infty & \text{if } a_i = -\infty \\ \beta_i & = (0, b_i) & \text{if } b_i \in \mathbf{L} \\ \beta_i & = \infty & \text{if } b_i = \infty \end{array}$$

The composition of a constructive isomorphism of $\langle \mathfrak{A}, \nu \rangle$ onto $\langle \mathfrak{B}_{\mathbf{L}}, \nu_{\mathbf{L}} \rangle$ and the constructive embedding φ define the required proper embedding in an obvious way. □

Theorem 3.8 *For any strongly constructive atomic Boolean algebra $\langle \mathfrak{A}, \nu \rangle$, there exists a constructive elementary extension φ to a strongly constructive Boolean algebra $\langle \mathfrak{B}, \mu \rangle$ such that there exists an element $b \in |\mathfrak{B}|$ with the following properties:*

(1) *for any* $a \in |\mathfrak{A}|$ *with* $\mathrm{ch}_2(a) = \infty$, *we have* $\mathrm{ch}_2(a \wedge b) = \infty$ *and* $\mathrm{ch}_2(a \wedge C(b)) = \infty$; *and*

(2) *if* $\mathfrak{A}$ *is infinite then* b *does not belong to* $\mathfrak{A}$.

Moreover, $\langle \mathfrak{B}, \mu \rangle$ *is uniformly constructed from* $\langle \mathfrak{A}, \nu \rangle$, *and for an infinite Boolean algebra this extension is proper.*

Proof. We consider the enrichment of the signature of a Boolean algebra $\mathfrak{A}$ by a unary predicate A distinguishing atoms. Let $\mathfrak{A}'$ be the Boolean algebra thus enriched. The theory of $\mathfrak{A}'$ is model-complete.

By results about trees generating Booelan algebras in [17, 29], there exists a tree $\mathcal{D}$ such that $\langle \mathfrak{B}_{\mathcal{D}}, \mu_{\mathcal{D}} \rangle \cong \langle \mathfrak{A}, \nu \rangle$. Since $\langle \mathfrak{A}, \nu \rangle$ is strongly constructive, the recursive and isomorphic Boolean algebra $\langle \mathfrak{B}_{\mathcal{D}}, \mu_{\mathcal{D}} \rangle$ is also strongly constructive. Therefore, the set of numbers of its atoms, as well as the set of end vertices of the tree $\mathcal{D}$, is recursive.

We identify $\mathcal{D}$ with a tree generating the Boolean algebra $\mathfrak{A}$ and consider elements of $\mathcal{D}$ as elements of $\mathfrak{A}$, keeping in mind the standard embedding. We introduce a new constant c and constants c_0 , c_1 , ... , c_n , ..., and consider the diagram of the Boolean algebra $\mathfrak{A}$ under the assumption that the value of the constant c_n is νn. In the course of the construction, we will define all relations between c and all the elements of the generating tree. In so doing, we define all relations between c and all elements of $\mathfrak{A}$. Let $c_{g(s)}$ denote the constant corresponding to an element s of $\mathcal{D}$.

With the help of an algorithm that decides all formulas, from any strongly constructive Boolean algebra $\langle \mathfrak{A}, \nu \rangle$ we can uniformly construct a tree $\mathcal{D}$ and a recursive embedding g of $\mathcal{D}$ into $\mathbb{N}$ so that the tree $\langle \mathcal{D}, \nu g \rangle$ generates $\mathfrak{A}$. In view of the strong constructibility of $\langle \mathfrak{A}, \nu \rangle$, the tree $\mathcal{D}$, as well as the set of endpoints of $\mathcal{D}$, are both recursive.

At the step t of the construction, we will define relations between c, its complement $C(c)$, and elements of $\mathcal{D}$ of level t if no relations are already defined for them. It is possible that such relations will already be defined for a finite number of points of higher level $s > t$, but this can happen only for those elements of level t under which at most four atoms are located. The set of relations constructed at step t will be denoted by S^t. In constructing a partition of c by elements of the Boolean algebra, we use counters $r(x, t)$ defining a part of the partition of c (or $C(c)$) the intersection of which with $\nu(x)$ is finite.

By a relation between c and an element n of the tree $\mathcal{D}$, we mean an expression of one of the following forms:

(a) $c \wedge c_{g(n)} = c_{g(n)}$, or $c \wedge c_{g(n)} \neq c_{g(n)}$, or $c \wedge c_{g(n)} = \mathbf{0}$, or $c \wedge c_{g(n)} \neq \mathbf{0}$,

(b) $c \wedge c_{g(n)} \neq c$,

(c) $A(c \wedge c_{g(n)})$, or $\neg A(c \wedge c_{g(n)})$.

Similar relations are defined for the complement $C(c)$ of an element c.

STEP 0:

We first check if at most four atoms are located under $\nu g(0)$. By the strong constructibility of $\langle \mathfrak{A}, \nu \rangle$, this condition is decidable. If the condition holds, then $\nu g(0) = \bigvee_{i=1}^{s} \nu x_i$, where $x_i \in \mathcal{D}$, x_i is an endpoint of $\mathcal{D}$, and $s \leqslant 4$. We set

$$S^0 = \{C(c) = c_{g(0)}\}$$

and $r(0,0) = 0$.

If there are more than four atoms, then

$$\begin{aligned} S^0 = \{ & c \wedge c_{g(0)} \neq c_{g(0)},\ C(c) \wedge c_{g(0)} \neq c_{g(0)},\ c \wedge c_{g(0)} \neq \mathbf{0}, \\ & C(c) \wedge c_{g(0)} \neq \mathbf{0},\ \neg A(c \wedge c_{g(0)}),\ \neg A(C(c) \wedge c_{g(0)})\} \end{aligned}$$

and $r(0,0) = 0$.

STEP $t+1$:

For all elements $x_1, \ldots, x_n$ of $\mathcal{D}$ of level $t+1$, we define all necessary relations if they are not already defined. If, for some $y \preccurlyeq x_i$, the relation $c \wedge c_{g(y)} = c_{g(y)}$ or $C(c) \wedge c_{g(y)} = c_{g(y)}$ is already defined, then the relations between $c_{g(x_i)}$ and c are defined or not, depending on what was added to S^t. We add similar equalities to the relations connecting $c_{g(x_i)}$ with c and $C(c)$. It is not necessary to define A on an intersection since it is already defined in the diagram of $\mathfrak{A}$. If the previous condition fails, then, for $y = H(x_i) \in \mathcal{D}$, the relations $c \wedge c_{g(y)} \neq c_{g(y)}$ and $C(c) \wedge c_{g(y)} \neq c_{g(y)}$ are already defined and both intersections are marked as not atoms. We also consider the neighboring element $S(x_i)$ that belongs to $\mathcal{D}$ and is located under y. We define all the relations and all the counters simultaneously. The following cases are possible.

CASE 1: There are at most four atoms located under $\nu g(x_i)$ and under $\nu g(S(x_i))$. Since at most four atoms of $\mathfrak{A}$ are located under $g(y)$, at least four atoms are located under $\nu g(x_i)$ and under $\nu g(S(x_i))$ as well. Let $y_1, \ldots, y_m$, $m \geqslant 4$, be those end vertices of $\mathcal{D}$ that are situated under x_i and under $S(x_i)$. We define $c \wedge c_{g(y_i)} = c_{g(y_i)}$ for $1 \leqslant i \leqslant 2$, and $C(c) \wedge c_{g(y_i)} = c_{g(y_i)}$ for $2 < i \leqslant m$. For the rest of the $z \in \mathcal{D}$ such that $z \preccurlyeq x_i$ or $z \preccurlyeq S(x_i)$, we can obtain relations for $c_{g(z)}$ from the above relations, since

$$\nu g(z) = \bigvee\{\nu g(y_i) \mid i \in K_z\},$$

where $K_z \subseteq \{1, \ldots, m\}$.

CASE 2: There are at least four atoms located under $\nu g(x_i)$ and under $\nu g(S(x_i))$. We define S^{t+1} by adding the following relations to S^t:

$$c_{g(x_i)} \wedge c \neq \mathbf{0} \quad \text{and} \quad c_{g(x_i)} \wedge c \neq c_{g(x_i)} \quad \text{and} \quad \neg A(c_{g(x_i)} \wedge c)$$

We add the same relations to S^t for $S(x_i)$:

$$c_{g(S(x_i))} \wedge c \neq \mathbf{0} \quad \text{and} \quad c_{g(S(x_i))} \wedge c \neq c_{g(S(x_i))} \quad \text{and} \quad \neg A(c_{g(S(x_i))} \wedge c)$$

For a complement, replace c by $C(c)$ in both these collections of relations. We set

$$r(x_i, t+1) = r(S(x_i), t+1) = r(H(x_i), t)$$

in the first two cases.

CASE 3: Under one of the elements of $\{\nu g(x_i), \nu g(S(x_i))\}$ there are at most four atoms while under another element there are more than four atoms. Let $a \in \{x_i, S(x_i)\}$ be such that at most four atoms are located under $\nu g(a)$, and more than four atoms are located under $\nu g(b)$,where $b \in \{x_i, S(x_i)\}$. If $r(H(x_i), t)$ is even, then we assume that in S^{t+1}, S^t is completed by the relations

$$c_{g(a)} \wedge c = c_{g(a)} \quad \text{and} \quad c_{g(a)} \wedge C(c) = \mathbf{0}.$$

If $r(H(x_i), t)$ is odd, then we assume that S^{t+1} contains

$$c_{g(a)} \wedge C(c) = c_{g(a)} \quad \text{and} \quad c_{g(a)} \wedge c = \mathbf{0}.$$

For b, we add the following relations to S^{t+1}:

$$c \wedge c_{g(b)} \neq c_{g(b)} \qquad c \wedge c_{g(b)} \neq \mathbf{0} \qquad \neg A(c \wedge c_{g(b)})$$

$$C(c) \wedge c_{g(b)} \neq c_{g(b)} \qquad C(c) \wedge c_{g(b)} \neq \mathbf{0} \qquad \neg A(C(c) \wedge c_{g(b)})$$

Naturally, we also add all the relations of S^t to S^{t+1}. We set

$$r(a\,,\,t+1) \;=\; r(H(a)\,,\,t) \quad \text{and} \quad r(b\,,\,t+1) \;=\; r(H(b)\,,\,t)+1.$$

This concludes our inductive construction of S^t.

Now consider the theory

$$T_c \leftrightharpoons \mathrm{At}_1^* \cup \mathcal{D}_1\langle\mathfrak{A},\nu\rangle \cup \bigcup_t S^t,$$

where $\mathcal{D}_1\langle\mathfrak{A},\nu\rangle$ is the diagram of the Boolean algebra $\mathfrak{A}$ in the signature $\sigma_1 = \sigma \cup \{I_0, A_1\}$, where the value of the constant c_n is assumed to be equal to νn andAt_1^* is the theory of atomic infinite Boolean algebras of the signature σ_1. For any t, we can choose $b \in \mathfrak{A}$ such that, provided that the value of c is assumed to be equal to b, all relations of S^t are satisfied by c in the enrichment $\langle\mathfrak{A}, b\rangle$ of $\mathfrak{A}$ by b for the value of the constant symbol c. By the Mal'tsev compactness theorem, the set of formulas T_c is consistent and there is an enrichment $\mathfrak{B}'$ of some Boolean algebra $\mathfrak{B}$ such that $\mathfrak{B}' \models T_c$. Hence $\mathfrak{B}$ is atomic. Since $\mathfrak{B}' \models \mathcal{D}_1\langle\mathfrak{A},\nu\rangle$, we can assume that $\mathfrak{B}'$ is an extension of the enrichment $\mathfrak{A}_1$ of $\mathfrak{A}$ to the signature σ_1. The theory of atomic Boolean algebras of the signature σ_1 is model-complete. Consequently, $\mathfrak{A}_1 \preccurlyeq \mathfrak{B}'$ and $\mathfrak{A} \preccurlyeq \mathfrak{B}$.

If there are infinitely many elements of $\mathcal{D}$ under $n \in \mathcal{D}$, then $c \wedge c_{g(n)} \neq c_{g(n)}$ and $C(c) \wedge c_{g(n)} \neq c_{g(n)}$. Consequently, for any element n such that $\mathrm{ch}_2(\nu(n)) = \infty$, we have $c \neq c_n$. Therefore, any element $\nu(n) \in \mathfrak{A}$ such that $\mathrm{ch}_2(\nu(n)) = \infty$ is divided by c into two parts in such a way that, under each of $c \wedge c_n$ and $C(c) \wedge c_n$, there are infinitely many c_m such that νm is an atom of $\mathfrak{A}$. (That is why we introduced the counters $r(x,t)$ in x.) Therefore, the value of c in the Boolean algebra $\mathfrak{B}'$ differs from the values of c_n and, consequently, it does not belong to $\mathfrak{A}$. Thus, $\mathfrak{B}$ is a proper extension of $\mathfrak{A}$ satisfying conditions (1) and (2), as required.

We consider the subalgebra $\mathfrak{B}_0$ of $\mathfrak{B}$ generated by

$$\{\nu g(n) \wedge (c)_{\mathfrak{B}} \mid n \in \mathcal{D}\} \;\cup\; \{\nu g(n) \wedge C((c)_{\mathfrak{B}}) \mid n \in \mathcal{D}\}$$

where $(c)_{\mathfrak{B}}$ denotes the value of the constant c in $\mathfrak{B}'$. It is obvious that $\mathfrak{A} \preccurlyeq \mathfrak{B}_0$, since atoms of $\mathfrak{A}$ go to atoms of $\mathfrak{B}_0$ and any element that is not an atom goes into such an element. The Boolean algebra $\mathfrak{B}_0$ is atomic, which follows from the construction of the relations S^t for atoms and the fact that $\mathfrak{A}$ is atomic. The elements of $\mathfrak{B}_0$ are as follows:

$$\bigvee_{i=1}^{k}(g(n_i)\wedge c)\ \vee\ \bigvee_{i=1}^{l}(g(m_i)\wedge C(c))$$

where

$$\{n_i \mid 1 \leqslant i \leqslant k\} \cup \{m_i \mid 1 \leqslant i \leqslant l\} \subseteq \mathcal{D}.$$

We regard x as the number of the pair $c(n,m)$ and n,m as numbers of finite sequences $n = \langle n_1, \ldots, n_k\rangle$ and $m = \langle m_1, \ldots, m_l\rangle$. Then

$$\nu_0(x) \leftrightharpoons (\bigvee_{i=1}^{k} g(n_i)\wedge c)\ \vee\ (\bigvee_{i=1}^{l} g(m_i)\wedge C(c))$$

defines a numbering of the Boolean algebra $\mathfrak{B}_0$. By the properties of the tree $\mathcal{D}$, it is easy to obtain the existence of recursive functions f, g, h, the construction of which depends only on the structure of the tree. Consequently, f, g and h are uniformly determined by the algebra $\langle\mathfrak{A},\nu\rangle$ so that

$$\begin{aligned}\nu_0(x)\vee\nu_0(y) &= \nu_0\, f(x,y)\\ \nu_0(x)\wedge\nu_0(y) &= \nu_0\, g(x,y)\\ C(\nu_0(x)) &= \nu_0\, h(x).\end{aligned}$$

By our construction of S^t, we are always able to determine if $\nu_0(x)$ is an atom. Given x and y, we can also determine whether the values $\nu_0(x)$ and $\nu_0(y)$ are equal. Thus, ν_0 is a strong constructivization of the atomic Boolean algebra $\mathfrak{B}_0$. The embedding φ of $\langle\mathfrak{A},\nu\rangle$ into $\langle\mathfrak{B}_0,\nu_0\rangle$ is recursive. Hence the element $\nu g(n)$, for $n \in \mathcal{D}$, is

$$(\nu g(n)\wedge c)\ \vee\ (\nu g(n)\wedge C(c)).$$

From n, we find $x(n)$ such that $\nu_0(x(n)) = \nu g(n)$. However, any element of $\mathfrak{A}$ can be effectively determined as a union of elements of the generating tree, and the images of such elements can also be effectively determined. Consequently, the embedding of $\langle\mathfrak{A},\nu\rangle$ into $\langle\mathfrak{B}_0,\nu_0\rangle$ is constructive and a function $f_{\mathfrak{A}}$ providing the constructibility of the embedding ($\varphi\nu = \nu_0 f_{\mathfrak{A}}$) is uniformly constructed from $\mathfrak{A}$. To construct this function, we need only information about the tree $\mathcal{D}$. □

Corollary 3.9 [15] *For any infinite strongly constructive Boolean algebra* $\langle \mathfrak{A}, \nu \rangle$*, there exists a proper strongly constructive elementary extension.*

Proof. If $\mathfrak{A}$ is atomic, Theorem 3.8 immediately yields the required conclusion. Otherwise, $\mathfrak{A}$ contains a nonzero atomless element a (i.e. $a \neq \mathbf{0}$ and has no atom below it) and we have the decomposition

$$\mathfrak{A} \cong \widehat{a} \times \widehat{C(a)}.$$

Since $\langle \mathfrak{A}, \nu \rangle$ is a constructive algebra, ν induces constructivizations ν_a and $\nu_{C(a)}$ such that

$$\langle \mathfrak{A}, \nu \rangle \cong \langle \widehat{a}, \nu_a \rangle \times \langle \widehat{C(a)}, \nu_{C(a)} \rangle.$$

In this case, ν_a and $\nu_{C(a)}$ are strong constructivizations because ν is a strong constructivization and all the predicates of the signature σ^* that are necessary for model completeness are defined on $x \leqslant a$ in $\widehat{a}$ in the same manner as in $\mathfrak{A}$. Using Theorem 3.8, we consider a proper elementary constructive extension $\langle \mathfrak{B}, \mu \rangle$ of $\langle \widehat{a}, \nu_a \rangle$. We note that $\langle \mathfrak{B}, \mu \rangle \times \langle \widehat{C(a)}, \nu_{C(a)} \rangle$ is a proper constructive elementary extension of $\langle \mathfrak{A}, \nu \rangle$. The strong constructibility of the product follows from the strong constructibility of its components. □

Corollary 3.10 (Constructive saturation of atomic Boolean algebras) *For any infinite atomic strongly constructive Boolean algebra* $\langle \mathfrak{A}, \nu \rangle$*, there exists a constructive elementary extension to a strongly constructive saturated Boolean algebra.*

Proof. Let $\langle \mathfrak{A}, \nu \rangle$ be a strongly constructive infinite Boolean algebra. We construct a uniformly computable sequence of strongly constructive Boolean algebras $\langle \mathfrak{A}_n, \nu_n \rangle$ and effective elementary embeddings $\varphi_n \colon \mathfrak{A}_n \preccurlyeq \mathfrak{A}_{n+1}$ expressed in terms of recursive functions f_n, i.e., $\varphi_n \nu_n = \nu_{n+1} f_n$, where $\langle \mathfrak{A}_0, \nu_0 \rangle = \langle \mathfrak{A}, \nu \rangle$. To construct this sequence, we iterate the construction of Theorem 3.8 on elementary proper extensions of strongly constructive atomic Boolean algebras. In this case, for any n and $a \in \mathfrak{A}_n$ such that $\mathrm{ch}_2(a) = \infty$, there is an element $b \in \mathfrak{A}_{n+1}$ such that

$$\mathrm{ch}_2(a \setminus b) = \mathrm{ch}_2(a \wedge b) = \infty.$$

We define a model $\mathfrak{A}_\infty$ as the union of models of this elementary chain and define a numbering ν_∞ by setting $\nu_\infty c(n, m) \leftrightharpoons \nu_n(m)$. In view of the computability of the elementary embeddings φ_n and the uniform strong constructibility of the Boolean algebras $\langle \mathfrak{A}_n, \nu_n \rangle$, we conclude that $\langle \mathfrak{A}_\infty, \nu_\infty \rangle$ is a strongly constructive Boolean algebra. Since $\mathfrak{A}_{n+1}$ is a condensation of $\mathfrak{A}_n$ for any n, we conclude that $\bigcup \mathfrak{A}_n = \mathfrak{A}_\infty$ is a dense Boolean algebra. Consequently, $\mathfrak{A}_\infty$ is a saturated Boolean algebra. Thus $\langle \mathfrak{A}_\infty, \nu_\infty \rangle$ is a strongly constructive saturated atomic Boolean algebra which is a constructive elementary extension of $\langle \mathfrak{A}, \nu \rangle = \langle \mathfrak{A}_0, \nu_0 \rangle$. □

Proposition 3.11 *Given any elementary characteristic α of a Boolean algebra, we can uniformly construct a strongly constructive prime model $\langle \mathfrak{A}_\alpha, \nu_\alpha \rangle$ and a strongly constructive saturated model $\langle \mathfrak{B}_\alpha, \nu_\alpha \rangle$ of the theory of Boolean algebras of characteristic α.*

Proof. We can obtain this assertion in view of the uniformity of constructing the corresponding families of types from characteristics and the uniformity of constructing a prime model and a saturated model from a family of types. □

Proposition 3.12 *If $\langle \mathfrak{A}, \nu \rangle$ and $\langle \mathfrak{B}, \mu \rangle$ are strongly constructive Boolean algebras with elementary characteristics*

$$\mathrm{ch}(\mathfrak{A}) \;=\; (n\,,\,\infty\,,\,0) \quad \textit{and} \quad \mathrm{ch}(\mathfrak{B}) \;=\; (n+1\,,\,1\,,\,0),$$

respectively, then a constructive elementary embedding of $\langle \mathfrak{B}, \mu \rangle$ into the product $\langle \mathfrak{A}, \nu \rangle \times \langle \mathfrak{B}, \mu \rangle$ can be uniformly constructed from $\langle \mathfrak{A}, \nu \rangle$, $\langle \mathfrak{B}, \mu \rangle$, and n.

Proof. Since $\langle \mathfrak{B}, \mu \rangle$ is strongly constructive, from a number m of the element of $\langle \mathfrak{B}, \mu \rangle$, we can effectively recognize if $\mathrm{ch}_1(\mu m) < n+1$ or $\mathrm{ch}_1(\mu m) = n+1$ holds. The embedding:

$$\varphi(\mu m) \;=\; \begin{cases} \langle 0\,,\,\mu m \rangle & \text{if } \mathrm{ch}_1(\mu m) < n+1 \\ \langle 1\,,\,\mu m \rangle & \text{if } \mathrm{ch}_1(\mu m) = n+1 \end{cases}$$

of $\mathfrak{B}$ into $\mathfrak{A} \times \mathfrak{B}$ becomes an isomorphic embedding in the enrichments of these algebras to the signature σ_{4n+1}. Since $\mathfrak{B}$ and $\mathfrak{A} \times \mathfrak{B}$ both have characteristics $(n+1\,,\,1\,,\,0)$, the model completeness of these enrichments means that φ is an elementary embedding; moreover, φ is clearly constructive. The strong constructibility of the product follows from that of its components. □

Almost the same method can be used to construct an embedding in the case of characteristic $(n+1\,,\,0\,,\,1)$.

Proposition 3.13 *Given $n \in \mathbb{N}$ and a strongly constructive Boolean algebra $\langle \mathfrak{B}, \mu \rangle$ with elementary characteristic*

$$\mathrm{ch}(\mathfrak{B}) \;=\; (n+1\,,\,0\,,\,1),$$

we can construct effectively a strongly constructive prime Boolean algebra $\langle \mathfrak{A}, \nu \rangle$ with $\mathrm{ch}(\mathfrak{A}) = (n\,,\,\infty\,,\,0)$ and a constructive elementary embedding of $\langle \mathfrak{B}, \mu \rangle$ into the product $\langle \mathfrak{A}, \nu \rangle \times \langle \mathfrak{B}, \mu \rangle$.

Proposition 3.14 *Given $n \in \mathbb{N}$ and a strongly constructive Boolean algebra $\langle \mathfrak{A}, \nu \rangle$ such that $\mathrm{ch}_1(\mathfrak{A}) = n$, we can construct effectively an elementary extension $\langle \mathfrak{B}, \mu \rangle$ and an elementary embedding $\varphi : \mathfrak{A} \preccurlyeq \mathfrak{B}$; moreover, if $\mathrm{ch}_2(\mathfrak{A}) = \infty$, then there exists an element $b \in |\mathfrak{B}|$ such that*

$$ch_1(b) = \mathrm{ch}_1(C(b)) = n \qquad \textit{and} \qquad \mathrm{ch}_2(b) = \mathrm{ch}_2(C(b)) = \infty.$$

Proof. The assertion is proved by a slight modification of the proof of Theorem 3.8 in the case $n = 1$. The difference comes in using counters not only for atoms of the nth quotient algebra by some Ershov–Tarski ideal I_n, but also for characteristics $k < n$, where elements of characteristic $\mathrm{ch}_1(a) = k$ are included in c or $C(c)$ respectively. □

Propositions 3.12–3.14 yield the following result.

Proposition 3.15 *Given $n \in \mathbb{N}$ and a strongly constructive Boolean algebra $\langle \mathfrak{A}, \nu \rangle$, we can construct effectively its constructive elementary extension to a strongly constructive Boolean algebra $\langle \mathfrak{B}, \mu \rangle$; moreover, if $\mathrm{ch}_1(\mathfrak{A}) = n$, then $\mathfrak{B}$ is a condensation of $\mathfrak{A}$.*

Proof. Given $n \in \mathbb{N}$ and $\langle \mathfrak{A}, \nu \rangle$, consider the set

$$\begin{aligned} A_n \;\leftrightharpoons\; & \big\{(m\,,\,k\,,\,(k+1,1,0)) \;\big|\; \mathrm{ch}(\nu m) = (k+1\,,\,1\,,\,0) \text{ and } k < n\big\} \\ & \cup \big\{(m\,,\,k\,,\,(k+1,0,1)) \;\big|\; \mathrm{ch}(\nu m) = (k+1\,,\,0\,,\,1) \text{ and } k < n\big\} \\ & \cup \big\{(m\,,\,k\,,\,(k,\infty,0)) \;\big|\; \mathrm{ch}_1(\nu m) = k \text{ and } k \leqslant n\big\} \end{aligned}$$

Since $\langle \mathfrak{A}, \nu \rangle$ is a strongly constructive structure, the set A_n is recursive. Let $\alpha_0\,,\,\dots\,,\,\alpha_n\,,\,\dots$ be an enumeration of the elements of A_n. We now uniformly construct a sequence of strongly constructive models and constructive embeddings

$$\langle\mathfrak{A},\nu\rangle = \langle\mathfrak{A}_0,\nu_0\rangle \preccurlyeq \langle\mathfrak{A}_1,\nu_1\rangle \preccurlyeq \cdots \preccurlyeq \langle\mathfrak{A}_n,\nu_n\rangle \preccurlyeq \cdots$$

so that the embeddings $\varphi : \mathfrak{A}_n \to \mathfrak{A}_{n+1}$ and $\varphi_n : \mathfrak{A}_0 \to \mathfrak{A}_n$ are represented in terms of the recursive functions g_n and f_n as follows. Let $\langle\mathfrak{A}_0,\nu_0\rangle \leftrightharpoons \langle\mathfrak{A},\nu\rangle$. By induction, assume that

$$\langle\mathfrak{A}_0,\nu_0\rangle \preccurlyeq \langle\mathfrak{A}_1,\nu_1\rangle \preccurlyeq \cdots \preccurlyeq \langle\mathfrak{A}_n,\nu_n\rangle$$

are already constructed, and consider the sequence α_n.

If $\alpha_n = (m\,,\,k\,,\,(k+1,1,0))$, then we apply the construction of Proposition 3.12 to the Boolean algebra

$$\langle \widehat{\nu_n f_n(m)}\,,\,\nu_n \restriction \nu f_n(m)\rangle,$$

and obtain the required elementary extension. Taking the product with

$$\langle \widehat{C(\nu m)}\,,\,\nu \restriction \widehat{C(\nu m)}\rangle,$$

we arrive at the desired extension.

If $\alpha_n = (m\,,\,k\,,\,(k+1,0,1))$, then the construction is similar to that in the previous case, applying Proposition 3.13 to $\widehat{\nu(m)}$.

If $\alpha_n = (m\,,\,k\,,\,(k,\infty,0))$, then we apply the construction of Proposition 3.14 to

$$\langle \widehat{\nu m}\,,\,\nu \restriction \widehat{\nu(m)}\rangle$$

and take the product of the strongly constructive model so obtained with

$$\langle \widehat{C(\nu m)}\,,\,\nu \restriction \widehat{C(\nu m)}\rangle.$$

Taking $\langle\mathfrak{A}_\infty,\nu_\infty\rangle$ as the union of this computable chain of elementary constructive extensions, we conclude that $\mathfrak{A}_\infty$ is a condensation of $\mathfrak{A}$ and $\langle\mathfrak{A}_\infty,\nu_\infty\rangle$ is a constructive elementary extension of $\mathfrak{A}$. □

Applying the construction of Proposition 3.15 ω times, from any strongly constructive Boolean algebra $\langle\mathfrak{A},\nu\rangle$, we obtain its constructive elementary extension to a dense strongly constructive Boolean algebra. Since dense Boolean algebras are countably saturated, we have the following result.

Theorem 3.16 *Any strongly constructive Boolean algebra* $\langle \mathfrak{A}, \nu \rangle$ *with* $\mathrm{ch}_1(\mathfrak{A}) < \infty$ *has a constructive elementary extension to a strongly constructive saturated Boolean algebra.* □

We now consider the case in which the first characteristic is infinite. For elements of a finite characteristic and Boolean algebras of an infinite characteristic, we can apply the propositions above to obtain constructive elementary extensions satisfying the density condition. In the case of infinite characteristics, the following proposition suffices.

Proposition 3.17 *Given a strongly constructive Boolean algebra* $\langle \mathfrak{A}, \nu \rangle$ *such that* $\mathrm{ch}_1(\mathfrak{A}) = \infty$, *we can construct effectively a constructive elementary extension* $\langle \mathfrak{B}, \mu \rangle$ *containing an element* $c \in |\mathfrak{B}|$ *such that for any* $a \in |\mathfrak{A}|$ *with* $\mathrm{ch}_1(a) = \infty$, *we have* $\mathrm{ch}(a \wedge c) = \mathrm{ch}(a \wedge C(c)) = (\infty\,,\,0\,,\,0)$.

Proof. We use a modification of the construction of Theorem 3.8 on elementary extensions of atomic Boolean algebras. But instead of the existence condition for four atoms, we verify the condition that $\mathrm{ch}_1(x) > t_n + 4$. In addition, the counter indicates the minimal characteristic of elements added under c and $C(c)$ but not the number of elements. □

Proposition 3.17 and Theorem 3.16 imply the following result.

Theorem 3.18 (Strongly constructive extensions) *Any strongly constructive Boolean algebra has a constructive elementary extension to a strongly constructive saturated Boolean algebra.*

Proof. We construct effectively a chain of condensations

$$\langle \mathfrak{A}_0, \nu_0 \rangle \preccurlyeq \langle \mathfrak{A}_1, \nu_1 \rangle \preccurlyeq \cdots \preccurlyeq \langle \mathfrak{A}_i, \nu_i \rangle \preccurlyeq \cdots$$

so that $\langle \mathfrak{A}_i, \nu_i \rangle$ is a computable sequence of constructive models, f_n is a computable sequence of recursive functions, and $\varphi_n : \mathfrak{A}_n \preccurlyeq \mathfrak{A}_{n+1}$ is an elementary embedding of $\mathfrak{A}_n$ into $\mathfrak{A}_{n+1}$ such that $\mathfrak{A}_{n+1}$ is a condensation of $\mathfrak{A}_n$ and $\varphi_n \nu_n(m) = \nu_{n+1} f_n(m)$ for any $m, n \in \mathbb{N}$. Defining $\langle \mathfrak{A}_\infty, \nu_\infty \rangle$ as the union of this chain, we obtain the required strongly constructive elementary extension of $\langle \mathfrak{A}_0, \nu_0 \rangle = \langle \mathfrak{A}, \nu \rangle$, where $\mathfrak{A}_\infty$ is the direct limit of the algebras $\mathfrak{A}_n$, and $\nu_\infty(c(n,m)) \leftrightharpoons \nu_n(m)$. The saturation of $\mathfrak{A}_\infty$ follows from its density. □

Corollary 3.19 *Any strongly constructive Boolean algebra* $\langle \mathfrak{A}, \nu \rangle$ *has a constructive elementary extension to a strongly constructive homogeneous Boolean algebra.*

References

[1] I. Bucur and A. Deleanu, Introduction to the Theory of Categories and Functors, Pure Appl. Math., **19** (1968).

[2] C. C. Chang and H. J. Keisler, Model Theory, 3rd. edn., Stud. Logic Found. Math., **73** (1990); [1st. edn. 1973, 2nd. edn. 1977].

[3] C. W. Day, Superatomic Boolean algebras, Pacific J. Math., **23** (1967) 479–489.

[4] H. Dobbertin, On Vaught's criterion for isomorphisms of countable Boolean algebras, Algebra Universalis, **15** (1982) 95–114.

[5] R. G. Downey, Every recursive Boolean algebra is isomorphic to one with incomplete atoms, Ann. Pure Appl. Logic, **60** (1993) 193–206.

[6] R. G. Downey and C. G. Jockusch Jr., Every low Boolean algebra is isomorphic to a recursive one, Proc. Amer. Math. Soc., **122** (1994) 871–880.

[7] B. N. Drobutun, Enumerations of simple models (Russian), Sibirsk. Math. Zh., **18** (1977) 1002–1014, 1205; [translated in: Siberian Math. J., **18** (1977) 707–716].

[8] Z. A. Dulatova, Constructivity of Boolean algebras with a distinguished subalgebra (Russian), Mat. Zametki**46** (1989) 53–56, 127; [translated in: Math. Notes, **46** (1989) 924–926].

[9] V. D. Dzgoev, On the constructivization of certain structures (Abstract in Russian), (Dep. VINITI, No. 1606–79, Moscow, 1979), Sibirsk. Math. Zh., **21** (1980) 231.

[10] V. D. Dzgoev, Constructive Boolean algebras (Russian), Mat. Zametki, **44** (1988) 750–757, 861; [translated in: Math. Notes, **44** (1988) 896–901].

[11] V. D. Dzgoev, Constructization of Boolean lattices (Russian), Algebra i Logika, **27** (1988) 641–648, 736; [translated in: Algebra and Logic, **27** (1988) 395–400].

[12] E. F. Eisenberg and J. B. Remmel, Effective isomorphisms of algebraic structures, in: Patras Logic Symposion, (Proc. Logic Sympos. Patras, Greece, Aug. 18–22, 1980) G. Metakides (ed.), Stud. Logic Found. Math., **109** (1982) 95–122.

[13] Yu. L. Ershov, Decidability of the elementary theory of relatively complemented distributive lattices, and of the theory of filters (Russian), Algebra i Logika, **3** (1964) 17–38.

[14] Yu. L. Ershov, Constructive models (Russian), in: Izbr. Vopr. Alg. i Log., [Selected Questions in Algebra and Logic], Sbornik posvjascen. pamjati A. I. Mal'ceva, [A collection dedicated to the memory of A. I. Mal'tsev], Izdat. Nauk Sibirsk. Otdel., Novosibirsk, (1973) 111–130.

[15] Yu. L. Ershov, Theory of Numerations III (Constructive Models) (Russian), Lib. Dept. Algebra and Math. Logic Novosibirsk Univ., **13**, (Novosibirsk Gosudarstr. Univ., Novosibirsk, 1974).

[16] Yu. L. Ershov, Theory of Numerations (Russian), Monographs in Math. Logic and Foundations of Math., (Nauka, Moskva, 1977).

[17] Yu. L. Ershov, Decision Problems and Constructivizable Models (Russian), (Mathematical Logic and Foundations of Mathematics, Nauka, Moskva, 1980).

[18] Yu. L. Ershov, I. A. Lavrov, A. D. Taĭmanov and M. A. Taĭtslin, Elementary theories (Russian), Uspekhi Mat. Nauk, **20** (1965) 37–108. [translated in: Russian Math. Surveys, **20** (1965) 35–105].

[19] Yu. L. Ershov and E. A. Palyutin, Mathematical Logic (Russian), 2nd. edn., (Nauka, Moskva, 1987). [English translation of 1st. edn., translated by V. Shokurov, (MIR Publishers, Moscow, 1984)].

[20] S. S. Goncharov, The constructivizability of superatomic Boolean algebras (Russian), Algebra i Logika, **12** (1973) 31–40, 120; [translated in: Algebra and Logic, **12** (1973) 17–22].

[21] S. S. Goncharov, Autostability and computable families of constructivizations (Russian), Algebra i Logika, **14** (1975) 647–680; [translated in: Algebra and Logic, **14** (1975) 392–409].

[22] S. S. Goncharov, Certain properties of the constructivization of Boolean algebras (Russian), Sibirsk. Math. Zh., **16** (1975) 264–278; [translated in: Siberian Math. J., **16** (1975) 203–214].

[23] S. S. Goncharov, Nonselfequivalent constructivizations of atomic Boolean algebras (Russian), Mat. Zametki, **19** (1976) 853–858; [translated in: Math. Notes, **19** (1976) 500–503].

[24] S. S. Goncharov, Restricted theories of constructive Boolean algebras (Russian), Sibirsk. Math. Zh., **17** (1976) 797–812; [translated in: Siberian Math. J., **17** (1976) 601–611].

[25] S. S. Goncharov, Strong constructivizability of homogeneous models (Russian), Algebra i Logika, **17** (1978) 363–388, 490; [translated in: Algebra and Logic, **17** (1978) 247–263].

[26] S. S. Goncharov, A totally transcendental decidable theory without constructivizable homogeneous models (Russian), Algebra i Logika, **19** (1980) 137–149, 250; [translated in: Algebra and Logic, **19** (1980) 85–93].

[27] S. S. Goncharov, Totally transcendental theory with a nonconstructivizable prime model (Russian), Sibirsk. Math. Zh., **21** (1980) 44–51; [translated in: Siberian Math. J., **21** (1980) 32–37].

[28] S. S. Goncharov, Constructive Models, Monash Univ., Clayton, Victoria, Australia, Monash Logic Papers, **56** (November 1984).

[29] S. S. Goncharov, Countable Boolean Algebras (Russian), (Nauka. Sibirsk. Otdel., Novosibirsk, 1988).

[30] S. S. Goncharov, Computable classes of constructivizations for models of finite constructivity type (Russian), Sibirsk. Math. Zh., **34** (1993) 23–37; [translated in: Siberian Math. J., **34** (1993) 812–824].

[31] S. S. Goncharov, Effectively infinite classes of weak constructivizations of models (Russian), Algebra i Logika, **32** (1993) 631–664, 712; [translated in: Algebra and Logic, **32** (1993) 342–360].

[32] S. S. Goncharov and B. N. Drobotun, Numerations of saturated and homogeneous models (Russian), Sibirsk. Math. Zh., **21** (1980) 25–41, 236; [translated in: Siberian Math. J., **21** (1980) 164–176].

[33] S. S. Goncharov and V. D. Dzgoev, Autostability of models (Russian), Algebra i Logika, **19** (1980) 45–58, 132; [translated in: Algebra and Logic, **19** (1980) 28–37].

[34] S. S. Goncharov and A. T. Nurtazin, Constructive models of complete decidable theories (Russian), Algebra i Logika, **12** (1973) 125–142, 243; [translated in: Algebra and Logic, **12** (1973) 67–77].

[35] P. Halmos, Lectures on Boolean algebras, (Van Nostrand, Toronto, New York, London, 1963).

[36] L. Harrington, Recursively presentable prime models, J. Symbolic Logic, **39** (1974) 305–309.

[37] N. G. Khisamiev, Strongly constructive models (Russian), Izv. Akad. Nauk Kaz. SSR, ser. Fiz.-Mat., **3** (1971) 59–63.

[38] N. G. Khisamiev, Strongly constructive models of a decidable theory (Russian), Izv. Akad. Nauk Kazakh. SSR, Ser. Fiz.-Mat., **1** (1974) 83–84.

[39] S. Koppelberg, Handbook of Boolean Algebras, Vols. 1–3, (text of Vol. 1 by S. Koppelberg), J. D. Monk, R. Bonnet and S. Koppelberg, (eds.), (North-Holland, Amsterdam, NY, 1989).

[40] P. La Roche, Recursively presented Boolean algebras, Notices Amer. Math. Soc., **24**:6 (1977) A 552–553 (abstract).

[41] A. I. Mal'tsev, Constructive algebras I (Russian), Uspekhi Mat. Nauk, 16 (1961) 3–60; [translated in: Constructive algebras I, Russian Math. Surveys, **16**:3 (1961) 77–129; also in: The Metamathematics of Algebraic Systems, Collected Papers: 1936–1967, translated and edited by B. F. Wells III, Stud. Logic Found. Math., **66** (1971), Ch. 18, 148–200].

[42] A. I. Mal'tsev, Untersuchungen aus dem Gebiete der mathematischen Logik, Mat. Sb. (N.S.), **1** (1936) 323–335; [Russian summary: ibid., 336; translated in: The Metamathematics of Algebraic Systems, Collected Papers: 1936–1967, translated and edited by B. F. Wells III, Stud. Logic Found. Math., **66** (1971), Ch. 1, 1–14.]

[43] A. I. Mal'tsev, Algorithms and Recursive Functions (Russian), (Izdat. Nauka, Moscow, 1965); [translated: by L. F. Boron, L. E. Sanchis, J. Stilwell and K. Iseki, (Wolters-Noordhoff Publishing, Groningen, 1970)].

[44] A. I. Mal'tsev, Algebraic Systems (Russian), (posth. edn.), V. D. Smirnov and M. Taĭclin, (eds.), (Izdat. Nauka, Moscow, 1970); [translated: by B. D. Seckler and A. P. Doohovskoy, Grundlehren Math. Wiss., **192** (1973).]

[45] D. A. Martin and M. B. Pour-El, Axiomatizable theories with few axiomatizable extensions, J. Symbolic Logic, **35** (1970) 205–209.

[46] J. Mead, Recursive prime models for Boolean algebras, Colloq. Math., **41** (1979) 25–33.

[47] G. Metakides and A. Nerode, Recursion theory and algebra, in: Algebra and Logic, (Proc. 14th. Summer Res. Inst. Austral. Math. Soc., Monash Univ., Clayton, Vic., Australia, Jan. 6 – Feb. 16, 1974), J. N. Crossley, (ed.), Lecture Notes in Math., **450** (1975) 209–219.

[48] G. Metakides and A. Nerode, Recursively enumerable vector spaces, Ann. Math. Logic, **11** (1977) 147–171.

[49] T. S. Millar, Foundations of recursive model theory, Ann. Math. Logic, **13** (1978) 305–320.

[50] M. Morley, Decidable models, Israel J. Math., **25** (1976) 233–240.

[51] A. S. Morozov, Countable homogeneous Boolean algebras (Russian), Algebra i Logika, **21** (1982) 269–282; [translated in: Algebra and Logic, **21** (1982) 181–190.

[52] A. S. Morozov, Strong constructivizability of countable saturated Boolean algebras (Russian), Algebra i Logika, **21** (1982) 193–203; [translated in: Algebra and Logic, **21** (1982) 130–137.

[53] M. Moses, Recursive properties of isomorphism types, J. Austral. Math. Soc., Ser. A, **34** (1983) 269–286.

[54] M. Moses, Recursive linear orders with recursive successivities, Ann. Pure Appl. Logic, **27** (1984) 253–264.

[55] A. T. Nurtazin, Strong and weak constructivizations and computable families (Russian), Algebra i Logika, **13** (1974) 311–323, 364; [translated in: Algebra and Logic, **13** (1974) 177–184].

[56] S. P. Odintsov, Atom-free ideals of constructive Boolean algebras (Russian), Algebra i Logika, **23** (1984) 278–295, 362; [translated in: Algebra and Logic, **23** (1984) 190–203].

[57] S. P. Odintsov, Lattice of a recursively enumerable subalgebras of recursive Boolean algebra (Russian), Algebra i Logika, **25** (1986) 631–642, 751; [translated in: Algebra and Logic, **25** (1986) 397–404].

[58] S. P. Odintsov, Restricted theories of constructive Boolean algebras of the lower stratum (Russian), Inst. of Math., Novosibirsk, (Preprint No. 21) (1986).

[59] S. P. Odintsov, Recursive Boolean algebras with a hyperhyperimmune set of atoms (Russian), Mat. Zametki, **44** (1988) 488–493, 557; [translated in: Math. Notes, **44** (1988) 747–749.]

[60] S. P. Odintsov, Hereditarily recursively enumerable subalgebras of recursive Boolean algebra (Russian), Algebra i Logika, **31** (1992) 38–46, 96; [translated in: Algebra and Logic, **31** (1992) 24–29].

[61] S. P. Odintsov and V. L. Selivanov, The arithmetical hierarchy and ideals of enumerated Boolean algebras (Russian), Sibirsk. Math. Zh., **30** (1989) 140–149; [translated in: Siberian Math. J., **30** (1989) 952–960].

[62] M. G. Peretyat'kin, Strongly constructive models and enumerations of the Boolean algebra of recursive sets (Russian), Algebra i Logika, **10** (1971) 535–557; [translated in: Algebra and Logic, **10** (1971) 332–345].

[63] M. G. Peretyat'kin, Complete theories with a finite number of countable models (Russian), Algebra i Logika, **12** (1973) 550–576, 618; [translated as: On complete theories with a finite number of denumerable models, Algebra and Logic, **12** (1973) 310–326].

[64] M. G. Peretyat'kin, A criterion for strong constructivizability of a homogeneous model (Russian), Algebra i Logika, **17** (1978) 436–454, 491; [translated in: Algebra and Logic, **17** (1978) 290–301.]

[65] M. G. Peretyat'kin, Calculations on Turing machines in finitely axiomatizable theories (Russian), Algebra i Logika, **21** (1982) 410–441; [translated as: Turing machine computations in finitely axiomatizable theories, Algebra and Logic, **21** (1982) 272–295.]

[66] R. C. Reed. A Decidable Ehrenfeucht Theory with Exactly Two Hyperarithmetic Models, Ph.D. Thesis, Univ. Wisconsin, Madison, WI, (1986).

[67] J. B. Remmel, Recursively enumerable Boolean algebras, Ann. Math. Logic, **15** (1978) 75–107.

[68] J. B. Remmel, r–maximal Boolean algebras, J. Symbolic Logic, **44** (1979) 533–548.

[69] J. B. Remmel, Complementation in the lattice of subalgebras of a Boolean algebra, Algebra Universalis, **10** (1980) 48–64.

[70] J. B. Remmel, Recursion theory on algebraic structures with independent sets, Ann. Math. Logic, **18** (1980) 153–191.

[71] J. B. Remmel, Recursive Boolean algebras with recursive atoms, J. Symbolic Logic, **46** (1981) 595–616.

[72] J. B. Remmel, Recursive isomorphism types of recursive Boolean algebras, J. Symbolic Logic, **46** (1981) 572–594.

[73] J. B. Remmel, Recursively categorical linear orderings, Proc. Amer. Math. Soc., **83** (1981) 387–391.

[74] J. B. Remmel, Recursively rigid Boolean algebras, Ann. Pure Appl. Logic, **36** (1987) 39–52.

[75] J. B. Remmel, Recursive Boolean algebras, in: Handbook of Boolean Algebras, Vol. 3, J. D. Monk, R. Bonnet and S. Koppelberg, (eds.), (North-Holland, Amsterdam, NY, 1989), Ch. 25, 1097–1165.

[76] H. Rogers Jr., Theory of Recursive Functions and Effective Computability, (1st. edn., McGraw-Hill, New York-Toronto, Ont.-London, 1967; 2nd. edn., MIT Press, Cambridge, Mass., London, 1987).

[77] R. Sikorski, Boolean Algebras, 2nd. edn., (Springer-Verlag, Berlin, NY, 1964).

[78] R. I. Soare, Recursively Enumerable Sets and Degrees. A Study of Computable Functions and Computably Generated Sets, Perspect. in Math. Logic, (1987).

[79] P. Štěpánek, Embeddings and automorphisms, in: Handbook of Boolean Algebras, Vol. 2, J. D. Monk, R. Bonnet and S. Koppelberg, (eds.), (North-Holland, Amsterdam, NY, 1989), Ch. 16, 607–635.

[80] M. N. Stone, The theory of representations for Boolean algebras, Trans. Amer. Math. Soc., **40** (1936) 37–111.

[81] A. Tarski, Arithmetical classes and types of Boolean algebras, Preliminary Report, Bull. Amer. Math. Soc. (N.S.), **55** (1949) 64.

[82] J. J. Thurber, Recursive and r.e. quotient Boolean algebras, Arch. Math. Logic, **33** (1994) 121–129.

[83] R. L. Vaught, Topics in the Theory of Arithmetical Classes and Boolean Algebras, Ph.D. Thesis, Univ. California, Berkeley, CA, (1954).

[84] V. N. Vlasov and S. S. Goncharov, Strong constructibility of Boolean algebras with elementary characteristic (1,1,0) (Russian), Algebra i Logika, **32** (1993) 618–630; [translated in: Algebra and Logic, **32** (1993) 334–341].

Yu. L. Ershov
Sobolev Institute of Mathematics
pr. Akademika Koptuga, 4
Novosibirsk, 630090
Russia

S. S. Goncharov
Sobolev Institute of Mathematics
pr. Akademika Koptuga, 4
Novosibirsk, 630090
Russia

Chapter 3

Isomorphic Recursive Structures

C. J. Ash*

Contents

An early result about recursive structures is the following, attributed to S. Tennenbaum. There is a recursive linear ordering of order type $\omega + \omega^*$ having no infinite recursive ascending or descending sequence. There are also, of course, more obvious recursive orderings of this type, for example $0, 2, 4, \ldots, 5, 3, 1$. An ordering of the first sort, which can be constructed by a priority argument, is classically isomorphic to the second, but cannot be *recursively* isomorphic since it has different recursive properties.

Some other examples are mentioned in [7]: a recursive linear ordering of type ω on which the successor relation is not recursive; a recursive vector space over any recursive infinite field having no recursive basis; a recursive algebraically closed field of any characteristic having no recursive transcendence basis.

*Due to the untimely death of the author, the final version of this paper was prepared by Professor J. Knight. The editors of the volume express their gratitude to Professor Knight for this valuable assistance.

HANDBOOK OF RECURSIVE MATHEMATICS
Edited by Yu. L. Ershov, S. S. Goncharov,
A. Nerode, and J. B. Remmel

In each of these cases it is clear that within the isomorphism type of a single structure we have recursive structures which are not recursively isomorphic. In this article we discuss general results which have been obtained concerning this and related matters.

1 Recursive structures

We consider structures of some fixed recursive similarity type (I, J, K, μ, ν) where I, J and K are recursive sets and $\mu : I \to \omega$ and $\nu : J \to \omega$ are recursive functions. A structure of this type is a quadruple of the form

$$\mathfrak{A} = (A, \langle R_i \rangle_{i \in I}, \langle f_j \rangle_{j \in J}, \langle a_k \rangle_{k \in K})$$

where each R_i is a $\mu(i)$–ary relation on A, each f_i is a $\nu(j)$–ary operation on A, and each $a_k \in A$.

Such a structure $\mathfrak{A}$ is a *recursive* structure if A is a recursive set, each R_i is a recursive relation on A, uniformly in i, each f_j is a recursive operation on A, uniformly in j, and the function $k \mapsto a_k$ is recursive.

An alternative definition is used by the Russian workers in this area. Define a *constructivization* α of an abstract structure $\mathfrak{A}$, to be an onto function $\alpha : \mathbb{N} \to A$, such that for atomic formulae φ the relations $\{\overline{a} \mid \mathfrak{A} \models \varphi[\alpha(\bar{a})]\}$ are recursive, uniformly in φ. Then such a pair $(\mathfrak{A}, \alpha)$ is essentially a recursive structure isomorphic to $\mathfrak{A}$ (together with a particular isomorphism). Indeed, this approach is closer to mathematical nature, where one thinks of a classical isomorphism type as being an abstract structure. But, in practice, it seems to reduce the number of symbols significantly to deal only with the recursive structures themselves, and we shall follow this line.

2 Recursive stability

A structure with which the least can be done by way of different recursive copies is $(\omega, <, S)$ where S is the successor relation on ω. This structure is *rigid*, that is, has no automorphisms other than the identity function, and it is easily seen that for any two recursive structures, $\mathfrak{A}$ and $\mathfrak{B}$, which are isomorphic to $(\omega, <, S)$, the unique isomorphism from $\mathfrak{A}$ to $\mathfrak{B}$ is recursive. This rather strong property, that *all* isomorphisms between recursive copies are recursive, has been called "recursive stability".

Definition 2.1 A recursive structure $\mathfrak{A}$ is said to be *recursively stable* if for every recursive structure $\mathfrak{B} \cong \mathfrak{A}$, every $f : \mathfrak{B} \cong \mathfrak{A}$ is recursive.

It seems plausible to expect that this phenomenon will occur only when the elements of the structure are suitably definable.

Definition 2.2 An *r.e. defining family* for a structure $\mathfrak{A}$ is an r.e. set S of existential formulae $\varphi(v)$ of one variable, such that each is satisfied by exactly one element of $\mathfrak{A}$, and each element of $\mathfrak{A}$ satisfies at least one of the formulae $\varphi(v)$ from S.

If a recursive structure $\mathfrak{A}$ is such that some expansion $(\mathfrak{A}, \bar{a})$ of $\mathfrak{A}$ by a sequence $\overline{a}$ of finitely many constants from $\mathfrak{A}$ has an r.e. defining family, then it is clear from the following that $\mathfrak{A}$ is recursively stable. For any recursive $\mathfrak{B} \cong \mathfrak{A}$ and any $f : \mathfrak{B} \cong \mathfrak{A}$, we have $f : (\mathfrak{B}, \overline{b}) \cong (\mathfrak{A}, \overline{a})$ for suitable $\overline{b}$. Then f is the unique isomorphism from $(\mathfrak{B}, \overline{b})$ to $(\mathfrak{A}, \overline{a})$, and its graph is r.e., since it is the set of pairs (d, c) such that, for some φ in the family, $(\mathfrak{B}, \overline{b}) \models \varphi[d]$ and $(\mathfrak{A}, \overline{a}) \models \varphi[c]$.

The converse statement, that a recursively stable structure $\mathfrak{A}$ has a finite expansion with an r.e. defining family, does not hold without some further assumption. Let us define the *existential diagram* of a recursive structure $\mathfrak{A}$ to be the set of existential sentences true in the structure $(\mathfrak{A}, \langle a \rangle_{a \in A})$. Since $\mathfrak{A}$ is a recursive structure, this set is clearly r.e..

Theorem 2.1 *Let $\mathfrak{A}$ be a recursive structure whose existential diagram is recursive. Then $\mathfrak{A}$ is recursively stable if and only if some finite expansion of $\mathfrak{A}$ by constants has an r.e. defining family.*

This result is implicit in [12]. The non-trivial direction is obtained by constructing simultaneously a recursive structure $\mathfrak{B}$ and an isomorphism $f : \mathfrak{B} \cong \mathfrak{A}$ which is not recursive, using a finite injury priority argument. This argument is also given in [7].

As with other results to come, Theorem 2.1 needs the assumption that at least one recursive isomorphic copy has some extra recursive properties. In [13], an example is constructed of a recursive structure $\mathfrak{A}$ which is recursively stable but has no r.e. defining family.

3 Intrinsically recursive relations

For a recursive structure $\mathfrak{A}$ and an additional relation R on A, a construction similar to that used to prove Theorem 2.1 can often be used to give a recursive structure $\mathfrak{B}$ and an isomorphism $f : \mathfrak{A} \cong \mathfrak{B}$ for which $f(R)$ is not recursive. If there are no such $\mathfrak{B}$ and f, then the relation R is said to be *intrinsically recursive* on $\mathfrak{A}$.

If a relation R on $\mathfrak{A}$ has a definition of the form

$$R(\overline{a}) \iff \mathfrak{A} \models \bigvee_n \varphi_n(\overline{a}, \overline{c})$$

for some sequence $\overline{c}$ from $\mathfrak{A}$ and some sequence $\varphi_n(\overline{x}, \overline{y})$ of existential formulae, then R will be called *formally r.e.* on $\mathfrak{A}$. Clearly, if both R and its complement are formally r.e., then R is intrinsically recursive. The converse can be shown provided that, again, $(\mathfrak{A}, R)$ satisfies further recursive conditions.

Theorem 3.1 (Ash [7]) *Let $(\mathfrak{A}, R)$ be a recursive structure whose existential diagram is recursive. Then R is intrinsically recursive on $\mathfrak{A}$ if and only if both R and its complement are formally r.e. on $\mathfrak{A}$*

The construction for this attempts to make both $f(R)$ and its complement not r.e.. A similar construction only attempting to make $f(R)$ not r.e. gives the more basic result:

Theorem 3.2 (Ash [7]) *Let $(\mathfrak{A}, R)$ be a recursive structure whose existential diagram is recursive. Then R is intrinsically r.e. on $\mathfrak{A}$ if and only if R is formally r.e. on $\mathfrak{A}$.*

The examples mentioned earlier can all be deduced from Theorem 3.1 without repeating the priority argument.

Some assumption is needed for the equivalence in Theorem 3.1. In [16], an example is given of a recursive $(\mathfrak{A}, R)$ for which R is intrinsically recursive on $\mathfrak{A}$ but is not formally r.e. on $\mathfrak{A}$.

If R is *not* intrinsically recursive on $\mathfrak{A}$, we may define the *spectrum* of R on $\mathfrak{A}$ to be the set of all possible Turing degrees of the relations $f(R)$ for recursive structures $\mathfrak{B}$ and isomorphisms $f : \mathfrak{A} \cong \mathfrak{B}$. The possible spectra obtained in this way are considered in detail by V. Harizanov in [15].

4 Recursive categoricity

A recursive structure $\mathfrak{A}$ is said to be *recursively categorical* if, for every recursive $\mathfrak{B} \cong \mathfrak{A}$, there is at least one recursive isomorphism $f : \mathfrak{B} \cong \mathfrak{A}$. For example, the ordering of the rationals $(\mathbb{Q}, <)$ has this property, as can be shown using an effective back-and-forth argument.

Clearly $(\mathbb{Q}, <)$ is not recursively stable, since it has $2^{\aleph_0}$ automorphisms, and so has $2^{\aleph_0}$ isomorphisms with any recursive copy. In fact, it follows fairly briefly from the definitions that a structure $\mathfrak{A}$ is recursively stable if and only if, for some finite sequence $\overline{a}$ from $\mathfrak{A}$, $(\mathfrak{A}, \overline{a})$ is both rigid and recursively categorical.

The example of $(\mathbb{Q}, <)$ suggests a sufficient condition for $\mathfrak{A}$ to be recursively categorical.

Definition 4.1 An *r.e. automorphism family* for $\mathfrak{A}$ is an r.e. family of existential formulae $\varphi_n(\overline{x}_n)$, in different numbers of variable, such that every finite sequence from $\mathfrak{A}$ satisfies at least one of these formulae while, if $\mathfrak{A} \models \varphi_n[\overline{b}]$ and $\mathfrak{A} \models \varphi_n[\overline{c}]$, then $(\mathfrak{A}, \overline{b}) \cong (\mathfrak{A}, \overline{c})$.

Clearly, if $\mathfrak{A}$, or more generally some $(\mathfrak{A}, \overline{a})$, has an r.e. automorphism family, then, by an effective back-and-forth argument, $\mathfrak{A}$ is recursively categorical. The converse can be shown subject to further assumptions on $\mathfrak{A}$. An $\exists\forall$ sentence means one of the form $\exists\, \overline{x}\, \forall\, \overline{y}\, \varphi(\overline{x}, \overline{y})$, where $\varphi(\overline{x}, \overline{y})$ is quantifier-free.

Theorem 4.1 (Goncharov [12]) *Let $\mathfrak{A}$ be a recursive structure whose $\exists\forall$ diagram is recursive. Then $\mathfrak{A}$ is recursively categorical if and only if some finite expansion $(\mathfrak{A}, \overline{a})$ has an r.e. automorphism family.*

In the case where there is no r.e. automorphism family, one can construct not just two, but infinitely many recursive copies of $\mathfrak{A}$ which are pairwise not recursively isomorphic. In other words:

Theorem 4.2 (Goncharov [13]) *Let $\mathfrak{A}$ be a recursive structure whose $\exists\forall$ diagram is recursive. Then either there is only one recursive isomorphism type of recursive copies of $\mathfrak{A}$, or there are infinitely many.*

In contrast, Goncharov in [14] has shown that there are recursive structures having any other number, 2, 3, 4, ..., of recursive isomorphism types of recursive copies, but of course not having recursive $\exists\forall$ diagrams.

It follows quickly from the definitions, that if a finite expansion $(\mathfrak{A}, \overline{a})$ of $\mathfrak{A}$ is recursively categorical, then so is $\mathfrak{A}$. The converse appears to be an open question: Is there an $\mathfrak{A}$ which is recursively categorical while some $(\mathfrak{A}, \overline{a})$ is not?[1]

From Theorem 4.1, there is no such $\mathfrak{A}$ whose $\exists\forall$ diagram is recursive. Millar in [17] shows, more generally, that there is no such $\mathfrak{A}$ whose existential diagram is recursive.

5 Δ^0_α–stability

While the structure $(\omega, <, S)$ is recursively stable, $(\omega, <)$ is not, as can be deduced from Theorem 2.1. On the other hand, for any recursive copy $(A, <)$, the successor relation is Π^0_1, so the unique isomorphism is Δ^0_2. We call such a structure Δ^0_2–stable, and in general:

Definition 5.1 A recursive structure $\mathfrak{A}$ is Δ^0_α*–stable* $(\alpha < \omega_1^{\text{CK}})$ if, for every recursive $\mathfrak{B} \cong \mathfrak{A}$, every isomorphism $f : \mathfrak{B} \cong \mathfrak{A}$ is Δ^0_α.

Thus, each $(\beta, <)$ with $\omega \leq \beta < \omega^2$ is Δ^0_2–stable, but not recursively stable.

For any recursive copy of $(\omega^2, <)$, there is a Δ^0_4 procedure which, given m and n, locates the element corresponding to the ordinal $\omega m + n$. It follows that $(\omega^2, <)$ is Δ^0_4–stable. It seems less obvious how to show that $(\omega^2, <)$ is not Δ^0_3–stable. Similarly, if

$$\omega^\alpha \leq \beta < \omega^{\alpha+1}$$

and $\alpha \geq 1$, then it is not difficult to show that $(\beta, <)$ is $\Delta^0_{2\alpha}$–stable. Theorem 6.1 below shows that $(\beta, <)$ is not Δ^0_γ–stable for any $\gamma < 2\alpha$.

6 Back-and-forth relations

Between structures $\mathfrak{A}$ and $\mathfrak{B}$ of the same type, we define the relations $\leq_\gamma$ for all ordinals $\gamma \geq 1$ as follows.

[1]Recently, the question has been answered affirmatively by Cholak, Goncharov, Khoussainov and Shore [11].

Let $\mathfrak{A} \leq_1 \mathfrak{B}$ if every existential sentence true in $\mathfrak{B}$ is also true in $\mathfrak{A}$. For $\gamma > 1$, let $\mathfrak{A} \leq_\gamma \mathfrak{B}$ if, for each β with $1 \leq \beta < \gamma$ and each finite sequence $\overline{b}$ from $\mathfrak{B}$, there is a finite sequence $\overline{a}$ from $\mathfrak{A}$ (of the same length as $\overline{b}$) such that $(\mathfrak{B}, \overline{b}) \leq_\beta (\mathfrak{A}, \overline{a})$. For sequences from the same structure $\mathfrak{A}$, we write $\overline{a} \leq_\gamma \overline{b}$ to mean $(\mathfrak{A}, \overline{a}) \leq_\gamma (\mathfrak{A}, \overline{b})$.

Define, for a finite sequence $\overline{a}$ from $\mathfrak{A}$ and $\alpha \geq 2$, the subset $\mathrm{cl}_\alpha(\overline{a})$ of A by $c \in \mathrm{cl}_\alpha(\overline{a})$ if there exists $\overline{b}$ from $\mathfrak{A}$ and $\beta < \alpha$ with $\beta \geq 1$ such that, for all c' and $\overline{b}'$ with $\overline{a}, c, \overline{b} \leq_\beta \overline{a}, c', \overline{b}'$, we have $c' = c$.

Theorem 6.1 (Ash [1]) *Suppose that $\alpha \geq 2$, and that $\mathfrak{A}$ is a recursive structure for which the relations $\leq_\beta$ on $\mathfrak{A}$ are r.e. uniformly in $\beta < \alpha$. Suppose also that there is a recursive procedure which yields for each $\overline{a}$ from $\mathfrak{A}$ an element of $\mathfrak{A}$ not in $\mathrm{cl}_\alpha(\overline{a})$. Then $\mathfrak{A}$ is not Δ^0_α-stable.*

To apply this result, one needs to consider these relations $\leq_\beta$ and the sets $\mathrm{cl}_\alpha(\overline{a})$ in some detail. It is at least true, as outlined in [1], that for all $\alpha, \gamma < \omega_1^{\mathrm{CK}}$ there is a recursive copy of $(\gamma\,, <)$ on which the relations $\leq_\beta$ for $\beta < \alpha$ and the sets $\mathrm{cl}_\alpha(\overline{a})$ are uniformly recursive.

In the case of such well-behaved structures $\mathfrak{A}$, Theorem 6.1 says that $\mathfrak{A}$ is not Δ^0_α-stable if there is no finite $\overline{a}$ for which $\mathrm{cl}_\alpha(\overline{a}) = A$. It seems desirable to try to obtain a result along the same lines as Theorem 2.1. For this purpose, one may consider recursive infinitary formulae.

7 Recursive infinitary formulae

The Σ_α and Π_α formulae of $L_{\omega_1\omega}$ are defined by transfinite induction: the Σ_0 and Π_0 formulae are the quantifier-free formulae, the Σ_α formulae are those of the form $\bigvee_n \exists \overline{y}_n \varphi_n$ for which each φ_n is a Π_β formula for some $\beta < \alpha$, and the Π_α formulae are those of the form $\bigwedge_n \forall \overline{y}_n \varphi_n$ for which each φ_n is a Σ_β formula for some $\beta < \alpha$.

One may then show, by transfinite induction on α, that $\mathfrak{A} \leq_\alpha \mathfrak{B}$ if and only if every Π_α sentence true in $\mathfrak{A}$ is also true in $\mathfrak{B}$, or equivalently, every Σ_α sentence true in $\mathfrak{B}$ is also true in $\mathfrak{A}$.

We define, for $\alpha < \omega_1^{\mathrm{CK}}$, the *recursive* Σ_α and Π_α formulae to be those in which all the infinitary disjunctions and conjunctions are recursively enumerable. For this to make sense, we need to define simultaneously indices for these formulae, in the style of Kleene's system of notations for the ordinals below ω_1^{CK}.

We can then define a Σ^0_α *defining family* for a structure $\mathfrak{A}$ to be a Σ^0_α set S of (indices for) recursive Σ_α formulae, $\varphi(v)$, each satisfied by exactly one element, such that each element of $\mathfrak{A}$ is satisfied by at least one $\varphi(v)$ from S.

Then it is clear that if a recursive structure $\mathfrak{A}$ has a finite expansion $(\mathfrak{A}, \overline{a})$ with a Σ^0_α defining family, then it is Δ^0_α-stable. For a converse, we need several assumptions.

Assume that the existential diagram of $\mathfrak{A}$ is recursive, and that the relations $\leq_\beta$ are r.e. uniformly in $\beta < \alpha$. Then, as shown in [1], and again, more neatly, in [5], there is a recursive procedure which obtains, for each $\beta < \alpha$ and each $\overline{a}$, a recursive Π_β formula $\varphi(\overline{v})$, such that for each $\overline{b}$ from $\mathfrak{A}$, we have $\overline{a} \leq_\beta \overline{b}$ if and only if $\mathfrak{A} \models \varphi[\overline{b}]$.

It follows that, under the same assumptions, if $c \in \mathrm{cl}_\alpha(\overline{a})$, then there is a recursive Σ_α formula $\varphi(v, \overline{a})$ satisfied in $\mathfrak{A}$ by c alone, and such a formula can be found recursively from c (and $\overline{a}$). Thus, if $A = \mathrm{cl}_\alpha(\overline{a})$, then $(\mathfrak{A}, \overline{a})$ has a Σ^0_α defining family. So Theorem 6.1 gives:

Theorem 7.1 (Ash [1]) *Suppose that $\mathfrak{A}$ is a recursive structure for which no finite expansion $(\mathfrak{A}, \overline{a})$ has a Σ^0_α defining family. Suppose also that the existential diagram of $\mathfrak{A}$ is recursive, that the relations $\leq_\beta$ are r.e. uniformly in $\beta < \alpha$, and that the relation $c \notin \mathrm{cl}_\alpha(\overline{a})$ is r.e.. Then $\mathfrak{A}$ is not Δ^0_α-stable.*

8 Δ^0_α–categoricity

In a similar way, we may define a recursive structure $\mathfrak{A}$ to be Δ^0_α*-categorical* if, for every recursive $\mathfrak{B} \cong \mathfrak{A}$, there exists a Δ^0_α-isomorphism $f : \mathfrak{B} \cong \mathfrak{A}$. Again, for a sufficiently well-behaved $\mathfrak{A}$, either Theorem 8.1 below can be used to show that $\mathfrak{A}$ is not Δ^0_α-categorical, or some $(\mathfrak{A}, \overline{a})$ has a Σ^0_α automorphism family – the obvious generalization of an r.e. automorphism family, above – and so $\mathfrak{A}$ is Δ^0_α-categorical.

Theorem 8.1 below gives the most straightforward result for non-Δ^0_α-categoricity. For a recursive structure $\mathfrak{A}$, $\alpha \geq 2$, and elements $\overline{a}$ of $\mathfrak{A}$, we define $C_\alpha(\overline{a})$ to be the set of sequences $\overline{c}$ from $\mathfrak{A}$ such that, for some $\overline{b}$ from $\mathfrak{A}$ and some $\beta < \alpha$, for all $\overline{c}'$ and $\overline{b}'$ of the same lengths as $\overline{c}$ and $\overline{b}$, if $\overline{a}, \overline{c}, \overline{b} \leq_\beta \overline{a}, \overline{c}', \overline{b}'$ then $(\mathfrak{A}, \overline{c}') \cong (\mathfrak{A}, \overline{c})$.

Theorem 8.1 (Ash [2]) *Suppose that $\alpha \geq 2$, and that $\mathfrak{A}$ is a recursive structure on which the relations $\leq_\beta$ are r.e. uniformly in $\beta < \alpha$, and the relation $(\mathfrak{A}, \overline{c}') \not\cong (\mathfrak{A}, \overline{c})$ is r.e.. Suppose also that there is a recursive procedure which gives, from each finite sequence $\overline{a}$, a sequence $\overline{c} \notin C_\alpha(\overline{a})$. Then $\mathfrak{A}$ is not Δ^0_α-categorical.*

Under similar recursive assumptions on $\mathfrak{A}$, including that its existential diagram is recursive, if every sequence is in $C_\alpha(\overline{a})$, then we can obtain a Σ^0_α automorphism family for $(\mathfrak{A}, \overline{a})$, so that a sufficiently well-behaved structure $\mathfrak{A}$ will be Δ^0_α-categorical if and only if there is some $\overline{a}$ such that $(\mathfrak{A}, \overline{a})$ has such a family.

A slight refinement of Theorem 8.1 is shown in [2], in which the definition of $C_\alpha(\overline{a})$ replaces $(\mathfrak{A}, \overline{c}') \cong (\mathfrak{A}, \overline{c})$ by $\overline{a}, \overline{c} \geq_\alpha \overline{a}, \overline{c}'$. This is an oversight, and should read $\overline{a}, \overline{c} \geq_\alpha \overline{a}, \overline{c}'$ *and* $\overline{a}, \overline{c} \leq_\alpha \overline{a}, \overline{c}'$.

By way of example, we consider in [2] certain Boolean algebras. We let $B(\beta)$ denote the Boolean algebra generated by a well-ordered set of type β. Then, for infinite β, $B(\beta)$ has $2^{\aleph_0}$ automorphisms, and so is not Δ^0_α-stable for any α. For $\omega^\alpha \leq \beta < \omega^{\alpha+1}$ and $\alpha \geq 1$, $B(\beta)$ is $\Delta^0_{2\alpha}$-categorical, and not Δ^0_γ-categorical for any $\gamma < 2\alpha$.

9 Intrinsically Σ^0_α relations

Say that a relation R on a recursive structure $\mathfrak{A}$ is *intrinsically* Σ^0_α if, for every recursive $\mathfrak{B} \cong \mathfrak{A}$ and every isomorphism $f : \mathfrak{A} \cong \mathfrak{B}$, $f(R)$ is Σ^0_α. If R is defined in some $(\mathfrak{A}, \overline{a})$ by a recursive Σ_α formula, then it is clear that R is intrinsically Σ^0_α. Otherwise, provided that $(\mathfrak{A}, R)$ has sufficiently many recursive properties, Theorem 9.1 below shows that R is not intrinsically Σ^0_α.

Assume that $\alpha \geq 2$. For $\overline{a}$ from $\mathfrak{A}$, define the relation $\mathrm{Rcl}_\alpha(\overline{a})$ on A by $\overline{a} \in \mathrm{Rcl}_\alpha(\overline{a})$ if there exist $\overline{b}$ and $\beta < \alpha$ such that, for all $\overline{c}'$ and $\overline{b}'$ with $\overline{a}, \overline{c}, \overline{b} \leq_\beta \overline{a}, \overline{c}', \overline{b}'$, we have $R(\overline{c}')$.

Theorem 9.1 (Barker [8]). *Suppose that $\alpha \geq 2$, and that the relations $\leq_\beta$ are r.e. uniformly in $\beta < \alpha$. Suppose further that there is a recursive procedure for finding, given $\bar{a}$, some $\bar{c} \notin \mathrm{Rcl}_\alpha(\overline{a})$. Then R is not intrinsically Σ^0_α on $\mathfrak{A}$.*

Again, assuming also that the existential diagram of $\mathfrak{A}$ is recursive, if for some $\overline{a}$ every sequence $\overline{C}$ of the relevant length is in $\mathrm{Rcl}_\alpha(\overline{a})$, then we can obtain a recursive Σ_α definition of R in $(\mathfrak{A}, \bar{a})$. So, for a sufficiently recursive $(\mathfrak{A}, R)$, R is intrinsically Σ^0_α if and only if it has such a recursive Σ_α definition.

By way of example, Barker shows in [8] that, if $\beta \geq \omega^{\alpha+1}$ and $\alpha \geq 1$, then the set of limit ordinals of the α–th kind in $(\beta\,, <)$ is intrinsically $\Pi^0_{2\alpha}$, but not intrinsically $\Sigma^0_{2\alpha}$. In [9], he also considers intrinsically Σ^0_α subsets of reduced abelian p–groups.

10 Non-recursive structures

By a structure here (in this section) we mean one which has a recursive similarity type, and has as universe a recursive set of natural numbers, but is not necessarily recursive. Instead, as a sequence of sequences of relations and operations, it has a well-defined Turing degree.

Let $\mathfrak{A}$ be a recursive structure, and let R be a relation on A. Define R to be *relatively* intrinsically Σ^0_α on $\mathfrak{A}$ if, for all $\mathfrak{B}$, not necessarily recursive, and all $f : \mathfrak{A} \cong \mathfrak{B}$, $f(R)$ is Σ^0_α relative to $\mathfrak{B}$. We have seen that, under the assumption of recursive properties of $(\mathfrak{A}, R)$, if R has no recursive Σ_α definition in any $(\mathfrak{A}, \bar{a})$, then R is not intrinsically recursive on $\mathfrak{A}$. By contrast, without any provisos, one can show the following, using forcing.

Theorem 10.1 *R is* relatively *intrinsically Σ^0_α on $\mathfrak{A}$ if and only if R has a recursive Σ_α definition in some $(\mathfrak{A}, \bar{a})$.*

Suitable definitions of relative Δ^0_α–stability and Δ^0_α–categoricity also have exact characterizations in terms of recursive infinitary formulae. These results are obtained in [10] and, independently, in [6]. In many ways, such results are more pleasing that our earlier ones. In this paper, however, we take the view that what matters is the construction of *recursive* structures.

11 α–systems

Theorems 6.1, 8.1 and 9.1 were obtained using the author's "metatheorems" from [1]. While relying on the same proof, these results have since then been improved and streamlined in [3] and [4]. We take the opportunity to state the present version from [4], and show how Theorem 6.1 can be obtained from it. This is a significant improvement on the argument in [1], since complete recursive metric spaces are no longer mentioned, and a single argument covers the cases where α is and is not a limit ordinal.

Let T be an r.e. tree having no terminal nodes. Let U and V denote the sets of nodes of T at even levels and at odd levels, respectively, so that the root of T is in U. Let E assign to each $v \in V$ an r.e. set $E(v)$, uniformly in v.

Consider the game, $\mathfrak{G}$, in which players I and II choose, alternately, nodes of T to form an infinite sequence $u_0, v_0, u_1, v_1, \ldots$, where I chooses the u_i and II chooses the v_i in such a way that u_0 is the root of T, each v_i is a successor of u_i in T, and each u_{i+1} is a successor of v_i. (By a successor of a node in a tree, we mean an *immediate* successor.) We say that such a sequence is a *play* of the game $\mathfrak{G}$, and that this play is *winning for* II if $\cup_i E(v_i)$ is r.e., and otherwise is *winning for* I.

A I–strategy is a function s which assigns to each $v \in V$ a successor, $s(v)$, of v in T. The play $u_0\ v_0\ u_1\ v_1\ \ldots$ is said to *follow* the I–strategy s if for each $i, u_{i+1} = s(v_i)$. A *winning strategy for player* I is an I–strategy s such that every play which follows s is winning for player I.

Proposition G Suppose that there exist uniformly r.e. relations $\langle \subseteq_\gamma \rangle_{\gamma<\alpha}$ on V satisfying the conditions below. Then there is no Δ^0_α winning strategy for player I in the game $\mathfrak{G}$.

(i) If $v \subseteq_0 v'$, then $E(v) \subseteq E(v')$.

(ii) If $\gamma < \beta < \alpha$ and $v \subseteq_\beta b'$, then $v \subseteq_\gamma v'$.

(iii) Each $\subseteq_\gamma$ is reflexive and transitive.

(iv) If $\alpha > \gamma_1 > \gamma_2 > \cdots > \gamma_k$ for $k = 1, 2, \ldots$, and $v_1 \subseteq_{\gamma_1} v_2 \subseteq_{\gamma_2} \cdots \subseteq_{\gamma_{k-1}} v_k$ with u any successor of v_1, then there is a successor v of u such that $v_i \subseteq_{\gamma_i} v$ for each $i = 1, 2, \ldots, k$. □

Thus, for any Δ^0_α I–strategy, there is a play which follows it, and is winning for II. Further information about such a play is given by:

Proposition U Under the supposition of Proposition G, from a recursive index for T, an r.e. index for E and a Δ^0_α index for a I–strategy s, we may recursively compute both a Δ^0_α index for a play $u_0, v_0, u_1, v_1, \ldots$ of $\mathfrak{G}$ such that $\cup_i E(v_i)$ is r.e., and also an r.e. index for $\cup_i E(v_i)$. □

Proof of Theorem 6.1. We may assume that A is infinite, since otherwise the second supposition cannot hold. Let B be an infinite recursive set. We wish to define a recursive structure $\mathfrak{B}$ having universe B, and an isomorphism $f : \mathfrak{B} \cong \mathfrak{A}$ which is not Δ^0_α.

Let $c^* \in A$ with $c^* \notin \mathrm{cl}_\alpha(\emptyset)$, and let $d^* \in B$ be arbitrary. We let the tree T have root $\big((\emptyset, \emptyset, c^*), d^*\big)$, and have as other nodes all longer sequences of the form $\big((f_0, \sigma_0, c_0)\,,\, d_0\,,\, (f_1, \sigma_1, c_1)\,,\, d_1\,,\, \ldots\big)$, satisfying the conditions:

(1) $f_0 = \emptyset$, $\sigma_0 = \emptyset$, $c_0 = c^*$, $d_0 = d^*$,

and, for each i, the conditions:

(2) f_i is a finite partial one-to-one function from B to A, $f_{i+1} \supseteq f_i$ and $\mathrm{ran}(f_{i+1})$ and $\mathrm{dom}(f_{i+1})$ contain at least the first i elements of A and B, respectively. Also, $c_i \in \mathrm{ran}(f_{i+1})$ and $f^{-1}_{i+1}(c_i) \neq d_i$.

(3) σ_i is a finite set of sentences $\varphi(\bar{b})$ for which $\varphi(\bar{v})$ is a quantifier-free formula of the language, $\bar{b} \in \mathrm{dom}(f_i)$ and $\varphi(f_i(\bar{b}))$ is true in $\mathfrak{A}$. Also, $\sigma_{i+1} \supseteq \sigma_i$, and σ_{i+1} contains each of the first i sentences $\varphi(\bar{b})$ which are induced in this sense by f_{i+1}.

(4) $c_i \in A$ and $c_i \notin \mathrm{cl}_\alpha(\mathrm{ran}(f_i))$.

(5) $d_i \in B$.

Thus, a play in $\mathfrak{G}$ consists of the initial segments of an infinite sequence

$$\big((f_0, \sigma_0, c_0)\,,\, d_0\,,\, (f_1, \sigma_1, c_1)\,,\, d_1\,,\, \ldots\big)$$

for which $\cup_i \sigma_i$ determines a structure $\mathfrak{B}$ with universe B, and $f = \cup_i f_i$ is an isomorphism from $\mathfrak{B}$ to $\mathfrak{A}$ such that for each i, $f^{-1}(c_i) \neq d_i$.

We consider a $\Delta^0_{\alpha+1}$ strategy for player I in which each

$$\big(\ldots\,,\, (f_{i+1}, \sigma_{i+1}, c_{i+1})\big)$$

is extended to a sequence

$$\big(\ldots\,,\, (f_{i+1}, \sigma_{i+1}, c_{i+1})\,,\, d_{i+1}\big),$$

where d_{i+1} is different from the i–th partial Δ^0_α function applied to c_{i+1}, if this is defined, and is arbitrary otherwise. Then, for a play which follows this strategy, $f = \cup_i f_i$ is different from each Δ^0_α function.

To use Proposition G, we arrange that, for a play which is winning for II, the diagram $\cup_i \sigma_i$ of $\mathfrak{B}$ is r.e., so that $\mathfrak{B}$ is a recursive structure. This follows if we define $E((\dots, (f, \sigma, c))$ to be σ. Of course, for α in Proposition G, we now read $\alpha + 1$.

To complete the argument, we need to define suitable relations $\subseteq_\gamma$ for $\gamma \leq \alpha$. We first define relations $\leq_\gamma$ for $\gamma \geq 1$ between finite partial one-to-one functions f and g from B to A by $f \leq_\gamma g$ if $\mathrm{dom}(g) \supseteq \mathrm{dom}(f)$ and $\overline{a} \leq_\gamma g(f^{-1}(\overline{a}))$, where $\overline{a} = \mathrm{ran}(f)$. Then for

$$v = (\dots, (f, \sigma, c)) \qquad \text{and} \qquad v' = (\dots, (f', \sigma', c')),$$

we define $v \subseteq_0 v'$ if $\sigma \subseteq \sigma'$, we define $v \subseteq_\alpha v'$ if $f \subseteq f'$ and $\sigma \subseteq \sigma'$, and, for $1 \leq \gamma < \alpha$, we define $v \subseteq_\gamma v'$ if $f \leq_\gamma f'$ and $\sigma \subseteq \sigma'$.

Conditions (i), (ii) and (iii) of Proposition G are immediate, so to prove the theorem it remains only to verify condition (iv).

Suppose, then, that

$$\alpha + 1 > \gamma_1 > \dots > \gamma_k,$$

and that

$$v \subseteq_{\gamma_1} v_2 \subseteq_{\gamma_2} \cdots \subseteq_{\gamma_{k-1}} v_k.$$

For simplicity we may assume, without loss of generality, that $\gamma_1 = \alpha$, $\gamma_k = 0$ and, since $\alpha \geq 2$, that $k > 2$, and so $\gamma_2 > 0$. Let $v_i = (\dots, (f_i, \sigma_i, c_i))$, and let $u = (\dots, (f_1, \sigma_1, c_1), d)$. Now we have $f_1 \subseteq f_2 \leq_{\gamma_2} \cdots \leq_{\gamma_{k-1}} f_k$ and $\sigma_0 \subseteq \sigma_1 \subseteq \cdots \subseteq \sigma_k$.

We appeal at this stage to the principal result about the relations $\leq_\gamma$ that, since $f_2 \leq_{\gamma_2} \cdots \leq_{\gamma_{k-1}} f_k$, there is a $g \supseteq f_2$ with $dom(f_k) \subseteq dom(g)$ and $f_i \leq_{\gamma_i} g$ for each $i = 3, \dots, k-1$. This fact is fairly easy to show by induction on k; it appears as Lemma 4 in [1], and is proved as Proposition 1.3 in [5].

By extending this g, we may assume that $c_1 \in \mathrm{ran}(g)$. If $g^{-1}(c_1) \neq d$, then take $h = g$. If $g^{-1}(c_1) = d$, then, since $c_1 \notin \mathrm{cl}_\alpha(\mathrm{ran}(f_1))$, there exists $h \supseteq f_1$ with $g \leq_{\gamma_2} h$ and $h(d) \neq c_1$. In either case, we may choose v to be $(\dots, (f_1, \sigma_1, c_1), d, (f, \sigma, c,))$, where $f \supseteq h$, $\sigma \supseteq \sigma_k$ and c are chosen to satisfy the conditions (2), (3) and (4). Condition (iv) is thus verified, and the theorem proved. □

References

[1] C. J. Ash, Recursive labelling systems and stability of recursive structures in hyperarithmetical degrees, Trans. Amer. Math. Soc., **298** (1986) 497–514; errata ibid., **310** (1988) 851.

[2] C. J. Ash, Categoricity in hyperarithmetical degrees, Ann. Pure Appl. Logic, **34** (1987) 1–14.

[3] C. J. Ash, Labelling systems and r.e. structures, Ann. Pure Appl. Logic, **47** (1990) 99–119.

[4] C. J. Ash, A construction for recursive linear orderings, J. Symbolic Logic, **56** (1991) 673–683.

[5] C. J. Ash and J. F. Knight, Pairs of recursive structures, Ann. Pure Appl. Logic, **46** (1990) 211–234.

[6] C. J. Ash, J. F. Knight, M. Manasse and T. A. Slaman, Generic copies of countable structures, Ann. Pure Appl. Logic, **42** (1989) 195–205.

[7] C. J. Ash and A. Nerode, Intrinsically recursive relations, in: Aspects of Effective Algebra, (Proc. Conf. Monash Univ., Clayton, Australia, Aug. 1–4, 1979), J. N. Crossley (ed.), (Upside Down A Book Co., Yarra Glen, Victoria, Australia, 1981), 26–41.

[8] E. J. Barker, Intrinsically Σ^0_α relations, Ann. Pure Appl. Logic, **39** (1988) 105–130.

[9] E. J. Barker, Hyperarithmetical Properties of Relations on Abelian p–Groups and Orderings, Ph.D. Thesis, School of Cognitive Science, Univ. Edinburgh, Edinburgh, Scotland, (1991).

[10] J. Chisholm, Effective model theory vs. recursive model theory, J. Symbolic Logic, **55** (1990) 1168–1191.

[11] P. A. Cholak, S. S. Goncharov, B. Khoussainov and R. A. Shore, On recursively categorical models, (in preparation).

[12] S. S. Goncharov, Autostability and computable families of constructivizations (Russian), Algebra i Logika, **14** (1975) 647–680; [translated in: Algebra and Logic, **14** (1975) 392–409].

[13] S. S. Goncharov, The quantity of nonautoequivalent constructivizations (Russian), Algebra i Logika, **16** (1977) 257–282, 377; [translated in: Algebra and Logic, **16** (1977) 169–185].

[14] S. S. Goncharov, The problem of the number of nonautoequivalent constructivizations (Russian), Dokl. Akad. Nauk SSSR, **251** (1980) 271–274; [translated in: Soviet Math. - Dokl., **21** (1980) 411–414].

[15] V. S. Harizanov, Some effects of Ash-Nerode and other decidability conditions on degree spectra, Ann. Pure Appl. Logic, **55** (1991) 51–65.

[16] M. S. Manasse, Notes on a theorem of Ash and Nerode, in: Recursion Theory, (Proc. AMS-ASL Summer Inst., Ithaca, NY, June 28 - July 16, 1982), A. Nerode and R. A. Shore, (eds.), Proc. Sympos. Pure Math., **42** (1985).

[17] T. S. Millar, Recursive categoricity and persistence, J. Symbolic Logic, **51** (1986) 430–434.

Chapter 4

Computable Classes of Constructive Models

V. P. Dobritsa

Contents

Introduction

The concept of computability for a collection of objects resulted from the development of the theory of algorithms and the theory of numerations. A general approach to the computable indexing of certain classes of objects was proposed by Mal'tsev [45]. From the mathematical point of view, computability of a collection of objects means that there exists an effective and uniform construction procedure for the objects, provided they are properly "numbered". When examining various particular cases of constructible objects, we should consider not only the objects themselves, but also collections of objects and uniform procedures for these collections. Such procedures lead to computable numerations of the objects. Various problems stem from the variable complexity of the objects under consideration. For example, in the theory of numerations [34], the computability of a system of recursively enumerable sets implies the possibility of the sets being numbered in such a way

HANDBOOK OF RECURSIVE MATHEMATICS
Edited by Yu. L. Ershov, S. S. Goncharov,
A. Nerode, and J. B. Remmel

that there exists a procedure which, given a set index, uniformly enumerates the set with that index. To make the analysis of constructive models and classes of constructs (effective numerations) consistent, it is necessary to study computable sequences and classes of constructive models. To perform this study, we must develop new methods and approaches. Along with the traditional tasks related to the numbered objects, specific new problems related to the particularities of the above mentioned constructions should be evolved as well. The concept of a computable class of constructive models was introduced by Ershov [32]. The construction of a strongly computable class of strongly constructive models was first performed by Ershov [32] in proving the existence theorem for strongly constructive models.

Theorem 0.1 *Let T be a decidable theory. Then there exists a sequence*

$$(\mathfrak{M}_0, \nu_0), \ldots, (\mathfrak{M}_n, \nu_n), \ldots$$

of strongly constructive models such that

(i) $T = \mathrm{Th}(\{\mathfrak{M}_0, \mathfrak{M}_1, \ldots\})$,

(ii) *the set* $\{\langle x, y\rangle \mid g(y) \in \mathrm{Th}_1(\mathfrak{M}_x, \nu_x)\}$ *is recursive.*

In this theorem, $T = \mathrm{Th}(\{\mathfrak{M}_0, \mathfrak{M}_1, \ldots\})$ denotes the elementary theory of the models for the main signature σ, and $\mathrm{Th}_1(\mathfrak{M}_n, \nu_n)$ is the elementary theory of the model $\mathfrak{M}_n$ for the extended signature $\sigma_1 = \sigma \cup \{c_k \mid k \in \omega\}$, the constants being interpreted as $c_k = \nu_n k$ in the constructive model $(\mathfrak{M}_n, \nu_n)$. Part (ii) of the theorem means exactly that ν–indexing, $n \longmapsto (\mathfrak{M}_n, \nu_n)$, is the strongly computable indexing of the class of strongly constructive models $\{(\mathfrak{M}_n, \nu_n) \mid n \in \omega\}$.

On the other hand, the concept of a computable class of constructive models is closely related to the problems of autostability dealt with by a number of authors. A well-known and fundamental result was obtained by Goncharov [43], and it may be formulated as follows.

Theorem 0.2 *For each $n \leq \omega$, a model exists with the property that its class of constructs is computable and contains exactly n elements, with the elements being different to within their autoequivalence.*

There is, however, a wide class of constructive models with non-computable classes of constructs. The following result by Nurtazin [49] illustrates this.

Theorem 0.3 *If a model contains non-equivalent strong constructs, then the class of all strong constructs of this model is not computable.*

Finally, it should be noted that the class of constructive models is a semantics for abstract data types. Indeed, a data bank may be considered to be a strongly constructive model. An axiomatic system defining a data bank can be used to naturally generate the computable class of constructive models, i.e., the class of possible states of the data bank, or the class of time-dependent states of the data bank, with the time being digitized as for computer processing. Interrelations between different axiomatic systems for the data bank, their equivalence, etc., are easily defined in terms of the corresponding tasks of computable indexing of the classes of constructive (even strongly constructive) models.

The above mentioned problems require theoretical investigations of the computable classes of constructive (or strongly constructive) models.

1 Non-computable, computable, and strictly computable classes of constructive models

Consider the algebraic system $\mathfrak{M} = \langle M; P_0, P_1, \dots; f_0, f_1, \dots \rangle$ for the signature $\sigma_0 = \langle P_0^{n_0}, P_1^{n_1}, \dots; f_0^{m_0}, f_1^{m_1}, \dots \rangle$. A pair $(\mathfrak{M}, \nu)$ is called a numbered model; here $\nu : \mathbb{N} \to M$ is a numeration of the domain M of the model $\mathfrak{M}$. Let $\mathfrak{M}_\nu$ denote an extension of the system $\mathfrak{M}$ with the signature $\sigma_1 = \sigma_0 \cup \{c_0, \dots, c_k, \dots\}$, with the element $\nu(k) \in \mathfrak{M}$ as the interpretation for the constant c_k.

The numbered system $(\mathfrak{M}, \nu)$ is constructive if the set $D(\mathfrak{M}_\nu)$, consisting of all atomic propositions and their negations that are true in the system $\mathfrak{M}_\nu$, is recursive. $(\mathfrak{M}, \nu)$ is strongly constructive if the elementary theory $\mathrm{Th}(\mathfrak{M}_\nu)$ of the model $\mathfrak{M}_\nu$ is decidable.

A mapping $\varphi : M_0 \to M_1$ of the domain M_0 of the constructive model $(\mathfrak{M}_0, \nu_0)$ into the domain M_1 of the constructive model $(\mathfrak{M}_1, \nu_1)$ is called a constructive homomorphism if the mapping φ is both an abstract homomorphism $\varphi : M_0 \to M_1$ of algebraic systems and a morphism of the numbered system $(\mathfrak{M}_0, \nu_0)$ into the numbered system $(\mathfrak{M}_1, \nu_1)$, i.e., there exists a general recursive function $\psi(x)$ such that $\varphi\nu_0 = \nu_1\psi$. The mapping φ is called an equivalence (or autoequivalence) if φ is both an isomorphism between the algebraic systems $\mathfrak{M}_0$, $\mathfrak{M}_1$ and a constructive homomorphism, as well as φ^{-1}

being a morphism of $(\mathfrak{M}_1, \nu_1)$ into $(\mathfrak{M}_0, \nu_0)$. In this case, the constructive models $(\mathfrak{M}_0, \nu_0)$ and $(\mathfrak{M}_1, \nu_1)$ are identified as equivalent (or autoequivalent). The respective constructive numerations ν_0 and ν_1 are also identified as autoequivalent, notated $\nu_0 \underset{a}{\equiv} \nu_1$.

If a model $\mathfrak{M}$ yields non-autoequivalent constructs (or strong constructs), then it is called non-autostable (or non-autostable with respect to the strong constructs).

A constructive homomorphism $\varphi : (\mathfrak{M}_0, \nu_0) \to (\mathfrak{M}_1, \nu_1)$ is called a constructive isomorphic injection if $\varphi : \mathfrak{M}_0 \hookrightarrow \mathfrak{M}_1$ is an isomorphic injection of $\mathfrak{M}_0$ into $\mathfrak{M}_1$.

It is quite natural to treat a collection of constructive models up to their autoequivalence. It is easy to show that in the definition of constructivity (or strong constructivity), the recursiveness condition for the set $D(\mathfrak{M}_\nu)$ (or $\mathrm{Th}(\mathfrak{M}_\nu)$) may be replaced by the equivalent condition of recursive enumerability of the set $D(\mathfrak{M}_\nu)$ (or $\mathrm{Th}(\mathfrak{M}_\nu)$). This makes it possible to determine whether a class of constructive or strongly constructive systems is computable or not.

A class $\mathbb{K}$ of constructive systems $(\mathfrak{M}, \nu)$ will be called computable (or strongly computable) if there exists a general recursive function $f(n, x)$ such that for each n the function $f(n, x)$ enumerates the set $D(\mathfrak{M}_\nu)$ (or $\mathrm{Th}(\mathfrak{M}_\nu)$), where the constructive system $(\mathfrak{M}_\nu, \nu)$ is autoequivalent to a certain system of $\mathbb{K}$, and each system from $\mathbb{K}$ contains the relevant n value. Let γ_n be a constructive numeration associated with n. In this manner, the effective numeration $\gamma : \mathbb{N} \to \mathbb{K}$ of the constructive models of $\mathbb{K}$ is represented, this numeration being called the computable indexing of the class of constructive models.

In the other case, the class $\mathbb{K}$ of constructive systems will be non-computable. $\mathbb{K}$ is called effectively infinite [41] if any computable subclass $\mathbb{K}_1 \subseteq \mathbb{K}$ may be used to effectively generate an element from the difference $\mathbb{K} \setminus \mathbb{K}_1$. It is clear that an effectively infinite class will be non-computable. These concepts, however, are not identical. This was first shown in [13].

Theorem 1.1 *There exists a non-computable class of constructive models with the property that it is not effectively infinite.*

The constructed class contains the constructs of a single abelian group $\sum_{n=1}^{\infty} \mathbb{Z}(p^n)$, but not all of those constructs. A stronger result was obtained by Ventsov [52].

Theorem 1.2 *There exists a constructible model such that the class of its constructs is infinite, yet it does not contain the computable infinite subclasses of the non-equivalent constructs.*

The class of all constructs of such a model is obviously non-computable, but is not effectively infinite.

It will be of interest to describe some of the conditions which allow an easy construction of effectively infinite classes, or the identification of non-computable classes.

Theorem 1.3 [38] *If a model $\mathfrak{A}$ is strongly constructible and yields a weak construct, then for any computable class $\mathbb{K}$ of constructs of $\mathfrak{A}$, it is possible to generate effectively a construct of $\mathfrak{A}$ whose theory is undecidable, and which is not autoequivalent to any element of $\mathbb{K}$.*

This theorem implies that the class of all constructs of a model which has both a strong and a weak construct, is effectively infinite. Another criterion for the effective infinity of the class of constructs of a model is given in [51].

Let $\mathfrak{A}_0$ be a Σ–subsystem of the algebraic system $\mathfrak{A}$, i.e., there exists a Σ_1–formula φ such that for any element a of $\mathfrak{A}$, a satisfies φ if and only if a is an element of $\mathfrak{A}_0$. If the system $\mathfrak{A}$ is constructible, then for any constructive numeration ν of $\mathfrak{A}$, the set $\nu^{-1}(\mathfrak{A}_0)$ of all numbers corresponding to the elements of the systems $\mathfrak{A}_0$ is recursively enumerable. If the model $\mathfrak{A}$ is constructible, then clearly the submodel $\mathfrak{A}_0$ will also be constructible.

The set $X = \{x_1, \ldots, x_k, \ldots\}$ will be called t–independent over subsystem $\mathfrak{A}_0$ of system $\mathfrak{A}$, if for every k, x_{k+1} is not a member of the subsystem generated by the set

$$\mathfrak{A}_0 \cup \{x_1, \ldots, x_k\}$$

in the algebraic system $\mathfrak{A}$, and for any finite submodel $\mathfrak{B}$ of the system generated by

$$\mathfrak{A}_0 \cup \{x_1, \ldots, x_k, x_{k+1}\}$$

in the system $\mathfrak{A}$, there exists a term $t(x_1, \ldots, x_k)$ with the property that the substitution of x_{k+1} by $t(x_1, \ldots, x_k)$ leads to the transformation of the submodel $\mathfrak{B}$ into an isomorphic submodel of the subsystem generated by the set

$$\mathfrak{A}_0 \cup \{x_1, \ldots, x_k\}$$

in the system $\mathfrak{A}$.

Theorem 1.4 [27] *Let the class $\mathbb{K}$ contain the constructive algebraic systems which extend the $\exists$-formulary system $\mathfrak{A}_0$. If $\mathbb{K}$ includes all the constructs of some system $\mathfrak{A}$ containing an infinite set X which is t–independent over $\mathfrak{A}_0$ and which is enumerable in the suitable construct of the system $\mathfrak{A}$, then $\mathbb{K}$ is effectively infinite.*

This theorem generalizes the result of [6].

In particular, the classes of all constructive abelian groups [7], constructive torsion-free abelian groups [47], constructive fields, constructive rings, etc., are of the kind described above.

A class of constructive systems is called hereditary if it is closed with respect to its enumerable subsystems. The least hereditary class containing a class $\mathbb{K}$ is called the hereditary closure of $\mathbb{K}$. The hereditary closure of a computable class of constructive models is also expected to be computable. A constructive system $(\mathfrak{A}, \alpha)$ from the hereditary class $\mathbb{K}$ is universal with respect to this class if any system from $\mathbb{K}$ may be constructively injected into $(\mathfrak{A}, \alpha)$. A hereditary class containing a universal constructive system will obviously be computable, because it represents the hereditary closure of this universal system.

A computable indexing γ of a class $\mathbb{K}$ of constructive algebraic systems is asserted to be reducible to the computable indexing η of the same class if there exists a general recursive function $f(x)$ such that $\gamma \equiv \eta_{f(n)}$. The reducibility is usually denoted by the common "less-than-or-equal-to" sign: $\gamma \leq \eta$. A computable indexing η of the class $\mathbb{K}$ is considered to be principal if any other computable indexing γ of the same class is reducible to η. If a class $\mathbb{K}$ of constructive algebraic systems has a computable principal indexing, then it is called strictly computable. Now it is possible to formulate a stronger outcome concerning hereditary classes than that stated above.

Theorem 1.5 *A hereditary class will be strictly computable if it contains a constructive universal system which has the property that the constructive models of the class can be effectively and uniformly injected into it.*

Corollary 1.6 [7] *For each $n \in \omega$, the class $\mathbb{A}$ of constructive abelian groups $(\mathfrak{A}, \nu)$ whose torsion-free rank $r_0(\mathfrak{A}) = n$ is strictly computable.*

If $n = 0$, then a strengthening of Khisamiev's result [44] is obtained.

Corollary 1.7 *The class of constructive abelian torsion-free groups with finite rank is strictly computable.*

Theorem 1.8 [7] *Let the class* $\mathbb{K}$ *of constructive algebraic systems have the property that, given systems* $(\mathfrak{A}, \alpha)$ *and* $(\mathfrak{B}, \beta)$ *in* $\mathbb{K}$, *there exists an algorithm to form a system* $(\mathfrak{M}_{\alpha\beta}, \mu_{\alpha\beta})$ *in* $\mathbb{K}$ *such that the system* $\mathfrak{A}$ *can be locally injected into* $\mathfrak{M}_{\alpha\beta}$, *and the system* $(\mathfrak{M}_{\alpha\beta}, \mu_{\alpha\beta})$ *can never be constructively isomorphic with respect to the system* $(\mathfrak{B}, \beta)$. *Then* $\mathbb{K}$ *is not strictly computable.*

Proof. Let γ represent the computable indexing of a class $\mathbb{K}$. Now we shall form a computable indexing ν of a certain subclass of $\mathbb{K}$ such that $\nu \leq \gamma$. Then the computable indexing $\eta = \gamma \oplus \nu$ of $\mathbb{K}$ will have the property that $\eta \leq \gamma$. The desired property is the immediate consequence.

The direct sum $\eta = \gamma \oplus \nu$ of the indexings γ and ν is determined by

$$\eta_n = \begin{cases} \gamma_n & \text{for even } n, \\ \nu_n & \text{for odd } n. \end{cases}$$

Construction of ν**:** The following algorithm will be used to construct the numeration ν_m for every $m \in \omega$. For each step in constructing ν_m it is necessary to advance a step further in computing $\varphi_m(m)$, and then proceed to enumerate the element such that the following conditions hold.

(i) We keep enumerating the elements of the system $(\mathfrak{A}_m, \gamma_m)$ in accordance with the constructive numeration γ_m until the value $\varphi_m(m)$ is computed.

(ii) As soon as $\varphi_m(m)$ is computed, we will proceed to enumerate the elements of the model

$$(\mathfrak{M}_{\gamma_m \gamma_{\varphi m(m)}}, \mu_{\gamma_m \gamma_{\varphi m(m)}}).$$

To do this it is necessary to reproduce the diagram of the elements enumerated by γ_m. Then the least $\mu_{\gamma_m \gamma_{\varphi m(m)}}$–numeral is found such that the said diagram is a sub-diagram of elements

$$\mu_{\gamma_m \gamma_{\varphi m(m)}}(0), \ldots, \mu_{\gamma_m \gamma_{\varphi m(m)}}(\ell).$$

Accordingly, the injection of the enumerated elements is accomplished, and then we proceed to enumerate the hitherto non-enumerated elements of the system $\mathfrak{M}_{\gamma_m \gamma_{\varphi m(m)}}$, the proper numeration being $\mu_{\gamma_m \gamma_{\varphi m(m)}}$.

The construction yields the numeration

$$\nu_m \underset{\alpha}{=} \begin{cases} \gamma_m & \text{if } \varphi_m(m) \text{ is not determined,} \\ \mu_{\gamma_m \gamma_{\varphi_m(m)}} & \text{if } \varphi_m(m) \text{ is determined.} \end{cases}$$

It is easy to show that none of the functions φ_m enables us to reduce ν to γ. □

Corollary 1.9 *The class of constructive torsion-free abelian groups with non-zero rank is computable, but not strictly computable.*

Proof. The computable indexing γ^n of the class of constructive torsion-free abelian groups with rank n is constructed uniformly for each n. In this case, a direct sum may be used as a computable indexing of the class of constructive torsion-free abelian groups with non-zero rank.

$$\gamma = \gamma^1 \oplus (\gamma^2 \oplus \cdots (\gamma^n \oplus (\gamma^{n+1} \oplus \cdots)) \cdots)$$

The second part of the corollary is the immediate consequence of Theorem 1.8. □

The fact given below strengthens the result of [47], and is obtained in the same manner.

Corollary 1.10 *The class of constructive torsion-free abelian groups with finite rank is computable, but not strictly computable.*

More detailed considerations lead to the following conclusion.

Theorem 1.11 *For any set $I \subseteq \omega$ such that $0 \in I$ and $I \setminus \{0\} \neq \emptyset$, the class $\mathbb{A}_I$ of constructive torsion-free abelian groups $(\mathfrak{A}, \nu)$ whose rank $r_0(\mathfrak{A})$ lies in the set I is not strictly computable.*

If this set I lies in the class Δ_2^0, then the class $\mathbb{A}_I$ is computable.

The above results may be used to easily construct computable, computable but not strictly computable, and non-computable classes of constructive Boolean algebras, distributive lattices, and fields [36].

The results obtained are given in the following table.

	Effectively infinite classes	*Non-computable classes*	*Computable but not strictly computable classes*	*Strictly computable classes*
1.	Class of constructive Boolean algebras with infinite sets of atoms	Class of all constructs [52] of Ventsov's model	Class of constructive Boolean algebras with finite sets of atoms	Class of constructive Boolean algebras
2.	Class of constructive torsion-free abelian groups	Special class of constructs [13] of the abelian group $\sum_{n=1}^{\infty} \mathbb{Z}(p^n)$	Class of constructive torsion-free abelian groups of finite Prüfer rank	Class of constructive abelian groups of fixed finite Prüfer rank
3.	Class of constructive fields (of a given characteristic)	Class $\mathbb{K}_A \setminus \{m_\omega\}$ of Theorem 2.15 for a simple set A	Class of finite dimensional transcendental extensions of a simple field	Class of constructive linearly ordered sets
4.	Class of all constructs of a model yielding strong and weak constructs		Class of finite dimensional vector spaces	Class of constructive distributive lattices

Now we will proceed to the problem of computability of classes of constructive extensions.

Let $(\mathfrak{A}, \alpha)$ and $(\mathfrak{B}, \beta)$ be two constructive algebraic systems such that $\mathfrak{A} \subseteq \mathfrak{B}$. If an identity mapping

$$\mathrm{id} : \mathfrak{A} \hookrightarrow \mathfrak{B}$$

represents a constructive isomorphic injection, then $(\mathfrak{A}, \alpha)$ is identified as a constructive subsystem of the system $(\mathfrak{B}, \beta)$, with the system $(\mathfrak{B}, \beta)$ being a constructive extension of the system $(\mathfrak{A}, \alpha)$. The constructive numeration β is called a continuation of the constructive numeration α of system $\mathfrak{A}$ into extension $\mathfrak{B}$. Two constructive systems $(\mathfrak{B}, \beta)$ and $(\mathfrak{G}, \gamma)$ constructively extending a system $(\mathfrak{A}, \alpha)$ will be equivalent over $\mathfrak{A}$ if there exists a constructive isomorphism

$$\varphi : (\mathfrak{B}, \beta) \longrightarrow (\mathfrak{G}, \gamma)$$

such that it is generally valid within the common subsystem $\mathfrak{A}$. In this case, the constructive numerations β and γ are identified as being equivalent over $(\mathfrak{A}, \alpha)$, with the following notation being used: $\beta \underset{a}{\equiv} \gamma$ (or $\beta \underset{a,\alpha}{\equiv} \gamma$).

Let $\mathbb{K}$ be the class of constructive extensions of the system $(\mathfrak{A}, \alpha)$. It will be computable if there exists a computable indexing γ of $\mathbb{K}$ and a general recursive function $f(n, x)$ such that for each fixed value of n, the function $f(n, x)$ specifies the constructive injection of the system $(\mathfrak{A}, \gamma)$ into the system $(\mathfrak{A}_n, \alpha_n)$, and for each constructive extension $(\mathfrak{B}, \beta)$ from $\mathbb{K}$ there exists a value m, for which the numerations β and γ_m are equivalent over $(\mathfrak{A}, \alpha)$. The ordered sequence $(\mathfrak{A}_0, \gamma_0)$, $(\mathfrak{A}_1, \gamma_1)$, ... will be identified as the recursively numbered class of constructive extensions of the algebra $(\mathfrak{A}, \alpha)$, and it will be denoted $F(\mathbb{K}, \alpha, \gamma)$. Thus, $\mathbb{K}$ will be the computable class of constructive extensions of the system $(\mathfrak{A}, \alpha)$ if there exists a corresponding recursively numbered class $F(\mathbb{K}, \alpha, \gamma)$.

A recursively numbered class $F(\mathbb{K}, \alpha, \gamma)$ is said to be reducible to the recursively numbered class $F(\mathbb{K}, \alpha, \delta)$ if there exists a general recursive function $g(x)$ such that $\gamma \underset{a,\alpha}{\equiv} \delta_{g(n)}$.

A class $F(\mathbb{K}, \alpha, \delta)$ is identified as principal if any other recursively numbered class $F(\mathbb{K}, \alpha, \gamma)$ is reducible to $F(\mathbb{K}, \alpha, \delta)$. If for the class $\mathbb{K}$ of constructive extensions of the system $(\mathfrak{A}, \alpha)$ there exists a principal recursively numbered class $F(\mathbb{K}, \alpha, \delta)$, then the class $\mathbb{K}$ is identified as being strongly computable.

Now we will give an example of a strictly computable class of constructive extensions. Let $(\mathfrak{P}, \nu)$ be a constructive field for which a variety of $\mathfrak{P}$–irreducible polynomials in one variable is recursive, and let $\mathbb{K}$ be the class of all separable normal extensions of the field $\mathfrak{P}$. Then there exists a principal recursively numbered class of all non-equivalent constructive extensions of the field $(\mathfrak{P}, \nu)$ in $\mathbb{K}$. To show this, we should remind the reader of two theorems concerning the continuation of a constructive numeration of a field into its extension [33].

Theorem 1.12 *To continue a construct of a field $\mathfrak{P}$ up to a construct of an algebraic extension $\mathfrak{P}'$ of the field $\mathfrak{P}$, it is necessary and sufficient that the family of all finite varieties of polynomials in denumerably many variables over the field $\mathfrak{P}$ which have roots in $\mathfrak{P}'$ should be recursively enumerable.*

If $\mathfrak{P}'$ is a separable extension, then the condition of recursive enumerability of the family of polynomials in one variable over the field $\mathfrak{P}$ having roots in $\mathfrak{P}'$ will suffice.

Theorem 1.13 *If a constructive field $(\mathfrak{P}, \nu)$ is such that the variety of $\mathfrak{P}$–irreducible polynomials in one variable over $\mathfrak{P}$ is recursive, then there exists at most one continuation of the construct ν into any normal extension of the field $\mathfrak{P}$.*

Let some single-valued Gödel-numbering η of all $\mathfrak{P}$–irreducible polynomials in one variable be fixed. By Theorem 1.12, a normal extension $\mathfrak{P}'$ from the class $\mathbb{K}$ of the field $\mathfrak{P}$ will be a constructible extension of the field $(\mathfrak{P}, \nu)$ if and only if the set of η–numbers which are irreducible over the field $\mathfrak{P}$ and have roots in the extension $\mathfrak{P}'$ is recursively enumerable. The recursively numbered class $F(\mathbb{K}, \nu, \gamma)$ associated with Post's numeration of recursively enumerable sets of η–numbers of $\mathfrak{P}$–irreducible polynomials will be principal. Any other class $F(\mathbb{K}, \nu, \delta)$ specifies a numeration μ of recursively enumerable sets of η–numbers. Since Post's numeration is principal, there exists a general recursive function $g(n)$ reducing μ to Post's numeration. Theorem 1.13 enables us to show that the same function $g(n)$ may be used to reduce the class $F(\mathbb{K}, \nu, \delta)$ to the class $F(\mathbb{K}, \nu, \gamma)$.

A constructive algebraic system $(\mathfrak{A}, \alpha)$ and the class $\mathbb{K}$ of constructive extensions of the system $(\mathfrak{A}, \alpha)$ are considered to satisfy the EF–condition if

(i) $(\mathfrak{A}, \alpha) \in \mathbb{K}$

(ii) there is an algorithm which uses the numeration β of each $(\mathfrak{B}, \beta) \in \mathbb{K}$ to construct the constructive numeration β' of $(\mathfrak{B}', \beta') \in \mathbb{K}$ such that $\beta \underset{a,\alpha}{\not\equiv} \beta'$, with the constructive injection

$$(\mathfrak{A}, \alpha) \hookrightarrow (\mathfrak{B}', \beta')$$

being fully specified by this algorithm.

Theorem 1.14 *If both the algebra* $(\mathfrak{A}, \alpha)$ *and the class* $\mathbb{K}$ *satisfy the* EF–*condition, then* $\mathbb{K}$ *is not strictly computable.*

Proof. Let $\mathbb{K}$ be a computable class of constructive extensions of the constructive system $(\mathfrak{A}, \alpha)$, then there exists a recursively numbered class $F(\mathbb{K}, \alpha, \gamma)$. Now we construct a recursively numbered class $F(\mathbb{K}, \alpha, \delta)$ such that $F(\mathbb{K}, \alpha, \delta) \not\leq F(\mathbb{K}, \alpha, \gamma)$. The desired proof is an immediate consequence of this irreducibility since the recursively numbered class $F(\mathbb{K}, \alpha, \gamma)$ was arbitrary.

Construction of $F(\mathbb{K}, \alpha, \delta)$**:**

I. Let $\delta_{2m+1} = \gamma_m$ for every $m \in \omega$.

II. A stepwise procedure will be used to construct the numeration δ_{2m} for each $m \in \omega$. γ'_m will designate the numeration defined in condition (ii) of the above description of the EF–condition for the γ_m–numbering of the system in the class $F(\mathbb{K}, \alpha, \gamma)$.

STEP $t \geq 0$: In computing $\varphi_m(2m)$, we proceed as far as step t.

(1) If the value of $\varphi_m(2m)$ has not yet been computed, then we take

$$\delta_{2m}(t) = \alpha(t).$$

(2) If this step gives the value of $\varphi_m(2m)$, then take

$$\delta_{2m}(t) = \gamma'_{\varphi_m(2m)}(0).$$

(3) If the value of $\varphi_m(2m)$ was computed before this step, then take

$$\delta_{2m}(t) = \gamma'_{\varphi_m(2m)}(\ell_t),$$

where ℓ_t is the least number such that $\gamma'_{\varphi_m(2m)}(\ell_t)$ has not yet been assigned a δ_{2m}–number.

The construction is completed.

Suppose that the recursive function φ_m performs the reduction of the recursively numbered class $F(\mathbb{K}, \alpha, \delta)$ to the class $F(\mathbb{K}, \alpha, \gamma)$. From the definition of reducibility, we can conclude that $\varphi_m(2m)$ is determined, and that $\delta_{2m} \underset{a,\alpha}{\equiv} \gamma'_{\varphi_m(2m)}$. This is inconsistent with condition (ii) in the description of the EF–condition. □

This result may be used to give an example of a computable class that is not strictly computable. Let $(\mathfrak{P}, \nu)$ be a constructive field, and $\mathbb{K}$ be the class of all purely transcendental extensions of the field $\mathfrak{P}$ having a finite degree of transcendence. It is quite obvious that there exists a recursively numbered class $F(\mathbb{K}, \nu, \gamma)$ of all non-equivalent constructive extensions of the field $(\mathfrak{P}, \nu)$ in $\mathbb{K}$, but a principal class does not exist. This last proposition is the consequence of Theorem 1.7. To verify the EF–condition for the $\mathfrak{P}(x_1, \dots, x_n)$–transcendental extension field of degree n, we should use a purely transcendental extension field $\mathfrak{P}'(x_1, \dots, x_n)$ of degree 1 to represent $\mathfrak{P}(x_1, \dots, x_n)$. Thus, the class of constructive purely transcendental extensions of a finite degree of transcendence over the given constructive field $(\mathfrak{P}, \nu)$ is computable rather than strictly computable.

It is possible to strengthen Theorem 1.4 for the class $\mathbb{K}$ of constructive extensions of the constructive system $(\mathfrak{A}, \alpha)$.

Theorem 1.15 *Let $\mathbb{K}$ be a class of constructive extensions of the constructive system $(\mathfrak{A}, \alpha)$, and let this class contain all continuations of the α–construct into some extension $\mathfrak{B}$ of the system $\mathfrak{A}$, with this extension including a set t–independent over $\mathfrak{A}$ and enumerable in some of the continuations of construct α. Then the class $\mathbb{K}$ will be effectively infinite.*

Corollary 1.16 *The class of all constructive extensions of a fixed constructive field is effectively infinite.*

Non-computability of this class was ascertained in an earlier study [46] as a consequence of the more general results of [3].

Thus the question posed in [32] has been answered.

2 Index sets of computable indexings of the classes of constructive systems

Let γ be a computable indexing of a class $\mathbb{K}$ of constructive algebraic systems, and let $\mathbb{K}_0 \subseteq \mathbb{K}$. Let the set

$$I^{\gamma}(\mathbb{K}_0) = \{n \mid (\mathfrak{M}_{\gamma_n}, \gamma_n) \in \mathbb{K}_0\}$$

be identified as the index set of the subclass $\mathbb{K}_0$ in the computable indexing γ of $\mathbb{K}$. If the subclass $\mathbb{K}_0 = \{(\mathfrak{M}, \nu)\}$ is composed of one system, then the relevant index set of the subclass $\mathbb{K}_0$ is classified as the index set of the constructive models, and it is designated as $I_\nu^\gamma = I^\gamma((\mathfrak{M}, \nu))$.

The concept of arithmetic complexity of index sets is concerned with the characteristics of computable indexings of a class of constructive algebraic systems. The most "simple" computable indexing has uniformly recursive index sets. A computable indexing having more complex index sets is not reducible to a computable indexing having less complex index sets. If a class is composed of models having simple algorithmic complexity, such as finite models, then the index sets of these models are also less complex in comparison with the maximal complexity of an index set. On the other hand, if in every computable indexing of a constructive model class the index sets have low arithmetic complexity, then both the class itself and its models have rather simple algorithmic complexity. These features call for a study of the index sets, their degree of complexity and the problems of reducibility to a computable indexing with less complex index sets. This section deals with just these issues. In the subsequent sections we will also obtain relationships between the complexity of index sets and the reducibility and the number of computable indexings, as well as between the index set complexity and the structure of a semilattice of computable indexings.

Now we shall estimate the complexity of an index set of a constructive model.

Theorem 2.1 *For each computable indexing γ of a class $\mathbb{K}$ of constructive systems and an arbitrary system $(\mathfrak{M}, \nu) \in \mathbb{K}$, the index set I_ν^γ is in the class Σ_3^0 of the arithmetical hierarchy of sets. For each set A of the class Σ_3^0 there exists $(\mathbb{K}, \gamma, \nu)$ such that $I_\nu^\gamma = A$, i.e., the given estimate of the arithmetical complexity of an index sets of a constructive system is exact.*

Proof. Let a universal recursive function be denoted by $\Phi(k, x)$. Then

$$\gamma_n \underset{a}{\equiv} \nu \leftrightarrow \exists k \Big((\forall x \exists y (\Phi(k,x) = y) \ \wedge \ (\forall y_1 \exists x_1 (\Phi(k, x_1) =_\nu y_1)) \ \wedge$$

$$\bigwedge_{P \in \sigma} (\forall r_1, \ldots, r_s (P_{\gamma_n}(r_1, \ldots, r_s) \leftrightarrow P_\nu(\Phi(k, r_1), \ldots, \Phi(k, r_s)))) \ \wedge$$

$$\bigwedge_{f \in \sigma} (\forall z_1, \ldots, z_m (\Phi(k, f_{\gamma_n}(z_1, \ldots, z_m)) =_\nu f_\nu(\Phi(k, z_1), \ldots, \Phi(k, z_m)))) \Big)$$

Here the model corresponding to the constructive numbering ν is considered to be a model in a factor set $\mathbb{N}/{=_\nu}$ where the inter-number equality predicate is:

$$\ell =_\nu m \leftrightarrow \nu(\ell) = \nu(m).$$

For each predicate P of the main signature σ, the term P_ν is chosen to denote the recursive analogue in the constructive numeration ν. Similarly, for each function f of the signature σ, the term f_ν denotes the recursive analogue in the constructive numeration ν. On the right-hand side of the equivalence under consideration, the condition is written in such a way that the function $\Phi(k, x)$ is a general recursive function and specifies the constructive isomorphism between the systems $(\mathbb{N}/{=_{\gamma_n}}, \gamma_n)$ and $(\mathbb{N}/{=_\nu}, \nu)$.

The Tarski-Kuratowski algorithm will be used to write out a prenex formula which specifies the least upper bound of the arithmetic complexity of an index set:

$$\exists k \forall x, y_1, r_1, \dots, r_s, z_1, \dots, z_m \exists y, x_1 \sim \exists\forall\exists$$

The proof of the second part of the theorem is found elsewhere [13]. In this reference, construction with priority is used to form a computable indexing γ of a class of constructive abelian groups

$$\sum_{n=1}^{\infty} \mathbb{Z}(p^n), \qquad \text{and} \qquad \sum_{n=1}^{\infty} \mathbb{Z}(p^n) + \sum^{\infty} \mathbb{Z}(p^\infty).$$

The second group in this class has only the constructive numeration ν for which the index set I_ν^γ is the same as the pre-chosen set A. □

One can succeed in proving a stronger result if the abelian group technique is used to modify the construction of Theorem 2.1 to form some additional constructions.

Theorem 2.2 [15, 17] *There exists a computable class $\mathbb{K}$ of constructive periodic abelian groups and a computable indexing γ such that all the index sets I_ν^γ of constructive systems in $\mathbb{K}$ have arithmetic complexity not less than Δ_3^0, with an infinite number of index sets lying in the class Σ_3^0 rather than belonging to the lower classes of the arithmetic hierarchy of sets.*

It is of interest to form a class of constructive models having a computable indexing such that all the index sets of constructive models be in the class Σ_3^0 but not in the lower classes of arithmetical hierarchy.

Complexity of an index set is a function of both the mode of system indexing and the properties of the constructive system class.

Theorem 2.3 *Let $(\mathfrak{M}, \nu)$ be a system in a class $\mathbb{K}$ of constructive systems, and the classes $\mathbb{K}$ and $\mathbb{K}/\{(\mathfrak{M}, \nu)\}$ be computable. Then the index set I_ν^γ will be Δ_3^0 for any computable indexing γ of $\mathbb{K}$, with this complexity estimate being exact.*

Proof. By Theorem 2.1, the index set I_ν^γ belongs to the class Σ_3^0 of the arithmetical hierarchy of sets for an arbitrary computable indexing γ. The index set $I^\gamma(\mathbb{K}_0)$ of the subclass $\mathbb{K}_0 = \mathbb{K} \setminus \{(\mathfrak{M}, \nu)\}$ will be a complement to the index set I_ν^γ, i.e.,

$$I^\gamma(\mathbb{K}_0) = \mathbb{N} \setminus I_\nu^\gamma = \overline{I_\nu^\gamma}.$$

Let η denote a computable indexing of $\mathbb{K}_0$ stipulated by the theorem.

Now we proceed to the estimation of the arithmetical complexity of the complement $\overline{I_\nu^\gamma}$:

$$n \in \overline{I_\nu^\gamma} \leftrightarrow n \in I^\gamma(\mathbb{K}_0) \leftrightarrow \exists m(\gamma_n \underset{a}{\equiv} \eta_m).$$

As in the proof of Theorem 2.1, the equivalence relation $\underset{a}{\equiv}$ specifies the complexity of $\exists\forall\exists$. Joining the $\exists m$ quantifier to the antecedent of the prenex prefix $\exists\forall\exists$ does not influence the arithmetical complexity. Thus, the sets I_ν^γ and $\overline{I_\nu^\gamma} = I^\gamma(\mathbb{K}_0)$ will be the Σ_3^0-sets, and therefore they will be Δ_3^0-sets.

As it was discovered [13], the exactness of this estimate results from the existence of a computable indexing μ for the class $\mathbb{K} = \{z, z+z, z+z+z\}$ of constructive autostable abelian groups, such that the index set I_{z+z}^μ will be a Δ_3^0-set which does not belong to the lower classes of the arithmetical hierarchy of sets. □

Theorem 2.3 and its proof lead to the result below.

Corollary 2.4 *For any computable indexing γ of a finite class of constructive systems $\mathbb{K} = \{(\mathfrak{M}_i, \nu_i) \mid i = 0, 1, \ldots, n\}$, the index sets I_ν^γ belong to the class Δ_3^0 of the arithmetical hierarchy of sets, and this estimate of the index set complexity is exact.*

It will be noted that regarding the accuracy of estimations, a stronger theorem holds.

Theorem 2.5 [17] *There exists a computable class*

$$\mathbb{K} = \{(\mathfrak{A}_{k_i+1}, \nu_i) \mid i = 0, 1, \ldots, n\}$$

of constructive periodic abelian groups

$$\mathfrak{A}_{k_i+1} = \sum_{\ell=1}^{\infty} \mathbb{Z}(p^{\ell}) + \sum_{s=1}^{k_i+1} \mathbb{Z}(p^{\infty})$$

and a computable indexing η of $\mathbb{K}$ such that all index sets $I^{\eta}_{\nu_i}$ belong to the class Δ^0_3 and do not belong to the lower classes of the arithmetic hierarchy of sets ($n \geq 1$).

The following result is a consequence of Theorem 2.3.

Corollary 2.6 *If a constructive system $(\mathfrak{M}, \nu)$ in a computable class $\mathbb{K}$ has an index set I^{γ}_{ν} for some computable indexing γ, with the complexity of this set being Σ^0_3 rather than Δ^0_3, then the class $\mathbb{K} \setminus \{(\mathfrak{M}, \nu)\}$ is not computable.*

Both this fact and the example used when proving Theorem 2.1 contribute to the proof of Theorem 1.1 by way of choosing a suitable set A in Theorem 2.1.

The following result is another example of the lower least upper bound of index set complexity which is a function of the properties of the class $\mathbb{K}$. Some preliminary definitions should be introduced here. A model $\mathfrak{M}$ will be called locally injectable into a model $\mathfrak{N}$, $\mathfrak{M} \hookrightarrow \mathfrak{N}$, if every finite submodel $\mathfrak{M}_0$ of the model $\mathfrak{M}$ can form an isomorphic injection into the model $\mathfrak{N}$, $\mathfrak{M}_0 \overset{\ell}{\hookrightarrow} \mathfrak{N}$. If neither of the models $\mathfrak{M}$, $\mathfrak{N}$ can have a local injection into the other, then those models are identified as finitely distinct ones.

Theorem 2.7 *If in a class $\mathbb{K}$ of constructive models all the models are pairwise finitely distinct, then for any computable indexing γ of $\mathbb{K}$ every index set belongs to the class Π^0_2 of the arithmetical hierarchy, with this estimate being exact.*

Proof. The signature σ of the models in $\mathbb{K}$ will be divided into ascending finite parts

$$\sigma_0 \subseteq \sigma_1 \subseteq \cdots \subseteq \sigma_i \subseteq \cdots \subseteq \sigma \qquad \text{such that} \qquad \bigcup_{s=0}^{\infty} \sigma_s = \sigma.$$

Let $D_s^{\mathfrak{M}}(x_1, \ldots, x_k)$ denote a diagram of a finite submodel of $\mathfrak{M}$ with the universal set $\{x_1, \ldots, x_n\}$ and the finite signature σ_s. Then the supremum estimate of the index set complexity is obtained from the following equivalence.

$$n \in I_\nu^\gamma \leftrightarrow \forall_s \forall \langle n_1, \ldots, n_k \rangle$$
$$\exists \langle t_1, \ldots, t_k \rangle D_s^{\mathfrak{M}_\nu}(\nu n_1, \ldots, \nu n_k) = D_s^{\mathfrak{M}_{\gamma_n}}(\gamma_n t_1, \ldots, \gamma_n t_k)$$

Here $\langle n_1, \ldots, n_k \rangle$ is the number of a suitable k–tuple of some pre-fixed Gödel numbering of all the natural number tuples. Because of the constructivity of $(\mathfrak{M}_\nu, \nu)$ and $(\mathfrak{M}_{\gamma_n}, \gamma_n)$, it is possible to effectively write out the diagrams $D_s^{\mathfrak{M}_\nu}(\nu n_1, \ldots, \nu n_k)$ and $D_s^{\mathfrak{M}_{\gamma_n}}(\gamma_n t_1, \ldots, \gamma_n t_k)$. This leads to effective verification of the equality:

$$D_s^{\mathfrak{M}_{\gamma_n}}(\gamma_n t_1, \ldots, \gamma_n t_k) = D_s^{\mathfrak{M}_\nu}(\nu n_1, \ldots, \nu n_k).$$

Prefixing $\forall\exists$ on the right-hand side of the equivalence gives an estimate of the supremum of the index set complexities for a class of finitely distinct constructive models.

An exactness proof for this estimate can be found in [20]. □

A finite model class is a particular case of a class of finitely distinct models. A constructive finite model will obviously be autostable. The signature being finite, all the finite models will be constructible. This is not so for a denumerable signature.

Let $I_n^\gamma = I^\gamma(\mathbb{K}_n)$ denote an index set of the subclass $\mathbb{K}_n$ of $\mathbb{K}$ which contains all models of cardinality n.

Theorem 2.8 *For any computable indexing γ of a class $\mathbb{K}$ of constructive finite models, the index set I_ν^γ is the difference of two recursively enumerable sets, i.e., I_ν^γ is a Σ_2^{-1}–set in the Ershov hierarchy. For any recursively enumerable sets A, B, there exists a class $\mathbb{K}$ of constructive finite models and a computable indexing α of $\mathbb{K}$ such that $I_n^\alpha = A \setminus B$.*

Proof. Indeed, a subclass of models having not less than k elements is defined by the formula

$$\exists x_1, \ldots, x_k \Phi(x_1, \ldots, x_k) \equiv \exists x_1, \ldots, x_k \bigwedge_{i \neq j} (x_i \neq x_j)$$

This leads to the equivalence

$$m \in I_n^\alpha \leftrightarrow \Big(\exists t_1, \ldots, t_n(\Phi(\gamma_m(t_1), \ldots, \gamma_m(t_n)))\Big)$$
$$\wedge \neg\Big(\exists \ell_1, \ldots, \ell_n, \ell_{n+1}(\Phi_{n+1}(\gamma_m(\ell_1), \ldots, \gamma_m(\ell_n), \gamma_m(\ell_{n+1})))\Big)$$

which is the proof of the first part of the theorem.

Now the exactness of the estimate will be shown. Let f and φ be general recursive functions enumerating the sets A and B respectively. The sign "$<$", denoting a strict linear order for the model elements, will be used as a signature symbol. Consider a model $\mathfrak{M} = \langle\{a_1, \ldots, a_n\}; <\rangle$, and suppose that $a_1 < a_2 < \cdots < a_n$. Models of the type $\mathfrak{M}_k$ with $k \in \{n, n+1, n+2\}$ will be treated as class $\mathbb{K}$.

The following algorithm is used to construct an α-indexing of $\mathbb{K}$. We begin to enumerate by α_m the elements of model $\mathfrak{M}_n$. Simultaneously, the values of $f(x)$ and $\varphi(x)$ are computed. If x is found such that $f(x) = m$, then the injection of $\mathfrak{M}_n$ into $\mathfrak{M}_{n+1}$ is carried out, and we proceed to further enumeration of the elements of $\mathfrak{M}_{n+1}$ by α_m. But if y is found such that $\varphi(y) = m$, then the injection of $\mathfrak{M}_n$ or $\mathfrak{M}_{n+1}$ into $\mathfrak{M}_{n+2}$ is carried out. The elements of this $\mathfrak{M}_{n+2}$ are enumerated by α_m to complete the construction.

Finally, it is easy to show that $I_{n+2}^\alpha = B$, $I_{n+1}^\alpha = A \setminus B$, and that $I_n^\alpha = \mathbb{N} \setminus (A \cup B)$. □

Observe that if for a given n the class $\mathbb{K}$ has a finite number of models of this cardinality, then the index set of each of these models will be a difference of two recursively enumerable sets. This condition is clearly satisfied as far as the models of finite signature are concerned. From this, we obtain the validity of the following result.

Corollary 2.9 *For any computable indexing γ of a class of finite models of finite signature, every index set I_ν^γ will be a Σ_2^{-1}-set. For any two recursively enumerable sets A, B, there exists a class of finite models of finite signature, a computable indexing γ of this class, and a model $\mathfrak{M}_{n+1}$ such that $I_{\mathfrak{M}_{n+1}}^\alpha = A \setminus B$.*

For a class $\mathbb{K}$ of constructive systems, let $L(\mathbb{K})$ denote a subclass consisting of all the systems of $\mathbb{K}$ which are locally injectable into all systems of $\mathbb{K}$.

Theorem 2.10 *Let γ be a computable indexing of a class $\mathbb{K}$ of constructive systems. Then $I^{\gamma}(L(\mathbb{K})) \in \Pi^0_2$. For an arbitrary set $A \in \Pi^0_2$, there exists a computable class $\mathbb{K}_A$ and a computable indexing γ_A of $\mathbb{K}_A$ such that $I^{\gamma_A}(L(\mathbb{K}_A)) = A$.*

Proof. The estimation of the complexity supremum is carried out in a fashion similar to that used above, using the prenex prefix of the proper formula. The example of [20] confirms the estimate's exactness. In this example, the subclass $L(\mathbb{K})$ contains a single model. □

Let $\mathbb{K}$ be a class of constructive systems, and let $(\mathfrak{M}_\nu, \nu) \in \mathbb{K}$. We shall define two subclasses which depend on $(\mathfrak{M}_\nu, \nu)$:

$$\mathbb{K}_\nu = \{(\mathfrak{M}_\mu, \mu) \mid (\mathfrak{M}_\mu, \mu) \in \mathbb{K},\ \mathfrak{M}_\mu \overset{\ell}{\hookrightarrow} \mathfrak{M}_\nu\}$$

$$\mathbb{L}_\nu = \{(\mathfrak{M}_\mu, \mu) \mid (\mathfrak{M}_\mu, \mu) \in \mathbb{K},\ \mathfrak{M}_\mu \overset{\ell}{\hookrightarrow} \mathfrak{M}_\nu,\ \mathfrak{M}_\nu \overset{\ell}{\hookrightarrow} \mathfrak{M}_\mu\}$$

Note that

$$(\mathfrak{M}_\mu \overset{\ell}{\hookrightarrow} \mathfrak{M}_\nu) \wedge (\mathfrak{M}_\nu \overset{\ell}{\hookrightarrow} \mathfrak{M}_\mu) \leftrightarrow \mathfrak{M}_\mu \equiv_1 \mathfrak{M}_\nu.$$

Hence the subclass $\mathbb{L}_\nu$, which will be called a local subclass having member $(\mathfrak{M}_\nu, \nu)$, may be defined in another way:

$$\mathbb{L}_\nu = \{(\mathfrak{M}_\mu, \mu) \mid (\mathfrak{M}_\mu, \mu) \in \mathbb{K},\ \mathfrak{M}_\mu \equiv_1 \mathfrak{M}_\nu\}.$$

The subclass $\mathbb{K}_\nu$ will be called a local cone specified by the model $(\mathfrak{M}_\nu, \nu)$.

Theorem 2.11 [28] *For computable indexings γ of classes of constructive models, the following supremum estimates of the arithmetic complexity of index sets are exact.*

$$I^{\gamma}(\mathbb{K}_\nu) \in \Pi^0_2, \qquad I^{\gamma}(\mathbb{L}_\nu) \in \Pi^0_2$$

It is important to lower the complexity of an index set of a constructive model. The exact formulation of the problem is the following: we are to clarify whether it is possible, having a computable indexing γ and a constructive numeration ν, to find or to prove the existence of a computable indexing α of the same class of constructive models such that the set I^{α}_ν will be of lower arithmetic complexity than I^{γ}_ν.

The simplest exercise in lowering the complexity of an index set leads to the proposition below.

Proposition 2.12 *If for some computable indexing γ of the class $\mathbb{K}$, the constructive model $(\mathfrak{M}, \nu)$ has index set I_ν^γ whose complexity is Π_1^0, then there exists a computable indexing η of the same class such that the index set I_ν^η is recursive.*

The class $\mathbb{L}$ of constructive systems, all locally injectable into each other, offers greater possibilities for lowering the complexity of the index set.

Theorem 2.13 [12, 20] *If, for some computable indexing γ of the class $\mathbb{L}$ of systems all locally injectable into each other, the index set I_ν^γ of the constructive model $(\mathfrak{M}, \nu)$ has complexity Π_2^0, then there exists a computable indexing α of $\mathbb{L}$ such that the index set I_ν^α is recursive.*

The proof of this is quite similar to the proof of Theorem 1 in [12]. This is evidence of the possibility of strengthening Proposition 2.12 in the case that additional conditions are imposed on the model $(\mathfrak{M}, \nu)$.

Theorem 2.14 *Let a computable class $\mathbb{K}$ have a system $(\mathfrak{M}, \nu)$ with the property that for every submodel of $\mathfrak{M}$, the isomorphic injection into some system of the class $\mathbb{K} \setminus \{(\mathfrak{M}, \nu)\}$ can be carried out, and suppose that a computable indexing γ of $\mathbb{K}$ exists such that $I_\nu^\gamma \in \Pi_1^0$. Then for an arbitrary $A \in \Pi_1^0$ having infinite complement, there exists a computable indexing α of $\mathbb{K}$ such that $I_\nu^\alpha = A$.*

However, in general the lowering of the index set complexity is not attainable.

Theorem 2.15 *There exists a computable class $\mathbb{K}$, a computable indexing γ of $\mathbb{K}$, and a constructive model $(\mathfrak{M}, \nu) \in \mathbb{K}$, such that $I_\nu^\gamma \in \Sigma_2^0$, and for none of the computable indexings of this class can the index set of $(\mathfrak{M}, \nu)$ be recursive.*

Proof. Let us consider a signature consisting of a single one-place function symbol $f(x)$. For each $n \in \mathbb{N}$ we shall define a finite model $\mathfrak{M}_n$ with domain $|\mathfrak{M}_n| = \{a_1, a_2, \ldots, a_{n+3}\}$. Within the model $\mathfrak{M}_n$, the function $f(x)$ is specified by equations $f(a_i) = a_{i+1}$ for $i < n+3$, and $f(a_{n+3}) = a_1$. We also define a model $\mathfrak{M}_\omega = \bigcup_{i=0}^{\infty} \mathfrak{M}_i$ as a free join of all finite models $\mathfrak{M}_i$.

On the basis of an arbitrary recursively enumerable set A, we define a class $\mathbb{K}_A = \{\mathfrak{M}_i \mid i \in A\} \cup \{\mathfrak{M}_\omega\}$. All the models $\mathfrak{M}_i$ $(i \in \omega)$, $\mathfrak{M}_\omega$ are obviously constructible (even strongly constructible) and autostable. The complement $\overline{A} = \mathbb{N} \setminus A$ to the set A may be considered non-empty. Let φ_A denote a general recursive function which enumerates the set A.

Construction of indexing γ. By γ_m we begin to enumerate the model $\mathfrak{M}_m$ and simultaneously compute the values of φ_A. If a value a of x has been found such that $\varphi(a) = m$, $m \in A$, then the isomorphic injection of $\mathfrak{M}_m$ into $\mathfrak{M}_\omega$ is performed, and we further proceed to enumerate the elements of $\mathfrak{M}_\omega$ by γ_m–numerals. If there is no x with the property that $\varphi(x) = m$, then γ_m will be a constructive numeration of the model $\mathfrak{M}_m$. The construction is completed.

Notice that if in a computable indexing α of $\mathbb{K}_A$, the index set $I^\alpha_{\mathfrak{M}_\omega}$ is recursive, then the set A will be recursive too. Indeed, for any construct α_m, with the index being $m \in \mathbb{N} \setminus I^\alpha_{\mathfrak{M}_\omega}$, it is easy to write out a number n_m which is 3 less than the cycle of the function $f(x)$ in a suitable model. This is a way to define an algorithm for the enumeration of the set $\overline{A} = \mathbb{N} \setminus I^\alpha_{\mathfrak{M}_\omega}$. Since the set A is recursively enumerable by definition, it will be recursive.

Thus, if when constructing γ, the set A is defined as recursively enumerable but not recursive, then the index set $I^\alpha_{\mathfrak{M}_\omega}$ can never be recursive for any computable indexing. □

A suitable choice of the set A in defining $\mathbb{K}_A$ may lead to the non-computability of the class $\mathbb{K}_A \setminus \{(\mathfrak{M}_\omega, \nu)\}$, to the exclusion of any infinite computable subclass in this class, etc.. If the complexity of an index set is still high, then it is impossible to lower its complexity.

Theorem 2.16 [20] *There exists a computable class $\mathbb{K}$ of constructive periodic abelian groups, a computable indexing γ of $\mathbb{K}$, and an abelian group $(\mathfrak{A}, \alpha) \in \mathbb{K}$, such that $I^\eta_\alpha \in \Sigma^0_3$ and the index set I^η_α is in the class $\Sigma^0_3 \setminus \Pi^0_2$ for any computable indexing η of $\mathbb{K}$.*

Proof. The appropriate objects from the proof of the second part of Theorem 2.1 will be chosen as $\mathbb{K}$ and γ. Let

$$\mathfrak{A} = \sum_{n=1}^{\infty} \mathbb{Z}(p^n) + \sum^{\infty} \mathbb{Z}(p^\infty)$$

with a suitable constructive numeration α. The abelian groups $\sum_{n=1}^{\infty} \mathbb{Z}(p^n)$ and $\sum_{n=1}^{\infty} \mathbb{Z}(p^n) + \sum^{\infty} \mathbb{Z}(p^\infty)$ will obviously be locally injectable into each other. The proper choice of the set A in Theorem 2.1 leads to $I^\eta_\alpha \in \Sigma^0_3 \setminus \Delta^0_3$ by construction. If the index set I^η_α were in the class Π^0_2 for some computable indexing η of $\mathbb{K}$, then Theorem 2.13 would specify a computable indexing μ

of the class $\mathbb{K}$ of locally mutually injectable systems, and for such an indexing the index set I^{μ}_{α} would be recursive. Thus, the index set $I^{\mu}(\mathbb{K}\setminus\{(\mathfrak{A},\alpha)\}) = I^{\mu}_{\alpha}$ would also be recursive. For the reasons given above, the class $\mathbb{K}\setminus\{(\mathfrak{A},\alpha)\}$ would be computable. By Theorem 2.3, this leads to the condition that $I^{\eta}_{\gamma} \in \Delta^0_3$, which is inconsistent with the choice of the set A. Therefore, there is no computable indexing η for which $I^{\eta}_{\alpha} \in \Pi^0_2$. □

It is important to find some other criteria to decide whether it is possible to lower the complexity of index sets.

3 Reducibility of computable indexing systems

The concept of reducibility of computable indexings was introduced in §2. In this section we shall consider different types of reducibility of computable numeral systems (or computable indexings) and their properties, as well as the possible number of different computable numeral systems.

Let α and β be two computable indexing systems of a class $\mathbb{K}$ of constructive algebraic systems. Indexing α is reducible to indexing β if there exists a general recursive function $f(x)$ such that for any constructive system $(\mathfrak{M},\nu)$, the inclusion $f(I^{\alpha}_{\nu}) \subseteq I^{\beta}_{\nu}$ holds. (The notation is $\alpha \leq \beta$.) Indexing α is reducible to indexing β in terms of local classes if there exists a general recursive function $\varphi(x)$ such that for any constructive system $(\mathfrak{M},\nu) \in \mathbb{K}$, the following is valid: $\varphi(I^{\alpha}(\mathbb{L}_{\nu})) \subseteq I^{\beta}(\mathbb{L}_{\nu})$, where $\mathbb{L}_{\nu}$ is the local class defined by the system $(\mathfrak{M},\nu)$. We shall denote reducibility in terms of the local classes by $\alpha \underset{\ell}{\leq} \beta$.

Reducibility $\alpha \leq \beta$ of indexing systems clearly implies the reducibility in terms of the local classes $\alpha \underset{\ell}{\leq} \beta$:

$$\alpha \leq \beta \;\rightarrow\; \alpha \underset{\ell}{\leq} \beta,$$

with the mapping function being the same. In general, the converse does not hold. One instance in which $\alpha \underset{\ell}{\leq} \beta$ but $\alpha \not\leq \beta$, is the example used to prove the second part of Theorem 2.1, wherein the constructive system class is a local class, and consequently any indexing systems of this class are mutually reducible in terms of local classes, and in addition, any general recursive function may be used as a mapping function. However, as will be shown below, such a class has infinitely many computable indexing systems which are non-computable in the sense of their reducibility.

There is a limit function to map α into β if there exists a Δ_2^0-function $f(x)$ such that $\alpha_n \underset{a}{\equiv} \beta_{f(n)}$ for each $n \in \omega$. In this definition, the existence condition for a Δ_2^0-function may be replaced by the existence condition for a general recursive function which has a limit, $\lim_{n\to\infty} \varphi(n,x)$, for each fixed value of x. Namely, there is a limit function for α to be reducible to β if there exists a general recursive function $\varphi(n,x)$ with the properties that $\lim_{n\to\infty} \varphi(n,x)$ exists, and

$$\alpha_x \underset{a}{\equiv} \beta_{\lim\limits_{n\to\infty} \varphi(n,x)}$$

for each $x \in \omega$. This kind of reducibility will be denoted $\alpha \underset{\lim}{\leq} \beta$, and the function $\varphi(n,x)$ will be called a limit mapping function.

If there exists a general recursive function $f(x)$ which bounds from above the number of value jumps in the sequence

$$\varphi(0,x)\,,\ \varphi(1,x)\,,\ \ldots\,,\ \varphi(n,x)\,,\ \ldots\,,$$

then the reducibility will be of limit bounded type (by means of $f(x)$). The function $f(x)$ will be identified as a bounding one for a limit mapping function (or a majorizing one for a mapping function), or simply a majorant of the function $\varphi(n,x)$. It will be noticed that if $f(x)$ is a majorant of a limit mapping function $\varphi(n,x)$ and $f(x) \leq h(x)$ holds for each $x \in \omega$, then $h(x)$ will also be a majorant of $\varphi(n,x)$. Bounded limit reducibility of α to β, with a majorant being $f(x)$, will be denoted $\alpha \underset{\lim, f(x)}{\leq} \beta$.

It is clear that the notion of common reducibility of α to β implies that the bounded limit reducibility of α to β can be carried out, with a majorant being $\mathbf{0}(x) \equiv 0$. Generally, the converse is not true.

Theorem 3.1 [25] *There exists a computable class $\mathbb{K}$ of constructive models and computable indexing systems α, β and γ of $\mathbb{K}$ such that $\alpha \leq \beta \leq \gamma$,*

$$\gamma \underset{\mathbf{lim,1}}{\leq} \beta, \qquad \beta \underset{\mathbf{lim,1}}{\leq} \alpha, \qquad \text{and} \qquad \gamma \underset{\mathbf{lim,1}}{\not\leq} \alpha,$$

with $\mathbb{K}$ having infinitely many different computable indexing systems and the reducibility $\kappa \underset{\mathbf{lim,2}}{\leq} \eta$ being valid for any two κ and η.

In this theorem, **1** and **2** are used to denote the general recursive identical functions $\mathbf{1}(x) \equiv 1$ and $\mathbf{2}(x) \equiv 2$.

By a similar method, it is possible to show that for each $n \geq 0$ there exists a computable class $\mathbb{K}^n$ of constructive models with the properties that the collection of different computable indexing systems of this class will be infinite, reducibility $\kappa \underset{\lim,n}{\leq} \eta$ will take place for any two κ and η, and there will exist an $(n+1)$–size chain of computable indexings of this class

$$\alpha^0 < \alpha^1 < \cdots < \alpha^n$$

for which $\alpha^{i+1} \underset{\lim,\mathbf{1}}{\leq} \alpha^i$ for each $i \leq n$, but $\alpha^{i+k+1} \underset{\lim,\mathbb{k}}{\not\leq} \alpha^i$. Here $\mathbb{k}(x) \equiv k$.

Finally, there is a bounded limit function to reduce α to β, i.e., $\alpha \underset{r,\lim}{\leq} \beta$, if there exists a general recursive function $f(x)$ such that $\alpha \underset{\lim,f(x)}{\leq} \beta$.

Taking into account all the above, and the given definitions, the satisfiability of the following implications is obvious:

$$\alpha \leq \beta \rightarrow \alpha \underset{\lim,f(x)}{\leq} \beta \rightarrow \alpha \underset{r,\lim}{\leq} \beta \rightarrow \alpha \underset{\lim}{\leq} \beta.$$

For each reducibility type ($\leq$, $\underset{\ell}{\leq}$, $\underset{\lim}{\leq}$, $\underset{r,\lim}{\leq}$ and $\underset{\lim,f(x)}{\leq}$) of computable indexing of the constructive model class, we shall define binary relations:

$$\alpha \approx \beta \leftrightarrow (\alpha \leq \beta) \wedge (\beta \leq \alpha)$$

$$\alpha \underset{\ell}{\approx} \beta \leftrightarrow (\alpha \underset{\ell}{\leq} \beta) \wedge (\beta \underset{\ell}{\leq} \alpha)$$

$$\alpha \underset{\lim}{\approx} \beta \leftrightarrow (\alpha \underset{\lim}{\leq} \beta) \wedge (\beta \underset{\lim}{\leq} \alpha)$$

$$\alpha \underset{r,\lim}{\approx} \beta \leftrightarrow (\alpha \underset{r,\lim}{\leq} \beta) \wedge (\beta \underset{r,\lim}{\leq} \alpha)$$

$$\alpha \underset{\lim,f(x)}{\approx} \beta \leftrightarrow (\alpha \underset{\lim,f(x)}{\leq} \beta) \wedge (\beta \underset{\lim,f(x)}{\leq} \alpha)$$

Theorem 3.2

(i) *The binary relations* $\approx$, $\underset{\ell}{\approx}$, $\underset{\lim}{\approx}$, $\underset{r,\lim}{\approx}$ *are equivalence relations.*

(ii) *The relation* $\underset{\lim,f(x)}{\approx}$ *is not an equivalence relation.*

Proof. Reflexivity and symmetry of all the relations under consideration are obvious from their definitions. We shall verify the transitivity of $\approx$, noting that the verification for $\underset{\ell}{\approx}$ is carried out similarly.

Let $\alpha \approx \beta$ and $\beta \approx \gamma$, with $\varphi_1(x)$ being used to reduce α to β, $\varphi_2(x)$ for β to α, $\psi_1(x)$ for β to γ, and $\psi_2(x)$ for γ to β. Then we have

$$\varphi_1(I_\nu^\alpha) \subseteq I_\nu^\beta, \quad \psi_1(I_\nu^\beta) \subseteq I_\nu^\gamma, \quad \varphi_2(I_\nu^\beta) \subseteq I_\nu^\alpha, \quad \text{and} \quad \psi_2(I_\nu^\gamma) \subseteq I_\nu^\beta.$$

Whence

$$\varphi_1{\circ}\psi_1(I_\nu^\alpha) \subseteq I_\nu^\gamma \qquad \text{and} \qquad \psi_2{\circ}\varphi_2(I_\nu^\gamma) \subseteq I_\nu^\alpha$$

for each constructive system $(\mathfrak{M}, \nu)$ in the constructive algebraic system class under consideration. Thus, the general recursive function $\varphi_1{\circ}\psi_1$ specifies the reducibility $\alpha \leq \gamma$, and the function $\psi_2{\circ}\varphi_2$ specifies the reducibility $\gamma \leq \alpha$, and hence $\alpha \approx \gamma$.

When proving the transitivity of the relation $\underset{\ell}{\approx}$, the index sets of type I_ν^α should be replaced by index sets of type $I^\alpha(\mathbb{L}_\nu)$.

Transitivity verification for the relations $\underset{\lim}{\approx}$ and $\underset{r,\lim}{\approx}$ is accomplished similarly to the above, but a proper general recursive majorant should be found for the second relation. Hence we will be content to verify the transitivity of the relation $\underset{r,\lim}{\approx}$.

Let $\alpha \underset{r,\lim}{\approx} \beta$ and $\beta \underset{r,\lim}{\approx} \gamma$, with

$$\alpha \underset{\lim,f_1}{\approx} \beta, \qquad \beta \underset{\lim,f_2}{\approx} \gamma, \qquad \gamma \underset{\lim,f_3}{\approx} \beta, \qquad \text{and} \qquad \beta \underset{\lim,f_4}{\approx} \alpha,$$

where f_1, f_2, f_3 and f_4 are the proper general recursive majorants. The notations $\psi_1(x,y)$, $\psi_2(x,y)$, $\psi_3(x,y)$ and $\psi_4(x,y)$ are used for the proper general recursive limit mapping functions. The algorithm $\varphi_1(n,x) = \psi_2(n, \psi_1(n,x))$ will be used to specify the function $\varphi_1(n,x)$ describing the limit reducibility of α to β. This function is obviously general recursive. We also have

$$\lim_{n\to\infty} \varphi_1(n,x) = \lim_{n\to\infty} \psi_2(n, \lim_{n\to\infty} \psi_1(n,x)).$$

Whence

$$\alpha_x \underset{a}{\equiv} \beta_{\lim\limits_{n\to\infty} \psi_1(n,x)} = \beta_{n_x} \underset{a}{\equiv} \gamma_{\lim\limits_{m\to\infty} \psi_2(m,n_x)} \underset{a}{\equiv} \gamma_{\lim\limits_{m\to\infty} \psi_2(m, \lim\limits_{n\to\infty} \psi_1(n,x))}$$

$$= \gamma_{\lim\limits_{n\to\infty} \psi_2(n, \lim\limits_{n\to\infty} \psi_1(n,x))} = \gamma_{\lim\limits_{n\to\infty} \varphi_1(n,x)}.$$

It is obvious that the function

$$h_1(x) = \sum_{i=0}^{f_1(x)} (f_2(i) + 1)$$

is a general recursive majorant of the function $\varphi_1(n, x)$. Thus, we have $\alpha \underset{\lim,h_1}{\approx} \gamma$, and hence $\alpha \underset{r,\overline{\lim}}{\leq} \gamma$.

A similar procedure is used to show that $\varphi_2(n, x) = \psi_4(n, \psi_3(n, x))$ is a general recursive function which accomplishes the limit reducibility $\gamma \underset{\lim,h_2}{\approx} \alpha$. Here,

$$h_2(x) = \sum_{i=0}^{f_3(x)} (f_4(i) + 1)$$

is a general recursive majorant of $\varphi_2(n, x)$. Therefore $\gamma \underset{r,\overline{\lim}}{\leq} \alpha$.

Thus it is established that

$$\alpha \underset{r,\lim}{\leq} \gamma \qquad \text{and} \qquad \gamma \underset{r,\lim}{\leq} \alpha,$$

and hence $\alpha \underset{r,\lim}{\approx} \gamma$.

The second part of Theorem 3.2 is a consequence of Theorem 3.1, since in the latter it was shown that the relation $\underset{\lim,1}{\approx}$ does not satisfy the transitivity condition. That is why the binary relation $\underset{\lim,f(x)}{\approx}$ will not in general be an equivalence relation. □

Two general recursive functions will be said to satisfy $f(x) \underline{\propto} h(x)$ if the set

$$\{x \mid h(x) < f(x)\}$$

is finite. Let a general recursive function $f(x)$ be a majorant of some general recursive function $\varphi(n, x)$, and let $f(x) \underline{\propto} h(x)$. Then $h(x)$ will obviously be a majorant of the general recursive function $\varphi_1(n, x)$ derived from $\varphi(n, x)$ by redefining values within the finite set of arguments.

The definition itself implies that the majorant is a non-negative function. If a function $f(x)$ is available to majorize the general recursive function $\varphi(n, x)$, then for each $x \in \omega$ there exists $\lim \varphi_1(n, x)$.

The notations given below will be used for some classes of functions defined in terms of computable indexing numerals α and β, provided that $\alpha \underset{\lim}{\leq} \beta$.

$$\Pi(\alpha,\beta) = \{\varphi(n,x) \mid \varphi(n,x) \text{ is a general recursive function to accomplish the limit reducibility of } \alpha \text{ to } \beta\}$$

$$\text{CI}(\alpha,\beta) = \{\varphi(n,x) \mid \varphi(n,x) \in \Pi(\alpha,\beta) \text{ and a general recursive majorant is available for } \varphi(n,x)\}$$

$$M(\alpha,\beta) = \{f(x) \mid f(x) \text{ is a majorant of some function } \varphi(n,x) \in \text{CI}(\alpha,\beta)\}$$

It should be noted that the condition $\alpha \underset{\lim}{\leq} \beta$ may be omitted, since for $\alpha \underset{\lim}{\not\leq} \beta$ it is quite natural that $\Pi(\alpha,\beta) = \emptyset$.

The relation $\underline{\propto}$ used above yields a partial ordering on the set $M(\alpha,\beta)$. Furthermore, if $f(x) \in M(\alpha,\beta)$ and $f(x) \underline{\propto} h(x)$, then $h(x) \in M(\alpha,\beta)$.

If $f(x) \underline{\propto} h(x)$ and $h(x) \underline{\propto} f(x)$ with $f(x), h(x) \in M(\alpha,\beta)$, then these majorants are considered to be equivalent, $f(x) \dot{\approx} h(x)$. This last relation is clearly an equivalence relation within $M(\alpha,\beta)$.

Theorem 3.3

(i) *Any finite subset in $M(\alpha,\beta)$ has an infimum and a supremum.*

(ii) *If $\mathbf{0} \notin M(\alpha,\beta)$, then there are no minimal elements in $M(\alpha,\beta)$.*

(iii) *If $\mathbf{0} \in M(\alpha,\beta)$, then the identity function $\mathbf{0}(x) \equiv 0$ is the least element of the set $M(\alpha,\beta)$.*

Proof. Part (iii) is obvious because of the non-negativity of majorant functions.

To prove part (i), it will be sufficient to consider a two-element subset. Let $f_1(x), f_2(x) \in M(\alpha,\beta)$. If $f_1(x) \dot{\approx} f_2(x)$, then the supremum and the infimum coincide with these functions.

Suppose $f_1(x) \dot{\not\approx} f_2(x)$. Let $\varphi_1(n,x)$ and $\varphi_2(n,x)$ denote limit mapping functions in $\text{CI}(\alpha,\beta)$ corresponding to the majorants $f_1(x)$ and $f_2(x)$. The two functions $\psi_1(n,x)$ and $\psi_2(n,x)$ will be specified as follows.

$$\psi_1(n,x) = \begin{cases} \varphi_1(n,x) & \text{if } f_1(x) \leq f_2(x) \\ \varphi_2(n,x) & \text{if } f_1(x) > f_2(x) \end{cases}$$

$$\psi_2(n,x) = \begin{cases} \varphi_1(n,x) & \text{if } f_1(x) \geq f_2(x) \\ \varphi_2(n,x) & \text{if } f_1(x) < f_2(x) \end{cases}$$

Each of these is a general recursive limit mapping function of $\mathfrak{M}(\alpha,\beta)$, since

$$\alpha_x \underset{a}{\equiv} \beta_{\lim_{n\to\infty} \varphi_1(n,x)} \underset{a}{\equiv} \beta_{\lim_{n\to\infty} \varphi_2(n,x)} \underset{a}{\equiv} \beta_{\lim_{n\to\infty} \psi_1(n,x)} \underset{a}{\equiv} \beta_{\lim_{n\to\infty} \psi_2(n,x)}.$$

The majorant $h_1(x)$ of $\varphi_1(n,x)$ is defined as follows.

$$h_1(x) = \begin{cases} f_1(x) & \text{if } f_1(x) \leq f_2(x) \\ f_2(x) & \text{if } f_1(x) > f_2(x) \end{cases}$$

Recursiveness of $h_1(x)$ is obvious.

We will now show $h_1(x)$ to be the infimum of the set $\{f_1(x), f_2(x)\}$. Let $g(x) \underline{\propto} f_1(x)$ and $g(x) \underline{\propto} f_2(x)$, with $g(x) \in M(\alpha,\beta)$. The sets

$$\{x \mid g(x) > f_1(x)\} \qquad \text{and} \qquad \{x \mid g(x) > f_2(x)\}$$

are finite by the definition of $\underline{\propto}$. Therefore their union

$$\{x \mid g(x) > f_1(x)\} \cup \{x \mid g(x) > f_2(x)\}$$
$$= \{x \mid (g(x) > f_1(x)) \vee (g(x) > f_2(x))\}$$

is finite. The equivalence $(g(x) > f_1(x)) \vee (g(x) > f_2(x)) \leftrightarrow (g(x) > h_1(x))$ is obviously valid. Therefore the set

$$\{x \mid g(x) > h_1(x)\} = \{x \mid g(x) > f_1(x)\} \cup \{x \mid g(x) > f_2(x)\}$$

will be finite. Consequently $g(x) \underline{\propto} h_1(x)$, i.e., $h_1(x)$ is the infimum of $f_1(x)$ and $f_2(x)$.

The procedure to show that $h_2(x) \in M(\alpha,\beta)$ and that $h_2(x)$ is the supremum of $\{f_1(x), f_2(x)\}$ is similar to the above procedure for $h_1(x)$.

The proof of part (ii) can be found in [24]. □

The majorant $f(x)$ will be called attainable if there exists $x \in \omega$ such that in the sequence $\varphi(0,x)\,,\ \ldots\ ,\ \varphi(n,x)\,,\ \ldots$ there are exactly $f(x)$ jumps of value.

Proposition 3.4 *There exists an attainable majorant for each $\varphi(n,x)$ in the set $\mathrm{CII}(\alpha,\beta)$.*

Proof. Indeed, thee exists some majorant $f(x)$ for the limit mapping function $\varphi(n,x) \in \mathrm{CII}(\alpha,\beta)$. If it is not attainable, then in the sequence

$$\varphi(0,x)\,,\ \ldots\ ,\ \varphi(n,x)\,,\ \ldots$$

the number of value jumps is strictly less that $f_0(x) = f(x)$ for each $x \in \omega$. Therefore $f_1(x) = f_0(x) - 1$ will also be a majorant. For each fixed $x_0 \in \omega$, the number of value jumps in the sequence $\varphi(0,x_0)\,,\ \ldots\ ,\ \varphi(n,x_0)\,,\ \ldots$ belongs to the set $\{0\,,\ \ldots\ ,\ f(x_0)\}$. That is why the described $f_0(x)$–to–$f_1(x)$–transition procedure is not more that $f(x_0)$–folded. Thus, for the proper $k \leq f(x_0)$, the function $f_k(x) = f(x) - k$ will be an attainable majorant for the limit mapping function $\varphi(n,x)$. □

If "complexity" of a limit mapping function is conceived of as the number of possible value jumps of this function, then there is a chance to lower the "complexity" of a limit mapping function when performing the transition to some equivalent computable indexing of the initial class of constructive models.

Theorem 3.5 *Let α and β be computable indexings of a class $\mathbb{K}$ of constructive algebraic systems, and $g(x)$ be a general recursive function such that $\sup\{g(x) \mid x \in \omega\} = \infty$. If $\alpha \underset{\lim, f(x)}{\leq} \beta$, with $f(x)$ being some general recursive majorant, then there exists a computable indexing γ of $\mathbb{K}$ such that $\gamma \equiv \alpha$ and $\gamma \underset{\lim, g(x)}{\leq} \beta$.*

Proof. The procedure to construct γ will be stepwise. The values of the subsidiary general recursive function $\psi(t)$ will be determined simultaneously.

STEP 0: Let $\gamma_0 = \alpha_0$, $\psi(0) = 0$. Proceed to the next step.

Suppose that, the step t having been completed, the value of $\psi(t)$ is obtained, and γ–images of constructive numerations $\alpha_0\,,\ \alpha_1\,,\ \ldots\ ,\ \alpha_{\psi(t)}$ are known.

STEP $t+1$: If $g(t+1) < f(\psi(t)+1)$, then

$$\gamma_{t+1} = \alpha_0 \qquad \text{and} \qquad \psi(t+1) = \psi(t).$$

However, if $g(t+1) \geq f(\psi(t)+1)$, then

$$\gamma_{t+1} = \alpha_{\psi(t)+1} \qquad \text{and} \qquad \psi(t+1) = \psi(t)+1.$$

Proceed to the next step of the construction procedure.

The construction is completed.

The construction implies that $\gamma \leq \alpha$. On the other hand, the number sequence in γ, consistent with the constructs $\alpha_0, \alpha_1, \ldots,$ will form a recursive set. This is a consequence of enumerability of the set elements in strictly ascending order. For each i, the numeration γ_{n_i}, consistent with and autoequivalent to α, may be found effectively. Thus, the reducibility $\alpha \leq \gamma$ is proven. Taking into account $g(n_i) \geq f(i)$ for $i \in \omega$, it is obvious from the construction that $\gamma \underset{\lim, g(x)}{\leq} \beta$. □

Corollary 3.6 *Let α and β be computable indexings of a class of constructive models such that $\alpha \underset{\lim, f(x)}{\leq} \beta$ for some general recursive majorant $f(x)$. Then there exists a computable indexing γ of this class such that $\alpha \equiv \gamma$, $\gamma \underset{\lim, (x-k)}{\leq} \beta$ for each fixed natural k.*

Proof. It is sufficient, in the proof of Theorem 3.5, to take $g(x) = [x - \ln x]$ as a new majorant, and to "correct" the proper limit mapping function within the finite set which is dependent of k. □

On the other hand, the results just below demonstrate the existence of computable indexings having highly "sophisticated" limit mapping functions.

Theorem 3.7 [24] *Let α be a computable indexing of a class $\mathbb{K}$ of constructive algebraic systems such that $I_\nu^\alpha \subseteq A \subseteq I^\alpha(\mathbb{K}_\nu)$, where $(\mathfrak{M}, \nu) \in \mathbb{K}$, $|\mathbb{L}_\nu| \geq 2$ and A is a Σ_2^0-set. Then for any general recursive $f(x)$, there exist computable indexings β and γ of $\mathbb{K}$ with the properties that $\gamma \underset{\lim, f(x)}{\not\leq} \beta$ and $\mathfrak{A}(\alpha, \beta) \neq \emptyset$.*

Modifying the proof of this theorem shows that in $\mathbb{K}$ it is possible to find a computably enumerable family of computable indexings which are pairwise incomparable with respect to the bounded reducibility $\underset{\lim, f(x)}{\leq}$.

Theorem 3.8 [24] *Let α be a computable indexing of a class $\mathbb{K}$ of constructive algebraic systems with the property that there exists a Δ_2^0-set A such that $I_\nu^\alpha \subseteq A$ and $I_\mu^\alpha \subseteq \overline{A}$, where $(\mathfrak{M}_\nu, \nu) \in \mathbb{K}$, $(\mathfrak{M}_\mu, \mu) \in \mathbb{K}$, $\mathfrak{M}_\mu \equiv_1 \mathfrak{M}_\nu$ and $\nu \underset{a}{\not\equiv} \mu$. Then for any general recursive $f(x)$, there exist computable indexings β and γ of $\mathbb{K}$ with the properties that $\gamma \underset{\lim, f(x)}{\not\leq} \beta$ and $\mathfrak{A}(\alpha, \beta) \neq \emptyset$.*

An analysis of computable indexings for a class of constructive systems show that when constructing a suitable equivalent computable indexing to a given one, it is possible to "raise the complexity" of a limit mapping function under certain conditions.

Theorem 3.9 *Let α and β be computable indexings of a class $\mathbb{K}$ of constructive algebraic systems, and let $g(x)$ be a general recursive function for which*

$$\sup\{g(x) \mid x \in \omega\} = \infty \qquad \textit{and} \qquad \alpha \underset{\lim, g(x)}{\not\leq} \beta.$$

Then for any general recursive $f(x)$ with the limit inferior $\underline{\lim}\, f(x) = k$ such that, if $\underline{\lim}\, g(x)$ is found, then $k \leq \underline{\lim}\, g(x)$ and there exists a computable indexing γ of $\mathbb{K}$ with the properties that $\gamma \equiv \alpha$ and $\gamma \underset{\lim, f(x)}{\not\leq} \beta$.

Proof. We shall determine the constructs γ_t (i.e., effective enumeration systems) of the indexing γ in a stepwise procedure.

STEP $t \geq 0$: Compute $f(t)$.

(i) If $f(t) \leq k$, then find the first numeral n_t such that there is no $i < t$ such that $\alpha_{n_t} = \gamma_i$. Set $\gamma_t = \alpha_{n_t}$. Proceed to the next step.

(ii) If $f(t) > k$, then find the least n_t such that $g(n_t) < f(t)$ and there is no $i < t$ such that $\alpha_{n_t} = \gamma_i$. Set $\gamma_t = \alpha_{n_t}$. Proceed to the following step.

The construction is completed.

Since $\underline{\lim}\, f(x) = k$, the set $\{x \mid f(x) < k\}$ is finite. The construction procedure does not appreciably depend on it. On the other hand, the set $\{x \mid f(x) = k\}$ is infinite. That is why part (i) of the procedure will be repeated infinitely. Then all the α_n-numerations will be included in the construction of γ as the numerations $\gamma_i = \alpha_n$, provided i is chosen properly. Therefore γ is a computable indexing of $\mathbb{K}$. The equivalence $\gamma \equiv \alpha$ is easily established.

Note that for all $t \in \omega$, the inequality $f(t) \leq g(n_t)$ holds. Here n_t denotes a number which is obtained in step t, and for which $\gamma_i \equiv \alpha_{n_t}$. Suppose that the reduction $\gamma \underset{\lim, f(x)}{\leq} \beta$ may be performed by using general recursive function $\psi(s,x)$, i.e., $\gamma_x \equiv \beta_{\lim\limits_{s\to\infty} \psi(s,x)}$ and $f(x)$ is a majorant for $\psi(s,x)$. We shall define $\varphi(s,x)$ in the following way. For the chosen value of y in the construction procedure, the step t is awaited such that $\gamma_t = \alpha_y$. For all s, let $\varphi(s,y) = \psi(s,t)$.

As has been noted above, each value of y has the corresponding value of t. The inequality $f(t) \leq g(y)$ is valid due to the construction procedure. Therefore $\varphi(s,y)$ is a general recursive function, and it can be majorized by a general recursive function $g(y)$ and can accomplish the limit reducibility $\alpha \underset{\lim, g(x)}{\leq} \beta$. But this cannot be so due to the restriction placed on computable α and β. Consequently, $\gamma \underset{\lim, f(x)}{\not\leq} \beta$. □

For any class $\mathbb{K}$, there exists a partition into maximal subclasses $\mathbb{L}_\nu$ of locally injectable systems which can be used to specify the reducibility by way of local subclasses. The content of this reducibility concept for constructive model classes is similar to the content of reducibility of computable numerations of families of recursively enumerable sets, because each local subclass is actually specified by a suitable set of finite submodels.

Without loss of generality, $\mathbb{K}$ may be considered to contain models of the signature σ which contains only predicate symbols. It may be subdivided into finite parts

$$\sigma_0 \subseteq \sigma_1 \subseteq \cdots \subseteq \sigma_n \subseteq \sigma_{n+1} \subseteq \cdots \subseteq \sigma \qquad \text{such that} \qquad \sigma = \bigcup_{n=0}^{\infty} \sigma_n.$$

It is obvious that there is a finite number of models of finite cardinality k for a finite signature σ_n. The notation $\mathfrak{M}^{nm}$ will be used for a finite submodel of the model $\mathfrak{M}$ which contains n elements and is restricted to the finite signature σ_m. If $\mathfrak{M} \equiv_1 \mathfrak{N}$, then for these models, the sets of finite submodels for finite signatures are equal. That is why the local subclass $\mathbb{L}_\nu$ of the class $\mathbb{K}$ having representative $(\mathfrak{M}, \nu)$ is uniquely specified by the set of finite models $\{\mathfrak{M}^{nm} \mid n \in \omega, m \in \omega\}$.

The number of non-equivalent computable indexings may be used as a characteristic of a computable class of constructive models.

Theorem 3.10 [21] *If for a class $\mathbb{K}$ of constructive systems, the computable indexings α and β are available with the property that $\alpha \underset{\ell}{\not\leq} \gamma$, then there exists a computable indexing τ of the same class such that*

$$\alpha \underset{\ell}{\not\leq} \tau, \qquad \gamma \leq \tau, \qquad \text{and} \qquad \tau \underset{\ell}{\not\leq} \gamma.$$

Corollary 3.11 *If for a constructive model class there exist two computable indexings α and γ such that $\alpha \underset{\ell}{\not\leq} \gamma$, then this class has infinitely many computable indexing systems which are non-equivalent with respect to local class reducibility.*

Proof. The proof consists of successive applications of the theorem establishing the sequence $\tau_0 = \gamma$, $\tau_1 = \tau$, τ_2, ..., τ_n, τ_{n+1}, ... for computable indexing systems of the initial class, with

$$\tau_0 < \tau_1 < \cdots < \tau_n < \tau_{n+1} < \cdots,$$

and $\tau_{n+1} < \tau_n$, $\alpha \underset{\ell}{\not\leq} \tau_n$, for $n \in \omega$. □

Corollary 3.12 *If for a class of pairwise finitely distinct constructive models there exist two non-equivalent computable indexing systems, then this class has infinitely many pairwise non-equivalent computable indexing systems.*

Note that for a class of finitely distinct constructive models, each subclass of locally injectable systems consists of a single model. Therefore, for computable indexing systems of a class of constructive finitely distinct models, two distinguishable reducibilities coincide, i.e., $\alpha \leq \beta \leftrightarrow \alpha \underset{\ell}{\leq} \beta$. This is an immediate consequence of Corollary 3.11.

As has been noted, a class of finite models is a particular case of a class of finitely distinct models. Therefore, the following corollary holds.

Corollary 3.13 *If for a class of finite distinct constructive models there exist two non-equivalent computable indexing systems, then this class has infinitely many non-equivalent computable indexing systems.*

The concept of a class of finitely distinct systems is inconsistent with the concept of a class of locally injectable systems, that being a class wherein all the models are locally mutually injectable. Although computable indexings of a class of locally injectable systems are mutually equivalent with respect to reducibility in terms of local classes, nevertheless, every such non-trivial computable class will always have infinitely many incomparable computable indexing systems in the sense of common reducibility. This fact is a consequence of a more general one.

Theorem 3.14 [12] *Let two models in a computable class $\mathbb{K}$ of constructive systems be $(\mathfrak{M}_\nu, \nu)$ and $(\mathfrak{M}_\mu, \mu)$, and let an effective sequence of constructive systems be $\{(\mathfrak{M}_i, \nu_i) \mid i \in I\}$ where I is an initial segment of ω, with the above satisfying the conditions*

(i) $\mu \underset{a}{\not\equiv} \nu$, *and* $\nu_0 \underset{a}{\equiv} \nu$,

(ii) $\mathfrak{M}_\mu \equiv_1 \mathfrak{M}_\nu$,

(iii) $\forall (\mathfrak{M}_\lambda, \lambda) \in \mathbb{K}\ \forall t\ \exists i \in I\ (\mathfrak{M}_\lambda^t \hookrightarrow \mathfrak{M}_i)$,

(iv) $\forall i \in I \setminus \{0\}\ (\nu_i \underset{a}{\not\equiv} \nu)$.

Then the class $\mathbb{K}$ has infinitely many computable indexing systems which are incomparable with respect to reducibility.

Corollary 3.15 *If a class $\mathbb{K}$ is a non-trivial computable class of locally injectable constructive systems, then this $\mathbb{K}$ has denumerably many computable indexing systems which are incomparable with respect to reducibility.*

Proof. Really, the existence of the constructive models $(\mathfrak{M}_\nu, \nu)$ and $(\mathfrak{M}_\mu, \mu)$ required by the theorem follows from the non-triviality of $\mathbb{K}$, and I should be a one-element set, $I = \{0\}$. □

Corollary 3.16 *If a computable class $\mathbb{K}$ of constructive systems has a non-trivial local subclass $\mathbb{L}_\nu$ and a model $(\mathfrak{M}_\alpha, \alpha)$ such that for any $(\mathfrak{M}_\beta, \beta) \in \mathbb{K}$ the injection $\mathfrak{M}_\beta \overset{\ell}{\hookrightarrow} \mathfrak{M}_\alpha$ exists, then $\mathbb{K}$ has denumerably many computable indexing systems which are incomparable with respect to reducibility.*

Proof. If $\mathfrak{M}_\alpha \overset{\ell}{\hookrightarrow} \mathfrak{M}_\nu$, then we can let $I = \{0\}$ in the theorem. However, if $\mathfrak{M}_\alpha \overset{\ell}{\not\hookrightarrow} \mathfrak{M}_\nu$, then we should assume $I = \{0, 1\}$ and $\nu_0 \underset{a}{\equiv} \nu$, $\nu_1 \underset{a}{\equiv} \alpha$. □

The strongly constructive models are quite special in a class of constructive models. The concept of effectivity is completely realizable in these models. In this light, it is of interest to consider computable and strongly computable classes of strongly constructive models.

Corollary 3.17 *If a computable class* $\mathbb{K}$ *of constructive systems contains the systems* $(\mathfrak{M}_\nu, \nu)$ *and* $(\mathfrak{M}_\mu, \mu)$ *such that*

(i) $\nu \underset{a}{\not\equiv} \mu$, $\mathfrak{M}_\nu \equiv_1 \mathfrak{M}_\mu$

(ii) $\exists$*–theory of the model* $\mathfrak{M}_\nu$ *is decidable,*

then $\mathbb{K}$ *has denumerably many computable indexing systems which are incomparable with respect to reducibility.*

Proof. Let $\mathfrak{D}$ be a finite model. The subclass

$$\mathbb{K}_{\mathfrak{D}} = \{(\mathfrak{M}_\alpha, \alpha) \mid (\mathfrak{M}_\alpha, \alpha) \in \mathbb{K} \ \wedge \ \mathfrak{D} \hookrightarrow \mathfrak{M}_\alpha\}$$

will be completely enumerable in computable indexings of $\mathbb{K}$, and therefore it will be computable. Due to the decidability of the $\exists$–theory of $\mathfrak{M}_\nu$, the finite model set will be effectively enumerable, with these finite models being non-injectable into $\mathfrak{M}_\nu$, so let $\mathcal{D} = \{\mathfrak{D} \mid \mathfrak{D} \not\hookrightarrow \mathfrak{M}_\nu\}$. Then the subclass $\mathbb{K}_{\mathcal{D}} = \cup\{\mathbb{K}_{\mathfrak{D}} \mid \mathfrak{D} \in \mathcal{D}\}$ will obviously be completely enumerable for each indexing of $\mathbb{K}$, and it will be a computable class. Any system $(\mathfrak{M}_\beta, \beta) \in \mathbb{K}$ which is locally non-injectable into $(\mathfrak{M}_\nu, \nu)$ belongs to $\mathbb{K}_{\mathcal{D}}$, because for it there exists a finite model $\mathfrak{D}$ such that $\mathfrak{D} \hookrightarrow \mathfrak{M}_\beta$ and $\mathfrak{D} \not\hookrightarrow \mathfrak{M}_\nu$.

The class $\mathbb{K}$, the constructive models $(\mathfrak{M}_\nu, \nu)$ and $(\mathfrak{M}_\mu, \mu)$, and the computable class $\mathbb{K}_{\mathcal{D}} \cup \{(\mathfrak{M}_\nu, \nu)\}$ satisfy the conditions of Theorem 3.14, and that gives the verification of Corollary 3.17, as required. □

Corollary 3.18 *Let* $\mathbb{K}$ *be a class of constructive systems having decidable* $\exists$*–theories. If* $\mathbb{K}$ *has two non-equivalent computable indexing systems, then it has denumerably many non-equivalent computable indexing systems.*

Proof. Indeed, if $\mathbb{K}$ includes a non-trivial local subclass $\mathbb{L}_\nu$, then $\mathbb{K}$ and the two constructively non-isomorphic models $(\mathfrak{M}_\nu, \nu)$ and $(\mathfrak{M}_\mu, \mu)$ in the subclass $\mathbb{L}_\nu$ satisfy the conditions of Corollary 3.17. Therefore, in this case, $\mathbb{K}$ has denumerably many incomparable computable indexing systems. □

However, if every local subclass in $\mathbb{K}$ is trivial, i.e., is composed of a single model, then $\mathbb{K}$ will be a class of finitely distinct constructive models. Then, from Corollary 3.12, it follows that such a class has infinitely many non-equivalent computable indexing systems.

Corollary 3.19 *Let $\mathbb{K}$ be a class of strongly constructive systems which has two non-equivalent computable indexing systems. Then $\mathbb{K}$ has denumerably many non-equivalent computable indexing systems.*

Now we shall proceed to address the problem of the number of non-equivalent strongly computable indexing systems of a class of strongly constructive models.

Let $\mathbb{K}$ be a class of strongly constructive models of a given signature σ. We shall introduce the predicate $P_\varphi(\overline{y})$ for each formula $\varphi(\overline{y})$ of the signature σ. The extended signature

$$\sigma' = \sigma \cup \{P_\varphi \mid \varphi \text{ is the formula for the signature } \sigma\}$$

is defined.

Each model $\mathfrak{M}$ of the signature σ may be extended to a model $\mathfrak{M}'$ of the signature σ' in the following way.

$$\mathfrak{M}' \models P_\varphi(a_1, \ldots, a_n) \leftrightarrow \mathfrak{M} \models \varphi(a_1, \ldots, a_n)$$

Proposition 3.20 *A model $(\mathfrak{M}_\nu, \nu)$ will be strongly constructive if and only if $(\mathfrak{M}'_\nu, \nu)$ is a constructive model having a decidable $\exists$–theory.*

Proof. Indeed, strong constructibility of $(\mathfrak{M}_\nu, \nu)$ follows immediately from the constructivity of $(\mathfrak{M}'_\nu, \nu)$ because of the method of extending $\mathfrak{M}_\nu$ up to $\mathfrak{M}'_\nu$, since the set of all the formulae of a given σ is effectively denumerable.

On the other hand, the constructibility of $(\mathfrak{M}'_\nu, \nu)$ is also an immediate consequence of the method of its construction from the strongly constructive model $(\mathfrak{M}_\nu, \nu)$. Each $\exists$–formula Ψ' of the signature σ' is easily and effectively transformable into a formula Ψ of the signature σ by way of replacing each predicate P_φ by its corresponding formula φ. It is easily seen that the equivalence

$$\mathfrak{M}'_\nu \models \Psi' \leftrightarrow \mathfrak{M}_\nu \models \psi$$

will hold. Whence, by reason of strong constructivity of $(\mathfrak{M}_\nu, \nu)$, we come to the decidability of $\exists$–theory of the model $(\mathfrak{M}'_\nu, \nu)$. □

Given a class $\mathbb{K}$, let $\mathbb{T} = \{(\mathfrak{M}'_\nu, \nu) \mid (\mathfrak{M}_\nu, \nu) \in \mathbb{K}\}$, and let some Gödel numbering of all formulae of signature σ be fixed. Then for every strongly computable indexing γ of $\mathbb{K}$, it is possible to construct a computable indexing γ' of $\mathbb{T}$, i.e., having $(\mathfrak{M}_{\gamma_n}, \gamma_n)$, we shall construct, uniformly with respect to n, the model $(\mathfrak{M}'_{\gamma'_n}, \gamma'_n)$. Here the equivalence $\gamma'_i \underset{a}{\equiv} \gamma'_j \leftrightarrow \gamma_i \underset{a}{\equiv} \gamma_j$ will obviously be valid.

Conversely, according to Corollary 3.19, for each computable indexing γ' of $\mathbb{T}$, it is possible to construct effectively a strongly computable indexing γ of $\mathbb{K}$. The general recursiveness of the characteristic function for the theory $\mathrm{Th}(\mathfrak{M}_{\gamma_n})$ follows from the general recursiveness of the characteristic function of the set of formulae $D(\mathfrak{M}'_{\gamma_n})$ and the equivalence below.

$$\varphi \in \mathrm{Th}(\mathfrak{M}_{\gamma_n}) \leftrightarrow \mathfrak{M}_{\gamma_n} \models \varphi \leftrightarrow \mathfrak{M}'_{\gamma'_n} \models P_\varphi \leftrightarrow P_\varphi \in D(\mathfrak{M}'_{\gamma'_n})$$

This establishes a one-to-one correspondence between the classes of equivalent strongly computable indexing systems of $\mathbb{K}$ and the classes of equivalent computable indexing systems of $\mathbb{T}$. Taking account of the fact that every model in $\mathbb{T}$ has a decidable $\exists$–theory, Corollary 3.18 leads to the following result.

Theorem 3.21 *Let $\mathbb{K}$ be a class of strongly constructive algebraic systems having two non-equivalent strongly computable indexing systems. Then $\mathbb{K}$ has denumerably many non-equivalent strongly computable indexing systems.*

Proof. The class $\mathbb{T}$ will be specified on the basis of $\mathbb{K}$ as described above. Then $\mathbb{T}$ will satisfy the conditions of Corollary 3.18 by reason of the one-to-one correspondence described above. Therefore, $\mathbb{T}$ will have denumerably many non-equivalent computable indexing systems which are related to the denumerably many non-equivalent strongly computable indexing systems for $\mathbb{K}$. □

The next result demonstrates the mutual relationship between the number of computable indexing systems and the complexity of index sets.

Theorem 3.22 [20] *Let $\mathbb{K}_0$ be a subclass of the class $\mathbb{K}$, and let $(\mathfrak{M}, \nu) \in \mathbb{K}_0$ be a locally injectable model for $\mathbb{K}$. If the index set $I^\gamma(\mathbb{K}_0)$ is not an m–complete Π^0_1–set for some computable γ in $\mathbb{K}$, then $\mathbb{K}$ has infinitely many non-equivalent computable indexing systems.*

Now we can conclude with results about the possible number of computable indexing systems within the finite classes of constructive algebraic systems.

Proposition 3.23 *Let the computable class* $\mathbb{K}$ *of constructive algebraic systems have two models* $(\mathfrak{A}, \alpha)$ *and* $(\mathfrak{B}, \beta)$*, such that* $\mathfrak{A} \hookrightarrow \mathfrak{B}$ *with* $\alpha \underset{a}{\not\equiv} \beta$*, and let the class* $\mathbb{K} \setminus \{(\mathfrak{A}, \alpha), (\mathfrak{B}, \beta)\}$ *be computable. Then* $\mathbb{K}$ *has infinitely many non-equivalent computable indexings.*

Proof. The computable indexing of $\mathbb{K} \setminus \{(\mathfrak{A}, \alpha), (\mathfrak{B}, \beta)\}$ will be denoted by η. Let γ and ν be some computable indexings of $\mathbb{K}_0 = \{(\mathfrak{A}, \alpha), (\mathfrak{B}, \beta)\}$. It will be noticed that $\eta \oplus \gamma$ and $\eta \oplus \nu$ will represent computable indexings in $\mathbb{K}$. The constructive models $(\mathfrak{A}, \alpha)$ and $(\mathfrak{B}, \beta)$ bear indices only in accordance with γ and ν. Therefore, the equivalence

$$(\eta \oplus \gamma \leq \eta \oplus \nu) \leftrightarrow (\gamma \leq \nu)$$

holds.

The computable indexing of $\mathbb{K}_0$ defined by $\gamma_{2n} \underset{a}{\equiv} \alpha$ and $\gamma_{2n+1} \underset{a}{\equiv} \beta$, for each $n \in \omega$, will be taken as an example. Both $\mathbb{K}_0$ and γ satisfy the conditions of Theorem 3.22. Therefore, there exist infinitely many non-equivalent computable indexings for $\mathbb{K}_0$, and consequently for $\mathbb{K}$. □

Theorem 3.24 *If there are two non-equivalent computable indexing systems for a finite class of constructive algebraic systems, then there are infinitely many non-equivalent computable indexing systems for this class.*

Proof. Let the class

$$\mathbb{K} = \{(\mathfrak{A}_i, \alpha_i) \mid i = 1, 2, \dots, n\}$$

of constructive systems with non-equivalent constructs $(i \neq j \to \alpha_i \underset{a}{\not\equiv} \alpha_j)$ contain two models, with one of them being locally injectable into the other. Then, according to Proposition 3.23, we obtain that $\mathbb{K}$ contains infinitely many non-equivalent computable indexing systems. □

The number of non-equivalent computable indexing systems of a constructive model class will depend upon the availability of computable indexing systems which are mutually irreducible in terms of a bounded limit function with a fixed majorant.

Theorem 3.25 *Let computable indexings α and β of a class $\mathbb{K}$ of constructive models satisfy the relations*

$$\alpha \leq \beta, \quad \beta \underset{\lim,\Bbbk}{\not\leq} \alpha, \quad \text{and} \quad \beta \underset{\lim,(\Bbbk+1)}{\leq} \alpha.$$

Then $\mathbb{K}$ has a computable indexing γ such that

$$\alpha \leq \gamma \underset{\lim,1}{\leq} \beta, \quad \beta \not\leq \gamma, \quad \gamma \underset{\lim,\Bbbk}{\leq} \alpha, \quad \text{and} \quad \gamma \underset{\lim,(\Bbbk-2)}{\not\leq} \alpha.$$

Corollary 3.26 *If a class $\mathbb{K}$ of constructive models has computable indexings α and β such that*

$$\alpha \leq \beta, \quad \beta \underset{\lim,(\Bbbk+1)}{\leq} \alpha, \quad \text{and} \quad \beta \underset{\lim,\Bbbk}{\not\leq} \alpha,$$

then the number of non-equivalent computable indexings of $\mathbb{K}$ is not less than $\left[\frac{k+1}{2}\right] + 2$.

It should be noted that the existence of a constructive model class with a finite number of non-equivalent computable indexing systems has not been established yet. That is why the question of the infimum for the number of non-equivalent computable indexing systems (considered in Corollary 3.26) remains open.

4 Semilattices of computable indexings of the classes of constructive models

As a computable indexing γ of a class $\mathbb{K}$ of constructive algebraic systems is specified by a binary general recursive function, the set of different computable indexings of $\mathbb{K}$ will be at most a computable set. It will be denoted by $I(\mathbb{K})$. In §4, it was shown that the relations $\approx$, $\underset{\ell}{\approx}$, $\underset{\lim}{\approx}$ and $\underset{r,\lim}{\approx}$ are the equivalence relations within the set $I(\mathbb{K})$. For a computable $\alpha \in I(\mathbb{K})$, the notation $[\alpha]$, $[\alpha]_\ell$, $[\alpha]_{\lim}$ and $[\alpha]^r_{\lim}$ will be used for the respective classes of equivalent elements.

The partial ordering $[\alpha] \leq [\gamma] \leftrightarrow \alpha \leq \gamma$ will be established in accordance with the reducibility of computable indexing systems in the factor set

$$\mathcal{L}(\mathbb{K}) = I(\mathbb{K})/\equiv = \{[\alpha] \mid \alpha \in I(\mathbb{K})\}.$$

As for any two computable indexings, there exists a supremum with respect to reducibility, then this same property holds for the order in the set $\mathcal{L}(\mathbb{K})$. Therefore, the order $\leq$ having been established, the set $\mathcal{L}(\mathbb{K})$ forms an upper semilattice. This semilattice will be called the (upper) semilattice of computable indexings of the class of constructive models, the notation $\mathcal{L}(\mathbb{K})$ being the same.

The definitions and notations for other types of semilattices are quite similar to the above. Reducibility $\underset{\ell}{\leq}$ by local classes will correspond to the upper semilattice $\mathcal{L}_\ell(\mathbb{K})$, limit reducibility $\underset{\lim}{\leq}$ corresponds to $\mathcal{L}_{\lim}(\mathbb{K})$, and bounded reducibility $\underset{r,\lim}{\leq}$ to $\mathcal{L}_r(\mathbb{K})$. If only the strongly computable indexing systems of a class $\mathbb{K}$ of strongly constructive models are considered, then an asterisk will be used in the notation: $\mathcal{L}^*(\mathbb{K})$, $\mathcal{L}^*_\ell(\mathbb{K})$, $\mathcal{L}^*_{\lim}(\mathbb{K})$ and $\mathcal{L}^*_r(\mathbb{K})$.

The relation $\leq$ is a more general one, whence all other ($\underset{\ell}{\leq}$, $\underset{\lim}{\leq}$, $\underset{r,\lim}{\leq}$) reducibilities follow. Therefore, the semilattices $\mathcal{L}_\ell(\mathbb{K})$, $\mathcal{L}_{\lim}(\mathbb{K})$ and $\mathcal{L}_r(\mathbb{K})$ are obviously factors of the semilattice $\mathcal{L}(\mathbb{K})$. Consequently, they have a number of properties in common. For example, the number of non-equivalent computable indexings of $\mathbb{K}$ may be used to characterize the cardinality $|\mathcal{L}(\mathbb{K})|$ of the semilattice. Cardinalities of the different semilattices under consideration satisfy the following inequalities.

$$|\mathcal{L}_{\lim}(\mathbb{K})| \leq |\mathcal{L}_r(\mathbb{K})|, \qquad |\mathcal{L}_\ell(\mathbb{K})| \leq |\mathcal{L}_r(\mathbb{K})| \quad \text{and} \quad |\mathcal{L}_\theta(\mathbb{K})| \leq |\mathcal{L}(\mathbb{K})|$$

where $\theta \in \{\ell, \lim, r\}$. The outcome of Theorem 4.1 corresponds to a reformulated Corollary 3.11.

Theorem 4.1 *If for a class $\mathbb{K}$ of constructive systems there exist computable indexings α and β such that $\alpha \underset{\ell}{\not\leq} \beta$, then $|\mathcal{L}_\ell(\mathbb{K})| = \aleph_0$.*

Proof. Taking into account that the semilattice cardinalities are bounded from above by $\aleph_0$, we come to the equality $|\mathcal{L}_\ell(\mathbb{K})| = |\mathcal{L}(\mathbb{K})| = \aleph_0$, which is valid if the conditions of Theorem 4.1 are satisfied. □

Corollary 4.2 *For any computable class $\mathbb{K}$ of constructive algebraic systems, the semilattice cardinality $|\mathcal{L}_\ell(\mathbb{K})|$ may take on one of only two possible values; 1 or $\aleph_0$.*

The fact below follows from the other results in §4.

Theorem 4.3 *There exists a computable class* $\mathbb{K}$ *on constructive models such that* $|\mathcal{L}_\ell(\mathbb{K})| = 1$ *and* $|\mathcal{L}(\mathbb{K})| = \aleph_0$.

Similarly, the following follows from the results of §4.

Theorem 4.4 *There exists a computable class* $\mathbb{K}$ *of constructive models such that* $|\mathcal{L}_{\lim}(\mathbb{K})| = |\mathcal{L}_r(\mathbb{K})| = 1$ *and* $|\mathcal{L}(\mathbb{K})| = \aleph_0$.

Some results in the above sections concerning the number of computable indexing systems for classes of strongly constructive models or constructive models with decidable ∃–theories may be formulated as follows.

Theorem 4.5 *Let* $\mathbb{K}$ *be a class of constructive models with decidable ∃–theories. The computable indexing semilattice cardinality* $|\mathcal{L}(\mathbb{K})|$ *will be equal either to* 1 *or to* $\aleph_0$.

Theorem 4.6 *Let* $\mathbb{K}$ *be a class of strongly constructive models. The semilattice cardinalities* $|\mathcal{L}(\mathbb{K})|$ *and* $|\mathcal{L}^*(\mathbb{K})|$ *will be equal either to* 1 *or to* $\aleph_0$.

Different dimension types are known for the semilattices. The semilattice may be

(i) of dimension n if it contains a sequence of n pairwise incomparable elements, yet any sequence of $(n+1)$ elements will contain comparable elements,

(ii) of denumerable dimension if for any natural n, the semilattice contains n mutually incomparable elements,

(iii) of effectively denumerable dimension (or enumerable dimension) if there exists a computable infinite sequence of semilattice elements which are mutually incomparable,

(iv) of effectively infinite (non-constructive [34]) dimension if in any computable sequence of the semilattice, incomparable elements may be used to effectively generate an element in the semilattice which is incomparable to the elements of the computable sequence.

Proposition 4.7 *The above listed possibilities for the dimension of a semilattice are mutually distinct.*

To distinguish between some particular classes of semilattices is a rather complicated problem. For example, the concepts of denumerable, enumerable, and effectively infinite dimensions coincide [34] for the semilattices of numerations of the families of recursively enumerable sets. For the semilattices of computable indexings of constructive model classes, the result below is true, which shows the relationship between bounded limit reducibility and semilattice properties.

Theorem 4.8 [26] *If a class* $\mathbb{K}$ *of constructive models has computable indexings* α *and* β *such that*

$$\alpha \underset{\lim, 2^{x^3}}{\leq} \beta,$$

then the semilattice $|\mathcal{L}(\mathbb{K})|$ *of computable indexings of* $\mathbb{K}$ *will be of enumerable dimension.*

Corollary 4.9 *Let a computable indexing* α *of a class* $\mathbb{K}$, *a constructive model* $(\mathfrak{M}, \nu) \in \mathbb{K}$, *and a* Σ^0_2*-set* A *satisfy* $|\mathbb{L}_\nu| \geq 2$ *and* $I^\alpha_\nu \subseteq A \subseteq I^\alpha(\mathbb{K}_\nu)$. *The semilattice* $\mathcal{L}(\mathbb{K})$ *of computable indexings of the class* $\mathbb{K}$ *will then be of enumerable (effectively denumerable) dimension.*

To verify the above, it is sufficient to assume $f(x)$ be equal to 2^{x^3} in Theorem 3.7, and to apply the theorem.

Corollary 4.10 *Let a computable indexing* α *of a class* $\mathbb{K}$ *of constructive systems, models* $(\mathfrak{M}_\nu, \nu) \in \mathbb{K}$ *and* $(\mathfrak{M}_\mu, \mu) \in \mathbb{K}$, *and the* Δ^0_2*-set* A *be such that* $\nu \underset{a}{\not\equiv} \mu$, $\mathfrak{M}_\nu \equiv_1 \mathfrak{M}_\mu$, $I^\alpha_\nu \subseteq A$ *and* $I^\alpha_\mu \subseteq \overline{A}$. *Then the semilattice* $\mathcal{L}(\mathbb{K})$ *of computable indexings of* $\mathbb{K}$ *will be of enumerable dimension.*

This result is an immediate consequence of Theorems 3.8 and 4.8.

Let $\theta \in \{\Delta^0_n, \Sigma^0_n, \Pi^0_n \mid n \in \omega\}$, and let W_x denote the x–th recursively enumerable set.

The set B will be called θ–simple in the set A if $B \subseteq A$ and $B \in \theta$, and if for each $x \in \omega$ the finiteness of W_x follows from $W_x \subseteq A \setminus B$. If B is a Σ^0_1–simple subset of A, then it will be identified as a simple subset of the set A.

It will be clear that a θ–simple subset always exists for an arithmetical set A belonging to the class θ. One example is the set A itself. However, a θ_1–simple subset with $\theta_1 \subsetneqq \theta$ does not always exist for the θ–set A. For example, a productive set does not contain a Σ^0_1–simple subset.

Theorem 4.11 [23] *Let $\mathbb{K}$ be a computable class of constructive models, let γ be a computable indexing of $\mathbb{K}$, $(\mathfrak{M}_\nu, \nu) \in \mathbb{K}$, $|\mathbb{L}_\nu| \geq 2$, and let the index subset $I^\gamma(\mathbb{K}_\nu)$ contain a Σ^0_2–simple subset. Then the semilattice $\mathcal{L}(\mathbb{K})$ of computable indexings of $\mathbb{K}$ will be of effectively denumerable dimension.*

Corollary 4.12 *Let $\mathbb{K}$ be a computable class of constructive systems, let γ be a computable indexing of $\mathbb{K}$, $(\mathfrak{M}_\nu, \nu) \in \mathbb{K}$, $|\mathbb{L}_\nu| \geq 2$, and $I^\gamma(\mathbb{K}_\nu) \in \Delta^0_2$. Then the semilattice $\mathcal{L}(\mathbb{K})$ of computable indexings of $\mathbb{K}$ will be of enumerable dimension.*

Proof. In this case, the index set $I^\gamma(\mathbb{K}_\nu)$ itself should be taken as the Σ^0_2–simple subset. □

Now, using a computable indexing α of the class $\mathbb{K}$, the semilattice $\mathcal{L}(\mathbb{K})$ will be analyzed to identify the sub-semilattice $\mathcal{L}(\mathbb{K}, \alpha)$, which includes classes $[\gamma]$ from $\mathcal{L}(\mathbb{K})$ such that $\gamma \leq \alpha$. If the semilattice $\mathcal{L}_\ell(\mathbb{K})$ is to be considered, then the sub-semilattice $\mathcal{L}_\ell(\mathbb{K}, \alpha)$ will be identified in the same way.

Now, let α and β be computable indexings of $\mathbb{K}$, and let $\alpha \leq \beta$. In the semilattice $\mathcal{L}(\mathbb{K})$ we define the sub-semilattice $\mathcal{L}(\mathbb{K}, \alpha, \beta)$ which includes the classes $[\gamma]$ from $\mathcal{L}(\mathbb{K})$ such that $\alpha \leq \gamma \leq \beta$. The sub-semilattice $\mathcal{L}_\ell(\mathbb{K}, \alpha, \beta)$ is specified in the same way.

Theorem 4.13 *There exists an effective functor which correlates each constructive model $\mathfrak{M}_\nu$ to a recursively enumerable set $\Phi(\mathfrak{M}_\nu)$, and transforms any computable class $\mathbb{K}$ of constructive models into a corresponding computable family $\mathcal{S} = \{\Phi(\mathfrak{M}_\nu) \mid (\mathfrak{M}_\nu, \nu) \in \mathbb{K}\}$ of recursively enumerable sets, and maps each computable indexing α of $\mathbb{K}$ into the corresponding computable indexing α' of the family $\mathcal{S}$ in such a way that*

(i) *if $\mathbb{K}$ is a computable class of constructive finite models, then the functor Φ generates the isomorphism $\mathcal{L}(\mathbb{K}) \cong \mathcal{L}(\mathcal{S})$,*

(ii) *if α is a computable indexing of a class $\mathbb{K}$ of finitely distinct constructive models, then the functor Φ generates the isomorphism $\mathcal{L}(\mathbb{K}, \alpha) \cong \mathcal{L}(\mathcal{S}, \alpha')$,*

(iii) *if α and β are computable indexings of a class $\mathbb{K}$ of constructive models, and $\alpha \underset{\ell}{\leq} \beta$, then the functor Φ generates the isomorphism $\mathcal{L}_\ell(\mathbb{K}, \alpha, \beta) \cong \mathcal{L}(\mathcal{S}, \alpha', \beta')$.*

Without loss of generality, the signature σ of the models under consideration may be restricted to consist of only predicate symbols. The notation $\sigma_n \subseteq \sigma$ is used for a sub-signature containing at most n predicate symbols from σ, with

$$\sigma_1 \subseteq \sigma_2 \subseteq \sigma_3 \subseteq \cdots \subseteq \sigma_n \subseteq \sigma_{n+1} \subseteq \cdots \subseteq \sigma \qquad \text{and} \qquad \sigma = \bigcup_{n=1}^{\infty} \sigma_n.$$

There exist a finite number of models of a given finite cardinality k of a finite signature σ_n. For such models, to check for isomorphism or, being the same thing, to check for coincidence of their diagrams, is completely effective.

It is possible to choose an injective Gödel numbering of the finite models of cardinality k of finite signatures σ_n such that all the models of the signature σ_1 are enumerated first, then the models of σ_2, etc.. The order of numerations corresponds to the increasing n of the signature σ_n. If beginning from some n, we have $\sigma_n = \sigma_{n+\ell}$ for $\ell \in \omega$, then for each finite cardinality k there will be a finite number of models of this signature.

Let

$$\bigcup_{i=1}^{\infty} R_i = \mathbb{N}$$

be an effective partition of the set $\mathbb{N}$ of natural numbers into an infinite sequence of infinite recursive sets R_i. Let the Gödel numbering of finite models of signatures σ_n be fixed in such a way that only the numbers from the set R_k are used to number the models of cardinality k is accord with the above procedure. The abbreviated notation $\mathfrak{D}_r$ will be used for the finite model having Gödel number r in this fixed numeration.

An arbitrary constructive model $(\mathfrak{M}_\nu, \nu)$ of the signature σ will be associated with the recursively enumerable set

$$\Phi(\mathfrak{M}_\nu) = \{r \mid \mathfrak{D}_r \hookrightarrow \mathfrak{M}_\nu\}.$$

The functor Φ is the desired one; the verification of this can be found in [23].

In part (iii) of Theorem 4.13, it was stated that the factor $\mathcal{L}_\ell(\mathbb{K})$ of the semilattice $\mathcal{L}(\mathbb{L})$ ("locally", within the class boundaries $[\alpha]$ and $[\beta]$) is similar to the corresponding segment of the semilattice $\mathcal{L}(\mathcal{S})$ of computable numerations of the proper family $\mathcal{S}$ of recursively enumerable sets. However, the structure of the semilattice $\mathcal{L}(\mathbb{K})$ itself may significantly differ from the structure of the semilattice $\mathcal{L}(\mathcal{S})$. For example, if $\mathbb{K}$ is a computable local class of constructive systems, then $\mathcal{S}$ includes only one set, and $|\mathcal{L}(\mathcal{S})| = 1$. However, for a non-trivial class $\mathbb{K}$, we will have $|\mathcal{L}(\mathbb{K})| = \aleph_0$, which follows from Theorem 4.15.

At the same time, all semilattices of computable numerations of recursively enumerable sets may be represented as semilattices of computable indexings of a constructive model class. This is a consequence of a stronger result.

Theorem 4.14 [20] *For each computable family $\mathcal{S}$ of recursively enumerable sets, there exists a computable class $\mathbb{K}$ of constructive algebraic systems $\mathfrak{M}_A$ which are constructed effectively on the basis of recursively enumerable sets $A \in \mathcal{S}$, and this class has the property that it establishes a one-to-one correspondence between the computable numerations γ of the family $\mathcal{S}$ and the computable indexings γ' of the class $\mathbb{K}$, and the following conditions hold.*

(i) *$\mathfrak{M}_A$ is an autostable constructive system,*

(ii) $I_A^\gamma = I^\gamma(\mathfrak{M}_A)$,

(iii) *the model $\mathfrak{M}_A$ may be used to effectively enumerate the corresponding set A,*

(iv) $\gamma \leq \mu \leftrightarrow \gamma' \leq \mu'$.

Corollary 4.15 *For each computable family $\mathcal{S}$ of recursively enumerable sets, there exists a computable class $\mathbb{K}$ of constructive algebraic systems such that $\mathcal{L}(\mathcal{S}) \cong \mathcal{L}(\mathbb{K}) \cong \mathcal{L}_\ell(\mathbb{K})$.*

To conclude, some problems for further investigation will be formulated.

(1) The spectra of possible cardinalities for various semilattices of constructive model classes should be found.

(2) It should be examined whether there is any possibility for the semilattices of computable indexings of constructive model classes to be represented as the semilattices of the numerations of proper families of recursively enumerable sets.

(3) The degree of distinctions in cardinality and structural properties between the named semilattices of computable indexings of a single class of constructive models should be analyzed.

References

[1] G. Birkhoff, Lattice theory, corr. reprint of the 1967 3rd. edn., Amer. Math. Soc. Colloq. Pub., **25** (1979).

[2] V. P. Dobritsa, Computable classes of constructive abelian groups (Russian), in: Proc. 13th. All-Union Students' Conf. (Math.), Novosibirsk, (1975) 25–26.

[3] V. P. Dobritsa, Recursively numbered classes of constructive extensions and autostability of algebras (Russian), Sibirsk. Math. Zh., **16** (1975) 1148–1154, 1369; [translated in: Siberian Math. J., **16** (1975) 879–883].

[4] V. P. Dobritsa, On computable and strictly computable classes of constructive algebras (Russian), Material Repub. Konf. Molod Uchen, Alma-Ata, Nauka, Kazakh. SSR, (1976) 187.

[5] V. P. Dobritsa, The single necessary condition for computability (Russian), in: Proc. 4th. All-Union Conf. Math., Kishinev, Shtiintsa, (1976) 44.

[6] V. P. Dobritsa, Theorem on non-computable classes (Russian), Mat. Nauk (Ped. Inst. Alma-Ata), **3** (1976) 3–8.

[7] V. P. Dobritsa, Computability of certain classes of constructive algebras (Russian), Sibirsk. Math. Zh., **18** (1977) 570–579; [translated in: Siberian Math. J., **18** (1977) 406–413].

[8] V. P. Dobritsa, A note on computable classes of algebraic systems (Russian), in: Proc. 15th. All-Union Conf. Algebra, Krasnoyarsk, part 2, (1979) 53.

[9] V. P. Dobritsa, On computable arithmetical classes of constructive algebraic systems (Russian), in: Proc. 16th. All-Union Conf. Extended Abstracts on Algebra, Leningrad, Part 2, (1981) 46.

[10] V. P. Dobritsa, Computable classes with locally injectable systems (Russian), in: Mathematics, (Proc. 5th. Kazakh. Intercollegiate Conf. Math. and Mech., Karaganda), Extended Abstracts, (1981) 132–133.

[11] V. P. Dobritsa, On the complexity of an index set of constructive numerations (Russian), in: Proc. 6th. All-Union Conf. Math. Logic, Tbilisi, (1981) 60.

[12] V. P. Dobritsa, On the number of computable indexings of a constructive model class (Russian), Bull. Investigations into the Constructive Models, Alma-Ata, (1982) 14–22.

[13] V. P. Dobritsa, Complexity of the index set of a constructive model (Russian), Algebra i Logika, **22** (1983) 372–381; [translated in: **24** (1983) 269–276].

[14] V. P. Dobritsa, On the finite classes of constructive systems (Russian), in: Proc. 8th. Kazakh. Intercollegiate Conf. Math. and Mech., Extended Abstracts, Alma-Ata, (1984) 175.

[15] V. P. Dobritsa, On computable classes of constructive systems having complete index sets (Russian), in: Proc. 7th. All-Union Conf., Inst. Math. Siberian Branch Acad. Sci. USSR, Extended Abstracts on Math. Logic, Novosibirsk, (1984) 61.

[16] V. P. Dobritsa, On computable classes of constructive periodic abelian groups (Russian), in: Proc. 17th. All-Union Conf. Algebra, Extended Abstracts, part 2, Kishinev, (1985) 169.

[17] V. P. Dobritsa, Complexity of index sets of computable classes with a finite number of constructive systems (Russian), Sibirsk. Math. Zh., **27** (1986) 68–74, 204; [translated in: Siberian Math. J., **27** (1986) 685–690].

[18] V. P. Dobritsa, The number of computable indexings of finite classes of constructive models (Russian), Mat. Zametki, **40** (1986) 93–97, 141; [translated in: Math. Notes, **40** (1986) 551–553].

[19] V. P. Dobritsa, Cardinality of the semilattice of computable indexings of a class of strongly constructive models (Russian), in: Proc. 8th. All-Union Conf. Math. Logic, Extended Abstracts, Moskva, (1986) 60.

[20] V. P. Dobritsa, Structural properties of computable classes of constructive models (Russian), Algebra i Logika, **26** (1987) 36–62, 125; [translated in: Algebra and Logic, **26** (1987) 24–41].

[21] V. P. Dobritsa, Local classes and computable indexings (Russian), Algebra i Logika, **26** (1987) 165–190, 271; [translated in: Algebra and Logic, **26** (1987) 100–114].

[22] V. P. Dobritsa, On computable classes of constructive systems under condition of mutual local injectability (Russian), in: Proc. 19th. All-Union Conf. Algebra, Extended Abstracts, part 1, L'vov, (1987) 84.

[23] V. P. Dobritsa, Semilattices of computable indexings of classes of constructive models (Russian), Algebra i Logika, **26** (1987) 558–576, 649; [translated in: Algebra and Logic, **26** (1987) 333–345].

[24] V. P. Dobritsa, Bounded limit reducibility of computable indexings of classes of constructive models (Russian), Vyčisl. Systemy, **120** Logich. Metody v Program., Novosibirsk, (1987) 143–158, 161.

[25] V. P. Dobritsa, Classes with boundedly equivalent indexings (Russian), Vyčisl. Systemy, **122** Prikl. Aspekty Mat. Logiki, Novosibirsk, (1987) 59–72, 164.

[26] V. P. Dobritsa, On the semilattices of effectively denumerable dimension (Russian), in: Proc. 2nd. All-Union Conf. Applied Logic, Extended Abstracts, Novosibirsk, (1988) 74.

[27] V. P. Dobritsa, On the effectively infinite classes (Russian), in: Proc. 9th. All-Union Conf. Math. Logic, Extended Abstracts, Leningrad, (1988) 53.

[28] V. P. Dobritsa, Computability of some subclasses of a computable class of constructive models (Russian), Sibirsk. Math. Zh., **30** (1989) 45–51, 217; [translated in: Siberian Math. J., **30** (1989) 382–386].

[29] B. N. Drobotun, On the non-computability of a class of strong constructs (Russian), 4th. Kazakh. Mezhvuzov. Konf. Mat. i Mekh., Alma-Ata, (1977) 130.

[30] Yu. L. Ershov, A hierarchy of sets I (Russian), Algebra i Logika, **7** (1968) 47–74 [translated in: Algebra and Logic, **7** (1968) 25–43].

[31] Yu. L. Ershov, A hierarchy of sets II (Russian), Algebra i Logika, **7** (1968) 15–47 [translated in: Algebra and Logic, **7** (1968) 212–232].

[32] Yu. L. Ershov, Numbered fields, in: Logic, Methodology and Philos. Sci. III, (Proc. 3rd. Internatl. Congr. Logic, Methodology and Philos. Sci., Amsterdam, 1967), B. van Rootselaar and J. F. Staal, (eds.), North-Holland, 1968) 31–34.

[33] Yu. L. Ershov, Existence of constructivizations (Russian), Dokl. Akad. Nauk SSSR, **204** (1972) 1041–1044 [translated in: Soviet Math. – Dokl., **13** (1972) 779–783].

[34] Yu. L. Ershov, Theory of Numerations (Russian), Monographs in Math. Logic and Foundations of Math., (Nauka, Moskva, 1977).

[35] Yu. L. Ershov, Constructive models (Russian), in: Izbr. Vopr. Alg. i Log., [Selected Questions in Algebra and Logic], Sbornik posvjascen. pamjati A. I. Mal'ceva, [A collection dedicated to the memory of A. I. Mal'tsev], Izdat. Nauk Sibirsk. Otdel., Novosibirsk, (1973) 111–130.

[36] Yu. L. Ershov, Algorithmic problems in the theory of fields (positive aspects) (Russian), (supplement added in translation to Russian of J. Barwise, (ed.), Handbook of Mathematical Logic, Stud. Logic Found. Math., **90** (1977)), Handbook of Math. Logic. Part III: Recursion Theory, (Nauka, Moskva, 1982).

[37] L. Fuchs, Infinite abelian groups, Vol. 1, Pure Appl. Math., **36** (1970).

[38] S. S. Goncharov, Autostability and computable families of constructivizations (Russian), Algebra i Logika, **14** (1975) 647–680; [translated in: Algebra and Logic, **14** (1975) 392–409].

[39] S. S. Goncharov, Computable classes of constructive models (Russian), All-Union Sympos. Algebra, Gomel, **2** (1975) 253–287.

[40] S. S. Goncharov, The quantity of nonautoequivalent constructivizations (Russian), Algebra i Logika, **16** (1977) 257–282, 377; [translated in: Algebra and Logic, **16** (1977) 169–185].

[41] S. S. Goncharov, Autostability of models and abelian groups (Russian), Algebra i Logika, **19** (1980) 23–44; [translated in: Algebra and Logic, **19** (1980) 13–27].

[42] S. S. Goncharov, The problem of the number of nonautoequivalent constructivizations (Russian), Dokl. Akad. Nauk SSSR, **251** (1980) 271–274; [translated in: Soviet Math. – Dokl., **21** (1980) 411–414].

[43] S. S. Goncharov, On the problem of the number of nonautoequivalent constructivizations (Russian), Algebra i Logika, **19** (1980) 621–639, 745; [translated in: Algebra and Logic, **19** (1980) 401–414].

[44] N. G. Khisamiev, Constructive periodic abelian groups (Russian), in: Proc. 5th. Kazakh. Conf. Math. Mech., Alma-Ata, Extended Abstracts, part 2, (1974) 253.

[45] A. I. Mal'tsev, Towards a theory of computable families of objects (Russian), Algebra i Logika, **3** (1964) 5–31; [translated in: The Metamathematics of Algebraic Systems, Collected Papers: 1936–1967, translated and edited by B. F. Wells III, Stud. Logic Found. Math., **66** (1971), Ch. 27, 353–378].

[46] G. Metakides and A. Nerode, Effective content of field theory, Ann. Math. Logic, **17** (1979) 289–320.

[47] A. T. Nurtazin Computable classes and algebraic criteria of autostability, summary of scientific thesis (Russian), (Math. Inst. Siberian Branch of SSSR Acad. Sci., Novosibirsk, 1974).

[48] A. T. Nurtazin, Strong and weak constructivizations of strongly constructible models (Russian), in: Proc. 3rd. All-Union Conf. Math. Logic, Novosibirsk, (1974) 156–158.

[49] A. T. Nurtazin, Strong and weak constructivizations and computable families (Russian), Algebra i Logika, **13** (1974) 311–323, 364; [translated in: Algebra and Logic, **13** (1974) 177–184].

[50] H. Rogers Jr., Theory of Recursive Functions and Effective Computability, (1st. edn., McGraw-Hill, New York-Toronto, Ont.-London, 1967; 2nd. edn., MIT Press, Cambridge, Mass., London, 1987).

[51] Yu. G. Ventsov, Algorithmic properties of branching models (Russian), Algebra i Logika, **25** (1986) 369–383, 494; [translated in: Algebra and Logic, **25** (1986) 229–238].

[52] Yu. G. Ventsov, A family of recursively enumerable sets with finite classes of nonequivalent univalent computable numerations (Russian), Vyčisl. Systemy, **120**, Logich. Metody v Program., (Novosibirsk, 1987) 105–142.

V. P. Dobritsa
Department of Mathematics
Alma-Ata State University
Alma-Ata, Kazakhstan.

Chapter 5

Σ–Definability of Algebraic Structures

Yu. L. Ershov

The notion of the Σ–definability of an algebraic structure in an admissible set was introduced in the author's article [4] as a generalization of the notion of constructivization.

Below, we use notions and facts from the theory of admissible sets or, generally speaking, KPU–models, following [5]. All the model-theoretic notions we use can be found in [2, 6]. Let $\mathbb{A}$ be a KPU–model and $\nu : B \to M$ be an $\mathbb{A}$–numbering, i.e., a mapping of a Σ–subset B of $\mathbb{A}$ onto M.

Definition 1 An n–ary predicate $P \subseteq M^n$ on M is called

(i) a Σ_ν*–predicate* if

$$\{\langle b_1, \ldots, b_n\rangle \mid \bar{b} \in B^n, \langle \nu b_1, \ldots, \nu b_n\rangle \in P\} = \nu^{-1}(P)$$

is a Σ–predicate;

(ii) a Δ_ν*–predicate* if P and $M^n \setminus P$ are Σ_ν–predicates.

Let $\mathfrak{M} = \langle M, P_0^{\mathfrak{M}}, \ldots, P_n^{\mathfrak{M}}\rangle$ be an algebraic structure with signature $\langle P_0^{m_0}, \ldots, P_n^{m_n}\rangle$ (for the sake of simplicity we consider here only the case of relational signatures).

Definition 2 An $\mathbb{A}$–numbering $\nu : B \to M$ of the universe of $\mathfrak{M}$ is called an $\mathbb{A}$–*constructivization* of the structure $\mathfrak{M}$ if the equality predicate and all the predicates $P_0^{\mathfrak{M}}, \ldots, P_n^{\mathfrak{M}}$ are Δ_ν–predicates.

A structure $\mathfrak{M}$ for which there exists at least one $\mathbb{A}$–constructivization is called $\mathbb{A}$–*constructivizable.*

A pair $(\mathfrak{M}, \nu)$, where $\nu : B \to M$ is an $\mathbb{A}$–constructivization of the structure $\mathfrak{M}$, is called an $\mathbb{A}$–*constructive* structure.

HANDBOOK OF RECURSIVE MATHEMATICS
Edited by Yu. L. Ershov, S. S. Goncharov,
A. Nerode, and J. B. Remmel

Remark 1 In the case where $\mathbb{A}$ is $\mathbb{HF}(\emptyset)$, the notion of an $\mathbb{A}$–constructivizable structure coincides with that of a constructivizable structure, while (if $\mathbb{HF}(\emptyset)$ and ω are identified as in Section 2.1 of [5]) the notions of an $\mathbb{A}$–constructivization and of an $\mathbb{A}$–constructive structure differ from the well-known ones only in the fact that an arbitrary nonempty recursively enumerable set (and not only ω) can be taken as the domain of the constructivization.

It is possible to give an equivalent definition of an $\mathbb{A}$–constructivizable structure which is based on the notion of "definability" well-known in model theory. Let $\mathfrak{M} = \langle M\,, P_0^{\mathfrak{M}}, \ldots, P_n^{\mathfrak{M}} \rangle$ be an algebraic structure with signature $\langle P_0^{m_0}, \ldots, P_n^{m_n} \rangle$.

Definition 3 $\mathfrak{M}$ is called *definable in* $\mathbb{A}$ if there exist formulas

$$\Psi_0(x_0)\,,\ \Psi_1(x_0\,, x_1)\,,\ \Phi_0(x_0\,, \ldots, x_{m_0-1})\,,\ \ldots,\ \Phi_n(x_0\,, \ldots, x_{m_n-1})$$

(with parameters in A), such that $M_0 \rightleftharpoons \Psi_0^{\mathbb{A}}[x_0] \neq \emptyset$ and $\eta \rightleftharpoons \Psi_1^{\mathbb{A}}[x_0, x_1] \cap M_0^2$ is a congruence on the algebraic structure

$$\mathfrak{M}_0 \ \rightleftharpoons\ \langle M_0\,, P_0^{\mathfrak{M}_0}, \ldots, P_n^{\mathfrak{M}_0} \rangle,$$

where

$$P_i^{\mathfrak{M}_0} \ \rightleftharpoons\ \Phi_i^{\mathbb{A}}[x_0\,, \ldots, x_{m_i-1}] \,\cap\, M_0^{m_i}, \qquad i \leqslant n,$$

and the structure $\mathfrak{M}$ is isomorphic to the quotient structure $\mathfrak{M}_0/\eta$. (In this case we say that the *system of formulas* $\Psi_0\,, \Psi_1\,, \Phi_0\,, \ldots, \Phi_n$ *defines* $\mathfrak{M}$ *in* $\mathbb{A}$.)

We recall that η is a congruence on $\mathfrak{M}_0$ if η is an equivalence on M_0 and, for any $i \leqslant n$ and $a_0\,, \ldots, a_{m_i-1}\,, b_0\,, \ldots, b_{m_i-1} \in M_0$ such that $\langle a_j, b_j \rangle \in \eta$ for $j < m_i$, we have

$$\langle \overline{a} \rangle \in P_i^{\mathfrak{M}_0} \ \Rightarrow\ \langle \overline{b} \rangle \in P_i^{\mathfrak{M}_0}.$$

Definition 4 A structure $\mathfrak{M}$ is called Σ–*definable in* $\mathbb{A}$ if there exists a system of Σ–formulas

$$\Psi_0\,, \Psi_1\,, \Phi_0\,, \ldots, \Phi_n\,, \Psi_1^*, \Phi_0^*, \ldots, \Phi_n^*$$

such that $\Psi_0\,, \Psi_1\,, \Phi_0\,, \ldots, \Phi_n$ defines $\mathfrak{M}$ in $\mathbb{A}$, and

$$\Psi_1^{*\mathbb{A}}[x_0\,, x_1] \,\cap\, M_0^2 \ =\ M_0^2 \setminus \Psi_1^{\mathbb{A}}[x_0\,, x_1], \qquad \text{and}$$

$$\Phi_i^{*\mathbb{A}}[x_0\,, \ldots, x_{m_i-1}] \,\cap\, M_0^{m_i} \ =\ M_0^{m_i} \setminus \Phi_i^{\mathbb{A}}[x_0\,, \ldots, x_{m_i-1}], \qquad i \leqslant n,$$

where $M_0 \rightleftharpoons \Psi_0^{\mathbb{A}}[x_0]$.

Remark 2 Let $M_0 \subseteq A$, and let $P \subseteq M_0^n$ be an n–ary predicate. If we refer to P as a Δ*–predicate on* M_0 in the case that there exist Σ–formulas $\Phi(x_0, \ldots, x_{n-1})$ and $\Phi^*(x_0, \ldots, x_{n-1})$ such that

$$P = \Phi^{\mathbb{A}}[\overline{x}] \cap M_0^n \qquad \text{and} \qquad M_0^n \setminus P = \Phi^{*\mathbb{A}}[\overline{x}] \cap M_0^n,$$

then it is possible to give another equivalent definition of Σ–definability. (The Σ–formulas $\Psi_1^*, \Phi_0^*, \ldots, \Phi_n^*$ in the above definition "confirm" the equivalence.)

Definition 5 A structure $\mathfrak{M}$ is called Σ*–definable in* $\mathbb{A}$ if there exists a sequence of Σ–formulas $\Psi_0, \Psi_1, \Phi_0, \ldots, \Phi_n$ which defines $\mathfrak{M}$ in $\mathbb{A}$, and for which the predicates $\Psi_1^{\mathbb{A}}[x_0, x_1], \ldots, \Phi_i^{\mathbb{A}}[\overline{x}], \ldots, i \leqslant n$, are Δ–predicates on $M_0 = \Psi_0^{\mathbb{A}}[x_0]$.

Proposition 1 *An algebraic structure* $\mathfrak{M}$ *is* $\mathbb{A}$*–constructivizable if and only if it is* Σ*–definable in* $\mathbb{A}$.

Proof. Let $\mathfrak{M}$ be Σ–definable in $\mathbb{A}$, and $\Psi_0, \Psi_1, \Phi_0, \ldots, \Phi_n$ be a sequence of Σ–formulas which defines $\mathfrak{M}$ in $\mathbb{A}$;

$$M_0 \rightleftharpoons \Psi_0^{\mathbb{A}}[x_0],$$

$$\eta \rightleftharpoons \Psi_1^{\mathbb{A}}[x_0, x_1] \cap M_0^2,$$

$$P_i^{\mathfrak{M}_0} \rightleftharpoons \Phi_i^{\mathbb{A}}[x_0, \ldots, x_{m_i-1}] \cap M_0^{m_i}, \qquad i \leqslant n,$$

$$\mathfrak{M}_0 \rightleftharpoons \langle M_0, P_0^{\mathfrak{M}_1}, \ldots, P_n^{\mathfrak{M}_n}\rangle.$$

Let $\varphi : \mathfrak{M}_0/\eta \to \mathfrak{M}$ be an isomorphism. We define an $\mathbb{A}$–numbering $\nu : M_0 \to M$ by setting $\nu(m_0) \rightleftharpoons \varphi([m_0])$, $m_0 \in M_0$, where $[m_0]$ is the class of elements of M_0 that are η–equivalent to the element m_0. A routine verification shows that the $\mathbb{A}$–numbering constructed in such a way is an $\mathbb{A}$–constructivization of the structure $\mathfrak{M}$.

Conversely, let $\mathfrak{M}$ be $\mathbb{A}$–constructivizable, and let $\nu : B \to M$ be some $\mathbb{A}$–constructivization of the structure $\mathfrak{M}$. Since B is a Σ–subset of $\mathbb{A}$, there exists a Σ–formula $\Psi_0(x_0)$ such that $B = \Psi_0^{\mathbb{A}}[x_0]$. Since $\nu : B \to M$ is an $\mathbb{A}$–constructivization of $\mathfrak{M}$, the relations

$$\eta_\nu \rightleftharpoons \{\langle b_0\,, b_1\rangle \mid b_0\,, b_1 \in B,\ \nu b_0 = \nu b_1\},$$

$$Q_0 \rightleftharpoons \{\langle b_0\,, \ldots\,, b_{m_0-1}\rangle \mid b_0\,, \ldots\,, b_{m_0-1} \in B,\ \langle \nu b_0\,, \ldots\,, \nu b_{m_0-1}\rangle \in P_0^{\mathfrak{M}}\},$$

$$\vdots$$

$$Q_n \rightleftharpoons \{\langle b_0\,, \ldots\,, b_{m_n-1}\rangle \mid b_0\,, \ldots\,, b_{m_0-1} \in B,\ \langle \nu b_0\,, \ldots\,, \nu b_{m_n-1}\rangle \in P_n^{\mathfrak{M}}\}$$

are Δ_ν–predicates. Consequently, there exist Σ–formulas

$$\Psi_1(x_0\,, x_1),$$

$$\Psi_1^*(x_0\,, x_1),$$

$$\Phi_0(x_0\,, \ldots\,, x_{m_0-1}),$$

$$\Phi_0^*(x_0\,, \ldots\,, x_{m_0-1}),$$

$$\vdots$$

$$\Phi_n(x_0\,, \ldots\,, x_{m_n-1}),$$

$$\Phi_n^*(x_0\,, \ldots\,, x_{m_n-1})$$

such that

$$\eta = \Psi_1^{\mathbb{A}}[x_0\,, x_1], \qquad B^2 \setminus \eta = \Psi_1^{*\mathbb{A}}[x_0\,, x_1],$$

$$Q_0 = \Phi_0^{\mathbb{A}}[x_0\,, \ldots\,, x_{m_0-1}], \qquad B^{m_0} \setminus Q_0 = \Phi_0^{*\mathbb{A}}[x_0\,, \ldots\,, x_{m_0-1}],$$

$$\vdots \qquad\qquad \vdots$$

$$Q_n = \Phi_n^{\mathbb{A}}[x_0\,, \ldots\,, x_{m_n-1}], \qquad B^{m_n} \setminus Q_n = \Phi_n^{*\mathbb{A}}[x_0\,, \ldots\,, x_{m_n-1}].$$

It is not hard to verify that the system of Σ–formulas $\Psi_0\,, \Psi_1\,, \Phi_0\,, \ldots\,, \Phi_n$ defines $\mathfrak{M}$ in $\mathbb{A}$. The existence of Σ–formulas $\Psi_1^*\,, \Phi_0^*\,, \ldots\,, \Phi_n^*$ shows that $\mathfrak{M}$ is Σ–definable in $\mathbb{A}$. □

The following claim is an interesting consequence of this (uninteresting) proposition. Let $\sigma_1 = \langle \emptyset, \in^2\rangle$.

Proposition 2 *If an algebraic structure $\mathfrak{M}$ of signature σ is $\mathbb{A}$-constructivizable, then there exists a* KPU*-model $\mathbb{B}$ of signature $\sigma_1 \cup \sigma$ with set of urelements M, such that $(\mathbb{B} \upharpoonright M) \upharpoonright \sigma = \mathfrak{M}$, and $\mathbb{B}$ is Σ-definable in $\mathbb{A}$ and "has the same ordinals" as $\mathbb{A}$ does. If $\mathbb{A}$ is an admissible set, then $\mathbb{B}$ is also an admissible set.*

For the proof of this proposition we refer the reader to Section 3.4 of [5].

Corollary 3 *If $\mathbb{A}$ is an admissible, set and the algebraic structure $\mathfrak{M}$ is $\mathbb{A}$-constructivizable, then the ordinal $o(\mathbb{A})$ is $\mathfrak{M}$-admissible.*

The definition of an $\mathfrak{M}$-admissible ordinal can be found in [1]. It is well-known that if $\mathfrak{M}$ is a constructivizable algebraic structure, then the $\exists$-theory $\mathrm{Th}(\mathfrak{M})_\exists$ of this structure is recursively enumerable. It is easy to see that if $\mathfrak{M}$ is constructivizable, then $\mathbb{HF}(\mathfrak{M})$ is also constructivizable. Hence the $\exists$-theory $\mathrm{Th}(\mathbb{HF}(\mathfrak{M}))_\exists$ (and moreover, the Σ-theory $\mathrm{Th}(\mathbb{HF}(\mathfrak{M}))_\Sigma$) of the admissible set $\mathbb{HF}(\mathfrak{M})$ is recursively enumerable.

If $\mathfrak{M}$ is $\mathbb{A}$-constructivizable, and $\mathbb{A}$ is a sufficiently "large" admissible set, it is possible to indicate, using infinite formulas, a larger fragment of the theory of the structure $\mathfrak{M}$ which is effectively "enumerable" in $\mathbb{A}$. We outline a description of the syntax and semantics of this Σ^*-logic of an admissible set $\mathbb{B}$ of signature $\langle \emptyset\,, \in^2\,; c_b\,, b \in B\,, P_0^{m_0}\,, \ldots\,, P_n^{m_n} \rangle$.

The variables $V_B = \{x_b \mid b \in B\}$ are coded as follows: $\llcorner x_b \lrcorner \leftrightharpoons \langle 0, b\rangle$, $b \in B$. The constants are coded as follows: $\llcorner c_b \lrcorner \leftrightharpoons \langle 1, b\rangle$, $b \in B$ (we assume that $c_\emptyset = \emptyset$).

We first introduce the notion of a Δ_0^*-formula. Simultaneously, we define the code of a Δ_0^*-formula.

(1) By elementary formulas we mean formulas of the forms $t_0 = t_1$, $t_0 \in t_1$, and $P_i(t_1\,, \ldots\,, t_{m_i})$ for $i \leqslant n$; they are Δ_0^*-formulas and have the following codes:

$$\llcorner t_0 = t_1 \lrcorner \;=\; \langle 2\,, 0\,, \llcorner t_0 \lrcorner\,, \llcorner t_1 \lrcorner \rangle,$$

$$\llcorner t_0 \in t_1 \lrcorner \;=\; \langle 2\,, 1\,, \llcorner t_0 \lrcorner\,, \llcorner t_1 \lrcorner \rangle,$$

$$\llcorner P_i(t_1\,, \ldots\,, t_{m_i}) \lrcorner \;=\; \langle 2\,, i+2\,, \llcorner t_1 \lrcorner\,, \ldots\,, \llcorner t_{m_i} \lrcorner \rangle, \qquad i \leqslant n.$$

(2) If Φ is a Δ_0^*-formula, then $\neg\Phi$ is a Δ_0^*-formula and its code is $\langle 3\,, \llcorner \Phi \lrcorner \rangle$.

(3) If $\boldsymbol{\Phi}$ is a family of Δ_0^*–formulas such that $\llcorner\boldsymbol{\Phi}\lrcorner \leftrightharpoons \{\llcorner\Phi\lrcorner \mid \Phi \in \boldsymbol{\Phi}\} \in B$, then $\wedge(\boldsymbol{\Phi})$ and $\vee(\boldsymbol{\Phi})$ are Δ_0^*–formulas, and their codes are defined as follows:

$$\llcorner\wedge(\boldsymbol{\Phi})\lrcorner \leftrightharpoons \langle 4\,,\,\llcorner\boldsymbol{\Phi}\lrcorner\rangle,$$

$$\llcorner\vee(\boldsymbol{\Phi})\lrcorner \leftrightharpoons \langle 5\,,\,\llcorner\boldsymbol{\Phi}\lrcorner\rangle.$$

(4) If Φ is a Δ_0^*–formula, x is a variable, and t is a term, then $\forall x \in t\,\Phi$ and $\exists x \in t\,\Phi$ are Δ_0^*–formulas, and their codes are defined as follows:

$$\llcorner\forall x \in t\,\Phi\lrcorner \leftrightharpoons \langle 6\,,\,\llcorner x\lrcorner\,,\,\llcorner t\lrcorner\,,\,\llcorner\Phi\lrcorner\rangle,$$

$$\llcorner\exists x \in t\,\Phi\lrcorner \leftrightharpoons \langle 7\,,\,\llcorner x\lrcorner\,,\,\llcorner t\lrcorner\,,\,\llcorner\Phi\lrcorner\rangle.$$

Something is a Δ_0^*–formula if and only if it is contained in the set generated by rules (1)–(4). It is easy to establish that the set $\boldsymbol{\Delta}_0^*$ of all codes of Δ_0^*–formulas is a Δ–subset of $\mathbb{B}$.

We now introduce the notion of a Σ^*–formula.

(1) Every Δ_0^*–formula is a Σ^*–formula, and its Σ^*–code coincides with its Δ_0^*–code.

(2) If $\boldsymbol{\Phi}$ is a family of Σ^*–formulas such that $\llcorner\boldsymbol{\Phi}\lrcorner \leftrightharpoons \{\llcorner\Phi\lrcorner \mid \Phi \in \boldsymbol{\Phi}\} \in B$, then $\wedge(\boldsymbol{\Phi})$ and $\vee(\boldsymbol{\Phi})$ are Σ^*–formulas and their Σ^*–codes are defined as follows:

$$\llcorner\wedge(\boldsymbol{\Phi})\lrcorner \leftrightharpoons \langle 4\,,\,\llcorner\boldsymbol{\Phi}\lrcorner\rangle,$$

$$\llcorner\vee(\boldsymbol{\Phi})\lrcorner \leftrightharpoons \langle 5\,,\,\llcorner\boldsymbol{\Phi}\lrcorner\rangle.$$

(3) If Φ is a Σ^*–formula, x is a variable, and t is a term, then $\forall x \in t\,\Phi$ and $\exists x \in t\,\Phi$ are Σ^*–formulas and their codes are defined as follows:

$$\llcorner\forall x \in t\,\Phi\lrcorner \leftrightharpoons \langle 6\,,\,\llcorner x\lrcorner\,,\,\llcorner t\lrcorner\,,\,\llcorner\Phi\lrcorner\rangle,$$

$$\llcorner\exists x \in t\,\Phi\lrcorner \leftrightharpoons \langle 7\,,\,\llcorner x\lrcorner\,,\,\llcorner t\lrcorner\,,\,\llcorner\Phi\lrcorner\rangle.$$

(4) If Φ is a Σ^*–formula, and X is a set of variables such that $\llcorner X\lrcorner \leftrightharpoons \{\llcorner x\lrcorner \mid x \in X\} \in B$, then $\langle\exists\,,\,X\,,\,\Phi\rangle$ is a Σ^*–formula, and its code is defined as follows:

$$\llcorner\langle\exists\,,\,X\,,\,\Phi\rangle\lrcorner \leftrightharpoons \langle 8\,,\,\llcorner X\lrcorner\,,\,\llcorner\Phi\lrcorner\rangle.$$

The set $\mathbf{\Sigma}^*$ of all codes of Σ^*–formulas is a Δ–subset of B. For any Σ^*–formula Φ, it is natural to define the set $\mathrm{FV}(\Phi)$ of free variables of Φ. For any such formula, $\llcorner\mathrm{FV}(\Phi)\lrcorner \in B$. To any element $a \in B$, we can assign a mapping $\gamma_a : V \to B$ from the set of all variables $\{x_b \mid b \in B\}$ into B by the following rule: given x_b, if there exists a unique element $c \in B$ such that $\langle \llcorner x_b \lrcorner , c\rangle = \langle\langle 0 , b\rangle , c\rangle \in a$, then $\gamma_a(x_b) \leftrightharpoons c$, otherwise $\gamma_a(x_b) \leftrightharpoons \emptyset$.

By induction on Σ^*–formulas, we define the relation $\mathbb{B} \models \Phi[\gamma_a]$ of the truth of a formula Φ in $\mathbb{B}$ under the interpretation γ_a. This is defined as usual. We must only specify the case in which $\Phi = \langle \exists , X , \Psi\rangle$. $\mathbb{B} \models \Phi[\gamma_a]$ is true if and only if there is a function $f : X \to B$ such that $f \in B$ and $\mathbb{B} \models \Psi[\gamma_{a'}]$ for $a' \leftrightharpoons (a \setminus X \times B) \cup f$.

The following theorem on the Σ–definability of the truth of Σ^*–formulas is not hard to prove.

Theorem 4 *The predicate*

$$\mathrm{Tr}_{\Sigma^*} \leftrightharpoons \{\langle \llcorner\Phi\lrcorner, a\rangle \mid \Phi \textit{ is a } \Sigma^*\textit{-formula}, \mathbb{B} \models \Phi[\gamma_a]\}$$

is a Σ–predicate in $\mathbb{B}$.

Corollary 5 *If $\mathfrak{M}$, $\mathbb{A}$, and $\mathbb{B} = \mathbb{B}(\mathfrak{M})$ are the same as in Proposition 2, then the Σ^*-theory of $\mathbb{B}(\mathfrak{M})$ is a "Σ-subset" of $\mathbb{A}$.*

In the sequel, our attention will be focused mainly on the Σ–definability of structures in admissible sets of the form $\mathbb{HF}(\mathfrak{M})$ for models $\mathfrak{M}$ of sufficiently simple theories T. It is useful to obtain beforehand information on definability in such structures. The following assertion turns out to be useful.

Proposition 6 *If $\mathfrak{H} \preccurlyeq \mathbb{HF}(\mathfrak{M})$, where $\mathfrak{H}$ is an elementary submodel of $\mathbb{HF}(\mathfrak{M})$, then $\mathfrak{H}$ has the form $\mathbb{HF}(\mathfrak{M}')$ for a suitable model $\mathfrak{M}' \preccurlyeq \mathfrak{M}$.*

We give a sketch of the proof of the fact that predicates of the system $\mathfrak{M}$ which are first-order definable in $\mathbb{HF}(\mathfrak{M})$ can be described by (infinite) formulas of the language $L_{\omega_1 , \omega}$.

Let $n \in \omega$. We set

$$\mathfrak{H}_n \rightleftharpoons \mathbb{HF}(n(= \{i \mid i \in n\})),$$

$$\mathfrak{H}_0 = \mathbb{HF}(\emptyset) \subseteq \mathfrak{H}_1 \subseteq \cdots ,$$

$$\bigcup_{n\in\omega} \mathfrak{H}_n = \mathbb{HF}(\omega).$$

For any $n \in \omega$, $\varkappa \in \mathfrak{H}_n$, $\overline{m} \in M^n$ we define an element $\varkappa(\overline{m}) \in \mathbb{HF}(M)$ in the following way. Let $\lambda_{\overline{m}} : n \to M$ be defined as follows: $\lambda_{\overline{m}}(i) = m_i$, $i < n$, where $\overline{m} = \langle m_0, \ldots, m_{n-1}\rangle$. The mapping $\lambda_{\overline{m}}$ can be uniquely extended to a mapping $\lambda^{\omega}_{\overline{m}} : \mathfrak{H}_n = \mathbb{HF}(n) \to \mathbb{HF}(M)$, so that

$$\lambda^{\omega}_{\overline{m}}(\{a_0, \ldots, a_k\} = \{\lambda^{\omega}_{\overline{m}}(a_0), \ldots, \lambda^{\omega}_{\overline{m}}(a_k)\}$$

for any set (not urelement) $\{a_0, \ldots, a_k\} \in \mathbb{HF}(n)$. Let

$$\varkappa(\overline{m}) \rightleftharpoons \lambda^{\omega}_{\overline{m}}(\varkappa).$$

Given any element $\varkappa \in \mathfrak{H}_n$, it is possible to define effectively a term

$$t_{\varkappa}(x_1, \ldots, x_n)$$

of the signature $\langle \emptyset, \{\ \}^1, \cup^2\rangle$ such that, for any $m_1, \ldots, m_n \in M$,

$$t^{\mathbb{HF}(M)}_{\varkappa}(m_1, \ldots, m_n) = \varkappa(\overline{m}).$$

Let $\varphi(\overline{x}, \overline{y})$ be a formula of the language of the signature $\sigma_1 \cup \sigma'$, where σ' is the signature of the structure $\mathfrak{M}$. We assume that the variables in the list $\overline{x}$ correspond to urelements of the structure $\mathbb{HF}(\mathfrak{M})$ (i.e., to elements of the structure $\mathfrak{M}$), and variables in the list $\overline{y}$ correspond to arbitrary elements of $\mathbb{HF}(\mathfrak{M})$. The formula $\varphi_*(\overline{x}, \overline{y})$ of the language $L_{\omega_1,\omega}$ of the signature $\sigma' \cup \langle \emptyset, \in^2, \{\ \}^1, \cup^2\rangle$ is constructed from the formula φ as follows:

(1) if φ does not contain quantifiers, then $\varphi_* \rightleftharpoons \varphi$,

(2) if $\varphi = \varphi_0\ q\ \varphi_1$, then $\varphi_* \rightleftharpoons (\varphi_0)_*\ q\ (\varphi_1)_*$, for $q \in \{\wedge, \vee, \to\}$,

(3) if $\varphi = \neg\varphi_0$, then $\varphi_* \rightleftharpoons \neg(\varphi_0)_*$,

(4) if $\varphi = Q\, x\varphi_0$, then $\varphi_* \rightleftharpoons Qx(\varphi_0)_*$, for $Q \in \{\forall, \exists\}$,

(5) if $\varphi = \exists\, y\varphi_0$, then $\varphi_* \rightleftharpoons \bigvee_{n\in\omega}\big(\bigvee_{\varkappa\in\mathfrak{H}_n} \exists x_1 \cdots \exists x_n\, ((\varphi_0)*)^{y}_{t_\varkappa(\overline{x})}\big)$,

(6) if $\varphi = \forall\, y\varphi_0$, then $\varphi_x \rightleftharpoons \bigwedge_{n\in\omega}\big(\bigwedge_{\varkappa\in\mathfrak{H}_n} \forall x_1 \cdots \forall x_n\, ((\varphi_0)_*)^{y}_{t_\varkappa(\overline{x})}\big)$.

If $\varphi(\overline{x})$ does not contain variables from the list $\overline{y}$, then neither does $\varphi_*(\overline{x})$. For any terms t_0 and t_1 of the signature $\sigma' \cup \langle \emptyset, \{\ \}^1, \cup^2\rangle$, with variables from the list $\overline{x}$, it is possible to write effectively quantifier-free (finite) formulas

$\Psi^*_{t_0,t_1}(\overline{x})$ and $\Phi_{t_0,t_1}(\overline{x})$ of the signature σ' such that $t_0 = t_1 \equiv_{\mathbb{HF}(\mathfrak{M})} \Psi^*_{t_0,t_1}$ and $t_0 \in t_1 \equiv_{\mathbb{HF}(\mathfrak{M})} \Phi_{t_0,t_1}$. For any pair of terms t_0 and t_1 of the signature $\langle \varnothing, \{\ \}^1, \cup^2 \rangle$, with variables from the list $\overline{x}$, we define formulas Φ_{t_0,t_1} and Ψ_{t_0,t_1} of the empty signature such that

$$\mathrm{FV}(\Phi_{t_0,t_1}) = \mathrm{FV}(\Psi_{t_0,t_1}) = \mathrm{FV}(t_0) \cup \mathrm{FV}(t_1)$$

and

$$t_0^{\langle \mathbb{HF}(\mathfrak{M}), \{\ \}, \cup \rangle}[\gamma] \in t_1^{\langle \mathbb{HF}(\mathfrak{M}), \{\ \}, \cup \rangle}[\gamma] \Leftrightarrow |\mathfrak{M}| \models \Phi_{t_0,t_1}[\gamma],$$

$$t_0^{\langle \mathbb{HF}(\mathfrak{M}), \{\ \}, \cup \rangle}[\gamma] \subseteq t_1^{\langle \mathbb{HF}(\mathfrak{M}), \{\ \}, \cup \rangle}[\gamma] \Leftrightarrow |\mathfrak{M}| \models \Psi_{t_0,t_1}[\gamma] \quad \text{or}$$

$$t_0^{\langle \mathbb{HF}(\mathfrak{M}), \{\ \}, \cup \rangle}[\gamma] = t_1^{\langle \mathbb{HF}(\mathfrak{M}), \{\ \}, \cup \rangle}[\gamma]$$

for any interpretation $\gamma : \mathrm{FV}(t_0 = t_1) \to |\mathfrak{M}|$. Let τ denote the sentence $\exists\, u(u = u)$, and let u_0 and u_1 represent variables from the list $\overline{x}$. We define

for $t_0 = u_0$:

$$\begin{array}{ll}
\text{if } t_1 = u_1, & \text{then } \Phi_{u_0,u_1} \rightleftharpoons \neg\tau,\ \Psi_{u_0,u_1} \rightleftharpoons (u_0 = u_1),\\
\text{if } t_1 = \varnothing, & \text{then } \Phi_{u_0,\varnothing} \rightleftharpoons \neg\tau,\ \Psi_{u_0,\varnothing} \rightleftharpoons \tau,\\
\text{if } t_1 = \{t_1'\}, & \text{then } \Phi_{u_0,t_1} \rightleftharpoons \Psi_{t_1',u_0},\ \Psi_{u_0,t_1} \rightleftharpoons \tau,\\
\text{if } t_1 = (t_1' \cup t_1''), & \text{then } \Phi_{u_0,t_1} \rightleftharpoons \Phi_{u_0,t_1'} \vee \Phi_{u_0,t_1''},\ \Psi_{u_0,t_1} \rightleftharpoons \tau;
\end{array}$$

for $t_0 = \varnothing$:

$$\begin{array}{ll}
\text{if } t_1 = u_1, & \text{then } \Phi_{\varnothing,u_1} \rightleftharpoons \neg\tau,\ \Psi_{\varnothing,u_1} \rightleftharpoons \neg\tau,\\
\text{if } t_1 = \varnothing, & \text{then } \Phi_{\varnothing,\varnothing} \rightleftharpoons \neg\tau,\ \Psi_{\varnothing,\varnothing} \rightleftharpoons \tau,\\
\text{if } t_1 = \{t_1'\}, & \text{then } \Phi_{\varnothing,t_1} \rightleftharpoons \Psi_{t_1',\varnothing} \wedge \Psi_{\varnothing,t_1'},\ \Psi_{\varnothing,t_1} \rightleftharpoons \tau,\\
\text{if } t_1 = (t_1' \cup t_1''), & \text{then } \Phi_{\varnothing,t_1} \rightleftharpoons \Phi_{\varnothing,t_1'} \vee \Phi_{\varnothing,t_1''},\ \Psi_{\varnothing,t_1} \rightleftharpoons \tau;
\end{array}$$

for $t_0 = \{t_0'\}$:

$$\begin{array}{ll}
\text{if } t_1 = u_1, & \text{then } \Phi_{t_0,u_1} \rightleftharpoons \neg\tau,\ \Psi_{t_0,u_1} \rightleftharpoons \neg\tau,\\
\text{if } t_1 = \varnothing, & \text{then } \Phi_{t_0,\varnothing} \rightleftharpoons \neg\tau,\ \Psi_{t_0,\varnothing} \rightleftharpoons \neg\tau,\\
\text{if } t_1 = \{t_1'\}, & \text{then } \Phi_{t_0,t_1} \rightleftharpoons \Psi_{t_0,t_1'} \wedge \Psi_{t_1',t_0},\ \Psi_{t_0,t_1} = \Phi_{t_0',t_1},\\
\text{if } t_1 = (t_1' \cup t_1''), & \text{then } \Phi_{t_0,t_1} \leftrightharpoons \Phi_{t_0,t_1'} \vee \Phi_{t_0,t_1''},\ \Psi_{t_0,t_1} \leftrightharpoons \Phi_{t_0',t_1};
\end{array}$$

for $t_0 = (t_0' \cup t_0'')$:

$$\begin{array}{lll} \text{if } t_1 = u_1, & \text{then } \Phi_{t_0,u_1} \rightleftharpoons \neg\tau, & \Psi_{t_0,u_1} \rightleftharpoons \neg\tau, \\ \text{if } t_1 = \emptyset, & \text{then } \Phi_{t_0,\emptyset} \rightleftharpoons \neg\tau, & \Psi_{t_0,\emptyset} \rightleftharpoons \Psi_{t_0',\emptyset} \wedge \Psi_{t_0'',\emptyset}, \\ \text{if } t_1 = \{t_1'\}, & \text{then } \Phi_{t_0,t_1} \rightleftharpoons \Psi_{t_0,t_1'}, & \Psi_{t_0,t_1} \leftrightharpoons \Psi_{t_0',t_1} \wedge \Psi_{t_0'',t_1}, \\ \text{if } t_1 = (t_1' \cup t_1''), & \text{then } \Phi_{t_0,t_1} \rightleftharpoons \Phi_{t_0,t_1'} \vee \Phi_{t_0,t_1''}, & \Psi_{t_0,t_1} \rightleftharpoons \Psi_{t_0',t_1} \wedge \Psi_{t_0'',t_1}. \end{array}$$

We set $\Psi^*_{t_0,t_1} \leftrightharpoons \Psi_{t_0,t_1} \wedge \Psi_{t_1,t_0}$. Using the formulas $\Psi^*_{t_0,t_1}$ and Φ_{t_0,t_1}, we transform the formula $\varphi_*(\overline{x})$ into a formula $\varphi^*(\overline{x})$ of the language $L_{\omega_1,\omega}$ of the signature σ' such that

$$\varphi_*(\overline{x}) \equiv_{\mathbb{HF}(\mathfrak{M})} \varphi^*(\overline{x}).$$

Then $\varphi^{\mathbb{HF}(\mathfrak{M})}[\overline{x}] = \varphi^{*\mathfrak{M}}[\overline{x}]$. Using this transformation and Karp's theorem [7, Theorem 9.10], we arrive at the following proposition.

Proposition 7 *Let $\mathfrak{M}_0$ and $\mathfrak{M}_1$ be ω–saturated models. If $\mathfrak{M}_0 \equiv \mathfrak{M}_1$, then $\mathbb{HF}(\mathfrak{M}_0) \equiv \mathbb{HF}(\mathfrak{M}_1)$; if $\mathfrak{M}_0 \preccurlyeq \mathfrak{M}_1$, then $\mathbb{HF}(\mathfrak{M}_0) \preccurlyeq \mathbb{HF}(\mathfrak{M}_1)$.*

Definition 6 A model $\mathfrak{M}_0$ is called *saturated enough* if there exists an ω–saturated model $\mathfrak{M}_1$ such that $\mathfrak{M}_0 \preccurlyeq \mathfrak{M}_1$ and $\mathbb{HF}(\mathfrak{M}_0) \preccurlyeq \mathbb{HF}(\mathfrak{M}_1)$.

Corollary 8 *Let $\mathfrak{M}_0$ and $\mathfrak{M}_1$ be saturated enough models. If $\mathfrak{M}_0 \equiv \mathfrak{M}_1$, then $\mathbb{HF}(\mathfrak{M}_0) \equiv \mathbb{HF}(\mathfrak{M}_1)$. If $\mathfrak{M}_0 \preccurlyeq \mathfrak{M}_1$, then $\mathbb{HF}(\mathfrak{M}_0) \preccurlyeq \mathbb{HF}(\mathfrak{M}_1)$.*

Proof. The assertion follows from the definition, Proposition 7, and the existence of ω–saturated models. □

Remark 3 In any saturated enough model $\mathfrak{M}$, any (not necessarily complete) arithmetic type of formulas with a finite number of quantifier alterations is realized over a finite $M_0 \subseteq |\mathfrak{M}|$. The author does not know whether this condition is equivalent to being saturated enough.

Remark 4 Any complete theory T (of a countable language) with infinite models has a countable saturated enough model. This fact follows from the Löwenheim–Skolem theorem, Proposition 6, and the existence of ω–saturated models of the theory T.

Any complete categorical theory T has sufficiently many ω–saturated models.

Proposition 9 *If a theory T is complete and ω-categorical, then any model of T is ω-saturated. If the theory T is complete and ω_1-categorical, then any uncountable model of T is saturated (and, in addition, ω-saturated).*

Proof. The first assertion follows from the Ryll-Nardzewski theorem on the characterization of ω-categorical theories (see [2, Theorem 2.3.13]), or more exactly, from the fact that there is a finite number of types over any finite set. The second assertion is well-known. (see, for example, [2, Corollary 7.1.15]). □

Remark 5 Not all countable models of ω_1-categorical theories are saturated enough. For example, the following claim is true.

> *Let T be the theory of algebraically closed fields of characteristic 0. If $F_0, F_1 \models T$, then $\mathbb{HF}(F_0) \equiv \mathbb{HF}(F_1)$ if and only if either F_0 and F_1 both have infinite transcendence degree over $\mathbb{Q}$, or the transcendence degrees of F_0 and F_1 over $\mathbb{Q}$ are finite and equal.*

Theorem 10 *Let $\mathfrak{M}$ be a saturated enough model. If $h_0, h_1 \in \mathbb{HF}(\mathfrak{M})$, then the types $t_{\mathbb{HF}(\mathfrak{M})}(h_0)$ and $t_{\mathbb{HF}(\mathfrak{M})}(h_1)$ coincide if and only if there exist $n \in \omega$, $\varkappa \in \mathfrak{H}_n$, $\overline{m}_0, \overline{m}_1 \in M^n$ such that $h_0 = \varkappa(\overline{m}_0)$, $h_1 = \varkappa(\overline{m}_1)$, and $t_{\mathfrak{M}}(\overline{m}_0) = t_{\mathfrak{M}}(\overline{m}_1)$.*

Proof.

Necessity. Let $h_0, h_1 \in \mathbb{HF}(\mathfrak{M})$ and

$$t_{\mathbb{HF}(\mathfrak{M})}(h_0) \;=\; t_{\mathbb{HF}(\mathfrak{M})}(h_1).$$

Let $\mathrm{sp}\, h_0$ (the support of h_0) have n elements: $\mathrm{sp}\, h_0 = \{m_0^0, \dots, m_{n-1}^0\}$. It is easy to see that the support of h_1 also has n elements, and there exists $\varkappa \in \mathfrak{H}_n$ such that $h_0 = \varkappa(\overline{m}^0)$, $\overline{m}^0 \rightleftharpoons \langle m_0^0, \dots, m_{n-1}^0\rangle$, and $h_1 = \varkappa(\overline{m}^1)$ for a suitable $\overline{m}^1 \rightleftharpoons \langle m_0^1, \dots, m_{n-1}^1\rangle$. Furthermore, for any formula $\varphi(\overline{x}) \in t_{\mathfrak{M}}(\overline{m}^0)$, there exists a permutation $\sigma \in S_n$ of the set n such that $\varphi(\overline{x}) \in t_{\mathfrak{M}}(\sigma(\overline{m}^1))$, where $\sigma(\overline{m}^1) \rightleftharpoons \langle m_{\sigma(0)}^1, \dots, m_{\sigma(n-1)}^1\rangle$ and $h_1 = \varkappa(\sigma(\overline{m}^1))$. But then there exists $\sigma \in S_n$ such that $h_1 = \varkappa(\sigma(\overline{m}^1))$ and $t_{\mathfrak{M}}(\overline{m}^0) = t_{\mathfrak{M}}(\sigma(\overline{m}^1))$. Indeed, if such σ does not exist, then for any $\sigma \in S_n$ such that $h_1 = \varkappa(\sigma(\overline{m}^1))$ there exists $\varphi_\sigma(\overline{x})$ belonging to $t_{\mathfrak{M}}(\overline{m}^0)$ but not to $t_{\mathfrak{M}}(\sigma(\overline{m}^1))$. Let $\varphi(\overline{x}) \rightleftharpoons \bigwedge_{\sigma \in S_n^\varkappa} \varphi_\sigma(\overline{x})$, where

$$S_n^\varkappa \;\rightleftharpoons\; \{\sigma \mid \sigma \in S_n,\ \varkappa(\sigma(\overline{m}^1)) = h_1\}.$$

Since $\varphi_\sigma(\overline{x}) \in t_{\mathfrak{M}}(\overline{m}^0)$ for all $\sigma \in S_n^\varkappa$, it follows that $\varphi(\overline{x}) \in t_{\mathfrak{M}}(\overline{m}^0)$. As already mentioned, there exists $\sigma_0 \in S_n^\varkappa$ such that $\varphi(\overline{x}) \in t_{\mathfrak{M}}(\sigma_0(\overline{m}^1))$, but $\varphi_{\sigma_0}(\overline{x}) \notin t_{\mathfrak{M}}(\sigma_0(\overline{m}^1))$. We arrive at a contradiction. Hence there exists $\sigma \in S_n^\varkappa$ such that

$$t_{\mathfrak{M}}(\overline{m}^0) = t_{\mathfrak{M}}(\sigma(\overline{m}^1))$$

and $h_1 = \varkappa(\sigma(\overline{m}^1))$.

Sufficiency. Let $\varkappa \in \mathfrak{H}_n$, $\overline{m}^0, \overline{m}^1 \in M^n$, and $t_{\mathfrak{M}}(\overline{m}^0) = t_{\mathfrak{M}}(\overline{m}^1)$. Using Propositions 5.1.7 (ii) and 5.1.8 of [2], we find an elementary extension $\mathfrak{M}'$ of the model $\mathfrak{M}$ which is a special model.

Since $\mathfrak{M}'$ is ω-saturated, it follows that $\mathbb{HF}(\mathfrak{M}) \preccurlyeq \mathbb{HF}(\mathfrak{M}')$. Then

$$t_{\mathfrak{M}'}(\overline{m}^0) = t_{\mathfrak{M}}(\overline{m}^0) = t_{\mathfrak{M}}(\overline{m}^1) = t_{\mathfrak{M}'}(\overline{m}^1)$$

and, since $\mathfrak{M}'$ is special, there exists an automorphism φ of $\mathfrak{M}'$ such that $\varphi(\overline{m}^0) = \overline{m}^1$. We extend φ to an automorphism φ^ω of the model $\mathfrak{H} = \mathbb{HF}(\mathfrak{M}')$. Then

$$\varphi^\omega(\varkappa(\overline{m}^0)) = \varkappa(\varphi^\omega(\overline{m}^0)) = \varkappa(\varphi(\overline{m}^0)) = \varkappa(\overline{m}^1).$$

Consequently, $t_{\mathfrak{H}}(\varkappa(\overline{m}^0)) = t_{\mathfrak{H}}(\varkappa(\overline{m}^1))$, and

$$t_{\mathbb{HF}(\mathfrak{M})}(\varkappa(\overline{m}^0)) = t_{\mathfrak{H}}(\varkappa(\overline{m}^0)) = t_{\mathfrak{H}}(\varkappa(\overline{m}^1)) = t_{\mathbb{HF}}(\varkappa(\overline{m}^1)).$$

The theorem is proved. □

By *special admissible sets* we mean admissible sets of the form $\mathbb{HF}(\mathfrak{M})$, where $\mathfrak{M}$ is an algebraic structure which is a model of a simple theory T. The preceding shows that many questions about admissible sets of the form $\mathbb{HF}(\mathfrak{M})$ can be reduced to the corresponding ones about the structure $\mathfrak{M}$. On the other hand, such admissible sets present examples of structures for which there is a natural notion of computability without conditions on the cardinality of structures. Thus, if $\mathbb{R}$ denotes the field of real numbers, then the notion of Σ-definability in $\mathbb{HF}(\mathbb{R})$ is a reasonable "computability over $\mathbb{R}$". However, the notion of "simplicity of a theory" requires refinement. One of the required features of a simple theory is its regularity in the sense of the following definition:

Definition 7 A theory T of the (finite) signature σ' is called *regular* if it is decidable and model complete.

Remark 6 In a model complete theory T any formula is equivalent to an $\exists$-formula (see [2, Proposition 3.17]). If T is also decidable, then such an $\exists$-formula can be found effectively.

The theories $\mathrm{Th}(\mathbb{R})$ and $\mathrm{Th}(\mathbb{Q}_p)$ of the fields of real and p-adic numbers, respectively, provide examples of regular theories. However, it is hard to consider such theories as simple ones because they have many nonconstructivizable countable models.

Definition 8 A theory T is called *simple* if it is a regular ω-categorical theory whose set of complete formulas is decidable.

The condition of ω-categoricity means the uniqueness (up to an isomorphism) of a countable model of the theory. The model completeness and the decidability of the set of complete formulas guarantee the autoequivalence of all constructivizations of this countable model (see [3]), i.e., the "uniqueness" of the computability theory for countable models of such theories.

We now turn to the question of the possibility of (Σ-)definition of classical objects, the fields $\mathbb{C}$ and $\mathbb{R}$ of complex and real numbers, in special admissible sets. We first consider the case of the simple theory T_0 of an infinite structure with empty signature, i.e., models of the theory T_0 are presented by infinite sets without any additional structure. We assume that some uncountable algebraic structure $\mathfrak{M} = \langle M\,, \ldots \rangle$ is defined in $\mathbb{HF}(S)$, where S is an infinite set. Let $a_0\,, \ldots\,, a_m \in S$ be all the urelements appearing in the definition of $\mathfrak{M}$ as parameters and let $\overline{a} \rightleftharpoons \langle a_0\,, \ldots\,, a_m\rangle$. Let $\Psi_0(x_0)$ and $\Psi_1(x_0\,, x_1)$ be formulas from the definition of $\mathfrak{M}$ in $\mathbb{HF}(S)$ such that $M_0 \rightleftharpoons \Psi_0^{\mathbb{HF}(S)}[x_0]$, $\eta \rightleftharpoons \Psi_1^{\mathbb{HF}(S)}[x_0\,, x_1] \cap M_0^2$ is an equivalence on M_0, and there exists a one-to-one correspondence (induced by an isomorphism between $\mathfrak{M}$ and $\mathfrak{M}_0/\eta$) between M and M_0/η.

With any $n \in \omega$ and $\varkappa \in \mathfrak{H}_n$, we associate the sets

$$T_\varkappa \rightleftharpoons \{\overline{b} \mid \overline{b} \in S^n,\ \mathbb{HF}(S) \models \Psi_0(\varkappa(\overline{a}, \overline{b}))\},$$

$$M_\varkappa \rightleftharpoons \{\varkappa(\overline{a}, \overline{b}) \mid \overline{b} \in T_\varkappa\} \,/\, (\delta \upharpoonright \{\varkappa(\overline{a}, \overline{b}) \mid \overline{b} \in T_\varkappa\}^2)$$

and the relation

$$\delta_\varkappa \rightleftharpoons \{\langle \overline{b}^0, \overline{b}^1\rangle \mid \overline{b}^0, \overline{b}^1 \in T_\varkappa,\ \mathbb{HF}(S) \models \Psi_1(\varkappa(\overline{a}, \overline{b}^0), \varkappa(\overline{a}, \overline{b}^1))\}.$$

Then $M_\varkappa \subseteq M_0$, $\delta_\varkappa$ is an equivalence on $T_\varkappa$, and $|M_\varkappa| = |T_\varkappa/\delta_\varkappa|$. We note that since S is saturated enough, Theorem 10 implies that the set $T_\varkappa$ and the relation $\delta_\varkappa$ are closed with respect to types over $\overline{a}$, i.e., if $\overline{b} \in T_\varkappa$, $\overline{b}' \in S^n$ and $t_{(S,\overline{a})}(\overline{b}) = t_{(S,\overline{a})}(\overline{b}')$, then $\overline{b}' \in T_\varkappa$, and if $\langle \overline{b}^0, \overline{b}^1\rangle \in \delta_\varkappa$ and $t_{(S,\overline{a})}(\overline{b}^0, \overline{b}^1) = t_{(S,\overline{a})}(\overline{d}^0, \overline{d}^1)$ for some $\overline{d}^0, \overline{d}^1 \in S^n$, then $\langle \overline{d}^0, \overline{d}^1\rangle \in \delta_\varkappa$. It should be noted that the following obvious relation is valid:

$$M_0 = \bigcup_n \bigcup_{\varkappa \in \mathfrak{H}_n} M_\varkappa = \bigcup_{\varkappa \in \mathbb{HF}(\omega)} M_\varkappa .$$

Since M is uncountable, we can find $n \in \omega$ and $\varkappa \in \mathfrak{H}_n$ such that $M_\varkappa$ is uncountable. Then there exist pairwise unequal

$$b_1^0, \ldots, b_k^0 \; (b_1^1, \ldots, b_k^1) \in S \setminus \{a_0, \ldots, a_m\}$$

so that $1 + m + k = n$, $\varkappa(\overline{a}, \overline{b}^0)$, $\varkappa(\overline{a}, \overline{b}^1) \in M_0$, and $\langle \varkappa(\overline{a}, \overline{b}^0), \varkappa(\overline{a}, \overline{b}^1)\rangle \notin \eta$, where $\overline{b}^0 \rightleftharpoons \langle b_1^0, \ldots, b_k^0\rangle$, $\overline{b}^1 \rightleftharpoons \langle b_1^1, \ldots, b_k^1\rangle$. Without loss of generality, we can assume that there exists a unique i, $1 \leqslant i \leqslant k$, such that $b_i^0 \neq b_i^1$. Indeed, let us consider the following sequence of collections of k elements.

$$\overline{b}^0 = b_1^0, \ldots, b_k^0;$$

$$b_1^1, b_1^0, \ldots, b_k^0;$$

$$\vdots$$

$$b_1^1, \ldots, b_i^1, b_{i+1}^0, \ldots, b_k^0;$$

$$\vdots$$

$$\overline{b}^1 = b_1^1, \ldots, b_k^1 .$$

Each collection differs from the neighboring one by at most one element, so if for each neighboring pair $\overline{b}, \overline{b}'$ the pair $\langle \varkappa(\overline{a}, \overline{b}), \varkappa(\overline{a}, \overline{b}')\rangle$ belongs to η, then the pair $\langle \varkappa(\overline{a}, \overline{b}^0), \varkappa(\overline{a}, \overline{b}^1)\rangle$ must also belong to η since η is an equivalence. We note that $\varkappa(\overline{a}, \overline{b}) \in M_0$ for any collection $\overline{b}$ from this sequence. This follows from the fact that the types of the elements $\varkappa(\overline{a}, \overline{b}^0)$ and $\varkappa(\overline{a}, \overline{b}^1)$ coincide.

Hence we can assume that there exist $b_1, \ldots, b_{k-1}, b_k, b'_k \in S \setminus \{a_0, \ldots, a_m\}$ such that

$$\langle \varkappa(\overline{a}, b_1, \ldots, b_{k-1}, b_k), \varkappa(\overline{a}, b_1, \ldots, b_{k-1}, b'_k) \rangle \notin \eta.$$

Now for any $b \neq b' \notin \{\overline{a}, b_1, \ldots, b_{k-1}\}$, the fact that

$$t_{(S,\overline{a})}(b_1, \ldots, b_{k-1}, b_k, b_1, \ldots, b_{k-1}, b'_k) = t_{(S,\overline{a})}(b_1, \ldots, b_{k-1}, b, b_1, \ldots, b_{k-1}, b')$$

implies (as mentioned above) that

$$\langle \varkappa(\overline{a}, b_1, \ldots, b_{k-1}, b), \varkappa(\overline{a}, b_1, \ldots, b_{k-1}, b') \rangle \notin \eta.$$

It should be noted that any permutation α of the set S uniquely defines (lifts up to) an automorphism α^* of the admissible set $\mathbb{HF}(S)$ where

$$\alpha^*(s) \rightleftharpoons \alpha(s), \quad s \in S,$$

$$\alpha^*(\{s_1, \ldots, s_n\}) \rightleftharpoons \{\alpha^*(s_1), \ldots, \alpha^*(s_n)\}, \quad \{s_1, \ldots, s_n\} \in \mathbb{HF}(S) \setminus S.$$

If $\alpha \upharpoonright \{\overline{a}\} = \mathrm{id}_{\{\overline{a}\}}$, then α^* induces an automorphism on the structures $\mathfrak{M}_0$ and $\mathfrak{M}_0/\eta$. The following proposition is a consequence of the above considerations.

Proposition 11 *If an uncountable algebraic structure $\mathfrak{M}$ is definable in $\mathbb{HF}(S)$, then the automorphism group $\mathrm{Aut}(\mathfrak{M})$ of the structure $\mathfrak{M}$ contains a subgroup G that is isomorphic to the symmetric group $\mathrm{Sym}(S)$ of all permutations of the set S.*

Proof. Indeed, if $\varkappa$, $\overline{a}$, $b_1, \ldots, b_{k-1}$ have the same meaning as above, then any nontrivial permutation α_0 of the set

$$S_0 \rightleftharpoons S \setminus \{\overline{a}, b_1, \ldots, b_{k-1}\}$$

extended to S by letting $\alpha_0(\overline{a}) = \overline{a}$, $\alpha_0(b_1) = b_1, \ldots, \alpha_0(b_{k-1}) = b_{k-1}$, induces a nontrivial automorphism on $\mathfrak{M}_0/\eta$. Thus

$$\mathrm{Aut}(\mathfrak{M}_0/\eta) \simeq \mathrm{Aut}(\mathfrak{M})$$

contains a subgroup isomorphic to $\mathrm{Sym}(S_0) \simeq \mathrm{Sym}(S)$. □

Corollary 12 *The fields $\mathbb{C}$ and $\mathbb{R}$ are not definable in $\mathbb{HF}(S)$ for any set S.*

Proof. Indeed, the field $\mathbb{R}$ has no automorphisms, and the field $\mathbb{C}$ has no automorphisms of finite order greater than 2. □

Remark 7 It is easy to see that if the field $\mathbb{R}$ is definable (Σ–definable) in the KPU–model $\mathbb{A}$, then $\mathbb{C}$ is also definable (Σ–definable) in $\mathbb{A}$.

Hence Proposition 11 is a serious obstacle to the definability of uncountable structures in admissible sets of the form $\mathbb{HF}(S)$, where S is an infinite set.

We now consider the simple theory T_1 — the theory of dense linear orders without endpoints. Let L be a dense linear order without endpoints. We describe a general model-theoretic construction in terms of which we state necessary conditions for the definability in $\mathbb{HF}(L)$ of an uncountable structure (see Proposition 16 below), and a necessary and sufficient condition for the Σ-definability in $\mathbb{HF}(L)$ of uncountable models of a given theory (see Theorem 17 below).

The category ${}^*\omega$ is defined as follows. Its objects are the sets of the form $[\boldsymbol{n}] \rightleftharpoons \{0, 1, \dots, n-1\}$, $n \in \omega$ ($[\mathbf{0}] \rightleftharpoons \emptyset$), and its morphisms are order-preserving embeddings. It should be noted that there is a unique morphism from $[\mathbf{0}]$ into $[\boldsymbol{n}]$ for any $n \in \omega$.

Definition 9 By a *${}^*\omega$–spectrum*, we mean any functor S from the category ${}^*\omega$ into the category Mod^*_σ of algebraic structures (of some fixed signature σ) whose morphisms are all possible embeddings.

To define a ${}^*\omega$–spectrum S, it is necessary to give an infinite sequence $\mathfrak{M}_0, \mathfrak{M}_1, \dots$ of algebraic structures of signature σ, and associate with each order-preserving embedding $\mu : [\boldsymbol{n}] \to [\boldsymbol{m}]$ an embedding $\mu_* : \mathfrak{M}_n \to \mathfrak{M}_m$ so that, if $\mu_0 : [\boldsymbol{n}] \to [\boldsymbol{m}]$ and $\mu_1 : [\boldsymbol{m}] \to [\boldsymbol{k}]$, $n \leqslant m \leqslant k \in \omega$, are morphisms of the category ${}^*\omega$, then $(\mu_1\mu_0)_* = \mu_{1*}\mu_{0*}$, and if $\mu : [\boldsymbol{n}] \to [\boldsymbol{n}]$ is the unique morphism from $[\boldsymbol{n}]$ into $[\boldsymbol{n}]$ ($= \mathrm{id}_{[\boldsymbol{n}]}$), then $\mu_* = \mathrm{id}_{\mu n} : \mathfrak{M}_n \to \mathfrak{M}_n$, $n \in \omega$.

If the ${}^*\omega$–spectrum $S = \{\mathfrak{M}_n, \mu_* \mid n \in \omega,\ \mu \in \mathrm{Mor}\,{}^*\omega\}$ has been defined, then for any linearly ordered set L, it is possible to define the algebraic structure $\mathfrak{M}_L$ ($\mathfrak{M}^S_L$) as a direct limit $\varinjlim_{L_0} \mathfrak{M}'_{L_0}$ of the spectrum

$$\{\mathfrak{M}'_{L_0}, \varphi_{L_0, L_1} \mid L_0 \subseteq L_1 \subseteq L,\ L_1 \text{ is finite}\},$$

where $\mathfrak{M}'_{L_0} \rightleftharpoons \mathfrak{M}_n$ if $L_0 \subseteq L$ is finite and $|L_0| = n$, and the embedding $\varphi_{L_0, L_1} : \mathfrak{M}'_{L_0} \to \mathfrak{M}'_{L_1}$ is defined for finite $L_0 \subseteq L_1$ ($\subseteq L$) as follows: if

$$L_1 = \{l_0 < l_1 < \cdots < l_{m-1}\} \quad \text{and} \quad L_0 = \{l_{i_0} < l_{i_1} < \cdots < l_{i_{n-1}}\}$$

(in which case $0 \leqslant i_0 < i_1 < \cdots < i_{n-1} \leqslant m$), and $\mu : [\boldsymbol{n}] \to [\boldsymbol{m}]$ is defined as $\mu(j) \rightleftharpoons i_j$ for $j < n$, then

$$\varphi_{L_0, L_1} \quad \rightleftharpoons \quad \mu_* : \mathfrak{M}'_{L_0} = \mathfrak{M}_n \quad \to \quad \mathfrak{M}_m = \mathfrak{M}'_{L_1}.$$

If $L \subseteq L'$ are linearly ordered sets, then the structure $\mathfrak{M}_L$ can be identified with a substructure of $\mathfrak{M}_{L'}$ in a natural way.

Remark 8 Any isomorphism between linearly ordered sets L and L' induces an isomorphism between $\mathfrak{M}_L$ and $\mathfrak{M}_{L'}$.

Proposition 13 *If $L \subseteq L'$ are dense linear orders without endpoints, then $\mathfrak{M}_L \preceq \mathfrak{M}_{L'}$.*

Proof. For countable L' the assertion follows from [6, Proposition 24.2]. The case of an arbitrary L' is proved by the application of the Tarski–Vaught theorem on elementary embeddings. □

Corollary 14 *If L and L' are dense linear orders without endpoints, then $\mathfrak{M}_L \equiv \mathfrak{M}_{L'}$.*

Let μ_0 and μ_1 be morphisms from $[\mathbf{1}]$ into $[\mathbf{2}]$ such that $\mu_0(0) \rightleftharpoons 0$ and $\mu_1(0) \rightleftharpoons 1$. The condition

$$\mu_{0*} \neq \mu_{1*} \tag{$*$}$$

is sufficient for $|\mathfrak{M}^S_L| \geqslant |L|$ to hold for any linearly ordered set L. Indeed, let $\xi \in M_1$ be such that $\mu_{0*}(\xi) \neq \mu_{1*}(\xi)$. For any linearly ordered set L, and $l \in L$, we define the corresponding embedding $\mu_l : \mathfrak{M}'_{\{l\}} = \mathfrak{M}_1 \to \mathfrak{M}_L$, and let $\xi_l \rightleftharpoons \mu_l(\xi)$. Then $l \neq l'$ implies $\xi_l \neq \xi_{l'}$ and $|L| = |\{\xi_l \mid l \in L\}| \leqslant |M_L|$.

The following model-theoretic claim is stated without proof.

Proposition 15 *Let S be a $^*\omega$-spectrum such that $\mu_{0*}(\xi) \neq \mu_{1*}(\xi)$ for some $\xi \in M_1$, and let L be a dense linear order without endpoints. Then*

$$\{\xi_l \mid l \in L\}$$

is a set of indiscernible elements of the structure $\mathfrak{M}_L$ (with order induced by L).

Proposition 16 *Let $\mathfrak{M}$ be an uncountable algebraic structure defined in $\mathbb{HF}(L)$ for some dense linear order L without endpoints. Then there exists a $^*\omega$-spectrum S such that $\mu_{0*} \neq \mu_{1*}$, all the models $\mathfrak{M}_0, \mathfrak{M}_1, \dots$ of this spectrum are countable and isomorphic, $\mathfrak{M}_0 \preccurlyeq \mathfrak{M}_1 \preccurlyeq \dots \preccurlyeq \mathfrak{M}$, and for any embedding $\mu : [\boldsymbol{n}] \to [\boldsymbol{m}]$, the embedding $\mu_* : \mathfrak{M}_n \to \mathfrak{M}_m$ is elementary.*

Proof. The preceding analysis of definable sets shows that there exist $k \in \omega$, $\varkappa \in \mathbb{HF}(k)$, and sequences $\overline{a}, \overline{b} \in L^k$ that have the same type over $(L, \overline{c})$, where $\overline{c}$ are constants appearing in the definition of $\mathfrak{M}$ in $\mathbb{HF}(L)$ (we can assume that $\{\overline{c}\} \subseteq \{\overline{a}\} \cap \{\overline{b}\}$), such that $\varkappa(\overline{a}), \varkappa(\overline{b}) \in M_0$ and $\langle \varkappa(\overline{a}), \varkappa(\overline{b}) \rangle \notin \eta_{\varkappa}$. Then $\overline{a} \neq \overline{b}$, and we can show (in a similar way to the proof of Proposition 11) that there exist such $\overline{a}$ and $\overline{b}$ so that $a_i \neq b_i$ for a single number $i < k$ ($a_j = b_j$ for $j < k$, $j \neq i$). For definiteness, we will assume that $a_i < b_i$. We suppose that $a_0 < a_1 < \dots < a_{k-1}$, and $0 < i < k-1$. Then $a_i < b_i < a_{i+1}$.

By the Löwenheim–Skolem theorem, we can find a countable elementary submodel $\mathbb{HF}(L^0) \preccurlyeq \mathbb{HF}(L)$ such that $\{\overline{c}\} \subseteq \{\overline{a}\} \cup \{b_i\} \subseteq L^0$. Since L^0 is a dense linear order, there are elements $d_2, d_3, \dots, d_n, \dots$ such that $b_i < d_2 < d_3 < \dots < d_n < \dots < a_{i+1}$. We set $d_0 \leftrightharpoons a_i$ and $d_1 \leftrightharpoons b_i$. Then

$$a_{i-1} < d_0 < d_1 < d_2 < \dots < d_n < \dots < a_{i+1}.$$

We set $\delta_k \leftrightharpoons [d_k, d_{k+1})$, $k \in \omega$, and

$$L_n \leftrightharpoons (-\infty, d_0) \cup \delta_0 \cup \dots \cup \delta_{n-1} \cup [a_{i+1}, \infty) = (-\infty, d_n) \cup [a_{i+1}, \infty)$$

(define $L_0 = (-\infty, d_0) \cup [a_{i+1}, \infty)$).

From the above definition, it is clear that L_n is a countable dense linear order without endpoints which is isomorphic to L^0 over $\{\overline{c}\}$. Let $\mathfrak{M}_n$ be an algebraic structure definable in $\mathbb{HF}(L_n)$ by the same formulas (and parameters $\overline{c}$) as $\mathfrak{M}$ in $\mathbb{HF}(L)$. For any $n \in \omega$ we fix an isomorphism between linearly ordered sets $\varphi_n : \delta_n \leftrightharpoons [d_n, d_{n+1}) \to \delta_0 \leftrightharpoons [d_0, d_1)$ (such an isomorphism exists because δ_n and δ_0 are countable dense linear orders with least elements but without largest elements). For $n = 0$ we set $\varphi_0 \leftrightharpoons \mathrm{id}_{\delta_0}$. For $n, k \in \omega$ let $\varphi_{n,k} : \delta_n \to \delta_k$ be an isomorphism between linearly ordered sets which is defined as follows: $\varphi_{n,k} = \varphi_k^{-1} \varphi_n$ (we note that $\varphi_{n,n} = \mathrm{id}_{\delta_n}$ and $\varphi_{n,l} = \varphi_{k,l}\, \varphi_{n,k}$ for $n, k, l \in \omega$).

Let $\mu : [\boldsymbol{n}] \to [\boldsymbol{m}]$ be a monotone embedding of $[\boldsymbol{n}] = \{0, \dots, n-1\}$ into $[\boldsymbol{m}] = \{0, \dots, m-1\}$ (i.e., μ is a morphism of the category $^*\omega$).

Corresponding to this embedding, we assign the monotone embedding $\overline{\mu} : L_n \to L_m$ uniquely defined by the following relations:

$$\overline{\mu} \restriction L_0 = \mathrm{id}_{L_0}, \qquad \overline{\mu} \restriction \delta_k = \varphi_{k,\mu(k)}, \quad \text{for } k < n$$

(we note that $L_n = L_0 \cup \delta_0 \cup \cdots \cup \delta_{n-1}$). It is easy to verify that $\mu \mapsto \overline{\mu}$ is a functor from $^*\omega$ into the category of linearly ordered sets with monotone embeddings (i.e., $\overline{\mathrm{id}_{[\boldsymbol{n}]}} = \mathrm{id}_{L_n}$ and $\overline{\mu\lambda} = \overline{\mu}\overline{\lambda}$ for $\mu : [\boldsymbol{n}] \to [\boldsymbol{m}]$, $\lambda : [l] \to [\boldsymbol{n}]$). For a monotone embedding $\mu : [\boldsymbol{n}] \to [\boldsymbol{m}]$, the embedding $\overline{\mu} : L_n \to L_m$ (identical on $\{\overline{c}\}$!) induces elementary embeddings $\mu' : \mathbb{HF}(L_n) \to \mathbb{HF}(L_m)$, and $\mu_* : \mathfrak{M}_n \to \mathfrak{M}_m$. It is easy to check that $\mu \mapsto \mu_*$ is a functor from $^*\omega$ into the category of algebraic structures that are isomorphic to $\mathfrak{M}_0$ with elementary embeddings as morphisms.

It remains to check that $\mu_{0*} \neq \mu_{1*}$. By definition ($d_0 = a_i$ and $d_1 = b_i$), we see that $\varkappa(\overline{a}) \in L_1$, $\mu_0'(\varkappa(\overline{a})) = \varkappa(\overline{a})$ and $\mu_1'(\varkappa(\overline{a})) = \varkappa(\overline{b})$ (since $\overline{\mu}_1(d_0) = d_1 = b_i$ and $\overline{\mu}_1(d_j) = a_j$ for $j < k$, $j \neq i$). Since $\langle \varkappa(\overline{a}), \varkappa(\overline{b}) \rangle \notin \eta_\varkappa$ for $\mathbb{HF}(L)$, this fact remains valid for $\mathbb{HF}(L_2) \preccurlyeq \mathbb{HF}(L)$. Thus, $\mu_{0*}([\varkappa(\overline{a})]) = [\varkappa(\overline{a})]_\eta \neq [\varkappa(\overline{b})]_\eta = \mu_{1*}([\varkappa(\overline{a})])$. □

Proposition 16 gives a necessary condition for definability in $\mathbb{HF}(L)$, where L is a dense linear order without endpoints. It turns out that the effectivization of this condition is already a necessary and sufficient condition for Σ–definability in $\mathbb{HF}(L)$.

Definition 10 A system of numberings $\nu_n : \omega \to M_n$, $n \in \omega$, is called a *computable sequence of constructivizations*

$$(\mathfrak{M}_0, \nu_0), (\mathfrak{M}_1, \nu_1), \ldots, (\mathfrak{M}_n, \nu_n), \ldots, \qquad n \in \omega,$$

if the following conditions hold (we assume that the signature σ of the structures $\mathfrak{M}_0$, $\mathfrak{M}_1$, ... is finite and without function symbols):

(1) $E \rightleftharpoons \{\langle n, m_0, m_1 \rangle \mid n, m_0, m_1 \in \omega,\ \nu_n(m_0) = \nu_n(m_1)\}$ is a Δ–predicate on ω,

(2) $N_P \rightleftharpoons \{\overline{n} = \langle n_0, n_1, \ldots, n_k \rangle \mid \overline{n} \in \omega^{k+1}, \langle \nu_{n_0}(n_1), \ldots, \nu_{n_0}(n_k) \rangle \in P^{\mathfrak{M}_{n_0}}\}$ is a Δ–predicate on ω for any (k–ary) predicate symbol $P \in \sigma$;

(3) for any constant symbol $c \in \sigma$ there exists a Σ–function $f_c : \omega \to \omega$ such that $c^{\mathfrak{M}_n} = \nu_n f_c(n)$.

Every morphism $\mu : [\boldsymbol{n}] \to [\boldsymbol{m}]$ of the category ${}^*\omega$ is uniquely defined by the number m and the subset $\mu([\boldsymbol{n}]) \subseteq [\boldsymbol{m}]$. This remark allows one to define a one-to-one correspondence $\mu^* : \Delta \to \mathrm{Mor}^*\omega$ between the subset $\Delta \rightleftharpoons \{n \mid n \in \omega,\ r(n) < 2^{l(n)}\} \subseteq \omega$ and the set $\mathrm{Mor}^*\omega$, provided that $n \in \Delta$ is assumed to code the morphism $\mu : [\boldsymbol{k}] \to [\boldsymbol{l}]$ such that $l = l(n)$ and $r(n)$ is the number of the subset $\mu([\boldsymbol{k}]) \subseteq [\boldsymbol{l}] = [\boldsymbol{l}(n)]$ in some standard listing of finite subsets of ω. (Here l and r are such that $n = \langle l(n), r(n)\rangle$.) It is evident that Δ is a Δ–subset of ω.

Let $S = \{\mathfrak{M}_n, \mu_* \mid n \in \omega, \mu \in \mathrm{Mor}^*\omega\}$ be a ${}^*\omega$–spectrum.

Definition 11 By a *constructivization* of S we mean any computable sequence of constructivizations

$$(\mathfrak{M}_0, \nu_0)\,,\ (\mathfrak{M}_1, \nu_1)\,,\ \ldots\ ,\ (\mathfrak{M}_n, \nu_n)\,,\ \ldots\ , \qquad n \in \omega,$$

together with a Σ–function $f : \Delta \times \omega \to \omega$ such that, for any $n, m, k \in \omega$ and $\mu : [\boldsymbol{n}] \to [\boldsymbol{m}] \in \mathrm{Mor}^*\omega$, if $n^* \in \Delta$ is such that $\mu^*(n^*) = \mu$, then $\mu_*\nu_n(k) = \nu_m f(n^*, k)$.

A ${}^*\omega$–spectrum S is called *constructivizable* if there exists a constructivization for it.

Theorem 17 *Let L be a dense linear order without endpoints. A theory T has an uncountable model Σ–definable in $\mathbb{HF}(L)$ if and only if there exists a constructivizable ${}^*\omega$–spectrum S satisfying condition $(*)$, and such that $\mathfrak{M}_L^S \models T$.*

Proof. Let $\mathfrak{M}$ be an uncountable model of the theory T which is Σ–definable in $\mathbb{HF}(L)$ for some dense linear order L without endpoints.

Acting as in the proof of Proposition 16, we find $\varkappa$, $\bar{a}$, $\bar{b}$, L^0, d_2, d_3, ..., and so on. Since L^0 is a countable dense linear order without endpoints, it is isomorphic to the set $\mathbb{Q}$ of rational numbers with the natural order. From this, the existence of a constructivization $\nu : \omega \to \mathbb{HF}(L^0)$ of the admissible set

$$\mathbb{HF}(L^0) \quad (\simeq\ \mathbb{HF}(\mathbb{Q}, \leqslant))$$

follows. Since M_0 is Σ–definable, the set $\nu^{-1}(M_0) = \{n \mid n \in \omega,\ \nu(n) \in M_0\}$ is a Σ–subset of ω. The predicate

$$\{\langle n\,,\ m\rangle \bigm| n, m \in \omega,\ \nu(n), \nu(m) \in M_0,\ \langle \nu(n)\,,\ \nu(m)\rangle \in \eta\}$$

is a Δ-predicate on $\nu^{-1}(M_0)$. The predicate

$$\nu^{-1}(P^{\mathfrak{M}_0}) \rightleftharpoons \{\langle n_1, \dots, n_k\rangle \mid n_i \in \omega, \ \langle \nu(n_1), \dots, \nu(n_k)\rangle \in P^{\mathfrak{M}_0}\}$$

is a Δ-predicate on $\nu^{-1}(M_0)$ for any (k-ary) predicate symbol $P \in \sigma$.

The elements $d_2 < \cdots < d_n < \cdots$, for $n \in \omega$, of the set L^0 (for which $b_i < d_n < a_{i+1}$ for all $n \in \omega$) can be chosen so that there exists a Σ-function $d : \omega \to \omega$ such that $\nu d(n) = d_n$ for all $n \in \omega$. If the substructures $\mathfrak{M}_n$, $n \in \omega$, of the structure $\mathfrak{M}$ are defined as in Proposition 16, then it is obvious that the corresponding $^*\omega$-spectrum has a constructivization. (We only need to choose uniformly effective isomorphisms $\varphi_n : \delta_n \to \delta_0$, $n \in \omega$.)

Conversely, let S be a $^*\omega$-spectrum such that $\mathfrak{M}_L^S \models T$ for (any) dense linear order L without endpoints. Let

$$(\mathfrak{M}_0, \nu_0), \dots, (\mathfrak{M}_n, \nu_n), \dots, \qquad n \in \omega,$$

and let $f : \Delta \times \omega \to \omega$ be a constructivization of this $^*\omega$-spectrum. We sketch the proof of the Σ-definability of the model $\mathfrak{M}_L^S$ in $\mathbb{HF}(L)$. Let $L_0 = \{l_0 < \cdots < l_{n-1}\} \subseteq L$ be a finite subset of the set L,

$$\overline{L_0} \rightleftharpoons \langle l_0, \dots, l_{n-1}\rangle \in L^n$$

(for $n = 0$ we set $\overline{L_0} \rightleftharpoons \emptyset$). Let

$$M^0_{L_0} \rightleftharpoons \{\langle \overline{L_0}, k\rangle \mid k \in \omega\},$$

$$\eta_{L_0} \rightleftharpoons \{\langle\langle \overline{L_0}, k_0\rangle, \langle \overline{L_0}, k_1\rangle\rangle \mid k_0, k_1 \in \omega, \ \nu_n(k_0) = \nu_n(k_1)\},$$

$$P^0_{L_0} \rightleftharpoons \{\langle\langle \overline{L_0}, k_1\rangle, \dots, \langle \overline{L_0}, k_m\rangle\rangle \mid k_1, \dots, k_m \in \omega, \ \langle \nu_n(k_1), \dots, \nu_n(k_m)\rangle \in P^{\mathfrak{M}_n}\}$$

for each (m-ary) predicate symbol $P \in \sigma$. For $c \in \sigma$ we choose $k \in \omega$ so that $c^{\mathfrak{M}_n} = \nu_n(k)$ and set $c^{\mathfrak{M}^0_{L_0}} \rightleftharpoons \langle \overline{L_0}, k\rangle$. Then η_{L_0} is a congruence on the structure

$$\mathfrak{M}^0_{L_0} \rightleftharpoons \langle M^0_{L_0}, \dots, P^{\mathfrak{M}^0_{L_0}}, \dots, c^{\mathfrak{M}^0_{L_0}}, \dots\rangle$$

and the quotient structure $\mathfrak{M}_{L_0} \rightleftharpoons \mathfrak{M}^0_{L_0} / \eta_{L_0}$ is isomorphic to $\mathfrak{M}_n$. Let $L_0 \subseteq L_1 \subseteq L$ be finite subsets of the set L,

$$L_0 = \{l_0 < \cdots < l_{n-1}\}, \qquad L_1 = \{l'_0 < \cdots < l'_{m-1}\}$$

and $0 \leqslant j_0 < \cdots < j_{n-1} < m$ be such that $l_i = l'_{j_i}$ for $i < n$. Now let $\mu \rightleftharpoons \mu_{L_0,L_1} : [\boldsymbol{n}] \to [\boldsymbol{m}]$ be the morphism of ${}^*\omega$ such that $\mu(i) = j_i$ for $i < n$. Then the embedding $\mu_* : \mathfrak{M}_n \to \mathfrak{M}_m$ defines the embedding $\overline{\mu}_* : \mathfrak{M}_{L_0} \to \mathfrak{M}_{L_1}$. On the set

$$M^0 \rightleftharpoons \bigcup\{M^0_{L_0} \mid L_0 \subseteq L \text{ is a finite subset of } L\},$$

we define the equivalence relation $\eta \subseteq (M^0)^2$ as follows. Then for $\langle \overline{L}_0, k\rangle$, $\langle \overline{L}_1, l\rangle \in M^0$, we set $\langle\langle \overline{L}_0, k\rangle, \langle \overline{L}_1, l\rangle\rangle \in \eta$ if

$$(\mu_{L_0,L_2})_*([\langle \overline{L}_0, k\rangle]_{\eta_{L_0}}) = (\mu_{L_1,L_2})_*([\langle \overline{L}_1, l\rangle]_{\eta_{L_1}})$$

for $L_2 \rightleftharpoons L_0 \cup L_1$. It is easy to verify that $\eta \cap (M^0_{L_0})^2 = \eta_{L_0}$ for finite $L_0 \subseteq L$. Taking into account that $L_0 \subseteq L_1 \subseteq L_2 \subseteq L$ implies $\mu_{L_0,L_2} = \mu_{L_1,L_2}\,\mu_{L_0,L_1}$ (for a finite L_2) and using the definition of a ${}^*\omega$–spectrum, we obtain the following equivalence:

$$\begin{aligned} \langle\langle \overline{L}_0, k\rangle, \langle \overline{L}_1, l\rangle\rangle \in \eta \Leftrightarrow {} & \text{there exists a finite subset } L_2 \subseteq L, \\ & \text{such that } L_0 \cup L_1 \subseteq L_2, \\ & \text{and } (\mu_{L_0,L_2})_*([\langle \overline{L}_0, k\rangle]_{\eta_{L_0}}) = (\mu_{L_1,L_2})_*([\langle \overline{L}_1, l\rangle]_{\eta_{L_1}}). \end{aligned}$$

Owing to this equivalence, it is not hard to check that η is an equivalence relation on M^0. Furthermore, η is a congruence on the algebraic structure

$$\mathfrak{M}^0 \rightleftharpoons \langle M^0, \ldots, \bigcup\{P^{\mathfrak{M}^0_{L_0}} \mid L_0 \subseteq L \text{ is finite}\}, \ldots, c^{\mathfrak{M}^0_{\varnothing}}, \ldots\rangle.$$

The quotient structure $\mathfrak{M}^0/\eta$ is isomorphic to $\mathfrak{M}^S_L$.

It is not hard to verify that the set M^0 is a Δ–subset of $\mathbb{HF}(L)$, and the relation η and predicates $\bigcup\{P^{\mathfrak{M}_{L_0}} \mid L_0 \subseteq L \text{ is finite}\}$, $P \in \sigma$, are Δ–predicates on $\mathbb{HF}(L)$. To this end, it is necessary to use the properties of constructivization of the ${}^*\omega$–spectrum S, and the remark that a Δ–subset (Σ–subset) and a Δ–predicate (Σ–predicate) in $\Omega \rightleftharpoons \langle \omega, 0, s, +, \cdot, \leqslant\rangle$ are the same in any admissible set. Theorem 17 is proved. □

Regarding this theorem, the following conjecture arises.

If a theory T has an uncountable model that is Σ–definable in $\mathbb{HF}(\mathfrak{M})$ for an algebraic structure $\mathfrak{M}$ with a simple theory, then the theory T also has an uncountable model that is Σ–definable in $\mathbb{HF}(L)$ for some dense linear order L.

We indicate an example of the application of Theorem 17. Let $\mathbb{Q}(a_0, a_1, \ldots)$ denote the purely transcendental extension of the field $\mathbb{Q}$ of rational numbers regarded as a linearly ordered field such that for $n \in \omega$, all elements of the field $\mathbb{Q}_n \rightleftharpoons \mathbb{Q}(a_0, \ldots, a_{n-1})$ (we set $\mathbb{Q}_0 \rightleftharpoons \mathbb{Q}$) are infinitesimal with respect to the element a_n. Such an ordering is unique, and there exists a single-valued constructivization

$$\nu : \omega \to \langle \mathbb{Q}(a_0, \ldots, a_n, \ldots \mid n \in \omega), \leqslant \rangle$$

such that $n \to \nu^{-1}(a_n)$, $n \in \omega$, is a Σ–function on Ω.

Let $\mathbb{R}^*$ be the real closure of the ordered field $\langle \mathbb{Q}(a_n \mid n \in \omega), \leqslant \rangle$. Then there exists a constructivization $\nu^* : \omega \to \mathbb{R}^*$ that "extends" ν. We note that the embedding $\mathbb{Q}(a_n \mid n \in \omega) \to \mathbb{R}^*$ is a morphism from $(\mathbb{Q}(a_n \mid n \in \omega), \nu)$ into $(\mathbb{R}^*, \nu^*)$.

If $\mathbb{R}^*_n$, $n \in \omega$, is the algebraic closure of $\mathbb{Q}_n$ in $\mathbb{R}^*$, then $n \to (\nu^*)^{-1}(\mathbb{R}^*_n)$, $n \in \omega$, is the computable numbering of a Σ–subset of ω, which allows us to construct a computable sequence of constructivizations

$$(\mathbb{R}^*_0, \nu_0), \ldots, (\mathbb{R}^*_n, \nu_n), \ldots$$

such that the embeddings $\mathbb{R}^*_n \subseteq \mathbb{R}^*$ are morphisms from $(\mathbb{R}^*_n, \nu_n)$ into $(\mathbb{R}^*, \nu^*)$.

Let $\mu : [\boldsymbol{n}] \to [\boldsymbol{m}]$ be a morphism of ${}^*\omega$. The embedding $\mu' : \mathbb{Q}_n \to \mathbb{Q}_m$ such that $\mu'(a_i) = a_{\mu(i)}$ can be uniquely determined from μ. It is an order-preserving embedding. Consequently, it induces the uniquely defined embedding $\overline{\mu} : \mathbb{R}^*_n \to \mathbb{R}^*_m$. Therefore, the structure

$$\{\mathbb{R}^*_n, \overline{\mu} \mid n \in \omega,\ \mu \in \mathrm{Mor}^*\omega\}$$

is a ${}^*\omega$–spectrum which is constructivizable. Indeed, the sequence of constructivizations

$$(\mathbb{R}^*_0, \nu_0), \ldots, (\mathbb{R}^*_n, \nu_n), \ldots, \qquad n \in \omega,$$

was given above. We can assume that all the constructivizations ν_n are single-valued. Then the functions f_μ such that $\overline{\mu}(\nu_n(l)) = \nu_m f_\mu(l)$ for any $l \in \omega$ are uniquely determined from $\mu \in \mathrm{Mor}^*\omega$. These functions are Σ–functions and the corresponding function $f : \Delta \times \omega \to \omega$ such that $f(n^*, k) = f_{\mu^*(n^*)}(k)$, $n^* \in \Delta$, $k \in \omega$, is a Σ–function. Hence the ${}^*\omega$–spectrum

$$\{\mathbb{R}^*_n, \overline{\mu} \mid n \in \omega,\ \mu \in \mathrm{Mor}^*\omega\}$$

is constructivizable and, by Theorem 17, the structure $\mathbb{R}^{*S}_L$, being the real-closed field of cardinality $|L|$, is Σ–definable in $\mathbb{HF}(L)$.

Corollary 18 *If L is a dense linear order without endpoints of the power of the continuum, then the field $\mathbb{C}$ of complex numbers is Σ-definable in $\mathbb{HF}(L)$.*

Proof. Indeed, in $\mathbb{HF}(L)$, some real-closed field $\overline{\mathbb{R}}$ of the power of the continuum is Σ-definable. Therefore, its finite extension $\overline{\mathbb{R}}(i)$, the algebraic closure of $\overline{\mathbb{R}}$, is also definable in $\mathbb{HF}(L)$, and $\overline{\mathbb{R}}(i)$ is isomorphic to $\mathbb{C}$. □

Neither Proposition 16 nor Theorem 17 can help in solving the question of the definability (Σ-definability) of a concrete uncountable model. By Corollary 18, the field $\mathbb{C}$ of complex numbers is Σ-definable in $\mathbb{HF}(L)$ for any dense linear order L of the power of the continuum. But what about the definability of the field $\mathbb{R}$ of real numbers?

Proposition 19 *The field $\mathbb{R}$ is not definable in $\mathbb{HF}(L)$ for any dense linear order L.*

Proof. If $\mathbb{R}$ is definable in $\mathbb{HF}(L)$ for some dense linear order L, then it is possible to choose two countable suborders $L_0 \prec L_1$ ($\prec L$) such that L_0 contains all the parameters from the definition of $\mathbb{R}$ in $\mathbb{HF}(L)$, and if $\mathbb{R}_0$ and $\mathbb{R}_1$ are subfields of $\mathbb{R}$ defined in $\mathbb{HF}(L_0)$ and $\mathbb{HF}(L_1)$ respectively (in the same way as $\mathbb{R}$ in $\mathbb{HF}(L)$), then $\mathbb{R}_0 < \mathbb{R}_1 < \mathbb{R}$. Since $\mathbb{R}_0$ and $\mathbb{R}_1$ are different countable subfields of $\mathbb{R}$, they are not isomorphic (since $\mathbb{R}$ is Archimedean). On the other hand, there exists an isomorphism between L_0 and L_1 (preserving parameters). Consequently, it must be that $\mathbb{R}_0 \simeq \mathbb{R}_1$. The contradiction thus obtained proves the proposition. □

Remark 9 Using a similar argument, we can prove that the field $\mathbb{Q}_p$ of p-adic numbers is not definable in $\mathbb{HF}(L)$ for any dense linear order L.

We establish a general fact which implies that $\mathbb{R}$ is not Σ-definable in $\mathbb{HF}(L)$ for any linear order L.

Definition 12 An algebraic structure $\mathfrak{M}$ of finite (recursive) signature is called *locally constructivizable* if for any finite family of elements $a_0, \ldots, a_n \in M$, the $\exists$-theory $\mathrm{Th}_\exists(\mathfrak{M}, \overline{a})$ of the structure $(\mathfrak{M}, a_0, \ldots, a_n)$ is recursively enumerable.

Remark 10 A structure $\mathfrak{M}$ is locally constructivizable if and only if, for any $a_0, \ldots, a_n \in M$, there exist a constructivizable structure $\mathfrak{N}$ and $b_0, \ldots, b_n \in N$ such that $\mathrm{Th}_\exists(\mathfrak{M}, \overline{a}) = \mathrm{Th}_\exists(\mathfrak{N}, \overline{b})$.

The decidability of the theory Th($\mathfrak{M}$) and its categoricity with respect to some cardinality is a sufficient condition for the local constructivizability of $\mathfrak{M}$. In particular, all models of simple theories are locally constructivizable.

Proposition 20 *If $\mathfrak{M}$ is locally constructivizable and $\mathfrak{N}$ is Σ-definable in $\mathbb{HF}(\mathfrak{M})$, then $\mathfrak{N}$ is locally constructivizable.*

Proof. The assertion is obtained from the above remark and the following general fact:

If $\mathfrak{M}$ is constructivizable, and $a_0, \ldots, a_{n-1} \in M$, then the Σ-theory of the structure $\langle \mathbb{HF}(\mathfrak{M}), \varkappa(\overline{a}) \rangle_{\varkappa \in \mathbb{HF}(n)}$ is recursively enumerable. □

Remark 11 Any two infinite linearly ordered sets have the same $\exists$-theory, whence any linearly ordered set is locally constructivizable.

Corollary 21 *The field $\mathbb{R}$ is not Σ-definable in $\mathbb{HF}(L)$ for any linear order L.*

In conclusion, we give an example of a complete decidable theory only the simple model of which is locally constructivizable. Let $E \rightleftharpoons \{0,1\}^*$ be the set of all finite sequences formed by 0 and 1. The signature $\sigma \rightleftharpoons \{P^1_\varepsilon \mid \varepsilon \in E\}$ consists only of unary predicates. The theory T is defined by the choice of an infinite recursive binary tree $D(\subseteq E)$ with no infinite recursive branches (an example of such a tree can be found, e.g., in [8]) and by the following system of axioms:

$$\forall x\, P_\Lambda(x),$$

$$\forall x\, \big(P_{\varepsilon 0}(x) \vee P_{\varepsilon 1}(x) \to P_\varepsilon(x)\big), \quad \varepsilon \in E,$$

$$\forall x\, \big((P_{\varepsilon 0}(x) \to \neg P_{\varepsilon 1}(x)) \wedge (P_{\varepsilon 1}(x) \to \neg P_{\varepsilon 0}(x))\big), \quad \varepsilon \in E,$$

$$\exists x\, \big(P_\varepsilon(x) \wedge \neg P_{\varepsilon 0}(x) \wedge \neg P_{\varepsilon 1}(x)\big), \quad \varepsilon \in D,$$

$$\forall x\, \neg P_\varepsilon(x), \quad \varepsilon \in E \setminus D,$$

$$\forall x\, \forall y\, \big(P_\varepsilon(x) \wedge \neg P_{\varepsilon 0}(x) \wedge \neg P_{\varepsilon 1}(x) \wedge P_\varepsilon(y) \wedge \neg P_{\varepsilon 0}(y) \wedge \neg P_{\varepsilon 1}(y) \to x = y\big), \quad \varepsilon \in E.$$

References

[1] J. Barwise, (ed.), Handbook of Mathematical Logic, Stud. Logic Found. Math., **90** (1977).

[2] C. C. Chang and H. J. Keisler, Model Theory, 3rd. edn., Stud. Logic Found. Math., **73** (1990); [1st. edn. 1973, 2nd. edn. 1977].

[3] Yu. L. Ershov, Decision Problems and Constructivizable Models (Russian), (Mathematical Logic and Foundations of Mathematics, Nauka, Moskva, 1980).

[4] Yu. L. Ershov, Σ–definability in admissible sets (Russian), Dokl. Akad. Nauk SSSR, **285** (1985) 259–262; [translated in: Soviet Math. – Dokl., **32** (1985) 767–770].

[5] Yu. L. Ershov, Definability and Computability (Russian), (Sibirsk. Shk. Alg. i Log., NII MIOONGU, Novosibirsk, 1996); [translated in: Siberian School of Algebra and Logic, (Consultants Bureau, Plenum, New York, 1996)].

[6] Yu. L. Ershov and E. A. Palyutin, Mathematical Logic (Russian), 2nd. edn., (Nauka, Moskva, 1987). [English translation of 1st. edn., translated by V. Shokurov, (MIR Publishers, Moscow, 1984)].

[7] H. J. Keisler, Fundamentals of Model Theory, in: J. Barwise, (ed.), Handbook of Mathematical Logic, Stud. Logic Found. Math., **90** (1977) 47–103.

[8] M. G. Peretyat'kin, Strongly constructive models and enumerations of the Boolean algebra of recursive sets (Russian), Algebra i Logika, **10** (1971) 535–557; [translated in: Algebra and Logic, **10** (1971) 332–345].

Yu. L. Ershov
Sobolev Institute of Mathematics
pr. Akademika Koptuga, 4
Novosibirsk, 630090
Russia

Chapter 6

Autostable Models and Algorithmic Dimensions

S. S. Goncharov*

In the theory of constructive models, the study of types of different effective representations of models is, together with the existence problem, one of the principal problems. In the basic notions of constructive models we will follow the paper of Ershov and Goncharov "Elementary Theories and Their Constructive Models" in this volume, and [11, 12, 14, 32]. Following Mal'tsev's approach to the definition of effective representations of models through enumerations, we arrive at the natural notion of recursive equivalence (Kolmogorov equivalence) between numberings [13, 44].

Let $(\mathfrak{N}, \nu)$ and $(\mathfrak{N}, \mu)$ be two enumerated algebraic structures. The numberings ν and μ of the algebraic structure $\mathfrak{N}$ are said to be *recursively equivalent* if there exist recursive functions f and g such that $\nu = \mu f$ and $\mu = \nu g$. It is obvious that for constructivizations ν and μ it suffices to require the existence of a recursive function f such that $\nu = \mu f$. We note that any constructivizable structure $\mathfrak{A}$, whose group of automorphisms has cardinality of the continuum, has continuum many non-recursively equivalent constructivizations. Thus, in spite of the simple algorithmic structure of a countable atomless Boolean algebra, we obtain continuum many non-recursively equivalent constructivizations.

Proposition 1 *Two constructivizations of an algebraic structure $\mathfrak{A}$ are recursively equivalent if and only if, for any $P \subseteq A^2$,*

$$\nu^{-1}(P) = \{\langle n, m\rangle \mid \langle \nu n, \nu m\rangle \in P\} \quad \textit{is recursive} \quad \Longleftrightarrow$$

$$\mu^{-1}(P) = \{\langle n, m\rangle \mid \langle \mu n, \mu m\rangle \in P\} \quad \textit{is recursive.}$$

---

*Supported by the Russian Foundation for Basic Research, Grant No. 96-01-01525.

HANDBOOK OF RECURSIVE MATHEMATICS
Edited by Yu. L. Ershov, S. S. Goncharov,
A. Nerode, and J. B. Remmel

However, we consider abstract structures up to isomorphism. Mal'tsev [12, 14] has introduced the suitable notion of autoequivalence in this case. Two numberings ν and μ of an algebraic structure $\mathfrak{A}$ are said to be *autoequivalent* if they are recursively equivalent up to automorphism, i.e., if there exists an automorphism φ of $\mathfrak{A}$ such that $\varphi\nu$ and μ are recursively equivalent. In this case, we obtain the following corollary.

Corollary 2 *Two constructivizations ν and μ of an algebraic structure $\mathfrak{A}$ are autoequivalent if and only if the same subsets of A^2 are recursive in $(\mathfrak{A}, \nu)$ and $(\mathfrak{A}, \mu)$ up to automorphism, i.e., if and only if, for any $P \subseteq A^2$,*

$$\nu^{-1}(P) = \{\langle n, m\rangle \mid \langle \nu n, \nu m\rangle \in P\} \quad \textit{is recursive} \qquad \Longleftrightarrow$$

there exists an automorphism φ such that

$$\mu^{-1}(\varphi(P)) = \{\langle n, m\rangle \mid \langle \varphi\mu n, \varphi\mu m\rangle \in P\} \quad \textit{is recursive.}$$

Uspenskiĭand Semenov [59] considered a more delicate notion of equivalence, involving not all subsets $P \subseteq A^{<\omega}$, but only algebraic ones. A subset $P \subseteq A^{<\omega}$ is called *algebraic*, or *stable*, if for any automorphism φ of $\mathfrak{A}$ and any n–tuple $\langle a_1, \ldots, a_n\rangle \in P$ the n–tuple $\langle \varphi(a_1), \ldots, \varphi(a_n)\rangle$ belongs to P.

Two constructivizations ν and μ of a structure $\mathfrak{A}$ are called *algebraically equivalent* if for any algebraic subset $P \subseteq A^{<\omega}$, $\nu^{-1}(P)$ is recursive if and only if $\mu^{-1}(P)$ is recursive.

In the case of autoequivalence, the decision procedure for $\nu^{-1}(P)$ allows us to define uniformly a decision procedure for $\mu^{-1}(\varphi(P))$ for an automorphism φ of the structure $\mathfrak{A}$ such that $\varphi\nu = \mu$. However, such a uniform construction is in general impossible in the case of algebraic equivalence.

On the class $\mathrm{Con}(\mathfrak{M})$ of constructivizations of the model $\mathfrak{M}$, we can introduce the following reducibilities [59]: $\nu \leqslant_K \mu$ (Kolmogorov reducibility) if there exists a recursive function f such that $\nu = \mu f$; and $\nu \leqslant \mu$ (autoreducibility) if there exists an automorphism φ of $\mathfrak{M}$ such that $\varphi\nu \leqslant_K \mu$. We say that ν is *uniformly* reduced to μ ($\nu \leqslant_U \mu$) if for some computable operator $\mathcal{F}$ we have $\mathcal{F}(\chi_{\mu^{-1}(S)}) = \chi_{\nu^{-1}(S)}$ for all $S \in \mathrm{St}(\mathfrak{M})$ (where $\mathrm{St}(\mathfrak{M})$ is the set of stable relations of $\mathfrak{M}$ and χ_A is the characteristic function of A); that ν is *programmedly* reduced to μ ($\nu \leqslant_P \mu$) if there exists a partially recursive function f such that if φ_n is the characteristic function of the set $\mu^{-1}(S)$ for

$S \in \mathrm{St}(\mathfrak{M})$, then $\varphi_{f(n)}$ is the characteristic function of the set $\nu^{-1}(S)$, where φ_m denotes the m–th partial recursive function; and that ν is *algebraically* reduced to μ ($\nu \leqslant_{\mathrm{Alg}} \mu$) if every stable relation of $\mathfrak{M}$ which is decidable under the constructivization μ is also decidable under the constructivization ν.

The above notions were defined in [18], together with that of restricted algebraic reducibility $\leqslant_{\mathrm{Alg}_n}$, $n \geqslant 1$, (the case in which only m–ary algebraic relations, $m \leqslant n$, are considered). In [18], a sufficient condition for the equivalence between algebraic reducibility and n–algebraic reducibility was obtained. With the help of this condition, models for which algebraic reducibility and n–algebraic reducibility are equivalent were indicated.

Let λ be one of the above reducibilities (algebraic, programmed, or uniform reducibility, autoreducibility or Kolmogorov reducibility). Two constructivizations of a model $\mathfrak{M}$ are called λ–equivalent if each is λ–reduced to the other. The number of classes of λ–equivalent constructivizations of $\mathfrak{M}$ is called the λ–*dimension* of $\mathfrak{M}$ (λ–dim($\mathfrak{M}$)). A model of λ–dimension 1 is called λ–*stable*.

Proposition 3 [18]

(1) *For anyn $\geqslant 1$ there exists a model $\mathfrak{M}$ such that algebraic reducibility and n–algebraic reducibility are equivalent on* Con($\mathfrak{M}$) *and such that, for any $m < n$, the following relation holds:*

$$m+1 = \mathrm{Alg}_m\text{-}\mathrm{dim}(\mathfrak{M}) < \mathrm{Alg}_n\text{-}\mathrm{dim}(\mathfrak{M}) = \mathrm{Alg}\text{-}\mathrm{dim}(\mathfrak{M}) = n+1.$$

(2) *There exists a model $\mathfrak{A}$ such that, for any n, we have*

$$\mathrm{Alg}_n\text{-}\mathrm{dim}(\mathfrak{A}) < \mathrm{Alg}\text{-}\mathrm{dim}(\mathfrak{A}) = \omega.$$

Proposition 4 *Let ν and μ be two constructivizations of the same algebraic structure.*

(1) *If ν is recursively equivalent to μ, then ν is autoequivalent to μ.*

(2) *If ν is autoequivalent to μ, then ν is uniformly equivalent to μ.*

(3) *If ν is uniformly equivalent to μ, then ν is programmedly equivalent to μ.*

(4) *If ν is programmedly equivalent to μ, then ν is algebraically equivalent to μ.*

In [16, 17], the concept of structures of λ–reducibilities of a model $\mathfrak{M}$ was introduced as follows. Each λ–reducibility relation defines a preorder on the set Con($\mathfrak{M}$) of constructivizations of $\mathfrak{M}$. Similarly, each λ–reducibility relation induces a partial order on the set of all classes of λ–equivalent constructivizations (or, in general, of numberings of a given class, e.g., positive numberings, constructive numberings, negative numberings, and so on). In [17], such a partially ordered set for a model $\mathfrak{M}$ was called the *structure of λ–reducibility* of $\mathfrak{M}$. We note the following obvious properties.

Proposition 5

(1) *The structure of autoreducibility of any model is an at most countable antichain.*

(2) *The structure of λ–reducibility of any rigid structure coincides with its structure of autoreducibility.*

Theorem 6 [17] *For any at most countable cardinal α, there exists a model $\mathfrak{M}$ whose structure of algebraic reducibility is a linear order of cardinality α.*

In [16, 19] it was shown that the following structures can be realized as structures of algebraic reducibility:

(i) finite Boolean algebras,

(ii) certain upper semi-lattices which are not lattices.

Question 1 *Describe all possible finite structures of algebraic reducibility.*

The following relations between algorithmic dimensions hold:

$$\text{Alg-}\dim(\mathfrak{M}) \leqslant P\text{-}\dim(\mathfrak{M}) \leqslant U\text{-}\dim(\mathfrak{M}) \leqslant A\text{-}\dim(\mathfrak{M}) \leqslant K\text{-}\dim(\mathfrak{M})$$

In 1979, Uspenskiĭand Semenov posed the following problems and suggested that they be studied both in the general case and in the case of classical classes of algebraic structures.

(1) The problem of relations between algorithmic dimensions:

(a) Are there constructivizable models for which one of the signs $\leqslant$ above can be replaced by the sign $<$?

(b) In which cases can the sign $\leqslant$ be replaced by the sign $=$?

(2) The problem of possible spectra of algorithmic dimensions: "What sets of numbers can serve as sets of algorithmic dimensions of algebraic structures (for each kind of dimension defined above: algebraic, programmed, and uniform dimensions, Kolmogorov dimension, and autodimension)?".

The first nontrivial example of relations between algorithmic dimensions and a proof that algebraic, programmed, and uniform dimension and autodimension coincide for Abelian groups of a certain class were obtained in [39, 40]. In [51], the relations $D_n(V_\infty)$ of linear dependence of n elements of an infinite-dimensional vector space V_∞ were studied. It was shown that

$$\text{Alg}-\dim(V_\infty) = P-\dim(V_\infty) = U-\dim(V_\infty) = A-\dim(V_\infty) = \omega.$$

In [39, 40], algebraic, programmed, and uniform dimension and autodimension were studied in the general case: for each pair λ_1, λ_2 of the above algorithmic dimensions, give a constructive algebraic structure $\mathfrak{M}$ for which $\lambda_1-\dim(\mathfrak{M}) \neq \lambda_2-\dim(\mathfrak{M})$.

Theorem 7 [40]

(1) *There exists an algebra $\mathcal{U}$ such that*

$$U-\dim(\mathcal{U}) = 1 < A-\dim(\mathcal{U}) = \omega.$$

(2) *There exists an algebra $\mathcal{B}$ such that*

$$\text{Alg}-\dim(\mathcal{B}) = 1 < P-\dim(\mathcal{B}) = U-\dim(\mathcal{B}) = A-\dim(\mathcal{B}) = \omega.$$

Problem 1 *Construct a programmedly stable but not uniformly stable constructive algebraic structure.*

In [16, 17], partially ordered sets were studied. It was shown that this class of structures is universal with respect to structure properties of all algorithmic reducibilities.

Theorem 8 [19] *Let $\lambda \in \{\text{Alg}, P, U, A, K\}$. The structure of λ-reducibility of any model $\mathfrak{M}$ of finite signature is isomorphic to the structure of λ-reducibility of a suitable partially ordered set $\mathfrak{A}$. Moreover, this isomorphism can be given by taking the quotient of a mapping $\mathcal{F} : H(\mathfrak{M}) \to H(\mathfrak{A})$ by the corresponding equivalence relation.*

From this theorem, and certain well-known facts, the solution of the problem of relations between algorithmic dimensions in the class of partially ordered sets follows.

In [20], so-called *independent families of algebraic relations* were considered. The Uspenskiĭ–Semenov problem and structures of algebraic reducibility were studied in the class of distributive lattices with relative complements and in the class of linear orders.

Let $\langle D, F\rangle$ be a distributive lattice with relative complements in the language with the Frechét ideal.

Theorem 9 [20] *For a constructivizable lattice* $\langle D, F\rangle$, *the following conditions are equivalent:*

(1) $\text{Alg-dim}\langle D, F\rangle \neq 1$,

(2) $\text{Alg-dim}\langle D, F\rangle = \omega$,

(3) *the structure of algebraic reducibility of* $\langle D, F\rangle$ *has countable width, and*

(4) F *is infinite.*

Corollary 10 *For distributive lattices* $\langle D, F\rangle$, *the algebraic, programmed, and uniform dimensions and the autodimension coincide. The spectrum of the above dimensions consists of* 0, 1, *and* ω, *and the spectrum of the Kolmogorov dimension consists of* 0, 1, *and* 2^ω.

Let $\langle \mathcal{U}, B\rangle$ be a linear order in the language with block-relation. The following theorem describes algorithmically stable linear orders $\langle \mathcal{U}, B\rangle$.

Theorem 11 [20] *For a constructivizable linear order* $\langle \mathcal{U}, \mathcal{B}\rangle$, *the following conditions are equivalent:*

(1) $\langle \mathcal{U}, \mathcal{B}\rangle$ *is algebraically stable,*

(2) *the linear order* $\mathcal{U}$ *has order type* $\sum_{i=1}^{n}(m_i \times \eta + k_i) + m_{n+1} \times \eta$ *for some* $m_i, k_i \in \omega$, $m_{n+1} \in \omega$, $k_i \neq 0$, $n \in \mathbb{N}$, *and*

(3) $\langle \mathcal{U}, \mathcal{B}\rangle$ *is autostable.*

The characterization thus obtained coincides with the Schwartz characterization of autostable linear orders $\langle \mathcal{U}, B\rangle$. Therefore, we obtain the following.

Corollary 12 *For a linear order $\langle \mathcal{U}, B \rangle$, algebraic, programmed, and uniform stability and autostability are equivalent.*

Let $\mathcal{D}$ be the class of distributive lattices with relative complements. In [22] recursively inconsistent algebraic relations on $\mathcal{U} \in \mathcal{D}$ were studied, i.e., decidable algebraic relations on $\mathcal{U} \in \mathcal{D}$ (for suitable constructivizations) which cannot be decidable in the same constructivization simultaneously.

An enumerated algebraic structure $(\mathfrak{A}, \nu)$ is called *n–constructive* and $\mathfrak{A}$ is called *n–constructivizable* if there exists an algorithm that checks the truth of Σ_n formulas on elements with given numbers.

Theorem 13 [22] *For any 1–constructivizable lattice $\mathcal{U} \in \mathcal{D}$ with an infinite set of atoms, there exists an infinite family Ω of algebraic relations on $\mathcal{U}$, each of which is decidable with respect to a suitable constructivization, but such that under any constructivization, no two relations from Ω are simultaneously decidable.*

The existence of such relations is closely connected with many problems in the theory of constructive algebraic structures. First of all, we recall the problem of relations between algorithmic dimensions and the problem of possible spectra of algorithmic dimensions.

Corollary 14 *Let $\lambda \in \{\mathrm{Alg}, P, U, A\}$. For an arbitrary 1–constructivizable lattice $\mathcal{U} \in \mathcal{D}$, the following conditions are equivalent:*

(1) λ–$\dim(\mathcal{U}) \neq 1$ *(respectively K–$\dim(\mathcal{U}) \neq 1$),*

(2) λ–$\dim(\mathcal{U}) = \omega$ *(resp. K–$\dim(\mathcal{U}) = 2^\omega$),*

(3) *$\mathcal{U}$ has infinitely many atoms (resp. $\mathcal{U}$ is infinite).*

From the above assertion the following consequences can be obtained: a criterion for the existence of a least element in the structure of algebraic reducibility of the lattice $\mathcal{U} \in \mathcal{D}$ (i.e., a best effective representation) as well as algebraic conditions under which the problem of effective choice of constructivizations of the lattice $\mathcal{U}$ with respect to specifications of algebraic relations on $\mathcal{U}$ has a solution. (For any computable class $\mathrm{Con} = \{\nu_i \mid i \geqslant 0\}$ of constructivizations of an algebraic structure $\mathfrak{A}$, and any computable class $\Omega = \{\Omega_i \mid i \geqslant 0\}$ of algebraic relations on $\mathfrak{A}$, the problem $\Sigma(\Omega, \mathrm{Con})$ of effective choice of constructivizations in Con with respect to Ω is said to have a solution provided there exists a recursive function f such that for all i, Ω_i is decidable with respect to the constructivization $\nu_{f(i)}$.)

The latter result is connected with the following problem stated by Uspenskiĭ: "Give an algebraic characterization of classes of constructive structures for which all the problems $\Sigma(\Omega, \mathrm{Con})$ of effective choice of constructivizations are trivial".

Corollary 15 *For a 1–constructivizable lattice $\mathcal{U} \in \mathcal{D}$, the following conditions are equivalent:*

(1) *the structure of algebraic reducibility of the lattice $\mathcal{U}$ has a least element,*

(2) *for any computable family Ω of algebraic relations on $\mathcal{U}$ which are decidable under suitable constructivizations of the lattice $\mathcal{U}$, there exists a computable class* Con *of constructivizations of $\mathcal{U}$ such that the problem* $\Sigma(\Omega, \mathrm{Con})$ *has a solution, and*

(3) *$\mathcal{U}$ has a finite number of atoms.*

We note that the structure of algebraic reducibility, say, of an algebraically closed field of countable transcendental degree has a least element. A linear order with linearly ordered 3–element structure of algebraic reducibility is defined in [20]; moreover, a method of constructing linear orders with finite structure of algebraic reducibility which have a least element is indicated there.

The cases of recursive stability and autostability have been well-studied.

A structure $\mathfrak{A}$ is called 1–*simple* if, for any $\forall$–formula $\psi(x_0, \dots, x_n)$ and any $a_0, \dots, a_n \in A$, there is an $\exists$–formula $\varphi(x_0, \dots, x_n)$ such that, if

$$\mathfrak{A} \models \psi(a_0, \dots, a_n),$$

then $\mathfrak{A} \models \varphi(a_0, \dots, a_n)$ and

$$\mathfrak{A} \models (\forall x_0 \cdots x_n)\big(\varphi(x_0, \dots, x_n) \to \psi(x_0, \dots, x_n)\big).$$

An $\exists$–formula $\varphi(\overline{x})$ is called 1–*complete* in $\mathfrak{A}$ if, for any $\overline{b}$ of $\mathfrak{A}$ and any $\exists$–formula $\psi(\overline{x})$, if $\mathfrak{A} \models \psi(\overline{b}) \wedge \varphi(\overline{b})$ then $\mathfrak{A} \models (\forall x)(\varphi(\overline{x}) \to \psi(\overline{x}))$.

A structure $\mathfrak{A}$ is called 1–*complete* if, for any $a_0, \dots, a_n \in A$, there is a 1–complete $\exists$–formula φ such that $\mathfrak{A} \models \varphi(a_0, \dots, a_n)$.

Theorem 16 [24] *A 2–constructivizable structure $\mathfrak{A}$ is autostable if and only if there exists a finite collection of elements $a_1, \ldots, a_k$ of $\mathfrak{A}$ such that the expansion $(\mathfrak{A}, a_1, \ldots, a_n)$ is 1–complete and there exists a computable family of 1–complete $\exists$–formulas $\{\varphi_i(\overline{x}, \overline{a}) \mid i \in \mathbb{N}\}$ such that for any tuple $\overline{b}$ from $\mathfrak{A}$ there is i such that $(\mathfrak{A}, a_1, \ldots, a_n) \models \varphi_i(\overline{b}, \overline{a})$.*

Such a criterion fails in the case of 1–constructivizable models [41]. However in that case, one can obtain a criterion for autostability in terms of orbits. Let $\mathfrak{M}$ be a model and $\overline{a}$ be a tuple from $\mathfrak{M}$. By the *orbit* $\mathrm{Orb}_{\overline{a}}(\overline{b})$ of the n–tuple $\overline{b}$, we mean the least subset of M^n containing $\overline{b}$ which is algebraic with respect to the expanded model $(\mathfrak{M}, \overline{a})$.

Theorem 17 [41] *If $(\mathfrak{M}, \nu)$ is a 1–constructive model, then the following conditions are equivalent:*

(1) *$\mathfrak{M}$ is autostable;*

(2) *there is a tuple $\overline{a}$ from $\mathfrak{M}$, such that the set*

$$S = \bigcup_{n \in \mathbb{N}} \mathrm{Orb}_{\overline{a}}(\langle \nu 0, \ldots, \nu n \rangle)$$

is intrinsically decidable, i.e., S is decidable for any constructivization.

The question of the number of nonequivalent representations and their classification is of great importance in the study of constructive structures, and is connected with computable classes of constructive models [24, 31, 32, 33, 34, 36, 60, 61, 62].

In [24, 26, 30, 36], sufficient criteria for an algebraic structure to have an infinite algorithmic dimension are considered. We now indicate their applications to some classes of algebras and models. These criteria allow us to show that for most classes of models and algebras the dimension is infinite. They are of a model-complete character and, in a number of natural classes of models and algebras, allow us to give a complete characterization of algebraic structures of infinite algorithmic dimension, while their negations characterize autostable structures.

The first criterion arose in the study of algorithmic dimensions of torsion-free Abelian groups and vector spaces. Its informal idea is to find an infinite number of limit points in the structure with respect to the topology defined by quantifier-free formulas.

Let σ be a signature, and

$$\sigma_0 \subseteq \sigma_1 \subseteq \cdots \subseteq \sigma_t \subseteq \cdots ;$$

$\sigma = \bigcup_{t\in\mathbb{N}} \sigma_n$ be a strongly computable chain of signatures. By a *representation* of a constructive model $(\mathfrak{A}, \nu)$ of signature σ we mean a pair $\Pi^\nu = \langle(\mathfrak{A}_t^\nu \mid t \in \mathbb{N}), f^\nu\rangle$, where $(\mathfrak{A}_t^\nu \mid t \in \mathbb{N})$ is a strictly computable chain of finite models such that $\mathfrak{A}_t^\nu$ has signature σ_t, and f^ν is a general recursive function, satisfying

(1) $A^\nu = \bigcup_{t\in\mathbb{N}} A_t^\nu$ is a recursive set,

(2) for every $t \in \mathbb{N}$, $\mathfrak{A}_t^\nu \subseteq \mathfrak{A}_{t+1}^\nu$, and for some $m_n \in \mathbb{N}$, $A_t^\nu = \{0, \ldots, m_t\}$, and

(3) $\nu f^\nu(A^\nu) = A$ and, for every $t \in \mathbb{N}$, $\nu f^\nu \restriction A_t^\nu : \mathfrak{A}_t^\nu \to \mathfrak{A}$ is an isomorphic embedding.

Let $\Psi = \{\psi_{i,j} \mid i, j \in \mathbb{N}\}$ be a computable set of quantifier-free formulas. A constructivizable model $\mathfrak{A}$ is called *unbounded with respect to the family of formulas* Ψ, if there exists a constructivization ν, a representation

$$\Pi^\nu = \langle(\mathfrak{A}_t^\nu \mid t \in \mathbb{N}), f^\nu\rangle$$

of $(\mathfrak{A}, \nu)$, and a general recursive function h, such that, for any $t_0 \in \mathbb{N}$ and any monotone increasing unbounded general recursive function g, there exist $t \geqslant t_0$ and an isomorphic embedding $\varphi : \mathfrak{A}_t^\nu \to \mathfrak{A}_{t+1}^\nu$ such that $\varphi \restriction \mathfrak{A}_t^\nu = \mathrm{id}_{\mathfrak{A}_t^\nu}$, and for some $i \leqslant h(t)$ and $n_1, \ldots, n_k \leqslant g(t)$ we have

$$\mathfrak{A}_{t+1}^\nu \models \bigwedge_{j=0}^{t} \psi_{i,j}(\nu n_1, \ldots, \nu n_k),$$

but

$$\mathfrak{A}_{t+1}^\nu \models \neg \bigwedge_{j=0}^{t+1} \psi_{i,j}(\varphi(\nu n_1), \ldots, \varphi(\nu n_k)).$$

In this case, we also say that the representation Π^ν *is unbounded with respect to* Ψ. If Φ is a set of formulas, then a homomorphism $\varphi : \mathfrak{A} \to \mathfrak{B}$ is called a *homomorphism with respect to predicates from* Φ if for all those relations that occur in formulas from Φ the homomorphism preserves not only the truth, but also the falsity of these relations.

A constructive model $(\mathfrak{A}, \nu)$ *is homomorphically embedded* into $(\mathfrak{B}, \mu)$ if there exists a homomorphism $\varphi : \mathfrak{A} \to \mathfrak{B}$ and a general recursive function f such that $\varphi\nu = \mu f$.

Theorem 18 [26] *If a model $\mathfrak{A}$ is unbounded with respect to a family of formulas Φ, then*

(1) *From any computable class S of constructive models of the same signature as $\mathfrak{A}$, it is possible to construct effectively a constructivization μ of the model $\mathfrak{A}$ such that $(\mathfrak{A}, \mu)$ is not isomorphically embedded with respect to Φ into any model of the class S.*

(2) *There exists a computable family of constructivizations $\{\mu_i \mid i \in \mathbb{N}\}$ of the model $\mathfrak{A}$ such that for any $i \neq j$ the model $(\mathfrak{A}, \mu_i)$ is not isomorphically embedded with respect to Φ into $(\mathfrak{A}, \mu_j)$.*

We give one more criterion for nonautostability, which is based on the special manner of definition of the truth of $\forall$–formulas on constructive models and the impossibility of their definitions via $\exists$–formulas. This criterion is a generalization of nonautostability conditions from [24, 32, 36, 52], and has wide application. We introduce the necessary notions and definitions.

Let $\Pi^\nu = \langle (\mathfrak{A}_t^\nu \mid t \in \mathbb{N}), f^\nu \rangle$ be some representation of a constructive model $(\mathfrak{A}, \nu)$. Let $\Psi_m \leftrightharpoons \bigwedge_{i \in \mathbb{N}} \psi_i^m$, $m \in \mathbb{N}$, be a computable sequence of infinite conjunctions of $\forall$–formulas, i.e., there exists an effective procedure for constructing the formula ψ_i^m from the indices i and m. (Ψ_m may contain constants for elements of $\mathfrak{A}_m^\nu$.)

For each m, we define the set of branching elements $B_{\Psi_m}^m$ of the formula Ψ_m, as the set of tuples $\langle m_0, \ldots, m_l \rangle$ such that

$$\mathfrak{A} \models \Psi_m(\nu f^\nu(m_0), \ldots, \nu f^\nu(m_l)),$$

for any finite sequence of tuples $\overline{m}_0, \ldots, \overline{m}_s$ such that $\overline{m}_s = \langle m_0, \ldots, m_l \rangle$, and for any $i \leqslant s$ the following relation holds:

$$\mathfrak{A} \models \Psi_m(\nu f^\nu(m_0^i), \ldots, \nu f^\nu(m_l^i)),$$

and such that there exists infinitely many t for which there is an isomorphic embedding $\varphi : \mathfrak{A}_t^\nu \to \mathfrak{A}_{t+1}^\nu$ and an $i \leqslant s$ such that

(1) $\mathfrak{A}_{t+1}^\nu \models \neg {\Psi_m}^{t+1}(\varphi(m_0^i), \ldots, \varphi(m_l^i))$, where ${\Psi_m}^{t+1} \leftrightharpoons \bigwedge_{j=0}^{t+1} \psi_{m,i}$, and

(2) the mapping φ is the identity on A_m^ν and on elements of the tuples $\overline{m}_j$ for $j < i$, where $\overline{m}_j \leftrightharpoons \langle m_0^j, \ldots, m_l^j \rangle$.

The representation $\Pi^\nu = \langle(\mathfrak{A}^\nu_t \mid t \in \mathbb{N}), f^\nu\rangle$ is *branching with respect to the sequence* Ψ_m, $m \in \mathbb{N}$, if for any m the set $B^m_{\Psi_m}$ is nonempty, and the set of elements $\{\overline{m} \mid \mathfrak{A} \models (\nu f^\nu(m_0), \ldots, \nu f^\nu(m_l))\} \setminus B^m_{\Psi_m}$ is finite. We say that a model $\mathfrak{A}$ *is branching* if there exists a computable sequence Ψ_m, $m \in \mathbb{N}$, of infinite conjunctions of $\forall$–formulas, and a constructivization ν, such that some representation $\Pi^\nu = \langle(\mathfrak{A}^\nu_t \mid t \in \mathbb{N}), f^\nu\rangle$ of $(\mathfrak{A}, \nu)$ is branching with respect to the sequence Ψ_m, $m \in \mathbb{N}$.

Theorem 19 *If* $\mathfrak{A}$ *is a branching constructivizable model, then the class of its constructivizations is effectively infinite, i.e., given any computable class* $\mathrm{Con}(\mathfrak{A})$ *of constructivizations of* $\mathfrak{A}$ *we can effectively find a constructivization of* $\mathfrak{A}$ *which is not autoequivalent to any element of* $\mathrm{Con}(\mathfrak{A})$.

This criterion turns out to be effective for many classes of groups, rings, Boolean algebras, and their expansions.

We give a recursion-theoretic criterion for an algorithmic dimension to be infinite. This criterion is of great interest due not only to its applications but also to the fact that it provides a new point of view on the question of the number of nonautoequivalent constructivizations. The above methods for proving that the algorithmic dimension of certain structures is infinite make it possible to construct nonautoequivalent constructivizations which are autoequivalent in limit. If there are no such constructivizations, none of these methods can be applied. Thus, we clarify the principal difficulty in studying models of finite algorithmic dimension.

We define the notion of autoequivalence in limit, and apply it to the study of nonautostable models. Let μ and ν be two constructivizations of a model $\mathfrak{A}$. We call them *autoequivalent in limit* if there exists an automorphism φ of the model $\mathfrak{A}$ and a Δ^0_2 function f such that $\varphi\nu = \mu f$. We recall that a function f is Δ^0_2 if there exists a recursive function g such that for any n the equality $f(n) = \lim_{m\to\infty} g(n,m)$ holds. Hence two constructivizations ν and μ of a model $\mathfrak{A}$ are autoequivalent in limit if there exists an automorphism φ of the model $\mathfrak{A}$ and a recursive function g such that for any n the equality $\varphi\nu(n) = \lim_{m\to\infty} \mu g(n,m)$ holds.

Theorem 20 [30] *If a model* $\mathfrak{A}$ *has two nonautoequivalent but equivalent in limit constructivizations, then*$\mathfrak{A}$ *has an infinite algorithmic dimension.*

By the Turing spectrum $Sp_T(\mathbf{M}, \mathbf{P})$ of the relation P with respect to constructivizations of the model $\mathbf{M}$ we mean the set of Turing degrees of sets of numbers for P under different constructivizations of the model. It is easy to see that the problem of the existence of nonautostable relations with a finite Turing spectrum is closely connected with the problem of the existence of algebraic structures of finite algorithmic dimension. As was shown in [28], the latter problem can be reduced to that of classical algorithm theory, namely, to the problem of the number of computable Friedberg numberings of a family of recursively enumerable sets, but with some additional properties. Ash and Nerode [5] were the first to begin to study characterizations of decidable and enumerable relations for different constructivizations. The most significant results on the definability of relations of restricted complexity in the arithmetic or hyperarithmetic hierarchies were obtained by Ash [1].

We now describe the construction in [28]. Given a family S of recursively enumerable sets, we define a category Un_S whose objects are single-valued computable numberings of S. If γ_1 and γ_2 are two single-valued computable numberings of S, then by a *morphism* from γ_1 into γ_2 we mean a recursive function f such that $\gamma_1(n) = \gamma_2(f(n))$ for all $n \in \mathbb{N}$.

For any family S of subsets from $\mathbb{N}$ we will define a model $\mathfrak{A}_S$ whose signature Σ contains one unary predicate A, and binary predicates L_n, P_n and Q_n, where $n \in \mathbb{N}$, and show that if the sets in S are recursively enumerable then the category Un_S is equivalent to the category $K^\Sigma(\mathfrak{A}_S)$ whose objects are all constructive models $(\mathfrak{A}, \nu)$ of signature Σ such that $\mathfrak{A}$ is isomorphic to $\mathfrak{A}_S$ and whose morphisms are all recursive isomorphisms between such constructive models.

We first consider some family of functions $F = \{f_s \mid s \in S\}$, where $f_s : \mathbb{N} \to \mathbb{N}$, and construct a model $\mathfrak{A}_S^F$ of signature Σ. The universe $|\mathfrak{A}_S|$ is independent of F, and we set it equal to $S \cup (S \times \mathbb{N}^2)$. We first define predicates which are also independent of the choice of F.

Let $A_{\mathfrak{A}_S^F} \leftrightharpoons S$ and $(P_n)_{\mathfrak{A}_S^F} \leftrightharpoons \{\langle s, \langle s, n, m\rangle\rangle \mid m \in \mathbb{N},\ s \in S\}$. The predicates L_n and Q_n for $n \in \mathbb{N}$, depend on the choice F. We define them as follows:

(1) $(Q_n)_{\mathfrak{A}_S} \leftrightharpoons \{\langle s, \langle s, n, f_s(n)\rangle\rangle \mid s \in S,\ n \in s\}$,

(2) If $f_s(n) \neq m$ and $f_s(n) \neq m+1$, then $\langle s, n, m\rangle\, L_n\, \langle s, n, m+1\rangle$,

(3) If $f_s(n) = m$ and $m \neq 0$, then $\langle s, n, m-1\rangle\, L_n\, \langle s, n, m+1\rangle$,

(4) If $f_s(n) = 0$, then $s\, L_n\, \langle s, n, 1\rangle$,

(5) If $f_s(n) \neq 0$, then $s\, L_n\, \langle s, n, 0\rangle$.

Thus we have constructed the model $\mathfrak{A}_S^F$. It depends on the choice of the family F. Let $\varphi_n(x) \leftrightharpoons (\exists y)Q_n(x,y)$.

Lemma 21 *If* $\mathfrak{A}_S^F \models A(a)$, *then* $\mathfrak{A}_S^F \models \varphi_n(a) \Leftrightarrow n \in a$.

Proof. Since $A = S$, we see that a is a subset of $\mathbb{N}$. We now prove the equivalence. If $n \in a$, then $f_a(n)$ is defined and $f_a(n) = m$. Consequently, $\mathfrak{A}_S^F \models Q_n(a, \langle a,n,m\rangle)$ and $\mathfrak{A}_S^F \models \varphi_n(a)$. If $\mathfrak{A}_S^F \models \varphi_n(a)$, then there exists b such that $\mathfrak{A}_S^F \models Q_n(a,b)$. However, by definition, if $Q_n(a,b)$ is true in $\mathfrak{A}_S^F$ then $a \in S$, $n \in a$, and $b = \langle a,n,f_a(n)\rangle$. □

Lemma 22 *For any* $b \in |\mathfrak{A}_S^F|$, *exactly one of the following conditions holds:*

(1) $\mathfrak{A}_S^F \models A(b)$,

(2) *there exista and* $n \in \mathbb{N}$ *such that* $\mathfrak{A}_S^F \models Q_n(a,b)$,

(3) *there exist* $a_1, \ldots, a_K$, *where* $K \geqslant 1$ *and* $n \in \mathbb{N}$, *such that*

$$\mathfrak{A}_S^F \models A(a_1) \wedge \bigwedge_{1=1}^{K-1} (a_i \, L_n \, a_{i+1}) \wedge a_K \, L_n \, b.$$

Proof. If $\mathfrak{A}_S^F \models A(b)$, then condition (1) holds, and conditions (2) and (3) fail. This follows directly from the definition of the model $\mathfrak{A}_S^F$. If $\mathfrak{A}_S^F \not\models A(b)$, then $b = \langle a,n,m\rangle$. In the case $n \in a$, we consider $f_a(n)$. If $f_a(n) = m$, then $\mathfrak{A}_S^F \models Q_n(a, \langle a,n,m\rangle)$ and condition (2) holds, while the rest of the conditions fail. If $f_a(n) \neq m$ or $n \notin a$, then, taking

$$a \, L_n \, \langle a,n,0\rangle \, L_n \, \langle a,n,1\rangle \, L_n \cdots$$
$$L_n \, \langle a,n,f(n)-1\rangle \, L_n \, \langle a,n,f_a(n)+1\rangle \, L_n \cdots L_n \, \langle a,n,m\rangle,$$

we obtain condition 3. □

Corollary 23 *The model* $\mathfrak{A}_S^F$ *is rigid, i.e., it has no nontrivial automorphisms.*

Lemma 24 *For any families of functions* F *and* G, *the models* $\mathfrak{A}_S^F$ *and* $\mathfrak{A}_S^G$ *are isomorphic.*

Proof. We construct the required isomorphism Ψ_{FG}. Let $a \in |\mathfrak{A}_S^F|$. We let $\Psi_{FG}(s) = s$ if $s \in S$. Now consider an element $a = \langle s, n, m\rangle$. If $n \notin s$ or $m < \min\{f_s(n)\,,\ g_s(n)\}$, then we set

$$\Psi_{FG}(\langle s,n,m\rangle) \;=\; \langle s,n,m\rangle.$$

If $m = f_s(n)$, we set

$$\Psi_{FG}(\langle s,n,m\rangle) \;=\; \langle s,n,g_s(n)\rangle.$$

If $f_s(n)\, L\, g_s(n)$, then we set

$$\Psi_{FG}(\langle s,n,m\rangle) \;=\; \begin{cases}\langle s,n,m-1\rangle & \text{if } f_s(n) < m \leqslant g_s(n),\\ \langle s,n,m\rangle & \text{if } m > g_s(n).\end{cases}$$

Finally, if $f_s(n) \geqslant g_s(n)$, then we set

$$\Psi_{FG}(\langle s,n,K\rangle) \;=\; \begin{cases}\langle s,n,K\rangle & (\forall K)\ (K < g_n(s) \ \text{ or } \ K > f_s(n))\\ \langle s,n,K+1\rangle & \text{if } f_s(n) > K \geqslant g_s(n)\end{cases}$$

It is easy to see that the mapping Ψ_{FG} is single-valued and "onto", and preserves the truth or falsity of all the predicates A, L_n, P_n and Q_n for $n \in \mathbb{N}$. Consequently, $\mathfrak{A}_S^F \cong \mathfrak{A}_S^G$. □

Thus, the model is independent of F up to an isomorphism. We denote this model by $\mathfrak{A}_S$.

Proposition 25 *For any family S of recursively enumerable sets, there exists a functor F_S from the category Un_S into the category $K^{\Sigma}(\mathfrak{A}_S)$ which realizes their equivalence.*

Proof. Let γ be a single-valued computable numbering of the family S. There exists a strictly computable family $\gamma^t(m)$, $m \in \mathbb{N}$, of finite sets such that $\gamma^t(m) \subseteq \gamma^{t+1}(m)$, $\bigcup_{t\geqslant 0} \gamma^t(m) = \gamma(m)$, and $\gamma^0(m) = \emptyset$ for all $m, t \in \mathbb{N}$. For each $s = \gamma(m) \in S$ we define the function

$$g_{\gamma(m)}(n) \;=\; \mu t(n \in \gamma^{t+1}(m) \smallsetminus \gamma^t(m)).$$

Informally, this function indicates the step at which n enters the set S for a given computable numbering γ. Let s_γ^t denote the set $\gamma^t(m)$, where $s = \gamma(m)$.

Let $G_\gamma \leftrightharpoons \{g_{\gamma(n)} \mid n \in \mathbb{N}\}$. It is easy to note that this is a computable sequence of partially recursive functions.

We introduce the model $\mathfrak{A}_S^{G_\gamma}$. Let c_{31}, c_{32}, c_{33} and c^3 be the Cantor functions of the numbering of triples of natural numbers. We define the numbering ν_γ of the model $\mathfrak{A}_S$ as follows:

$$\nu_\gamma(2n) = \gamma(n), \qquad \nu_\gamma(2n+1) = \langle \gamma(c_{31}(n)), c_{32}(n), c_{33}(n)\rangle.$$

It is clear that the predicates A and P_n, $n \in \mathbb{N}$, are recursive with respect to this numbering. We note that

$$\begin{aligned} &Q_n(\gamma(k), \gamma_\nu(m)) \\ &\Leftrightarrow\ k \text{ is even } \wedge\ c_{31}(m) = \frac{k}{2}\ \wedge\ c_{32}(m) \leftrightharpoons n\ \wedge\ c_{33}(n) = g_{\gamma(\frac{k}{2})}(n) \\ &\Leftrightarrow\ k \text{ is even } \wedge\ c_{31}(m) = \frac{k}{2}\ \wedge\ c_{32}(m) = n\ \wedge\ n \in \gamma^{c_{33}(m)+1}\Big(\frac{k}{2}\Big) \setminus \gamma^{c_{33}(m)}\Big(\frac{k}{2}\Big) \end{aligned}$$

However, all these conditions are recursive. Consequently, the predicates Q_n, $n \in \mathbb{N}$, are recursive. The binary predicates $L_n, n \in \mathbb{N}$, are also recursive, which can be verified as in the case of Q_n directly from the definition.

We set $F(\gamma) \leftrightharpoons (\mathfrak{A}_S^{G_\gamma}, \nu_\gamma)$. We now need a lemma.

Lemma 26 *If γ_1 and γ_2 are nonequivalent single-valued computable numberings of the family S, then the constructive models $(\mathfrak{A}_S^{G_{\gamma_1}}, \nu_{\gamma_1})$ and $(\mathfrak{A}_S^{G_{\gamma_2}}, \nu_{\gamma_2})$ are not isomorphic.*

Proof. Assume for a contradiction that there exists a recursive function g and an isomorphism Ψ of the model $\mathfrak{A}_S^{G_{\gamma_1}}$ onto $\mathfrak{A}_S^{G_{\gamma_2}}$ such that $\Psi\nu_{\gamma_1} = \nu_{\gamma_2} g$. We consider the function h such that $h(n) = \left[\frac{g(2n)}{2}\right]$. For any n the value of $g(2n)$ is even, since $\mathfrak{A}_S^{G_{\gamma_1}} \models A(\nu_{\gamma_1}(2n))$ and consequently, $\mathfrak{A}_S^{G_{\gamma_2}} \models A(\nu_{\gamma_2}(g(2n)))$. By Lemma 21, the following equivalences hold:

$$\begin{aligned} m \in \gamma_1(n) \ &\Leftrightarrow\ \mathfrak{A}_S^{G_{\gamma_1}} \models \varphi_m(\gamma_1(n)) \ \Leftrightarrow\ \mathfrak{A}_S^{G_{\gamma_1}} \models \varphi_m(\nu_{\gamma_1}(2n)) \\ &\Leftrightarrow\ \mathfrak{A}_S^{G_{\gamma_2}} \models \varphi_m(\Psi\nu_{\gamma_1}(2n)) \ \Leftrightarrow\ \mathfrak{A}_S^{G_{\gamma_2}} \models \varphi_m \nu_{\gamma_2}(g(2n)) \\ &\Leftrightarrow\ m \in \nu_{\gamma_2}(g(2n)) = \gamma_2\Big[\frac{g(2n)}{2}\Big], \end{aligned}$$

which means that $\gamma_1(n) = \gamma_2(h(n))$ and $\gamma_1 \leqslant \gamma_2$. However, γ_2 is a single-valued numbering. Hence $\gamma_1 \equiv \gamma_2$. We arrive at a contradiction. □

We now resume the proof of Proposition 25. Let γ_1 and γ_2 be objects of Un_S, and let f be a morphism of the category Un_S from γ_1 into γ_2. By definition, this is a recursive function f such that $\gamma_1(n) = \gamma_2(f(n))$. We define the value of $\varphi = F(\gamma_1, \gamma_2)(f)$, by setting

$$\varphi(x) = \begin{cases} x & \text{if } x \in S \\ & \text{or } x = \langle s, n, m\rangle \wedge \big[(\bigwedge_{i=0}^{m+1} n \notin S_{\gamma_1}^i \wedge \bigwedge_{i=0}^{m+1} n \notin S_{\gamma_2}^i) \\ & \qquad \vee (\bigvee_{i=0}^{m} n \in S_{\gamma_1}^i \wedge \bigvee_{i=0}^{m} n \in S_{\gamma_2}^i)\big], \\ \langle s, n, m'\rangle & \text{if } x = \langle s, n, m\rangle \wedge (n \in S_{\gamma_1}^{m+1} \setminus S_{\gamma_1}^{m}) \wedge (n \in S_{\gamma_2}^{m'+1} \setminus S_{\gamma_2}^{m'}), \\ \langle s, n, m+1\rangle & \text{if } x = \langle s, n, m\rangle \wedge (\bigwedge_{i=0}^{m+1} n \notin S_{\gamma_1}^i) \wedge (\bigvee_{i=0}^{m} n \in S_{\gamma_2}^i), \\ \langle s, n, m-1\rangle & \text{if } x = \langle s, n, m\rangle \wedge (\bigvee_{i=0}^{m} n \in S_{\gamma_1}^i) \wedge (\bigwedge_{i=0}^{m} n \notin S_{\gamma_2}^i). \end{cases}$$

Now let

$$h(k) = \begin{cases} 2f\left(\frac{k}{2}\right) & \text{if } k = 2l, \\ 2c^3(fc_{31}(l), c_{32}(l), c_{33}(l)) + 1 & \text{if } k = 2l+1 \\ & \quad \wedge\big[(\bigwedge_{i=0}^{m+1} n \notin \gamma_1^i(c_{31}(l)) \\ & \qquad\qquad \wedge \bigwedge_{i=0}^{m+1} n \notin \gamma_2^i(c_{31}(l))) \\ & \qquad \vee (\bigvee_{i=0}^{m} n \in \gamma_1^i(c_{31}(l)) \\ & \qquad\qquad \wedge \bigvee_{i=0}^{m} n \in \gamma_2^i(f(c_{31}(l))))\big], \\ 2c^3(fc_{31}(l), c_{32}(l), m') + 1 & \text{if } k = 2l+1 \\ & \quad \wedge \big(n \in \gamma_1^{m+1}(c_{31}(l)) \setminus \gamma_1^{m}(c_{31}(l))\big) \\ & \quad \wedge \big(n \in \gamma_2^{m'+1}(fc_{31}(l)) \setminus \gamma_2^{m'}(fc_{31}(l))\big), \\ 2c^3(fc_{31}(l), c_{32}(l), c_{33}(l)+1) & \text{if } k = 2l+1 \\ & \quad \wedge \big(\bigwedge_{i=0}^{m+1} n \notin \gamma_1^i(c_{31}(l))\big) \\ & \quad \wedge \big(\bigvee_{i=0}^{m} n \in \gamma_2^i(fc_{31}(l))\big), \\ 2c^3(fc_{31}(l), c_{32}(l), c_{33}(l)-1) & \text{if } k = 2l+1 \\ & \quad \wedge \big(\bigvee_{i=0}^{m} n \in \gamma_1^i(c_{31}(l))\big) \\ & \quad \wedge \big(\bigwedge_{i=0}^{m} n \notin \gamma_2^i(fc_{31}(l))\big), \end{cases}$$

where $n = c_{32}(l)$ and $m = c_{33}(l)$. It is not hard to verify that $\varphi\nu_{\gamma_1} = \nu_{\gamma_2}h$. Consequently, φ is an isomorphism of the constructive model $(\mathfrak{A}_S^{G_{\gamma_1}}, \nu_{\gamma_1})$ onto $(\mathfrak{A}_S^{G_{\gamma_2}}, \nu_{\gamma_2})$.

Let $(\mathfrak{A}, \nu)$ be a constructive model which is an object of the category $k^{\Sigma}(\mathfrak{A}_S)$. Since any constructivization is recursively equivalent to a single-valued constructivization, we can regard ν as a single-valued constructivization. We consider the recursive set $\widehat{A}$ of ν–numbers of those elements of the model $\mathfrak{A}$ for which the predicate $A(x)$ is true. These are exactly the ν–numbers of elements of S. Since $\widehat{A}$ is an infinite recursive set, by Theorem III in section 5.1 of [58], there exists a single-valued recursive function $f : \mathbb{N} \xrightarrow{1-1} \widehat{A}$. We introduce $\gamma(n) = \{m \mid \mathfrak{A} \models \varphi_m(\nu f(n))\}$. Since $(\mathfrak{A}, \nu)$ is a constructive model and $\varphi_m, m \in \mathbb{N}$, is an effective sequence of $\exists$–formulas, γ is a computable numbering. In view of Lemma 21, since ν and f are single-valued, γ is also a single-valued numbering of the family S.

We consider $F_S(\gamma) = (\mathfrak{A}_S^{G_\gamma}, \nu_\gamma)$, and show that $(\mathfrak{A}_S^{G_\gamma}, \nu_\gamma)$ is isomorphic to $(\mathfrak{A}, \nu)$. We define the function h as follows: $h(2k) = f(2k)$, if

$$\mathfrak{A}_S^{G_\gamma} \models Q_{c_{32}(k)}(\nu_\gamma(2c_{31}(k)), \nu_\gamma(2k+1))$$

and

$$l = \mu n(\mathfrak{A} \models Q_{c_{32}(k)}(\nu f c_{31}(k), \nu(n))$$

then $h(2k+1) = l$; and if

$$\mathfrak{A}_S^{G_\gamma} \models \neg Q_{c_{32}(k)}(\nu_\gamma(2c_{31}(k)), \nu_\gamma(2k+1))$$

and there exist $l_1, \ldots, l_d, \widehat{l_i}, \ldots, \widehat{l_d}$ such that

$$\mathfrak{A}_S^{G_\gamma} \models (\nu_\gamma(l_1)\ L_{c_{32}(k)}\ \nu_\gamma(l_2)\ L_{c_{32}(k)} \cdots \\ L_{c_{32}(k)}\ \nu_\gamma(l_d)\ L_{c_{32}(k)}\ \nu_\gamma(2k+1)) \wedge A(\nu_\gamma(l_1)),$$

and

$$\mathfrak{A} \models (\nu(\widehat{l_1})\ L_{c_{32}(k)} \cdots L_{c_{32}(k)}\ \nu(\widehat{l_d})\ L_{c_{32}(k)}\ \nu(l)) \wedge A(\nu(\widehat{l_1})),$$

and $\widehat{l_1} = f\left[\frac{l_1}{2}\right]$, then $h(2k+1) = l$.

We show that h is everywhere defined. If $n = 2k$, then $h(n)$ is defined, and

$$\mathfrak{A}_S^{G_\gamma} \models \varphi_m(\nu_\gamma(n)) \Leftrightarrow m \in \gamma(k) \\ \Leftrightarrow m \in \{m \mid \mathfrak{A} \models \varphi_m(\nu f(k))\} \Leftrightarrow \mathfrak{A} \models \varphi_m(\nu f(k)). \quad (*)$$

If $n = 2k+1$ then, by Lemma 22, exactly one of the conditions (2) and (3) in that lemma holds for $\nu_\gamma(n)$. If (2) holds, then

$$\mathfrak{A}_S^{G_\gamma} \models Q_{c_{32}(k)}(\nu_\gamma(2c_{31}(k)), \nu_\gamma(n)).$$

But in this case we have $\mathfrak{A}_S^{G_\gamma} \models \varphi_{c_{32}(k)}(\nu_\gamma(2c_{31}(k)))$, and by condition (*), $\mathfrak{A} \models \varphi_{c_{32}(k)}(\nu f c_{31}(k))$. Consequently, there exists a unique l such that

$$\mathfrak{A} \models Q_{c_{32}(k)}(\nu f c_{31}(k), \nu(l)),$$

and $h(n) = l$. On the other hand, if

$$\mathfrak{A}_S^{G_\gamma} \models \neg Q_{c_{32}(k)}(\nu_\gamma(2c_{31}(k)), \nu_\gamma(n)),$$

then condition (3) holds, and there is a unique sequence $l_1, \ldots, l_d$ such that

$$\mathfrak{A}_S^{G_\gamma} \models A(\nu_\gamma(l_1)) \wedge \left(\bigwedge_{i=1}^{d-1} \nu_\gamma(l_i)\ L_{c_{32}(k)}\ \nu_\gamma(l_{i+1})\right) \wedge \left(\nu_\gamma(l_d)\ L_{c_{32}(k)}\ \nu_\gamma(n)\right)$$

and a unique sequence $\widehat{l}_1, \ldots, \widehat{l}_d, l$ such that

$$\mathfrak{A}_S \models A(\nu(\widehat{l}_1)) \wedge \left(\nu(\widehat{l}_1)\ L_{c_{32}(k)}\ \nu(\widehat{l}_2)\ L_{c_{32}(k)}\ \cdots\ L_{c_{32}(k)}\ \nu(\widehat{l}_d)\ L_{c_{32}(k)}\ \nu(l)\right)$$

and $\widehat{l}_1 = f\left[\frac{l_1}{2}\right]$. The uniqueness of the first sequence follows from the fact that the $L_{c_{32}(k)}$–predecessor is uniquely defined, and that of the second from the facts that $\widehat{l}_1$ can be uniquely found and that the $L_{c_{32}(k)}$–successor is uniquely defined. Thus, in this case also, there is a unique value for $h(n)$.

Let Ψ be an isomorphism of the model $\mathfrak{A}$ onto $\mathfrak{A}_S^{G_\gamma}$. Since $(\mathfrak{A}, \nu) \in k^\Sigma(\mathfrak{A}_S)$, such an isomorphism exists, and since the model $\mathfrak{A}_S$ is rigid, it is unique. We show that

$$\Psi\nu_\gamma(n) = \nu h(n) \tag{**}$$

for all $n \in \mathbb{N}$. If n is even, then the assertion is true by Lemma 21. If n is odd, then condition (**) holds provided that $h(n)$ was defined in accordance with the first case, because $\Psi\nu_\gamma(2c_{31}(k)) = \nu f(c_{31}(k))$, and for all a and k there exists at least one b such that $\mathfrak{A}_S \models Q_k(a, b)$. But if $h(n)$ was defined by the second case then, because for every element the L_k–successor and the L_k–predecessor are uniquely defined, and $\Psi\nu_\gamma(l) = \nu(\widehat{l}_1)$, condition (**) also holds. This completes the proof of Proposition 25. □

In view of the above proposition, the existence problem is reduced to the construction of families of recursively enumerable sets with a finite number of nonequivalent computable Friedberg numberings.

Theorem 27 [27, 29] *For any $n \in \mathbb{N}$, there exists a computable family of recursively enumerable sets that has exactly n nonequivalent computable Friedberg numberings.*

The construction from the proof of this theorem was used by many mathematicians.

The following theorem was proved by Goncharov and Khoussainov on the basis of a modification of the same method as in [27].

Theorem 28 *There exists a computable family of recursively enumerable sets S that possesses exactly 2 nonequivalent computable Friedberg numberings ν and μ. Moreover, it is possible to choose two subfamilies A and B of S so that $\nu^{-1}(A)$ and $\mu^{-1}(B)$ are recursive while $\nu^{-1}(B)$ and $\mu^{-1}(A)$ are recursively enumerable, but not recursive.*

From this theorem we directly obtain the solution of the Harizanov problem on two-element Turing spectra.

Theorem 29 *There exists a constructive model $\mathbf{M}$ with exactly two constructivizations, and an algebraic subset A of its domain such that the Turing spectrum $Sp_T(\mathbf{M}, A)$ consists of exactly two degrees, one of which is recursive and the other recursively enumerable.*

It can be seen that, for any finite number of elements $a_1, \ldots, a_n$ of a model $\mathcal{A}$, the dimension of $\mathcal{A}$ does not exceed the dimension of the expanded model $(\mathcal{A}, a_1, \ldots, a_n)$. It is not obvious what effect the expansion by a finite number of constants has on the number of recursive isomorphism types of a model. Kudinov has noted that if the $\exists$–diagram of a recursively categorical model is decidable, then the expansion of this model by a finite number of constants is also recursively categorical. But without the assumption of the decidability of the $\exists$–diagram, the problem remained open. This question is formulated in Novosibirsk's "Logic Notebook" as the Goncharov–Millar problem. The problem was solved by P. A. Cholak, S. S. Goncharov, R. A. Shore, and B. M. Khoussainov by considering Friedberg enumerations of symmetric families of pairs of recursively enumerated sets.

Theorem 30 *There exists a recursively categorical model* $\boldsymbol{A}$*, such that for each element* $a \in A$*, the expanded model* $(\boldsymbol{A}, a)$ *has dimension* 2.

Let S be a family of recursively enumerable sets, and let A and B be two subfamilies of S such that $\nu^{-1}(A)$ and $\mu^{-1}(B)$ are recursive, and $\nu^{-1}(B)$ and $\mu^{-1}(A)$ are recursively enumerable but not recursive, as in Theorem 28. Starting from a single-valued numbering γ, we can define the model $\mathfrak{A}_S^{G_\gamma}$. Now we can construct a new model with two equivalence classes by some predicate η and each equivalence class is exactly the model $\mathfrak{A}_S^{G_\gamma}$. Now we consider a signature $\Sigma^* = \Sigma \cup \{\eta^2, F^2\}$, where Σ is a signature of the model $\mathfrak{A}_S^{G_\gamma}$ and η and F are new predicate symbols. Each of these classes is isomorphic to $\mathfrak{A}_S^{G_\nu}$ and $\mathfrak{A}_S^{G_\mu}$ respectively, where ν and μ two nonequivalent single-valued computable numberings of the family S. It is easy to prove that there exist partial recursive functions f from $\nu^{-1}(A)$ onto $\mu^{-1}(A)$ and g from $\mu^{-1}(B)$ onto $\nu^{-1}(B)$ such that for x, y from their domains, the equalities $\nu(x) = \mu(f(x))$ and $\mu(y) = \nu(g(y))$ hold. Adding the graphs of these functions to the model which are distinguished by the predicate symbol F, we can conclude that the constructed model has only one constructivization up to autoequivalence. Indeed, it is impossible to distinguish equivalence classes and, in these classes, nonautoequivalent submodels are necessarily realized. By adding a constant, we thereby fix equivalence classes and two nonautoequivalent constructivizations of the model appear.

References

[1] C. J. Ash, Recursive labelling systems and stability of recursive structures in hyperarithmetical degrees, Trans. Amer. Math. Soc., **298** (1986) 497–514; errata ibid., **310** (1988) 851.

[2] C. J. Ash and J. F. Knight, Pairs of recursive structures, Ann. Pure Appl. Logic, **46** (1990) 211–234.

[3] C. J. Ash, J. F. Knight, M. Manasse and T. A. Slaman, Generic copies of countable structures, Ann. Pure Appl. Logic, **42** (1989) 195–205.

[4] C. J. Ash, J. F. Knight and T. A. Slaman, Relatively recursive expansions II, Fund. Math., **142** (1993) 147–161.

[5] C. J. Ash and A. Nerode, Intrinsically recursive relations, in: Aspects of Effective Algebra, (Proc. Conf. Monash Univ., Clayton, Australia, Aug. 1–4, 1979), J. N. Crossley (ed.), (Upside Down A Book Co., Yarra Glen, Victoria, Australia, 1981), 26–41.

[6] C. C. Chang and H. J. Keisler, Model Theory, 3rd. edn., Stud. Logic Found. Math., **73** (1990); [1st. edn. 1973, 2nd. edn. 1977].

[7] J. N. Crossley, A. B. Manaster and M. F. Moses, Recursive categoricity and recursive stability, Logic Paper 49, (Monash Univ., Clayton, Victoria, Australia, 1983).

[8] J. N. Crossley and A. Nerode, Combinatorial Functors, Ergeb. Math. Grenzgeb. Ser. 3, **81** (1974).

[9] V. D. Dzgoev, On the constructivization of certain structures (Abstract in Russian), (Dep. VINITI, No. 1606–79, Moscow, 1979), Sibirsk. Math. Zh., **21** (1980) 231.

[10] E. F. Eisenberg and J. B. Remmel, Effective isomorphisms of algebraic structures, in: Patras Logic Symposion, (Proc. Logic Sympos. Patras, Greece, Aug. 18–22, 1980) G. Metakides (ed.), Stud. Logic Found. Math., **109** (1982) 95–122.

[11] Yu. L. Ershov, Constructive models (Russian), in: Izbr. Vopr. Alg. i Log., [Selected Questions in Algebra and Logic], Sbornik posvjascen. pamjati A. I. Mal'ceva, [A collection dedicated to the memory of A. I. Mal'tsev], Izdat. Nauk Sibirsk. Otdel., Novosibirsk, (1973) 111–130.

[12] Yu. L. Ershov, Theory of Numerations III (Constructive Models) (Russian), Lib. Dept. Algebra and Math. Logic Novosibirsk Univ., **13**, (Novosibirsk Gosudarstr. Univ., Novosibirsk, 1974).

[13] Yu. L. Ershov, Theory of Numerations (Russian), Monographs in Math. Logic and Foundations of Math., (Nauka, Moskva, 1977).

[14] Yu. L. Ershov, Decision Problems and Constructivizable Models (Russian), (Mathematical Logic and Foundations of Mathematics, Nauka, Moskva, 1980).

[15] Yu. L. Ershov and E. A. Palyutin, Mathematical Logic (Russian), 2nd. edn., (Nauka, Moskva, 1987). [English translation of 1st. edn., translated by V. Shokurov, (MIR Publishers, Moscow, 1984)].

[16] S. T. Fedoryaev, Structures of algebraic reducibility of positive numerations (Russian), Teor. Algoritm i ee Prilozhen, Vyčisl. Systemy, **129** (1989) 144–151, 197.

[17] S. T. Fedoryaev, Constructivizable models with linear structure of algebraic reducibility, Mat. Zametki, **48** (1990) 106–111; [translated in: Math. Notes, **48** (1990) 1245–1249].

[18] S. T. Fedoryaev, Some properties of algebraic reducibility of constructivizations, in: Kleene '90, (Proc. 3rd. Biennial Logic Conf., Sofia, Bulgaria, June, 1990), 24–25.

[19] S. T. Fedoryaev, Some properties of algebraic reducibility of constructivizations (Russian), Algebra i Logika, **29** (1990) 597–612, 627; [translated in: Algebra and Logic, **29** (1990) 395–405].

[20] S. T. Fedoryaev, Countability of widths of algebraic reducibility structures for models in some classes, Siberian Adv. Math., **3** (1993) 81–102.

[21] S. T. Fedoryaev, Decidable algorithmic problems on relatively complemented distributive lattices which cannot be simultaneously decidable, Bulletin of Symbolic Logic, **1** (1995) 109 (abstract).

[22] S. T. Fedoryaev, Recursively incompatible algorithmic problems on 1–constructivizable distributive lattices with relative complements (Russian), Algebra i Logika, **34** (1995) 667–680, 729; [translated in: Algebra and Logic, **34** (1995) 371–378].

[23] A. Fröhlich and J. C. Shepherdson, Effective procedures in field theory, Philos. Trans. Roy. Soc. London, Ser. A, **248** (1956) 407–432.

[24] S. S. Goncharov, Autostability and computable families of constructivizations (Russian), Algebra i Logika, **14** (1975) 647–680; [translated in: Algebra and Logic, **14** (1975) 392–409].

[25] S. S. Goncharov, Nonselfequivalent constructivizations of atomic Boolean algebras (Russian), Mat. Zametki, **19** (1976) 853–858; [translated in: Math. Notes, **19** (1976) 500–503].

[26] S. S. Goncharov, Autostability of models and abelian groups (Russian), Algebra i Logika, **19** (1980) 23–44; [translated in: Algebra and Logic, **19** (1980) 13–27].

[27] S. S. Goncharov, Computable numerations (Russian), Algebra i Logika, **19** (1980) 507–551, 617; [translated in: Algebra and Logic, **19** (1980) 325–356].

[28] S. S. Goncharov, On the problem of the number of nonautoequivalent constructivizations (Russian), Algebra i Logika, **19** (1980) 621–639, 745; [translated in: Algebra and Logic, **19** (1980) 401–414].

[29] S. S. Goncharov, The problem of the number of nonautoequivalent constructivizations (Russian), Dokl. Akad. Nauk SSSR, **251** (1980) 271–274; [translated in: Soviet Math. - Dokl., **21** (1980) 411–414].

[30] S. S. Goncharov, Nonequivalent constructivizations (Russian), Proc. Math. Inst. Sib. Branch Acad. Sci., (Nauka, Novosibirsk, 1982).

[31] S. S. Goncharov, Constructive Models, Monash Univ., Clayton, Victoria, Australia, Monash Logic Papers, **56** (November 1984).

[32] S. S. Goncharov, Countable Boolean Algebras (Russian), (Nauka. Sibirsk. Otdel., Novosibirsk, 1988).

[33] S. S. Goncharov, Computable classes of constructivizations for models of finite constructivity type (Russian), Sibirsk. Math. Zh., **34** (1993) 23–37; [translated in: Siberian Math. J., **34** (1993) 812–824].

[34] S. S. Goncharov, Effectively infinite classes of weak constructivizations of models (Russian), Algebra i Logika, **32** (1993) 631–664, 712; [translated in: Algebra and Logic, **32** (1993) 342–360].

[35] S. S. Goncharov and B. N. Drobotun, The algorithmic dimension of nilpotent groups (Russian), Sibirsk. Math. Zh., **30** (1989) 52–60, 225; [translated in: Siberian Math. J., **30** (1989) 210–217].

[36] S. S. Goncharov and V. D. Dzgoev, Autostability of models (Russian), Algebra i Logika, **19** (1980) 45–58, 132; [translated in: Algebra and Logic, **19** (1980) 28–37].

[37] S. S. Goncharov, A. V. Molokov and N. S. Romanovskii, Nilpotent groups of finite algorithmic dimension, Sibirsk. Math. Zh., **30** (1989) 82–88; [translated in: Siberian Math. J., **30** (1989) 63–68].

[38] V. S. Harizanov, The possible Turing degree of the nonzero member in a two element degree spectrum, Ann. Pure Appl. Logic, **60** (1993) 1–30.

[39] B. M. Khoussainov, Algorithmic degree of unars (Russian), Algebra i Logika, **27** (1988) 479–494, 499; [translated in: Algebra and Logic, **27** (1988) 301–312].

[40] B. M. Khoussainov, Recursive unary algebras and trees, (Special issue: A selection of papers presented at the Sympos. Logic at Tvev '92, July 20–24, 1992), A. Nerode and M. A. Taĭtslin, (eds.), Ann. Pure Appl. Logic, 67 (1994) 213–268.

[41] O. V. Kudinov, A criterion for the autostability of 1–decidable models (Russian), Algebra i Logika, **31** (1992) 479–492, 562; [translated in: Algebra and Logic, 31 (1992) 284–292].

[42] O. V. Kudinov, Algebraic dependencies and reducibilities of constructivizations in universal domains, Siberian Adv. Math., **3** (1993) 121–128.

[43] P. La Roche, Recursively presented Boolean algebras, Notices Amer. Math. Soc., **24**:6 (1977) A 552–553 (abstract).

[44] A. I. Mal'tsev, Constructive algebras I (Russian), Uspekhi Mat. Nauk, 16 (1961) 3–60; [translated in: Constructive algebras I, Russian Math. Surveys, **16**:3 (1961) 77–129; also in: The Metamathematics of Algebraic Systems, Collected Papers: 1936–1967, translated and edited by B. F. Wells III, Stud. Logic Found. Math., **66** (1971), Ch. 18, 148–200].

[45] A. I. Mal'tsev, Untersuchungen aus dem Gebiete der mathematischen Logik, Mat. Sb. (N.S.), **1** (1936) 323–335; [Russian summary: ibid., 336; translated in: The Metamathematics of Algebraic Systems, Collected Papers: 1936–1967, translated and edited by B. F. Wells III, Stud. Logic Found. Math., **66** (1971), Ch. 1, 1–14.]

[46] A. I. Mal'tsev, Algorithms and Recursive Functions (Russian), (Izdat. Nauka, Moscow, 1965); [translated: by L. F. Boron, L. E. Sanchis, J. Stilwell and K. Iseki, (Wolters-Noordhoff Publishing, Groningen, 1970)].

[47] A. I. Mal'tsev, Algebraic Systems (Russian), (posth. edn.), V. D. Smirnov and M. Taĭclin, (eds.), (Izdat. Nauka, Moscow, 1970); [translated: by B. D. Seckler and A. P. Doohovskoy, Grundlehren Math. Wiss., **192** (1973).]

[48] G. Metakides and A. Nerode, Recursion theory and algebra, in: Algebra and Logic, (Proc. 14th. Summer Res. Inst. Austral. Math. Soc., Monash Univ., Clayton, Vic., Australia, Jan. 6 – Feb. 16, 1974), J. N. Crossley, (ed.), Lecture Notes in Math., **450** (1975) 209–219.

[49] M. Moses, Recursive properties of isomorphism types, J. Austral. Math. Soc., Ser. A, **34** (1983) 269–286.

[50] M. Moses, Recursive linear orders with recursive successivities, Ann. Pure Appl. Logic, **27** (1984) 253–264.

[51] A. Nerode and J. B. Remmel, Recursion theory on matroids. in: Patras Logic Symposion, (Proc. Logic Sympos. Patras, Greece, Aug. 18–22, 1980) G. Metakides (ed.), Stud. Logic Found. Math., **109** (1982) 41–65.

[52] A. T. Nurtazin, Strong and weak constructivizations and computable families (Russian), Algebra i Logika, **13** (1974) 311–323, 364; [translated in: Algebra and Logic, **13** (1974) 177–184].

[53] S. P. Odintsov, Recursive Boolean algebras with a hyperhyperimmune set of atoms (Russian), Mat. Zametki, **44** (1988) 488–493, 557; [translated in: Math. Notes, **44** (1988) 747–749.]

[54] M. G. Peretyat'kin, Strongly constructive models and enumerations of the Boolean algebra of recursive sets (Russian), Algebra i Logika, **10** (1971) 535–557; [translated in: Algebra and Logic, **10** (1971) 332–345].

[55] M. O. Rabin, Computable algebra, general theory and theory of computable fields, Trans. Amer. Math. Soc., **95** (1960) 341–360.

[56] J. B. Remmel, Recursive isomorphism types of recursive Boolean algebras, J. Symbolic Logic, **46** (1981) 572–594.

[57] J. B. Remmel, Recursively categorical linear orderings, Proc. Amer. Math. Soc., **83** (1981) 387–391.

[58] H. Rogers Jr., Theory of Recursive Functions and Effective Computability, (1st. edn., McGraw-Hill, New York-Toronto, Ont.-London, 1967; 2nd. edn., MIT Press, Cambridge, Mass., London, 1987).

[59] V. A. Uspenskiĭ and A. L. Semenov, Theory of Algorithms: Fundamental Principles and Applications (Russian), (Nauka, Moscow, 1987).

[60] Yu. G. Ventsov, Algorithmic dimension of models (Russian), Dokl. Akad. Nauk, **305** (1989) 21–24; [translated in: Soviet Math. – Dokl., **39** (1989) 237–239].

[61] Yu. G. Ventsov, A problem on the effective choice of constructivizations, and recursive consistency of problems on constructive models (Russian), Algebra i Logika, **31** (1992) 3–20, 96; [translated in: Algebra and Logic, **31** (1992) 1–11].

[62] Yu. G. Ventsov, The effective choice problem for relations and reducibilities in classes of constructive and positive models (Russian), Algebra i Logika, **31** (1992) 101–118, 220; [translated in: Algebra and Logic, **31** (1992) 63–73].

S. S. Goncharov
Sobolev Institute of Mathematics
pr. Akademika Koptuga, 4
Novosibirsk, 630090
Russia

Chapter 7

Degrees of Models

J. F. Knight*

Contents

Introduction

The languages to be considered here are recursive, and the structures have universe ω. Theories, atomic and complete diagrams of structures, etc., are all identified with subsets of ω through Gödel numbering. The Turing degree of a set X will be denoted by $\deg(X)$. We shall often identify a structure $\mathcal{A}$ with its atomic diagram, $D(\mathcal{A})$. Thus, we may write $\deg(\mathcal{A})$ for $\deg(D(\mathcal{A}))$, and we say that $\mathcal{A}$ is *recursive* if $D(\mathcal{A})$ is recursive.

It is natural to ask questions such as the following:

(1) What are the degrees of isomorphic copies of a particular structure?

(2) What are the degrees of models of a particular theory?

*Partially supported by NSF Grant DMS 90 01513

HANDBOOK OF RECURSIVE MATHEMATICS
Edited by Yu. L. Ershov, S. S. Goncharov,
A. Nerode, and J. B. Remmel

We consider the set of degrees of isomorphic copies of a given structure. Let

$$DI(\mathcal{A}) = \{\deg(\mathcal{B}) : \mathcal{B} \cong \mathcal{A}\}.$$

For most structures $\mathcal{A}$, $DI(\mathcal{A})$ is closed upwards. Thus, if T is a consistent theory, and the models of T are like most structures, the set of degrees of models of T will be closed upwards. For a complete, consistent theory T, the Henkin construction yields a model $\mathcal{A}$ which is recursive in T — even the complete diagram, $D^c(\mathcal{A})$, is recursive in T. However, for some theories T, there are models of degree well below $\deg(T)$. Constructing such models, and determining which degrees are possible, can be delicate.

Some general facts about $DI(\mathcal{A})$ are stated in Section 1. In Section 2, there is a discussion of structures having "α–th jump degree", where α is a recursive ordinal. Many of the results on jumps involve structures $\mathcal{A}$ which encode a set S in a special way, so that, for example,

$$DI(\mathcal{A}) = \{\deg(X) : S \text{ is r.e. relative to } X\}, \quad \text{or}$$

$$DI(\mathcal{A}) = \{\deg(X) : S \text{ is } \Sigma^0_\alpha \text{ relative to } X\}.$$

In Section 3 there are some results on degrees of models of arithmetic, including the result of Harrington for which he introduced the method of "workers", and results of Solovay characterizing the degrees of models of various completions of PA. In Section 4, there are some results on existence of recursive models, including a result of Lerman and Schmerl on models of $\aleph_0$-categorical theories.

For some results, the proof is sketched, possibly in a special case, or there is a hint about the main ingredients of the proof. For other results, the proof is not described at all.

1 Properties of the set $DI(\mathcal{A})$

In this section, we consider the general behavior of the sets $DI(\mathcal{A})$. For some structures $\mathcal{A}$, $DI(\mathcal{A})$ is a single degree.

Proposition 1.1 *For each degree* $\mathbf{d}$, *there is a structure* $\mathcal{A}$ *such that* $DI(\mathcal{A}) = \{\mathbf{d}\}$.

Proof. Let S be a set of degree $\mathbf{d}$, and let $\mathcal{A} = (\omega\,,\, U_n\,{}_{n\in\omega})$, where for all n, $U_n = \omega$ if $n \in S$ and $U_n = \varphi$ if $n \notin S$. □

A structure $\mathcal{A}$ is *trivial* if there is a finite set $\alpha \subseteq \omega$ such that the automorphism group of $\mathcal{A}$ includes all permutations of ω which fix the elements of α. The structures from Proposition 1.1 are trivial since all permutations of ω are automorphisms. It was shown in [16] that $DI(\mathcal{A})$ consists of a single degree iff $\mathcal{A}$ is trivial.

The result below was proved by Solovay and by Marker [25] for models of arithmetic. The proof of the general case is in [16].

Theorem 1.2 *If $\mathcal{A}$ is a non-trivial structure, then $DI(\mathcal{A})$ is closed upwards.*

We sketch the proof in the special case where $\mathcal{A}$ is a linear ordering. Suppose $\mathbf{d} >$ deg $(\mathcal{A})$. We must produce $\mathcal{B} \cong \mathcal{A}$ such that deg$(\mathcal{B}) = \mathbf{d}$. Let S be a set of degree $\mathbf{d}$, and let π be a permutation of ω with the following feature: for each n, π either fixes $2n$ and $2n+1$ or else switches them, such that

$$\mathcal{A} \models \pi^{-1}(2n) < \pi^{-1}(2n+1) \quad \text{iff} \quad n \in S.$$

Let $\mathcal{B}$ be the structure such that $\mathcal{A} \cong_\pi \mathcal{B}$.

For some non-trivial structures $\mathcal{A}$, $DI(\mathcal{A})$ has a least element. Richter [27] gave conditions under which this is so, and other conditions under which it is not so.

Theorem 1.3 (Richter). *For any degree* $\mathbf{c}$, *there is a group G with* $DI(G) = \{\mathbf{d} : \mathbf{d} \geq \mathbf{c}\}$.

Sketch Proof. Let $S \subseteq \omega$ and let $G(S)$ be the direct sum of $\mathbb{Z}_{p_n}$ for $n \in S$, where p_n is the n–th prime. Then

$$DI(G(S)) \;=\; \{\deg(X) : S \text{ is } \Sigma^0_1 \text{ relative to } X\}.$$

If $S = C \oplus (\omega - C)$, where $\deg(C) = \mathbf{c}$, then $\mathbf{c}$ is least in $DI(G(S))$. Here

$$A \oplus B \;=\; \{2n : n \in A\} \cup \{2n+1 : n \in B\}.$$

□

The next result characterizes the degrees $\mathbf{b}$ such that for all $\mathbf{d} \in DI(\mathcal{A})$, $\mathbf{b} \leq \mathbf{d}$.

Theorem 1.4 *For any structure $\mathcal{A}$ and any $S \subseteq \omega$, the following are equivalent:*

(1) *for all $\mathcal{B} \cong \mathcal{A}$, $S \leq_T \mathcal{B}$,*

(2) *there is a finite sequence $\bar{a}$ in $\mathcal{A}$ such that $S \leq_T f$ for all functions f enumerating the existential type of $\bar{a}$ in $\mathcal{A}$.*

It is easy to see that (2) implies (1). The fact that (1) implies (2) can be proved by a forcing argument (see [16] or [6]).

Theorem 1.5 (Richter). *If $\mathcal{A}$ is a linear ordering such that $DI(\mathcal{A})$ has a least element* $\mathbf{d}$, *then* $\mathbf{d} = \mathbf{0}$.

Sketch Proof. Note that in a linear ordering, the existential types are all recursive. Then the result follows from Theorem 1.4. □

The orderings $\mathcal{A}$ such that $DI(\mathcal{A})$ has no least element are those with no recursive copy. Such orderings can be produced either by diagonalizing or by coding. Here we describe one method for coding a set in an ordering. If F is a countable family of orderings, then the *shuffle* of F, denoted by $\sigma(F)$, consists of densely many copies of each ordering in F. (To construct a copy of $\sigma(F)$, partition the rationals into dense subsets $Q_{\mathcal{A}}$, one for each ordering $\mathcal{A}$ in F, and replace each rational q in $Q_{\mathcal{A}}$ by a copy of $\mathcal{A}$.) For each $S \subseteq \omega$, let $\sigma^*(S) = \sigma(F)$, where F consists of ω and $n+1$ for $n \in S$. By the following result from [3], if S is not Σ^0_3, then $\sigma^*(S)$ has no recursive copy.

Proposition 1.6 *For all $S \subseteq \omega$,*

$$DI(\sigma^*(S)) = \{\deg(X) : S \text{ is } \Sigma^0_3 \text{ relative to } X\}.$$

The next result generalizes Theorem 1.4, characterizing the sets $S \subseteq \omega$ which are Δ^0_α with respect to $\mathcal{B}$ for all $\mathcal{B} \cong \mathcal{A}$. The characterization involves "recursive" infinitary Σ_α formulas. Roughly speaking, the recursive infinitary formulas are the formulas of $L\omega_1\omega$ in which the disjunctions and conjunctions are taken over r.e. sets. In a formal treatment, the recursive infinitary Σ_α and Π_α formulas are defined, together with their indices, in terms of ordinal notations (see [5]).

Theorem 1.7 *Let $\alpha \geq 1$ be a recursive ordinal. For any structure $\mathcal{A}$ and any $S \subseteq \omega$, the following are equivalent:*

(1) *for all* $\mathcal{B} \cong \mathcal{A}$, *$S$ is* Δ^0_α *relative to* $\mathcal{B}$,

(2) *there is a finite sequence* $\bar{a}$ *in* $\mathcal{A}$ *such that* $S \leq_T f$ *for all functions* f *enumerating the recursive infinitary* Σ_α *type of* $\bar{a}$ *in* $\mathcal{A}$.

The proof of Theorem 1.4 also yields the following result.

Theorem 1.8 *Let C be a countable set of Turing degrees, and let $\mathcal{A}$ be a structure such that for all* $\mathbf{d} \in DI(\mathcal{A})$, *there exists* $\mathbf{c} \in C$ *such that* $\mathbf{c} \leq \mathbf{d}$. *Then there is some* $\mathbf{c} \in C$ *such that for all* $\mathbf{d} \in DI(\mathcal{A})$, $\mathbf{c} \leq \mathbf{d}$.

We have seen that for all structures $\mathcal{A}$, either $DI(\mathcal{A})$ has the form $\{\mathbf{c}\}$, or else it is closed upwards. If $DI(\mathcal{A})$ is closed upwards, then it may have the form $\{\mathbf{d} : \mathbf{d} \geq \mathbf{c}\}$ or it may not. Lempp posed the following.

Problem 1 *Is there a structure $\mathcal{A}$ for which* $DI(\mathcal{A}) = \{\mathbf{d} : \mathbf{d} > \mathbf{0}\}$?

The author, believing that the answer to Problem 1 should be negative, posed some related problems on enumerations. If $\mathcal{S}$ is a countable family of subsets of ω, then an *enumeration* of $\mathcal{S}$ is a relation $R \subseteq \omega \times \omega$ such that $\mathcal{S} = \{R_n : n \in \omega\}$, where $R_n = \{k : (n,k) \in R\}$.

Problem 2 *If for each non-recursive set X, $\mathcal{S}$ has an enumeration recursive in X, must $\mathcal{S}$ have a recursive enumeration?*

Problem 3 *If for each non-recursive set X, $\mathcal{S}$ has an enumeration r.e. relative to X, must* **S** *have an r.e. enumeration?*

The author thought the answers to Problems 2 and 3 should be negative. However, Wehner [36] answered both Problems 2 and 3 affirmatively, showing that there is a family $\mathcal{S}$ such that for all non-recursive sets X, $\mathcal{S}$ has an enumeration recursive in X, but $\mathcal{S}$ has no r.e. enumeration. A positive solution to Problem 2 yields a positive solution to Problem 1. We may take $\mathcal{A}$ to be a structure with an equivalence relation and a 1-1 function. The equivalence classes, all infinite, correspond to sets $A \in \mathcal{S}$. The function divides each equivalence class into finite cycles in such a way that the equivalence class corresponding to A has a cycle of size $2n$ for $n \in A$ and $2n+1$ for $n \notin A$ (and these are the only sizes). Slaman [29] gave a different solution to Problem 1, not using Wehner's results (this was after Wehner had solved Problems 2 and 3 but before he was aware of Problem 1).

Our picture of the possible sets $DI(\mathcal{A})$ is not complete. Some further patterns will be exhibited in the next section.

2 Jump degrees

It would be pleasant to have a single degree to assign to the isomorphism type of a structure $\mathcal{A}$. When $DI(\mathcal{A})$ has a least element, this is the obvious choice. However, for many interesting structures $\mathcal{A}$, $DI(\mathcal{A})$ has no least element. Jockusch suggested a family of possible measures of complexity of isomorphism types, involving jumps.

For any set $X \subseteq \omega$, the *jump* of X is the set $X' = \{e : \varphi_e^X(e) \downarrow\}$. It is possible to iterate the jump operation through the recursive ordinals and define $X^{(\alpha)}$ for all recursive ordinals α. To be precise, we should define $X^{(a)}$ for all ordinal notations a, but we shall identify the ordinal α with a notation a and write $X^{(\alpha)}$ while thinking of $X^{(a)}$. Let $X^{(\alpha+1)} = (X^{(\alpha)})'$. For α a recursive limit ordinal, each notation for α picks out (notations for) an increasing sequence of ordinals $(\alpha_n)_{n\in\omega}$ with limit α. Let $X^{(\alpha)} = \{(n,k) : k \in X^{(\alpha_n)}\}$. An infinite recursive ordinal α has more than one notation, so the set $X^{(\alpha)}$ is not unique. However, the Turing degree is independent of the notation. We write $0^{(\alpha)}$ for $\varphi^{(\alpha)}$ and $\mathbf{0}^{(\alpha)}$ for deg $(0^{(\alpha)})$.

Jockusch suggested the following. Let α be a recursive ordinal and let $\mathcal{A}$ be a structure. Consider the set $\{\mathbf{b}^{(\alpha)} : \mathbf{b} \in DI(\mathcal{A})\}$. If this set has a least element, say $\mathbf{d} = \mathbf{b}^{(\alpha)}$, then $\mathcal{A}$ is said to have α–th *jump degree* $\mathbf{d}$. Note that $\mathcal{A}$ has 0–th jump degree $\mathbf{d}$ iff $\mathbf{d}$ is least in $DI(\mathcal{A})$. For a given structure $\mathcal{A}$, if there is some α such that $\mathcal{A}$ has α–th jump degree, then there is at least such α. Suppose $\mathcal{A}$ has α–th jump degree $\mathbf{d}$, and for $\beta < \alpha$, $\mathcal{A}$ does not have β–th jump degree. Then $\mathcal{A}$ is said to have α–th jump degree $\mathbf{d}$ *sharply*. It is easy to see that if $\mathcal{A}$ has α–th jump degree, then for all recursive ordinals $\beta > \alpha$, $\mathcal{A}$ has β–th jump degree. As we shall see, for each recursive ordinal α, there exist structures having α–th jump degree sharply. There also exist structures not having α–th jump degree for any recursive ordinal α.

Below, we give some results on jumps for orderings, groups, and Boolean algebras. The first result resembles Theorem 1.5.

Theorem 2.1 *If $\mathcal{A}$ is a linear ordering and $\mathcal{A}$ has* 1*–st jump degree* $\mathbf{d}$, *then* $\mathbf{d} = \mathbf{0}'$.

See [16] for the proof.

In view of Theorems 1.5 and 2.1, it is natural to ask whether a linear ordering can have 1–st jump degree sharply. Jockusch and Soare [13] showed that it can.

Theorem 2.2 (Jockusch-Soare) *For any r.e. set S which is non-recursive, there is an ordering $\mathcal{A} \leq_T S$ such that $\mathcal{A}$ has no recursive copy.*

The proof is a "permitting" construction. Recursive copies of $\mathcal{A}$ are avoided by diagonalizing, rather than by encoding S.

A set S is *low* if $S' \leq 0'$. If, in Theorem 2.2, S is taken to be low, the resulting structure $\mathcal{A}$ has 1–st jump degree $\mathbf{0}'$. By Theorem 1.5, since $\mathcal{A}$ has no recursive copy, it does not have 0–th jump degree. Seetapun (in unpublished work) extended Theorem 2.2 to Δ_2^0 sets S which are non-recursive.

Theorem 2.3 *For each recursive ordinal $\alpha \geq 2$, and each $\mathbf{d} \geq \mathbf{0}^{(\alpha)}$, there is an ordering $\mathcal{A}$ which has α–th jump degree $\mathbf{d}$ sharply.*

In [3], there are constructions of orderings having α–th jump degree $\mathbf{d}$ sharply for all recursive ordinals $\alpha \geq 2$ and all $\mathbf{d} > \mathbf{0}^{(\alpha)}$. For certain α, the same constructions serve for $\mathbf{d} = \mathbf{0}^{(\alpha)}$. In [8], the remaining cases are filled in, showing that for each $\alpha \geq 1$, there is an ordering $\mathcal{A}$ which has α–th jump degree $\mathbf{0}^{(\alpha)}$ sharply.

Most of the cases in the proof of Theorem 2.3 involve coding a set or family of sets in an ordering $\mathcal{A}$, and characterizing $DI(\mathcal{A})$ in terms of the set or family of sets. These lemmas may be of interest apart from Theorem 2.3, as they illustrate some possible patterns for $DI(\mathcal{A})$. The first lemma extends Proposition 1.6.

Lemma 2.4 *Let α be a recursive ordinal, at least 3, and not a limit ordinal. Then for each $S \subseteq \omega$, there is an ordering $\mathcal{A}$ such that*

$$DI(\mathcal{A}) = \{\deg(X) : S \text{ is } \Sigma_\alpha^0 \text{ relative to } X\}.$$

Lemma 2.4 is a composite of results proved in [3] and [8].

If $\mathcal{A}$ and $\mathcal{B}$ are orderings, then $\mathcal{A} + \mathcal{B}$ is the result of putting a copy of $\mathcal{B}$ after a copy of $\mathcal{A}$, and $\mathcal{A} \cdot \mathcal{B}$ is the result of replacing each element of $\mathcal{B}$ by a copy of $\mathcal{A}$. Let Z be the usual ordering of the integers. We define Z^β for countable ordinals β. For finite β, we may proceed by induction, letting $Z^{n+1} = Z^n \cdot Z$. More generally, we obtain an ordering of type Z^β as follows. Let F consist of the functions $f : \beta \to Z$ such that $f(\gamma) = 0$ for all but finitely many $\gamma < \beta$. For f and g distinct elements of F, let γ be greatest such that $f(\gamma) \neq g(\gamma)$. Then $f < g$ iff $f(\gamma) < g(\gamma)$.

One ingredient in the proof of Lemma 2.4 is the result below.

Lemma 2.5 *Let $\mathcal{A}$ be an ordering.*

(a) $\mathbf{d} \in DI(Z \cdot \mathcal{A})$ *iff* $\mathbf{d}'' \in DI(\mathcal{A})$,

(b) *if β is a recursive* <u>*limit*</u> *ordinal, then* $\mathbf{d} \in DI(Z^{\beta} \cdot \mathcal{A})$ *iff* $\mathbf{d}^{(\beta+1)} \in DI(\mathcal{A})$.

Watnik [35] proved (a). The statement of (b) is in [3], with a sketch of a proof using workers. In [2] Ash gave another proof, using his "α–systems" (introduced in [1]).

In Lemma 2.4, each ordering codes a single set. In the next two lemmas, the orderings code a family of sets. In the first, we code a fixed enumeration of the family. Say $R \subseteq \omega \times \omega$, and let $\mathcal{A}$ be the sum of orderings of type $\eta + Z^n \cdot \sigma^*(R_n)$, for $n \in \omega$. For this $\mathcal{A}$, we have

$$DI(\mathcal{A}) \;=\; \{\deg(X) : R_n \text{ is } \Sigma^0_{2n+3} \text{ relative to } X, \text{ uniformly in } n\}.$$

Extending this idea, we obtain the following result, used in proving Theorem 2.3 for the case of limit ordinals.

Lemma 2.6 *Let α be a recursive limit ordinal. There is a recursive increasing sequence of ordinals $(\alpha_n)_{n\in\omega}$, with limit α, such that for $R \subseteq \omega \times \omega$, there is an ordering $\mathcal{A}$ for which*

$$DI(\mathcal{A}) \;=\; \{\deg(X) : R_n \textit{ is } \Sigma^0_{\alpha_n} \textit{ relative to } X, \textit{ uniformly in } n\}.$$

The next result involves coding a family $\mathcal{S}$ of sets, but not coding a specific enumeration of $\mathcal{S}$ as in Lemma 2.6. Moreover, each set in the family is coded slowly, through infinitely many levels. This result is used in proving Theorem 2.3 for the case of successors of limit ordinals. We let $\mathcal{E}(\mathcal{S})$ denote the set of all enumerations of the family $\mathcal{S}$.

Lemma 2.7 *Let α be a recursive limit ordinal. For any notation for α, we can find another notation which picks out a sequence $(\alpha_n)_{n\in\omega}$ such that for any countable family $\mathcal{S}$ of subsets of ω, there is an ordering $\mathcal{A}$ such that*

$$DI(\mathcal{A}) \;=\; \Big\{ \deg(X) : (\exists R \in \mathcal{E}(\mathcal{S}))\,(R_k \cap n \textit{ is } \Sigma^0_{\alpha_n} \textit{ relative to } X, \textit{ uniformly in } k \textit{ and } n)\Big\}.$$

Here is a sketch of the proof of Theorem 2.3 for $\alpha = 3$. Let $\mathbf{d} \geq \mathbf{0}^{(3)}$. First, choose a set S such that $\deg(S^{(3)}) = \mathbf{d}$, $S^{(3)} \equiv_T S \oplus 0^{(3)}$, and

$$\{\deg(X)'' : S \text{ is } \Sigma^0_3 \text{ relative to } X\}$$

has no least element. Let $\mathcal{A} = \sigma^*(S)$, where $\mathcal{A} \equiv_T S$. By Proposition 1.6, $DI(\mathcal{A}) = \{\deg(X) : S \text{ is } \Sigma^0_3 \text{ relative to } X\}$. The choice of S guarantees that $\mathcal{A}$ does not have 2–nd jump degree. For $\mathcal{B} \cong \mathcal{A}$,

$$\mathcal{A}^{(3)} \equiv_T S^{(3)} \equiv_T 0^{(3)} \oplus S \leq_T \mathcal{B}^{(3)}.$$

Therefore, $\mathcal{A}$ has 3–rd jump degree $\mathbf{d}$.

Question 2.1 *Suppose α is a recursive limit ordinal. Is there a scheme for coding an arbitrary set S in an ordering $\mathcal{A}(S)$ such that $DI(\mathcal{A}(S)) = \{\deg(X) : S$ is Σ^0_α relative to $X\}$?*

Lemmas 2.4, 2.6 and 2.7 provide the orderings needed for all but one case of Theorem 2.3. The case not covered by these coding lemmas is where $\alpha = 2$ and $\mathbf{d} = \mathbf{0}''$. Here the proof combines Theorem 2.2 (or a relatization of it) with the following coding trick of Downey, which is related to Lemma 2.5 (see [8]).

Lemma 2.8 *Let $\mathcal{A}$ be an ordering, and let $\varphi = \eta+2+\eta$. Then $\mathbf{d} \in DI(\varphi\cdot\mathcal{A})$ iff $\mathbf{d}' \in DI(\mathcal{A})$.*

The following result is stated in [3].

Theorem 2.9 *The ordering ω_1^{CK} does not have α–th jump degree for any recursive ordinal α.*

Sketch Proof. Suppose that ω_1^{CK} has α–th jump degree $\mathbf{d}$, say $\mathcal{A}$ is a copy such that $\mathcal{A}^{(\alpha)}$ has degree $\mathbf{d}$. By Theorem 1.7, there is a finite sequence $\bar{a}$ in $\mathcal{A}$ such that for all functions f enumerating the recursive infinitary Σ_α type of $\bar{a}$ in $\mathcal{A}$, $\mathcal{A} \leq_T f$. (Ordinal notation is being suppressed here.) There is a recursive ordinal β with a finite sequence $\bar{b}$ realizing this type. The type has a hyperarithmetical enumeration, so $\mathcal{A}$ is hyperarithmetical. Since ω_1^{CK} is known not to have a hyperarithmetical copy, this is a contradiction. □

Before leaving linear orderings, we state the following curious fact.

Theorem 2.10 *For each $\mathbf{d} > \mathbf{0}$, there is a linear ordering of degree $\mathbf{d}$ with no recursive copy.*

This is proved in several cases, using results mentioned above, Seetapun's extension of Theorem 2.2, Lemma 2.8, and the results from [3].

Oates [26] obtained results on jumps of groups, similar to those on jumps of orderings.

Theorem 2.11 (Oates) *For each recursive ordinal $\alpha \geq 1$, and each* $\mathbf{d} > \mathbf{0}^{(\alpha)}$, *there is a group G which has α–th jump degree* $\mathbf{d}$ *sharply.*

Oates used coding lemmas like the ones for orderings. Below are two of these coding lemmas. The first was actually stated above, in the proof of Theorem 1.3.

Lemma 2.12 *Let $S \subseteq \omega$ and let G be the direct sum of $\mathbb{Z}_{p_n}$ for $n \in S$. Then $DI(G) = \{\deg(X) : S$ is r.e. relative to $X\}$.*

Lemma 2.13 *Let $S \subseteq \omega$, and let G be a reduced Abelian p–group of length ω, with $u_{2n}(G) = 1$, $u_{2n+1}(G) = 1$ if $n \in S$, and $u_{2n+1}(G) = 0$ if $n \notin S$. Then $DI(G) = \{\deg(X) : S$ is Σ^0_2 relative to $X\}$.*

Now, we turn to Boolean algebras. The results differ from those for orderings more than might be expected. The phenomenon in Theorem 2.1 not only extends but holds at higher levels [14].

Theorem 2.14 (Jockusch-Soare) *If a Boolean algebra has n–th jump degree* $\mathbf{d}$, *then* $\mathbf{d} = \mathbf{0}^{(n)}$.

Feiner [9] showed that there is a Δ^0_2 Boolean algebra with no recursive copy. This has ω–th jump degree $\mathbf{0}^{(\omega)}$ sharply. Thurber [33] clarified and extended Feiner's ideas, showing the following.

Proposition 2.15 (Thurber)

(a) *For any set S, there is a Boolean algebra $\mathcal{A}$ for which*

$$DI(\mathcal{A}) = \{\deg(X) : S \cap (n+1) \text{ is } \Sigma^0_{2n+4} \text{ relative to } X, \text{ uniformly in } n\}.$$

(b) *For any recursive limit ordinal α, there is a recursive increasing sequence of ordinals $(\alpha_n)_{n\in\omega}$ with limit α, such that for any S, there is a Boolean algebra $\mathcal{A}$ for which*

$$DI(\mathcal{A}) = \{\deg(X) : S \cap (n+1) \text{ is } \Sigma^0_{\alpha_n} \text{ relative to } X, \text{ uniformly in } n\}.$$

We end the discussion of Boolean algebras by mentioning results which contrast with Theorem 2.10. Recall that a set S is *low* if $S' \leq_T 0'$. More generally, S is low_n if $S^{(n)} \leq_T 0^{(n)}$. Downey and Jockush [7] showed that every low Boolean algebra has a recursive copy. Thurber [34] improved upon this, showing that every low_2 Boolean algebra has a recursive copy. It is conjectured that for all finite n, every low_n Boolean algebra has a recursive copy.

3 Models of Arithmetic

For all structures $\mathcal{A}$, $\text{Th}(\mathcal{A}) \leq_T D^c(\mathcal{A}) \leq_T D(\mathcal{A})^{(\omega)}$. Turing equivalence is possible. For example, if $\mathcal{N} = (\omega\,, +\,, \cdot\,, S\,, 0\,, <)$, then $\mathcal{N}$ is recursive, while $\text{Th}(\mathcal{N})$ has degree $\mathbf{0}^{(\omega)}$. By "arithmetic", we mean a theory including some part of $\text{Th}(\mathcal{N})$. *True arithmetic*, or TA, is $\text{Th}(\mathcal{N})$. *First order Peano arithmetic*, or PA, is the fragment with a recursive set of axioms consisting of those of Peano's axioms which are first order, together with the *induction* axioms

$$\forall \bar{u} \left[\varphi(u,0) \ \& \ \forall x\,(\varphi(\bar{u},x) \rightarrow \varphi(u,S(x))) \rightarrow \forall x\, \varphi(u,x)\right].$$

A model of arithmetic is called *standard* if it is isomorphic to $\mathcal{N}$ and *non-standard* otherwise. Let DPA denote the set of degrees of non-standard models of PA. The following result, due to Tennenbaum [32], says that $Zero \notin$ DPA.

Theorem 3.1 *If $\mathcal{A}$ is a non-standard model of* PA, *then $\mathcal{A}$ is not recursive.*

Sketch Proof. Let A, B be a recursively inseparable pair of r.e. sets, $A = W_a$ and $B = W_b$. There is a natural formula $\psi(u,x)$ saying that for $c, k \leq u$, if c is a computation for $\varphi_a(k)$, then $p_k \mid x$, and if c is a computation for $\varphi_b(k)$, then $p_k \nmid x$. For each $n \in \omega$, $\mathcal{A} \models \exists x\, \psi(S^{(n)}(0), x)$. Therefore, for some infinite a, $\mathcal{A} \models \exists x\, \psi(a,x)$. Let $\mathcal{A} \models \psi(a,b)$, and let $C = \{k : \mathcal{A} \models p_k \mid b\}$. Then C separates A and B. We can determine whether $\mathcal{A} \models p_k \mid b$ by searching the atomic diagram for a sentence which says $b = p_k d + r$, for some $r < p_k$. Therefore, $C \leq_T \mathcal{A}$, and $\mathcal{A}$ cannot be recursive. □

A set $X \subseteq \omega$ is said to be *arithmetical* if X is definable in $\mathcal{N}$. Equivalently, X is arithmetical iff $X \leq_T 0^{(n)}$ for some $n \in \omega$. McAloon asked whether there was a non-standard model $\mathcal{A}$ of PA such that $\mathcal{A}$ is arithmetical while $\text{Th}(\mathcal{A})$ is not. Harrington [10] produced such a model.

Theorem 3.2 (Harrington). *There is a non-standard model $\mathcal{A}$ of* PA *such that $\mathcal{A}$ has degree $\mathbf{0}'$ and* Th($\mathcal{A}$) *has degree $\mathbf{0}^{(\omega)}$.*

The proof requires some kind of infinitely nested priority construction. It was for this result that Harrington developed his method of "workers". There is also a proof using the method Ash introduced in [1]. In order to code a set of degree $\mathbf{0}^{(\omega)}$ in Th($\mathcal{A}$), Harrington used the fact that there is a recursive sequence of sentences $(\varphi_n)_{n\in\omega}$ such that for all completions T of PA, φ_n is independent over PA and the set of sentences of T which are either Σ_n or Π_n (i.e., the set is consistent with φ_n and also with $\neg\varphi_n$).

A *Scott set* is a family $\mathcal{S}$ of subsets of ω such that

(1) if $A \in \mathcal{S}$ and $B \leq_T A$, then $B \in \mathcal{S}$,

(2) if $A, B \in \mathcal{S}$, then $A \oplus B \in \mathcal{S}$,

(3) if $\mathcal{T} \in \mathcal{S}$ is an infinite subtree of $2^{<\omega}$, then there exists $p \in \mathcal{S}$ such that p is a path, or infinite branch, through $\mathcal{T}$; equivalently, if $A \in \mathcal{S}$ is a consistent set of axioms, then there is a complete theory $T \in \mathcal{S}$ such that $T \supseteq A$.

The arithmetical sets form one Scott set. Scott [28] originally isolated these families of sets in studying "representability". Let T be a theory in the language of arithmetic. A set $X \subseteq \omega$ is said to be *representable with respect to* T if there is a formula $\varphi(x)$ such that for all $n \in \omega$, if $n \in X$, then $T \vdash \varphi(S^n(0))$ and if $n \notin X$, then $T \vdash \neg\varphi(S^n(0))$. Let Rep($T$) denote the family of sets representable with respect to T. If $T =$ PA, then Rep(T) is the family of recursive sets. The same is true for any recursively axiomatizable extension of PA. Scott was interested in completions of PA, which are not recursively axiomatizable.

Theorem 3.3 (Scott) *For any countable family $\mathcal{S}$ of subsets of ω, the following are equivalent:*

(1) *there is a completion T of* PA *such that $\mathcal{S} =$* Rep(T),

(2) *$\mathcal{S}$ is a Scott set.*

Let $\mathcal{A}$ be a non-standard model of PA. For any element a, let $S_a = \{n \in \omega : \mathcal{A} \models p_n \mid a\}$. The *Scott set of* $\mathcal{A}$ is $\mathcal{SS}(\mathcal{A}) = \{S_a : a \in \omega\}$. The proof of Theorem 3.1 suggests the proofs of some other useful facts.

Remarks 3.1

(1) If $\mathcal{S}$ is a Scott set, then $\{\deg(A) : A \in \mathcal{S}\}$ has no maximum element.

(2) If $\mathcal{A}$ is a non-standard model of PA and $A \in \mathcal{SS}(\mathcal{A})$, then $A \leq_T \mathcal{A}$.

(3) If $\mathcal{A}$ is a non-standard model of PA and $\bar{a}$ is a tuple in $\mathcal{A}$, then for all n, the finitary Σ_n type realized by $\bar{a}$ in $\mathcal{A}$ is an element of $\mathcal{SS}(\mathcal{A})$.

Proposition 3.4 *If $\mathcal{A}$ is a non-standard model of* PA, *then $DI(\mathcal{A})$ has no least element.*

Proof. Suppose $\mathbf{d}$ is least in $DI(\mathcal{A})$, and let S have degree $\mathbf{d}$. By Theorem 1.4, there is a complete existential type Γ realized in $\mathcal{A}$ such that S is recursive in all enumerations of Γ. Then $S \leq_T \Gamma$, and by (3) above, $\Gamma \in \mathcal{SS}(\mathcal{A})$. By (1), there exists $A \in \mathcal{SS}(\mathcal{A})$ such that $A \not\leq_T \Gamma$. Therefore, $A \not\leq_T \mathcal{S}$, contradicting (2). □

The following is due to Scott [28] (see also [21] and [15]).

Theorem 3.5 *If T is a completion of* PA, *and $\mathcal{S}$ is a countable family of subsets of ω, then the following are equivalent:*

(1) *there is a non-standard model $\mathcal{A}$ of T such that $\mathcal{S} = \mathcal{SS}(\mathcal{A})$.*

(2) *$\mathcal{S}$ is a Scott set including* $\mathrm{Rep}(T)$.

Remark 3.2 If $\mathcal{A}$ is a non-standard model of PA, and given $R = \{(a, n) : \mathcal{A} \models p_n \mid a\}$, then R is an enumeration of $\mathcal{SS}(\mathcal{A})$ such that $R \leq_T \mathcal{A}$.

Earlier, we defined DPA as the set of degrees of non-standard models of PA. It is also the set of degrees of enumerations of countable Scott sets, and of completions of PA (see [12] for further equivalences). The set DPA has no least element, and it contains low degrees.

Now, we turn to true arithmetic. Let $\mathcal{A}$ be a non-standard model of TA. Since Rep(TA) consists of the arithmetical sets, the degree of $\mathcal{A}$ is an upper bound for the sequence $(\mathbf{0}^{(n)})_{n\in\omega}$. The converse is not true.

Theorem 3.6 (Lachlan-Soare) *There is a degree $\mathbf{d}$ such that for all n, $\mathbf{d} > \mathbf{0}^{(n)}$ but $\mathbf{d}$ is not the degree of any non-standard model of* TA.

Theorem 3.6 is proved by constructing a set X such that for all n, $0^{(n)} \leq_T X$, but if $R \leq_T X$, then R is not an enumeration of a family which includes all of the sets $0^{(n)}$, see [20]. Much more is known, see [22].

Using workers, Marker [25] showed that if $\mathbf{d} > \mathbf{0}^{(n)}$ for all n, and $\mathbf{d}' \geq \mathbf{0}^{(\omega)}$, then there is a non-standard model of TA of degree $\mathbf{d}$. Marker asked whether the condition $\mathbf{d}' \geq \mathbf{0}^{(\omega)}$ is necessary. The next result, from [20], implies that it is not.

Theorem 3.7 *There exists a non-standard model $\mathcal{A}$ of* TA *such that* $\deg(\mathcal{A})'' = 0^{(\omega)}$.

Solovay [30] characterized the degrees of non-standard models of TA in terms of "effective" enumerations of Scott sets. Suppose $\mathcal{S}$ is a countable Scott set. An *effective enumeration of $\mathcal{S}$* is an enumeration R equipped with functions f, g, h which witness the fact that $\mathcal{S}$ is a Scott set:

(1) if $A = R_n$ and $\varphi_e^A = X_B$, then $B = R_{f(n,e)}$,

(2) if $A = R_m$ and $B = R_n$, then $A \oplus B = R_{g(m,n)}$,

(3) if $\mathcal{T} = R_m$ where $\mathcal{T}$ is an infinite subtree of $2^{<\omega}$, and $P = R_{h(m)}$, then X_P is a path through $\mathcal{T}$.

The *degree* of the effective enumeration R, f, g, h, is the join of the degrees of R, f, g and h, and the effective enumeration is said to be *recursive* in X if R, f, g and h are all recursive in X. If $\mathcal{A}$ is a non-standard model of PA, and R is the enumeration of $\mathcal{SS}(\mathcal{A})$ described above, then R can be equipped with functions f, g and h to form an effective enumeration which is recursive in $D^c(\mathcal{A})$.

Here is Solovay's characterization of the degrees of non-standard models of TA.

Theorem 3.8

(a) *If $\mathcal{S}$ is a countable Scott set containing the arithmetical sets, then the degrees of non-standard models $\mathcal{A}$ of* TA *such that $\mathcal{SS}(\mathcal{A}) = \mathcal{S}$ are just the degrees of effective enumerations of $\mathcal{S}$.*

(b) *The degrees of non-standard models of* TA *are the degrees of effective enumerations of Scott sets which contain the arithmetical sets.*

Macintyre and Marker [24] investigated the degrees of complete diagrams of recursively saturated models. Using this work, Marker obtained the following.

Theorem 3.9 *If R is an enumeration of a Scott set $\mathcal{S}$, then there is an effective enumeration of $\mathcal{S}$ recursive in R.*

The proof of Theorem 3.9 is not at all direct. Here is the outline. There is a completion T of PA such that $T \in \mathcal{S}$. By one of the main results in [24], T has a model $\mathcal{A}$ such that $D^c(\mathcal{A}) \leq_T R$ and $\mathcal{SS}(\mathcal{A}) = \mathcal{S}$. Then $D^c(\mathcal{A})$ yields the desired effective enumeration of $\mathcal{SS}(\mathcal{A})$.

Using Theorem 3.9, Marker simplified Theorem 3.8 as follows.

Corollary 3.10 *The degrees of non-standard models of* TA *are precisely the degrees of enumerations of Scott sets which contain the arithmetical sets.*

Theorem 3.9 also yields information related to Scott's original result.

Corollary 3.11 *For a given countable Scott set $\mathcal{S}$, the degrees of completions T of* PA *such that $\mathcal{S} = \mathrm{Rep}(T)$ are precisely the degrees of enumerations of $\mathcal{S}$.*

The characterization of the degrees of non-standard models of TA leaves certain questions unanswered.

Problem 4 *Is there a non-standard model $\mathcal{A}$ of* TA *whose degree is minimal over the degrees $\mathbf{0}^{(n)}$?*

Some time after characterizing the degrees of non-standard models of TA, Solovay also characterized the degrees of models of completions of other completions of PA [31].

Theorem 3.12 *Let T be a completion of* PA *such that $T \neq$* TA *(so the models are all non-standard), and let $T_n = T \cap \Sigma_n$. For any set X, the following are equivalent:*

(1) *T has a model $\mathcal{A} \leq_T X$,*

(2) *there is a Scott set $\mathcal{S}$ such that $T_n \in \mathcal{S}$ for all n, and $\mathcal{S}$ has an enumeration $R \leq_T X$ with a family of functions t_n, Δ^0_n relative to X, uniformly in n, such that $\lim_{s\to\infty} t_n(s)$ is an R-index for T_n, and for all s, $t_n(s)$ is an R-index for a subset of T_n.*

In the next section, we state a general result from which this follows.

4 Models for highly non-recursive theories

There are a number of results on existence of recursive models for non-recursive theories. We mention only a few. We begin with a result on $\aleph_0$–categorical theories, by Lerman and Schmerl [23].

Theorem 4.1 *Let T be an arithmetical $\aleph_0$–categorical theory such that the set of finitary Σ_{n+2} sentences of T is Σ^0_{n+1}. Then T has a recursive model.*

The proof of this result is inductive, based on the lemma below. For a structure $\mathcal{A}$, let $D_n(\mathcal{A})$ denote the set of sentences in $D^c(\mathcal{A})$ which are Boolean combinations of Σ_n sentences.

Lemma 4.2 *Suppose T is an $\aleph_0$–categorical elementary first order theory, and let $\mathcal{A}$ be a model of T. If the set of finitary Σ_{n+2} sentences of T is Σ^0_{n+1}, and $D_{n+1}(\mathcal{A})$ is Δ^0_{n+2}, then T has a model $\mathcal{B}$ such that $D_n(\mathcal{B})$ is Δ^0_{n+1}.*

It is possible to improve Theorem 4.1 somewhat, dropping the assumption that T is arithmetical (see [18]).

Theorem 4.3 *Let T be an $\aleph_0$–categorical theory, and for each n, let T_n be the set of finitary Σ_n sentences of T. If T_{n+2} is Σ^0_{n+1} uniformly in n, then T has a recursive model.*

Hurlburt [11] constructed recursive models for certain complicated infinitary theories by taking a limit of recursive models for various fragments of the theory. For a structure $\mathcal{A}$, let $\Sigma_\alpha(\mathcal{A})$ be the set of all Σ_α sentences of $L_{\omega_1\omega}$ which are true in $\mathcal{A}$. Similarly, let $\Pi_\alpha(\mathcal{A})$ be the set of Π_α sentences of $L_{\omega_1\omega}$ true in $\mathcal{A}$. Hurlburt's result involves the notion of an "α–friendly" family of structures, from [4]. Fix a language L, and let α be a recursive ordinal. Consider a family $(\mathcal{A}_i)_{i\in\omega}$ of L–structures, and form the relations $\leq_\beta$ such that $(i,\bar{a}) \leq_\beta (j,\bar{b})$ iff $\Pi_\beta((\mathcal{A}_i,\bar{a})) \subseteq \Pi_\beta((\mathcal{A}_j,\bar{b}))$. The family $(\mathcal{A}_i)_{i\in\omega}$ is said to be *α–friendly* if the structures are uniformly recursive and the relations $\leq_\beta$, for $\beta < \alpha$, are uniformly r.e.. (Ordinal notation is being suppressed again.)

Let α be a recursive limit ordinal, and let $(\alpha_n)_{n\in\omega}$ be the increasing sequence picked out by a notation for α. We call f an *(α_n)–function* if $f(n)$ is $\Delta^0_{\alpha_n}$, uniformly in n (i.e., there is a recursive function σ such that for each $n, \sigma(n)$ is an index for a procedure for computing $f(n)$ using a $\Delta^0_{\alpha_n}$ oracle). We also speak of *relatively* (α_n)–functions, where f is (α_n) relative to X if $f(n)$ is $\Delta^0_{\alpha_n}$ relative to X, uniformly in n.

Here is a version of Hurlburt's theorem.

Theorem 4.4 *Let α be a recursive limit ordinal and let $(\alpha_n)_{n\in\omega}$ be a recursive increasing sequence of successor ordinals with limit α, say $\alpha_n = \beta_n + 1$. Suppose $(\mathcal{A}_i)_{i\in\omega}$ is an α-friendly family of structures, and i is an (α_{n-1})-function such that $\Sigma_{\alpha_n}(\mathcal{A}_{i(n)}) \subseteq \Sigma_{\alpha_n}(\mathcal{A}_{i(n+1)})$ for all n. Then there is a recursive structure $\mathcal{B}$ such that*

(a) $\Sigma_{\alpha_n}(\mathcal{A}_{i(n)}) \subseteq \Sigma_{\alpha_n}(\mathcal{B})$ *for all $n \in \omega$,*

(b) *for each tuple $\bar{b}$ in $\mathcal{B}$, there exists m such that for all $n \geq m$, there exists $\bar{a}$ in $\mathcal{A}_{i(n)}$ such that $\Pi_{\beta_n}(\mathcal{A}_{i(n)}, \bar{a}) \subseteq \Pi_{\beta_n}(\mathcal{B}, \bar{b})$.*

Here is a companion result for successor ordinals.

Theorem 4.5 *Let α be a recursive successor ordinal, say $\alpha = \beta + 1$. Suppose $(\mathcal{A}_i)_{i\in\omega}$ is an α-friendly family, and i is a Δ^0_α-function such that $\Sigma_\alpha(\mathcal{A}_{i(n)}) \subseteq \Sigma_\alpha(\mathcal{A}_{i(n+1)})$ for all n. Then there is a recursive structure $\mathcal{B}$ such that*

(a) $\Sigma_\alpha(\mathcal{A}_{i(n)}) \subseteq \Sigma_\alpha(\mathcal{B})$ *for all n, and*

(b) *for each tuple $\bar{b}$ in $\mathcal{B}$, there exists m such that for all $n \geq m$, there exists $\bar{a}$ in $\mathcal{A}_{i(n)}$ such that $\Pi_\beta(\mathcal{A}_{i(n)}, \bar{a}) \subseteq \Pi_\beta(\mathcal{B}, \bar{b})$.*

The proofs of Theorems 4.4 and 4.5 use Ash's methods [1].

The next result yields both Theorem 3.8 and a slight strengthening of Theorem 3.2 as consequences. The statement involves Scott sets, but the result does not apply just to models of arithmetic. A structure $\mathcal{A}$ is said to *represent* a Scott set $\mathcal{S}$ if for all n and all complete finitary Σ_n types Γ consistent with $\mathrm{Th}(\mathcal{A})$, Γ is realized in $\mathcal{A}$ iff $\Gamma \in \mathcal{S}$. If $\mathcal{A}$ is a non-standard model of PA, then $\mathcal{SS}(\mathcal{A})$ is the unique Scott set represented by $\mathcal{A}$. However, some structures do not represent any Scott set, and some represent all Scott sets. We give some examples below.

Example 4.1 The structure $\mathcal{N}$ does not represent any Scott set.

To see why this is so, let $\mathcal{S}$ be a Scott set. If, for all n and all complete finitary Σ_n types Γ realized in $\mathcal{N}$, we have $\Gamma \in \mathcal{S}$, then $\mathcal{S}$ must contain all arithmetical sets. There are arithmetical complete finitary Σ_1 types which are realized in non-standard models of TA but are omitted in $\mathcal{N}$.

Example 4.2 The structure $(Q, <)$ represents all Scott sets.

Let $\mathcal{S}$ be a Scott set. For any n and any complete Σ_n type Γ which is consistent with $\text{Th}(Q, <), \Gamma \in \mathcal{S}$, and Γ is realized in $(Q, <)$. (The same is true for any structure $\mathcal{A}$ such that $\mathcal{A}$ is saturated and for all n, the Σ_n types realized in $\mathcal{A}$ are all recursive.)

Theorem 4.6 *Let T be a complete theory, and for each n, let T_n be the set of finitary Σ_n sentences of T. Let $\mathcal{S}$ be a countable Scott set such that for all n, $T_n \in \mathcal{S}$. Let R be an enumeration of $\mathcal{S}$ and let f be a function such that for all n, $R_{f(n)} = T_n$. If $R \leq_T X$ and f is (n) relative to X, then T has a model $\mathcal{A}$ recursive in X. Moreover, $\mathcal{A}$ can be taken to represent $\mathcal{S}$.*

This is proved in [17] using workers.

Here is the improved version of Theorem 3.2 which follows from Theorem 4.6. Recall that DPA is the set of degrees of non-standard models of PA.

Corollary 4.7 *If $\mathbf{d} \in$ DPA, then there is a non-standard model $\mathcal{A}$ of PA of degree $\mathbf{d}$ such that $\text{Th}(\mathcal{A})$ has degree $\mathbf{d}^{(\omega)}$.*

The following is a strengthening of Theorem 4.6, which can be used in proving Theorem 3.12. In fact, the statement was suggested by Theorem 3.12.

Theorem 4.8 *Let T be a complete theory, and for each n, let T_n be the set of finitary Σ_n sentences of T. Let $\mathcal{S}$ be a countable Scott set such that for all n, $T_n \in \mathcal{S}$. Let R be an enumeration of $\mathcal{S}$ such that $R \leq_T X$, and let $(t_n)_{n\in\omega}$ be a family of functions, such that t_n is Δ^0_n relative to X, uniformly in n, for all n, $\lim_{s\to\infty} t_n(s)$ is an R-index for T_n, and for all s, $R_{t_n(s)} \subseteq T_n$. Then T has a model $\mathcal{A} \leq_T X$ such that $\mathcal{A}$ represents $\mathcal{S}$.*

Theorem 4.8 is proved in [19].

References

[1] C. J. Ash, Recursive labelling systems and stability of recursive structures in hyperarithmetical degrees, Trans. Amer. Math. Soc., **298** (1986) 497–514; errata ibid., **310** (1988) 851.

[2] C. J. Ash, A construction for recursive linear orderings, J. Symbolic Logic, **56** (1991) 673–683.

[3] C. J. Ash, C. G. Jockusch Jr. and J. F. Knight, Jumps of orderings, Trans. Amer. Math. Soc., **319** (1990) 573–599.

[4] C. J. Ash and J. F. Knight, Pairs of recursive structures, Ann. Pure Appl. Logic, **46** (1990) 211–234.

[5] C. J. Ash and J. F. Knight, Relatively recursive expansions, Fund. Math., **140** (1992) 137–155.

[6] C. J. Ash, J. F. Knight, M. Manasse and T. A. Slaman, Generic copies of countable structures, Ann. Pure Appl. Logic, **42** (1989) 195–205.

[7] R. G. Downey and C. G. Jockusch Jr., Every low Boolean algebra is isomorphic to a recursive one, Proc. Amer. Math. Soc., **122** (1994) 871–880.

[8] R. G. Downey and J. F. Knight, Orderings with α–th jump degree $\mathbf{0}^{\alpha}$, Proc. Amer. Math. Soc., **114** (1992) 545–552.

[9] L. J. Feiner, Hierarchies of Boolean algebras, J. Symbolic Logic, **35** (1970) 365–374.

[10] L. Harrington, Building non-standard models of Peano arithmetic, (handwritten notes, 1979).

[11] K. R. Hurlburt, Sufficiency conditions for theories with recursive models, Ann. Pure Appl. Logic, **55** (1992) 305–320.

[12] C. G. Jockusch Jr. and R. I. Soare, Π_1^0 classes and degrees of theories, Trans. Amer. Math. Soc., **173** (1972) 33–56.

[13] C. G. Jockusch Jr. and R. I. Soare, Degrees of orderings not isomorphic to recursive linear orderings, (Special issue: Internatl. Sympos. Math. Logic and its Applications, Nagoya, 1988), Ann. Pure Appl. Logic, **52** (1991) 39–64.

[14] C. G. Jockusch Jr. and R. I. Soare, Boolean algebras, Stone spaces, and the iterated Turing jump, J. Symbolic Logic, **59** (1994) 1121–1138.

[15] J. F. Knight, Additive structure in uncountable models for a fixed completion of P, J. Symbolic Logic, **48** (1983) 623–628.

[16] J. F. Knight, Degrees coded into jumps of orderings, J. Symbolic Logic, **51** (1986) 1034–1042.

[17] J. F. Knight, Degrees of models with prescribed Scott set, in: Classification Theory, (Proc. Joint U.S.–Israel Workshop on Model Theory in Math. Logic, Chicago, Dec. 15–19, 1985), J. T. Baldwin, (ed.), Lecture Notes in Math., **1292** (1987) 182–191.

[18] J. F. Knight, Nonarithmetical $\aleph_0$–categorical theories with recursive models, J. Symbolic Logic, **59** (1994) 106–112.

[19] J. F. Knight, Requirement systems, J. Symbolic Logic, **60** (1995) 222–245.

[20] J. F. Knight, A. H. Lachlan and R. I. Soare, Two theorems on degrees of models of true arithmetic, J. Symbolic Logic, **49** (1984) 425–436.

[21] J. F. Knight and M. Nadel, Models of arithmetic and closed ideals, J. Symbolic Logic, **47** (1982) 833–840.

[22] A. H. Lachlan and R. J. Soare, Models of arithmetic and upper bounds for arithmetic sets, J. Symbolic Logic, **59** (1994) 977–983.

[23] M. Lerman and J. H. Schmerl, Theories with recursive models, J. Symbolic Logic, **44** (1979) 59–76.

[24] A. Macintyre and D. Marker, Degrees of recursively saturated models, Trans. Amer. Math. Soc., **282** (1984) 539–554.

[25] D. Marker, Degrees of models of true arithmetic, in: Logic Colloq. '81. (Proc. Herbrand Sympos., Marseilles, July, 16–24, 1981), J. Stern (ed.), Stud. Logic Found. Math., **107** (1982) 233–242.

[26] S. Oates, Jump Degrees of Groups, Ph.D. Thesis, Univ. Notre Dame, IN, (1989).

[27] L. J. Richter, Degrees of structures, J. Symbolic Logic, **46** (1981) 723–731.

[28] D. Scott, Algebra of sets binumerable in complete extensions of arithmetic, Proc. Sympos. Pure Math., **5** (1962) 117–121.

[29] T. A. Slaman, Relative to any non-recursive set, Proc. Amer. Math. Soc., (to appear).

[30] R. M. Solovay, Degrees of models of true arithmetic, (manuscript written for Colloquium '84, but not actually published there).

[31] R. M. Solovay, (personal correspondenc, 1991).

[32] S. Tennenbaum, Non-archimedean models for arithmetic, Notices Amer. Math. Soc., **6** (1959) 270.

[33] J. J. Thurber, Recursive and r.e. quotient Boolean algebras, Arch. Math. Logic, **33** (1994) 121–129.

[34] J. J. Thurber, Every low$_2$ Boolean algebra has a recursive copy, Proc. Amer. Math. Soc., **123** (1995) 3859–3866.

[35] R. Watnick, A generalization of Tennenbaum's Theorem on effectively finite recursive linear orderings, J. Symbolic Logic, **49** (1984) 563–569.

[36] S. Wehner, Enumerations, countable structures, and Turing degrees, Proc. Amer. Math. Soc., (to appear).

J. F. Knight
Department of Mathematics
University of Notre Dame
Notre Dame, IN 46556, USA

Chapter 8

Groups of Computable Automorphisms

A. S. Morozov*

Contents

Introduction

The notion of algorithm has been extensively investigated from various points of view, such as, for instance, algebraic, topological, set-theoretic, etc.. Here we study symmetries related to the notion of algorithm; this, undoubtedly, will help us to understand this notion better.

This paper is a survey of the results on computable automorphisms. The term "computable" is understood here in a broad sense ("recursive", "computable with respect to an oracle", "arithmetical", "hyperarithmetical", etc.).

In this paper we do not consider automorphisms of the lattices of recursively enumerable sets and Turing degrees. This subject is rather advanced now and merits an individual survey.

Generally, two related topics arise in the investigations of automorphisms. The first topic concerns investigations of properties of automorphisms for a

*Partially supported by the Soros Foundation.

HANDBOOK OF RECURSIVE MATHEMATICS
Edited by Yu. L. Ershov, S. S. Goncharov,
A. Nerode, and J. B. Remmel

given class of objects. The second topic is about finding, or more specifically reconstructing some properties of the class of objects from the automorphisms. We will discuss both topics in various situations.

We follow commonly used notions and definitions from Recursion Theory [54, 59, 69], Model Theory [5], Constructive Model Theory [11], and Admissible Sets [2].

A main definition of this paper is that of constructive models. We consider here only *effective languages*, i.e., languages whose symbols are naturally (may be, implicitly) coded by an initial segment of natural numbers; and given a code of a symbol one can effectively find its number of arguments. Consider a pair (M, ν) where M is a model and ν is an *enumeration* of $|M|$, that is, a mapping from ω onto the domain $|M|$ of the model M. Using effectiveness of our language and this enumeration ν, we can build some Gödel enumeration of the set of all quantifier-free sentences (as well as the set of all sentences) in the language extended by constants $c_{\nu(n)}$, $n \in \omega$, whose interpretations in M are $\nu(n)$. If the set of all quantifier-free sentences (respectively all sentences) which are true in M is recursive, then the pair (M, ν) is called a *constructive* (resp. *strongly constructive*) *model.* A model M is called *constructivizable* (resp. *strongly constructivizable*) if there exists an enumeration ν such that (M, ν) is a constructive (resp. strongly constructive) model. The notion of recursive model is close to that of constructive model. A model M is said to be *recursive* (resp. *decidable*) if its universe is a recursive set and there exists a procedure which, given a quantifier–free formula (resp. any formula) $\varphi(\overline{x})$ and an n–tuple $\overline{m}$ of elements, answers whether $M \models \varphi(\overline{m})$ or not. In fact, these two notions of constructive model and recursive model (as well as decidable model and strongly constructive model) are, respectively, the eastern and western versions of the same intuitive concept. We use both of them; in each case we select a variant which can show discussed results more clearly; sometimes we even reformulate original results or definitions in the other form.

The central notion of this paper is the definition of *recursive automorphism*. Below we give two definitions of this notion, for recursive and constructive models.

Definition 0.1

(1) Let M be a recursive model. An *automorphism* φ of M is called *recursive* if φ is a partial recursive function. All such automorphisms form a group $\mathrm{Aut}_r M$.

(2) Let (M, ν) be a constructive model. An *automorphism* φ of the model M is called *recursive* if there exists a recursive function f such that $\varphi\nu = \nu f$. All such automorphisms form a group $\mathrm{Aut}_r(M, \nu)$.

1 General results on permutations

It is natural to ask the question of which classes of recursive functions are closed under composition and inversion, i.e., form a group. Clearly, the set of all permutations recursive in an oracle is a group. The following result shows that the primitive recursive functions fail to form a group.

Theorem 1.1 (A. V. Kuznetsov, [27]) *There exists a primitive recursive permutation f whose inverse f^{-1} is not a primitive recursive function.*

A sharper version of this and other results which take into account the Grzegorczyk hierarchy are in [4].

It would be interesting to know algorithmic reducibilities with the property that all permutations whose degrees are less than or equal to some degree with respect to one of those reducibilities form a group (a list of most known reducibilities are in [54, 59, 69], for instance.)

Let $\mathrm{Sym}(\omega)$ be the group of all permutations of the set of natural numbers ω and let d be a Turing degree. We denote the Turing degree of a set or a function f by $\deg(f)$. Define

$$G_d \rightleftharpoons \{f \in \mathrm{Sym}(\omega) \mid \deg(f) \leqslant d\}.$$

As noted above, this is a group. Our main question here is the following: *What is the relationship between the group G_d and the Turing degree d?*

A basic tool which we use in investigations of this question is an interpretation of the action of the group $\mathrm{Sym}(\omega)$ or its subgroups on natural numbers. We first distinguish the so-called *transpositions*, that is, those permutations which permute exactly two elements. The following result by R. McKenzie [29] plays a key role here.

Theorem 1.2 *Let G be a group of permutations of natural numbers which contains all finite permutations. Then a permutation $x \in G$ is a transposition if and only if*

$$G \models \neg(x = 1) \;\&\; (x^2 = 1) \;\&\; \forall y((x^{-1}y^{-1}xy)^6 = 1).$$

This result allows us to associate with each pair $\langle x, y\rangle$ of transpositions such that $xy \neq yx$ a unique number α which is moved by both of them; in this case we say that the pair $\langle x, y\rangle$ *holds* the unique number α. Now we can distinguish the pairs $\langle x_0, x_1\rangle$ and $\langle y_0, y_1\rangle$ which hold the same element. If we denote this property by $E(x_0, x_1, y_0, y_1)$, it can be defined as follows:

$$E(x_0, x_1, y_0, y_1) = \Big(\bigwedge_{i,j=0,1} (x_i y_j)^3 = 1 \Big)$$

$$\& \bigwedge_{i,j,=0,1} \big((x_i = y_j \ \& \ x_{1-i} \neq y_{1-j}) \rightarrow x_{1-i} \neq x_i^{-1}\, y_{1-j}\, x_i\big).$$

Thus, each natural number can be identified with the class of such equivalent pairs of transpositions.

Note that if a pair of transpositions $\langle x, y\rangle$ holds a natural number n and $g \in G$, then the pair of transpositions $\langle g\,x\,g^{-1}\,,\ g\,y\,g^{-1}\rangle$ holds $g(n)$. Therefore the action of G on ω is also interpretable in the group G.

Further steps consist of interpretation of the structure of natural numbers — addition, multiplication, Turing reducibility, etc.. Therefore we can translate arithmetical statements into statements of the first order language of groups and vice versa.

It turns out that every group G_d contains all the information about the Turing degree d.

Theorem 1.3 [49] *For any Turing degrees s and d, the group G_s is embeddable into G_d if and only if $s \leqslant d$. In particular, $G_s \cong G_d$ if and only if $s = d$.*

A family $\mathcal{A}$ of sets of natural numbers is *arithmetical* if there exists a formula $\varphi(P)$ of the standard language of arithmetics extended by the unary predicate symbol P such that, for each set X of natural numbers,

$$\langle \omega\,,+\,,\cdot\,,0\,,s\,,<\,,X\rangle \models \varphi(P) \ \Leftrightarrow\ X \in \mathcal{A} \ \text{ holds.}$$

The following two theorems show the relationship between the definability of families of sets, and the first order definability in the class $\{G_s \mid s \text{ is a Turing degree}\}$.

Theorem 1.4 [39] *Let $\mathcal{A}$ be an arithmetical family of sets. Then there exists a formula φ of the first order group-theoretic language, such that for all Turing degrees d*

$$G_d \models \varphi \ \Leftrightarrow\ \mathcal{A} \cap d \neq \emptyset \ \text{ holds.}$$

Theorem 1.5 [39] *Let φ be a first order group-theoretic formula. Then the set $\{X \mid G_{\deg(X)} \models \varphi\}$ is arithmetical.*

Corollary 1.6 [39] *If $\mathcal{A}$ is an arithmetical family of sets, then*

$$\{X \subseteq \omega \mid \deg(X) \cap \mathcal{A} \neq \emptyset\}$$

is arithmetical.

We can also describe all Turing degrees d whose group G_d can be characterized within the class $\{G_s \mid s \text{ is a Turing degree}\}$ by a single sentence.

Corollary 1.7 [39] *A Turing degree d is an arithmetical class of sets if and only if there exists a formula φ_d of the group-theoretic language such that for all Turing degrees s*

$$G_s \models \varphi_d \Leftrightarrow d = s \quad \textit{holds},$$

i.e., G_d can be characterized within the class

$$\{G_s \mid s \textit{ is a Turing degree}\}$$

by a single sentence.

Corollary 1.8 [39] *The following statements are valid:*

(1) *Any arithmetical degree is an arithmetical class of sets.*

(2) *For any degree d which contains an arithmetical set, there exists a group-theoretic sentence φ_d such that for each Turing degree s*

$$G_s \models \varphi_d \Leftrightarrow d = s \quad \textit{holds}.$$

(3) *Any degree $0^{(\alpha)}$, where α is a constructive ordinal, is an arithmetical family.*

(4) *For any constructive ordinal α, there exists a group-theoretic sentence φ_α such that for each degree d*

$$G_d \models \varphi_\alpha \Leftrightarrow d = 0^{(\alpha)} \quad \textit{holds}.$$

Each degree d can be distinguished by the theory of its group G_d only up to some level. Indeed, using the result on determinacy of Borel sets [50] and its consequence concerning degrees [68], as well as using the technique of [39] and [68], we can establish the following result:

Corollary 1.9 *There exists a Turing degree d such that G_a is elementarily equivalent to G_b for all degrees $a, b \geqslant d$.*

This is not surprising since the higher a Turing degree is, the fewer individual features it possesses. The highest degrees become more and more similar and starting from some level they cannot be distinguished by elementary first order tools. Thus, there exists a "most typical" theory of the group of permutations which are computable with an oracle. This is $\mathrm{Th}(G_d)$, where d satisfies Corollary 1.9. It would be interesting to investigate the properties of this theory.

Here we have considered only the first order properties of the groups G_d. It would be very interesting to answer similar questions for some other languages, for instance, fragments of $L_{\omega_1 \omega}$. The whole language $L_{\omega_1 \omega}$ is too strong for this purpose, since it is possible to construct for each G_d a formula θ_d which defines the group G_d up to isomorphism [2].

C. F. Kent [20] studied the normal structure of the groups G_d; in fact, he investigated a more general problem setting. He proved that the normal series

$$1 \triangleleft A \triangleleft \mathrm{Fin} \triangleleft G_d$$

is a unique normal series for each group G_d, where Fin is the group of all permutations of natural numbers which move only a finite number of elements, and A is the group of all even permutations of natural numbers, i.e., finite permutations which can be decomposed into a product of an even number of two-element cycles. This is also valid for the group of all arithmetical permutations, which is defined as follows:

$$\mathrm{Ar} = \{f \in S(\omega) \mid \exists n \deg(f) \leqslant \emptyset^{(n)}\}.$$

C. F. Kent [20] also investigated automorphisms of such groups of permutations. In order to formulate the next two results we need the notion of an *effectively closed group*:

Definition 1.10 A subgroup G in the group of all permutations of natural numbers $\mathrm{Sym}(\omega)$ is *effectively closed* if, for all n and $f_1, \ldots, f_n \in G$, each permutation recursive in $f_1, \ldots, f_n$ belongs to G.

Theorem 1.11 (C. F. Kent, [20]) *Let G be an effectively closed group. Suppose also that $G \supset \mathrm{Fin}$, $G \neq \mathrm{Fin}$, and each normal subgroup $G' \leqslant G$ which contains an infinite permutation satisfies $G' = G$. Then all automorphisms of the group G are inner, i.e., of the form $x \mapsto \alpha^{-1} x \alpha$, for $\alpha \in G$.*

Corollary 1.12 (C. F. Kent, [20]) *Each automorphism of a group G_d, or of the group of all arithmetical permutations, is inner.*

This shows that in these groups the situation is just the same as in the group Sym(ω) [63, 64].

C. F. Kent also studied in [20] conjugacy classes in the groups G_d and Ar. He proved that in the group Ar, the conjugacy class of an element is completely defined by its cyclic structure, just as in the group Sym(ω), but this fails to be true in the groups G_d, i.e., there exist two permutations which are conjugate in the group Sym(ω), but not in G_d.

Now we will discuss complexity questions for the recursive automorphism groups of recursive models.

We say that a model M *has Turing degree* d if d is the least element in the set

$$\{\deg(H) \mid M \text{ is isomorphic to a model recursive in } H\}.$$

We denote this fact by $\deg M = d$. If there is no such minimal degree, then we say that the model M is *has no degree*.

It turns out that each group G_d has a degree.

Theorem 1.13 [39] *For each Turing degree d,* $\deg G_d = d''$.

Other questions concerning computable permutations, their decomposition and related problems are in [4, 20, 27].

Hereafter we denote the group G_0, i.e., the group of all recursive permutations of the set of natural numbers, by $\mathrm{Aut}_r\omega$.

2 Some general results on recursive automorphism groups

In this section we will discuss some general results about recursive automorphism groups of recursive models. The following results demonstrate how properties of usual automorphisms can be far from properties of recursive automorphisms. V. D. Dzgoev, A. B. Manaster and J. B. Remmel independently proved the first result of this kind

Theorem 2.1 (V. D. Dzgoev [9, 15], A. B. Manaster and J. B. Remmel [28]) *There exists a recursive model M such that $|\mathrm{Aut}M| = 2^\omega$, but $|\mathrm{Aut}_r M| = 1$.*

It is often the case in the theory of recursive automorphisms that when the whole automorphism group is sufficiently large but the recursive automorphism group is almost trivial, as we will see later.

In order to formulate the next result we need two definitions. The first one is a classical definition of homogeneity for countable models. A countable model M is called *homogeneous* provided that, for all $n < \omega$, and for all pairs of n–tuples of elements $\overline{a}, \overline{b} \in M$, $(M, \overline{a}) \equiv (M, \overline{b})$ implies that they are isomorphic, i.e., there exists $f \in \mathrm{Aut}M$ such that $f(\overline{a}) = \overline{b}$. The next definition is an effective version of the preceding one: a constructive model (M, ν) is called *effectively homogeneous* provided that, for all pairs of n–tuples of elements $\overline{a}, \overline{b} \in M$, $(M, \overline{a}) \equiv (M, \overline{b})$ implies that there exists an $f \in \mathrm{Aut}_r M$ such that $f(\overline{a}) = \overline{b}$.

Theorem 2.2 (K. Z. Kudaibergenov, [22]) *There exists a homogeneous strongly constructivizable model whose automorphism group is of the cardinality of the continuum, but in any constructivization its recursive automorphism group is trivial. In particular, this model is not effectively homogeneous with respect to all constructivizations.*

This theorem implies that two isomorphic elements, i.e., elements which can be transformed into each other by an automorphism, cannot be transformed into each other by a recursive automorphism in all constructivizations.

The theorems cited may be strengthened up to hyperarithmetics. A model M is called *hyperarithmetical* if its universe is hyperarithmetical, and the set of all quantifier-free sentences of its signature, extended by constants c_i, $i \in M$ (so that the value of c_i is i) which are true on M, is hyperarithmetical.

We will prove the following theorem later in this paper.

Theorem 2.3 [46] *There exists a recursive model M such that each hyperarithmetical model isomorphic to M has no nontrivial hyperarithmetical automorphisms, and $|\mathrm{Aut}M| = 2^\omega$.*

This theorem follows from the construction for trees below.

It should be noted that a further extension of this result is impossible, since, obviously, each model possessing a nontrivial automorphism also possesses an automorphism which is recursive in a Π^1_1–complete set.

In the course of the proof of this theorem, the following statement will be used, which amplifies a well-known result of Kueker [23] on the equivalence between the property $|\mathrm{Aut}M| < 2^\omega$ and the existence of an n-tuple with $\mathrm{Aut}(M, \overline{p}) = 1$.

Theorem 2.4 *Let M be a countable model. The following three statements are equivalent:*

(1) $|\mathrm{Aut}M| < 2^\omega$,

(2) $\mathrm{Aut}M \subset \mathrm{HYP}_M$, *and*

(3) $\mathrm{Aut}M \in \mathrm{HYP}_M$.

Here HYP_M is the least admissible set over M containing the model M [2]. From here we deduce that if a recursive model M has at most ω automorphisms, then all these automorphisms are hyperarithmetical.

The following result, due to Goncharov, is a direct consequence of categorical considerations from [13]. This theorem is proved with the use of a standard interpretation technique.

Theorem 2.5 (S. S. Goncharov, [13]) *For each recursive model M, there exist a recursive binary predicate R and a recursive partial order L such that*

$$\mathrm{Aut}_r M \;\cong\; \mathrm{Aut}_r L \;\cong\; \mathrm{Aut}_r \langle \omega, R\rangle .$$

Now we consider a natural correspondence between functional trees and automorphisms of models. Consider a recursive model $\mathcal{M}$ and an effective enumeration of its signature $\{=\} = \sigma_0 \subseteq \sigma_1 \subseteq \cdots \bigcup_{i<\omega} \sigma_i = \sigma$. Without lost of generality we may assume that the signature of $\mathcal{M}$ doesn't contain operation symbols.

Let the recursive functions $\ell(x)$ and $r(x)$ give us the left and right coordinates of the pair coded by a number x, and let $\ln(s)$ be the length of s. We may define the recursive tree $T_\mathcal{M}$ as follows:

$$\begin{aligned} T_\mathcal{M} \;=\; \big\{ s \in \omega^{<\omega} \mid\ & \ell s(0) \neq rs(0), \text{ and elements of tuples } \langle \ell s(0), \ldots, \ell s(k-1)\rangle \\ & \text{and } \langle rs(0), \ldots, rs(k-1)\rangle \text{ satisfy the same formulas of kind} \\ & P(x_{i_1}, \ldots, x_{i_n}),\ P \in \sigma_k,\ k = \ln(s), \text{ and } \{0, \ldots, [\tfrac{k-4}{2}]\} \\ & \subseteq \{\ell s(0), \ldots, \ell s(k-1)\} \cap \{rs(0), \ldots, rs(k-1)\} \text{ if } k \geqslant 4 \big\}. \end{aligned}$$

The definition of this tree reflects an informal attempt to construct all possible automorphisms of the model as follows: at each step we have finite parts of potential automorphisms which correspond to elements of the tree. Trying to extend a finite part of a potential automorphism, we add some elements to its domain and to its range so that they satisfy the same quantifier-free formulas. Since each automorphism is onto, we should also take care that each element with number n must be added to the domain and to the range at some step.

It is not difficult to verify that $T_{\mathcal{M}}$ possesses an infinite branch iff $\mathcal{M}$ has a nontrivial automorphism. Moreover, one can ascertain that, for any infinite branch t of T, there is an automorphism of $\mathcal{M}$ of the same Turing degree, and for any automorphism φ of $\mathcal{M}$, there is an infinite branch of $T_{\mathcal{M}}$ of the same Turing degree.

Now we consider a converse construction, which given an arbitrary tree builds a model that possesses a nontrivial automorphism iff this tree possesses an infinite branch.

Assume $T \subseteq \omega^{<\omega}$ is a tree. Let

$$T_n = \{t \in T \mid \ln(t) = n\}.$$

Consider for each $n < \omega$ the Abelian group G_n of order 2 with the basis T_n, i.e., $G \models \forall x\,(x + x = 0)$ and each element $a \in G_n$ possesses a unique representation as a sum $a = t_1 + \ldots + t_k$ of pairwise distinct $t_1, \ldots, t_k \in T_n$ up to the order of summands.

We call this representation *canonical.* Let 0_n be the zero element of G_n. Denote by $t \restriction m$ the initial m–element segment of $t \in T$. An immediate check shows that for each $m, n < \omega$, $m \geqslant n$ the mapping $h_n^m : G_m \to G_n$ defined as

$$h_n^m(t_1 + \cdots + t_k) = t_1 \restriction n + \cdots + t_k \restriction n$$

is a homomorphism. Clearly, $h_k^n h_n^m = h_k^m$, for all $k \leqslant n \leqslant m$.

Now we impose the structure of commutative semi-group on the set $S = \bigcup_{n<\omega} G_n$. If $a, b \in S$, $a \in G_m$, $b \in G_n$, $k = \min(m, n)$, we let

$$a + b = h_k^m(a) + h_k^n(b).$$

One can easily ascertain that this operation is commutative and associative. Note also that $0_m + 0_n = 0_{\min(m,n)}$ and $h_n^m(x) = x + 0_n(x)$ if $\ln(x) = m$, $m \geqslant n$.

Consider a polygon $P(S)$ over S defined as follows: $P(S) = \langle S, f_s\rangle_{s\in S}$ and

$$f_s(x) = s + x, \text{ for all } s, x \in S.$$

Now we define a natural ordering on S:

$$x \leqslant y \Leftrightarrow x = h_n^m(y), \quad \text{if } m = \ln(y),\ n = \ln(x),\ m \geqslant n.$$

To prove that the ordering $\leqslant$ on S is preserved under automorphisms of $P(S)$, take an arbitrary automorphism φ. Note that $x \leqslant y$ is equivalent to $x = 0_n + y$, for some $n < \omega$. We have

$$\begin{aligned} x \leqslant y &\Leftrightarrow \exists n < \omega\,(x = 0_n + y) \Leftrightarrow \exists n < \omega\,(x = f_{0_n}(y)) \\ &\Leftrightarrow \exists n < \omega\,(\varphi(x) = f_{0_n}(\varphi(y))) \Leftrightarrow \exists n < \omega\,(\varphi(x) = 0_n + \varphi(y)) \\ &\Leftrightarrow \varphi(x) \leqslant \varphi(y). \end{aligned}$$

Assume $\varphi \in \mathrm{Aut}\,P(S)$. We have

$$f_a(\varphi(x)) = \varphi(f_a(x)), \quad \text{for all } a, x \in S.$$

For $x = 0_n$, $n \geqslant \ln(a)$, we obtain

$$\varphi(a) = a + \varphi(0_n).$$

Since elements $(0_n)_{n<\omega}$ form an infinite increasing chain $0_0 < 0_1 < 0_2 < \cdots$, so do the elements $(\varphi(0_n))_{n<\omega}$, i.e., $\varphi(0_0) < \varphi(0_1) < \varphi(0_2) < \cdots$. Therefore, there is a one–to–one correspondence between all automorphisms φ of $P(S)$ and the chains $\varphi(0_0) < \varphi(0_1) < \varphi(0_2) < \cdots$, where $\ln(\varphi(0_n)) = n$, for all $n < \omega$.

Assume φ is a nontrivial automorphism of $P(S)$. Then for some $n < \omega$ $\varphi(0_n) \neq 0_n$, $\varphi_n(0_n) \in T_n$. There is $t \in T_n$ such that $\varphi(0_n) = t + \cdots$, where the right hand sum is canonical. This t occurs only once in this sum. Since $\varphi(0_{n+1}) > \varphi(0_n)$ and $\varphi(0_{n+1}) \in T_{n+1}$, $\varphi(0_{n+1})$ contains some $t' > t$ in its canonical representation. We can take the lexicographically minimal element t' with this property. By definition, t' extends t as an element of the tree. Repeating this process, we obtain an infinite branch in T.

On the other hand, if $(t_i)_{i<\omega}$, $t_i \in T_i$, $t_i < t_{i+1}$ is an infinite branch in T, the automorphism φ defined by $\varphi(0_n) = t_i$ is nontrivial, since $\varphi(0_n) \neq 0_n$.

Thus, T possesses an infinite branch iff $P(S)$ possesses a nontrivial automorphism.

Let T be a recursive tree. Consider an arbitrary constructivization ν of $P(S)$ for which, given an element $\nu(n)$ in $P(S)$, we can effectively write down its canonical representation. Since T is a recursive tree, this constructivization can be easily constructed.

Based on the considerations above, we may see that if an automorphism φ is nontrivial and H–recursive (that is, recursive in H for some oracle $H \subseteq \omega$), then T possesses an H–recursive infinite branch. On the other hand, if T possesses an H–recursive infinite branch $(t_i)_{i<\omega}$, $t_i \in T_i$, then this branch defines a nontrivial automorphism $\varphi(x) = x + t_k$, where $k = \ln(x)$. We have proved the following:

Lemma 2.6

$$\begin{aligned}\{\deg(f) \mid f \textit{ is a nontrivial automorphism of } P(S)\} \\ = \{\deg(t) \mid t \textit{ is an infinite branch in } T\}\,.\end{aligned}$$

Thus, we reduce the study of nontrivial automorphisms of models to the study of infinite branches of trees.

Let R_n be $\{\langle \ell x, rx\rangle \mid x \in W_n\}$, and let M_n be the model $\langle \omega, R_n\rangle$. Using the usual uniform interpretation of our polygons in binary relations, we obtain the following result:

Theorem 2.7 *The set* $\{n \mid \mathrm{Aut} M_n \neq 1\}$ *is a* Π^1_1*–complete set.*

Now we can transfer some results for trees to nontrivial automorphisms. For instance, consider the following result

Proposition 2.8 [59] *There exists a recursive tree having* 2^ω *infinite branches, but having no infinite hyperarithmetical branches.*

This result, together with some properties of $P(S)$, proves Theorem 2.3.

As a result of this correspondence between trees and automorphisms, we can also prove

Theorem 2.9 *A family of Turing degrees is the family of Turing degrees of all infinite branches of some recursive tree if and only if it is the family of all Turing degrees of automorphisms of an appropriate recursive model.*

Without loss of generality we may consider only the trees whose branches all contain some fixed initial segment $a = \{\langle 0, 0\rangle\}$. In this case the resulting polygon 0_1 is isomorphic to a if and only if the polygon possesses a nontrivial automorphism, since all nontrivial automorphisms are multiplications by an infinite branch, which always contains a, hence each nontrivial automorphism takes a to 0_1 and vice versa. Consider two models, which are obtained from the polygon by adding some new predicate which is true only on a in the first model, and only on 0_1 in the second one. These two models are isomorphic if and only if the original polygon possesses a nontrivial automorphism. This yields, as above

Theorem 2.10 *The set* $\{\langle m, n\rangle \mid M_n \cong M_n\}$ *is a* Π_1^1*-complete set.*

With slight modification of this construction, we could demonstrate that isomorphic embeddability is also a Π_1^1-complete property.

The following results deal with the notion of *computability* for recursive automorphism groups. Informally, one may say that a computable automorphism group is one which has a good system of coordinates.

Definition 2.11 A group $G \leqslant \mathrm{Aut}_r\omega$ is said to be *computable* if it possesses a computable enumeration ν as a set of recursive functions [10], i.e., if there exists a binary recursive function $f(i, x)$ such that

$$G = \{\lambda\, x\, f(i, x) \mid i < \omega\}.$$

We say that the number of a function $\lambda x f(i, x)$ is i, $\nu(i) = \lambda\, x\, f(i, x)$.

It turns out that the class of all computable recursive automorphism groups possesses a rather simple characterization. To formulate the result, we need one more definition. Two n-tuples $\overline{a}$ and $\overline{b}$ are called *recursively isomorphic* if $f(\overline{a}) = \overline{b}$ for some recursive automorphism f. We denote this by $\overline{a} \cong_r \overline{b}$.

Theorem 2.12 (Computability criterion, [38]) *Suppose M is a recursive model. Then the following statements are equivalent:*

(1) $\mathrm{Aut}_r M$ *is computable.*

(2) *There exists an n-tuple $\overline{p} \in M$ such that* $\mathrm{Aut}_r(M, \overline{p}) = 1$, *and the set* $\{\langle \overline{m}, \overline{n}\rangle \mid \overline{m}, \overline{n} \in |M|$ *and* $\overline{m} \cong_r \overline{n}\}$ *is recursively enumerable.*

(3) *There exists an n-tuple $\overline{p} \in M$ such that* $\mathrm{Aut}_r(M, \overline{p}) = 1$, *and the set* $\{\langle \overline{q}, x, y\rangle \mid \overline{q}, x, y \in |M|$ *and* $\overline{q}x \cong_r \overline{p}y\}$ *is recursively enumerable.*

The proof uses a diagonalization–like argument. This theorem implies the following corollaries.

Corollary 2.13 *Let* $\mathrm{Aut}_r M$ *be computable. Then* $\mathrm{Aut}_r M$ *possesses exactly one computable enumeration up to recursive equivalence; in other words, for any two computable enumerations* ν_0 *and* ν_1 *of* $\mathrm{Aut}_r M$ *there exists a recursive function* f *such that* $f\nu_0 = \nu_1$. *Moreover, any computable enumeration of* $\mathrm{Aut}_r M$ *is decidable.*

Corollary 2.14 *If a group* $\mathrm{Aut}_r M$ *is computable then it is constructivizable; moreover, each computable enumeration of* $\mathrm{Aut}_r M$ *is, in fact, a constructivization of this group. Conversely, for each constructive group* (G, ν), *there exists a recursive model* M *whose group of all recursive automorphisms is computable and, for each computable enumeration* μ *of this group,* $(G, \nu) \cong_r (\mathrm{Aut}_r M, \mu)$.

Corollary 2.15 *Any finitely generated group is isomorphic to a group of all recursive automorphisms of some recursive model if and only if this is a constructivizable group (i.e., the word problem for this group is decidable).*

Corollary 2.16 *The group* $\mathrm{Aut}_r B$ *is not computable for all infinite recursive Boolean algebras* B.

Corollary 2.17 *Let* M *be a decidable infinite model of an* ω*–categorical theory* T *with a recursive set of atomic formulas. Then the group* $\mathrm{Aut}_r M$ *is not computable.*

If we do not state that the set of atomic formulas is decidable, then the result may be the opposite.

Theorem 2.18 [40, 48] *There exists a countably–categorical decidable model without nontrivial recursive automorphisms.*

The problem of existence of a decidable countably–categorical theory T such that all its decidable models have no nontrivial recursive automorphisms is still unsolved.

It would be very desirable to have a general and sufficiently clear description of all recursive automorphism groups which would give us a tool for recognizing groups which are isomorphic to recursive automorphism groups.

It is a pity, but the results so far obtained make us regard this as hopeless. However, there are certain characterizations of this class of groups. The only case for which we currently have a satisfactory characterization is the finitely generated case (see Corollary 2.15).

One might suppose that it is possible to characterize the class of all recursive automorphism groups as the class of all H–constructivizable groups for a suitable oracle H. The results below demonstrate this to be impossible. In addition, we obtain a complete description of Turing degrees that can be realized as degrees of recursive automorphism groups, and of which groups may be contained in various Turing degrees. First, with the use of universal partial recursive functions, one can easily prove that each group of recursive automorphisms is isomorphic to a $0''$–recursive group. The rest of the cases are given by the following theorem.

Theorem 2.19 [47] *The following statements are true:*

(1) *For each Turing degree $d \neq 0$, there is a group whose degree is d, and which is not isomorphic to any group of all recursive automorphisms of a recursive model.*

(2) *For each Turing degree d, $d \leqslant 0''$, there exists a recursive model M such that* $\deg \mathrm{Aut}_r M = d$.

(3) *There exists a recursive model whose group of recursive automorphisms has no degree.*

One might hope for a simple description of finitely generated subgroups of the group $\mathrm{Aut}_r\omega$; this would be a description of finitely generated subgroups in groups of recursive automorphisms. It is not difficult to see that each group $G \leqslant \mathrm{Aut}_r\omega$ finitely generated by $g_0, \ldots, g_k$ satisfies the following property:

> "the set $\{t \mid t \text{ is a group term from } g_0, \ldots, g_k \text{ and } t \neq 1\}$ is recursively enumerable".

In other words, the equality problem for G is co–enumerable.

A natural question arises about the truth of the converse statement, that is, if any finitely generated group with a co–r.e. equality relation is isomorphic to a group generated by finitely many recursive permutations. Higman stated this question in a conversation with Ershov and Belegradek in 1989. The answer turned out to be negative:

Theorem 2.20 [44] *There exists a 4–generated group G whose equality problem is co–enumerable such that G is not embeddable into the group of all recursive permutations* $\mathrm{Aut}_r\omega$.

The question about the minimal possible number of generators for such groups remains open.

One of the main objects of our study is the group of all recursive permutations of the natural numbers. This group is a group of all recursive automorphisms of the infinite recursive model with the empty signature; on the other hand, this group contains all recursive automorphism groups.

Nurtazin was the first to prove non–constructivizability of this group [53]. Nowadays, due to Theorem 1.13, we have the exact complexity level of this group:

$$\deg \mathrm{Aut}_r\omega \;=\; 0''.$$

This group possesses two characterizations. The first one asserts categoricity and the finite axiomatizability of the group $\mathrm{Aut}_r\omega$ in the class of all recursive permutation groups:

Theorem 2.21 [45] *There exists a sentence φ of the group-theoretic language such that, for all groups $G \leqslant \mathrm{Aut}_r\omega$,*

$$G \models \varphi \;\Leftrightarrow\; G \cong \mathrm{Aut}_r\omega.$$

This group can be also characterized in the class of all groups:

Theorem 2.22 [45] *There exists a sentence φ of the group-theoretic language such that the group $\mathrm{Aut}_r\omega$ is the only model of φ, up to isomorphism, which contains exactly two nontrivial normal subgroups. All other models of φ contain at least three distinct nontrivial normal subgroups. In other words, $\mathrm{Aut}_r\omega$ is the only model of φ with the minimal possible number of normal subgroups.*

Using methods developed in the proofs of the above theorems, one can prove similar results for other groups, for example, for the group of all *almost periodic permutations.*

A permutation β is called an *almost periodic* provided that there exists $t > 0$ such that for all sufficiently large m, $\beta(m + t) = \beta(m) + t$ holds.

Corollary 2.23 [45] *There exists a sentence* φ_ω *of the group-theoretic language such that, for all* $G \leqslant \mathrm{Aut}_r\omega$,

$$G \models \varphi_\omega \;\Leftrightarrow\; G \text{ is isomorphic to the group of all almost periodic permutations of } \omega.$$

It would be interesting to investigate the question of categoricity and finite axiomatizability for some other recursive automorphism groups, for instance, for the group $\mathrm{Aut}_r\langle Q, \leqslant\rangle$ of all recursive order–preserving permutations of the rational numbers.

Having obtained the characterization of the group $\mathrm{Aut}_r\omega$ we may pass to the characterization of the class of all groups of recursive automorphisms:

Theorem 2.24 [45] *There exists a sentence* $\varphi(R)$ *of the group-theoretic language extended by a unary predicate* R, *such that the set of all subsets* $X \subseteq \mathrm{Aut}_r\omega$ *which can be realized as a group of all recursive automorphisms for a suitable recursive model, coincides with the set*

$$\{X \subseteq \mathrm{Aut}_r\omega \mid (\mathrm{Aut}_r\omega, X) \models \varphi(R)\}.$$

Here, X *is the interpretation of* R *in* $\mathrm{Aut}_r\omega$.

In other words one can say that the class of all recursive automorphism groups of recursive models is definable in the monadic second order language of Aut_r.

The following results are concerned with complexities of elementary theories of classes of groups of recursive permutations.

Let $\mathcal{K}_r$ be the class of all groups isomorphic to groups of all recursive automorphisms of suitable recursive models, let $\mathcal{K}_0$ be the class of all groups isomorphic to recursive groups, and let $\mathcal{K}$ be the class of all groups isomorphic to subgroups of $\mathrm{Aut}_r\omega$. The following results are about theories of these classes. In particular, the results below give exact levels of the undecidability degrees for the theories of $\mathcal{K}_r$, $\mathcal{K}_0$, and $\mathcal{K}$.

Theorem 2.25 [45] *The theories of the classes* $\mathcal{K}_0$, $\mathcal{K}_r$, *and* $\mathcal{K}$ *are pairwise distinct.*

Theorem 2.26 [45] *The theory of the class* $\mathcal{K}$ *is a* Π^1_1*-complete set.*

Theorem 2.27 [45] *The theories of the classes $\mathcal{K}_0$ and $\mathcal{K}_r$ are recursively isomorphic to the theory of the standard model of arithmetic.*

In what follows, we denote by $\mathbf{S}(G)$ the class of all groups which are isomorphic to subgroups of a group G. We also call two subsets $A, B \subseteq \omega$ *arithmetically inseparable* if there is no an arithmetical subset $R \subseteq \omega$ with $A \subseteq R$ and $B \subseteq \overline{R}$.

Theorem 2.28 [45] *There exists a sentence φ of the group–theoretic language such that*

(1) *φ is consistent with the theory of constructivizable groups.*

(2) *for each class $\mathcal{C} \subseteq \mathbf{S}(\mathrm{Aut}_r\omega)$, if $\mathrm{Th}(\mathcal{C}) \cup \{\varphi\}$ is consistent, then $\emptyset^{(\omega)}$ is 1–reducible to the set $\mathrm{Th}(\mathcal{C})$. Moreover, the set $\mathrm{Th}(\mathcal{C})$, and the set of sentences refutable in the class $\mathcal{C}$, are arithmetically inseparable.*

Corollary 2.29 *The theories of the following classes of groups:*

(1) *groups of recursive permutations;*

(2) *groups of recursive automorphisms of recursive models;*

(3) *recursive groups (constructivizable groups)*

are distinct, and differ from the theory of all groups. The first cannot be axiomatized by a hyperarithmetical set of axioms, the others cannot be axiomatized by an arithmetical set of axioms.

Corollary 2.30 *There exists a sentence which is consistent with the theory of groups, and is not true in all groups of recursive permutations, in particular, in all groups of recursive automorphisms and recursive (constructivizable) groups.*

This strengthens the results by Kreisel, Mostowski, and Vaught. Kreisel [21] and Mostowski [51, 52] constructed consistent sentences which are not true in all constructive and recursively enumerable models. Vaught [73, 74, 75] studied complexities of theories of some classes of models. Corollaries 2.29 and 2.30 show similar statements to be true even in the class of groups.

Note that the class $\mathcal{K}_0$ is not closed under subgroups, since the class is countable and the group $\mathrm{Aut}_r\omega$ has 2^ω many nonisomorphic subgroups. Also, this class is not closed under homomorphisms, since the free group of infinite rank is in $\mathcal{K}_0$ and the free group has all countable groups as its homomorphic images. In addition, we have the opportunity to see that even the class of finitely generated subgroups of $\mathrm{Aut}_r\omega$ is not as simple as it might seem. However, one can easily prove the class $\mathcal{K}_0$ to be closed under finite direct products.

The following question, formulated by Goncharov, seems to be rather interesting: under which algebraic constructions is the class of recursive automorphism groups closed? (In particular, it would be interesting to find the answer for free products.)

The problem of how to distinguish recursive and decidable models by their automorphism groups was studied in [48]. The first result shows it to be impossible if we consider these groups as abstract groups.

Theorem 2.31 *For each recursive model $\mathcal{M}$, there exists a decidable model $\mathcal{M}'$, and an isomorphism $\varphi : \mathrm{Aut}\mathcal{M} \longrightarrow \mathrm{Aut}\mathcal{M}'$, such that $\varphi(\mathrm{Aut}_r\mathcal{M}) = \mathrm{Aut}_r\mathcal{M}'$.*

However, if we take into account the action of the automorphism group, the situation changes.

Theorem 2.32 *There exists a recursive model $\mathcal{M}$ with the universe ω such that, for any decidable model $\mathcal{N}$ (of any language), $\mathrm{Aut}\mathcal{M}\cap\mathrm{Fin} \neq \mathrm{Aut}\mathcal{N}\cap\mathrm{Fin}$ holds. In particular, $\mathrm{Aut}\mathcal{M} \neq \mathrm{Aut}\mathcal{N}$ and $\mathrm{Aut}_r\mathcal{M} \neq \mathrm{Aut}_r\mathcal{N}$.*

The following theorem contains an exact level for the algorithmic complexity of orbits.

Theorem 2.33 [38] *For each recursive model M, the set*

$$\{\langle \overline{m}, \overline{n}\rangle \;\big|\; \overline{m}, \overline{n} \in |M| \;\&\; \overline{m} \cong_r \overline{n}\}$$

is in Σ^3_0. This set is Σ^3_0–complete for some model M.

The next area of investigation is the study of automorphisms which are recursive under all constructivizations. It seems to be natural to study computability of such automorphisms in connection with its definability in HF_M [2]. In fact, Ventsov [76] proved that there exists a recursive model M and a recursive automorphism thereof which is recursive with respect to all constructivizations of M, but not Σ–definable in HF_M.

3 Recursive automorphisms of Boolean algebras

The investigation of usual automorphisms of Boolean algebras has been rather successful (see, for example, [3, 14, 17, 19, 30, 31, 32, 33, 58, 60, 61, 62, 67, 72, 7, 71]). The reader also can find a lot of information about recursive Boolean algebras in [57].

A basic method in investigations of recursive automorphisms of atomic Boolean algebras is based on an interpretation of the action of these groups on the sets of atoms. This can be done as in Section 1 for the groups of permutations of natural numbers. As in Section1, we first can define *transpositions* as those automorphisms of a Boolean algebra which permute exactly two atoms and don't move atomless elements. The transpositions are definable by the same formula:

Theorem 3.1 (McKenzie, [30]) *An automorphism of a Boolean algebra B is a transposition if and only if it satisfies the formula*

$$\neg(x = 1) \ \& \ (x^2 = 1) \ \& \ \forall y((x^{-1} y^{-1} x \, y)^6 = 1)$$

in the group of all (recursive) automorphisms of B.

Hence, we again can consider the pairs $\langle x, y\rangle$ such that $xy \neq yx$, define in the same way the equivalence relation between the pairs, etc.. Thus, we can interpret the action of the (recursive) automorphism group on atoms, and so on.

A natural question which arises here is about constructivizability of the groups of recursive automorphisms of Boolean algebras. The following theorems answer this question for many cases.

Theorem 3.2 [42] *If the set of atoms of an infinite recursive Boolean algebra is not immune, i.e., contains a nontrivial r.e. set, then the group* $\mathrm{Aut}_r B$ *is not constructivizable.*

Theorem 3.3 [42] *Suppose that a Boolean algebra B possesses a finite family $\{\varphi_1, \ldots, \varphi_k\}$ of recursive automorphisms such that the set of atoms of B contains an infinite $(\varphi_1, \ldots, \varphi_k)$–orbit, i.e., the orbit of the group generated by $\varphi_1, \ldots, \varphi_k$. Then the group* $\mathrm{Aut}_r B$ *cannot be embedded into a constructivizable group.*

Using some automorphisms which we can construct under the assumption that the set of atoms is recursive, we can prove the following result:

Corollary 3.4 [42] *Let B be an infinite recursive Boolean algebra with a recursive set of atoms. Then the group* $\mathrm{Aut}_r B$ *cannot be embedded into a constructive group. In particular, this is so for the group of all recursive automorphisms of any infinite decidable (strongly constructive) Boolean algebra.*

In addition, using the fact that the group of all recursive permutations cannot be embedded into a recursive group [53], we can also construct an embedding of this group into the group of all recursive automorphisms of an atomless recursive Boolean algebra and prove the following result:

Theorem 3.5 *Suppose that a recursive Boolean algebra contains a nonzero atomless element. Then the recursive automorphism group of this Boolean algebra cannot be embedded into a recursive group.*

The next area of inquiry concerns the problem of how properties of Boolean algebras interact with properties of their recursive automorphism groups. One could also say that we are interested in the reconstruction of recursive Boolean algebras from their recursive automorphism groups. There are many ways to investigate this problem. Here we will discuss a few of them.

We begin with results which show a relationship between the isomorphism types of recursive Boolean Algebras and the isomorphism types of their recursive automorphism groups. When we do not impose recursiveness considerations on Boolean algebras and their automorphisms, the following result of McKenzie is exemplary.

Theorem 3.6 (Mckenzie, [30]) *Let B_0 be a countable Boolean algebra with a maximal atomic element, and let B_1 be an arbitrary countable Boolean algebra. Also, let the numbers of atoms in B_0 and in B_1 be different from* 1. *Then* $\mathrm{Aut} B_0 \cong \mathrm{Aut} B_1$ *implies* $B_0 \cong B_1$.

It is the analogous situation in the theory of recursive Boolean algebras we wish to consider. It turns out that for atomic decidable Boolean algebras a (classical) isomorphism of recursive automorphism groups implies that the original algebras are isomorphic, in fact, recursively isomorphic:

Theorem 3.7 [34] *Let B_0 be an atomic decidable Boolean algebra, and let B_1 be an arbitrary recursive Boolean algebra. Then* $\mathrm{Aut}_r B_0 \cong \mathrm{Aut}_r B_1$ *implies* $B_0 \cong_r B_1$.

In addition, the proof of this theorem gives us

Corollary 3.8 *Let B be an infinite recursive atomic Boolean algebra with a recursive set of atoms. Then the group* $\mathrm{Aut}_r B$ *is a complete group, i.e., it has trivial center, and each automorphism of the group is a conjugation.*

However, the following results serve as counterexamples to the preceding ones, which show that the restrictions in Theorem 3.7 above cannot be completely omitted.

Theorem 3.9 (Remmel, [56]) *For each recursive Boolean algebra there exists a Boolean algebra isomorphic to it such that each recursive automorphism of it moves only finite number of atoms.*

For atomic Boolean algebras, this result was independently obtained by the author in [37].

This theorem implies that for atomic Boolean algebras the group $\mathrm{Aut}_r B$ can be isomorphic to the group of all finite permutations of the natural numbers. Therefore without additional restrictions, the group of all recursive automorphisms of an atomic Boolean algebra may be constructivizable. One even can select a constructivization with respect to which this group is isomorphic to some fixed group, namely Fin.

We also can prove that each recursive Boolean algebra whose automorphism group is isomorphic to Fin is atomic and has an immune set of atoms. The converse is not true. That is, immunity of the set of atoms of an atomic Boolean algebra does not imply that its recursive automorphism group is isomorphic to Fin [36].

Corollary 3.10 *For each pair of infinite recursive atomic Boolean algebras B_0 and B_1, there exist recursive Boolean algebras B_0^* and B_1^* such that*

$$B_0 \cong B_0^*, \quad B_1 \cong B_1^*, \quad \text{and} \quad \mathrm{Aut}_r B_0^* \cong \mathrm{Aut}_r B_1^*$$

This (in some sense) negative effect is caused by the possibility of choosing "bad" enough constructivizations (recursive presentations). However, sometimes even the isomorphism types of recursive Boolean algebras cannot be recognized from the isomorphism types of their recursive automorphism groups. More exactly:

Theorem 3.11 [34] *There exist two nonisomorphic strongly constructive Boolean algebras $\mathcal{C}_0$ and $\mathcal{C}_1$ whose recursive automorphism groups are isomorphic.*

We should also note that the constructivizability of a Boolean algebra cannot be determined by its automorphism group [34]. That is, there exists a constructive Boolean algebra whose automorphism group is isomorphic to the automorphism group of a Boolean algebra without constructivizations.

Remmel proved one more result in this direction:

Theorem 3.12 (Remmel, [56]) *There exist atomic recursive Boolean algebras B_0 and B_1 such that $\mathrm{Aut}_r B_0 \cong \mathrm{Aut}_r B_1$, where B_1 is isomorphic to a decidable Boolean algebra, but B_2 is not isomorphic to any decidable Boolean algebra.*

Informally this result says that recursive automorphism groups of Boolean algebras cannot distinguish the isomorphism types of recursive Boolean algebras from the isomorphism types of decidable ones.

Now we will briefly be interested in the reconstruction of recursive Boolean algebras from the elementary types of their recursive automorphism groups. It turns out that the case of atomic and decidable Boolean algebras is especially interesting. Indeed, we can prove that the group of all recursive automorphisms of any such algebra is finitely axiomatizable, and categorical in the class of all recursive permutation groups:

Theorem 3.13 [42] *For each atomic decidable Boolean algebra B, there exists a sentence φ of the first order group-theoretic language such that, for each group of recursive permutations,*

$$G \models \varphi \Leftrightarrow G \cong \mathrm{Aut}_r B.$$

For comparison with this and the result below, we cite a well-known result of Rubin on usual automorphisms of Boolean algebras.

Theorem 3.14 (Rubin, [61]) *For every complete theory T of Boolean algebras, there is a sentence φ_T in the language of groups such that, for every countable or finite Boolean algebra whose number of atoms is not one,*

$$B \models T \quad \textit{if and only if} \quad \mathrm{Aut} B \models \varphi_T.$$

Thus, taking into account Theorem 3.7, we immediately obtain that for a complete reconstruction of an atomic decidable Boolean algebra from its recursive automorphism group a single sentence is enough. More exactly,

Corollary 3.15 [42] *Let* B_0 *be an atomic decidable Boolean algebra, and let* B_1 *be an arbitrary recursive Boolean algebra. Then* $\mathrm{Aut}_r B_0 \equiv \mathrm{Aut}_r B_1$ *implies* $B_0 \cong_r B_1$.

This result has a nonrecursive counterpart. Rubin in [61] proved that under some natural conditions, a Boolean algebra can be reconstructed from the elementary type of its automorphism group.

Using the proof of Theorem 3.13 we now can prove:

Corollary 3.16 *The theory of the class of all recursive automorphism groups of Boolean algebras, and the theory of each recursive automorphism group of an infinite decidable atomic Boolean algebra, are recursively isomorphic to* $\emptyset^{(\omega)}$.

Let us now turn to another formulation of the problem of the reconstruction of recursive Boolean algebras from their recursive automorphism groups. We will now show how to reconstruct a Boolean algebra from the action of its recursive automorphism group on the domain of the Boolean algebra.

Recall that two enumerations ν_0 and ν_1 are called *recursively equivalent* if there exists a partial recursive function f such that $\nu_0 = \nu f$ [10]. We denote this by $\nu_0 \approx_r \nu_1$.

The following theorem was proved in [36]:

Theorem 3.17 *If the sets of atoms, atomic elements, and atomless elements of a constructive Boolean algebra* (B, ν_0) *are decidable, and for some constructivization* ν_1 *of* B

$$\mathrm{Aut}_r(B, \nu_0) = \mathrm{Aut}_r(B, \nu_1) \quad \textit{holds},$$

then $\nu_0 \approx_r \nu_1$.

Corollary 3.18 *If* ν_0 *is a strong constructivization of a Boolean algebra* B, *and for some other constructivization* ν_1, $\mathrm{Aut}_r(B, \nu_0) = \mathrm{Aut}_r(B, \nu_1)$, *then* $\nu_0 \approx_r \nu_1$.

It would be interesting to know whether it is it possible to prove the result of Theorem 3.17 under weaker conditions.

Consider another setting for this problem. Let B_0 and B_1 be recursive Boolean algebras whose universe is ω and let the group $\mathrm{Aut}_r B_0 = \mathrm{Aut}_r B_1 \leqslant \mathrm{Sym}(\omega)$ be given. What can we say about B_0 and B_1? Dauletbaev investigated this question.

Let B be a Boolean algebra. A one-to-one mapping $\varphi : B \to B$ is called *two-valued* if the following two statements are true:

(1) $\forall x \in B\,(\varphi(x) \in \{x, \overline{x}\})$;

(2) $\forall xy \in B\,\big(\exists \psi \in \mathrm{Aut}_r B\, \psi(x) = y \ \&\ \varphi(x) = x\big) \Rightarrow \varphi(y) = y$.

Suppose that $B = \langle |B|\,;\cap, \cup, {}^{-}\rangle$ is a Boolean algebra, and φ is a two-valued mapping of B. One can easily see that, if we define operations $\cup^\varphi$, $\cap^\varphi$, and ${}^{-\varphi}$, and a Boolean algebra B^φ as follows:

$$x \cap^\varphi y \rightleftharpoons \varphi(\varphi^{-1}(x) \cap \varphi^{-1}(y)), \qquad \overline{x}^{\varphi} \rightleftharpoons \varphi(\overline{\varphi^{-1}(x)}),$$

$$x \cup^\varphi y \rightleftharpoons \varphi(\varphi^{-1}(x) \cup \varphi^{-1}(y)), \qquad B^\varphi \rightleftharpoons \langle |B|\,; \cap^\varphi, \cup^\varphi, {}^{-\varphi}\rangle$$

then $\mathrm{Aut} B = \mathrm{Aut} B^\varphi$. Dauletbaev proved that this is the only case when $\mathrm{Aut}_r B_0 = \mathrm{Aut}_r B_1$ holds.

Theorem 3.19 (Dauletbaev, [6])

(1) *Let B_0 be an infinite Boolean algebra, and B_1 be an arbitrary Boolean algebra with $|B_0| = |B_1|$ and $\mathrm{Aut} B_0 = \mathrm{Aut} B_1$. Then there is a two-valued isomorphism $\varphi : B_0 \to B_1$ such that $B_0 = B_1^\varphi$.*

(2) *Let B_0 be an infinite Boolean recursive algebra, and B_1 be an arbitrary recursive Boolean algebra with $|B_0| = |B_1|$ and $\mathrm{Aut}_r B_0 = \mathrm{Aut}_r B_1$. Then there is a recursive two-valued isomorphism $\varphi : B_0 \to B_1$ such that $B_0 = B_1^\varphi$.*

Using the method of Anderson [1], Dulatova proved the simplicity of the group of all recursive automorphisms of a recursive atomless Boolean algebra [8].

The structure of normal subgroups of the (recursive) automorphism groups of superatomic Boolean algebras is unknown. It would be interesting to know whether all normal subgroups of a such group form a chain.

We also would like to mention a result of Guichard about the automorphisms of the lattice $\mathcal{L}(B)$ of recursively enumerable subalgebras of a Boolean algebra B.

Theorem 3.20 (Guichard, [16]) *Let B be a free Boolean algebra on a countable set of generators. Then each automorphism of the lattice $\mathcal{L}(B)$ is induced by some recursive automorphism of the algebra B.*

4 Recursive automorphisms of modules and vector spaces

Let P be a recursive ring. A module M over P is *recursive* if M is a recursive subset of ω and, given $m, n \in M$ and $\alpha \in P$, one can effectively find $\alpha\, n$, and the sum of elements m and n in M.

Recursive endomorphisms of modules are not well understood yet. We mention some papers by Puninskaya [25, 26, 24, 55].

In all reasonable cases one can find a recursive presentation of a free module of infinite rank whose recursive automorphisms are only those which cannot be avoided, i.e., the multiplications by scalars:

Theorem 4.1 [43] *Let K be a recursive division ring. There exists a recursive module M over K such that*

(1) $M \cong K^\omega = K \oplus K \oplus K \oplus \cdots$

(2) *each recursive automorphism φ of the module M is a multiplication by a scalar, i.e., $\varphi(x) = \alpha\, x$, for some fixed invertible element α of K.*

Corollary 4.2 *Let P be an infinite recursive field. There exists an infinite dimensional recursive vector space V over P such that each recursive automorphism of V is exactly multiplication by a nonzero scalar.*

Of course, the vector space in this corollary has infinite dimension. Moreover, this result is not valid for a finite P, since the dependency problem over a finite field is always decidable. Indeed, the linear dependency condition of vectors $v_1, \ldots, v_n$ can be written as

$$\exists\, \alpha_1 \cdots \alpha_n \in P \,\Big(\bigvee_{i=0}^{n} \alpha_i \neq 0 \ \ \& \ \sum_{i=1}^{n} \alpha_i v_i = 0 \Big),$$

where the quantifier $\exists$ is taken over the finite field.

Corollary 4.3 *There exists a recursive Abelian group isomorphic to*

$$\mathbb{Z}^{\omega} = \mathbb{Z} \oplus \mathbb{Z} \oplus \mathbb{Z} \oplus \cdots$$

whose only nontrivial automorphism is $x \mapsto -x$.

Corollary 4.4 *For each recursively enumerable sub-ring K with unity in the field of rational numbers $\mathbb{Q}$, there exists a recursive abelian group whose group of recursive automorphisms is isomorphic to K^* (the group of all invertible elements of K).*

Karp obtained some results in this direction. He studied properties of constructivizations of an infinite dimensional vector space which can be reconstructed from their automorphism groups.

Theorem 4.5 (Karp, [18])

(1) *Suppose ν_1 and ν_2 are constructivizations of a space V, and linear dependency over V is decidable with respect to ν_1. If*

$$\mathrm{Aut}_r(V, \nu_1) = \mathrm{Aut}_r(V, \nu_2),$$

then linear dependency over V with respect to ν_2 is also decidable.

(2) *Let (V, ν) be a recursive vector space with a decidable linear dependency problem. Then there exists a countable family of elements in $\mathrm{Aut}_r(V, \nu)$ which are conjugate in $\mathrm{Aut}(V, \nu)$ but not in $\mathrm{Aut}_r(V, \nu)$.*

The problem of reconstruction of properties of vector spaces from their recursive automorphism groups has not been deeply studied yet. It would be interesting to answer the following questions:

(1) Let V_0 and V_1 be recursive vector spaces over an infinite recursive field and

$$\mathrm{Aut}_r V_0 = \mathrm{Aut}_r V_1.$$

Is it true that $V_0 \cong_r V_1$?

(2) Let V_0 and V_1 be recursive vector spaces over an infinite recursive field, let the linear dependency problem for V_0 be decidable, and let

$$\mathrm{Aut}_r V_0 \cong \mathrm{Aut}_r V_1.$$

Is it true that $V_0 \cong_r V_1$?

5 Other settings

Recursive automorphisms of linear orders have also been studied. Here we mention only two results:

Theorem 5.1 (Schwartz, [65]) *Let* $L = (\omega\,,\, <_L)$ *be a recursive linear ordering. Then the following conditions are equivalent:*

(1) *L contains a dense interval, and*

(2) *every recursive linear ordering* $R = (\omega\,,\, <_R)$ *which is (classically) isomorphic to L has a non-identity recursive automorphism.*

If a recursive ordering contains an interval isomorphic to the rational numbers, then its recursive automorphism group is rather complicated, more exactly:

Theorem 5.2 *Let* (L, ν) *be a recursive order which is isomorphic to the natural ordering of the rational numbers. Then the group* $\mathrm{Aut}_r(L, \nu)$ *is not positively presentable, that is, it can not be isomorphic to a group whose positive atomic diagram is r.e.. Moreover,* $\mathrm{Aut}_r(L, \nu)$ *cannot be embedded into a recursive group.*

We can prove this result in the framework of methods developed in the book of Glass [12], or with a technique from [35].

Selivanov studied automorphisms of enumerated sets. Let (S, ν) be an enumerated set [10], i.e., the set S together with an enumeration $\nu : \omega \xrightarrow{\text{onto}} S$. The map $\varphi : S \to S$ is called a *morphism* from (S, ν) to (S, ν) if there exists a recursive function f such that $\varphi\nu = \nu f$. A morphism φ from (S, ν) to (S, ν) is called an *automorphism* of (S, ν) if there exists a morphism ψ from (S, ν) onto (S, ν) such that $\varphi \circ \psi = \psi \circ \varphi = \mathrm{id}_S$. This is just the same as the usual notion of automorphism in a category. Selivanov obtained an unexpected result:

Theorem 5.3 (Selivanov, [66]) *Each at most countable group is isomorphic to the group of all automorphisms of a suitable enumerated set.*

Selivanov in [66] also asks what are the classes of automorphism groups for known classes: positive, negative enumerations, sub-objects of the Post enumeration, etc.. Interested readers can find these notions in [10].

In this survey we should also mention the paper by Soloviev [70]. This paper considers the notion of computational theory and studies their automorphisms.

I wish to thank Bakhadyr Khoussainov who read the manuscript and made suggestions on improvements of this paper.

References

[1] R. D. Anderson, The algebraic simplicity of certain groups of homeomorphisms, Amer. J. Math., **80** (1958) 955–963.

[2] J. Barwise, Admissible Sets and Structures, Perspect. in Math. Logic, (1975).

[3] M. Bekkali and R. Bonnet. Rigid Boolean algebras, Chapter 17, in: Handbook of Boolean Algebras, Vol. 2, J. D. Monk, R. Bonnet and S. Koppelberg, (eds.), (North-Holland, Amsterdam, NY, 1989), 637–678.

[4] F. B. Cannonito and M. Finkelstein, On primitive recursive permutations and their universes, J. Symbolic Logic, **34** (1969) 634–638.

[5] C. C. Chang and H. J. Keisler, Model Theory, 3rd. edn., Stud. Logic Found. Math., **73** (1990); [1st. edn. 1973, 2nd. edn. 1977].

[6] B. K. Dauletbaev, Reconstruction of an atomic Boolean algebra from the action of the automorphism group (Russian), Sibirsk. Math. Zh., **34** (1993) 49–51; [translated in: Siberian Math. J., **34** (1993) 1041–1043].

[7] E. K. van Douwen, J. D. Monk, and M. Rubin, Some questions about Boolean algebras, Algebra Universalis, **11** (1980) 220–243.

[8] Z. A. Dulatova, On recursive automorphisms of atomless Boolean algebras (Russian), in: Proc. 9th. All-Union Conf. Math. Logic, Leningrad, (1988) 55.

[9] V. D. Dzgoev, Recursive automorphisms of constructive models (Russian), in: Proc. 15th. All-Union Algebraic Conf., (Novosibirsk, 1979), Part 2, (1979) 52.

[10] Yu. L. Ershov, Theory of Numerations (Russian), Monographs in Math. Logic and Foundations of Math., (Nauka, Moskva, 1977).

[11] Yu. L. Ershov, Decision Problems and Constructivizable Models (Russian), (Mathematical Logic and Foundations of Mathematics, Nauka, Moskva, 1980).

[12] A. M. W. Glass, Ordered Permutation Groups, London Math. Soc. Lecture Note Ser., **55** (1981).

[13] S. S. Goncharov, On the problem of the number of nonautoequivalent constructivizations (Russian), Algebra i Logika, **19** (1980) 621–639, 745; [translated in: Algebra and Logic, **19** (1980) 401–414].

[14] S. S. Goncharov, Countable Boolean Algebras (Russian), (Nauka. Sibirsk. Otdel., Novosibirsk, 1988).

[15] S. S. Goncharov and V. D. Dzgoev, Autostability of models (Russian), Algebra i Logika, **19** (1980) 45–58, 132; [translated in: Algebra and Logic, **19** (1980) 28–37].

[16] D. Guichard, Automorphisms of substructure lattices in recursive algebra, Ann. Pure Appl. Logic, **25** (1983) 47–58.

[17] B. Jónsson, A Boolean algebra without proper automorphisms, Proc. Amer. Math. Soc., **2** (1951) 766–770.

[18] A. L. Karp, Recursive vector spaces (Russian), in: Proc. 11th. Interrepublic Conf. Math. Logic, (Kazan', 1993), (1993) 68.

[19] M. Katětov, Remarks on Boolean algebras, Colloq. Math., **2** (1951) 229–235.

[20] C. F. Kent, Constructive analogues of the group of permutations of the natural numbers, Trans. Amer. Math. Soc., **104** (1962) 347–362.

[21] G. Kreisel, Note on arithmetic models for consistent formulæ of the predicate calculus, Fund. Math., **37** (1950) 265–285.

[22] K. Zh. Kudaibergenov, Effectively homogenous models (Russian), Sibirsk. Math. Zh., 27 (1986) 180–182, 200.

[23] D. W. Kueker, Definability, automorphisms and infinitary languages, in: The Syntax and Semantics of Infinitary Languages, J. Barwise, (ed.), Lecture Notes in Math., **72** (1968) 152–165.

[24] V. A. Kuzicheva, Recursive endomorphisms of countable vector spaces with recursive operations (Russian), Dep. VINITI AN SSSR, (dep. No. 4175–84, 24 pages, June 1984).

[25] V. A. Kuzicheva, Minimal modules over the rings of recursive endomorphisms (Russian), in: Proc. 18th. Algebra Conf., (Kishinev, 1985), Part 1, (1985) 295.

[26] V. A. Kuzicheva, Inverse isomorphisms of rings of recursive endomorphisms (Russian), Vestnik Moskov. Univ. Ser. I Mat. Mekh., **41**:3 (1986) 91–93, 121; [translated in: Moscow Univ. Math. Bull., **41**:3 (1986) 82–84].

[27] A. V. Kuznetsov, On primitive recursive functions of large oscillation (Russian), Dokl. Akad. Nauk SSSR, **71** (1950) 233–236.

[28] A. B. Manaster and J. B. Remmel, Some recursion theoretic aspects of dense two-dimensional partial orderings, in: Aspects of Effective Algebra, (Proc. Conf. Monash Univ., Clayton, Australia, Aug. 1–4, 1979), J. N. Crossley (ed.), (Upside Down A Book Co., Yarra Glen, Victoria, Australia, 1981) 161–188.

[29] R. McKenzie, On elementary types of symmetric groups, Algebra Universalis, **1** (1971) 13–20.

[30] R. McKenzie, Automorphism groups of denumerable Boolean algebras, Canad. J. Math., **29** (1977) 466–471.

[31] J. D. Monk, On the automorphism groups of denumerable Boolean algebras, Math. Ann., **216** (1975) 5–10.

[32] J. D. Monk, Automorphism groups, in: Handbook of Boolean Algebras, Vol. 2, J. D. Monk, R. Bonnet and S. Koppelberg, (eds.), (North-Holland, Amsterdam, NY, 1989), Ch. 14, 517–546.

[33] J. D. Monk and R. McKenzie, On automorphism groups of Boolean algebras, in: Infinite and Finite Sets, Vol II, (to P. Erdös on his 60th. birthday, Keszthely, 1973), A. Hajnal, R. Rado and V. T. Sós, (eds.), Coll. Math. Soc. János Bolyai, **10** (1975), Vol. 2, 951–988.

[34] A. S. Morozov, Groups of recursive automorphisms of constructive Boolean algebras (Russian), Algebra i Logika, **22** (1983) 138–158; [translated in: Algebra and Logic, **22** (1983) 95–112].

[35] A. S. Morozov, Group $\mathrm{Aut}_r \langle Q, \leqslant \rangle$ is not constructivizable (Russian), Mat. Zametki, **36** (1984) 473–478; [translated in: Math. Notes, **36** (1984) 733–736].

[36] A. S. Morozov, Automorphisms of constructivizations of Boolean algebras (Russian), Sibirsk. Math. Zh., **26** (1985) 98–110; [translated in: Siberian Math. J., **26** (1985) 555–565].

[37] A. S. Morozov, Constructive Boolean algebras with almost-identical automorphisms (Russian), Mat. Zametki, **37** (1985) 478–482; [translated in: Math. Notes, **37** (1985) 266–268].

[38] A. S. Morozov, Computable groups of automorphisms of models (Russian), Algebra i Logika, **25** (1986) 415–424; [translated in: Algebra and Logic, **25** (1986) 261–266].

[39] A. S. Morozov, Permutations and implicit definability (Russian), Algebra i Logika, **27** (1988) 19–36; [translated in: Algebra and Logic, **27** (1988) 12–24].

[40] A. S. Morozov, A countably categorical decidable model without nontrivial recursive automorphisms (Russian), **30** (1989) 221–224; [translated in: Siberian Math. J., 30 (1989) 346–348].

[41] A. S. Morozov, Elementary properties of groups of recursive permutations (Russian), Dokl. Akad. Nauk, **305** (1989) 274–276; [translated in: Soviet Math. - Dokl., **39** (1989) 282–284].

[42] A. S. Morozov, Recursive automorphisms of atomic Boolean algebras (Russian), Algebra i Logika, **29** (1990) 464–490; [translated in: Algebra and Logic, **29** (1990) 310–330].

[43] A. S. Morozov, Rigid constructive modules (Russian), Algebra i Logika, **28** (1989) 570–583; [translated in: Algebra and Logic, **28** (1989) 379–387].

[44] A. S. Morozov, A question of Higman (Russian), Algebra i Logika, **29** (1990) 29–34; [translated in: Algebra and Logic, **29** (1990) 22–26].

[45] A. S. Morozov, On theories of classes of groups of recursive permutations (Russian), in: Proc. Inst. Math. Acad. Sibirsk. SSSR, Trudy Inst. Mat., Novosibirsk, **12** (1989) 91–104, 189; [translated in: Siberian Adv. Math., **1** (1991) 138–153].

[46] A. S. Morozov, Functional trees and automorphisms of models (Russian), Algebra i Logika, 32 (1993) 54–72; [translated in: Algebra and Logic, 32 (1993) 28–38].

[47] A. S. Morozov, On degrees of the recursive automorphism groups (Russian), in: Algebra, Logic, and Applications, in memoriam of A. I. Kokorin, (Irkutsk Univ., Irkutsk, 1994), 79–85.

[48] A. S. Morozov, Automorphism groups of decidable models (Russian), Algebra i Logika, **34** (1995) 437–447; [translated in: Algebra and Logic, **34** (1995) 242–248].

[49] A. S. Morozov, Turing reducibility as algebraic embeddability (Russian), Sibirsk. Math. Zh., **38** (1997) 362–364; [translated in: Siberian Math. J., **38** (1997) 312–313.

[50] Y. N. Moschovakis, Descriptive Set Theory, Stud. Logic Found. Math., **100** (1980).

[51] A. Mostowski, On a system of axioms which has no recursively enumerable arithmetic model, Fund. Math., **40** (1953) 56–61.

[52] A. Mostowski, A formula with no recursively enumerable model, Fund. Math., **42** (1955) 125–140.

[53] A. T. Nurtazin, On constructive groups (Russian), in: Proc. 4th. All-Union Conf. Math. Logic, (Kishinev, 1976), (1976) 106.

[54] P. Odifreddi, Classical Recursion Theory. The Theory of Functions and Sets of Natural Numbers, Stud. Logic Found. Math., **125** (1989).

[55] V. A. Puninskaya, On certain ring close to regular, in: Proc. 9th. All-Union Conf. Math. Logic, (Leningrad, 1988), (1988) 134.

[56] J. B. Remmel, Recursively rigid Boolean algebras, Ann. Pure Appl. Logic, **36** (1987) 39–52.

[57] J. B. Remmel, Recursive Boolean algebras, in: Handbook of Boolean Algebras, Vol. 3, J. D. Monk, R. Bonnet and S. Koppelberg, (eds.), (North-Holland, Amsterdam, NY, 1989), Ch. 25, 1097–1165.

[58] L. Rieger, Some remarks on automorphisms of Boolean algebras, Fund. Math., **38** (1951) 209–216.

[59] H. Rogers Jr., Theory of Recursive Functions and Effective Computability, (1st. edn., McGraw-Hill, New York-Toronto, Ont.-London, 1967; 2nd. edn., MIT Press, Cambridge, Mass., London, 1987).

[60] M. Rubin, On the automorphism groups of homogeneous and saturated Boolean algebras, Algebra Universalis, **9** (1979) 54–86.

[61] M. Rubin, On the automorphism groups of countable Boolean algebras, Israel J. Math., **35** (1980) 151–170.

[62] M. Rubin, On the reconstruction of Boolean algebras from their automorphism groups, in: Handbook of Boolean Algebras, Vol. 2, J. D. Monk, R. Bonnet and S. Koppelberg, (eds.), (North-Holland, Amsterdam, NY, 1989), Ch. 15, 547–606.

[63] J. Schreier and S. Ulam, Über die Permutationsgruppe der natürlichen Zahlenfolge, Studia Math., **4** (1933) 134–141.

[64] J. Schreier and S. Ulam, Über die Automorphismen der Permutationsgruppe der natürlichen Zahlenfolge, Fund. Math., **28** (1937) 258–260.

[65] S. Schwarz, Recursive automorphisms of recursive linear orderings, Ann. Pure Appl. Logic, **26** (1984) 69–73.

[66] V. L. Selivanov, Automorphism groups of numbered sets (Russian), Mat. Zametki, **41** (1987) 592–597, 622; [translated in: Math. Notes, **41** (1987) 330–333].

[67] S. Shelah, Boolean algebras with few endomorphisms, Proc. Amer. Math. Soc., **74** (1979) 135–142.

[68] S. G. Simpson, Degrees of unsolvability: a survey of results, in: J. Barwise, (ed.), Handbook of Mathematical Logic, Stud. Logic Found. Math., **90** (1977) 631–652.

[69] R. I. Soare, Recursively Enumerable Sets and Degrees. A Study of Computable Functions and Computably Generated Sets, Perspect. in Math. Logic, (1987).

[70] V. D. Soloviev, Automorphisms of the structure of computational theories (Russian), in: Proc. 9th. All-Union Conf. Math. Logic, (Leningrad, 1988), (1988) 153.

[71] P. Štěpánek, Embeddings and automorphisms, in: Handbook of Boolean Algebras, Vol. 2, J. D. Monk, R. Bonnet and S. Koppelberg, (eds.), (North-Holland, Amsterdam, NY, 1989), Ch. 16, 607–635.

[72] S. Todorčević, Rigid Boolean algebras, Publ. Inst. Math. (Beograd) (N.S.), **25(39)** (1979) 219–224.

[73] R. L. Vaught, Non-recursive-enumerability of the set of sentences true in all constructive models, Bull. Amer. Math. Soc. (N.S.), **63** (1957) 230.

[74] R. L. Vaught, Sentences true in all constructive models, Summaries Summer Inst. Symbolic Logic, Cornell Univ., (1957) 341–343 (1960).

[75] R. L. Vaught, Sentences true in all constructive models, J. Symbolic Logic, **25** (1960) 39–58.

[76] Yu. G. Ventsov, Nonuniform autostability of models (Russian), Algebra i Logika, **26** (1987) 684–714; [translated in: Algebra and Logic, **26** (1987) 422–440].

A. S. Morozov
Sobolev Institute of Mathematics,
pr. Akademika Koptuga, 4,
Novosibirsk, 630090,
Russia

Chapter 9

Constructive Models of Finitely Axiomatizable Theories

M. G. Peretyat'kin*

Contents

Introduction

Finitely axiomatizable theories form a subclass of the class of recursively axiomatizable theories. The latter can be defined as the class of recursively enumerable theories, and also as the class of recursively enumerably axiomatizable theories. As usual, we use the term *axiomatizable theory* instead of *recursively axiomatizable theory*. In fact, such theories are defined by recursively enumerable systems of axioms. The key idea is to study the expressiveness of finitely axiomatizable theories by comparing them with the

*The research described in this publication was made possible in part by Grant RUQ000 from the International Science Foundation.

HANDBOOK OF RECURSIVE MATHEMATICS
Edited by Yu. L. Ershov, S. S. Goncharov,
A. Nerode, and J. B. Remmel

well known class of recursively axiomatizable theories. Recently, significant progress has been made in this direction, establishing a theorem on the similarity of expressiveness of these two classes of theories for a large list of model-theoretic properties [29, 32, 33]. (This allows a wide variety of applications of finitely axiomatizable theories to the study of various algorithmic aspects of logic.)

In this work, we give a review of the applications of finitely axiomatizable theories to constructive models and related objects. We consider the algorithmic complexity of models, the complexity of Lindenbaum algebras and the complexity of semantic classes of sentences connected with constructive models and decidability. Our focus lies on a detailed account of the basic concepts, as well as an exposition of all available constructions of finitely axiomatizable theories. This enables us to exhibit various aspects of applications of finitely axiomatizable theories to constructive models.

Preliminaries

We consider theories in first order predicate logic with equality. In the basic concepts of model theory, algorithm theory, Boolean algebras and constructive models, we follow the monographs of Chang and Keisler [2], Rogers [36], Goncharov [11] and Ershov [7]. A detailed account of results on finitely axiomatizable theories is contained in [33].

The superscripts of a signature specify the arities of the appropriate symbols. A finite signature is called *rich* if it contains an n–ary predicate or function symbol for $n > 1$, or two unary function symbols. We consider only signatures that allow a Gödel numbering of their formulae. Such signatures are called *enumerable*. The set of all sentences of a signature σ is denoted by $\mathrm{SL}(\sigma)$. A set $E \subseteq \mathrm{SL}(\sigma)$ is *closed under deducibility* if for any $\Phi, \Psi \in \mathrm{SL}(\sigma)$, $\vdash (\Phi \leftrightarrow \Psi)$ implies $\Phi \in E \leftrightarrow \Psi \in E$. For $A, B \subseteq \mathbb{N}$, $A \approx B$ denotes a recursive isomorphism between the sets A and B, i.e., there exists a bijection $\pi : \mathbb{N} \to \mathbb{N}$ that is a generally recursive function and satisfies $\pi(A) = B$. For $A \subseteq \mathbb{N}$ we introduce the following notation:

$$\Sigma_n^0 \leqslant_m A \;\Leftrightarrow\; (\forall X \in \Sigma_n^0)\,(X \leqslant_m A),$$

$$A \approx \Sigma_n^0 \;\Leftrightarrow\; A \in \Sigma_n^0 \,\wedge\, (\forall X \in \Sigma_n^0)\,(X \leqslant_m A).$$

Similar notations exist for other hierarchy classes.

The *algorithmic dimension* of a model $\mathfrak{M}$ is the number $d_{sc}(\mathfrak{M})$ of strong constructivizations of this model up to autoequivalence. The dimension $d_c(\mathfrak{M})$ of the model under (ordinary) constructivizations is similarly defined. It is known that $d_{sc}(\mathfrak{M})$ can be only one of the three values $0, 1$ and ω, see [25].

Based on a standard Post numbering of recursively enumerable sets W_n, $n \in \mathbb{N}$, we construct an effective numbering for the class of all axiomatizable theories as follows. Let σ be an enumerable signature. We fix a Gödel numbering Φ_i, $i \in \mathbb{N}$, for all sentences in the signature σ. If a theory T in the signature σ is defined by the set of axioms $\{\Phi_i \mid i \in W_m\}$, then m is called the *recursively enumerable index* of the theory T. There is an alternative definition: Using the same set of axioms $\{\Phi_i \mid i \in W_m\}$ we construct a theory T' in the signature $\sigma' \subseteq \sigma$, where σ' contains only those symbols from σ that occur in formulae of the sequence Φ_i, $i \in W_m$. The number m is called a *weak recursively enumerable index* of the theory T'.

1 Similarity of theories

Whenever we describe an axiomatizable theory T having a certain set of properties, we naturally ask whether there is a finitely axiomatizable theory with the same set of properties. In these situations we try to use a notion of "similarity" or "likeness" of the two theories in terms of the properties in question. It appears that for axiomatizable theories this notion can be formalized precisely.

Our definition of similarity for two theories is based on a division of the properties of a theory T into two classes. The first class contains properties related to the structure of the Lindenbaum algebra $\mathcal{L}(T)$ (without free variables), including some properties of a more algorithmic nature. This class includes, for example, such properties as consistency, completeness, the existence of decidable extensions, etc.. All these properties are uniquely determined by the recursive isomorphism type of the Lindenbaum algebra. The second class contains model-theoretic properties related to any complete extension T' of the theory T (including some properties of an algorithmic nature), such as stability, existence of a prime model, its strong constructivizability, etc..

We will specify the concept of a model-theoretic property.

Two theories T_0 and T_1 in the signatures σ_0 and σ_1 are called *isomorphic* (written as $T_0 \approx T_1$) if T_1 can be obtained from T_0 by a finite number of the following operations: renaming some signature symbols and adding or eliminating those symbols that are first order definable under other signature symbols.

If two theories are isomorphic, the signature symbols of one can be represented by formulae in the other. Thus, the two theories share any properties that are determined by the structure of first order definable relations. For model-theoretic applications, two isomorphic theories are regarded as two copies of the same theory.

A *model-theoretic* or *semantic* property is a class p of complete theories in enumerable signatures such that

$$T_0 \approx T_1 \Rightarrow (T_0 \in p \Leftrightarrow T_1 \in p)$$

for all complete theories T_0 and T_1 in enumerable signatures. Let ML denote the set of all possible model-theoretic properties. Any subset $L \subseteq ML$ is called *a list of model-theoretic properties*, or simply *a list*.

Now let T_1 and T_2 be axiomatizable theories, and L a list of model-theoretic properties. The theories T_1 and T_2 are *semantically similar* with respect to the list L (written as $T_1 \equiv_L T_2$) if there exists a recursive isomorphism $\mu : \mathcal{L}(T_1) \to \mathcal{L}(T_2)$ between their Lindenbaum algebras, such that for any completion T_1' of T_1 and the corresponding completion $T_2' = \mu(T_1')$ of T_2, T_1' and T_2' have the same properties from the list L. In this case, we say that the isomorphism μ *preserves the properties of* L.

2 Predicate Expressiveness Theorem

The expressiveness of finitely axiomatizable theories is described in the following general theorem [33].

Theorem 2.1 (Main Theorem). *Let T be a recursively axiomatizable theory without finite models, and σ a finite rich signature. Then effectively in a recursively enumerable index of T one can construct a finitely axiomatizable model complete theory $F = \mathbb{F}_\sigma(T)$ in the signature σ, and a recursive isomorphism $\mu : \mathcal{L}(T) \to \mathcal{L}(F)$ between their Lindenbaum algebras, such that any completion T' of the theory T and the corresponding completion $F' = \mu(T')$ of the theory F have the same model-theoretic properties from the following list.*

(a) *Stability, superstability, ω–stability, stability in cardinality α.*

(b) *The existence of a prime model and its algorithmic dimension (with respect to strong constructivizations); the number of atomic models of cardinality $\alpha > \omega$.*

(c) *The number of countable minimal models (Jònsson models) and their algorithmic dimensions; the number of minimal models of cardinality $\alpha > \omega$.*

(d) *The existence of countable strongly constructivizable homogeneous models; the existence of a countable strongly constructivizable ω^+–homogeneous model; the existence of an α^+–homogeneous model of cardinality $\alpha \geqslant \omega$.*

(e) *The existence and strong constructivizability of a countable saturated model; the existence of a saturated model of cardinality $\alpha > \omega$.*

(f) *The existence of a model with only first order definable elements and its strong constructivizability; the existence of a model whose elements are almost all first order definable (algebraic), and its algorithmic dimension.*

(g) *The number of countable rigid models (those with only the trivial automorphism) and their algorithmic dimensions; the number of rigid models of cardinality $\alpha > \omega$.*

(h) *The non-maximality of the spectrum function.*

We use the notation $\mathbb{F}(T)$ whenever the signature is either not important or clear from the context. The list of model-theoretic properties in Theorem 2.1 is called *universal* and denoted by *MQL*.

Note 2.1 Theorem 2.1 is also applicable if T has finite models. Starting from T, we can construct a finitely axiomatizable theory $F = \mathbb{F}_\sigma(T)$ and a recursive isomorphism μ. The *MQL*–properties of every complete extension $T' \supseteq T$ with infinite models (as in the statement of the theorem) are then transferred to the corresponding completion $F' \supseteq F$. If T' is a theory of a finite model then the corresponding theory F' has only infinite models and has exactly the same properties from the list *MQL* as the theory $T' \oplus SI$, where SI is the axiomatizable theory in the signature $\sigma = \{\lhd^2, c\}$ whose axioms state that $\lhd$ is the successor relation, c has no $\lhd$–predecessor, all other elements have $\lhd$–predecessors, and there are no cycles (this theory is ω_1–categorical and complete).

Note 2.2 Theorem 2.1 also holds for weak recursively enumerable indexes of the theory T. This allows a complete effectivization of the construction of the main theorem. By taking a very large enumerable signature (with infinitely many predicates of each arity) and using weak recursively enumerable indexes, we can determine a uniform numbering of all possible axiomatizable theories in any enumerable signature. The main construction $T \mapsto \mathbb{F}_\sigma(T)$ becomes an effective operator from the class of all axiomatizable theories into the class of finitely axiomatizable theories in a given finite rich signature σ.

Note 2.3 With respect to the model completeness of the theory $F = \mathbb{F}_\sigma(T)$, the concepts "constructive model" and "strongly constructive model" are equivalent. In particular, if some model of this theory is not strongly constructivizable, it is not constructivizable at all.

The following general statement follows from the main theorem.

Theorem 2.2 *Let T be an axiomatizable theory without finite models and with some given properties from the list MQL. Then there is a finitely axiomatizable theory with the same properties.*

Yet another variation of Theorem 2.1 is the following.

Theorem 2.3 *Let T be a complete decidable theory without finite models. Then there is a complete and model complete finitely axiomatizable theory F in a given finite rich signature σ, such that the theories T and F have exactly the same properties from the list MQL.*

3 Constructions

In this section we introduce a series of constructions for finitely axiomatizable theories, and one construction for axiomatizable theories. Each of them represents a general method of constructing theories of a given class. It produces a series of theories depending on one or more parameters. By choosing suitable parameters one can control the properties of the resulting theory. Typical parameters are natural numbers, recursively enumerable trees, and axiomatizable theories. The 'strength' of a construction is determined by the number of properties involved, as well as the simplicity of the relationship between these properties and the parameters that are necessary to obtain them.

Figure 1 represents the general relation between the constructions. The arrows are supposed to indicate the historical development that led to the ideas behind the constructions, as well as their relative 'strength', pointing from 'weaker' to 'stronger' constructions.

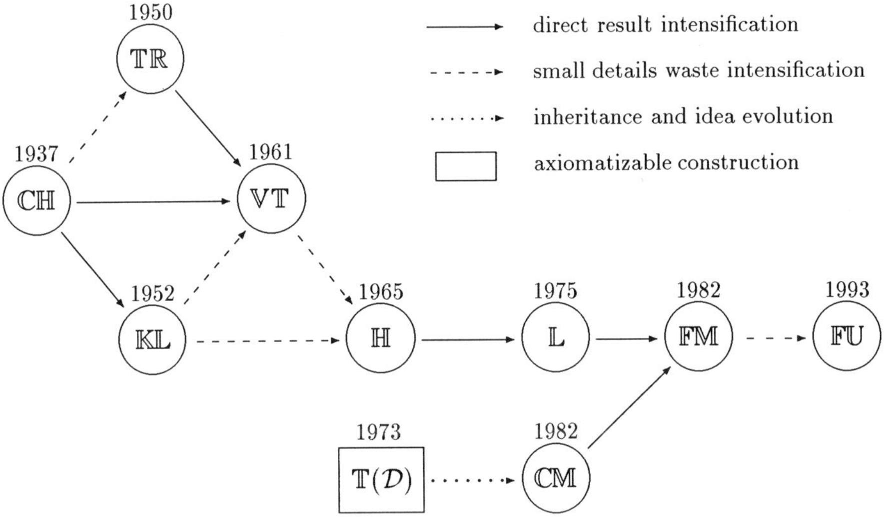

Figure 1: Constructions of finitely axiomatizable theories

Let $\mathbb{FU}$ denote the universal construction from Theorem 2.1, but in concrete applications we often write $\mathbb{F}(T)$ instead of $\mathbb{FU}(T)$. Note that each construction allows us to build a finitely axiomatizable theory in any given finite rich signature σ. If it is necessary to indicate the signature, we write $\mathbb{VT}_\sigma(n)$, $\mathbb{F}_\sigma(T)$, etc.. For the sake of completeness we include the construction $\mathbb{T}(\mathcal{D})$, which produces axiomatizable theories in a signature with countably many unary predicates [13, 33]. This construction precedes the construction $\mathbb{CM}$, see [26, 27].

We start with the Church construction [3]. In the following, let K denote the set $\{n \mid n \in W_n\}$.

Theorem 3.1 *On a natural number n, one can effectively construct a finitely axiomatizable theory $F = \mathbb{CH}(n)$, such that:*

(a) *For $n \in K$, the theory F is inconsistent.*

(b) *For $n \notin K$, the theory F is consistent.*

The Trakhtenbrot construction [38] gives a similar statement for finite models.

Theorem 3.2 *On a natural number n, one can effectively construct a finitely axiomatizable theory $F = \mathbb{TR}(n)$, such that:*

(a) *For $n \in K$, the theory F has a finite model.*

(b) *For $n \notin K$, the theory F has no finite models.*

The Vaught construction [39] generalizes the constructions of Church and Trakhtenbrot.

Theorem 3.3 *On a natural number n, one can effectively construct a finitely axiomatizable theory $F = \mathbb{VT}(n)$, and a 'halting sentence' Φ (not depending on n), such that:*

(a) *If $n \in K$, then $\vdash (F \to \Phi)$, and the theory F has a unique (up to isomorphism) model $\mathfrak{M}$, which is finite.*

(b) *If $n \notin K$, then $\nvdash (F \to \Phi)$, and the theory F is essentially and hereditary undecidable, and has no recursively enumerable models.*

It is possible to regard K as the halting problem for the universal Turing machine, where the halting sentence causes the machine to stop. It seems useful to specify the properties of the halting sentence in the statement of Theorem 3.3. In this form, Theorem 3.3 immediately implies Theorem 3.1 and Theorem 3.2. To see this, one simply defines

$$\mathbb{CH}(n) = \mathbb{VT}(n) \wedge \neg\Phi \qquad \text{and} \qquad \mathbb{TR}(n) = \mathbb{VT}(n) \wedge \Phi.$$

The most important among the early results is Kleene's construction [19], which demonstrates the expressiveness of finitely axiomatizable theories.

Theorem 3.4 *Let T be an axiomatizable theory in a finite signature σ without finite models. Then effectively on a recursively enumerable index for T, one can construct a finitely axiomatizable theory $F = \mathbb{KL}(T)$ in the signature $\sigma' = \sigma \cup \{R^2\}$, such that $T = F \upharpoonright \sigma$.*

Note that $\mathbb{KL}$ is the first construction which uses an axiomatizable theory as a parameter.

Next we consider the Hanf construction [15]. We fix a decidable theory H whose Lindenbaum algebra $\mathcal{L}(H)$ is atomless. We could for example choose the finitely axiomatizable theory of the successor relation with one initial element in the signature $\{\triangleleft^2, c\}$, where cycles of length $\geqslant 1$ are permitted.

Theorem 3.5 *For any recursively axiomatizable theory T there is a finitely axiomatizable theory $F = \mathbb{H}(T)$ such that the Lindenbaum algebras of F and $T \oplus H$ are recursively isomorphic.*

This completes our characterization of the early constructions.

Before we can describe the intermediate construction $\mathbb{FM}$, and its analogue $\mathbb{CM}$ for complete theories [26, 27, 28], we need to introduce some preliminary concepts. Although the constructions seem to be quite complicated, they are very convenient in concrete applications. They make use of the fact that the algorithmic construction of a binary recursively enumerable tree yields as an intermediate step a finitely axiomatizable theory whose properties depend on the properties of the tree in a well known way.

We give the basic definitions.

The *full binary tree* is the partially ordered set $\mathcal{D}_0 = \langle \mathbb{N}, \preceq \rangle$, where $\preceq$ is generated by the rules $n \preceq 2n+1$ and $n \preceq 2n+2$, $n \in \mathbb{N}$.

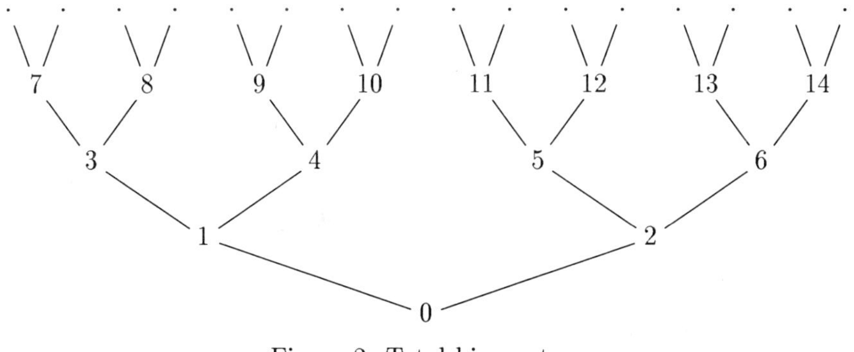

Figure 2: Total binary tree

We define two natural operations:

$L(n) = 2n + 1$ is the left successor of the element n,

$R(n) = 2n + 2$ is the right successor of the element n.

A *tree* is an arbitrary set $\mathcal{D} \subseteq \mathbb{N}$ such that:

(a) $(m \preceq n) \wedge (n \in \mathcal{D}) \longrightarrow m \in \mathcal{D}$, for all $m, n \in \mathbb{N}$,

(b) $L(n) \in \mathcal{D} \leftrightarrow R(n) \in \mathcal{D}$, for all $n \in \mathbb{N}$.

An element $n \in \mathcal{D}$ with $L(n) \notin \mathcal{D}$ is called a *terminal node* of the tree. The set of all terminal nodes of $\mathcal{D}$ is denoted by $Term(\mathcal{D})$. A tree is *atomic* if above each of its element there is at least one terminal node.

A *chain* is a set $\pi \subseteq \mathbb{N}$ such that:

(a) $m, n \in \pi \rightarrow (m \preceq n) \vee (n \preceq m)$, for all $m, n \in \mathbb{N}$.

(b) $(m \preceq n) \wedge (n \in \pi) \rightarrow m \in \pi$, for all $m, n \in \mathbb{N}$.

Let $\Pi(\mathcal{D})$ be the set of all maximal chains of the tree $\mathcal{D}$, and $\Pi^{\text{fin}}(\mathcal{D})$ the set of all finite maximal chains of $\mathcal{D}$.

Consider the family of all maximal chains of $\mathcal{D}$, and let G be a subset of $\Pi(\mathcal{D})$. A chain $\pi \in G$ is *isolated* in G if there is an element $t \in \pi$ such that π is the unique chain in G that contains t. Let G' be the set of all chains $\pi \in G$ that are not isolated in G. By transfinite induction we define the following subsets $\Pi_\alpha(\mathcal{D}) \subseteq \Pi(\mathcal{D})$:

(a) $\Pi_0(\mathcal{D}) = \Pi(\mathcal{D})$,

(b) $\Pi_{\alpha+1}(\mathcal{D}) = (\Pi_\alpha(\mathcal{D}))'$,

(c) $\Pi_\gamma(\mathcal{D}) = \bigcap\{\Pi_\beta(\mathcal{D}) \mid \beta < \gamma\}$, if γ is a limit ordinal.

The least ordinal α such that $\Pi_{\alpha+1}(\mathcal{D}) = \Pi_\alpha(\mathcal{D})$ is called the *rank* of $\mathcal{D}$, and denoted by $Rank(\mathcal{D})$. The *rank of a chain* $\pi \in \Pi(\mathcal{D})$, $Rank(\pi)$, is an ordinal α such that $\pi \in \Pi_\alpha(\mathcal{D}) \setminus \Pi_{\alpha+1}(\mathcal{D})$. It is clear that in general the rank function is only partially defined on the set $\Pi(\mathcal{D})$. We call $\mathcal{D}$ *superatomic* if the rank function is totally defined on $\Pi(\mathcal{D})$.

Let $[W]_\mathcal{D}$ be the naturally defined closure of a set $W \subseteq \mathbb{N}$ with respect to the tree $\mathcal{D}$. Define

$$\mathcal{D}_n = [W_n]_\mathcal{D},\ n \in \mathbb{N}, \qquad \text{and} \qquad \mathcal{D}_n^A = [W_n^A]_\mathcal{D},\ n \in \mathbb{N},\ A \subseteq \mathbb{N},$$

where W_n is the r.e. set with Post index n, and W_n^A is the n–th r.e. set in some standard numbering of the sets that are enumerable with oracle $A \subseteq \mathbb{N}$.

For $k \in \mathbb{N}$, let ε_k be the condition 'the set contains the element k'. A *truth-table condition*, or briefly tt–*condition*, is a propositional formula τ constructed from the elementary expressions ε_k, $k \in \mathbb{N}$. For $A \subseteq \mathbb{N}$, $A \models \tau$ means that the tt–condition τ is true of A.

We fix a Gödel numbering τ_k, $k \in \mathbb{N}$, of all tt–conditions, and define for $m \in \mathbb{N}$ the following class of sets:

$$\mathcal{R}_m = \{A \subseteq \mathbb{N} \mid (\forall k \in W_m) A \models \tau_k\}.$$

Now we can state the theorem on the version of the intermediate construction that deals with the completeness condition. Here we use the notation $\mathbb{F}(\mathcal{D})$ instead of $\mathbb{CM}(\mathcal{D})$.

Theorem 3.6 (C-version of the intermediate construction). *Let σ be a finite rich signature. Effectively on σ and on a recursively enumerable index of a tree $\mathcal{D}$, one can construct a finitely axiomatizable model complete theory $F = \mathbb{F}(\mathcal{D})$ in the signature σ with the following properties:*

(a) *The theory $\mathbb{F}(\mathcal{D})$ has a prime model if and only if the tree $\mathcal{D}$ is atomic.*

(b) *If a prime model of $\mathbb{F}(\mathcal{D})$ exists, then it is strongly constructivizable if and only the family of chains $\Pi^{\text{fin}}(\mathcal{D})$ is computable.*

(c) *If a prime model of $\mathbb{F}(\mathcal{D})$ exists and is strongly constructivizable, then it is autostable with respect to strong constructivizations if and only if $\mathcal{D}$ is a recursive tree.*

(d) *$\mathbb{F}(\mathcal{D})$ has a countable saturated model if and only if $\mathcal{D}$ is superatomic.*

(e) *The countable saturated model of $\mathbb{F}(\mathcal{D})$ is strongly constructivizable if and only if the family of chains $\Pi(\mathcal{D})$ is computable.*

(f) *$\mathbb{F}(\mathcal{D})$ is ω–stable if and only if $\mathcal{D}$ is superatomic.*

(g) *The Morley rank of $\mathbb{F}(\mathcal{D})$ is equal to*

$$\max\{12\,,\, 1 + \mathit{Rank}(\mathcal{D}) + \gamma\},$$

where $\gamma = 2$ if $\mathcal{D}$ is superatomic and $\gamma = 0$ otherwise.

Next we state a general version of the intermediate construction, where we use the notation $\mathbb{F}(m,s)$ instead of $\mathbb{FM}(m,s)$.

Theorem 3.7 (The intermediate construction). *Let σ be a finite rich signature. Effectively on σ and on a pair of natural numbers (m,s), one can construct a finitely axiomatizable model complete theory $F = \mathbb{F}(m,s)$ in the signature σ, and a recursive sequence of sentences Ψ_n, $n \in \mathbb{N}$, in this signature, such that:*

(1) *The sentences Ψ_n, $n \in \mathbb{N}$, generate the Lindenbaum algebra of the theory $\mathbb{F}(m,s)$.*

(2) *The theory*

$$\mathbb{F}(m,s)[A] = \mathbb{F}(m,s) \cup \{\Psi_i \mid i \in A\} \cup \{\neg\Psi_j \mid j \in \mathbb{N} \setminus A\},$$

for $A \subseteq \mathbb{N}$, is consistent if and only if $A \in \mathcal{R}_m$.

(3) *For any set $A \in \mathcal{R}_m$ the following statements are true:*

(a) *$\mathbb{F}(m,s)[A]$ has a prime model if and only if $\mathcal{D}_s^A$ is atomic.*

(b) *If a prime model of $\mathbb{F}(m,s)[A]$ exists, then it is strongly constructivizable if and only if the set A is recursive and the family of chains $\Pi^{\text{fin}}(\mathcal{D}_s^A)$ is computable.*

(c) *If a prime model of $\mathbb{F}(m,s)[A]$ exists and is strongly constructivizable, then it is autostable with respect to strong constructivizations if and only if $\mathcal{D}_s^A$ is recursive.*

(d) *$\mathbb{F}(m,s)[A]$ has a countable saturated model if and only if $\mathcal{D}_s^A$ is superatomic.*

(e) *A countable saturated model of $\mathbb{F}(m,s)[A]$ is strongly constructivizable if and only if A is recursive and $\Pi(\mathcal{D}_s^A)$ is computable.*

(f) *$\mathbb{F}(m,s)[A]$ is ω-stable if and only if $\mathcal{D}_s^A$ is superatomic.*

(g) *The Morley rank of $\mathbb{F}(m,s)[A]$ is equal to*

$$\max\{12\,,\,1 + \mathit{Rank}(\mathcal{D}_s^A) + \gamma\},$$

where $\gamma = 2$ if $\mathcal{D}_s^A$ is superatomic and $\gamma = 0$ otherwise.

We will use the notations $\mathbb{F}(T)$, $\mathbb{F}(m,s)$ and $\mathbb{F}(\mathcal{D})$ for the constructions $\mathbb{FU}$, $\mathbb{FM}$ and $\mathbb{CM}$, respectively, provided this simplification does not lead to any confusion.

4 Existence theorems

We start with a characterization of the Lindenbaum algebras of finitely axiomatizable theories.

Theorem 4.1 *Let σ be a finite rich signature. A numerated Boolean algebra $(\mathcal{B},\nu)$ is equivalent to the Lindenbaum algebra of a finitely axiomatizable theory F in the signature σ if and only if $(\mathcal{B},\nu)$ is a positively numerated Boolean algebra.*

Proof. If $(\mathcal{B},\nu)$ is equivalent to the Lindenbaum algebra of some finitely axiomatizable theory, then $(\mathcal{B},\nu)$ is obviously a positively numerated algebra.

Now suppose that $(\mathcal{B},\nu)$ is a positively numerated Boolean algebra. We will construct a finitely axiomatizable theory F such that $(\mathcal{B},\nu)$ is equivalent to the Lindenbaum algebra of F. Let $\mathcal{B}^*$ be the Boolean algebra whose elements are the classes of equivalent tt–conditions and whose operations are induced by the appropriate propositional operations. It is obvious that $\mathcal{B}^*$ is a countable atomless Boolean algebra, freely generated by the elementary tt–conditions ε_k, $k \in \mathbb{N}$. The algebra $\mathcal{B}^*$ has a natural constructivization μ, determined by a Gödel numbering of the tt–conditions. By Theorem 3.7, (1) and (2), the Lindenbaum algebra of the theory $\mathbb{F}(m,s)$ is isomorphic to the numerated quotient algebra

$$(\mathcal{B}^*/\mathcal{F}_m, \mu^*), \tag{4.1}$$

where $\mathcal{F}_m$ is the filter generated by the set $\{\tau_k \mid k \in W_m\}$, and μ^* is the numeration of the quotient algebra induced by the numeration μ.

It is easy to see that each recursively enumerable filter of the algebra $\mathcal{B}^*$ coincides with $\mathcal{F}_m$ for some $m \in \mathbb{N}$. Any positively numerated Boolean algebra is (up to isomorphism) representable as a quotient of the form (4.1), and this concludes our proof. □

As an immediate consequence, we obtain the following statement for the construction $\mathbb{L}$, see [16, 26]:

Theorem 4.2 *For any recursively axiomatizable theory T there is a finitely axiomatizable theory $F = \mathbb{L}_\sigma(T)$ in a given finite rich signature σ, such that the Lindenbaum algebras of the theories T and F are recursively isomorphic.*

Another consequence of Theorem 4.1 is the following important result.

Theorem 4.3

(a) *The theory $PC(\sigma)$ in a rich finite signature σ that is defined by the empty set of axioms has a non-constructivizable Lindenbaum algebra.*

(b) *There is a decidable finitely axiomatizable theory whose Lindenbaum algebra is constructivizable but not strongly constructivizable.*

Proof.

(a) Feiner [8] constructed a positively numerated Boolean algebra that is not constructivizable. Thus, by Theorem 4.1, there is a finitely axiomatizable theory F in a finite rich signature σ that has a non-constructivizable Lindenbaum algebra. F is finitely axiomatizable, and hence the algebra $\mathcal{L}(F)$ is isomorphic to a quotient algebra

$$\mathcal{L}(PC(\sigma))/_{\mathcal{F}}$$

over a principal filter $\mathcal{F}$. It follows that the algebra $\mathcal{L}(PC(\sigma))$ is not constructivizable.

(b) Goncharov [9] constructed a constructivizable but not strongly constructivizable Boolean algebra for each elementary type, except those categorical in a cardinality $\alpha \leqslant \omega$. An easy application of Theorem 4.1 completes the proof.

□

Combining Theorem 3.6 and Theorem 3.7, we find a method of constructing finitely axiomatizable theories whose prime models, countable saturated models and homogeneous models have certain prescribed properties.

Theorem 4.4

(a) *There is a complete finitely axiomatizable ω–stable theory whose prime model and countable saturated model are not constructivizable.*

(b) *There is a complete finitely axiomatizable ω–stable theory whose prime model is strongly constructivizable but not autostable with respect to strong constructivizations.*

Proof.

(a) Consider a recursively enumerable superatomic tree $\mathcal{D}$ such that the families of chains $\Pi(\mathcal{D})$ and $\Pi^{\text{fin}}(\mathcal{D})$ are not computable; a construction of such a tree can be found in [13]. Let σ be a finite rich signature. Then the theory $\mathbb{F}_\sigma(\mathcal{D})$ has the required properties.

(b) Let $\mathcal{D}$ be a recursively enumerable non-recursive superatomic tree such that $\Pi(\mathcal{D})$ is computable; the construction of such a tree is obvious. Then the theory $\mathbb{F}_\sigma(\mathcal{D})$ has the desired properties.

□

Theorem 4.4 solves two problems posed by Harrington [17].

Theorem 4.5 *There is a complete finitely axiomatizable ω–stable theory T that does not have constructive homogeneous models.*

Proof. A construction of an axiomatizable theory with the desired properties can be found in [10]. We simply have to apply the universal construction to this theory. □

The following three theorems characterize the Morley rank of a finitely axiomatizable theory.

Theorem 4.6 *For any constructive ordinal $\beta \geqslant 0$ there is a complete finitely axiomatizable ω–stable theory T with Morley rank $\alpha_T = \beta + 3$.*

Proof. If $\beta < \omega$, we choose T to be ω_1–categorical. Note that T can easily be constructed from the ω_1–categorical quasi-succession theory QS with Morley rank 3, which can be found in [31, 33].

If $\beta \geqslant \omega$, let T be the theory $\mathbb{F}(\mathcal{D})$, where $\mathcal{D}$ is a recursive superatomic tree of rank $\beta + 1$, as constructed in [33]. □

Let λ be the first non-constructive ordinal.

Theorem 4.7 *There is a complete finitely axiomatizable theory T with Morley rank $\alpha_T = \lambda$.*

Proof. Let $\mathcal{D}$ be an r.e. tree with $Rank(\mathcal{D}) = \lambda$; a construction of $\mathcal{D}$ can be found in [33]. Then we can choose the theory T as $T(\mathcal{D})$. □

Theorem 4.8 *There is a complete finitely axiomatizable theory F with Morley rank $\alpha_F = \omega_1$.*

Proof. In [20], Lachlan gave a construction of a decidable theory T with Morley rank ω_1. (Note that decidability is not mentioned in [20].) We apply the universal construction to T to get the finitely axiomatizable theory $F = \mathbb{F}(T)$. There is an interpretation of the theory T in the theory F, hence the Morley rank of F is at least ω_1. But, as shown by Lachlan, the Morley rank of a countable theory is at most ω_1, and therefore $\alpha_F = \omega_1$. □

Note that the Morley rank of an ω–stable theory cannot be a limit ordinal. Therefore, the theories from Theorem 4.7 and Theorem 4.8 are not ω–stable. As Sacks has shown [37], the Morley rank of an ω-stable decidable theory is a constructive ordinal. The above theorems thus describe complete finitely axiomatizable theories for most possible values of the Morley rank.

5 Complexity of models

Let $AD(\mathfrak{M}, \nu)$ be the atomic diagram of a numerated model $(\mathfrak{M}, \nu)$, and $FD(\mathfrak{M}, \nu)$ its full first order formula diagram.

We first mention a few results on estimates of the algorithmic complexity of a model for a consistent formula.

Theorem 5.1 *There is a consistent finitely axiomatizable theory without constructive models.*

This result first appeared in [20], but the proof was based on the consistency of axiomatic set theory. The direct proof of a slightly stronger result is due to Mostowski [23].

Theorem 5.2 *There is a consistent finitely axiomatizable theory without a recursively enumerable model.*

Mostowski asked, in [22], how to characterize the complexity of the models for a consistent finitely axiomatizable theory. In particular, he asked the following question: Is it true that an arbitrary consistent formula of predicate logic has a model of complexity Σ_1^*, where Σ_1^* is the smallest class of sets that contains all r.e. sets and is closed under Boolean operations?

In [34] and [18], Putnam and Hensel give a positive answer to this question.

Theorem 5.3 *For any consistent formula of predicate logic there is a numerated model* $(\mathfrak{M}, \nu)$ *of complexity* $AD(\mathfrak{M}, \nu) \in \Sigma_1^*$, *and this model is normal, i.e., equality is recursive in the numeration* ν.

As noted in [18], the same statement holds for any axiomatizable theory in a finite signature.

As described by Ershov [6], there is a natural difference hierarchy Σ_n^{-1} for the class Σ_1^*. Following his ideas, we see that a set $A \subseteq \mathbb{N}$ belongs to the class Σ_n^{-1} if and only if it can be represented in the form

$$A = (B_0 \setminus B_1) \cup (B_2 \setminus B_3) \cup \cdots ,$$

where B_0, B_1, B_2, ... are r.e. sets with

$$B_0 \supseteq B_1 \supseteq \cdots \supseteq B_{n-1} \supseteq B_n = \emptyset.$$

The proof in [18] actually gives a model $(\mathfrak{M}, \nu)$ whose atomic diagram is of complexity $AD(\mathfrak{M}, \nu) \in \Sigma_n^{-1}$, provided that the signature of the formula Φ contains only predicates that are at most n–ary.

In [4], Denisov established a lower bound for the complexity of a model for a consistent formula in Ershov's hierarchy.

Theorem 5.4 *For any* $n \geqslant 1$ *there is a consistent formula* Φ *in a signature with* $2n$*–ary predicates that does not have a model of complexity* Σ_n^{-1}.

Next we state a few results on the algorithmic complexity of certain models for a complete finitely axiomatizable theory.

We begin with prime models.

Theorem 5.5

(a) *Let* $\mathfrak{M}$ *be the prime model of a complete decidable theory* T. *Then there is a numeration* ν *of* $\mathfrak{M}$ *such that* $FD(\mathfrak{M}, \nu) \in \Delta_2^0$.

(b) *For any set* $A \in \Delta_2^0$ *there is a complete finitely axiomatizable* ω*–stable theory* F *such that, for any numeration* ν *of its prime model* $\mathfrak{M}$, *the set* $FD(\mathfrak{M}, \nu)$ *is not* m*–reducible to* A.

Proof.

(a) The standard construction of a prime model for a complete decidable theory is recursive in the set of the halting problem. Therefore, it gives a model $(\mathfrak{M}, \nu)$ with $FD(\mathfrak{M}, \nu) \in \Delta_2^0$.

(b) Fix $A \in \Delta_2^0$, and consider a superatomic recursively enumerable tree $\mathcal{D}$ such that the family of chains $\Pi^{\mathrm{fin}}(\mathcal{D})$ is not recursively enumerable with oracle A, see [5] for a construction of $\mathcal{D}$. We want to show that we can choose $F = \mathbb{F}(\mathcal{D})$, where $\mathbb{F}(\mathcal{D})$ is the construction from Theorem 3.6. Assume that the diagram $FD(\mathfrak{M}, \nu)$ is m–reducible to the set A. Then this diagram is obviously recursive in A. Enumerating the components of the model $\mathfrak{M}$ starting with U, see [33], we can compute the corresponding finite chains of the tree $\mathcal{D}$. Hence we can enumerate the family $\Pi^{\mathrm{fin}}(\mathcal{D})$ by a function recursive in A, in contradiction to the properties of $\mathcal{D}$.

□

The next result deals with countable saturated models.

Theorem 5.6

(a) *Let $\mathfrak{M}$ be a countable saturated model of a countable decidable theory T. Then there exists a numeration ν of $\mathfrak{M}$ such that $FD(\mathfrak{M}, \nu) \in \Delta_1^1$.*

(b) *For any $A \in \Delta_1^1$ there is a complete finitely axiomatizable ω–stable theory F such that $A \leqslant_m FD(\mathfrak{M}, \nu)$ for any numeration ν of its countable saturated model $\mathfrak{M}$.*

Proof.

(a) The standard construction of a saturated model for a complete decidable theory has the complexity of a hyperarithmetical set, which proves our statement.

(b) As in [12], we construct a recursively enumerable superatomic tree $\mathcal{D}$ on a given set $A \in \Delta_1^1$, such that $A \leqslant_m \pi^*$ for some chain π^* of $\mathcal{D}$. Then we can define $F = \mathbb{F}(\mathcal{D})$. In fact, let $c \in U(\mathfrak{M})$ be an element of the model $\mathfrak{M}$ that generates the component corresponding to the chain π^*, see [33]. If we enumerate this component in a numerated model $(\mathfrak{M}, \nu)$, then we can enumerate both the chain π^* and its complement with oracle $FD(\mathfrak{M}, \nu)$. Thus π^* is recursive in $FD(\mathfrak{M}, \nu)$ and we have $A \leqslant_m FD(\mathfrak{M}, \nu)$. □

The last result in this section is about homogeneous models.

Theorem 5.7

(a) *Let T be a complete decidable theory. Then there is a homogeneous model $\mathfrak{M}$ of this theory, together with a numeration ν, such that $FD(\mathfrak{M}, \nu) \in \Delta_2^0$.*

(b) *For any $A \in \Delta_2^0$ there a complete finitely axiomatizable ω-stable theory F such that for any numeration ν of a homogeneous model $\mathfrak{M}$ of F the diagram $FD(\mathfrak{M}, \nu)$ is not m-reducible to A.*

Proof.

(a) The statement follows immediately from the standard construction of a homogeneous model for a decidable theory.

(b) Apply the universal construction $\mathbb{FU}$ to the main construction in [10]. □

6 Complexity of semantic classes

Many important applications of the constructions of finitely axiomatizable theories are based on their effectiveness. We give estimates for the algorithmic complexity of concrete semantic classes of sentences that play an important role in Logic. For the following, we assume that σ is a finite rich signature, and we fix a Gödel numbering Φ_i, $i \in \mathbb{N}$, of all sentences in this signature. For $E \subseteq \mathrm{SL}(\sigma)$ let $\mathrm{Nom}(E)$ be the set of Gödel numbers $\{i \mid \Phi_i \in E\}$.

On the family of all sets $E \subseteq \mathrm{SL}(\sigma)$ that are closed under deducibility, we define two natural operations in the following way:

$$C(E) = \mathrm{Compl}(E) = \mathrm{SL}(\sigma) \setminus E,$$

$$N(E) = \mathrm{Neg}(E) = \{\Psi \in \mathrm{SL}(\sigma) \mid (\exists \Phi \in E) \vdash \neg\Phi \leftrightarrow \Psi\}.$$

Also, we denote $I(E) = E$, and $CN(E) = C(N(E))$. These operations have order 2 and commute with each other.

Lemma 6.1 *Let $E_0 \subseteq \mathrm{SL}(\sigma)$ be a set that is closed under deducibility, and define $E_1 = N(E_0)$, $E_2 = C(E_0)$, and $E_3 = CN(E_0)$. Then*

(a) $\mathrm{Nom}(E_i) \approx \mathrm{Nom}(E_{i+1})$, $i \in \{0, 2\}$,

(b) $\mathrm{Nom}(E_i) \approx \mathbb{N} \setminus \mathrm{Nom}(E_j)$, $i \in \{0, 1\}$, $j \in \{2, 3\}$.

We list some of the basic results.

Theorem 6.2 $\{n \mid \Phi_n$ *is true in all models*$\} \approx \Sigma_1^0$.

Proof. The upper estimate follows from the existence of a recursive axiomatization for classical predicate logic, while the lower estimate follows from the Church construction. □

Theorem 6.3 $\{n \mid \Phi_n$ *is true in all finite models* $\} \approx \Pi_1^0$.

Proof. The upper estimate is obvious, and the lower estimate follows from the Trakhtenbrot construction. □

Let $M_{\text{fin}}(\sigma)$ be the class of all finite models, $M_{\text{p.r.}}(\sigma)$ the class of all primitive recursive models, $M_{\text{c}}(\sigma)$ the class of all constructivizable models, $M_{\text{dec}}(\sigma)$ the class of all models with a decidable theory, $M_{\text{f.a.}}(\sigma)$ the class of all models with a finitely axiomatizable theory, $M_{\text{s.c.}}(\sigma)$ the class of all strongly constructivizable models, and $M_{\text{r.e.}}(\sigma)$ the class of all r.e. models.

The following two results are due to Mostowski [34].

Theorem 6.4 *Let K be a class of models of signature σ such that*

$$M_{\text{p.r.}}(\sigma) \subseteq K \subseteq M_{\text{c}}(\sigma).$$

Then $\varnothing^{(\omega)} \leqslant_m \mathrm{Th}(K)$.

Proof. It suffices to consider the signature $\sigma = \{+, -, 0, 1\}$, since all other cases can be obtained by the standard signature reduction [33]. In [34], Mostowski constructed a formula Ψ in the signature σ with a unique constructive model $\mathfrak{N}$ that is isomorphic to the standard model of arithmetic. Therefore

$$\mathfrak{N} \models \Phi_i \;\Leftrightarrow\; K_c(\sigma) \models \Psi \to \Phi_i.$$

This gives the necessary lower estimate for $\mathrm{Th}(K_c(\sigma))$. □

Theorem 6.5 $\mathrm{Th}(M_{\mathrm{c}}(\sigma)) \approx \emptyset^{(\omega)}$, *i.e., the first order theory of the class* $M_{\mathrm{c}}(\sigma)$ *is recursively isomorphic to the elementary theory of the standard model of arithmetic.*

Proof. The upper estimate is obtained by standard methods of arithmetical representation of constructive models. The lower estimate follows from Theorem 6.4. □

Theorem 6.6 *Let* K *be a class of models of signature* σ *such that*

$$M_{\mathrm{fin}}(\sigma) \subseteq K \subseteq M_{\mathrm{r.e.}}(\sigma).$$

Then $\Pi_1^0 \leqslant_m \mathrm{Th}(K)$.

Proof. The result follows immediately from the Vaught construction. □

The last result is of fundamental significance. It shows that a logic based on the traditional principles of axiomatization, whose semantics is based on some variation of the concept of constructive models, cannot exist.

Theorem 6.7 $\mathrm{Th}(M_{\mathrm{dec}}(\sigma)) = \mathrm{Th}(M_{\mathrm{s.c.}}(\sigma))$ *and* $\mathrm{Th}(M_{\mathrm{dec}}(\sigma)) \approx \Pi_3^0$, *so that* $\mathrm{Th}(M_{\mathrm{s.c.}}(\sigma)) \approx \Pi_3^0$ *as well.*

Proof. The equality of the theories follows from the fact that each decidable theory has a strongly constructive model. Therefore, it is enough to prove the estimate for the complexity of $\mathrm{Th}(M_{\mathrm{dec}}(\sigma))$. By virtue of Lemma 6.1, it suffices to show that

$$\{n \mid \Phi_n \text{ has a decidable completion }\} \approx \Sigma_3^0.$$

The upper estimate is obvious. For the lower estimate, we consider the standard set

$$A = \{n \mid W_{l(n)} \text{ and } W_{r(n)} \text{ are recursively separable }\}$$

for Σ_3^0 from [36]. Using this set, it is possible to construct effectively in the parameter $n \in \mathbb{N}$ an axiomatizable theory T_n such that

$$n \in A \Rightarrow T_n \text{ has a decidable extension},$$

$$n \notin A \Rightarrow T_n \text{ has no decidable extensions}.$$

Using the construction $\mathbb{H}$ (or any stronger construction), one can define a general recursive function $f(n)$ such that:

$$n \in A \Rightarrow \Phi_{f(n)} \text{ has a decidable extension,}$$

$$n \notin A \Rightarrow \Phi_{f(n)} \text{ has no decidable extensions.}$$

The function f realizes the necessary reduction. □

The following result is due to Boone and Rogers [1].

Theorem 6.8 $\{n \mid \Phi_n$ *determines a decidable theory* $\} \approx \Sigma_3^0$.

Proof. The upper estimate can be established immediately. For the lower estimate, consider the standard set

$$A = \{n \mid W_n \text{ is recursive }\}$$

for Σ_3^0, as described in [36]. Using this set, one can, effectively in a parameter n, construct an axiomatizable theory T_n, such that:

$$n \in A \Rightarrow T_n \text{ is decidable,}$$

$$n \notin A \Rightarrow T_n \text{ is not decidable.}$$

Using the construction $\mathbb{H}$, we define a general recursive function $f(n)$ such that:

$$n \in A \Leftrightarrow \Phi_{f(n)} \text{ is decidable.}$$

The function f realizes the necessary reduction. □

Theorem 6.9 $\{n \mid \Phi_n$ *determines a complete theory* $\} \approx \Pi_2^0$.

Proof. The upper estimate can be proved easily.

For the lower estimate we consider the standard Π_2^0–set defined in [36]:

$$I = \{n \mid W_n \text{ is infinite }\}.$$

Let σ' be the signature consisting only of the equality predicate. Let Δ_k denote a sentence stating that there exist at least k different elements. Consider the theory T_n in the signature σ' determined by the set of axioms

$\{\Delta_k \mid k \in W_n\}$. It is easy to see that this theory is complete if and only if $n \in I$. Applying Theorem 2.1, we effectively in n construct a finitely axiomatizable theory F_n in the signature σ such that $\mathcal{L}(F_n) \cong \mathcal{L}(T_n)$. Then

$$n \in I \leftrightarrow T_n \text{ is complete } \leftrightarrow F_n \text{ is complete.}$$

The theory F_n is defined effectively in T_n, and therefore there exists a general recursive function $f(n)$ such that the sentence $\Phi_{f(n)}$ is the axiom of this theory. Finally we have

$$n \in I \leftrightarrow \Phi_{f(n)} \text{ determines a complete theory,}$$

and this is the necessary reduction. □

The set defined in Theorem 6.9 is not recursively enumerable. This fact gives the answer to a problem posed in [14].

Theorem 6.10 $\mathrm{Nom}(\mathrm{Th}(M_{\mathrm{f.a.}}(\sigma))) \approx \Pi^0_3$.

Proof. In view of Lemma 6.1, it suffices to show the estimate

$$\{n \mid \Phi_n \text{ has a finitely axiomatizable extension }\} \approx \Sigma^0_3.$$

The upper estimate is obvious. For the lower estimate we use the standard Σ^0_3-set, as defined in [36]:

$$J = \{n \mid \text{ the set } \mathbb{N} \setminus W_n \text{ is finite }\}.$$

Consider the signature σ' and the sentences Δ_n, $n \in \mathbb{N}$, as defined in the proof of Theorem 6.8. Let T_n be the theory in the signature σ' defined by the set of axioms

$$\{\Delta_k \longrightarrow \Delta_{k+1} \mid k \in W_n\}.$$

Apply the intermediate construction to T_n to obtain a finitely axiomatizable theory F_n. Then $\mathcal{L}(F_n) \cong \mathcal{L}(T_n)$. Let $f(n)$ be a general recursive function such that $\Phi_{f(n)}$ is an axiom of the theory F_n. From our construction we have

$$n \in J \leftrightarrow F_n \text{ has a finitely axiomatizable extension.}$$

This proves the necessary lower estimate. □

The following table summarizes the results on the complexity estimates of semantic classes of sentences [26, 27]. Here σ is a fixed finite rich signature. The following notations for classes of models of this signature are used:

$$SC = M_{\text{s.c.}}(\sigma),$$
$$D = M_{\text{dec}}(\sigma),$$
$$F = M_{\text{fin}}(\sigma),$$

P is the class of prime models,

S is the class of countable saturated models,

T is the class of models with ω–stable theories.

Class of sentences	Estimate
$\{n \mid \Phi_n$ determines a complete stable theory $\}$	Π^0_2
$\{n \mid \Phi_n$ has a finite number of completions $\}$	Σ^0_3
$\{n \mid \Phi_n$ is complete and has a prime model $\}$	Π^0_3
$\{n \mid \Phi_n$ is complete and has no prime model $\}$	Σ^0_3
$\{n \mid \Phi_n$ is complete and has an s.c. prime model $\}$	Σ^0_4
$\{n \mid \Phi_n$ is complete and has a prime model that is not s.c. $\}$	Π^0_4
$\{n \mid \Phi_n$ is complete and has a countable saturated model $\}$	Π^1_1
$\{n \mid \Phi_n$ is complete and ω–stable $\}$	Π^1_1
$\text{Th}(P)$	Π^1_1
$\text{Th}(S)$, $\text{Th}(T)$	Π^1_2
$\text{Th}(P \cap D)$, $\text{Th}(P \cap F)$, $\text{Th}(P \cap SC)$, $\text{Th}(P \cap SC \cap F)$	Π^0_4
$\text{Th}(S \cap D)$, $\text{Th}(S \cap F)$, $\text{Th}(S \cap SC)$, $\text{Th}(S \cap SC \cap F)$	Σ^1_1
$\text{Th}(T \cap D)$, $\text{Th}(T \cap F)$, $\text{Th}(T \cap SC)$, $\text{Th}(T \cap SC \cap F)$	Σ^1_1

Table 1: Complexity estimates

7 Analogues of Rice's Theorem

The concept of semantic similarity that we used in Theorem 2.1 gives rise to the important concept of semantic classes of sentences. They are in some sense analogous to the index sets of the Post numbering. In this section we give lower estimates for the complexity of semantic classes of sentences.

We fix a finite rich signature σ and a Gödel numbering Φ_i, $i \in \mathbb{N}$, for the set of all sentences of σ. For $E \subseteq \mathrm{SL}(\sigma)$ let $\mathrm{Nom}(E)$ be the set of Gödel numbers $\{i \mid \Phi_i \in E\}$. Let $[\Phi]_\sigma$ denote the theory in the signature σ that is generated by the axiom Φ. Consider a list of properties $L \subseteq MQL$. The formulae Φ and Ψ in the signature σ are *similar with respect to* L if the theories $[\Phi]_\sigma \oplus SI$ and $[\Psi]_\sigma \oplus SI$ are similar over L, where SI is the ω_1-categorical theory of the successor relation, as defined in Note 2.1. For theories without finite models, this 'generalized' similarity coincides with the ordinary similarity; for arbitrary theories it is more appropriate than the ordinary similarity. In this section we only talk about generalized semantic similarity.

A set $E \subseteq \mathrm{SL}(\sigma)$ is called *semantically closed with respect to* L if for all sentences $\Phi, \Psi \in \mathrm{SL}(\sigma)$,

$$[\Phi] \equiv_L [\Psi] \;\Rightarrow\; (\Phi \in E \Leftrightarrow \Psi \in E).$$

Let $\Lambda \in \{I, C, N, CN\}$. The set $E \subseteq \mathrm{SL}(\sigma)$ is *semantically* Λ*-closed with respect to* L if there exists a set $E' \subseteq \mathrm{SL}(\sigma)$ that is semantically closed with respect to L such that $E = \Lambda(E')$. The set $E \subseteq \mathrm{SL}(\sigma)$ is *semantic with respect to* L if it is semantically Λ-closed with respect to L for some $\Lambda \in \{I, C, N, CN\}$. It is easy to show that if a set E is semantic with respect to L, then it is also semantic with respect to any bigger list $L' \supseteq L$. A set $E \subseteq \mathrm{SL}(\sigma)$ that is semantic with respect to the empty list $L = \emptyset$ is called *absolutely semantic*.

The following statement can be regarded as the analogue of the first level of Rice's Theorem for semantic classes.

Theorem 7.1 *Let L be a list such that $L \subseteq MQL$, and let $E \subseteq \mathrm{SL}(\sigma)$ be semantic with respect to L.*

(a) $\mathrm{Nom}(E)$ *is recursive* $\Leftrightarrow E = \emptyset$ *or* $E = \mathrm{SL}(\sigma)$.

(b) *If $E \neq \emptyset$ and $E \neq \mathrm{SL}(\sigma)$, then $\Sigma^0_1 \leqslant_m \mathrm{Nom}(E)$ or $\Pi^0_1 \leqslant_m \mathrm{Nom}(E)$.*

Proof. It is enough to prove Part (b), and in view of Lemma 6.1, it will suffice to consider a semantically closed set E. We consider the following two cases.

CASE 1: The set E contains an inconsistent formula. Then all inconsistent formulae from $\mathrm{SL}(\sigma)$ belong to E, since E is semantically closed. Let Ψ be a (consistent) formula from $\mathrm{SL}(\sigma) \setminus E$.

For a natural number n, consider the axiomatizable theory T_n that is determined by the following system of axioms:

$$T_n = \begin{cases} [\Psi]_\sigma, & \text{if } n \notin K, \\ [\{\Psi, \neg\Psi\}]_\sigma, & \text{if } n \in K, \end{cases}$$

where K is the set of the halting problem. It is obvious that the set of axioms of T_n can be enumerated effectively in n.

Applying the s–m–n Theorem [36], we define a general recursive function $e(x)$ such that for any n the number $e(n)$ is a recursively enumerable index of the theory T_n. We apply Theorem 2.1 to the theory T_n to obtain a general recursive function $h(x)$ such that $T_n \equiv_L [\Phi_{h(n)}]$. Then we have:

$$n \notin K \Rightarrow [\Phi_{h(n)}] \equiv_L [\Psi],$$

$$n \in K \Rightarrow \Phi_{h(n)} \text{ is contradictory.}$$

This implies $K \leqslant_m \mathrm{Nom}(E)$.

CASE 2: The set E does not contain an inconsistent formula. Then we apply similar arguments as in Case 1 to the complement of E to obtain

$$K \leqslant_m \mathbb{N} \setminus \mathrm{Nom}(E). \qquad \square$$

We now consider an analogue of the second level of Rice's Theorem, but only for a smaller list of properties. Let MQL' be the part of the list MQL that includes the following properties:

(a) The existence of a prime model and its algorithmic dimension;

(b) the existence and strong constructivizability of a countable saturated model.

Theorem 7.2 *Let L be a sublist of MQL', and let*

$$E_0 = \{\Phi \in \mathrm{SL}(\sigma) \mid\ \vdash \neg\Phi\}.$$

Let $E \subseteq \mathrm{SL}(\sigma)$ be semantic with respect to L.

(a) $\mathrm{Nom}(E)$ *is recursive if and only if* $E = \emptyset$ *or* $E = \mathrm{SL}(\sigma)$.

(b) $\mathrm{Nom}(E) \approx \Sigma_1^0$ *if and only if* $E = E_0$ *or* $E = N(E_0)$.

(c) $\mathrm{Nom}(E) \approx \Pi_1^0$ *if and only if* $E = C(E_0)$ *or* $E = CN(E_0)$.

(d) *In all other cases we have* $\Sigma_2^0 \leqslant_m \mathrm{Nom}(E)$ *or* $\Pi_2^0 \leqslant_m \mathrm{Nom}(E)$.

Next we present possible analogues for the remaining levels of Rice's Theorem. We will use the thin hierarchy of sets above Σ_2^0, similar to the hierarchy in [1]. For $A \subseteq \mathbb{N}$ we define

$$D(A) = \{\Phi \in \mathrm{SL}(\sigma) \mid (\exists k \in A)\, \mathcal{L}([\Phi]) \cong 2^k\}$$

where 2^k is a finite Boolean algebra with k atoms.

Theorem 7.3 *There is a family $\mathcal{A}^*$ of sets $A \subseteq \mathbb{N} \setminus \{0\}$ which is of the cardinality of the continuum, such that for all $A \in \mathcal{A}^*$:*

(a) *Both A and $\mathbb{N} \setminus A$ are infinite.*

(b) $(\forall$ *arithmetical* $X)\, X \leqslant_m \mathrm{Nom}(D(A)) \Leftrightarrow X \in \Sigma_{2,\omega}^0$.

Statement 7.4 *There is no effectively formulated analogue of the third level of Rice's Theorem for semantic classes of sentences with respect to any list of model-theoretic properties.*

Substantiation. In view of Theorem 7.3, any analogue of the third level of Rice's Theorem with respect to L has to include the special cases $D(A)$, $A \in \mathcal{A}^*$, that represent the sets that are semantic with respect to $\emptyset$, and therefore with respect to L. In particular, such an analogue should include the following statement: *In the remaining cases, we have* $\Sigma_3^0 \leqslant_m \mathrm{Nom}(E)$ *or* $\Pi_3^0 \leqslant_m \mathrm{Nom}(E)$. But this continuum of special cases is nonconstructive and therefore not suitable for an effective formulation of an analogue of Rice's Theorem. □

There is an analogue for the 'level two plus' of Rice's Theorem [30], but we won't include it since it is a technically very complicated statement.

Nevertheless we can look for 'restricted' analogues of Rice's theorem, bypassing the difficulties mentioned in Statement 7.4.

Theorem 7.5 (Zaurbekov [40]). *Let $E \subseteq \mathrm{SL}(\sigma)$ be a set that is semantic with respect to the empty list $L = \emptyset$, such that $E \neq D(A)$ for all $A \in \mathcal{A}^*$.*

(a) $\mathrm{Nom}(E)$ *is recursive if and only if* $E = \emptyset$ *or* $E = \mathrm{SL}(\sigma)$.

(b) $\mathrm{Nom}(E) \approx \Sigma_1^0$ *if and only if* $E = E_0$ *or* $E = N(E_0)$.

(c) $\mathrm{Nom}(E) \approx \Pi_1^0$ *if and only if* $E = C(E_0)$ *or* $E = CN(E_0)$.

(d) *In the remaining cases we have* $\Sigma_2^0 \leqslant_m \mathrm{Nom}(E)$ *or* $\Pi_2^0 \leqslant_m \mathrm{Nom}(E)$.

8 Open questions

Question 8.1 Is there a complete finitely axiomatizable ω–stable theory T with Morley rank $\alpha_T = \omega + 1$?

Question 8.2 Is there a complete finitely axiomatizable theory T whose Morley rank α_T satisfies the inequalities $\lambda < \alpha_T < \omega_1$, where λ is the first nonconstructive ordinal? (*Conjecture*: no).

Question 8.3 Describe the class of all finitely axiomatizable theories whose constructive models are precisely their strongly constructive models.

Question 8.4

(a) Describe all finitely axiomatizable theories T such that the class of all strongly constructive models of T is computable, and the theory of this class coincides with T.

(b) Describe all finitely axiomatizable theories T such that the class of all strongly constructive models of T is computable and has a principal numbering, and the theory of this class coincides with T.

Question 8.5 Find an exact estimate for the algorithmic complexity of the realization of a consistent formula in a given signature (see Theorem 5.3 and Theorem 5.4 for upper and lower bounds).

Question 8.6 Find exact estimates for the algorithmic complexity of the elementary theories of the following classes of countable models:

(a) $\{\mathfrak{M} \mid d_{\text{s.c.}}(\mathfrak{M}) = 1\}$,

(b) $\{\mathfrak{M} \mid d_{\text{s.c.}}(\mathfrak{M}) = \omega\}$,

(c) $\{\mathfrak{M} \mid d_{\text{c}}(\mathfrak{M}) = 1\}$,

(d) $\{\mathfrak{M} \mid d_{\text{c}}(\mathfrak{M}) = n,\ 1 < n < \omega\}$,

(e) $\{\mathfrak{M} \mid d_{\text{c}}(\mathfrak{M}) = \omega\}$,

(f) $\{\mathfrak{M} \mid d_{\text{s.c.}}(\mathfrak{M}) = 1,\ d_{\text{c}}(\mathfrak{M}) > 1\}$,

(g) $\{\mathfrak{M} \mid d_{\text{s.c.}}(\mathfrak{M}) = 1,\ d_{\text{c}}(\mathfrak{M}) = \omega\}$.

Question 8.7 Find exact estimates for the algorithmic complexity of the following classes of sentences:

(a) $\{n \mid \mathcal{L}(\Phi_n)$ is strongly constructivizable $\}$,

(b) $\{n \mid \mathcal{L}(\Phi_n)$ is constructivizable $\}$,

(c) $\{n \mid \mathcal{L}(\Phi_n)$ is not constructivizable $\}$.

References

[1] W. W. Boone and H. Rogers Jr., On a problem of J. H. C. Whitehead and a problem of Alonzo Church, Math. Scand., **19** (1966) 185–192.

[2] C. C. Chang and H. J. Keisler, Model Theory, 3rd. edn., Stud. Logic Found. Math., **73** (1990); [1st. edn. 1973, 2nd. edn. 1977].

[3] A. Church, A note on the Entscheidungsproblem, J. Symbolic Logic, **11** (1937) 40–41, corr. ibid., 101–102.

[4] S. D. Denisov, Models of non-contradictory formulas and the Ershov hierarchy (Russian), Algebra i Logika, **11** (1972) 648–655, 736; [translated in: Algebra and Logic, **11** (1972) 359–362].

[5] B. N. Drobutun, Enumerations of simple models (Russian), Sibirsk. Math. Zh., **18** (1977) 1002–1014, 1205; [translated in: Siberian Math. J., **18** (1977) 707–716].

[6] Yu. L. Ershov, A hierarchy of sets I (Russian), Algebra i Logika, **7** (1968) 47–74 [translated in: Algebra and Logic, **7** (1968) 25–43].

[7] Yu. L. Ershov, Decision Problems and Constructivizable Models (Russian), (Mathematical Logic and Foundations of Mathematics, Nauka, Moskva, 1980).

[8] L. J. Feiner, Hierarchies of Boolean algebras, J. Symbolic Logic, **35** (1970) 365–374.

[9] S. S. Goncharov, Certain properties of the constructivization of Boolean algebras (Russian), Sibirsk. Math. Zh., **16** (1975) 264–278; [translated in: Siberian Math. J., **16** (1975) 203–214].

[10] S. S. Goncharov, Totally transcendental theory with a nonconstructivizable prime model (Russian), Sibirsk. Math. Zh., **21** (1980) 44–51; [translated in: Siberian Math. J., **21** (1980) 32–37].

[11] S. S. Goncharov, Countable Boolean algebras and decidability (Russian), Sibirsk. Skh. Alg. i Log. NII MIOONGU, Novosibirsk, (1996); [translated in: Siberian School of Algebra and Logic, (Consultants Bureau, Plenum, New York, 1997)].

[12] S. S. Goncharov and B. N. Drobotun, Numerations of saturated and homogeneous models (Russian), Sibirsk. Math. Zh., **21** (1980) 25–41, 236; [translated in: Siberian Math. J., **21** (1980) 164–176].

[13] S. S. Goncharov and A. T. Nurtazin, Constructive models of complete decidable theories (Russian), Algebra i Logika, **12** (1973) 125–142, 243; [translated in: Algebra and Logic, **12** (1973) 67–77].

[14] Yu. Sh. Gurevich, The decision problem for the logic of predicates and of operations, Algebra i Logika, **8** (1969) 284–308; [translated in: Algebra and Logic, **8** (1969) 160–174].

[15] W. Hanf, Model-theoretic methods in the study of elementary logic, in: Theory of Models, (Proc. Internatl. Sympos., Berkeley, 1963), (North-Holland, Amsterdam, 1965), 132–145.

[16] W. Hanf, The Boolean algebra of logic, Bull. Amer. Math. Soc., **81** (1975) 587–589.

[17] L. Harrington, Recursively presentable prime models, J. Symbolic Logic, **39** (1974) 305–309.

[18] G. Hensel and H. Putnam, Normal models and the field Σ_1^*, Fund. Math., **64** (1969) 231–240.

[19] S. C. Kleene, Finite axiomatizability of theories in the predicate calculus using additional predicate symbols, in: Two Papers on the Predicate Calculus, Mem. Amer. Math. Soc., **10** (1952) 27–68.

[20] G. Kreisel, Note on arithmetic models for consistent formulæ of the predicate calculus, Fund. Math., **37** (1950) 265–285.

[21] A. H. Lachlan, The transcendental rank of a theory, Pacific J. Math., **27** (1971) 119–122.

[22] A. Mostowski, The present state of investigations in the foundations of mathematics (Russian), Uspekhi Mat. Nauk, **9** (1954) 3–38, [in Polish in: Prace Mat., **1** (1955) 13–55].

[23] A. Mostowski, A formula with no recursively enumerable model, Fund. Math., **42** (1955) 125–140.

[24] A. Mostowski, On recursive models of formalized arithmetic, Bull. Acad. Polon. Sci. Cl. III, **5** (1957) 705–710.

[25] A. T. Nurtazin, Strong and weak constructivizations and computable families (Russian), Algebra i Logika, **13** (1974) 311–323, 364; [translated in: Algebra and Logic, **13** (1974) 177–184].

[26] M. G. Peretyat'kin, Calculations on Turing machines in finitely axiomatizable theories (Russian), Algebra i Logika, **21** (1982) 410–441; [translated as: Turing machine computations in finitely axiomatizable theories, Algebra and Logic, **21** (1982) 272–295.]

[27] M. G. Peretyat'kin, Finitely axiomatizable totally transcendental theories (Russian), Trudy Inst. Mat., **2** ("Nauka" Sibirsk. Otdel., Novosibirsk) (1982) 88–135.

[28] M. G. Peretyat'kin, Finitely axiomatizable theories (Russian), in: Proc. Int. Congr. Math., Berkeley, CA, USA, 1986, **1** (1987) 322–330; [translated in: Nine Papers from the International Congress of Mathematicians, Aug. 3–11, 1986, Berkeley, CA, USA, B. Silver, (ed.), Amer. Math. Soc. Transl. Ser. 2, **147** (1990) 11–19.]

[29] M. G. Peretyat'kin, The similarity of properties of recursively enumerable and finitely axiomatizable theories (Russian), Dokl. Akad. NaukSSSR, **308** (1989) 788–791; [translated in: Soviet Math. - Dokl., **40** (1990) 372–375].

[30] M. G. Peretyat'kin, Analogues of Rice's theorem for semantic classes of propositions (Russian), Algebra i Logika, **30** (1991) 517–539, 626; [translated in: Algebra and Logic, **30** (1991) 332–348].

[31] M. G. Peretyat'kin, Uncountably categorical quasisuccession of Morley rank 3 (Russian), Algebra i Logika, **30** (1991) 74–89; [translated in: Algebra and Logic, **30** (1991) 51–61 (1992)].

[32] M. G. Peretyat'kin, Imbedding recursively enumerable theories in finitely axiomatizable theories (Russian), Trudy Inst. Sib. Otdel. Akad. Nauk SSSR, (1990).

[33] M. G. Peretyat'kin, Finitely axiomatizable theories (Russian), Sibirsk. Shk. Alg. i Log., (NII MIOONGU, Novosibisk, 1997); [translated in: Siberian School of Algebra and Logic. (Consultants Bureau, New York, 1997).

[34] H. Putnam, Trial and error predicates and the solution to a problem of Mostowski, J. Symbolic Logic, **30** (1965) 49–57.

[35] H. G. Rice, Classes of recursively enumerable sets and their decision problems, Trans. Amer. Math. Soc., **74** (1953) 358–366.

[36] H. Rogers Jr., Theory of Recursive Functions and Effective Computability, (1st. edn., McGraw-Hill, New York-Toronto, Ont.-London, 1967; 2nd. edn., MIT Press, Cambridge, Mass., London, 1987).

[37] G. Sacks, Effective bounds on the Morley rank, Fund. Math., **103** (1979) 111–121.

[38] B. A. Trakhtenbrot, The impossibility of an algorithm for the decision problem on finite domains (Russian), Dokl. Akad. Nauk SSSR, **70** (1950) 569–572.

[39] R. L. Vaught, Non-recursive-enumerability of the set of sentences true in all constructive models, Bull. Amer. Math. Soc. (N.S.), **63** (1957) 230.

[40] S. S. Zaurbekov, A bounded analogue of Rice's theorem at the third level for semantic classes of propositions (Russian), Algebra i Logika, **32** (1993) 131–138; [translated in: Algebra and Logic, **32** (1993) 71–75.]

M. G. Peretyat'kin
Institute of Pure and Applied Mathematics,
Academy of Sciences of Kazakhstan,
Alma-Ata, Kazakhstan.

Chapter 10

Complexity Theoretic Model Theory and Algebra

D. Cenzer and J. B. Remmel*

Contents

Introduction

In this paper, we will survey some recent results on complexity theoretic model theory and algebra. Essentially there are two major themes in this work. The first, which we call complexity theoretic model theory, deals with model existence questions. For example, given a recursive model $\mathcal{A}$, is there there a polynomial time (exponential time, polynomial space, etc.) model $\mathcal{B}$ which is isomorphic to $\mathcal{A}$. The second theme, which we call complexity

*The work of the second author was partly supported by Dept. of Commerce Agreement 70-NANB5H1164 and NSF grant DMS-9306427.

HANDBOOK OF RECURSIVE MATHEMATICS
Edited by Yu. L. Ershov, S. S. Goncharov,
A. Nerode, and J. B. Remmel

theoretic algebra, fixes a given polynomial time structure and explores the properties of that structure. For example, we can ask whether every polynomial time ideal of a given polynomial time representation of the free Boolean algebra can be extended to a maximal polynomial time ideal. In both cases, one uses the rich theory of recursive model theory and algebra as a reference but looks at resource bounded versions of the results in those areas. It turns out that not only are there a number of contrasts between results in recursive model theory and algebra and complexity theoretic model theory and algebra, but some new and interesting phenomena occur in the study of complexity theoretic model theory and algebra. That is, there are results in recursive model theory and algebra for which the natural complexity theoretic analogue is true but requires a more delicate proof which incorporates the resource bounds. There are also results in recursive model theory and algebra for which the natural complexity theoretic analogue is false because the proof of the recursive result uses the unbounded resources allowed in recursive constructions in a crucial way. However, there are a number of interesting new phenomena which arise due to the fact that not all infinite polynomial time sets are polynomial time isomorphic or due to the fact that complexity theoretic results do not relativize as is the case for most recursion theoretic results. For example, in recursive model theory any two infinite recursive sets are recursively isomorphic, so that one can restrict one's attention to models whose universe is the set of natural numbers. It is not the case that any two infinite polynomial time sets are polynomial time isomorphic so that the choice of a particular universe, say the tally representation of the natural numbers versus the binary representation of the natural numbers, makes a difference. Also, it is well known that the question of whether $\mathrm{P} = \mathrm{NP}$ is oracle dependent. That is, Baker, Gill and Solovay [3] proved that there are recursive oracles X and Y such that $\mathrm{P}^X = \mathrm{NP}^X$ and $\mathrm{P}^Y \neq \mathrm{NP}^Y$. We shall see that some of the natural complexity theoretic analogues of results in recursive algebra are oracle dependent as well.

There are several other areas of complexity theoretic model theory and algebra which will not be covered in this survey. There is the work of Friedman and Ko (see for example, [27], [42], and [43]) on polynomial time analysis, where complexity theoretic versions of various theorems of analysis are studied. Some of these results are oracle dependent and some are shown to be equivalent to $\mathrm{P} = \mathrm{NP}$. There is the work of Crossley, Nerode and Remmel on p–time equivalence types and p–time isols, as developed in [56], [57], [22] and [23]. We will present some results from [56] on p–time equivalence types in Section 3. There is the work of Khoussainov and Nerode [41] on automatic, or automata presentable structures, which is a further restriction of polynomial time structures.

We will start with a survey of complexity theoretic model theory. In Section 1, we will provide a general introduction to complexity theoretic model theory. In Section 2, we shall give our basic complexity theoretic definitions and establish notation. In Section 3, we will give a series of lemmas which are useful for building models with standard universes such as the binary representation of the natural numbers, $Bin(\omega)$, and the tally representation of the natural numbers, $Tal(\omega)$. In Section 4 we provide a survey of the main existence theorems for feasible models. In Section 5, we survey various feasible categoricity results. In Section 6, we give an introduction to complexity theoretic algebra. Then in Section 7, we focus on the structure of the binary and tally representation of an infinite dimensional vector space over a polynomial time field. In Section 8, we look at the semilattice of NP ideals of the binary and tally representation of the free Boolean algebra. Finally in Section 9, we give conclusions as well as some directions for further work.

1 Complexity Theoretic Model Theory

Complexity theoretic or feasible model theory is the study of resource-bounded structures and isomorphisms and their relation to computable structures and computable isomorphisms. The focus of complexity theoretic model theory in this paper is very different from classical complexity theory. A primary focus in classical complexity theory has been to determine the complexity of certain classes of finite models encoded as a decision problem. That is, one is interested in classifying decision problems as being in P, NP, PSPACE, etc.. A typical example is the graph-coloring problem, where it is known that the family of finite graphs which can be 3–colored is NP–complete. Complexity theoretic model theory is more concerned with infinite models whose universe, functions, and relations are in some well known complexity class such as polynomial time, exponential time, polynomial space, etc.. Thus if one studies graph colorings from this point of view, one would study the complexity of graph colorings in an infinite polynomial time graph as was done by Cenzer and Remmel in [12]. However complexity theoretic model theory has been more concerned with the complexity of the model itself. Thus one can pick any complexity class and ask questions about what structures can be represented by models in that complexity class. By far, the complexity class that has received the most attention is polynomial time. The basic questions that have been consider are to classify which recursive models are isomorphic or recursively isomorphic to a polynomial time model.

To establish some notation, let $\omega = \{0\,,\,1\,,\,\ldots\}$ denote the set of natural numbers. Let $[\ ,\ ]$ denote the usual quadratic-time pairing function $[m,n] = m + \frac{1}{2}(m+n)(m+n+1)$, which maps $\omega \times \omega$ onto ω. Let $\varphi_{e,n}$ denote the n–ary partial function on $(\{0,1\}^*)^n$ computed by the e–th Turing machine. Then we say that a structure

$$\mathcal{A} = \left(A\,,\,\{R_i^{\mathcal{A}}\}_{i\in S}\,,\,\{f_i^{\mathcal{A}}\}_{i\in T}\,,\,\{c_i^{\mathcal{A}}\}_{i\in U}\right),$$

(where the universe A of $\mathcal{A}$ is a subset of $\{0,1\}^*$) is *recursive* if A is a recursive subset of $\{0,1\}^*$, S, T, and U are initial segments of ω, the set of relations $\{R_i^{\mathcal{A}}\}_{i\in S}$ is uniformly recursive in the sense that there is a recursive function G such that for all $i \in S$, $G(i) = [n_i, e_i]$ where $R_i^{\mathcal{A}}$ is an n_i–ary relation and φ_{e_i,n_i} computes the characteristic function of $R_i^{\mathcal{A}}$, the set of functions $\{f_i^{\mathcal{A}}\}_{i\in T}$ is uniformly recursive in the sense that there is a recursive function F such that for all $i \in T$, $F(i) = [n_i, e_i]$ where $f_i^{\mathcal{A}}$ is an n_i–ary function and φ_{e_i,n_i} restricted to A^{n_i} computes $f_i^{\mathcal{A}}$, and there is a recursive function interpreting the constant symbols in the sense that there is a recursive function H such that for all $i \in U$, $H(i) = c_i^{\mathcal{A}}$. Note that if $\mathcal{A}$ is a recursive structure, then the atomic diagram of $\mathcal{A}$ is recursive. We say that a recursive structure $\mathcal{A} = (A, \{R_i^{\mathcal{A}}\}_{i\in S}, \{f_i^{\mathcal{A}}\}_{i\in T}, \{c_i^{\mathcal{A}}\}_{i\in U})$, is *polynomial time* if A is a polynomial time subset of $\{0,1\}^*$ and the set of relations $\{R_i^{\mathcal{A}}\}_{i\in S}$ and the set of functions $\{f_i^{\mathcal{A}}\}_{i\in T}$ are uniformly polynomial time in the sense that, in addition to the functions G and F defined above, there are recursive functions G' and F' such that for $i \in S$, $G'(i) = m_i$ where for all $(x_1\,,\,\ldots\,,\,x_{n_i})$ in $(\{0,1\}^*)^{n_i}$, it takes at most $(\max\{2\,,\,|x_1|\,,\,\ldots\,,\,|x_{n_i}|\})^{m_i}$ steps to compute $\varphi_{e_i,n_i}(x_1\,,\,\ldots\,,\,x_{n_i})$ and for all $i \in T$, $F'(i) = q_i$ where for all $(x_1\,,\,\ldots\,,\,x_{n_i})$ in $(\{0,1\}^*)^{n_i}$, it takes at most $(\max\{2\,,\,|x_1|\,,\,\ldots\,,\,|x_{n_i}|\})^{q_i}$ steps to compute $\varphi_{e_i,n_i}(x_1\,,\,\ldots\,,\,x_{n_i})$. Note that if $\mathcal{A}$ is a polynomial time structure with infinitely many relation symbols or with infinitely many function symbols, then our definition of a polynomial time structure does not ensure that the atomic diagram of $\mathcal{A}$ is polynomial time. Thus we say $\mathcal{A}$ is *uniformly polynomial time* if the atomic diagram of $\mathcal{A}$ is polynomial time. Note that the fact that $\mathcal{A}$ is uniformly polynomial time implies, among other things, that the sequence of run times $\{x^{m_i} : i \in S\}$ and $\{x^{q_i} : i \in T\}$ are bounded by some fixed polynomial. Of course, if $\mathcal{A}$ is a structure over a finite language, then $\mathcal{A}$ is a polynomial time structure if and only if $\mathcal{A}$ is a uniformly polynomial time structure. Similar definitions may be given for other resource-bounded classes.

There are two basic types of questions which have been studied in polynomial time model theory. First, as discussed above, there is the basic existence problem, i.e., whether a given infinite recursive structure $\mathcal{A}$ isomorphic or recursively isomorphic to a polynomial time model. For example, the authors showed in [10, p. 24] that every recursive relational structure is recursively isomorphic to a polynomial time model and that the standard model of arithmetic $(\omega, +, -, \cdot, <, 2^x)$ with addition, subtraction, multiplication,order and the 1–place exponential function is isomorphic to a polynomial time model. The fundamental effective completeness theorem says that any decidable theory has a decidable model. It follows that any decidable relational theory has a polynomial time model. However, one is naturally led to ask more refined existence questions in complexity theoretic algebra than one asks in recursive algebra. That is, since all infinite recursive sets are recursively isomorphic, it is easy to see that any infinite recursive structure is recursively isomorphic to a recursive structure whose universe is ω. It is certainly not the case that any two infinite polynomial-time sets are polynomial-time isomorphic. For example, the tally representation of the natural numbers is not polynomial time isomorphic to the binary representation of the natural numbers. Hence it no longer the case that any infinite polynomial-time structure can be identified with a polynomial-time structure whose universe is $\{0,1\}^*$. Thus a more refined existence questions is to take a fixed universe, such as the tally representation of the natural numbers or the binary representation of the natural numbers, and ask if a recursive model is isomorphic or recursively isomorphic to a polynomial time model with that given universe.

Here are two examples which illustrate both the negative and positive outcomes to the simplest existence type question, i.e., whether a given recursive model is isomorphic to a polynomial time model.

Example 1.1 Let $\mathcal{A} = (A, 0, S, R)$ where $A = \{1\}^*$ (that is, the set of natural numbers in unary representation), S is the successor function, (that is, $S(1^n) = 1^{n+1}$), and R is a unary relation, (that is, a subset of $\{1\}^*$). Now if $\mathcal{A}$ is isomorphic to a polynomial-time structure $\mathcal{B} = (B, 0^B, S^B, R^B)$, then we can test for membership in R as follows. Given 1^n, compute $(S^B)^n(0^B) = y_n$ and then test whether y_n is in R^B. Now if we assume that we can compute

$S^B(x)$ in $|x|^k$ steps for $|x| \geqslant 2$, then it takes at most $\sum_{i=1}^{n} |0^B|^{k^i} \leqslant |0^B|^{k^n+1}$ steps to compute y_n. Next we may assume that testing whether $x \in R^B$ takes $|x|^r$ steps if $|x| \geqslant 2$, so that it takes at most $|0^B|^{r(k^n+1)}$ steps to test whether 0^n is in R. This means that R is a doubly-exponential-time set. Thus if we start with any recursive structure $\mathcal{A} = (A, 0, S, R)$ where R is a recursive set but is not doubly exponential-time, then $\mathcal{A}$ is not even isomorphic, much less recursively isomorphic, to a polynomial-time structure.

Despite this example, there are lots of recursive structures which are recursively isomorphic to polynomial-time structures.

Example 1.2 Let $\mathcal{A} = (A, f)$, where $A = \{1\}^*$ and f is a unary function. We say that 1^m and 1^n are in the same f–orbit if, for some $k \geqslant 0$, either $f^k(1^m) = 1^n$ or $f^k(1^n) = 1^m$. If f is length-increasing, then it is clear that each f–orbit is isomorphic to (A, S). Now let f and g be any two recursive length- increasing functions from $\{1\}^*$ into $\{1\}^*$. Then the structures (A, f) and (A, g) are recursively isomorphic if and only if they have the same number of orbits. Thus, for example, we can let $f(1^n) = 1^{a(n)}$ where a is Ackermann's function and still be guaranteed that (A, f) is recursively isomorphic to a polynomial-time structure.

Next consider the more restricted kind of existence question, i.e., whether a given recursive model is isomorphic or recursively isomorphic to a polynomial time model which has a standard universe such as the binary representation of the natural numbers, *Bin* (ω), or the tally representation of the natural numbers, *Tal* $(\omega) = \{1^n : n \in \omega\}$. Grigorieff [32] proved that every recursive linear ordering is isomorphic to a linear time linear ordering which has universe $Bin(\omega)$. However Grigorieff's result can not be improved to the result that every recursive linear ordering is *recursively isomorphic* to a linear time linear ordering over *Bin* (ω). That is, Cenzer and Remmel [10, p. 25] showed that for any infinite polynomial time set $A \subseteq \{0,1\}^*$, there exists a recursive copy of the linear ordering $\omega + \omega^*$ which is not recursively isomorphic to any polynomial time linear ordering which has universe A. Here $\omega + \omega^*$ is the ordering obtained by taking a copy of $\omega = \{0, 1, 2, \ldots\}$ under the usual ordering followed by a copy of the negative integers under the usual ordering.

The general problem of determining which recursive models are isomorphic or recursively isomorphic to feasible models has been studied by the authors in [10], [13], and [11]. For example, it was shown in [13, pp. 343–348] that any recursive torsion Abelian group G is isomorphic to a polynomial time group A and that if the orders of the elements of G are bounded, then A may be taken to have a standard universe, i.e., either $Bin(\omega)$ or $Tal(\omega)$. It was also shown in [13, p. 357] that there exists a recursive torsion Abelian group which is not even isomorphic to any polynomial time (or any primitive recursive) group with a standard universe. Feasible linear orderings were studied by Grigorieff [32], by Cenzer and Remmel [10], and by Remmel [68, 69]. Feasible vector spaces were studied by Nerode and Remmel in [53] and [55]. Feasible Boolean algebras were studied by Cenzer and Remmel in [10] and by Nerode and Remmel in [54]. Feasible permutation structures and feasible Abelian groups were studied by Cenzer and Remmel in [13] and [11]. By a *permutation structure* $\mathcal{A} = (A, f)$, we mean a set A together with a unary function f which maps A one-to-one and onto A. Similarly an *equivalence structure* $\mathcal{A} = (A, R^{\mathcal{A}})$ consists of a set A together with an equivalence relation.

The second basic type of problem studied in polynomial time model theory is the problem of feasible categoricity. Here we say that a recursive model $\mathcal{A}$ is *recursively categorical* if any other recursive model isomorphic to $\mathcal{A}$ is in fact recursively isomorphic to $\mathcal{A}$. The notion of recursive categoricity was first defined by Mal'cev [46] and is referred to in the Russian literature as *autostability*. Recursively categorical structures have been widely studied in the literature of recursive algebra and recursive model theory.

The recursively categorical structures for various theories have been classified, including Boolean algebras independently by Goncharov [30] and LaRoche [44], Abelian groups by Smith [74] and linear orderings independently by Dzgoev [31] and Remmel [66]. For example, Remmel showed in [66] that a recursive linear ordering $L = (D, <)$ is recursively categorical if and only if L has only finitely many successivities, where a pair $a < b$ is a successivity if there is no c with $a < c < b$.

Defining a natural analogue of feasible categoricity is complicated by the fact that unlike the case of infinite recursive models, where any two infinite recursive universes are recursively isomorphic, it is not the case that any two polynomial time universes are polynomial time isomorphic. It turns out to be more natural to define polynomial categorical structures with respect to a fixed universe. Thus we say that a p–time structure $\mathcal{A}$ with universe $D \subseteq \{0,1\}^*$ is *p–time categorical with respect to D* if every p–time structure

$\mathcal{B}$ with universe D which is isomorphic to $\mathcal{A}$ is necessarily p–time isomorphic to $\mathcal{A}$, i.e., there exist polynomial time functions f, g such that f restricted to D is an isomorphism from $\mathcal{A}$ onto $\mathcal{B}$ and g restricted to D is an isomorphism from $\mathcal{B}$ onto $\mathcal{A}$.

Remmel showed in [66] that there are no p-time categorical linear orderings with respect to the standard universes $Bin(\omega)$ and $Tal(\omega)$. There are two parts to this strongly negative statement. For any p–time linear ordering L with universe B (either $Bin(\omega)$ or $Tal(\omega)$), there is a p–time linear ordering L' with universe B which is not primitive recursively isomorphic to L. Furthermore, if L is not recursively categorical, then L' is not even recursively isomorphic to L. Similar results will be shown for other structures. The problem of feasible categoricity for permutation structures and torsion Abelian groups was studied by Cenzer and Remmel in [11]. Here there are some limited positive results. In particular, a permutation structure (A, f) such that all orbits of f have the same finite size is p–time categorical over $Tal(\omega)$. There are also structures $\mathcal{A}$ which are not p–time categorical over B, but such that any p–time structure $\mathcal{D}$ with universe B which is isomorphic to $\mathcal{A}$ must be exponential time isomorphic or double exponential time isomorphic to $\mathcal{A}$. More generally, we can define a larger notion of feasibility, *q–time* or iterated exponential time computability, and show that there are many natural structures which are q–time categorical over $Bin(\omega)$ and $Tal(\omega)$.

General semantic conditions for when a decidable model is recursively categorical were given by Nurtazin [60] and Goncharov [29] and similar results were found by Ash and Nerode [2] for models in which one can effectively decide all Σ_1 formulas. These methods are based on the existence of a so-called *Scott family* of formulas. We discuss in Section 5 various notions from [15] of a feasible Scott family of formulas for a feasible model and show that any two families which possess a common Scott family and have the same universe B are feasibly isomorphic. Structures considered in [15] include linear orderings, permutation structures, Abelian groups and equivalence structures.

2 Preliminaries

In this section, we will give the basic definitions from complexity theory which will be needed for the rest of the paper.

Let Σ be a finite alphabet. Then Σ^* denotes the set of finite strings of letters from Σ and Σ^ω denotes the set of infinite strings of letters from Σ where $\omega = \{0, 1, 2, \ldots\}$ is the set of natural numbers. For any natural number $n \neq 0$, $tal(n) = 1^n$ is the tally representation of n and $bin(n) = i_0 i_1 \ldots i_e \in \{0,1\}^*$ is the (reverse) binary representation of n if $n = i_0 + 2 \cdot i_1 + \cdots + 2^e \cdot i_e$ and $i_e \neq 0$. In general, the k–ary representation $b_k(n) = i_0 i_1 \ldots i_e$ if $n = i_0 + i_1 \cdot k + \cdots i_e \cdot k^e$ and $i_e \neq 0$. We let $tal(0) = bin(0) = b_k(0) = 0$. Then we let $Tal(\omega) = \{tal(n) : n \in \omega\}$, $Bin(\omega) = \{bin(n) : n \in \omega\}$ and, for each $k \geqslant 3$, $B_k(\omega) = \{b_k(n) : n \in \omega\}$. Occasionally, we will want to say that $B_2(\omega) = Bin(\omega)$ and that $B_1(\omega) = Tal(\omega)$.

For a string $\sigma = (\sigma(0), \sigma(1), \ldots, \sigma(n-1))$, $|\sigma|$ denotes the length n of σ. The empty string has length 0 and will be denoted by $\emptyset$. A constant string σ of length n will be denoted by k^n. For $m < |\sigma|$, $\sigma \lceil m$ is the string $(\sigma(0), \ldots, \sigma(m-1))$; σ is an *initial segment* of τ (written $\sigma \prec \tau$) if $\sigma = \tau \lceil m$ for some m. The *concatenation* $\sigma^\frown\tau$ (or sometimes just $\sigma\tau$) is defined by

$$\sigma^\frown\tau = (\sigma(0), \sigma(1), \ldots, \sigma(m-1), \tau(0), \tau(1), \ldots, \tau(n-1)),$$

where $|\sigma| = m$ and $|\tau| = n$; in particular we write $\sigma^\frown a$ for $\sigma^\frown(a)$ and $a^\frown\sigma$ for $(a)^\frown\sigma$.

Our basic computation model is the standard multitape Turing machine of Hopcroft and Ullman [5]. Note that there are different heads on each tape and that the heads are allowed to move independently. This implies that a string σ can be copied in linear time. An oracle machine is a multitape Turing machine M with a distinguished work tape, a query tape, and three distinguished states QUERY, YES, and NO. At some step of a computation on an input string σ, M may transfer into the state QUERY. In state QUERY, M transfers into the state YES if the string currently appearing on the query tape is in an oracle set A. Otherwise, M transfers into the state NO. In either case, the query tape is instantly erased. The set of strings accepted by M relative to the oracle set A is

$$L(M, A) = \{\sigma \mid \text{there is an accepting computation of } M \text{ on input } \sigma \text{ when the oracle set is } A\}.$$

If $A = \emptyset$, we write $L(M)$ instead of $L(M, \emptyset)$.

Let $t(n)$ be a function on natural numbers. A Turing machine M is said to be *$t(n)$–time bounded* if each computation of M on inputs of length n where $n \geqslant 2$ requires at most $t(n)$ steps. A function $f(x)$ on strings is said to be in DTIME(t) if there is a $t(n)$–time bounded deterministic Turing machine M which computes $f(x)$. For a function f of several variables, we let the length of $(x_1, \ldots, x_n)$ be $|x_1| + \cdots + |x_n|$. A set of strings or a relation on strings is in DTIME(t) if its characteristic function is in DTIME(t). We let

$$\text{R} = \bigcup_c \{\text{DTIME}(n+c) : c \geqslant 0\},$$

$$\text{LIN} = \bigcup_c \{\text{DTIME}(cn) : c \geqslant 0\},$$

$$\text{P} = \bigcup_i \{\text{DTIME}(n^i) : i \geqslant 0\},$$

$$\text{DEXT} = \bigcup_{c \geqslant 0} \{\text{DTIME}(2^{c \cdot n})\}, \text{ and}$$

$$\text{DOUBEXT} = \bigcup_{c \geqslant 0} \{\text{DTIME}(2^{2^{c \cdot n}})\},$$

$$\text{EXPTIME} = \bigcup_{c \geqslant 0} \{\text{DTIME}(2^{n^c})\}, \text{ and in general,}$$

$$\text{DEX}(S) = \bigcup_{t(n) \in S} \{\text{DTIME}(2^{t(n)})\}.$$

A function $f(x)$ on strings is said to be in NTIME(t) if there is a $t(n)$–time bounded nondeterministic Turing machine M which computes $f(x)$. A set of strings or a relation on strings is in NTIME(t) if its characteristic function is in NTIME(t). We let

$$\text{NP} = \bigcup_i \{\text{NTIME}(n^i) : i \geqslant 0\},$$

$$\text{NEXT} = \bigcup_{c \geqslant 0} \{\text{NTIME}(2^{c \cdot n})\},$$

$$\text{NEXPTIME} = \bigcup_{c \geqslant 0} \{\text{NTIME}(2^{n^c})\},$$

$$\text{DOUBNEXT} = \bigcup_{c \geqslant 0} \{\text{NTIME}(2^{2^{c \cdot n}})\}, \text{ and in general,}$$

$$\text{NEX}(S) = \bigcup_{t(n) \in S} \{\text{NTIME}(2^{t(n)})\}.$$

We fix enumerations $\{P_i\}_{i \in \omega}$ and $\{N_i\}_{i \in \omega}$ of the polynomial time bounded deterministic oracle Turing machines and the polynomial time bounded nondeterministic oracle Turing machines respectively. We may assume that $p_i(n) = \max(2, n)^i$ is a strict upper bound on the length of any computation by P_i or N_i with any oracle X on inputs of length n. P_i^X and N_i^X denote the oracle Turing machine using oracle X and in an abuse of notation we shall denote $L(P_i, X)$ by simply P_i^X and $L(N_i, X)$ by N_i^X. This given, $\mathrm{P}^X = \{P_i^X : i \in \omega\}$ and $\mathrm{NP}^X = \{N_i^X : i \in \omega\}$.

For A, $B \subset \Sigma^*$, we shall write $A \leqslant_m^P B$ if there is a polynomial-time function f such that for all $x \in \Sigma^*$, $x \in A$ iff $f(x) \in B$. We shall write $A \leqslant_T^P B$ if A is polynomial time Turing reducible to B. For r equal to m or T, we write $A \equiv_r^P B$ if $A \leqslant_r^P B$ and $B \leqslant_r^P A$ and we write $A \,|_r^P B$ if not $A \leqslant_r^P B$ and not $B \leqslant_r^P A$.

We define the standard notions of feasibility as follows. We say that a function $f(x)$ is *quasi-real-time* if $f(x) \in R$. (This is slightly more general than the usual notion of real-time as computable by a Turing machine which simply reads the input one symbol at a time from left to right (or right to left) and simultaneously leaves the output in its place on the tape. In particular, a real-time function is always in $\mathrm{DTIME}(n)$.) The function $f(x)$ is *linear time* if $f(x) \in L$, *polynomial time* if $f(x) \in \mathrm{P}$, *nondeterministic polynomial time* if $f(x) \in \mathrm{NP}$, *exponential time* if $f(x) \in \mathrm{DEXT}$, *nondeterministic exponential time* if $f(x) \in \mathrm{NEXT}$, and is *double exponential time* if $f(x) \in \mathrm{DOUBEXT}$. We say that $f(x)$ is exponentially feasible if $f(x) \in \mathrm{DEX}(T)$ for a notion T of feasibility. In particular, if $f(x) \in \mathrm{DEX}(\mathrm{DOUBEXT})$, then $f(x)$ is said to be *triple exponential time.*

The smallest class including P and closed under DEX can be defined by iterating DEX. That is, let

$$\mathrm{P}^0 = \mathrm{P}, \quad \mathrm{P}^{n+1} = \mathrm{DEX}(\mathrm{P}) \text{ for each } n, \quad \text{and } \mathrm{Q} = \bigcup_{n<\omega} \mathrm{P}^n.$$

A function $f(x) \in \mathrm{Q}$ is said to be *iterated exponential time* or *q–time.* The iterated exponential functions $E_n(x)$ can be defined recursively by $E_0(x) = x$ and $E_{n+1}(x) = 2^{E_n(x)}$ for all n and x. It is easy to see that $x^r \leqslant E_r(x)$ for all $r > 0$, from which it follows that $\mathrm{Q} = \bigcup_{m<\omega} \mathrm{DEX}^m(\mathrm{DTIME}(n))$.

We observe that the classes R, LIN, P, NP, and Q are all closed under composition, whereas the other classes defined above are not. In addition, *Tal* (ω) and *Bin* (ω) are q–time isomorphic.

Observe that for a function $f(x_1, \ldots, x_k)$ of several variables, the above definitions are equivalent, if we declare the size of the input $(x_1, \ldots, x_k)$ to be the maximum of the sizes $|x_1|, \ldots, |x_k|$ since we allow multiple tapes. This occasionally simplifies the computation of the complexity of various functions.

We refer the reader to Odifreddi [61] for the basic definitions of recursion theory. Let $\varphi_{i,n}$ be the partial recursive function of n variables computed by the i–th Turing machine M_i. If $n = 1$, we will write φ_i instead of $\varphi_{i,1}$. Given a string $\sigma \in \{0,1\}^*$, we write $\varphi_i^s(\sigma)\downarrow$ if M_i gives an output in s or fewer steps when started on input string σ. Thus the function φ_i^s is uniformly polynomial time. We write $\varphi_e(\sigma)\downarrow$ if $(\exists s)\,(\varphi_e^s(\sigma)\downarrow)$ and $\varphi_e(\sigma)\uparrow$ if not $\varphi_e(\sigma)\downarrow$.

The notion of a p–time structure was defined in Section 1. We need a few refinements of that definition.

Definition 2.1

(i) A p–time function f is *honest p–time* if there is a polynomial function q such that for all $x_1, \dots, x_n$,

$$y = f(x_1, \dots, x_n) \to (\forall i \leqslant n)(|x_i| \leqslant q(|y|)).$$

(ii) A p–time structure $\mathcal{A}$ is *honest p–time* if all of its functions are honest p–time.

(iii) A structure $\mathcal{A}$ has *honest witnesses* if for any quantifier-free formula $\varphi(y, x_1, \dots, x_n)$, there is a polynomial q such that for any $a_1, \dots, a_n \in A$, if $\mathcal{A} \models (\exists y)\,\varphi(y, a_1, \dots, a_n)$, then there is a $z \in A$ with $|z| \leqslant q(|a_1| + \cdots + |a_n|)$ such that $\mathcal{A} \models \varphi(z, a_1, \dots, a_n)$.

Note that for an honest p-time function mapping *Tal* (ω) into *Tal* (ω), Nerode and Remmel showed in [56] that f^{-1} is also honest p–time.

For a group, we will distinguish two types of computability. The structure of a group $\mathcal{G}$ is determined by the binary operation which we will denote by the addition sign $+^G$, since we are interested in Abelian groups. We let e^G denote the additive identity of $\mathcal{G}$. However, the inverse operation, denoted by inv^G, may also be included as an inherent part of the group. Thus we have the following distinction.

Definition 2.2 A group $\mathcal{G}$ is Γ*–computable* if $(G, +^G, e^G)$ is Γ–computable, and is *fully* Γ*–computable* if $(G, +^G, inv^G, e^G)$ is Γ–computable.

It is easy to see that any recursive group is also fully recursive, since $inv^G(a)$ can be computed as the least member b of G such that $a +^G b = e^G$, where the elements of G are ordered first by length and then lexicographically for elements of the same length.

On the other hand, the fully p-time groups make up a proper subclass of the p–time groups, as shown by Proposition 1.1 of [13].

Definition 2.3 For any complexity class Γ and any structures

$$\mathcal{A} = (A, \{R_i^A\}_{i\in S}, \{f_i^A\}_{i\in T}, \{c_i\}_{i\in U}),$$

and

$$\mathcal{B} = (B, \{R_i^B\}_{i\in S}, \{f_i^B\}_{i\in T}, \{c_i\}_{i\in U}),$$

we say that $\mathcal{A}$ and $\mathcal{B}$ are Γ–*isomorphic* if there is an isomorphism f from $\mathcal{A}$ onto $\mathcal{B}$ and Γ–computable functions F and G such that $f = F\lceil A$ (the restriction of F to A) and $f^{-1} = G\lceil B$.

3 Polynomial-Time Sets and Isomorphisms

In this section we shall give a number of useful lemmas about the relations between the various standard universes that we will consider in our study of feasible structures. The most basic standard universe is the set Σ^* where Σ is a finite alphabet and in particular where $\Sigma = \{0,1\}$, $\{1\}$, or $\{0\}$. Other standard universes include the set $Tal\,(\omega)$ of tally representations of natural numbers, the set $Bin\,(\omega)$ of binary representations of natural numbers, and, for any k, the set $B_k(\omega)$ of k–ary representations of natural numbers. In recursion theory, all of these sets are recursively isomorphic and therefore interchangeable. For our purposes, we must consider carefully which of these isomorphisms are polynomial time or even polynomial time in one direction.

First we need to explicitly define a polynomial time pairing function. For any finite alphabet Σ, there is a natural embedding ρ of Σ^* into $Bin\,(\omega)$ given as follows. We may suppose that $\Sigma \subset \{0\,, 1\,, 2\,, \ldots\,, n\}$ for some n. Let $\rho(\emptyset) = 0$ and, for $\sigma = (i_1\,, \ldots\,, i_k)$, let

$$\rho(\sigma) = 0^{i_1}{}^\frown 1^\frown 0^{i_2}{}^\frown 1^\frown \cdots {}^\frown 0^{i_k}{}^\frown 1.$$

The function ρ is actually an isomorphism from ω^* onto $Bin\,(\omega)$ and has an inverse ρ^{-1}. It is also clear that, for each n, the set $\rho(\{0\,, 1\,, \ldots\,, n\}^*)$ is linear time (uniformly in n). Thus we can normally assume that an arbitrary structure has universe a subset of $Bin\,(\omega)$.

The coding function $\langle \sigma_1, \sigma_2, \dots, \sigma_k \rangle_k$ for $\sigma_1, \dots, \sigma_k \in \{0,1\}^*$ is now defined by

$$\langle \sigma_1, \sigma_2, \dots, \sigma_k \rangle_k = \rho(\sigma_1{}^\frown 2^\frown \sigma_2{}^\frown 2^\frown \dots{}^\frown \sigma_{k-1}{}^\frown 2^\frown \sigma_k).$$

Let $Q_k = \{\langle \sigma_1, \sigma_2, \dots, \sigma_k \rangle_k : \sigma_i \in \{0,1\}^* \text{ for each i}\}$. For $i = 1, \dots, k$, the projection functions π_i^k from Q_k onto $\{0,1\}^*$ are implicitly defined by the equation

$$\sigma = \langle \pi_1^k(\sigma), \pi_2^k(\sigma), \dots, \pi_k^k(\sigma) \rangle_k.$$

The subscript k will normally be omitted. It is easy to see that the sets Q_k and $B_k(\omega)$ are all linear time and that the functions π_i^k, and $\langle\ ,\ \dots\ ,\ \rangle_k$ are all computable in linear time.

Given two subsets A and B of $\{0,1\}^*$, define

$$A \otimes B = \{\langle a, b \rangle : a \in A, b \in B\}$$

and

$$A \oplus B = \{\langle 0, a \rangle : a \in A\} \cup \{\langle 1, b \rangle : b \in B\}.$$

It is clear that if A and B are p–time, then both $A \oplus B$ and $A \otimes B$ will also be p–time.

Now, for each $k \geqslant 2$, a natural number in (reverse) k–ary form is simply a string $\sigma \in \{0, 1, \dots, k-1\}^*$ which is either 0 or else ends with an element of $\{1, \dots, k-1\}$. Thus the set $B_k(\omega)$ of k–ary representations of natural numbers is a linear time subset of $\{0, 1, \dots, k-1\}^*$. $Tal\,(\omega) = \{0\} \cup \{1\}^* 1$ is of course a linear time subset of $\{0,1\}^*$, but is not polynomial time isomorphic to the whole set. (This will follow immediately from Lemma 3.6 below.) For a string $\sigma = b_k(n) \in B_k(\omega)$ with $n > 0$, let $\mu_k(\sigma)$ be the unary representation, 1^n, of n and let $\mu_k(0) = 0$. We now state a sequence of lemmas which will be useful for our main results. Most of these lemmas are proved in [13] or [11].

Lemma 3.1

(a) *Suppose that $\mathcal{A}$ is a polynomial time structure and that φ is a polynomial time set isomorphism from A onto a set B. Then $\mathcal{B}$ is a polynomial time structure, where the functions and relations on B are defined to make φ an isomorphism of the structures.*

(b) *Suppose that $\mathcal{A}$ is an* EXPTIME *structure and that φ is a polynomial time set isomorphism from A onto a set B. Then $\mathcal{B}$ is an* EXPTIME *structure, where the functions and relations on B are defined to make φ an isomorphism of the structures.*

(c) *Suppose that $\mathcal{A}$ is a q–time structure and that φ is a q–time set isomorphism from A onto a set B. Then $\mathcal{B}$ is a q–time structure, where the functions and relations on B are defined to make φ an isomorphism of the structures.*

Proof. We sketch the proof for the p–time case. The other cases are similar. To simplify the proof, let us suppose that A has one function f^A and one relation R^A. Observe first that B is a polynomial time set, since $b \in B \iff \varphi^{-1}(b) \in A$. The function f^B is polynomial time, since $f^B(b_1, \ldots, b_n) = \varphi(f^A(\varphi^{-1}(b_1), \ldots, \varphi^{-1}(b_n)))$. The relation R^B is polynomial time, since $R^B(b_1, \ldots, b_n) \iff R^A(\varphi^{-1}(b_1), \ldots, \varphi^{-1}(b_n))$. □

Next we state two lemmas which relate tally representation of a structure to its binary representation and, more generally, to its k–ary representation. Part (b) of the first lemma is an improvement of Lemma 2.2 of [13] where the computation was bounded in polynomial time.

Lemma 3.2 *For each $k > 1$,*

(a) *$Bin(\omega)$ is linear time isomorphic to $\{0,1\}^*$.*

(b) *There is a linear time function p such that, for all n, both the computation of $\mu_k(b_k(n)) = 1^n$ and the inverse computation of $\mu_k^{-1}(1^n) = b_k(n)$ can be computed in time $p(n)$;*

(c) *For each $n > 0$ and $\sigma = b_k(n)$, $k^{|\sigma|-1} < n + 1 \leqslant k^{|\sigma|}$*

Proof. We sketch the proof of (b) for $k = 2$. The basic computation in either direction consists of enumerating the binary numbers from 1 to n on one tape while either writing or reading n 1's on the other tape. The enumeration of the binary numbers is done by repeatedly adding 1 by the usual algorithm, which consists of replacing 1's with 0's while looking for the first 0, and then replacing the first 0 with a 1. If we define $h(n)$ to be the total number of symbols written by this procedure for the numbers from $k = 1$ up to $k = 2^n - 1$, then we observe that $h(1) = 1$ and that, in general, $h(n+1) = n+1+2h(n)$. This is because the numbers from 2^n up to $2^{n+1} - 1$ are obtained by writing 2^n, which takes $n + 1$ steps and then essentially writing the numbers from 1 up to $2^n - 1$ again, while leaving the final 1 on the end of each. Now it is easy to see by induction that $h(n) \leqslant 2^{n+1} - n - 2$ for each n. Counting a slightly smaller number of steps for returning to the beginning of the string, we see that all of the binary numbers from 1 up to $k = 2^n - 1$ may be written in total time $\leqslant 2h(n) \leqslant 2^{n+1} - 2 = 4k$. For any number k with $2^{n-1} < k \leqslant 2^n$, the binary numbers from 1 to k may be written in total time $\leqslant 2^{n+1} - 2 \leqslant 8k$. □

For any subset M of ω, let $tal\,(M) = \{tal(n) : n \in M\}$, let $bin\,(M) = \{bin\,(n) : n \in M\}$, and for any finite $k > 1$, let $b_k(M) = \{b_k(n) : n \in M\}$.

Lemma 3.3 *For any finite $k > 1$, any $M \subset \omega$ and any oracle X:*

(a) $tal(M) \in \mathrm{P}^X \iff b_k(M) \in \mathrm{DEXT}^X$;

(b) $tal(M) \in \mathrm{NP}^X \iff b_k(M) \in \mathrm{NEXT}^X$.

(c) $b_k(M) \in \mathrm{P}^X \to tal\,(M) \in \mathrm{LIN}^X$.

Proof. We give the proof for $k = 2$, where $b_k(M) = bin\,(M)$. The proof of (b) is the same as the proof of (a).

(a) Suppose first that $tal\,(M) \in P^X$. Then there is a procedure with oracle X which tests whether $tal\,(n) \in tal\,(M)$ in time $\leqslant n^c$ for some fixed c and all $n \geqslant 2$. To test whether $\sigma = bin\,(n) \in bin\,(M)$, we first compute $tal\,(n)$, which requires time $\leqslant 8n$ by the proof of Lemma 3.2. Then we test whether $tal\,(n) \in tal(M)$, which requires $\leqslant n^c$ steps by assumption. Now by part (c) of Lemma 3.2, $n^c < (2^{|\sigma|})^c = 2^{c|\sigma|}$ so that $bin\,(M) \in \mathrm{DEXT}^X$.

Next suppose that $bin\,(M) \in \mathrm{DEXT}^X$. Then there is a procedure with oracle X which tests whether $bin\,(n) \in bin\,(M)$ in time $\leqslant 2^{c|bin(n)|}$ for some fixed c and all n. To test whether $tal\,(n) \in tal\,(M)$, we first compute $bin\,(n)$, which requires time $\leqslant 8n$ by the proof of Lemma 3.2. Then we test whether $bin\,(n) \in bin\,(M)$, which requires $\leqslant 2^{c|bin(n)|}$ steps by assumption. Now by part (c) of Lemma 3.2, $2^{c|bin\,(n)|} = (2^{|bin\,(n)|})^c \leqslant (2n)^c \leqslant \min(2,n)^{2c}$. Hence $tal(M) \in P^X$.

(c) Suppose that $bin\,(M) \in P^X$, so that we can test $\sigma = bin\,(n) \in bin\,(M)$ in time $\leqslant |\sigma|^c$. Now let $|\sigma| = r$, so that $2^{r-1} \leqslant n < 2^r$. Then as in part (a) above, we see that we can test $tal\,(n) \in tal\,(M)$ in time $\leqslant r^c$. Now it is clear that for sufficiently large r, $r^c \leqslant 2^{r-1} \leqslant n$. Thus $tal\,(M) \in \mathrm{LIN}^X$. □

We note that the assumption that $tal\,(M) \in \mathrm{LIN}^X$ actually implies that we can test $\sigma \in Bin\,(M)$ in time $\leqslant c\,2^{|\sigma|}$ for some fixed c and almost all σ.

For any structure $\mathcal{M}$ with universe $M \subseteq \omega$, let $tal\,(\mathcal{M})$ be the tally representation of $\mathcal{M}$ with universe $tal\,(\omega)$ and relations and functions defined so that the mapping taking n to $tal\,(n)$ is an isomorphism from $\mathcal{M}$ onto $tal\,(\mathcal{M})$; $bin\,(\mathcal{M})$ and $b_k(\mathcal{M})$ are similarly defined. Lemma 3.3 is easily extended to tally and binary representations of relational structures $\mathcal{M}$ and one direction extends to structures with functions.

Lemma 3.4 *Let $k \geqslant 2$, let $\mathcal{M}$ be a structure with universe $M \subseteq \omega$ and let $\mathcal{A} = tal\,(\mathcal{M})$ and $\mathcal{B} = b_k(\mathcal{M})$. Then*

(a) *If $\mathcal{A}$ is p–time, then $\mathcal{B}$ is exponential time.*

(b) *If $\mathcal{B}$ is exponential time, then $\mathcal{A}$ is* EXPTIME.

(c) *If $\mathcal{B}$ is exponential time and, for all functions $f^{\mathcal{M}}$, $f^{\mathcal{M}}(n_1, \ldots, n_k) \leqslant 2^{c(n_1+\cdots+n_k)}$ for some fixed constant c and all but finitely many k–tuples, then $tal\,(\mathcal{M})$ is exponential time.*

(d) *If $\mathcal{B}$ is exponential time and, for all functions $f^{\mathcal{M}}$, $f^{\mathcal{M}}(n_1, \ldots, n_k) \leqslant (n_1 + \cdots + n_k)^c$ for some fixed constant c and all but finitely many k–tuples, then $\mathcal{A}$ is p–time.*

(e) *If $\mathcal{B}$ is polynomial time and, for all functions $f^{\mathcal{M}}$, $f^{\mathcal{M}}(n_1, \ldots, n_k) \leqslant c(n_1 + \cdots + n_k)$ for some fixed constant c and all but finitely many k–tuples, then $\mathcal{A}$ is linear time.*

Proof. We sketch the proofs for $k = 2$.

(a) Suppose that $\mathcal{A}$ is p–time. It follows from Lemma 3.3 that $\mathcal{B}$ has a exponential time universe and it is easy to see that the relations of $\mathcal{B}$ are also exponential time. For simplicity, suppose that $f^{\mathcal{M}}$ is a unary function. (The general proof can be found in [13, p. 320].) Suppose $tal\,(m) = f^{\mathcal{A}}(tal\,(n))$. By assumption, $tal\,(m)$ may be computed from $tal\,(n)$ in time $\leqslant n^c$ for some fixed c and all $n \geqslant 2$, so that $m \leqslant n^c$. To compute $f^{\mathcal{B}}(bin\,(n))$, we first compute $tal\,(n)$ from $bin\,(n)$, which takes exponential time by Lemma 3.2. Then we compute $tal\,(m) = f^{\mathcal{A}}(tal\,(n))$ in $\mathcal{A}$, which takes time $\leqslant n^c \leqslant 2^{c|bin\,(n)|}$. Finally, we must compute $bin\,(m)$ from $tal\,(m)$. This final computation takes time $\leqslant 8m \leqslant 8n^c \leqslant 8(2^{|bin(n)|})^c$. Thus $\mathcal{B}$ is exponential time.

For parts (b), (c) and (d), suppose that $\mathcal{B}$ is exponential time. Then we easily see that the universe of $\mathcal{A}$ and the relations of $\mathcal{A}$ are polynomial time. For simplicity, let $f^{\mathcal{M}}$ be a unary function. The procedure for computing

$f^{\mathcal{A}}(tal\,(n))$ has three parts as above. First, compute $bin\,(n)$ from $tal\,(n)$, which takes time $\leqslant 8n$ by Lemma 3.2. Next, compute $bin\,(m) = f^{\mathcal{B}}(bin\,(n))$ which takes time $\leqslant 2^{c|bin(n)|} \leqslant (2n)^c$. The final and most time-consuming part of the computation is to compute $tal\,(m)$ from $bin\,(m)$, which takes time $\leqslant 8m$. In general, we only know that $|bin\,(m)| \leqslant 2^{c|bin\,(n)|} \leqslant (2n)^c$ so that $m \leqslant 2^{(2n)^c}$ and hence we can only conclude that $\mathcal{A}$ is EXPTIME. For part (c), we have $m \leqslant 2^{cn}$ and hence $\mathcal{A}$ is exponential time. For part (d), we have $m \leqslant n^c$ so that $\mathcal{A}$ is p–time.

(e) It follows easily from the hypothesis and Lemma 3.2 that the universe of $\mathcal{A}$ and the relations of $\mathcal{A}$ are linear time. Now let $tal\,(m) = f^{\mathcal{A}}(tal\,(n)$ and observe that by the hypothesis, $m \leqslant nc$. The computation of $bin\,(n)$ from $tal\,(n)$ takes time $\leqslant 8n$ by Lemma 3.2. Since $\mathcal{B}$ is polynomial time, the computation of $bin\,(m) = f^{\mathcal{B}}(bin\,(n))$ takes time $\leqslant |bin\,(n)|^d$ for some fixed d. It follows as in the proof of Lemma 3.2(c) that this computation can almost always be done in time $\leqslant n$. Finally, the computation of $tal\,(m)$ from $bin\,(m)$ takes time $\leqslant 8m \leqslant 8nc$. Thus $\mathcal{A}$ is a linear time structure. □

Note that the hypothesis needed for part (d) follows from the assumption that, for some fixed constant c and all but finitely many k–tuples,

$$|f^{\mathcal{B}}(\sigma_1\,, \ldots\,, \sigma_k)| \leqslant c(|\sigma_1| + \ldots + |\sigma_k|).$$

Nerode and Remmel define in [56] the notion of a *p–time equivalence type* (PET) by saying that two subsets A and B of $Tal\,(\omega)$ are p–time equivalent if there is a partial one-to-one honest p–time function f with domain including A such that $f(A) = B$. It is natural to extend this notion to p–time subsets of $Bin(\omega)$ by defining two sets A and B to be *p–time isomorphic* if there is a one-to-one p–time function mapping A onto B whose inverse is also p–time. (We may assume that f and f^{-1} have domain $Bin\,(\omega)$ since A and B assumed to be p–time.)

It follows from Theorem 3 of [56] that for any p–time subset A of $Tal\,(\omega)$, there are infinitely many p–time subsets of $Tal\,(\omega)$ which are recursively isomorphic to A but not ptime isomorphic to A. We can now characterize those subsets of $\{1\}^*$ which are polynomial time isomorphic to $Tal\,(\omega)$, and put conditions on those subsets of $\{0,1\}^*$ which are polynomial time isomorphic to $Tal\,(\omega)$. The following was proved in [13].

Lemma 3.5

(a) *Let A be a p–time subset of $Bin(\omega)$ which is polynomial time isomorphic to $Tal(\omega)$, and let $a_0, a_1, \ldots$ list the elements of A in the standard ordering, first by length and then lexicographically. Then for some j, k and all $n \geqslant 3$, $n \leqslant |a_n|^j$ and $|a_n| \leqslant n^k$.*

(b) *Let A be a p–time subset of $Tal(\omega)$ and let $a_0, a_1, \ldots$ list the elements of A in the standard ordering, Then the following are equivalent.*

  (i) *A is p–time isomorphic to $Tal(\omega)$.*

  (ii) *For some k and all $n \geqslant 2$, $|a_n| \leqslant n^k$.*

  (iii) *The canonical map taking 1^n to a_n is p–time.*

Lemma 3.6 *For any infinite set M of natural numbers,*

$$tal(M) = \{tal(n) : n \in M\} \qquad \text{and} \qquad bin(M) = \{bin(n) : n \in M\}$$

are not p–time isomorphic.

Lemma 3.7 *Let $B_k(\omega)$ be the set of k–ary representations of natural numbers. Then*

(a) *The addition, subtraction, multiplication and division (with remainder) functions from $B_k(\omega) \otimes B_k(\omega)$ to $B_k(\omega)$, the order relation on $B_k(\omega)$, and the length function from $B_k(\omega)$ to $B_k(\omega)$ are all p–time. (As usual, $m - n$ is set to 0 if $m < n$.)*

(b) *$Bin(\omega) \setminus \{1\}^*$ is p–time isomorphic to $Bin(\omega)$.*

(c) *For each $k \geqslant 2$ and for A equal to either the set $B_k(\omega)$ or the set $\{0, 1, \ldots, k-1\}^*$, there is a polynomial time isomorphism φ from A to $Bin(\omega)$ and a constant c such that, for all but finitely many $a \in A$, $|a| \leqslant |\varphi(a)| \leqslant c|a|$.*

We will frequently want to combine structures using disjoint unions and direct sums. Nerode and Remmel proved in [56] that there exist p–time subsets A, B and C of $Tal(\omega)$ such that A is not p–time isomorphic to B but $A \oplus C$ is p–time isomorphic to $B \oplus C$, and similarly there exist subsets X, Y and Z of $Tal(\omega)$ such that X is not p–time isomorphic to Y but $X \otimes Z$ is p–time isomorphic to $Y \otimes Z$. This result of Nerode and Remmel also easily follows from our next results due to Cenzer and Remmel [13].

Lemma 3.8

(a) *Let A be a p–time subset of $Tal(\omega)$. Then $A \oplus Tal(\omega)$ is p–time isomorphic to $Tal(\omega)$ and $A \oplus Bin(\omega)$ is p–time isomorphic to $Bin(\omega)$.*

(b) *Let A be a nonempty p–time subset of $Tal(\omega)$. Then $A \otimes Tal(\omega)$ is p–time isomorphic to $Tal(\omega)$ and $A \otimes Bin(\omega)$ is p–time isomorphic to $Bin(\omega)$.*

Proof.

(a) First observe that $A \oplus Tal(\omega)$ is p–time isomorphic to the set

$$C = \{2a : a \in A\} \cup \{2n+1 : n \in Tal(\omega)\}$$

by the obvious isomorphism. Now let $c_0, c_1, \ldots$ enumerate C in increasing order. Since C contains every odd number, it is clear that $c_n \leqslant 2n+1$. It follows from Lemma 3.5 that C is p–time isomorphic to $Tal(\omega)$.

Next, $A \oplus Bin(\omega)$ is certainly p–time isomorphic to

$$(A \oplus Tal(\omega)) \oplus (Bin(\omega) \setminus Tal(\omega)).$$

Now by the preceding discussion, $A \oplus Tal(\omega)$ is p–time isomorphic to $Tal(\omega)$. It follows that $A \oplus Bin(\omega)$ is p–time isomorphic to

$$Tal(\omega) \oplus (Bin(\omega) \setminus Tal(\omega))$$

which is clearly p–time isomorphic to $Bin(\omega)$.

(b) If A has only one element, this is obvious. If A has at least two elements, let a be one of them. Then $A \times Tal(\omega)$ is p–time isomorphic to

$$(\{a\} \times Tal(\omega)) \oplus ((A \setminus \{a\}) \times Tal(\omega)).$$

Now the first part of this sum is obviously p–time isomorphic to $Tal(\omega)$ and the second part is p–time isomorphic to some p–time subset of $Tal(\omega)$. It now follows from part (a) that the sum is p–time isomorphic to $Tal(\omega)$.

For the case of $Bin(\omega)$, again if A consists of single element, the result is trivial. Otherwise, let a be the least element of A and consider the following mapping. If $b \in A$ and $b \neq a$, then let $f(\langle b, \sigma \rangle) = \sigma^\frown 1\, 0^b\, 1$. The idea is to define f so that the image of $\{a\} \times Bin(\omega)$ under f is

$$Bin(\omega) \setminus f((A \setminus \{a\}) \times Bin(\omega)) = S$$

by letting $f(\langle a, bin(n)\rangle)$ be the n–th element of S. Observe that for any $\tau = bin\,(q) \in Bin\,(\omega)$, which is not of the form $\sigma^\frown 1\,0^b\,1$ for some $b \in A \smallsetminus \{a\}$, we can find in polynomial time in $|\tau|$ all strings of the form $1^{k_b\frown} 1\,0^b\,1$ such that

(i) $b > 0$,

(ii) $1^b \in A$, and

(iii) under the usual ordering on $Bin\,(\omega)$, $1^{k_b\frown} 1\,0^b\,1 < \tau$, but $1^{k_b+1\frown} 1\,0^b\,1 \nleqslant \tau$.

That is, since A is a polynomial time subset of $Tal\,(\omega)$, we can test each element of the 1^b with $0 < b \leqslant |\tau|$ for membership in A in b^k steps for some fixed k and hence find all strings of the form 1^b where $q \geqslant b > 0$ and $1^b \in A$ in at most $\sum_{j=1}^{|\tau|} j^k \leqslant (|\tau|)^{k+1}$ steps. Thus in polynomial time in $|\tau|$, we can find all the strings

$$1^{k_{b_1}\frown} 1\,0^{b_1}\,1\,, \ldots\,, 1^{k_{b_p}\frown} 1\,0^{b_p}\,1$$

satisfying properties (1)–(3) above. This means that exactly $\sum_{i=1}^{p} 2^{k_i}$ elements of $Bin\,(\omega)$ which are less than or equal to τ are in the image of

$$f((A \smallsetminus \{a\}) \times Bin\,(\omega)).$$

Thus we let

$$f(\langle a, bin\,(q - \sum_{i=1}^{p} 2^{k_i}\rangle) \;=\; \tau.$$

It follows that f^{-1} is polynomial time. To see that f is polynomial time, note that

$$f(\langle a, bin\,(n)\rangle) \;\leqslant\; bin\,(n)^\frown 1\,1$$

so a similar computation starting with

$$\tau \;=\; bin\,(n)^\frown 1\,1$$

will allow us to find the n–th element of S in polynomial time in $|bin\,(n)|$. □

It follows easily from the lemmas above that any p–time relational structure $\mathcal{A}$ is recursively isomorphic to a p–time structure $\mathcal{B}$ such that $\mathcal{A}$ and $\mathcal{B}$ are not p–time isomorphic. (We assume that $\mathcal{A}$ has universe $Bin\,(A) = \{bin\,(a) : a \in A\}$ for some $A \subseteq \omega$ and let $\mathcal{B}$ have universe $Tal\,(A) = \{tal\,(a) : a \in A\}$.) Now these structures are actually exponential-time isomorphic. However, there is a stronger result.

Lemma 3.9 *For any p–time set*

$$A = \{bin\,(a_0) < bin\,(a_1) < \cdots\},$$

there is a set

$$M = M(A) = \{bin\,(m_0) < bin\,(m_1) < \cdots\}$$

such that M is p–time and the map which takes $bin\,(m_i)$ to $bin\,(a_i)$ is p–time, but there is no primitive recursive map from A into M which maps at most k elements of A to any element of M' where k is any fixed finite number. Furthermore, M may be taken to be a subset of $Tal\,(\omega)$.

Proof. Let φ_e be the e–th primitive recursive function mapping $Bin\,(\omega)$ into $Bin\,(\omega)$ and, for each e, let t_e be the total time required to test all numbers up to a_e for membership in A and to compute $\varphi_i(bin(a_j))$ for all $j, i \leqslant a_e$; clearly $t_e < t_{e+1}$. For each e, let $m_e = 2^{t_e}$, so that

$$bin\,(m_e) = 0^{t_e}1$$

and $|bin\,(m_e)| = t_e + 1$. It follows that

$$\varphi_e(bin(a_i)) < bin\,(m_i)$$

for all $i \leqslant e$, since by convention it takes at least k steps to compute an output of length k. Let $A^* = \{bin\,(m_e) : e < \omega\}$.

Here is the p–time algorithm for testing whether $x \in A^*$. First check to see that $x = 0^t1$ for some n. Then start to test $bin\,(0)\,,\, bin\,(1)\,, \ldots$ for membership in A. As soon as we find that $bin\,(n)$ is the e–th member of A so that $bin\,(n) = bin\,(a_e)$, then compute in order $\varphi_e(a_0)\,, \cdots, \varphi_e(a_{e-1})$ and $\varphi_0(a_e)\,, \cdots, \varphi_e(a_e)$ and then return to testing whether $bin\,(n+1)$, $bin\,(n+2)\,, \cdots$ are in A. If the total number of steps reaches t exactly when the computation of some $\varphi_e(a_e)$ has just been completed, then $t = t_e$ so that $x = bin\,(m_e)$ belongs to A^*. Otherwise, $x \notin A^*$. This argument also shows that the map which takes $bin\,(m_e)$ to $bin\,(a_e)$ is p–time. Finally, let

$$M = M(A) = \{bin\,(m_{i^2}) : i < \omega\}.$$

Then M is also p–time since we can clearly check in polynomial time whether i is a square if we discover that $x = bin\,(m_e)$ at the end of the computation.

The map taking $bin\,(m_{i^2})$ to $bin\,(a_i)$ is p–time since we can test all numbers $\leqslant a_{i^2}$ for membership in A in time $\leqslant |bin\,(m_{i^2})|$ and therefore determine a_i from $bin\,(m_{i^2})$.

Suppose now by way of contradiction that φ_e were a map from A into M' which were at most k to 1. Since each primitive recursive function has infinitely many indices, we may assume that $e \geqslant k$. Then

$$\{\varphi_e(bin(a_0)), \ldots, \varphi_e(bin(a_{e^2}))\}$$

must contain at least e distinct elements, so that at least one of them is $\geqslant bin\,(m_{e^2})$, which contradicts the observation above that

$$\varphi_e(bin\,(a_i)) \;<\; bin\,(m_e)$$

for all $i \leqslant e$, and thus establishes the result.

Since $Bin\,(\omega)$ and $Tal\,(\omega)$ are primitive recursively isomorphic via the standard map μ_2, it follows that we could replace M in this argument with the set

$$M^* \;=\; \{tal\,(m_0),\, tal\,(m_1),\, \ldots\}. \qquad \square$$

Remarks 3.1 The only properties of the set of primitive recursive functions needed for this result is that the primitive recursive functions is a class of total recursive functions which can be effectively listed. Thus in the statement of Theorem 3.9 we can replace the primitive recursive functions by any class of total recursive functions which can be effectively listed.

Letting $k = 1$, we see that there is no primitive recursive embedding of A into $M(A)$. Furthermore, there is no primitive recursive embedding φ of a cofinite subset C of A into $M(A)$, since if $|A \setminus C| = k - 1$, then we could extend φ to a map from A into $M(A)$ which is at most k to 1 by mapping all elements of $A \setminus C$ to some fixed element of $M(A)$.

4 The Existence of Feasible Structures

In this section, we shall survey a number of model existence results for polynomial time structures. In particular, we will consider four existence questions for any class **C** of structures.

(1) Is every recursive structure in **C** isomorphic to a polynomial time structure?

(2) Is every recursive structure in **C** recursively isomorphic to a polynomial time structure?

(3) Is every recursive structure in **C** isomorphic to a polynomial time structure with a specified universe such as the binary or tally representation of the natural numbers?

(4) Is every recursive structure in **C** recursively isomorphic to a polynomial time structure with a specified universe such as the binary or tally representation of the natural numbers?

The fundamental result for relational structures is due to Grigorieff [32] and (for structures with infinitely many relations) to Cenzer-Remmel [10]. Recall that a structure with no function is said to be relational.

Theorem 4.1 *Every relational structure is recursively isomorphic to a real time structure with universe a subset of* $Bin(\omega)$ *and to a linear time structure with universe a subset of* $Tal(\omega)$.

Sketch Proof. We may assume without loss of generality that the structure $\mathcal{A}$ has universe ω. The element a is represented in the binary real time model by $\psi(a) = 1^{a+1}\,0\,1\,0^t$, where t is the time required to compute whether $R_j(x_1, \ldots, x_{t(j)})$ for all $j \leqslant a$ and all tuples $(x_1, \ldots, x_{t(j)})$ from $\{0, \ldots, a\}^{t(j)}$.

The set $B = \{\psi(a) : a \in \omega\}$ is real time by the following algorithm.

Given $bin(n)$, start to read $bin(n)$ from left to right.

If at any time we discover that $bin(n)$ is not of the form $1^{a+1}\,0\,1\,0^t$ for some a and t, then $bin(n) \notin B$.

Otherwise, having read the first 0 in $bin(n)$ so that we have found a such that $bin(n) = 1^{a+1}\,0\,1\,0^t$, start the computation which tests for all $j \leqslant a$ and for all tuples $(x_1, \ldots, x_{t(j)})$ from $\{0, \ldots, a\}^{t(j)}$, whether $R_j(x_1, \ldots, x_{t(j)})$.

If the total computation finishes in exactly t steps, then $bin(n) \in B$. If the computation either finishes in fewer steps, or has not finished by t steps, then $bin(n) \notin B$.

Each relation R_j is real time, by the following algorithm.

> Given $(bin(n_0), \ldots, bin(n_{t(j)})) \in B^{t(j)}$, first compute a_k and t_k for each $k \leqslant j$ so that $bin(n_k) = 1^{a_k+1}, 0, 1\,0^{t_k}$.
>
> Then let $t = \max\{t_0, \ldots, t_{t(j)}\}$.
>
> Now by the construction we can test whether $R_j(a_0, \ldots, a_{t(j)})$ in time $\leqslant t$.

The tally representation of $\mathcal{A}$ has universe $\{tal(n) : bin(n) \in B\}$ and is linear time by Lemma 3.4 (e). □

Note that Theorem 4.1 allows us to conclude that if G is a recursive graph, i.e., $G = (V, E)$ where V, the vertex set of G, is a recursive set of natural numbers and the edge relation E is also recursive, then G is recursively isomorphic to a polynomial time graph G'. However if G has a recursive k–coloring, then to conclude that G is recursively isomorphic to a polynomial time graph with a polynomial time k–coloring requires a stronger version of Theorem 4.1. We will present an improved version of Theorem 4.1 due to Cenzer and Remmel [16] which is their primary tool in their analysis of polynomial time combinatorial structures and Π^0_1–classes. The improved version of the theorem presented applies to structures with two distinct types of objects, the first type being the normal universe of the structure, and with functions which map the first type into the second type. The type of example that we have in mind is a function from the vertices of a graph into the natural numbers which computes the degree of a vertex or the color assigned to a vertex. The universe of the graph is now expanded by adding a p–time set which represents the natural numbers and the degree function or coloring now becomes part of the structure. Naturally, the new objects are not vertices and therefore are not joined to any other objects by edges.

Theorem 4.2 *Let*

$$\mathcal{C} = (C, A, B, \{R_i^{\mathcal{C}}\}_{i \in S}, \{f_i^{\mathcal{C}}\}_{i \in T}),$$

be a recursive structure such that

(i) *A and B are disjoint subsets of C with $C = A \cup B$ and B is a polynomial time set;*

(ii) *there is a recursive isomorphism from* $Bin(\omega)$ *onto a subset of* $Bin(\omega) \setminus B$ *with a p–time inverse;*

(iii) *for each* $i \in T$, f_i *maps* C *into* B;

(iv) *for each* $i \in S$, *the relation* R_i *is independent of* B, *that is, for any* $(x_1, \ldots, x_n) \in C^n$, *where* $n = s(i)$, *any* $j \leqslant n$ *such that* $x_i \in B$, *and any* $b \in B$, $R_i^{\mathcal{C}}(x_1, \ldots, x_n)$ *holds if and only if* $R_i^{\mathcal{C}}(x_1, \ldots, x_{j-1}, b, x_{j+1}, \ldots, x_n)$ *holds;*

(v) *for each* $i \in T$, *the function* f_i *is independent of* B, *that is, for any* $(x_1, \ldots, x_n) \in C^n$, *where* $n = t(i)$, *any* $j \leqslant n$ *such that* $x_i \in B$, *and any* $b \in B$,

$$f_i^{\mathcal{C}}(x_1, \ldots, x_n) = f_i^{\mathcal{C}}(x_1, \ldots, x_{j-1}, b, x_{j+1}, \ldots, x_n).$$

Then there is a recursive isomorphism φ *of* $\mathcal{C}$ *onto a p–time structure* $\mathcal{M}$ *such that* $\varphi(b) = b$ *for all* $b \in B$.

4.1 Structures with Functions

In contrast to purely relational structures, recursive structures with functions need not be effectively isomorphic to feasible structures. The following results are Theorems 3.1, 3.2 and 3.3 of [10].

Theorem 4.3 *Let* $\mathcal{L}_0$ *be the language with exactly one function symbol* f *which is unary.*

(a) *There is a recursive structure* $\mathcal{A} = (A, f^{\mathcal{A}})$ *which is not recursively isomorphic to any primitive recursive structure.*

(b) *There is an exponential time structure* $\mathcal{D} = (D, f^{\mathcal{D}})$ *which is not recursively isomorphic to any polynomial time structure.*

Proof.

(a) Let (A_0, f_0), (A_1, f_1), ... be an effective list of all primitive recursive structures over $\mathcal{L}_0$ and let $\varphi_0, \varphi_1, \ldots$ be a list of all one-to-one partial recursive functions. We must meet the following set of requirements in our construction of $\mathcal{A}$.

$R_{i,j}$: φ_j is not a recursive isomorphism from $\mathcal{A}$ to (A_i, f_i).

To meet the requirements $R_{i,j}$, recursively partition $\{0,1\}^*$ into infinitely many disjoint infinite recursive sets $S_{i,j}$. We then define $\mathcal{A}$ so that $A = \bigcup_{i,j} S_{i,j} = \{0,1\}^*$ and for all i, j, $f^{\mathcal{A}}$ maps $S_{i,j}$ into $S_{i,j}$.

We now fix i, j and then we define $f = f^{\mathcal{A}}$ on $S_{i,j}$ in stages. We let $a_0, a_1, \ldots$ be some effective listing of $S_{i,j}$. At stage s, we shall define $f(a_s)$. We start by defining $f(a_0) = a_1$ at stage 0. At stage $s+1$, compute $\varphi_j^s(a_0)$. If $\varphi_j^s(a_0) \uparrow$ or if $\varphi_j^{s-1}(a_0) \downarrow$, then we define $f(a_s) = a_{s+1}$. Otherwise, that is, if $\varphi_j^s(a_0) \downarrow$ but $\varphi_j^{s-1}(a_0) \uparrow$, let $x = \varphi_j^s(a_0)$ and do the following. Compute the sequence $x, f_i(x), f_i(f_i(x)), \ldots, f_i^{(s+1)}(x)$, where here $f_i^{(k)}$ denotes f_i composed with itself k times. Note that if φ_j were an isomorphism, then it must be the case that $\varphi_j(a_k) = f_i^{(k)}(x)$ for all k. Thus if φ_j were an isomorphism, then it must be the case that

$$f(a_s) = a_0 \iff f_i^{(s+1)}(x) = x.$$

Thus if $f_i^{(s+1)}(x) = x$, then we define $f(a_s) = a_{s+1}$. If $f_i^{(s+1)}(x) \neq x$, then we define $f(a_s) = a_0$. Note that in either case, we will have ensured that φ_j cannot be an isomorphism from $\mathcal{A}$ onto (A_i, f_i).

(b) The proof of part (a) must be modified in several ways. First, let $(E_0, f_0), (E_1, f_1), \ldots$ be an effective list of all p–time structures whose universe is contained in $\{0,1\}^8$ over $\mathcal{L}_0$. Let $\varphi_0, \varphi_1, \ldots$ be a list of all one-to-one partial recursive functions which map $\{0,1\}^*$ into $\{0,1\}^*$. We shall build our structure (D, f^D) so that $D \subseteq \mathit{Tal}(\omega)$ and we meet the following set of requirements.

$$R_{i,j}: \quad \varphi_j \text{ is not a recursive isomorphism from } \mathcal{D} \text{ to } (E_i, f_i).$$

To meet the requirements $R_{i,j}$, we construct D as a disjoint polynomial-time union of infinite p-time sets $D = \bigcup_{i,j} T_{i,j}$. Define the function $\psi : \mathit{Tal}(\omega) \times \mathit{Tal}(\omega) \times \mathit{Tal}(\omega) \to \mathit{Tal}(\omega)$ by

$$\psi(0, \mathit{tal}(i), \mathit{tal}(j)) = \mathit{tal}(2[i,j]) + 3), \text{ and}$$

$$\psi(\mathit{tal}(n+1), \mathit{tal}(i), \mathit{tal}(j)) = \mathit{tal}(2^p), \text{ if } \psi(\mathit{tal}(n), \mathit{tal}(i), \mathit{tal}(j)) = \mathit{tal}(p).$$

Note that $\mathit{tal}(x) = \mathit{tal}(2[i,j]+3)$ can be computed from input $(\mathit{tal}(i), \mathit{tal}(j))$ in time $a \cdot x$ for some fixed constant a and that the computation of $\mathit{tal}(y) = \mathit{tal}(2^x)$ from input $\mathit{tal}(x)$ can be computed in time at most $b \cdot y$ for some fixed constant $b \geqslant a$. Thus the computation of $tal(z) = \psi(tal(n), tal(i), tal(j))$ from input $(tal(n), tal(i), tal(j))$ takes at most the following number of steps:

$$b[i,j] + bx + b2^x + b2^{2^x} + \cdots + bz < b(1 + 2 + \cdots + z) < bz^2.$$

For each i, j, we let $T_{i,j} = \{\psi(n,i,j) : n < \omega\}$. Then we can test whether $tal(z) = \psi(tal(n), tal(i), tal(j))$, perform the computation of

$$\psi(0, tal(i), tal(j)) \ , \ \psi(1, tal(i), tal(j)) \ , \ \ldots \ , \ \psi(tal(n), tal(i), tal(j))$$

for bz^2 steps and see if the computation converges to z by that time. It follows that the sets $T_{i,j}$ are uniformly p–time and that D is also p–time, since

$$tal(z) \in T_{i,j} \iff (\exists n < z)\, tal(z) = \psi(tal(n), tal(i), tal(j))$$

and

$$tal(z) \in D \iff (\exists i, j < z)\, tal(z) \in T_{i,j}.$$

We now fix i, j and define $f = f^{\mathcal{D}}$ on $T_{i,j} = \{tal(a_0)\, , \, tal(a_1)\, , \ldots\}$, where $tal(a_n) = \psi(tal(n), tal(i), tal(j))$. For each m, perform the following series of computations for at most 2^{am} steps.

(i) Start to compute $\varphi_j(tal(a_0))$. If this converges in less than 2^{am} steps, let $b_0 = \varphi_j(tal(a_0))$.

(ii) Check that $b_0 \in E_i$.

(iii) Compute the sequence $b_1 = f_i(b_0)\, , \, b_2 = f_i(b_1)\, , \ldots \, , \, b_{m+1} = f_i(b_m)$.

Let s be the least m such that the computations can be successfully completed in at most 2^{am} steps. Assuming the existence of $b_0 = \varphi_j(a_0)$, we can show that such an s must exist. That is, it takes some constant amount c_0 of time to compute b_0. Since f_i is p–time, $f_i(y)$ can be computed in time bounded by $|y|^k$ for some fixed integer $k > 1$ and any $y \in \{0,1\}^*$ with $|y| > 1$. Let c_1 be the time required to compute $f_i(\emptyset)$, $f_i(0)$ and $f_i(1)$, if needed, and let $c = c_0 + c_1$. Then to compute the sequence $b_0 = \varphi_j(a_0)$, $b_1 = f_i(b_0)\, , \ldots \, , \, b_{m+1} = f_i(b_m)$ takes time at most

$$t(m) = c + |b_0|^k + (|b_0|^k)^k + \cdots \quad = c + |b_0|^k + |b_0|^{k^2} + \cdots + |b_0|^{k^m}.$$

We need to show that this sequence is eventually dominated by the sequence $a_0, a_1 = 2^{a_0}, \ldots, a_m = 2^{a_{m-1}}$. We may assume without loss of generality that $|b_0| > 1$. Now if m is large enough so that both c and m are $< |b_0|^{k^m}$, then

$$t(m) < (m+1)|b_0|^{k^m} \leqslant |b_0|^{2k^m}.$$

Now let m be large enough so that $|b_0|^2 < 2^{2^m}$, $k < 2^m$, and $m^2 + m < 2^m$. Then $k^m < 2^{m^2}$ and $t(m) < 2^{2^m \cdot 2^{m^2}} = 2^{2^{m^2+m}} < 2^{2^{2^m}} = exp_3(m)$. To show that the latter is dominated by a_m, note first that $a_0 \geqslant 3$ and that, for any m, $a_m \geqslant m+3$; it follows that $a_{m+3} \geqslant \exp_3(m+3)$.

The definition of f now proceeds in stages, as in part (a). Let s be the least m such that the computations described above can be successfully completed. Now for $t \neq s$, we let $f(a_t) = a_{t+1}$. To compute $f(a_s)$, we let $x = \varphi_j(a_0)$ and compute $f_i^{(s+1)}(x)$. Then if $f_i^{(s+1)}(x) = x$, we define $f(a_s) = a_{s+1}$. If $f_i^{(s+1)}(x) \neq x$, then we define $f(a_s) = a_0$. Note that in either case, we will have ensured that φ_j cannot be an isomorphism from $\mathcal{D}$ onto (E_i, f_i).

It remains to be seen that the computation of f can be done in exponential time. Given $tal\,(x) \in D$, we first compute the unique triple (n, i, j) such that $tal\,(x) = \psi(tal\,(n), tal\,(i), tal\,(j))$. This can be done in polynomial time since n, i and j are all less than x. Next we perform the computations (i), (ii) and (iii) for $m = 0, 1, \ldots, n$ in turn. This can be done in exponential time since each series of computations is bounded by time 2^{a_m}. The remainder of the computation of $f(x)$ takes little time. We look for the least n, if any, such that the n-th series of computations has been successfully completed. If $m = n$, then we check to see if $f_i^{(s+1)}(x) = x$ and let $f(tal(a_m)) = tal(2^{a_m})$ if so; otherwise, $f(tal(a_m)) = tal(a_0)$. □

We note here that the functions in the previous theorem may be taken to be permutations of the sets A and D. Then next two results show that we can diagonalize over Δ_2^0 isomorphisms if the underlying language has at least two unary function symbol or at least one n-ary function symbol with $n \geqslant 2$.

Theorem 4.4 *Let $\mathcal{L}_0$ be the language with exactly two function symbols f and g which are unary.*

(a) *There is a recursive structure $\mathcal{A} = (A, f^{\mathcal{A}}, g^{\mathcal{A}})$ which is not Δ_2^0 isomorphic to any primitive recursive structure.*

(b) *There is an exponential time structure* $\mathcal{D} = (D, f^{\mathcal{D}}, g^{\mathcal{D}})$ *which is not* Δ_2^0 *isomorphic to any polynomial time structure.*

Theorem 4.5 *Let* $\mathcal{L}_0$ *be the language with exactly one function symbol* h *which is binary.*

(a) *There is a recursive structure* $\mathcal{A} = (A, h^{\mathcal{A}})$ *which is not* Δ_2^0 *isomorphic to any primitive recursive structure.*

(b) *There is an exponential time structure* $\mathcal{D} = (D, h^{\mathcal{D}})$ *which is not* Δ_2^0 *isomorphic to any polynomial time structure.*

Two natural types of structures with functions will be considered below in more detail. These are permutation structures (A, f^A), where f^A is a permutation of the set A, and Abelian groups.

Next we state an unpublished theorem due to H. Freidman and J. Remmel which characterizes when structures which are finitely generated are isomorphic to polynomial time structures.

Definition 4.6

(a) We say that $\mathcal{A} = (A, \{R_i^A\}_{i=1,\dots,k}, \{f_i^A\}_{i=1,\dots,n}, \{c_i^A\}_{i=1,\dots,m})$ is *finitely generated* if A equals the closure of $\{c_i^A\}_{i=1,\dots m}$ under the set of functions $\{f_i^A\}_{i=1,\dots,n}$.

(b) A finitely generated structure $\mathcal{A}$ as above has a *double exponential time decision procedure* if

(i) $A \subseteq \{0,1\}^*$ is double-exponential-time.

(ii) There is an algorithm which given any two terms $t_1(c_1^A, \dots, c_m^A)$ and $t_2(c_1^A, \dots, c_m^A)$ in the free term algebra generated by $c_1^A, \dots, c_m^A$ and $f_1^A, \dots, f_n^A$, decides in $2^{2^{dk_1}}$ steps for some constant k_1 if $t_1(c_1^A, \dots, c_m^A) = t_2(c_1^A, \dots, c_m^A)$ where d is equal to the maximum of the depth of $t_1(c_1^A, \dots, c_m^A)$ and the depth of $t_2(c_1^A, \dots, c_m^A)$.

(iii) There is an algorithm which given a relation R_i^A and terms $t_1(c_1^A, \dots, c_m^A), \dots, t_p(c_1^A, \dots, c_m^A)$ in the free term algebra generated by $c_1^A, \dots, c_m^A$ and $f_1^A, \dots, f_n^A$, decides in $2^{2^{dk_2}}$ steps for some constant k_2 if $R_i^A(t_1(c_1^A, \dots, c_m^A), \dots, t_p(c_1^A, \dots, c_m^A))$ holds where d is equal to the maximum of the depths of $t_j(c_1^A, \dots, c_m^A)$ for $j = 1, \dots, p$.

Theorem 4.7 *Let* $\mathcal{A} = (A, \{R_i^A\}_{i=1,\ldots,k}, \{f_i^A\}_{i=1,\ldots,n}, \{c_i^A\}_{i=1,\ldots,m})$ *be a finitely generated structure. Then* $\mathcal{A}$ *is isomorphic to a polynomial time model iff* $\mathcal{A}$ *has a double exponential time decision procedure.*

4.2 Linear Orderings

There are several results on the existence of p–time linear orderings.

Theorem 4.8 (Grigorieff 1989) *Every recursive linear ordering* $\mathcal{L}$ *is isomorphic to a real time linear ordering* $\mathcal{L}' = (Bin(\omega), <_{\mathcal{L}'})$.

Sketch Proof. We sketch a proof showing that $\mathcal{L}$ is isomorphic to a p–time ordering. There are two cases.

The first case is where $\mathcal{L}$ has either a recursive increasing sequence

$$S = (s_0 <_{\mathcal{L}} s_1 <_{\mathcal{L}} s_2 <_{\mathcal{L}} \cdots)$$

such that S is cofinal or S has a limit or $\mathcal{L}$ has a recursive decreasing sequence

$$D = (d_0 >_{\mathcal{L}} d_1 >_{\mathcal{L}} d_2 >_{\mathcal{L}} \cdots)$$

such that D is cofinal or D has a limit.

In this case, we have a p–time copy of either S or D with universe $Bin(\omega)$ and we can make a p–time copy of $\mathcal{L} \setminus S$ (or $\mathcal{L} \setminus D$) with universe a subset of $Tal(\omega)$ by Theorem 4.1. Then we can apply Lemma 3.8 to combine the two orderings into one p–time ordering with universe $Bin(\omega)$ which is recursively isomorphic to $\mathcal{L}$.

The second case is where no such sequences exist. In this case $\mathcal{L}$ is isomorphic to $\omega + \mathbb{Z} \cdot \lambda + \omega^*$ for some recursive ordinal λ. Here $\mathbb{Z}$ is the order type of the integers. Then there are two subcases. First if λ has a first or last element or has a pair of elements $x <_L y$ such y is an immediate successor of x, then $\mathcal{L}$ contains an explicit recursive copy of $D = \omega + \omega^*$ which is isomorphic to a p–time linear ordering with universe $Bin(\omega)$. We can make a p–time copy of $\mathcal{L} \setminus D$ with universe a subset of $Tal(\omega)$ by Theorem 4.1. Then we can apply Lemma 3.8 to combine the two orderings into one p–time ordering with universe $Bin(\omega)$. The only other subcase is that λ is a dense linear ordering without endpoints so that λ is isomorphic to the rationals $\mathbb{Q}$. But in this case it is easy to construct a p–time linear ordering with universe $Bin(\omega)$ which is isomorphic to $\mathcal{L}$. □

The natural question is whether the isomorphism in the previous theorem is effective. It should be observed that the proof is not uniform. Indeed in case 2, we only constructed an isomorphic copy and not a recursively isomorphic copy of $\mathcal{L}$. The next result due to Remmel is Theorem 2.2 of [68] which shows that in case 2 the isomorphism cannot be replaced by a recursive isomorphism.

Theorem 4.9 *Let $A \subseteq \{0,1\}^*$ be any infinite p–time set and let $\mathcal{L}$ be a recursive linear ordering which is isomorphic to $\omega + \mathbb{Z} \cdot \lambda + \omega^*$ for some linear ordering λ. Then there exists a recursive linear ordering $\mathcal{K}$ which is isomorphic to $\mathcal{L}$ but which is not recursively isomorphic to any p–time linear ordering whose universe is A.*

Sketch Proof. We will not give the full proof as it requires an infinite injury priority argument. However we will give the proof in the case where the $\mathcal{L}$ is isomorphic to $\omega + \omega^*$ since in that case, a simple finite injury priority argument suffices.

Recall that φ_i is the i–th partial recursive function and let $R_0, R_1, \ldots$ be an effective list of all polynomial time binary relations on $\{0,1\}^*$. For simplicity, we let $\langle A, R_e \rangle$ denote the structure with universe A and relation R which is the restriction of R_e to $A \times A$.

Let $\tau_0, \tau_1, \ldots$ be an effective enumeration of A in the usual order (first by length and then by lexicographic order.)

We shall construct our desired recursive linear ordering $\mathcal{L}$ in stages. Let $\sigma_0, \sigma_1, \ldots$ be an effective listing of $\{0,1\}^*$. At any given stage s, we shall specify two sequences $a_0^s, a_1^s, \ldots, a_{n_s}^s$ and $b_0^s, b_1^s, \ldots, b_{n_s}^s$ for some $n_s \geqslant s$ such that $B_s = \{\sigma_0, \ldots, \sigma_{2n_s+1}\} = \{a_0^s, b_0^s, a_1^s, b_1^s, \ldots, a_{n_s}^s, b_{n_s}^s\}$. Moreover, at stage s we shall define the ordering $< = <_{\mathcal{L}}$ on $B_s \times B_s$ so that

$$a_0^s < a_1^s < \cdots < a_{n_s}^s < b_{n_s}^s < b_{n_s-1}^s < \cdots < b_0^s.$$

Our construction will ensure that for all i, $\lim_s a_i^s = a_i$ and $\lim_s b_i^s = b_i$ exist. Moreover, our construction will ensure that $\{0,1\}^* = \{a_0, b_0, a_1, b_1, \ldots\}$ and that

(a) for all i, $a_i < a_{i+1}$ and $b_{i+1} < b_i$, and

(b) for all i and j, $a_i < b_j$.

Thus $(\{0,1\}^*, <_{\mathcal{L}})$ will have order type $\omega + \omega^*$. To ensure that $\mathcal{L}$ is not recursively isomorphic to (A, R_e) for any e, we shall meet the following set of requirements.

$P_{[e,k]}$: There exist n and m such that one of the following four conditions holds.

(i) $\varphi_k(a_n) \uparrow$ or $\varphi_k(a_n) = x \notin A$.

(ii) $\varphi_k(b_m) \uparrow$ or $\varphi_k(b_m) = x \notin A$.

(iii) $\varphi_k(a_n) = x \in A$ and there exist $n+1$ elements $v_0, \ldots, v_n$ of A such that $(v_i, x) \in R_e$ for $i = 0, \ldots, n$.

(iv) $\varphi_k(b_m) = y \in A$ and there exist $m+2$ elements $w_0, \ldots, w_{m+1}$ of A such that $(w_i, y) \notin R_e$ for $i = 0, \ldots, m+1$.

We write $(w_i, y) \notin R_e$ rather than $(y, w_i) \in R_e$ in clause (iv) to allow for the possibility that R_e is not actually a linear ordering.

It is easy to see that if requirement $P_{[e,k]}$ is satisfied, then φ_e is not a recursive isomorphism from $\mathcal{L} = (\{0,1\}^*, <_{\mathcal{L}})$ onto (A, R_e). Thus meeting all the requirements $P_{[e,k]}$ ensures that $\mathcal{L}$ is not recursively isomorphic to any p–time linear ordering with universe A.

Our basic strategy for meeting a requirement P_z, where $z = [e,k]$, is as follows. Let us assume that $s > z$ is a stage large enough so that requirements $P_0, \ldots, P_{z-1}$ no longer require action at any stage $t \geqslant s$. Then at stage s, we consider a_z^s. Our construction will then ensure that $a_j^s = a_j^t$ for all $j \leqslant z$ and $t \geqslant s$ unless there is a stage $u \geqslant s$ such that $\varphi_k^u(a_z^s) \downarrow$. Of course if there is no such u, then $a_z^s = a_z$ and a_z will witness that requirement P_z is satisfied (by virtue of clause (i)).

Now if there is such a stage u, then let $x = \varphi_k^u(a_z^s)$. If $x \notin A$, then again we will simply ensure $a_z^s = a_z$ so that once again a_z will witness that requirement P_z is satisfied. If $x \in A$, then we will compare x to the first $4n_{u-1}+4$ elements of A (in the fixed order $\tau_0, \tau_1, \ldots$ prescribed above) with respect to the binary relation R_e. Note that since A and R_e are polynomial time, we can effectively make these $4n_{u-1}+4$ comparisons. There are two possibilities.

(1) There are $h = 2n_{u-1}+2$ of these elements v of A such that $(v, x) \in R_e$. Denote these elements by $v_0, \ldots, v_{h-1}$.

(2) There are $h = 2n_{u-1}+3$ of these elements w of A such that $(w, x) \notin R_e$. Denote these elements by $w_0, \ldots, w_{h-1}$.

In case (1), we will simply ensure $a_z^s = a_z$. But then a_z is preceded by exactly z elements in $\mathcal{L}$, where $z \leqslant n_{u-1}$, whereas $x = \varphi_k(a_z)$ is preceded by at least $2n_{u-1} + 2$ elements in $\langle A, R_e \rangle$. Thus φ_k is not an isomorphism from $\mathcal{L}$ onto (A, R_e).

In case (2), we will switch $a_z^s = a_z^{u-1}$ from the ω side of $\mathcal{L}$ to the ω^* side of $\mathcal{L}$. That is, we shall let $n_u = 2n_{u-1} - z + 1$ and let $b^u_{n_{u-1}+i} = a_{n_{u-1}-i+1}$ for $i = 1, \ldots, n_u - n_{u-1}$. We also let $b_i^u = b_i^{u-1}$ for $i \leqslant n_{u-1}$. Then our construction will ensure that for all $t \geqslant u$, $b^t_{n_u} = b^u_{n_u} = a_z^s$. Thus in this case, there will be precisely $n_u + 1$ elements w (namely $b_0, \ldots, b_{n_u}$) such that $(w, b_{n_u}) \notin <_{\mathcal{L}}$. However, in (A, R_e) there are at least $2n_{u-1} + 3$ elements x such that $(x, \varphi_k(b_{n_u})) \notin R_e$. But $n_u + 1 = 2n_{u-1} + 2 - z < 2n_{u-1} + 3$, so that φ_u cannot be an isomorphism from $\mathcal{L}$ onto (A, R_e). Our construction will ensure that a_z^s can switch from the ω side to the ω^* side of $\mathcal{L}$ only for the sake of requirements $P_0, \ldots, P_z$. The usual priority argument will then show that a_z^s "switches sides" for at most finitely many s.

We shall employ a set of movable markers Γ_e to help us keep track of which requirements we have acted on. The idea is that if we have taken an action as described above which ensures a_z^u will witness that requirement P_z is satisfied, then we will place a Γ_z marker on a_z^u. Thus at any given stage s, either Γ_z is *inactive*, i.e., Γ_z does not rest on any element at stage s, or Γ_z is *active*, i.e., Γ_z rests on some element $x \in \{a_0^s, b_0^s, \ldots, a_{n_s}^s, b_{n_s}^s\}$. If Γ_z is active, we let $\Gamma_z(s) = x$, where x is the element on which Γ_z is placed.

Construction

STAGE 0:

Let $a_0^0 = \sigma_0$, $b_0^0 = \sigma_1$, and declare $a_0^0 < b_0^0$. We let Γ_z be inactive for all z at stage 0.

STAGE $s + 1$:

Assume we have defined $n = n_s$, $a_0^s, b_0^s, \ldots, a_n^s, b_n^s$ so that $n \geqslant s$ and

$$\{a_0^s, b_0^s, \ldots, a_n^s, b_n^s\} = \{\sigma_0, \ldots, \sigma_{2n+1}\} = B_s.$$

Moreover, assume we have defined a linear order $< = <_{\mathcal{L}}$ on $B_s \times B_s$ so that

$$a_0^s < a_1^s < \cdots < a_n^s < b_n^s < b_{n-1}^s < \cdots < b_0^s.$$

Look for a $p \leqslant s$ such that Γ_p is inactive at stage s and $\varphi_k^s(a_p^s) \downarrow$, where $p = [e, k]$.

If there is no such p, then for all z, let Γ_z be inactive at stage $s+1$ if and only if Γ_z is inactive at stage s. If Γ_z is active, let $\Gamma_z(s+1) = \Gamma_z(s)$. In addition, let $a_i^{s+1} = a_i^s$ and $b_i^{s+1} = b_i^s$ for all $i \leqslant n = n_s$. Finally, let $n_{s+1} = n+1$, $\sigma_{2n+2} = a_{n+1}^{s+1}$ and $\sigma_{2n+3} = b_{n+1}^{s+1}$ and extend our definition of $<=<_{\mathcal{L}}$ to $B_{s+1} \times B_{s+1}$ by declaring

$$a_0^{s+1} < \cdots < a_{n+1}^{s+1} < b_{n+1}^{s+1} < \cdots < b_0^{s+1}.$$

If there is such a p, let $p = p(s+1) = [e(s+1), k(s+1)] = [e, k]$ be the least such p and $x = x(s+1) = \varphi_k(a_p^s)$. If $x(s+1) \notin A$, then proceed exactly as in the case where $p(s+1)$ is not defined, except declare Γ_p active and let $\Gamma_p(s+1) = a_p^{s+1}$. If $x(s+1) \in A$, then find the first $4n_s + 4$ elements of A. Now compare these elements to x with respect to R_e. We will then be in one of two cases.

Case 1: There are h $= 2n_s + 2$ elements $v_0, \ldots, v_{h-1}$ among the first $4n_s + 4$ elements of A such that $(v_i, x) \in R_e$ for $i = 0, 1, \ldots, h-1$. In this case, we proceed exactly as in the case where $x(s+1) \notin A$.

Case 2: Otherwise, there must be $h+1$ elements $w_0, \ldots, w_{h+1}$, among the first $4n_s + 4$ elements of A, such that $(w_i, x) \notin R_e$ for all i. Then we let $a_i^{s+1} = a_i^s$ for $i < p$ and $b_j^{s+1} = b_j^s$ for all $j \leqslant n = n_s$. Set $n_{s+1} = 2n + 1 - p$. Let $b_{n+i}^{s+1} = a_{n-i+1}^s$ for $i = 1, \ldots, n+1-p$ and let $a_{p+j}^{s+1} = \sigma_{2n+2+j}$ for $j = 0, \ldots, 2n+1-2p$. Activate the Γ_p marker and place it on $b_{n_{s+1}}^{s+1} = a_p^s$. Remove any markers Γ_z that were on elements among $a_p^s, \ldots, a_n^s$ and make them inactive. Any marker Γ_z which was active at stage s where $\Gamma_z(s) \in \{a_0^s, \ldots, a_{p-1}^s, b_0^s, \ldots, b_n^s\}$ is still active at stage $s+1$ and $\Gamma_z(s+1) = \Gamma_z(s)$. Markers Γ_z where $z \neq p$ which were inactive at stage s remain inactive at stage $s+1$. Finally, extend $< = <_{\mathcal{L}}$ to $B_{s+1} \times B_{s+1}$ by declaring

$$a_0^{s+1} < \cdots < a_{n_{s+1}}^{s+1} < b_{n_{s+1}}^{s+1} < \cdots < b_0^{s+1}.$$

This completes our construction. Because A is a polynomial time set and each R_e is a polynomial time relation, it easily follows that each stage is completely effective. The following facts are easily proved by induction.

(a) For all s, $n_s \geqslant s$.

(b) For all s, $\{a_0^s, b_0^s, \ldots, a_{n_s}^s, b_{n_s}^s\} = \{\sigma_0, \ldots, \sigma_{2n_s+1}\}$.

(c) Our definition of $<_{\mathcal{L}}$ is consistent, that is, if $i, j \leqslant 2n_s + 1$ and stage s, we declare $\sigma_i <_{\mathcal{L}} \sigma_j$, then for all $t \geqslant s$, we declare $\sigma_i <_{\mathcal{L}} \sigma_j$ at stage t.

Note that these facts imply that $\mathcal{L} = (\{0,1\}^*, <^{\mathcal{L}})$ is a recursive linear ordering, because to decide if $\sigma_i < \sigma_j$, we simply go to stage $s = \max\{i,j\}$ and then $\sigma_i < \sigma_j$ if and only if at stage s, we declare $\sigma_i < \sigma_j$.

Next we prove two lemmas which will complete the proof that $\mathcal{L}$ has the desired properties.

Lemma 4.10 *For each z, $\lim_s a_z^s = a_z$ and $\lim_s b_z^s = b_z$ exist and there is a stage t_z such that either Γ_z is inactive at stage s for all $s \geqslant t_z$ or Γ_z is active at stage s and $\Gamma_z(s) = \Gamma_z(t_z)$ for all $s \geqslant t_z$.*

Proof. We proceed by induction on $z = [e,k]$. By induction, we can assume that there is a stage $u > z$ large enough so that

(i) $a_j^s = a_j^u$ and $b_j^s = b_j^u$ for all $j < z$ and $s \geqslant u$ and

(ii) for each $j < z$, either Γ_j is inactive at stage s for all $s > u$ or for all $s \geqslant u$, Γ_j is active at stage s and $\Gamma_j(s) = \Gamma_j(u)$.

Note that our construction ensures that a Γ_j marker can be removed from an element at stage s only if $\Gamma_j(s-1) = a_k^s$ for some k and we take action to meet a requirement $\Gamma_{p(s)}$ at stage s where $p(s) < k$. Similarly, the only way $a_k^s \neq a_k^{s+1}$ is if $p(s+1) \leqslant k$ and we act according to Case 2 at stage $s+1$. Moreover, our construction ensures that if $j \leqslant n_s$, then $b_j^t = b_j^s$ for all $s \geqslant t$. It follows that $\lim_s b_z^s = b_z^u$. Now if Γ_z is active at stage s, then our choice of u ensures $p(s) > z$ for all $s > u$ so that $\lim_s a_z^s = a_z^u$ and Γ_z is active for all $s > u$. If Γ_z is not active at stage u, then either

(i) $\varphi_k^s(a_z^u) \uparrow$ for all $s \geqslant u$, in which case, for all $s \geqslant u$, Γ_z is inactive at stage s, $p(s) > z$ and $a_z^s = a_z^u$, or

(ii) There is an $s > u$ such that $\varphi_k^s(a_z^u) \downarrow$.

In case (ii), let t be the least $s > u$ such that $\varphi_k^t(a_z^u) \downarrow$. Then our choice of u ensures that, for all $u \leqslant s \leqslant t$, $p(s) > z$ and Γ_z is inactive at stage s, so that $p(t+1)$ will be defined and $p(t+1) = z$. But then at stage $t+1$, Γ_z becomes active and is placed on either a_z^{t+1} or $b_{n_{t+1}}^{t+1}$. If Γ_z is placed on a_z^{t+1}, then $a_z^{t+1} = a_z^t$ and Γ_z will never be removed from a_z^{t+1}. This is because Γ_z can be removed from a_z^{t+1} only if $p(s) < z$ for some $s > t+1$, which is ruled out by our choice of u. If Γ_z is placed on $b_{n_{t+1}}^{t+1}$, then again Γ_z can never be removed from $b_{n_{t+1}}^{t+1}$. Thus in either case Γ_z will remain active for all $s \geqslant t+1$. But this means $p(s) > z$ for all $s \geqslant t+1$, so that $a_z^s = a_z^{s+1}$ for all $s \geqslant t+1$. □

Lemma 4.11 *Requirement P_p is satisfied for all p.*

Proof. Let $p = [e, k]$ and let s_p be a stage such that $s_p > p$ and

(i) $(\forall s \geqslant s_p)(\forall j \leqslant p)\,[a_j^s = a_j^{s_p}$ and $b_j^s = b_j^{s_p}]$, and

(ii) $s_p \geqslant \max\{t_o, \ldots, t_p\}$, where t_z is a stage such that either

(a) for all $s \geqslant t_z$, Γ_z is active at s, or

(b) for all $s \geqslant t_z$, Γ_z is inactive at s.

It then easily follows from our construction that if Γ_p is inactive at stage s_p, then $\varphi_k^s(a_p^s)\uparrow$ and $a_p^s = a_p^{s_p}$ for all $s \geqslant s_p$. Thus $\varphi_k(a_p)\uparrow$, where $a_p = \lim_s(a_p^s)$. Hence, the requirement P_p is automatically satisfied. If Γ_p is active at stage s_p, then there are two possibilities. The first is that $\Gamma_p(s) = a_p^s = a_p$, in which case our construction guarantees that either $\varphi_k^s(a_p) \notin A$ or $\varphi_k^s(a_p) \in A$ but there are at least $p+1$ elements $v_0, \ldots, v_p \in A$ such that $(v_i, \varphi_k(a_p)) \in R_e$ for $i = 0, \ldots, p$. The second possibility is that $\Gamma_p(s) = b_m^s = b_m$ for some $m \leqslant n_{s_p}$. In this case, our construction ensures that $\varphi_k^s(b_m^s) \in A$ and there are at least $m+2$ elements $w_0, \ldots, w_{m+1} \in A$ such that $(w_i, \varphi_k(b_m)) \notin R_e$ for $i = 0, \ldots, m$. Thus in any case, P_p is satisfied. □

It now follows from Lemma 4.10 that $a_i <_{\mathcal{L}} a_{i+1}$ and $b_{i+1} <_{\mathcal{L}} b_i$ for all i and that $a_i <_{\mathcal{L}} b_j$ for all i and j, so that $\mathcal{L}$ is isomorphic to $\omega + \omega^*$. By our remarks preceding the construction of $\mathcal{L}$, it follows that meeting all the requirements P_p ensures that $\mathcal{L}$ is not recursively isomorphic to any polynomial time linear ordering over A.

In the general case, if λ is a recursive linear ordering, then a finite injury priority argument very similar to the one given above for $\omega+\omega^*$ will prove the theorem. However, the only thing that we can conclude from the fact that $\omega + \mathbb{Z}\cdot\eta + \omega^*$ is a recursive linear ordering is that λ is a Π_2^0 linear ordering and in that case, a more complicated infinite injury priority argument is required. □

We note that in the special case of $\omega+\omega^*$, one can make K have universe *Tal* (ω) when $A =$ *Bin* (ω). (This is Theorem 2.2 of [69].)

Cenzer and Remmel [18] gave a general condition which implies that a relational structure is recursively isomorphic to a p–time structure with universe *Bin* (ω). This condition can be thought of as a generalization of the argument in case 1 of Gregorieff's Theorem (5.8) on linear orderings.

Definition 4.12 Let $\mathcal{A}$ be a recursive substructure with universe A of the recursive relational structure $\mathcal{M}$ with universe M. Then $\mathcal{A}$ is said to be a *highly recursive relatively indiscernible binary substructure* of $\mathcal{M}$ if

(i) There is a recursive map f from $\mathcal{A}$ to $Bin\,(\omega)$ which induces a p–time model $\widetilde{\mathcal{M}}$. (Let $a_i = f^{-1}(bin\,(i))$.)

(ii) For any m–ary relation $R(x_1, \ldots, x_m)$ of $\mathcal{M}$, any fixed sequence $b_1, \ldots, b_k \in M \setminus A$ with $k \leqslant m$ and any fixed sequence $1 \leqslant I_1 < \ldots < I_k \leqslant m$, let $R^{b_1,\ldots,b_k}_{I_1,\ldots,I_k}$ denote the $(m-k)$–ary relation on $M - A$ which results by substituting b_{I_j} for x_{I_j} for $j = 1, \ldots, k$.

Then, for any such R, $1 \leqslant I_1 < \ldots < I_k \leqslant m$ and $b_1, \ldots, b_k \in M \setminus A$, either $R^{b_1,\ldots,b_k}_{I_1,\ldots,I_k}$ holds for all but finitely many elements in $(M - A)^{m-k}$, or $\neg R^{b_1,\ldots,b_k}_{I_1,\ldots,I_k}$ holds for all but finitely many elements in $(M - A)^{m-k}$.

Furthermore, there is a uniform effective algorithm which, given an index for R and sequences $b_1, \ldots, b_k \in M \setminus A$ and $I_1, \ldots, I_k$, will compute whether $R^{b_1,\ldots,b_k}_{I_1,\ldots,I_k}$ holds for all but finitely many elements in $(M - A)^{m-k}$ or $\neg R^{b_1,\ldots,b_k}_{I_1,\ldots,I_k}$ holds for all but finitely many elements in $(M - A)^{m-k}$, along with with a complete list of the finitely many sequences of $(M - A)^{m-k}$ which are exceptions.

Theorem 4.13 *Suppose that $\mathcal{A}$ is a highly recursive relatively indiscernible binary substructure of the relational structure $\mathcal{M}$. Then $\mathcal{M}$ is recursively isomorphic to a p–time structure with universe $Bin\,(\omega)$ (and therefore also recursively isomorphic to a p–time structure with universe $Tal\,(\omega)$).*

Proof. There is no loss in generality in assuming the universe of $\mathcal{M}$ is the set of natural numbers ω. Let $S_0, S_1, \ldots$ be an effective list of all relations of $\mathcal{A}$ and assume that there is a recursive function F such that S_i is an $F(i)$–ary relation. Let $\mathcal{A}$ with universe $A = \{a_0, a_1, \ldots\}$ be isomorphic to $\widetilde{\mathcal{M}}$ as described above and let $\omega \setminus A = \{b_0 < b_1 < \cdots\}$. For any $x \in \omega \setminus A$, let $t(x)$ be the total time needed to search all elements $y \leqslant x$ and determine if $y \in A$ and to compute the common value of all relations of the form $R^{b_1,\ldots,b_k}_{I_1,\ldots,I_k}$ and the finite list of exceptions where $R = S_i$ for some $i \leqslant x$ and $b_1, \ldots, b_k$ are elements of $M - A$ which are less than or equal to x. Let $\varphi(x) = 1^{[x,t(x)]}$ and let $P = \{\varphi(x) : x \in M \setminus A\}$. It is clear that P is a p–time subset of $Tal\,(\omega)$.

We define a polynomial time model $\mathcal{D}$ with universe $Bin\,(\omega)$ by defining the relation $S_q^{\mathcal{D}}$ for $\mathcal{D}$ as follows. Given an element $bin\,(i)$, search the strings of the form 1^k for $k \leqslant |bin\,(i)|$ and determine whether each such string is in P. If $bin\,(i) \in P$, then let $r(bin\,(i)) = bin\,(n)$ where $bin\,(i) = 1^{[b_n, t(b_n)]}$. If $bin\,(i) \notin P$, then let $r(bin\,(i)) = bin\,(m)$ where $bin\,(i)$ is the m–th element of $Bin\,(\omega) \setminus P$. Note that because P is a polynomial time set, there is a fixed polynomial p such that we can compute whether $x \in P$ in $p(|x|)$ steps. It thus takes at most $\sum_{j=0}^{|bin\,(i)|} p(j)$ steps to search the strings of the form 1^k for $k \leqslant |bin\,(i)|$ for membership in P so that the function r is polynomial time. Then $S_q^{\mathcal{D}}(s_1, \ldots, s_{F(q)})$ holds if either

(A) no $s_i \in P$ and $S_q(r(s_1), \ldots, r(s_{F(q)}))$ holds in $\widetilde{\mathcal{M}}$, or

(B) there is some s_i in P and $S_q(Z_1, \ldots, Z_{F(q)})$ holds in $\mathcal{M}$ where $Z_j = b_k$, if $s_j \in P$ and $r(s_j) = bin\,(k)$, and $Z_j = a_k$, if $s_j \notin P$ and $r(s_j) = bin\,(k)$.

Note that in case (A) we can compute whether $S_q^{\mathcal{D}}(s_1, \ldots, s_{F(q)})$ holds in time polynomial in $|s_1| + \cdots + |s_{F(q)}|$ since $\widetilde{\mathcal{M}}$ is a polynomial time model.

In case (B), let n be the maximum value such that there is an $s_j \in P$ with $j \leqslant F(q)$ and $r(s_j) = bin\,(n)$. If $n \geqslant q$, then $s_j = 1^{[b_n, t(b_n)]}$ and in $t(b_n)$ steps we can compute whether $R_q(s_1, \ldots, s_{F(q)})$ holds by the definition of t. Thus in case (B), the only cases in which we can not directly compute in linear time whether $S_q^{\mathcal{D}}(s_1, \ldots, s_{F(q)})$ holds is if $n \leqslant q$. However it is easy to see that our assumptions ensure that it takes only a finite amount of information to determine whether $S_q^{\mathcal{D}}(s_1, \ldots, s_{F(q)})$ holds in all such cases. Thus $S_q^{\mathcal{D}}$ is a polynomial time relation.

Finally it is easy to see that our definitions ensure that we have defined the map g where $g(bin\,(i)) = b_n$, if $bin\,(i) \in P$ and $r(bin\,(i)) = bin\,(n)$, and $g(bin\,(i)) = a_n$, if $bin\,(i) \notin P$ and $r(bin\,(i)) = bin\,(n)$, is an isomorphism from $\mathcal{D}$ onto $\mathcal{M}$.

The model $\mathcal{C}$ with universe $Tal\,(\omega)$ simply has relations $R^{\mathcal{C}}$ defined by

$$R^{\mathcal{C}}(tal\,(i_1, \ldots, tal(i_n)) \iff R^{\mathcal{D}}(bin\,(i_1), \ldots, bin\,(i_n)).$$

The relation $R^{\mathcal{C}}$ is p-time since $bin(i)$ can be computed from $tal\,(i)$ in polynomial time and $|bin\,(i)| \leqslant |tal\,(i)|$. □

For a simple example, consider a well-ordering $\mathcal{M} = (M, <^M)$ of type $> \omega$ and let $\mathcal{A} = (A, <^M)$ be the first ω elements of $\mathcal{M}$. A is a recursive set since $x \in A \iff x <^M w$ where w is the ω–th element of $\mathcal{M}$. $\mathcal{A}$ is certainly recursively isomorphic to the standard ordering on $Bin(\omega)$ and is indiscernible since for all $a \in A$ and all $x \in M \setminus A$, $a <^M x$. Thus by Theorem 4.13, $\mathcal{M}$ is recursively isomorphic to a p–time model with standard universe $Bin(\omega)$. This is a special case of Theorem 4.9 above.

4.3 Boolean Algebras

There are some cases of structures with function symbols where we can get some positive results. For example, every recursive Boolean algebra is recursively isomorphic to a polynomial time Boolean algebra.

Definition 4.14

(i) The *language* $\mathcal{L}$ of Boolean algebras consists of two binary function symbols $\wedge$ (meet) and $\vee$ (join), one unary function symbol $\neg$ (complement) and two constant symbols 0 (zero) and 1 (unity). A Boolean algebra $\mathcal{B}$ is a structure $(B, \wedge^B, \vee^B, \neg^B, 0^B, 1^B)$ for this language which satisfies the usual axioms.

(ii) Given a linear ordering $\mathcal{M} = (\mathcal{M}, <)$ with a first element, the *interval algebra* Intalg($\mathcal{M}$) is the Boolean algebra of subsets of M generated by the left-closed, right-open intervals of M, $[a, b) = \{x : a \leqslant x < b\}$.

The partial ordering $\leqslant^B$ of a Boolean algebra $\mathcal{B}$ is defined by $a \leqslant^B b$ if and only if $b = a \vee c$ for some $c \neq 0$. An element a of a Boolean algebra is said to be an *atom* if whenever $b \leqslant a$, either $b = 0$ or $b = a$. The Boolean algebra $\mathcal{A}$ is said to be *atomic* if for any $b \neq 0$, there exists some atom a with $a \leqslant b$. An element $x \in \mathcal{B}$ is *atomless* if x is not the zero of $\mathcal{B}$ and there is no atom a of $\mathcal{B}$ such that $a <^B x$. $\mathcal{B}$ is said to be *non-atomic* if $\mathcal{B}$ contains an atomless element.

The following is Lemma 2.5 of [10].

Lemma 4.15 *For any p–time linear ordering $\mathcal{L}$ with a first element, the interval algebra $Intalg(\mathcal{L})$ is a p–time Boolean algebra.*

Sketch Proof. The nonzero elements of Intalg($\mathcal{L}$) are given the natural representation $[a_1, a_2) \cup [a_3, a_4) \cup \cdots \cup [a_{2n-1}, a_{2n})$, where $a_1 <^L a_2 <^L \cdots <^L a_{2n-1}$ and either $a_{2n} = \infty$ or $a_{2n} \in L$ and $a_{2n-1} < a_{2n}$. □

This lemma is used to prove the following result from [10].

Theorem 4.16 *Every recursive Boolean algebra $\mathcal{B}$ is recursively isomorphic to a p–time Boolean algebra.*

Sketch Proof. First observe that the classical proof that every countable Boolean algebra is isomorphic to the interval algebra of a countable linear ordering is effective (see Remmel [67]). Thus every recursive Boolean algebra is recursively isomorphic to Intalg($\mathcal{M}$) where $\mathcal{M}$ is a recursive linear ordering. However by Theorem 4.8, $\mathcal{M}$ is recursively isomorphic to a polynomial time linear ordering $\mathcal{P}$. The interval algebra of $\mathcal{P}$ is thus recursively isomorphic to $\mathcal{B}$ and is polynomial time by Lemma 4.15. □

The next two theorems are unpublished results due to Cenzer and Remmel.

Theorem 4.17 *Every infinite* non-atomic *recursive Boolean algebra is recursively isomorphic to p–time Boolean algebras*

(a) *with universe Bin* (ω) *and*

(b) *with universe Tal* (ω).

Theorem 4.18 *Let $A \subseteq \{0,1\}^*$ be an infinite p–time set and let B be an infinite* atomic *recursive Boolean algebra. Then there is a recursive Boolean algebra $\mathcal{D}$ which is isomorphic to $\mathcal{B}$ but is not recursively isomorphic to any p–time Boolean algebra with universe A.*

4.4 Graphs

Next we give two applications of Theorem 4.13 to recursive graphs due to Cenzer and Remmel in [18].

Definition 4.19

(i) A graph $G = (V, E)$ is *locally finite* (respectively *locally cofinite*) if for every $v \in V$, the set $\text{NB}(v)$ of neighbors of v is finite (resp. cofinite).

(ii) A graph $G = (V, E)$ is *locally finite/cofinite* if for every $v \in V$, either the set $\mathrm{NB}(v)$ of neighbors of v is finite or $V \setminus \mathrm{NB}(v)$ is finite.

(iii) A locally finite/cofinite recursive graph G is *highly recursive* if there are algorithms for deciding whether a given $v \in V$ has finite degree and for computing $NB(v)$ is the degree is finite and $V \setminus \mathrm{NB}(v)$ if not.

Theorem 4.20 *Every infinite highly recursive locally finite/cofinite graph G is recursively isomorphic to a p–time graph H with universe $Bin(\omega)$.*

Sketch Proof. If there are infinitely many vertices of G whose degree is finite, then we can construct a recursive subset of such vertices U such that G restricted to U is the empty graph. Moreover we can construct U so that it will be the highly recursive relatively indiscernible binary substructure required to apply Theorem 4.13. If there are infinitely many vertices of finite co-degree. We can construct a complete subgraph C which will be the highly recursive relatively indiscernible binary substructure needed for Theorem 4.13. □

Theorem 4.21 *Every infinite recursive graph G which is either locally finite or locally cofinite, is recursively isomorphic to a p–time graph H with universe $Bin(\omega)$.*

Sketch Proof. Suppose that all vertices have finite co-degree. Then we again pick out a complete subgraph

$$C = (\{v_0, v_1, \dots\}, E),$$

now with the additional property that v_i is not joined to any vertex from $\{0, 1, \dots, i-1\}$. □

The next result shows that the hypotheses of the two preceding theorems are needed.

Theorem 4.22 *Let A be any infinite polynomial time subset of $\{0,1\}^*$. Then there is a recursive graph, having every vertex of either finite degree or finite co-degree, which is not recursively isomorphic to any p–time graph with universe A.*

4.5 Trees

A connected graph G with no cycles is said to be a *tree*. The vertices of T are called *nodes*. We will assume that any tree T has a designated *root* ε. Then any node v of T can be reached from the root by a *path*, that is, a sequence of edges $(\varepsilon, v_1), (v_1, v_2), \ldots, (v_{n-1}, v)$. We say that v is a *successor* of v_{n-1}. It is clear that the successor relation is recursive, since the path from the root to a node v may be computed from v in a uniform fashion. The partial ordering $u \prec^T v$, which says that there is a path from the root to v which passes through u, is also recursive. On the other hand, if T is a p–time tree, then this computation of the path from ε to v might not be in polynomial time in $|v|$, so that the successor relation and the relation $\prec^T$ need not be p–time. Thus we say that T is *fully* p–time if both the successor relation and the relation $\prec^T$ are p–time. Similar notions may be defined for any bounded resource class.

T is said to be *highly recursive* if there is a recursive function which computes from any node v a list of successors of v. The corresponding notion of a highly feasible tree (or more generally, of a highly feasible graph) is more difficult to formulate. Several inequivalent notions are studied in [14, 12, 18]. In particular, a graph is said to be *locally p–time*, if there is a p–time function which computes from any node v a list of successors of v and is said to be *highly p–time* if there is a p–time function which computes from a vertex v and a tally number $tal(n)$ a list of all vertices at distance n from v.

The following results are Theorems 9, 10 and 12 of [18].

Theorem 4.23 *Any infinite recursive tree T is recursively isomorphic to a p–time tree with universe $Bin(\omega)$.*

Sketch Proof. There are two cases. First, every node of T may have only finitely many successors. In this case, Theorem 4.20 may be applied. Second, some node v of T may have an infinite set A of successors. In this case, the set A plays the role of the highly recursive relatively indiscernible binary substructure, so that Theorem 4.13 may be applied. □

Theorem 4.24 *There is a highly recursive tree T which is not recursively isomorphic to any fully primitive recursive tree P with a standard universe $Bin(\omega)$.*

Theorem 4.25 *There is a highly recursive binary tree which is not isomorphic as a directed graph to any highly primitive recursive tree with universe Bin* (ω).

A similar but stronger result for graphs was given in Theorem 2.3 of [12].

Theorem 4.26 *There is a highly recursive graph* G *which is not isomorphic to any locally primitive recursive graph.*

4.6 Equivalence Structures

Another type of relational structure is an *equivalence structure*, (A, R^A), where R^A is an equivalence relation on A. A recursive equivalence structure (A, R^A) is said to be *highly recursive* if the set of elements that belong to infinite equivalence classes is recursive, and there is a recursive function f such that $f(a)$ is the cardinality of $[a]^R$ when $[a]^R$ is finite (so that the equivalence class $[a]^R$ can be computed from a). We say that (A, R^A) is *highly p–time* if A is a p–time subset of *Bin* (ω), the set of elements that belong to infinite equivalence classes is p–time, and there is a p–time function f such that $f(a)$ codes $[a]^R$ when it is finite. The *full spectrum* of (A, R^A) is the set of pairs $(0, n)$ such that there are at least $n + 1$ infinite equivalence classes in (A, R^A) and pairs (q, n) such that $q > 0$ and there are at least $n + 1$ equivalence classes of size q in (A, R^A).

The following results are Theorems 13, 14 and 16 of [18].

Theorem 4.27 *Any recursive equivalence structure* (A, R^A) *is recursively isomorphic to a p–time model with universe Bin* (ω).

Sketch Proof. There are two cases. If all equivalence classes of (A, R^A) are finite, then Theorem 4.21 can be applied to (A, R^A) viewed as a graph. If (A, R^A) has an infinite equivalence class, then this class is a highly recursive relatively indiscernible substructure so that Theorem 4.13 may be applied. □

Theorem 4.28 *For any equivalence structure* (A, R^A) *with full spectrum* S *such that* $S^* = \{\langle tal(q), tal(r)\rangle : (q, r) \in S\}$ *is p–time, there is a highly recursive, p–time equivalence structure* $(Bin\,(\omega), R)$ *isomorphic to* (A, R^A).

Theorem 4.29 *There is an infinite recursive full spectrum S of ω such that no highly primitive recursive equivalence structure with universe $Bin\,(\omega)$ has full spectrum a subset of S.*

We now turn to the study of some structures with functions. There are three basic models which we have been considered: first, models of some fragment of arithmetic; second, Abelian groups; and, third, permutation structures.

4.7 Models of Arithmetic

Our first result, taken from [10], demonstrates that the unary exponential function 2^x may be adjoined to the standard model of arithmetic while being represented by a p–time function.

Theorem 4.30 $\mathbb{N} = (\omega\,,\,S\,,\,+\,,\,-\,,\,\cdot\,,\,2^x\,,\,<\,,\,0)$ *is isomorphic to a p–time structure $\mathcal{M}$.*

Sketch Proof. The elements of $\mathcal{M}$ are terms in the language $\{0, A, I, E\}$. The natural number n is represented by the expression $\sigma(n)$ defined as follows.

$$\begin{aligned}\sigma(0) &= 0,\\ \sigma(2^k) &= E\sigma(k),\\ \sigma(2^k+m) &= AE\sigma(k)\sigma(m),\text{ for } 0<2m\leqslant 2^k,\\ \sigma(2^k-m) &= IE\sigma(k)\sigma(m),\text{ for } 0<2m<2^k.\end{aligned}$$

It is easy to see that the universe of $\mathcal{M}$ is a linear time set and that the term $\sigma(n)$ which represents n can be computed from $bin\,(n)$ in polynomial time. It can be shown by induction that $|\sigma(n)| \leqslant 2|bin\,(n)|^2$ and that $|\sigma \pm^M \tau| \leqslant |\sigma| + |\tau| + 1$. □

We note that it is an open question whether $(\omega\,,\,S\,,\,+\,,\,-\,,\,\cdot\,,\,\exp\,,\,<\,,\,0)$ is isomorphic to a polynomial time model where $\exp(m,n) = m^n$ is the general exponential, or even whether $(\omega\,,\,S\,,\,+\,,\,-\,,\,\cdot\,,\,2^x\,,\,3^x\,,\,<\,,\,0)$ is isomorphic to a polynomial time. Bäuerle [5] proved that in the model of Theorem 4.30 the function 3^x is not polynomial time.

We also note that the model of Theorem 4.30 can be used to build a EXPTIME group isomorphic to the integers under $+$, $\mathbb{Z} = (Z, +)$, which is not q–time isomorphic to the standard model of $\mathbb{Z}$ where the positive integer n is coded as $bin\,(2n)$ and a negative integer $-n$ is coded as $bin\,(2n-1)$; see Section 5.3.

4.8 Injection Structures

The simplest type of structure with a function is an *injection structure* (A, f^A) where f^A is a one-to-one mapping from A into itself. If f^A maps A onto A, then we say that (A, f^A) is a *permutation structure.* The *orbit* $\mathcal{O}(a)$ of an element a of A is $\{b \in A : (\exists n \in \omega)(f^n(a) = b \vee f^n(b) = a)\}$. There are two types of infinite orbits, one of type ω which is isomorphic to (ω, S) and the other of type $\mathbb{Z}$ which is isomorphic to $(\mathbb{Z}, S)$. The *order* $|a|$ of a is card$(\mathcal{O}(a))$ if $\mathcal{O}(a)$ is finite and is either ω or $\mathbb{Z}$ otherwise. The *full spectrum* of (A, f^A) is the set of pairs $(0, n)$ such that there are at least $n+1$ orbits of type ω, pairs $(1, n)$ such that there are at least $n+1$ orbits of type $\mathbb{Z}$, and pairs (q, n) such that $q > 1$ and there are at least $n+1$ orbits of size $q-1$ in (A, f^A).

It is easy to see that the full spectrum of a recursive injection structure is always a recursively enumerable set. It is shown in Theorem 3.4 of [11] that any r.e. spectrum can be realized by a p–time injection structure. Thus every recursive injection structure is isomorphic to a p–time structure. However, we know by Theorem 4.3 that this isomorphism need not be recursive. Now we consider the question of when we can obtain a p–time injection structure with a standard universe $Bin\,(\omega)$ or $Tal\,(\omega)$.

The basic result here is Theorem 3.2 of [13].

Theorem 4.31 *Any recursive permutation structure (A, f^A) with all finite orbits is recursively isomorphic to an honest p–time structure with universe a subset of $Tal\,(\omega)$.*

Sketch Proof. The element $a \in A$ is represented by $tal\,(n)$, where $bin\,(n) = 1^a\,0^t$ and t is the total time required to compute the orbit of a. The details follow as in the proof of Theorem 4.1 above. □

Theorems 3.4 and 3.6 of [13] give two cases in which we can improve this result to obtain a standard universe.

Theorem 4.32 *Let $B = Bin\,(\omega)$ or $Tal\,(\omega)$. Any recursive injection structure (A, f) with at least one but only finitely many infinite orbits is recursively isomorphic to a p–time structure with universe B.*

Sketch Proof. Let $F = \{a \in A : |a| \text{ is finite}\}$ and let $I = A \setminus F$. Since there are only finitely many orbits in I, both F and I are recursive. It is easy to see that (I, f) is recursively isomorphic to a p–time structure with universe B. By Theorem 4.31, (F, f) is recursively isomorphic to a p–time structure with universe a subset of $Tal\,(\omega)$. The result now follows from Lemma 3.8. □

Theorem 4.33 *Let $B = Bin\,(\omega)$ or $Tal\,(\omega)$. Any recursive injection structure (A, f) with infinitely many orbits of size q, for some finite q, is recursively isomorphic to a p–time structure with universe B.*

Sketch Proof. Let $C = \{a \in A : |a| = q\}$ and let $D = A \setminus C$. It follows from Theorem 4.31 that (D, f) is recursively isomorphic to a p–time structure with universe a subset of $Tal\,(\omega)$. It therefore suffices by Lemma 3.8 to show that C is recursively isomorphic to a p–time structure (B, g). For $B = Bin\,(\omega)$, the permutation g is simply defined by $g(bin\,(nq + i)) = bin\,(nq + i + 1)$, if $i + 1 < q$, and $g(bin\,(nq + q - 1) = bin\,(nq)$. The tally definition is similar. □

A general result on the existence of p–time injection structures is given by Theorems 3.4 and 3.8 of [11].

Theorem 4.34

(a) *For any r.e. full spectrum S, there is a p–time injection structure (A, f) with full spectrum S.*

(b) *If $\{\langle tal\,(q), tal\,(r)\rangle : (q, r) \in S\}$ is p–time, then A may be taken to be either $Bin\,(\omega)$ or $Tal\,(\omega)$.*

Sketch Proof. We sketch the proof of part (b) for $B = Tal\,(\omega)$, assuming that all orbits will be finite. Let $q_0, q_1, \ldots$ enumerate in non-decreasing order the set of orbit sizes (with repetitions). Then the permutation f may be defined by

$$f(tal(q_0+q_1+\cdots+q_{k-1}+r)) = \begin{cases} (tal(q_0 + q_1 + \cdots + q_{k-1} + r + 1)) & \text{if } r < q_k \\ (tal(q_0 + q_1 + \cdots + q_{k-1} + q_k)) & \text{if } r = q_k \end{cases}$$

An infinite orbit of type ω is given by the standard successor function on $Tal\,(\omega)$ and an orbit of type $\mathbb{Z}$ is given by $f(tal\,(2n)) = tal\,(2n+2)$, $f(1) = 0$ and $f(tal\,(2n+3)) = tal\,(2n+1)$. Multiple infinite orbits and a combination of finite and infinite orbits may then be obtained by Lemma 3.8. □

Corollary 4.35 *Any recursive injection structure is isomorphic to a p–time injection structure.*

Finally, we consider some negative results. The first is Theorem 3.13 of [13] and deals with structures with a fixed universe.

Theorem 4.36 *There is a recursive set M such that no injection structure with full spectrum M is isomorphic to any primitive recursive structure with universe $Bin\,(\omega)$ or $Tal\,(\omega)$.*

Sketch Proof. Let f_e enumerate all primitive recursive unary functions and let $\mathcal{B}_e = (Bin\,(\omega), f_e)$. Construct a set $R = \{r_0 < r_1 < \cdots\}$ by a diagonal argument so that, for all e, either

(1) f_e is not one-to-one, or

(2) $\mathcal{B}_e$ has an infinite orbit, or

(3) $\mathcal{B}_e$ has two disjoint orbits of the same finite size, or

(4) $\mathcal{B}_e$ has an orbit of size $q \notin R$.

We establish this requirement, given $r_0 < \cdots < r_{e-1}$ by computing enough of $\mathcal{B}_e$ to either find two orbits of the same size, or an orbit (perhaps incomplete) of size $r > r_{e-1}$. If the orbit is complete, we let $r_e = r + 1$, thus ensuring that $\mathcal{B}_e$ has an orbit of size $r \notin R$. If the orbit is incomplete, we continue to build the orbit at later stages and take a similar action when the orbit becomes complete. If this never happens, then the orbit is infinite, so that (2) is satisfied.

It follows that no primitive recursive permutation structure with all finite orbits can have $M = \{(r, 1) : r \in R\}$ as a subset of its full spectrum. □

The final result here is Corollary 3.18 of [13] and deals with the question of recursive isomorphism. The proof is omitted.

Theorem 4.37 *For any recursive injection structure (C, f) with infinitely many infinite orbits, there is a recursive structure (A, f^A) which is isomorphic to (C, f) but is not recursively isomorphic to any primitive recursive structure.*

4.9 Abelian Groups

We now turn to the study of feasible versus recursive Abelian groups. The results here are parallel to those for permutation structures. We begin by recalling some basic notation. For any natural number $n > 1$, $\mathbb{Z}(n)$ is the cyclic group of order n. For a prime number p, the group $\mathbb{Z}(p^\infty)$ is the inverse limit of the sequence $\mathbb{Z}(p^n)$, or more concretely, the set of rational numbers with denominator equal to a power of p and where the group operation is addition modulo 1. The group $\mathbb{Z}(p^\infty)$ is said to be *quasi-cyclic.* The additive group of rational numbers is denoted by $\mathbb{Q}$.

For any element a of an Abelian group $\mathcal{A} = (A\,, +^A\,, -^A\,, 0^A)$ and any integer n, $n{\cdot}a$ is defined recursively by $0{\cdot}a = 0$ and $(n+1){\cdot}a = a+^A n{\cdot}a$. Then $(-n) \cdot a = 0^A -^A n \cdot a$. The *order* $|a|_\mathcal{A}$ of a is the least n such that $n \cdot a = 0$. $\mathcal{A}$ is said to be *torsion* if all elements have finite order and *torsion-free* if all elements (except the identity) have infinite order.

We will frequently be concerned with products of Abelian groups.

Definition 4.38 For any sequence $\mathcal{A}_0\,, \mathcal{A}_1\,, \ldots$ of Abelian groups, where $\mathcal{A}_i = (A_i\,, +_i\,, -_i\,, e_i)$ and $A_i \subseteq \{0,1\}^*$, the *direct product* $\mathcal{A} = \oplus_n \mathcal{A}_n$ is defined to have domain

$$A = \{\langle \sigma_1, \sigma_2, \ldots, \sigma_k\rangle : k \in \omega,\ \sigma_i \in A_i \text{ for } 1 \leqslant i \leqslant k, \text{ and } \sigma_k \neq e_k\}.$$

The identity of $\mathcal{A}$ is $e^A = \emptyset$, and addition $+^A$ and subtraction $-^A$ are defined as follows: for $\sigma = \langle \sigma_1, \sigma_2, \ldots, \sigma_m\rangle$ and $\tau = \langle \tau_1, \tau_2, \ldots, \tau_n\rangle$, $\sigma +^A / -^A \tau = \rho = \langle \rho_1, \rho_2, \ldots, \rho_k\rangle$, where

$$k = \max\{i : (i \leqslant m \wedge i \leqslant n \wedge \sigma_i +_i / -_i \tau_i \neq e_i) \vee m < i \leqslant n \vee n < i \leqslant m\}$$

and, for $i \leqslant k$,

$$\rho_i = \begin{cases} \sigma_i +_i / -_i \tau_i & \text{for } i \leqslant \min(m, n) \\ \sigma_i & \text{for } n < i \leqslant k \\ \tau_i & \text{for } m < i \leqslant k. \end{cases}$$

In particular, we write $\oplus_{i<\omega}\mathcal{G}$ to be the direct product of a countably infinite number of copies of $\mathcal{G}$.

Definition 4.39 Let B be either $Bin(\omega)$ or $Tal(\omega)$. For any complexity class Γ, the sequence $\mathcal{A}_0, \mathcal{A}_1, \dots$ of groups, where $\mathcal{A}_n = (A_n, +_n, -_n, e_n)$ is said to be *uniformly Γ-computable over B* if

(i) $\{\langle b(n), a\rangle : a \in A_n\}$ is a Γ-computable subset of $B \otimes B$, where $b(n) = bin(n)$ if $B = Bin(\omega)$ and $b(n) = tal(n)$ if $B = Tal(\omega)$.

(ii) The functions $F(b(n), a, b) = a +_n b$ and $G(b(n), a, b) = a -_n b$ are both the restrictions of Γ-computable functions from B^3 into B where we set $F(b(n), a, b) = G(b(n), a, b) = \emptyset$ if either a or b is not in A_n.

(iii) The function from $Tal(\omega)$ into B given by $e(tal(i)) = e_i$ is Γ-computable.

The following is Lemma 4.2 of [13].

Lemma 4.40 *Let B be either $Bin(\omega)$ or $Tal(\omega)$ and let Γ be one of the following complexity classes: recursive, primitive recursive, exponential time, polynomial time. Suppose that the sequence $\mathcal{A}_i = (A_i, +_i, -_i, e_i)$ of Abelian groups is Γ-computable over B. Then*

(a) *The direct product $\mathcal{A}$ of the sequence is recursively isomorphic to a Γ-computable group with universe contained in B.*

(b) *If A_i is a subgroup of A_{i+1} for all i and if there is a Γ-computable function f such that, for all $a \in \cup_i A_i$, $a \in A_{f(a)}$, then the union $\cup_i \mathcal{A}_i$ is recursively isomorphic to a Γ-computable group with universe contained in B.*

(c) *If the sequence is finite and one of the components has universe B, then the product is recursively isomorphic to a Γ-computable group with universe B.*

(d) *If the sequence is infinite and if each component has universe B, then the product is recursively isomorphic to a Γ-computable group with universe $Bin(\omega)$.*

(e) *If each component has universe $Tal(\omega)$ and there is a uniform constant c such that for any i and any $a, b \in A_i$, $|a +_i b|$ and $|a -_i b|$ are both $\leqslant c(|a| + |b|)$, then the product is recursively isomorphic to a Γ-computable group with universe $Tal(\omega)$.*

If a torsion Abelian group $\mathcal{G}$ is isomorphic to a direct sum $\oplus_i \mathbb{Z}(q_i^{n_i})$ of prime power cyclic groups, then we define the *characteristic* $\eta(\mathcal{G})$ to be

$$\{(p^m, k) : q_i^{n_i} = p^m \text{ for at least } k+1 \text{ distinct values of } i\}.$$

Khisamiev shows in Corollary 3.4 of [40] that for any $k \leqslant \omega$ and any torsion Abelian p-group $\mathcal{G}$, $\mathcal{G} \oplus (\oplus_{i<k}\mathbb{Z}(p^\infty))$ is isomorphic to a recursive group if and only if $\eta(G)$ is a Σ_2^0 set. We will say that a subset Q of $\omega \times \omega$ is *hereditary* if $(m, k+1) \in Q$ implies $(m, k) \in Q$ for all m, k. It is clear that a subset Q of $\omega \times \omega$ is the characteristic $\eta(\mathcal{G})$ for some Abelian torsion group $\mathcal{G}$ if and only if Q is hereditary and $(m, k) \in Q$ implies m is a prime power. Therefore we will say that any such set Q is a *characteristic*.

The following are results 4.3 and 4.4 of [13].

Theorem 4.41 *Each of the groups $\mathbb{Z}$, $\oplus_{i<\omega}\mathbb{Z}(k)$, $\mathbb{Z}(p^\infty)$ and $\mathbb{Q}$ are isomorphic to polynomial time groups*

(a) *with universe Bin (ω) and*

(b) *with universe Tal (ω).*

Theorem 4.42 *Any finitely generated recursive Abelian group is recursively isomorphic to p–time Abelian groups*

(a) *with universe Bin (ω) and*

(b) *with universe Tal (ω).*

The simplest torsion groups are *primary* groups, or p-groups, in which every element has order a power of p where p is a prime. In [74], Smith characterized the recursively categorical p-groups as follows.

Theorem 4.43 (Smith) *A recursive p–group $\mathcal{G}$ is recursively categorical iff either*

(1) $\mathcal{G} \approx \oplus_{i<n}\mathbb{Z}(p^\infty) \bigoplus \mathcal{F}$, *or*

(2) $\mathcal{G} \approx \oplus_{i<n}\mathbb{Z}(p^\infty) \bigoplus \oplus_{i<\omega}\mathbb{Z}(p^m) \bigoplus \mathcal{F}$ *where $\mathcal{F}$ is a finite p–group and m and n are nonnegative integers.*

Corollary 4.44 *Any recursively categorical p–group is recursively isomorphic to a polynomial time group*

(a) *with universe Bin (ω) and*

(b) *with universe Tal (ω).*

Note that not every product of cyclic groups is recursively categorical. For example, consider $\oplus_{i<\omega}\mathbb{Z}(2) \bigoplus \oplus_{i<\omega}\mathbb{Z}(4)$. The following fundamental result from [13] shows that this group is recursively isomorphic to a p–time group.

Theorem 4.45 *Any recursive Abelian torsion group $\mathcal{G} = (G, +^G, -^G, e^G)$ is recursively isomorphic to a polynomial time group $\mathcal{H}$ with universe a subset of Tal (ω).*

Sketch Proof. It suffices, by the remarks following Lemma 3.4, to define a p–time group $\mathcal{H}$ with universe a subset of *Bin* (ω) such that both $a +^{\mathcal{H}} b$ and $a -^{\mathcal{H}} b$ have length bounded by some constant multiple of $|a| + |b|$.

Let $\mathcal{A}_k$ be the subgroup generated by $\{1, 2, \ldots, k\}$. Renumber the elements of $\mathcal{A}$ as $a_0, a_1, \ldots$ so that the elements of $\mathcal{A}_k$ precede the elements of $\mathcal{A}_{k+1} \setminus \mathcal{A}_k$. This can be done so that the map taking i to a_i is a recursive isomorphism. Now map $a_k \in A$ to $\varphi(a_k) = 1^k\, 0^{t(k)}$, where $t(k)$ is the total time required to compute the operation table for $\mathcal{A}_k$. The key to the proof is that whenever $a_i \pm^{\mathcal{G}} a_j = a_k$, where $i \leqslant j$, then we have $t(k) \leqslant t(j)$, since a_k will be in the group generated by $\{a_0, \ldots, a_j\}$. Furthermore, $|\varphi(a_k)| = k + t(k) \leqslant 2t(j) \leqslant 2(|\varphi(a_i) + \varphi(j)|)$, since $k \leqslant t(k)$. □

We need an effective version of the Fundamental Theorem of Abelian groups, which states that every torsion Abelian group is a direct product of primary groups. This is Lemma 4.8 of [13].

Theorem 4.46 *Any recursive Abelian torsion group $\mathcal{G}$ is recursively isomorphic to a p–time direct product of primary groups over B where B may be taken to be either Tal (ω) or Bin (ω).*

The main result on the existence of feasible groups with standard universe is the following theorem from [13].

Theorem 4.47 *Let $\mathcal{G}$ be an infinite recursive Abelian group with bounded order. Then $\mathcal{G}$ is recursively isomorphic to a polynomial time group with universe Tal (ω) and to a polynomial time group with universe Bin (ω).*

Sketch Proof. Let B be either *Bin* (ω) or *Tal* (ω). We may assume that $\mathcal{G}$ is p–time by Theorem 4.45. Since the orders are bounded, there is no loss of generality in assuming that $\mathcal{G}$ is a p–group for some prime p. Let p^m be the largest order of an element of $\mathcal{G}$. The proof is by induction on m. We can express $\mathcal{G}$ as a product $\mathcal{H} \oplus \mathcal{K}$, where $\mathcal{H}$ is generated by some independent set of elements of order p^m and $\mathcal{K}$ is maximal independent of $\mathcal{H}$ with no elements of order p^m. There are two cases.

Case (1). If $\mathcal{H}$ is finite, then $\mathcal{K}$ is infinite and may be assumed to have universe B by induction. The result now follows from Lemma 3.8.

Case (2). If $\mathcal{H}$ is infinite, then $\mathcal{H}$ is isomorphic to $\oplus_{i<\omega}\mathbb{Z}(p^m)$ and is therefore recursively isomorphic to a p–time group with universe B by Theorem 4.44. Since $\mathcal{K}$ is recursively isomorphic to a p–time group with universe a subset of $Tal\,(\omega)$ by Theorem 4.45, the result again follows from Lemma 3.8. □

Next we state some results on characteristics. The first result here follows from Theorem 4.45 together with the theorem of Khisamiev cited above.

Theorem 4.48 *For any Σ_2^0 characteristic Q, there is a p–time Abelian group with characteristic Q.*

We will show in Theorem 4.50 that not all recursive characteristics can be realized by p–time groups. The next result shows that any p–time characteristic can be so realized.

Theorem 4.49 ([11]) *Let Q be a nonempty, infinite characteristic such that $tal\,(Q) = \{tal\,([p^m, k]) : (p^m, k) \in Q\}$ is a p–time set. Then there exists a p–time Abelian group with characteristic Q and universe B where*

(a) $B = Bin\,(\omega)$ *or*

(b) $B = Tal\,(\omega)$.

Theorem 4.50 ([13]) *There is a recursive characteristic M such that no recursive group $\mathcal{G}$ with characteristic Q can be isomorphic to any primitive recursive group with universe $Bin\,(\omega)$ or $Tal\,(\omega)$.*

Proof. Let $\mathcal{A}_0, \mathcal{A}_1, \ldots$ be an effective enumeration of the primitive recursive structures (ω, f_e) with one binary operation f_e. Let

$$Q_e = \{p : (p, 0) \in \eta(\mathcal{A}_e)\},$$

and for any $a \in \mathcal{A}_e$, let $|a|_e$ be the order of a in $\mathcal{A}_e$. Construct a set $R = \{r_0 < r_1 < \cdots\}$ of prime numbers such that, for any e, $\mathcal{A}_e$ either

(i) is not an Abelian group with identity 0, or

(ii) has an element of infinite order, or

(iii) has an element of order p^2 for some prime p, or

(iv) has two subgroups of the same prime order, or

(v) has an element of prime order $p \notin Q$.

Now the order $|a|_e$ of an element a may not be a prime. Therefore we need some way to control the prime factors of $|a|_e$. We will now define a recursive function ν such that for any q and any $r > \nu(q)$, r either is divisible by p^2 for some prime p or r has a prime factor bigger than q. Given q, $\nu(q)$ is simply the product of all of the prime numbers $p \leqslant q$.

We will define the set Q in stages. At stage s, we will have $s+1$ elements $q_0 < q_1 < \cdots q_s$ in Q^s along with a certain finite subset I^s of $\omega \times \omega$ of restraints which will prevent numbers from coming into Q at stage s or at any later stage. Let $q_0 = 1$, $Q^0 = \{1\}$ and $I^0 = \emptyset$.

The initial stage of the construction proceeds as follows.

Compute $f_0(1,1)$. There are then two cases.

Case (1). If $f_0(1,1) = 0$, then we have $2 \in Q_0$ or else $\mathcal{A}_0$ is not an Abelian group with identity 0. Thus we can ensure that Q is not the characteristic of $\mathcal{A}_0$ by setting $q_1 = 3$ and restraining 2 from ever coming into Q. We let $I^0 = \emptyset$.

Case (2). If $f_0(1,1) \neq 0$, then we know that either $\mathcal{A}_0$ is not an Abelian group with identity 0 or $|1|_0 > 2$, and therefore either has a prime factor $q > 2$ or is divisible by 4. Now let $q_1 = 2$ and let $I_0 = \{(1,0)\}$. This means that either $\mathcal{A}_0$ will have an element of order 4, thus satisfying part (iii) of the requirement, or we will eventually restrain at least one of the prime factors of $|1|_0$ from ever coming into Q.

At stage $s+1$, we are given $q_0 < \cdots < q_s$ and the set I^s of previous restraints. Moreover, assume by induction that for every $(a,e) \in I^s$, either

(a) $|a|_e \leqslant \nu(q_s)$ and there is a prime factor $q < q_s$ of $|a|_e$ such that $q \neq q_i$ for any $i \leqslant s$, or

(b) $|a|_e \leqslant \nu(q_s)$ and $|a|_e$ is divisible by the square of a prime, or

(c) $|a|_e > \nu(q_s)$.

Now let $k = 1 + \nu(q_s)^2$ and compute $i \cdot a$ in $\mathcal{A}_s$ for each $a \leqslant k$ and each $i \leqslant k$. This will produce a set of equivalence classes $[a]$, where a and b are equivalent if either $b = i \cdot a$ or $a = i \cdot b$ for some $i \leqslant k$. Note that every number $a \leqslant k$ belongs to some equivalence class, but numbers greater than k can also belong. Now all we need is that the computation of $i \cdot a$ in $\mathcal{A}_s$ produces a sequence of distinct elements up until it produces 0. If this is ever violated, then $\mathcal{A}_s$ is not an Abelian group, so that we will have satisfied the e–th requirement. In this case, we let $I^{s+1} = I^s$ and we choose q_{s+1} to be the least prime $p > q_s$ which does not violate any of the restraints $(t, b) \in I^{s+1}$. That is, q_{s+1} is the least q such that (a), (b) or (c) above are satisfied for all restraints in I^{s+1}. Otherwise, there are two further cases.

Case (1). There is some equivalence class which has more than $\nu(q_s)$ elements. In this case, let a be the least such that $[a]$ has more than $\nu(q_s)$ elements. It follows that $|a|_s > \nu(q_s)$. Now put (a, s) into the set of restraints, so that $I^{s+1} = I^s \cup \{(a, s)\}$. Since we will keep this restraint active throughout the construction, it will be the case that either $|a|_s| = \omega$ or else it is finite and has a prime factor p such that either $p \notin Q$ or p^2 divides $|a|_s$.

Then let q_{s+1} be the least prime $p > q_s$ which does not violate any of the restraints $(t, b) \in I^{s+1}$. That is, q_{s+1} is the least q such that (a), (b) or (c) above are satisfied for all restraints in I^{s+1}.

Case (2). Each class has $\nu(q_s)$ or fewer elements. In this case, $|a|_s \leqslant \nu(q_s)$ for all $a \leqslant k$ and each equivalence class is a cyclic subgroup of $\mathcal{A}_s$. Now, since $k = 1 + \nu(q_s^2)$, there must be at least $\nu(q_s)$ different subgroups among the classes. Since there are no more than $\nu(q_s)$ possible orders (that is, numbers between 2 and $\nu(q_s)$) for these subgroups, there must be two distinct subgroups of the same order in $\mathcal{A}_s$. It follows from this that $\mathcal{A}_s$ has two distinct subgroups of some prime order and hence part (iv) of the s–th requirement will be satisfied.

Then we again let q_{s+1} be the least $q > q_s$ which does not violate any of the restraints $(t, b) \in I^{s+1}$.

This completes the construction. The set $Q = \{q_0, q_1, \ldots\}$ is recursive since $q \in Q \iff (\exists s < q)(q = q_s)$. Let $M = Q \times \{0\}$.

Now suppose that $\eta(\mathcal{A}_s) = M$ for some s. Then $\mathcal{A}_s$ does not have two distinct subgroups of the same order so that Case (2) does not apply at stage $s + 1$. Thus Case (1) must apply and hence there is an element a with finite order $|a|_s = q > \nu(q_s)$ such that (a, s) is in I^t for all $t > s$. Thus there must be a stage $t > s$ such that either condition (a) or (b) is satisfied. That is, $|a|_s \leqslant \nu(q_t)$ and there is a prime factor $q < q_t$ of $|a|_s$ such that either

(i) $q \neq q_i$ for any $i \leqslant t$, or

(ii) $|a|_e$ is divisible by q^2.

In case (i), we have $(q, 0) \in \eta(\mathcal{A}_s)$ but $q \notin Q$. In case (ii), we have $(q^2, 0) \in \eta(\mathcal{A}_s)$ but $(q^2, 0) \notin M$. In either case, we see that $\eta(\mathcal{A}_s)$ is not a subset of M. □

Corollary 4.51 *There is a recursive torsion Abelian group $\mathcal{A}$ which is not isomorphic to any primitive recursive Abelian group with universe Bin (ω) or Tal (ω).*

For groups of unbounded order, Theorem 3.6 of [10] and Theorem 4.21 of [13] give different results.

Theorem 4.52 ([10])

(a) *There is a recursive Abelian group $\mathcal{A}$ which is not recursively isomorphic to any primitive recursive group.*

(b) *There is an exponential-time Abelian group $\mathcal{B}$ which is not recursively isomorphic to any polynomial-time group.*

Theorem 4.53 ([13]) *There is a recursive torsion-free Abelian group which cannot be embedded into any p–time Abelian group.*

5 Uniqueness of Feasible Structures

In this section, we shall survey results on feasible categoricity. Again we shall concentrate mainly on polynomial time structures. As we shall see, unlike recursive model theory where there are many beautiful classification results on recursively categorical structures, there are very few examples of polynomial time categorical structures even if we restrict the universe. Thus most of the results on polynomial time categoricity are negative. Recall that a structure $\mathcal{A}$ with universe B is said to be *p–time categorical over B* if any structure $\mathcal{D}$ with universe B which is isomorphic to $\mathcal{A}$ is in fact p–time isomorphic to $\mathcal{A}$. A similar definition can be given for other notions of feasibility. We note that restricting the universe is crucial if we are to have any positive results due to the following general theorem of Cenzer and Remmel [11].

Theorem 5.1 *For any p–time relational structure* $\mathcal{A} = (A, \{R_i^A\}_{i \in S})$, *there are infinitely many p–time structures* $\mathcal{B}_0 = \mathcal{A}$, $\mathcal{B}_1 = (B_1, \{R_i^1\}_{i \in S})$, $\mathcal{B}_2 = (B_2, \{R_i^2\}_{i \in S})$, ... *which are each recursively isomorphic to* $\mathcal{A}$ *and such that, for each* $m < n$, *there is a p–time map taking* B_n *one-to-one and onto* B_m, *but there is no primitive recursive map from* B_m *into* B_n *which is at most c to 1, for some finite number c. Furthermore, the universes* B_n *may be taken to be subsets of* $Tal(\omega)$ *for each* $n \geqslant 1$.

Sketch Proof. Let $B_0 = A$ and $\mathcal{B}_0 = \mathcal{A}$. Given

$$B_n = \{bin(b_0) < bin(b_1) < \cdots\},$$

let $B_{n+1} = M(B_n) = \{bin(m_0) < bin(m_1) < \cdots\}$ as defined above in the proof of Lemma 3.9 and define the relations R_i on B_{n+1} to make the map taking $bin(b_e)$ to $bin(m_e)$ an isomorphism. □

5.1 Linear Orderings

In this subsection, we survey results of Remmel [69], which was the first paper on polynomial time categoricity. Remmel essentially showed that there are no polynomial categorical linear orderings over either $Tal(\omega)$ or $Bin(\omega)$.

The classic back-and-forth method of Cantor which shows that any two dense linear orderings without end points are isomorphic is crucial to the study of categoricity in linear orderings. The key step in defining an isomorphism between two structures requires a way to select, given two elements $a < b$ of one structures, an element $c < a$, an element $d > b$ and an element e with $a < e < b$. Thus we are led to the following effective notion of density functions in the effort of finding conditions which will provide some form of feasible categoricity.

Definition 5.2 A Γ–computable dense linear ordering $L = (D, <)$ without end points is said to have *Γ–computable density functions* if there are Γ–computable functions f_a, f_b and f_i such that for any x and y in D, $f_b(x) < x < f_a(x)$ and $x < f_i(x, y) < y$.

By carefully following the back-and-forth argument and keeping track of the number of steps required, we obtain the following, Theorems 3.1, 3.2 and 3.3 of [69].

Theorem 5.3 *Suppose* $L_1 = (B, <_1)$ *and* $L_2 = (B, <_2)$ *are polynomial-time dense linear orderings without endpoints with polynomial-time density functions. Then*

(a) *if* $B = Tal(\omega)$, L_1 *and* L_2 *are double-exponential-time isomorphic, and*

(b) *if* $B = Bin(\omega)$, L_1 *and* L_2 *are triple-exponential-time isomorphic.*

Theorem 5.4 *Suppose* $L_1 = (B, <_1)$ *and* $L_2 = (B, <_2)$ *are polynomial-time dense linear orderings without endpoints with linear-time density functions. Then*

(a) *if* $B = Tal(\omega)$, L_1 *and* L_2 *are exponential-time isomorphic, and*

(b) *if* $B = Bin(\omega)$, L_1 *and* L_2 *are double-exponential-time isomorphic.*

Theorem 5.5 *Suppose* $L_1 = (Bin(\omega), <_1)$ *and* $L_2 = (Bin(\omega), <_2)$ *are polynomial-time dense linear orderings without endpoints with quasi-real-time density functions. Then* L_1 *and* L_2 *are exponential-time isomorphic.*

Note that the standard ordering on the dyadic rationals in the interval $(0, 1)$ is in fact a p–time linear ordering with quasi-real density functions and has universe p–time isomorphic to $Bin(\omega)$. Details are given in Theorem 3.4 of [69]. On the other hand, there are p–time structures without nice density functions, as shown by Corollary 3.6 of [69].

Theorem 5.6 *There exist p–time dense linear orderings without end points with universe* B *for* $B = Bin(\omega)$ *and* $B = Tal(\omega)$ *which have no primitive recursive density functions.*

We note that there is a possible positive result, namely one can show that any two p–time linear orderings with universe $Tal(\omega)$ which have quasi-real-time density functions are polynomial time isomorphic. However Ash showed that there are no p–time linear orderings with universe $Tal(\omega)$ which have quasi-real-time density functions, see [69].

Examination of the previous theorems shows that the complexity of the back-and-forth isomorphism falls within the scope of exponential iteration. Thus we have the following.

Theorem 5.7 *Suppose $L_1 = (B, <_1)$ and $L_2 = (B, <_2)$ are polynomial-time dense linear orderings without endpoints with q–time density functions. Then for $B = Bin(\omega)$ or $B = Tal(\omega)$, L_1 and L_2 are q–time isomorphic.*

The main result of [69] improves Theorem 5.1 above by obtaining models with a fixed universe. This result shows that there really are no categorical linear orderings.

Theorem 5.8 *Let L be a p–time linear ordering with universe B, either $Tal(\omega)$ or $Bin(\omega)$. Then*

(a) *There exists a p–time linear ordering L' with universe B which is not primitive recursively isomorphic to L.*

(b) *If L is not recursively categorical, then there exists a p–time linear ordering L'' with universe B which is not recursively isomorphic to L.*

Sketch Proof. We just sketch the proof of part (a). If L is recursively categorical, then L contains a copy of a dense linear ordering without end points. Then by Theorems 5.6 and 5.7, there exist p–time orderings L_1 and L_2 with universe B, one having p–time density functions and one without primitive recursive density functions. Thus L may not be primitive recursively isomorphic to both structures. □

5.2 Injection Structures

For injection structures, Cenzer and Remmel classified in Theorem 3.2 of [11] the recursively categorical injection structures.

Theorem 5.9 *A recursive injection structure (A, f) is recursively categorical if and only if it has only finitely many infinite orbits.*

The feasible categoricity results for injection structures depend on the spectrum of orbits. For example, there is one very nice positive result, Theorem 3.7 of [15].

Theorem 5.10 *Let $\mathcal{A} = (A, f)$ and $\mathcal{B} = (B, g)$ be two finitary permutation structures such that all but finitely many orbits have the same size q for some finite q.*

(a) *If $\mathcal{A}$ and $\mathcal{B}$ are both p–time over Tal (ω), then $\mathcal{A}$ is p–time isomorphic to $\mathcal{B}$.*

(b) *If $\mathcal{A}$ and $\mathcal{B}$ are both p–time over Bin (ω), then $\mathcal{A}$ is exponential-time isomorphic to $\mathcal{B}$.*

(c) *If $\mathcal{A}$ and $\mathcal{B}$ are both q–time over either Bin (ω) or Tal (ω), then $\mathcal{A}$ is q–time isomorphic to $\mathcal{B}$.*

Sketch Proof. We sketch the argument for *Tal* (ω). We may assume without loss of generality that all orbits have the same size q. The desired isomorphism φ is defined in stages φ^s, in which we enumerate s orbits $A_1, A_2, \dots, A_s$ and $B_1, B_2, \dots, B_s$ of each structure, by defining a sequence of elements $a_1, \dots, a_s$ and $b_1, \dots, b_s$ so that $A_i = \{a_i, f(a_i), \dots, f^{q-1}(a_i)\}$ and similarly for B_i. Then we let $\varphi_k(f^n(a_i)) = f^n(b_i)$. The key to measuring the complexity of this mapping is that since each orbit has q members, $a_k = tal\,(m)$ for some $m \leqslant kq$. □

The general negative result is analogous to Theorem 5.8 above for linear orderings, except that we cannot specify the universe for the non-recursively categorical structures, since as seen by Theorem 5.10 there actually are some p–time categorical structures. Our next result combines Corollaries 3.3 and 3.5 of [11].

Theorem 5.11 *Let $\mathcal{A}$ be a p–time injection structure with universe B where B is either Tal (ω) or Bin (ω). Then*

(a) *There exists an infinite family $\mathcal{A}_i$ of p–time structures each recursively isomorphic to $\mathcal{A}$ which are pairwise not primitive recursively isomorphic.*

(b) *If $\mathcal{A}$ is not recursively categorical, then there exists a p–time structure $\mathcal{A}''$ with universe B which is not recursively isomorphic to $\mathcal{A}$.*

The most general result for recursively categorical structures is the following. This combines Theorems 3.6 and 3.10 of [11].

Theorem 5.12 *Let B be either $Bin(\omega)$ or $Tal(\omega)$ and let $\mathcal{A}$ be an injection structure such that either*

(a) *$\mathcal{A}$ has an infinite orbit, or*

(b) *$\mathcal{A}$ has infinitely many orbits of size q for some finite q and has infinitely many other orbits.*

Then there is an infinite family $\mathcal{A}_i$ of p–time structures each with universe B and isomorphic to $\mathcal{A}$ which are pairwise not primitive recursively isomorphic.

Sketch Proof. There are two distinct arguments. We first sketch the proof in the case that $\mathcal{A}$ has either an infinite orbit or infinitely many orbits of finite size q, together with an infinite set of other elements. We partition the structure into two parts. The first part $\mathcal{B}$ is either the infinite orbit or the infinitely many orbits of size q and may be assumed to have universe B by Theorem 4.31 and 4.32. The second part $\mathcal{C}$ has an infinite family of copies $\mathcal{C}_i$ with universe C_i such that B cannot be primitive recursively embedded in any C_i and such that, by Theorem 5.11, and for any $i \neq j$, $\mathcal{C}_i$ is not primitive recursively isomorphic to $\mathcal{C}_j$. Now just let $\mathcal{A}_i = \mathcal{B} \oplus \mathcal{C}_i$.

For the case of a single infinite orbit, we appeal directly to Lemma 3.9. Here is the construction of a copy of (ω, S) with universe $Bin(\omega)$, but not primitive recursively isomorphic to the standard structure. Let $m_0 < m_1 < \ldots$ be the set from Lemma 3.9, where $A = Bin(\omega)$, and assume $m_0 = 0$.

We define $0\,,\ f^B(0)\,,\ f^B(f^B(0))\,,\ \ldots$ in blocks so that the k–th block is in three parts:

$$3m_k\,,\ 3m_k + 3\,,\ \ldots\,,\ 3m_{k+1} - 3\,,$$

followed by

$$3m_{k+1} - 2\,,\ 3m_{k+1} - 5\,,\ \ldots\,,\ 3m_k + 1\,,$$

and then

$$3m_k + 2\,,\ 3m_k + 5\,,\ \ldots\,,\ 3m_{k+1} - 1\,.$$

The unique isomorphism φ from (ω, S) to (B, f^B) maps $3m_k + 1$ to $2m_{k+1} - m_k - 1$ and is not primitive recursive by Lemma 3.9. □

Finally, we note that if $\mathcal{A}$ has no infinite orbits and the spectrum of $\mathcal{A}$ is p–time in tally as in Theorem 4.34, then the conclusion of Theorem 5.12 also applies by Theorem 3.8 of [11].

5.3 Models of Arithmetic

Before focusing on the categoricity of torsion Abelian groups, we briefly present two results for the group $\mathbb{Z}$ of integers. These are Theorems 4.28, 4.29 and 4.30 of [11].

Theorem 5.13 *Let B be Tal (ω) or Bin (ω). There is a p–time structure $(B, S^B, +^B)$ isomorphic to $(\mathbb{Z}, S, +)$, but not exponential-time isomorphic.*

Sketch Proof (Binary case). Let $\langle 0, bin(n) \rangle$ represent $n \geqslant 0$ and let $\langle 1, bin(2^{n^2}) \rangle$ represent $-n < 0$. □

Theorem 5.14 *There is a fully p–time group $\mathcal{A}$ isomorphic to $\mathbb{Z}$ but not q–time isomorphic.*

Sketch Proof. This is a corollary of 4.30, since the model defined there is not q–time isomorphic to $\mathcal{N}$. To see this, observe that the term $E^n 0$, which has length $n+1$ is mapped to the iterated exponential $2^{2^{\cdots}}$. □

Let $Bin(\mathbb{Z})$ be the standard structure of $\mathbb{Z}$ with universe p–time isomorphic to $Bin(\omega)$, and similarly for $Tal(\mathbb{Z})$.

Theorem 5.15 *There is an* EXPTIME *(respectively, exponential-time) group $\mathcal{A}$ with universe Bin (ω) (resp. Tal (ω)) which is isomorphic to Bin $(\mathbb{Z})$ (resp. Tal $(\mathbb{Z})$), but not q–time isomorphic.*

5.4 Torsion Abelian Groups

The results for Abelian groups are parallel to those given above for injection structures. We begin with the positive results, Theorems 4.24 and 4.25 of [11].

Theorem 5.16 *Let p be a prime, and let $\mathcal{A}$ and $\mathcal{B}$ be two groups with universe B, where $B = Tal(\omega)$ or $B = Bin(\omega)$, both isomorphic to $\oplus_{n<\omega}\mathbb{Z}(p)$.*

(a) *If $\mathcal{A}$ and $\mathcal{B}$ are p–time, then $\mathcal{A}$ and $\mathcal{B}$ are* EXPTIME *isomorphic if $B = Tal(\omega)$ and double-exponential-time isomorphic if $B = Bin(\omega)$.*

(b) *If $\mathcal{A}$ and $\mathcal{B}$ are q–time, then $\mathcal{A}$ and $\mathcal{B}$ are q–time isomorphic.*

Sketch Proof. We sketch the proof of (a) for universe $Tal(\omega)$. The standard structure $\mathcal{B}$ may be viewed as an infinite dimensional vector space over $\mathbb{Z}(p)$, where the general element $(c_1, \ldots, c_n)$ is represented by

$$tal(c_1 + c_2 \cdot p + \cdots + c_n \cdot p^{n-1}).$$

The arbitrary structure $\mathcal{A}$ will have a basis defined recursively by letting a_n be the least element independent of $\{a_1, \ldots, a_{n-1}\}$. It can then be seen that the map taking $tal(c_1 + c_2 \cdot p + \cdots + c_n \cdot p^{n-1})$ to $c_1 \cdot a_1 + \cdots + c_n \cdot a_n$ is exponential time and its inverse is EXPTIME. □

The next result, Theorem 4.26 of [11] shows that a p–time isomorphism is not always possible in Theorem 5.16.

Theorem 5.17 *For any prime p and for $B = Tal(\omega)$ or $B = Bin(\omega)$, there exist two p–time groups with universe B and which are isomorphic to $\oplus_{i<\omega}\mathbb{Z}(p)$ but which are not p–time isomorphic to each other.*

The case of $\oplus_\omega \mathbb{Z}(p^m)$ where $m > 1$ requires a stronger hypothesis. The difficulty is that only elements not divisible by p, can be used for the generators $a_1, a_2, \ldots$ and these may all be very large. (This is the basis for the proof of Theorem 5.17 above.) What is needed is the ability to compute a divisor of an element x which is divisible by p. Let us say that the group $\mathcal{A}$ has *recursive divisors* if there is an algorithm which, for any $a \in \mathcal{A}$, determines whether a is divisible and which computes a divisor of a if there is one; if the algorithm runs in polynomial time, then we say that $\mathcal{A}$ has *p–time divisors.* Note that the standard models of the recursively categorical groups all have p-time divisors.

Theorem 5.18 *Let p be a prime, let $m > 1$ be finite, and let $\mathcal{A}$ and $\mathcal{B}$ be two groups with universe B, $B = Tal(\omega)$ or $B = Bin(\omega)$, both isomorphic to $\oplus_\omega \mathbb{Z}(p^m)$.*

(a) *If $\mathcal{A}$ and $\mathcal{B}$ are p–time and have p–time divisors, then $\mathcal{A}$ and $\mathcal{B}$ are* EXPTIME *isomorphic if $B = Tal(\omega)$ and double-exponential-time isomorphic if $B = Bin(\omega)$.*

(b) *If $\mathcal{A}$ and $\mathcal{B}$ are q–time and have q–time divisors, then $\mathcal{A}$ and $\mathcal{B}$ are q–time isomorphic.*

The next result, Theorem 4.19 of [11] shows that the hypothesis of p–time divisibility is needed in Theorem 5.18.

Theorem 5.19 *Let B be either Bin (ω) or Tal (ω), let p be a prime and let $m > 1$ be finite. Then there exists an infinite family of p–time groups $\mathcal{G}_i$ each recursively isomorphic to $\oplus_{i<\omega}\mathbb{Z}(p^m)$ and having universe B such that there is no primitive recursive structure preserving embedding from $\mathcal{G}_i$ into $\mathcal{G}_j$ for any $i < j$.*

The basic non-categoricity result for torsion groups is Theorem 4.11 of [11].

Theorem 5.20 *For any infinite recursive Abelian torsion group $\mathcal{A}$, there is an infinite family $\mathcal{A}_i$ of p–time groups each recursively isomorphic to $\mathcal{A}$ and having universe a subset of Tal (ω) which are pairwise not primitive recursively isomorphic.*

It is also the case that if some p–primary component of $\mathcal{A}$ is infinite and has bounded order, or is isomorphic to $\mathbb{Z}(p^\infty)$, then each $\mathcal{A}_i$ in Theorem 5.20 may be taken to have standard universe.

Next we give two results for p–groups, Theorems 4.9 and 4.23 of [11]. The first is the fundamental result for non-recursively categorical p–groups and the second is a summary of results for products of basic p–groups.

Theorem 5.21 *Let $\mathcal{G}$ be a recursive p–group which is not recursively categorical. Then there exist p–time groups $\mathcal{H}_1$ and $\mathcal{H}_2$ both isomorphic to $\mathcal{G}$ but not recursively isomorphic to each other. If $\mathcal{G}$ has bounded order, then we may take $\mathcal{H}_1$ and $\mathcal{H}_2$ to have universe B where either $B =$ Tal (ω) or $B =$ Bin (ω).*

Theorem 5.22 *Let p be a prime number, let $B =$ Tal (ω) or Bin (ω), and let $\mathcal{C}$ be an infinite recursive group which is a product of cyclic and quasi-cyclic p–groups and which is not isomorphic to $(\oplus_{i<\omega}\mathbb{Z}(p)) \bigoplus \mathcal{F}$ for any finite group $\mathcal{F}$, and either $\mathcal{C}$ has a quasi-cyclic factor or is a product of cyclic groups such that $\eta(\mathcal{C})$ is p–time in tally.*

(a) *Then there exists an infinite family $\mathcal{A}_i$ of p–time groups with universe B and isomorphic to $\mathcal{C}$ which are pairwise not primitive recursively isomorphic.*

(b) *If $\mathcal{C}$ is not recursively categorical, then there exist p–time groups $\mathcal{A}_1$ and $\mathcal{A}_2$, each with universe B and isomorphic to $\mathcal{C}$, which are not recursively isomorphic to each other.*

The group $\mathbb{Q}$ of rationals is closely related to the quasi-cyclic groups $\mathbb{Z}(p^\infty)$, since $\mathbb{Q}\cap[0,1]$ is isomorphic to the product of the quasi-cyclic groups. We use this to obtain the following result, Theorem 4.31 of [11].

Theorem 5.23 *Let B be either Bin (ω) or Tal (ω). Then there is an infinite family of p–time groups $\mathcal{H}_i$ each with universe B and isomorphic to $\mathbb{Q}$ but not pairwise primitive recursively isomorphic.*

5.5 Scott Families

We now consider some general, syntactic conditions which lead to some feasible categoricity results. Nurtazin [60] and Goncharov [29] provided sufficient conditions to ensure that a model $\mathcal{A}$ with universe A is recursively categorical, namely if there is a finite sequence $(c_0, \ldots, c_{k-1})$ of elements of A and a recursive sequence (called a *Scott family*) of recursive existential formulas $\{\varphi_n(x_1,\ldots,x_m,c_0,\ldots,c_{k-1}) : n < \omega\}$ in the extended language with names for $c_0, \ldots, c_{k-1}$ satisfying the following two conditions:

(1) Every $a_1, \ldots, a_m \in A$ satisfies one of the formulas φ_n.

(2) For each n and for any $(a_1, \ldots, a_m)$ and $(d_1, \ldots, d_m)$, if $\mathcal{A}$ satisfies $\varphi_n(a_1,\ldots,a_m,c_0,\ldots,c_{k-1})$ and $\varphi_n(d_1,\ldots,d_m,c_0,\ldots,c_{k-1})$, then $(A,a_1,\ldots,a_m,c_0,\ldots,c_{k-1})$ is isomorphic to $(A,d_1,\ldots,d_m,c_0,\ldots,c_{k-1})$ via the map which sends a_i to d_i for $i = 1$ to m and c_i to c_i for $i < k$.

Several notions of feasible Scott families were developed in [15] and applied to the feasible structures we have studied. We will present one such formulation here.

A Scott family $\{\varphi_n(x_1,\ldots,x_m,c_0,c_1,\ldots,c_{k-1}) : n < \omega\}$ of p–time existential formulas, for a p–time model $\mathcal{A}$ with universe A, satisfying (1) and (2) as described above is said to be *strongly p–time* if there is some fixed integer $r > 1$ such that the following conditions are satisfied, for each $m \geqslant 0$.

(3) For any finite sequence $a_1, \ldots, a_m$ of elements of $\mathcal{A}$, we can compute in time $\leqslant (\max\{2, m, |a_1|, \ldots, |a_m|\})^r$ a formula φ_t from the list such that $\varphi_t(a_1,\ldots,a_m,c_0,c_1,\ldots,c_{k-1})$ holds in $\mathcal{A}$.

(4) For each formula $\varphi_t(x_1, \ldots, x_m, c_0, \ldots, c_{k-1})$ and each $a_1, \ldots, a_m \in A$, if there exists a such that $\mathcal{A}$ satisfies $\varphi_t(a_1, \ldots, a_m, a, c_0, c_1, \ldots, c_{k-1})$, then there exists such an a with $|a| \leqslant (m+2)^r + \max\{|a_1|, \ldots, |a_m|\}$.

(5) For each $\varphi_t(x_1, \ldots, x_m, c_0, c_1, \ldots, c_{k-1})$ and each $a_1, \ldots, a_m \in A$, if there exists a such that $\mathcal{A}$ satisfies $\varphi_t(a_1, \ldots, a_m, a, c_0, c_1, \ldots, c_{k-1})$, then we can compute an a as described in (4) in time less than or equal to $(\max\{2, m, |a_1|, \ldots, |a_m|\})^r$.

Note that clause (4) above implies that the structure $\mathcal{A}$ has only finitely many types of each arity. The following theorem is proved by a careful analysis of the back-and-forth method.

Theorem 5.24 *If $\mathcal{A}$ and $\mathcal{B}$ possess a common strongly p–time Scott family, then $\mathcal{A}$ and $\mathcal{B}$ are p–time isomorphic if both have universe Tal (ω), and are exponential-time isomorphic if both have universe Bin (ω).*

Theorem 5.5 can be proved directly from this general result. We give one other corollary here which provides some additional feasible categoricity for permutation structures.

Corollary 5.25 *Let $\mathcal{A} = (Tal\,(\omega), f)$ and $\mathcal{B} = (Tal\,(\omega), g)$ be two isomorphic p–time permutation structures such that for some fixed integer k,*

(i) *for any a and a' in the same orbit,*

$$|a'| \leqslant |a| + k,$$

and

(ii) *for any $a_0, a_1, \ldots, a_{m-1} \in B$ and any finite q, if there is an orbit of size q not containing any of $a_0, \ldots, a_{m-1}$, then there is such an orbit containing an element a of size*

$$|a| \leqslant \max\{|a_0|, \ldots, |a_{m-1}|\} + (m+2)^k.$$

Then $\mathcal{A}$ and $\mathcal{B}$ are p–time isomorphic.

Weaker notions of Scott families defined in [15] include the *strongly exponential-time Scott family*, which leads to exponential-time isomorphism for universe *Tal* (ω) and double-exponential-time isomorphism for universe *Bin* (ω), and the *strongly* EXPTIME *Scott family*, which leads to EXPTIME isomorphism for universe *Tal* (ω) and double-exponential-time isomorphism for universe *Bin* (ω). The following applications are given in [15].

Corollary 5.26 *Let $\mathcal{A} = (B, \equiv^A)$ and $\mathcal{B} = (B, \equiv^B)$ be two polynomial-time models of an equivalence relation $\equiv$ such that, for some fixed integer k, both models satisfy the following:*

(i) *for any a and a' in the same equivalence class,*

$$|a'| \leqslant k \cdot |a| \quad \textit{if } B = \textit{Tal}\,(\omega)$$

or, (where $a = \textit{bin}\,(n)$ and $a' = \textit{bin}\,(n')$)

$$n - k|a| \leqslant n' \leqslant n + k|a| \quad \textit{if } B = \textit{Bin}\,(\omega),$$

and

(ii) *for any $a_0, \ldots, a_{m-1} \in B$ and any finite q, if there is an equivalence class of size q not containing any of $a_0, \ldots, a_{m-1}$, then there is such a class containing an element b of size*

$$|b| \leqslant k \cdot \max\{k^m, |a_0|, \ldots, |a_{m-1}|\} \quad \textit{if } B = \textit{Tal}\,(\omega),$$

or

$$|b| \leqslant k \cdot \max\{2, m\} \quad \textit{if } B = \textit{Bin}\,(\omega).$$

Then $\mathcal{A}$ and $\mathcal{B}$ are exponential-time isomorphic if $B = \textit{Tal}\,(\omega)$, and double-exponential-time isomorphic if $B = \textit{Bin}\,(\omega)$.

Corollary 5.27 *Let $\mathcal{A}$ and $\mathcal{B}$ be two isomorphic p–time torsion Abelian groups with the same universe $\textit{Tal}\,(\omega)$ such that for some fixed integer k:*

(i) *for any a, b,*

$$|a + b| \leqslant k \cdot \max\{|a|, |b|\},$$

and

(ii) *for any $a_0, \ldots, a_{m-1}$ in either $\mathcal{A}$ or $\mathcal{B}$ and any finite q, if there is an element of order q not in $G(a_0, \ldots, a_{m-1})$ (that is, the subgroup generated by $\{a_0, a_1, \ldots, a_{m-1}\}$), then there is such an element b of size*

$$|b| \leqslant k^m \cdot \max\{|a_0|, \ldots, |a_{m-1}|\}.$$

Then $\mathcal{A}$ and $\mathcal{B}$ are EXPTIME *isomorphic if $B = \textit{Tal}\,(\omega)$ and are double-exponential-time isomorphic if $B = \textit{Bin}\,(\omega)$.*

6 Complexity Theoretic Algebra

In this section, we introduce the second theme of our survey. That is, instead of focusing on problems of comparing polynomial-time versus recursive models, we will fix a given polynomial-time model such as an infinite dimensional vector space over a polynomial-time field or a polynomial-time atomless Boolean algebra and consider the internal structure of that model. Once again we shall use established results from recursive algebra as a guide.

In recursive algebra, one studies the effective content of results like the fact that every independent subset of a vector space V can be extended to a basis. If the vector space V is infinite dimensional, then all known proofs of this fact use some version of the axiom of choice, e.g., Zorn's Lemma, which is known to be non-constructive. Thus one would expect that it is not the case that every recursive independent set can be extended to a recursive basis in infinite dimensional recursive vector space. Indeed, Metakides and Nerode [49] proved that not every recursive independent set of a recursively presented infinite dimensional vector space over a recursive field could be extended to a recursive basis. Another theme in the study of recursive algebra has been to study the lattice of r.e. substructures of various recursive structures structures; see the survey article by Nerode and Remmel [50]. Nerode and Remmel began the study of complexity theoretic algebra in a series of papers, [53, 54, 55, 58]. We survey their results as well as results by Bäuerle [5] in the next two sections.

The overriding paradigm of Nerode and Remmel's study of complexity theoretic algebra was to use the admittedly flawed analogy that "recursive is to r.e." as "P is to NP" to formulate natural complexity theoretic analogues of theorems in recursive algebra. For example, Dekker [24] proved that every r.e. subspace of a recursively presented infinite dimensional vector space over a recursive field with a dependence algorithm has a recursive basis. The natural complexity theoretic analogue of Dekker's Theorem is that in a suitable polynomial-time infinite dimensional vector space V over a polynomial-time field with a polynomial-time dependence algorithm, every NP subspace of V has a basis in P. It turns out that the proof of Dekker's Theorem is not uniform in that the proof breaks up into two cases depending on whether the underlying field of V is finite or infinite. The complexity theoretic analogue of Dekker's Theorem behaves very differently in these two cases. That is, Nerode and Remmel [55] proved that if the underlying field is infinite and has a polynomial-time representation with certain nice properties, then every

NP subspace of V has a basis in P. However if the underlying field is finite, then Dekker's Theorem is oracle dependent. That is, there is an oracle X such that $\mathrm{P}^X \neq \mathrm{NP}^X$ and every NP^X subspace of V has a basis in P^X and there is an oracle Y such that $\mathrm{P}^Y \neq \mathrm{NP}^Y$ and there is a subspace W of V which is NP^Y but has no basis in P^Y. This presents us with two general themes. Sometimes the complexity theoretic analogue of a theorem of recursive algebra is true but must be proved by more delicate methods which take into account the bounds of the resources used in a computation. Sometimes the complexity theoretic analogue is false or oracle dependent because the proof of the recursive algebra result uses unbounded resources available in a recursive construction in a crucial way. Thus complexity theoretic algebra is not just a mere translation of the results of recursion theoretic algebra.

Another problem that complicates the study of complexity theoretic algebra is that fact that not all polynomial-time models are equivalent, as we have seen in the previous sections. That is, Metakides and Nerode showed that all infinite dimensional recursive vector spaces with an effective dependence algorithm are recursively isomorphic. Similarly, Cantor's basic back-and-forth argument which shows that all countable free Boolean algebras are isomorphic is effective so that all recursive free Boolean algebras are recursively isomorphic. As we have seen in the previous section, it is certainly not the case that all polynomial-time free Boolean algebras are polynomial-time isomorphic. Thus in complexity theoretic algebra, one fixes a polynomial-time presented structure over a natural universe such as the tally representation of the natural numbers or the binary representation of the natural numbers, and studies that particular structure. Indeed, Nerode and Remmel studied two basic models of vector spaces, the *tally representation* of an infinite dimensional vector space of a polynomial-time field with a polynomial-time dependence algorithm, $Tal\,(V_\infty)$, where the underlying universe is the tally representation of the natural numbers, and the *binary representation* of an infinite dimensional vector space of a polynomial-time field with a polynomial-time dependence algorithm, $Bin\,(V_\infty)$, where the underlying universe is the binary representation of the natural numbers. Similarly they consider a tally representation and a binary representation of the free Boolean algebra. The results for the tally representation and standard representation of a structure are not always the same.

Another basic question in the study of complexity theoretic algebra is whether the priority method which was so useful in the study of recursive algebra would again play a central role. In 1975, Metakides and Nerode [48] initiated the systematic study of recursion theoretic algebra and introduced the use of the finite injury priority method from recursion theory as a uniform tool to meet algebraic requirements. Prior to that time the priority method has been limited primarily to internal applications within recursion theory in the theory of recursively enumerable sets and in the theory of degrees of unsolvability and their generalizations. Recursion theoretic algebra has been developed since, in depth, by many authors in such subjects as commutative fields, vector spaces, orderings, and Boolean algebras (see Crossley [21] for references and a cross-section of results before 1980). Recursion theoretic algebra yielded as a byproduct a theory of recursively enumerable substructures (see the survey article Nerode-Remmel [50] for references).

Simultaneously in computer science there was a vast development of P and NP problems in complexity theory. This subject started out as a tool for measuring the relative difficulties of classes of computational problems (see Cobham [19], Cook [20], Hartmanis and Stearns [35]). Many papers in this area have dealt with coding a given problem M into a calibrated problem to find an upper bound on the complexity of M, and coding a calibrated problem into a given problem M to find a lower bound the complexity of M (see Hopcroft and Ullman [37] and Garey and Johnson [28]). Due to the intractability of the fundamental problem $\mathrm{P} = \mathrm{NP}$, Baker-Gill-Solovay [3] began a line of inquiry using diagonal arguments to produce sets ("oracles") R_1, R_2 such that $\mathrm{P}^{R_1} = \mathrm{NP}^{R_1}$, $\mathrm{P}^{R_2} \neq \mathrm{NP}^{R_2}$. Typical of recent work in this direction is the construction by Yao [76] of oracles relative to which none of the polynomial time hierarchy collapses, and the result of Cai [8] that this holds for oracles with probability 1. The Baker-Gill-Solovay, Yao, and Cai results are fundamental, but they do not use the priority method which was used systematically with success in recursion-theoretic algebra.

Priority arguments have been used by many authors in the study of P^A and NP^A sets for recursive or recursively enumerable oracles A. For example, Homer and Maass [36], used priority arguments to investigate the lattice of NP^A sets. Shinota and Slaman [72] and Shore and Slaman [73] have used priority arguments to study the structure of the polynomial time Turing degrees relative to a recursive oracle. Downey and Fellows [25] used priority arguments to study the density of their fixed parameter complexity classes. Nerode and Remmel [53, 54, 55, 58] showed that indeed priority methods play a central role in the study of complexity theoretic algebras as we will bring out in the following sections.

We will start by surveying results of Nerode, Remmel and Bäuerle on polynomial time vector spaces.

7 Polynomial-Time Vector Spaces

In this section, we shall study the structure of an infinite dimensional vector space V_∞ over a polynomial time field. We will start by giving some basic definitions and defining the binary or standard representation of V_∞ and the tally representation of V_∞. Our definitions of the standard and tally representation of V_∞ will be broken down into two cases depending on whether the underlying field F is finite or infinite.

A recursive field $F = \langle U_F, +_F, \cdot_F, AI_F, MI_F \rangle$ consists of a recursive subset U_F of the natural numbers ω and partial recursive functions $+_F$ (field addition), $\cdot_F$ (field multiplication), AI_F (field additive inverse), and MI_F (field multiplicative inverse) such that these operations restricted to F turn U_F into a field. A *recursively presented vector space* $V = \langle U_V, +_V, \cdot_V \rangle$ consists of a recursive subset U_V of the natural numbers and partial recursive functions $+_V$ (vector space addition) and $\cdot_V : U_F \times U_V \to U_V$ (scalar multiplication) which turn U_V into a vector space. V is said to have a *dependence algorithm* if there is a uniform effective procedure which given any n-tuple $v_0, \ldots, v_{n-1}$ will determine if $v_0, \ldots, v_{n-1}$ are dependent.

We say that a recursive field $F = \langle U_F, +_F, \cdot_F, AI_F, MI_F \rangle$ is a *polynomial-time field* if U_F is a polynomial-time subset of $\{0,1\}^*$, and the operations $+_F, \cdot_F, AI_F, MI_F$ are the restrictions of total polynomial-time functions. We will always assume that $0, 1 \in U_F$ and that 0 is the zero of F and that 1 is the multiplicative identity of F.

Let V_∞ be the infinite dimensional vector space over a polynomial-time field F which consists of all finite sequences $\langle a_1, \ldots, a_n \rangle$ of elements of F where $a_n \neq 0$ together with the empty sequence $\emptyset$ which is the zero of the vector space. The operations on V_∞ are induced by coordinate-wise addition and scalar multiplication. Finally we say that a vector $v = \langle a_1, \ldots, a_n \rangle$ of V_∞ where $a_i \in F$ for $1 \leqslant i \leqslant n$ and $a_n \neq 0$ has *height* n. We say that the zero vector of V_∞ has height 0.

Case 1: F is finite. Suppose that $F = \{0, 1, \ldots, k-1\}$ is a finite field where 0 is the zero of F and 1 is the multiplicative identity of F. The space V_∞ can be coded into the natural numbers $\omega = \{0, 1, 2, \ldots\}$ as a polynomial-time vector space in many ways. We refer to $e_1, e_2, \ldots$ as the standard basis of V_∞ where e_n is the sequence of the length n, $\langle 0, \ldots, 0, 1\rangle$ with $n-1$ zeros and 1 denotes the unit of F. Now the question of whether V_∞ is polynomial-time, recursive, etc., depends on how we code the sequences $\langle a_1, \ldots, a_n\rangle$. Following [53, 55], we will distinguish two specific polynomial-time representations of V_∞ which we call the *tally* and *binary* (or *standard*) representations of V_∞. We identify each vector $v \in V_\infty$ with a natural number $R(v)$ by $R(\vec{0}) = 0$ and

$$R(\langle a_1, \ldots, a_n\rangle) = a_1 + a_2 k + \ldots a_n k^{n-1} \quad \text{if } a_n \neq 0.$$

Next, with a slight abuse of notation, we define maps $b_k : V_\infty \to B_k(\omega)$, $bin : V_\infty \to Bin(\omega)$ and $tal : V_\infty \to Tal(\omega)$ by $b_k(v) = b_k(R(v))$, $bin(v) = bin(R(v))$ and $tal(v) = tal(R(v))$.

Then $B_k(V_\infty)$ consists of the set $B_k(\omega)$ with the operations of vector addition $+_{B_k}$ and scalar multiplication $\cdot_{B_k}$ induced by the corresponding operations from V_∞. Similarly, $bin(V_\infty)$ consists of the set $Bin(\omega) = \{bin(v) : v \in V_\infty\}$ with corresponding induced operations $+_{bin}$ and $\cdot_{bin}$, and $tal(V_\infty)$ consists of the set $Tal(\omega)$ with the induced operations $+_{tal}$ and scalar multiplication $\cdot_{tal}$. It is easy to see that $B_k(V_\infty)$, $bin(V_\infty)$ and $tal(V_\infty)$ are polynomial-time structures and it follows from Lemma 3.4 that $B_k(V_\infty)$ and $bin(V_\infty)$ are p–time isomorphic. We shall normally refer to either of these two structures as the *standard* representation $st(V_\infty)$ of V_∞ and write the operations as $+_{st}$ and $\cdot_{st}$.

Case 2: F is infinite. Recall the p–time coding functions $\langle \sigma_1, \ldots, \sigma_k\rangle_k$ defined in Section 3. Now suppose that $F = \langle U_F, +_F, \cdot_F, AI_f, MI_F\rangle$ is an infinite polynomial-time field of characteristic 0. Let $\mathbf{0}$ and $\mathbf{1}$ denote the zero and 1 of F respectively. For any positive integer n, let $\mathbf{n} = \mathbf{1} + \cdots + \mathbf{1}$ where there are n summands and let $-\mathbf{n} = AI_F(\mathbf{n})$. For any integers n and $m \neq 0$, let $\mathbf{n}/\mathbf{m} = \mathbf{n} \cdot_F MI(\mathbf{m})$. Then set

$$Q^+ = \{\mathbf{n}/\mathbf{m} : n \in \omega, m \in \omega \setminus \{0\}\}.$$

Thus Q^+ is a copy of the nonnegative rationals inside of F. We say that Q is *properly embedded* in F if

(i) Q^+ is a polynomial-time subset of $\{0,1\}^*$, and

(ii) the map $f : Q^+ \to \{0,1\}^*$ given by $f(\mathbf{n}/\mathbf{m}) = [bin(n), bin(m)] = bin([n, m])$ is the restriction of a polynomial-time function from $\{0,1\}^*$ to $\{0,1\}^*$.

Now suppose that $F = \langle U_F, +_F, \cdot_F, AI_f, MI_F \rangle$ is a polynomial-time field where Q^+ is properly embedded and $U_F = \{0,1\}^*$, and then define $bin : V_\infty \to Bin(\omega)$ by

$$bin(\vec{0}) = 0, \quad \text{and}$$

$$bin(\langle a_1, \ldots, a_n \rangle) = \langle a_1, \ldots, a_n \rangle_n \quad \text{for } a_1, \ldots, a_n \text{ in } F \text{ with } a_n \neq 0.$$

In this case, we let $st(V_\infty) = (U_b, +_b, \cdot_b)$, where $U_b = \{bin(v) : v \in V_\infty\}$ and where the operations $+_b$ and $\cdot_b$ are defined so that *bin* is an isomorphism from V_∞ onto $st(V_\infty)$. It is easy to see that the operations $+_b$ and $\cdot_b$ are the restrictions of polynomial-time functions and that U_b is polynomial-time isomorphic to $\{0,1\}^*$. We call $st(V_\infty)$ the *binary representation* (or the *standard representation*) of V_∞ in this case.

The tally representation of V_∞ is defined by observing that that if $\sigma = \sigma_0 \cdots \sigma_n$ is any string of U_{st} other than the empty string, then σ ends in a 1. Hence there is an integer n_σ such that $bin(n_\sigma) = \sigma_n \cdots \sigma_1$.

Now define a map $tal : V_\infty \to Tal(\omega)$ by $tal(v) = tal(n_v)$, where n_v is the natural number such that $bin(v) = bin(n)$ and let $tal(V_\infty) = (U_t, +_t, \cdot_t)$, where $U_t = \{tal(v) : v \in V_\infty\}$ and the operations $+_t$ and $\cdot_t$ are defined so that *tal* is an isomorphism from V_∞ onto $tal(V_\infty)$. It is easy to see that the operations $+_t$ and $\cdot_t$ are the restrictions of polynomial-time functions, and that U_t is polynomial-time isomorphic to $Tal(\omega)$. We call $tal(V_\infty)$ the *tally representation* of V_∞ in this case.

Finally we argue that both the standard and tally representation of V_∞ have polynomial time dependence algorithms. First the decoding functions π_i^k defined in Section 3 allow us to recover the coefficients $a_1, \ldots, a_k$ from any vector $bin(v) = \langle a_1, \ldots, a_k \rangle_k \in st(V_\infty)$. We can then similarly recover the coefficients from $tal(v)$ by first computing $bin(v)$. It follows that, given any set $v_1, \ldots, v_n$ of vectors in either of our representations of V_∞, we can recover the matrix of coefficients of $v_1, \ldots, v_n$ corresponding to the expansions of those vectors in terms of the standard basis $e_1, e_2, \ldots$ of V_∞ in polynomial time in the sums of the lengths $|v_1| + \cdots + |v_n|$. We can then use Gaussian elimination on the matrix of coefficients to determine whether or not $\{v_1, \ldots, v_n\}$ is an independent set. Since Gaussian elimination is polynomial-time over the coefficients (since the operations of F are polynomial-time), it follows that in each of our representations, there is a polynomial p such that we can decide if $\{v_1, \ldots, v_n\}$ is dependent in $p(|v_1| + \cdots + |v_n|)$ steps.

We end this section with some basic definitions and notations for vector spaces. Let V be either V_∞, $st(V_\infty)$ or $tal\,(V_\infty)$. We shall abuse notation and let $\vec{0}$ denote the zero vector for V_∞, $st(V_\infty)$, and $tal\,(V_\infty)$ even though technically the zero vectors of the three vector spaces are distinct objects. Given a subset A of V, we let $space(A)$ denote the subspace of V generated by A. Given two subspaces U and W of V, we let $U+W$ denote the subspace generated by $U \cup W$. We shall write $W = U_1 \oplus U_2$ if W, U_1 and U_2 are subspaces of V such that $W = U_1 + U_2$ and $U_1 \cap U_2 = \{\vec{0}\}$. We say U is a complementary subspace of W if $U \oplus W = V$. Given $x \in V$, we let $ht(x)$ denote the height of x. We note that if $x \in st(V_\infty)$, then in polynomial-time in $|x|$, we can produce the binary representations of the integers $a_1, \ldots, a_n$ such that $x = bin\,(\langle a_1, \ldots, a_n\rangle)$ with $a_n \neq 0$ so that we can find the height of x in polynomial-time in $|x|$. Similarly if $x \in tal\,(V_\infty)$, then in polynomial-time in $|x|$, we can produce the tally representations of the integers $a_1, \ldots, a_n$ such that $x = tal\,(\langle a_1, \ldots, a_n\rangle)$ with $a_n \neq 0$ so that we can find the height of x in polynomial-time in $|x|$.

7.1 Subspaces and Bases over Infinite Polynomial-Time Fields

We shall see that there is a vast difference between the theory of bases and subspaces of $st(V_\infty)$ or $tal\,(V_\infty)$ when the underlying field is infinite as opposed to when the underlying field is finite. For example, Nerode and Remmel proved the following strengthening of Dekker's Theorem that every r.e. subspace of a recursively presented vector space over a recursive field with a dependence algorithm has a recursive basis.

Theorem 7.1 ([55]) *Let F be a polynomial-time field where Q is properly embedded. Then*

(a) *every r.e. subspace V of $tal\,(V_\infty)$ has a basis in* P, *and*

(b) *every r.e. subspace W of $st(V_\infty)$ has a basis in* P.

Bäuerle [5] proved the existence of simple and maximal subspaces of $tal\,(V_\infty)$ which are in P. To properly state Bäuerle's results, we first need some definitions.

In the lattice, $\mathcal{E}$, of recursively enumerable (r.e.) sets of natural numbers, an r.e. set S is *simple* if $\omega \setminus S$ is infinite and for any infinite r.e. set W, $W \cap S \neq \emptyset$. An r.e. set M is *maximal* if $\omega \setminus M$ is infinite and for any r.e. set $W \supseteq M$, either $\omega \setminus W$ or $W \setminus M$ is finite.

The analogues of these notions in the lattice of NP^A sets, $\mathcal{E}_{\mathrm{NP}^A}$, for any oracle A are the following. An NP^A set $S \subseteq \{0,1\}^*$ is NP^A*–simple* if $\{0,1\}^* \setminus S$ is infinite and for any infinite NP^A set $W \subseteq \{0,1\}^*$, $W \cap S \neq \emptyset$. An NP^A set $M \subseteq \{0,1\}^*$ is NP^A*–maximal* if $\{0,1\}^* \setminus M$ is infinite and for any NP^A set $W \supseteq M$, either $\{0,1\}^* \setminus W$ or $W \setminus M$ is finite.

It was shown by Homer and Maass [36], that there exist oracles A and B such that $\mathrm{NP}^A \neq \mathrm{P}^A$ and no NP^A–simple sets exist, and $\mathrm{NP}^B \neq \mathrm{P}^B$ and there exist NP^B–simple sets. It follows from a result of Briedbart [7] that there are no NP^A-maximal sets for any A.

In the lattice, $\mathcal{L}(V_\infty)$, of r.e. subspaces of a recursively presented copy of V_∞, an r.e. subspace S of V_∞ is *simple* if the dimension of the quotient space V_∞/S is infinite and for any infinite dimensional r.e. subspace W of V_∞, $W \cap S \neq \{\vec{0}\}$. An r.e. subspace M is *maximal* if the dimension of V_∞/M is infinite and for any r.e. subspace $W \supseteq M$, either the dimension of V_∞/W or the dimension of W/M is finite. An r.e. subspace M is *supermaximal* if the dimension of V_∞/M is infinite and for any r.e. subspace $W \supseteq M$, either $V_\infty = W$ or the dimension of W/M is finite.

The NP analogues of these notions in $st(V_\infty)$ and $tal\,(V_\infty)$ are the following. Let A be an oracle, then an NP^A subspace S of $st(V_\infty)$ (respectively $tal\,(V_\infty)$) is NP^A*–simple* if the dimension of $st(V_\infty)/S$ (resp. $tal\,(V_\infty)/S$) is infinite and for any infinite NP^A subspace W of $st(V_\infty)$ (resp. $tal\,(V_\infty)$), $W \cap S \neq \{bin\,(\vec{0})\}$ (resp. $W \cap S \neq \{tal\,(\vec{0})\}$). An NP^A subspace M is NP^A*–maximal* if the dimension of $st(V_\infty)/M$ (resp. $tal(V_\infty)/M$) is infinite and for any NP^A subspace W of $st(V_\infty)$ (resp. $tal\,(V_\infty)$), either the dimension of $st(V_\infty)/W$ (resp. $tal\,(V_\infty)/W$) or the dimension of W/M is finite. An NP^A subspace M is NP^A*–supermaximal* if the dimension of $st(V_\infty)/M$ (resp. $tal\,(V_\infty)/M$) is infinite and for any NP^A subspace W of $st(V_\infty)$ (resp. $tal\,(V_\infty)$), either $st(V_\infty) = W$ (resp. $tal\,(V_\infty) = W$) or the dimension of W/M is finite.

Nerode and Remmel [58] introduced a slightly weaker notion than an NP^X–simple subspace which they called a P^X–simple subspace. Note that in the case of simple sets or simple subspaces, we can replace the infinite r.e. set W or the infinite dimensional r.e. subspace W by an infinite recursive set W or an infinite dimensional recursive subspace. That is, every infinite r.e. set W contains an infinite recursive set and every infinite dimensional r.e. subspace V of V_∞ contains an infinite dimensional recursive subspace. Thus an r.e. set S is simple iff $\omega \setminus S$ is infinite and for any infinite recursive set W, $W \cap S \neq \emptyset$. Similarly an r.e. subspace S of V_∞ is simple iff the dimension of V_∞/S is infinite and for any infinite dimensional recursive subspace W of V_∞, $W \cap S \neq \{\vec{0}\}$. Thus we make the following definition. Let A be an oracle, then an NP^A subspace S of $st(V_\infty)$ (respectively $tal\,(V_\infty)$) is P^A*-simple* if the dimension of $st(V_\infty)/S$ (resp. $tal\,(V_\infty)/S$) is infinite and for any infinite dimensional P^A subspace W of $st(V_\infty)$ (resp. $tal\,(V_\infty)$), $W \cap S \neq \{bin\,(\vec{0})\}$ (resp. $W \cap S \neq \{tal\,(\vec{0})\}$). It follows from results of Nerode and Remmel [55] that there exist oracles A such that there exists an infinite dimensional NP^A subspace V of $tal\,(V_\infty)$ such that V has no infinite dimensional subspace $W \in \mathrm{P}^A$. Thus while a subspace W which is NP^A–simple is certainly P^A–simple, it is not clear that every P^A–simple subspace of $tal\,(V_\infty)$ is NP^A–simple.

Given a subspace V of $st(V_\infty)$ $(tal(V_\infty))$, we let

$$D_n(V) = \{\langle v_1, \ldots, v_n\rangle_n : v_1, \ldots, v_n \text{ are dependent}\}$$

$$D(V) = \bigcup_{n \geqslant 1} D_k(V).$$

The Turing degree of $D_n(V)$ is called the n–th dependence degree and the Turing degree of $D(V)$ is called the dependence degree of V. (The sets $D_n(V)$ and $D(V)$ can be defined for any subspace of a recursively presented vector space over a recursive field using a suitable coding of the finite sequences of ω.) Nerode and Remmel [51] proved the following.

Theorem 7.2 *Assume the underlying field F of $tal\,(V_\infty)$ is an infinite recursive field. Let $A_0, A_1, A_2, \ldots$ be any effective sequence of r.e. sets such that $A_1 \leqslant_T A_2 \leqslant_T \cdots \leqslant_T A_0$ and A_0 is not recursive. Then there is a supermaximal subspace V in $tal\,(V_\infty)$ such that $D(V) \equiv_T A_0$ and $D_k(V) \equiv_T A_k$.*

There is a nice application of Theorem 7.2 in the case where we pick $A_1, A_2, \ldots$ to be recursive and A_0 to be nonrecursive. In that case, the supermaximal space V of Theorem 7.2 is recursive, so the quotient space $W = tal\,(V_\infty)/V$ is a recursively presented vector space such that

(i) every r.e. independent set I of W is finite,

(ii) for any fixed n, there is an effective procedure which given an n-tuple $w_1, \ldots, w_n$ will determine if $w_1, \ldots, w_n$ are dependent, but

(iii) W has no dependence algorithm.

Bäuerle [5] proved the following result for $tal(V_\infty)$.

Theorem 7.3 ([5]) *Let F be a polynomial-time field where Q is properly embedded and δ be any nonzero r.e. degree. Then for any finite $k \geqslant 1$, there is a supermaximal subspace V of $tal(V_\infty)$ such that*

(i) *$D_1(V), \ldots, D_k(V)$ are polynomial-time,*

(ii) *for all j, $D_j(V) \in$ PSPACE $\cap$ DEXT, and*

(iii) *$D(V) \in \delta$.*

Thus in particular, there exists a polynomial-time supermaximal subspace W of $tal(V_\infty)$ which is of course automatically NP–simple and NP–maximal. Moreover, if we consider the quotient space $U = tal(V_\infty/W)$, then it is easy to see that U is a polynomial-time vector space. That is, if we identify U with the set of minimal elements in each equivalence class of $tal(V_\infty/W)$, the Q will be a polynomial-time set, and the operations of $tal(V_\infty)$ will induce polynomial-time operations on U which will make it isomorphic to $tal(V_\infty/W)$. Thus we have the following

Theorem 7.4 *There exists a polynomial-time presented vector space U such that the only r.e. independent sets of U are finite.*

As we shall see in the next section, the analogues of Theorems 7.1 and 7.3 are oracle dependent.

7.2 Subspaces and Bases over finite fields

In this section, we shall state several results on the relation between the complexity of a subspace V of either $st(V_\infty)$ or $tal(V_\infty)$ and the complexity of a basis of that subspace when the underlying field is finite. These results turn out to be essential for many of the more complicated results and constructions in polynomial-time vector spaces.

Note that since the universe of $st(V_\infty)$ is $Bin\,(\omega)$, there is a natural order $<$ on the elements of $st(V_\infty)$ inherited from the standard ordering of the natural numbers. Similarly, since the universe of $tal\,(V_\infty)$ is $Tal\,(\omega)$, there is a natural order $<$ on the elements of $st(V_\infty)$ inherited from the standard ordering of the natural numbers. This given, we can now state some very useful definitions for our purposes. Recall that $e_1, e_2, \ldots$ is the standard basis for V_∞. Thus $R(e_n) = k^{n-1}$.

We start with the definition of a height increasing basis.

Definition 7.5 Let V be a subspace of $st(V_\infty)$ or $tal\,(V_\infty)$.

(1) Call B a *height increasing basis* of V if B is a basis for V and for all $n \geqslant 1$, B has at most one element of height n.

(2) The *standard height increasing basis* of V, B_V, is defined by declaring that $x \in B_V$ iff $x \in V$ and there is no $y \in V$ such that $y < x$ and $ht(y) = ht(x)$.

(3) The *standard height increasing complementary basis* of $V \subseteq tal\,(V_\infty)$, $B_{\overline{V}}$, is defined in $tal\,(V_\infty)$ by declaring that $tal\,(e_n) \in B_{\overline{V}}$ iff $tal\,(e_n) \notin V$ and there is no $y \in V$ such that $ht(y) = n$. Similarly the *standard height increasing complementary basis* of $V \subseteq st(V_\infty)$, $B_{\overline{V}}$, is defined in $st(V_\infty)$ by declaring that $bin\,(e_n) \in B_{\overline{V}}$ iff $bin\,(e_n) \notin V$ and there is no $y \in V$ such that $ht(y) = n$.

(4) We call the $space(B_{\overline{V}})$, the *standard complement* of V.

There is a crucial difference between $st(V_\infty)$ and $tal\,(V_\infty)$ with respect to searches. That is, the vector of height n with the smallest R value is e_n and $R(e_n) = k^{n-1}$. The vector of height n with the largest R value is $(k-1)e_1 + \cdots + (k-1)e_n$ and

$$R((k-1)e_1 + \cdots + (k-1)e_n) = \sum_{i=1}^{N} (k-1)k^{i-1} = k^n - 1\,.$$

Thus in $tal\,(V_\infty)$, given a vector v of height n, we can produce in polynomial time in $|v|$, a list of all vectors of height n in $tal\,(V_\infty)$. However in $st(V_\infty)$, given a vector v of height n, it takes exponential time in $|v|$ to produce a list of all vectors of height n in $st(V_\infty)$. For this reason, the relation between the complexity of V, B_V, $B_{\overline{V}}$, and $space(B_{\overline{V}})$ is very different in $tal\,(V_\infty)$ than in $st(V_\infty)$. For this reason, we shall divide this subsection into two parts, one for $tal\,(V_\infty)$ and one for $st(V_\infty)$, and discuss the relation between the complexity of bases and subspaces for each case separately.

7.2.1 Bases and Subspaces for $tal\,(V_\infty)$.

Nerode and Remmel in [55] studied bases of NP–subspaces of $tal\,(V_\infty)$, so we start by listing a number of results from that paper.

Theorem 7.6 ([55]) *Let V be a subspace of $tal\,(V_\infty)$.*

(a) *If B is a height increasing basis of V, then $V \leqslant_T^{\mathrm{P}} B$.*

(b) *$B_V \leqslant_T^{\mathrm{P}} V$ and $B_{\overline{V}} \leqslant_T^{\mathrm{P}} V$.*

Proof. The key point here is that in our tally representation,

$$\mathrm{card}\,(\{x \in V_\infty : ht(x) \leqslant n\}) = k - 1 + (k-1)k + \cdots + (k-1)k^{n-1} = k^n - 1.$$

Moreover, if $ht(y) \leqslant n$, then $|y| < k^n$. Given $x \in V_\infty$ such that $ht(x) = n$, we know that $|x| \geqslant k^{n-1}$. So there are at most $k|x|$ elements of $tal\,(V_\infty)$ with height less than or equal to $ht(x)$. For q fixed we can run any (uniform) computation which takes at most n^q steps on strings of length n for all the elements of $tal\,(V_\infty)$ of height less than or equal to $ht(x)$ in polynomial time. This is because

$$\sum_{\substack{y \in tal(V_\infty)\\ ht(y) \leqslant ht(x)}} |y|^q \;\leqslant\; \sum_{i=0}^{(k\,|x|)^q} i \;=\; \frac{1}{2}\,(k\,|x|)^q\,[(k\,|x|)^q - 1] \;\leqslant\; k^2\,|x|^{2q}\,.$$

Given these observations it is immediate from our definitions of B_V and $B_{\overline{V}}$ that $B_V \leqslant_T^{\mathrm{P}} V$ and $B_{\overline{V}} \leqslant_T^{\mathrm{P}} V$.

To prove Theorem 7.6 (a), note that if B is a height increasing basis for V, then $x \in V$ iff $x \in space(\{y \in B : ht(y) \leqslant ht(x)\})$. Thus to decide if $x \in V$, we simply search all the elements y in V_∞ with $ht(y) \leqslant ht(x)$ and produce all vectors $y_1, \ldots, y_k$ in $\{y \in B : ht(y) \leqslant ht(x)\}$. We can then use the polynomial-time dependence algorithm to determine if

$$y \in space(\{y \in B : ht(y) \leqslant ht(x)\})$$

in polynomial-time in $|y|$. Thus $V \leqslant_T^{\mathrm{P}} B$. □

An immediate corollary of Theorem 7.6 is the following.

Corollary 7.7 ([55])

(i) *A subspace V of tal (V_∞) is in* P *iff V has a height increasing basis B in* P.

(ii) *If V is a subspace of tal (V_∞) and $V \in$ P, then V has a complementary subspace W in* P.

We note that one cannot replace $\leqslant_T^{\mathrm{P}}$ by $\leqslant_m^{\mathrm{P}}$ in the statement of Theorem 7.6 due to the following result of Nerode and Remmel.

Theorem 7.8 ([55]) *There exists a subspace V of tal (V_∞) such that neither $B_V \leqslant_m^{\mathrm{P}} V$ nor $V \leqslant_m^{\mathrm{P}} B_V$.*

Next we observe that height increasing bases in NP generate NP spaces.

Theorem 7.9 ([58]) *Suppose that A is a height increasing independent set of tal (V_∞) in* NP. *Then space$(A) \in$* NP.

Proof. Note that if A is a height increasing independent set, then $x \in$ *space*(A) iff $x \in$ *space*$(\{y \in A : ht(y) \leqslant ht(x)\})$. Thus $x \in$ *space*(A) iff there are elements $b_1, \ldots, b_n$ of height $\leqslant ht(x)$ and $\lambda_1, \ldots, \lambda_n \in F$ such that $x = \sum_{i=1}^{n} \lambda_i b_i$. Moreover, if $ht(x) = m$, then $k^{m-1} \leqslant |x| \leqslant k^m - 1$ so that each b_i must have length $\leqslant k|x|$. Thus in nondeterministic polynomial time, we can guess $\lambda_1, \ldots, \lambda_n, b_1, \ldots, b_n$, and computations which show that $b_i \in A$, and then verify that $x = \sum_{i=1}^{n} \lambda_i b_i$. Thus *space*$(A)$ is in NP if $A \in$ NP. □

Similarly one can show that if $\mathrm{NP}^X = \text{co–}\mathrm{NP}^X$, then we have the following.

Theorem 7.10 ([55]) *Suppose $\mathrm{NP}^X = co\text{–}\mathrm{NP}^X$, and V is a subspace of tal (V_∞). Then*

(i) *$V \in \mathrm{NP}^X$ iff V has a height increasing basis in NP^X,*

(ii) *$V \in \mathrm{NP}^X$ implies V has a complementary subspace W in NP^X.*

Our next result will allow us to show that the property of a subspace V of $tal(V_\infty)$ having a basis in P does not necessarily tell us anything about the complexity of V other than that V is recursively enumerable.

Theorem 7.11 ([55]) *Let V be a recursively enumerable infinite dimensional subspace of $tal(V_\infty)$. Then the following are equivalent:*

(1) *V has a basis C in* P,

(2) *V contains an infinite dimensional subspace W in* P,

(3) *V contains an infinite height increasing independent subset S in* P.

Another consequence of a subspace containing an infinite independent subset in P is the following.

Theorem 7.12 *Let V be a recursive subspace of $tal(V_\infty)$ such that V contains an infinite height increasing independent set C in* P. *Then if the dimension of $tal(V_\infty)/V$ is infinite, there is an infinite height increasing independent set D in* P *such that $V \cap space(D) = \{\vec{0}\}$.*

Proof. Note that $B_{\overline{V}}$ is recursive. Let $b_0, b_1, \ldots$ be a list of the elements of $B_{\overline{V}}$ such that

$$h(b_0) < h(b_1) < \cdots .$$

Let f be a recursive function such that $f(0^n) = b_n$. Similarly let $c_0, c_1, \ldots$ be a list of the elements of C such that $h(c_0) < h(c_1) < \cdots$. Then let $d_s = b_s +_{tal} c_{r(s)}$ where

$$r(s) = 1 + \sum_{i=0}^{s} h(b_i) + \text{the number of steps to compute } f(0), \ldots, f(s).$$

Then we claim that $D = \{d_0, d_1, \ldots\}$ is our required height increasing independent set. First observe that by our definition of $r(s)$, $r(s) > h(b_s)$ so that $h(d_s) = h(c_{r(s)})$. Also it is clear that $r(0) < r(1) < \cdots$ so that $h(d_0) < h(d_1) < \cdots$. Thus D is a height increasing basis. Moreover it is easy to see that D is independent over V. Thus we need only show that D is p–time. To decide whether a given $x \in tal(V_\infty)$ is in D, we first compute which elements y with $h(y) \leqslant h(x)$ are in C. Now C is a p–time set, so that for all z we can determine whether $z \in C$ in $\max(2, |z|)^m$ steps for some fixed m. Moreover, if $h(x) = n$, then $x = 1^{|x|}$ where $k^{n-1} \leqslant |x| \leqslant k^n - 1$ so that it requires at most

$$2^m + 2^m + \sum_{j=2}^{k^n-1} j^m \leqslant \sum_{j=0}^{k|x|+1} j^m < ((k|x|+1)^m)^2 = (k|x|+1)^{2m}$$

steps to find the elements of C of height less than or equal to $h(x)$. If no element of height $h(x)$ is in C, then clearly $x \notin D$. If there is an element of height $h(x)$ in C, then in polynomial time in $|x|$, we can find r such that $h(c_r) = h(x)$. At this point, we start to compute the sequence of elements $f(0), f(1), \ldots$ in order, for r steps. Suppose that end the end of r steps, we have successfully computed $f(0), \ldots, f(t)$. Note that if we are not successful in computing $f(0)$ by the end of r steps, then $x \notin D$. Otherwise, see if there is some $s \leqslant t$ such that

$$r = 1 + \sum_{i=0}^{s} h(b_i) + \text{the number of steps to compute } f(0), \ldots, f(s).$$

If there is no such s, then $x \notin D$ and if there is such an s, then $x \in D$ iff $x = f(s) +_{tal} c_r$. It follows that we can decide if $x \in D$ in polynomial time in $|x|$, so that D is a p–time height increasing independent set which is independent over V. □

Next we show that having a basis in P does not restrict the degree of a subspace other than ensuring the subspace is recursively enumerable.

Theorem 7.13 *Let δ be any r.e. degree. Then there there exists an r.e. subspace V in tal (V_∞) such that V has a basis in* P.

Proof. Let B_1 be an infinite subset of $\{e_{2n} : n \geqslant 1\}$ in P, and for any given r.e. degree δ, let B_δ be an infinite r.e. subset of $\{e_{2n+1} : n \geqslant 0\}$ of degree δ. Then it is easy to see that the Turing degree of $V_\delta = space(B_1 \cup B_2)$ is δ. By Theorem 8.11, V_δ has a basis in P since $space(B_1)$ is an infinite dimensional subspace of V_δ which is in P. □

It is also easy to construct spaces with no basis in P. In fact, Nerode and Remmel [55] gave a general construction which, given any effective list of r.e. independent sets of *tal* (V_∞), $A_0, A_1, \ldots$, produced a subspace V of *tal* (V_∞) such that $V \cap A_i$ is finite for all. Their construction can be specialized to prove the following results.

Theorem 7.14

(1) *There is a subspace V of $tal\,(V_\infty)$ in* DEXT *such that V has no basis in* P.

(2) *There is a recursive subspace V of $tal\,(V_\infty)$ such that V has no primitive recursive basis.*

(3) *There is a subspace V of $tal\,(V_\infty)$ which is recursive in $\mathbf{0}'$ such that for any r.e. independent set I, $I \cap V$ is finite.*

We should also note that every r.e. subspace has a basis which has high complexity.

Theorem 7.15 ([55]) *Let V be an r.e. subspace of either $tal\,(V_\infty)$ or $st(V_\infty)$. Then V has a recursive basis B which is not primitive recursive.*

All of the results so far do not settle the question of whether every subspace V of $tal\,(V_\infty)$ which is in NP has a basis in P. In fact, this question is oracle dependent. To prove the existence of an oracle B such that every subspace V of $tal\,(V_\infty)$ which is in NP^B has a basis in P^B, Nerode and Remmel proved the following result which strengthens a similar result of Homer and Maass [36].

Theorem 7.16 ([55]) *There is a recursive oracle B such that $\mathrm{P}^B \neq \mathrm{NP}^B$ and such that every infinite set X which is p–time Turing reducible to a set Y in NP^B contains an infinite subset in P^B.*

We note that in light of Theorem 7.11, it also follows that for the oracle B of Theorem 7.16, every NP^B subspace V of $tal\,(V_\infty)$ has a basis in P^B. Thus we have the following.

Theorem 7.17 ([55]) *There is a recursive oracle B such that $\mathrm{P}^B \neq \mathrm{NP}^B$ and every NP^B subspace V of $tal\,(V_\infty)$ has a basis in P^B.*

Via a delayed diagonal argument, Nerode and Remmel also proved the following.

Theorem 7.18 ([55]) *There is a recursive oracle A such that*

(a) *there is an infinite dimensional subspace V in NP^A such that V has no basis in P^A (and hence $\mathrm{NP}^A \neq \mathrm{P}^A$), and*

(b) $\mathrm{NP}^A = co\text{–}\mathrm{NP}^A$.

Combining Theorems 7.17 and 7.18, we have the following.

Theorem 7.19 ([55]) *Arguments valid under relativization are not sufficient to prove*

(1) $\mathrm{P} \neq \mathrm{NP} \implies$ *every subspace of $tal\,(V_\infty)$ in NP has a basis in P, and*

(2) $\mathrm{P} \neq \mathrm{NP} \implies$ *there is a subspace V of $tal\,(V_\infty)$ in NP which has no basis in P.*

We end this section with some results of Bäuerle [5]. We say a set $A \subseteq \{0,1\}^*$ is P^X–immune if there are no infinite subsets of A in P^X. The next results show that a subspace $V \subseteq tal\,(V_\infty)$ can have a basis in P without the standard basis being in P.

Theorem 7.20 ([5]) *There exists an exponential-time subspace V of $tal\,(V_\infty)$ which has a basis in P, but for which the standard height increasing basis of V, B_V, is P–immune.*

Theorem 7.21 ([5]) *There exists a recursive oracle A such that there exists a subspace V of $tal\,(V_\infty)$ which is in $\mathrm{NP}^A \setminus \mathrm{P}^A$, has a basis in P^A, and yet the standard height increasing basis of V, B_V, is P^A–immune.*

Theorem 7.22 ([5]) *There exists a recursive oracle B such that there exists a subspace V of $tal\,(V_\infty)$ in $\mathrm{NP}^B \setminus \mathrm{P}^B$ and such that, for all $\mathrm{NP}^B \setminus \mathrm{P}^B$ subspaces V of $tal\,(V_\infty)$, the standard height increasing basis B_V has an infinite subset in P^B.*

Theorem 7.23 *Let $\mathcal{F}$ be finite and $V \subset tal\,(V_\infty)$. If V has an infinite dimensional subspace in P, then V has a height increasing basis D with a subset in P such that $B_V \equiv_T^{\mathrm{P}} D$.*

Theorem 7.24 ([5]) *Let A be an oracle such that $(\mathrm{NP}^A \setminus \mathrm{P}^A)$–subspaces of $tal(V_\infty)$ exist. Then if V has an infinite dimensional subspace in P^A, then V has a height increasing basis D with a subset in P^A such that $B_V \equiv_T^{\mathrm{P}} D$.*

7.2.2 Bases and Subspaces of $st(V_\infty)$.

It will be convenient to think of $st(V_\infty)$ via the representation $B_k(V_\infty)$ defined above. The advantage is that for nonzero $x \in B_k(V_\infty)$, $ht(x) = |x|$. The standard basis for $B_k(V_\infty)$ is given by $e_n = b_k(k^{n+1}) = 0^n 1$.

As pointed out in the introduction to this section, there is a significant difference between $st(V_\infty)$ and $tal(V_\infty)$ with regard to searches. Indeed many of the proofs of the propositions and theorems in the previous subsection relied on the fact that given an $x \in tal(V_\infty)$, we could produce a list of all elements $tal(V_\infty)$ of height $\leqslant ht(x)$ in polynomial time in $|x|$. This is no longer the case in $st(V_\infty)$. That is, if $x \in tal(V_\infty)$ and $ht(x) = n$, then $k^{n-1} \leqslant |x| \leqslant k^n - 1$ while if $x \in st(V_\infty)$, then $ht(x) = |x|$ so that there are $k^{|x|} - 1$ elements of height less than or equal to $ht(x)$ in $st(V_\infty)$. Thus in $st(V_\infty)$, we cannot find all the elements of height less than or equal to $ht(x)$ in a p–time height increasing set S in polynomial time in $|x|$. However there is a special class of p–time independent sets of $st(V_\infty)$, which we call strongly p–time independent sets, which do have most of the useful properties possessed by p–time height increasing bases of $tal(V_\infty)$.

Definition 7.25 An independent set $B \subseteq st(V_\infty)$ is called *strongly p–time* if

(i) B is a p–time set,

(ii) B is height increasing, and

(iii) if $B = \{b_0, b_1, \dots\}$ where $ht(b_0) < ht(b_1) < \cdots$, then there is a polynomial-time function f such that for all $n > 0$

(a) $f(1^n) = b_k$ if $ht(b_k) = n$ and B has an element of height n,

(b) $f(1^n) = 0$ if B has no element of height n.

We note that condition (iii) allows us to find, for any $x \in st(V_\infty)$, all elements of b of height $\leqslant ht(x)$ in polynomial time in $|x|$. That is, given $x \in st(V_\infty)$, $ht(x) = |x|$ and we can compute $f(1), f(1^2), \dots, f(1^{ht(x)})$ in polynomial time in $|x|$. Then

$$\{b : b \in B \wedge ht(b) \leqslant ht(x)\} = \{f(1^n) : n \leqslant |x| \wedge f(1^n) \neq 0\}.$$

As noted above, any p–time height increasing independent set B in $tal(V_\infty)$ also has the property that, for any x, we can find all elements of B of height $\leqslant ht(x)$ in polynomial time in $|x|$. Thus condition (iii) is specifically designed to give us this property which holds for all p–time height increasing bases in $tal(V_\infty)$ automatically. It is easy to see that our standard basis $\{e_1, e_2, e_3, \ldots\}$ of $st(V_\infty)$ is strongly p–time.

Our next proposition lists several basic properties of subspaces generated by subsets of a strongly p–time basis.

Theorem 7.26 *Let B be a strongly p–time basis of $st(V_\infty)$ and suppose that $S \subseteq B$. Then*

(i) *$S \in$ P iff $space(S) \in$ P.*

(ii) *$S \in$ NP iff $space(S) \in$ NP.*

(iii) *$S \in$ co–NP iff $space(S) \in$ co–NP.*

(iv) *$S \equiv_T^{\mathrm{P}} space(S)$.*

Proof. Since $S = space(S) \cap B$, it follows that $S \leqslant_T^{\mathrm{P}} space(S)$ and S is in P (respectively NP, co–NP) if $space(S)$ is in P (resp. NP, co–NP).

Let f be the p–time function such that $f(1^n) = b_n$, where b_n is the element of height n in B. Then, given an $x \in st(V_\infty)$ of height n, we can compute $f(1) = b_1, \ldots, f(1^n) = b_n$ and test $b_1, \ldots, b_n$ for membership in S, all in time polynomial in $|x|$. Thus in polynomial time in $|x|$, we can find $\{s_1, \ldots, s_k\}$, where $\{b_{s_1}, \ldots, b_{s_k}\} = \{y \in S : ht(y) \leqslant ht(x)\}$. Moreover the fact that B is a height increasing basis means that $x = \sum_{i=1}^{|x|} \lambda_i b_i$ for some $\lambda_1, \ldots, \lambda_{|x|}$ in F. Now suppose that $|x| = n$, then we can write $x = x_1 \cdots x_n$ where all $x_i \in F$ and and each $b_i = b_{i,1} \cdots b_{i,n}$ where $b_{i,j} \in F$. Then we can solve the matrix equation over F

$$BY = X$$

where $B = (b_{i,j})$, Y is a column vector of unknowns, and X is the column vector $(x_1, \ldots, x_n)$ in polynomial time in $n = |n|$. Thus in polynomial time in $|x|$, we can find $\lambda_1, \ldots, \lambda_k$ such that $x = \sum_{i=1}^{|x|} \lambda_i b_i$. This given,

$$x \in space(S) \text{ iff } \{i : \lambda_i \neq 0\} \subseteq \{s_1, \ldots, s_k\}.$$

It then easily follows that $space(S) \leqslant_T^{\mathrm{P}} S$ and $space(S)$ is in P (resp. NP, co–NP) if S is in P (resp. NP, co–NP). □

Our next result is a weak analogue for $st(V_\infty)$ of Theorem 7.7.

Theorem 7.27 *Let V be a subspace of $st(V_\infty)$ with strongly p–time basis R. Then $R \cup B_{\bar{V}}$ is a strongly p–time basis for $st(V_\infty)$, and both V and $space(B_{\bar{V}})$ are in* P.

Our next theorem shows that no extra condition on a height increasing basis, such as condition (iii), is required to generate subspaces of $st(V_\infty)$ in NP.

Theorem 7.28 *Let B be a height increasing independent set of $st(V_\infty)$ which is in* NP. *Then $space(B)$ is in* NP.

Proof. The key property of a height increasing basis is that if $x \in space(B)$, then $x \in space(\{b \in B : ht(b) \leqslant ht(x)\})$. That is, x must be generated by the elements of height $\leqslant ht(x)$ in B if $x \in space(B)$. Thus to see that $space(B) \in$ NP, we simply guess the elements of B of height $\leqslant ht(x)$, say $\{b_1, \ldots, b_k\} = \{b \in B : ht(b) \leqslant ht(x)\}$, where $ht(b_1) < \cdots < ht(b_k)$. Now, for all nonzero $y \in st(V_\infty)$, $ht(y) = |y|$ so $|b_i| \leqslant |x|$ for all i and $k \leqslant |x|$. Then we perform a nondeterministic polynomial-time computation to check if $b_1, \ldots, b_k$ are all in B. Finally, we use our polynomial-time dependence algorithm to check whether $x \in space(\{b_1, \ldots, b_k\})$. Thus $space(B)$ is in NP. □

Theorem 7.29 *Suppose* $\text{NP}^X = co\text{–}\text{NP}^X$ *and V is a subspace of $st(V_\infty)$. Then*

(i) *$V \in \text{NP}^X$ iff V has a height increasing basis in NP^X.*

(ii) *$V \in \text{NP}^X$ implies V has a complementary subspace W in NP^X.*

Our next result is a weak analogue of Theorem 7.11 of [55] for $st(V_\infty)$.

Theorem 7.30 *Let V be an r.e. infinite dimensional subspace of $st(V_\infty)$. Suppose that there exists an infinite strongly p–time independent subset $I \subseteq V$. Then V has a basis in* P.

Our next next result is the analogue of Theorem 7.12 for $st(V_\infty)$.

Theorem 7.31 *Let V be a recursive co-infinite dimensional subspace of $st(V_\infty)$ such that V contains an infinite strongly p–time height increasing independent set C. Then there is an infinite strongly p–time height increasing independent set D such that $V \cap space(D) = \{\vec{0}\}$.*

Theorem 7.32 *Given any r.e. Turing degree δ, there exists an r.e. subspace V of $st(V_\infty)$ such that V has degree δ and V has a basis in P.*

Again one can show that there exists an exponential-time subspace of $st(V_\infty)$ which has no basis in P.

Theorem 7.33 *There is a subspace V of $st(V_\infty)$ such that $V \in$ DEXT and V has no basis in* P.

7.2.3 The semilattice of NP^X subspaces

In this section we shall study various properties of the lower semilattice of NP^X subspaces of $tal(V_\infty)$ and $st(V_\infty)$ for various oracles X. Our first result shows that in contrast to the collection of r.e. subspaces which is closed under both intersection ($\cap$) and sum ($+$) and hence forms a lattice, the collection of NP^X subspaces of either $tal(V_\infty)$ or $st(V_\infty)$ is only closed under intersection and hence only forms a lower semilattice.

Theorem 7.34 *There exist two polynomial-time subspaces W and V of $tal(V_\infty)$ (respectively $st(V_\infty)$) such that $W \cap V = \{\vec{0}\}$ and $W + V$ is not recursive.*

Proof. The proof that we present below works equally well for both $tal(V_\infty)$ and $st(V_\infty)$. Thus we shall write a generic proof where V_∞ may be interpreted as either $tal(V_\infty)$ or $st(V_\infty)$ and the standard basis $e_1, e_2, \ldots$ may be interpreted as either the standard basis $tal(e_1), tal(e_2), \ldots$ of $tal(V_\infty)$ or the standard basis $st(e_1), st(e_2), \ldots$ of $st(V_\infty)$, as appropriate.

By a result of Metakides and Nerode [48], a subspace V of V_∞ is recursive iff V is r.e. and V has an r.e. complementary space. It is easy to see that we can form an effective list $(A_0, B_0), (A_1, B_1), \ldots$ of all pairs of r.e. subspaces W_i and W_j of V_∞ such that $W_i \cap W_j = \{\vec{0}\}$. That is, if $W_0, W_1, \ldots$ is an effective list of all r.e. subspaces of V_∞, and W_i^n denotes the set of elements enumerated into W_i after n steps, then (A_i, B_i) is the pair of r.e. subspaces given by letting (A_i, B_i) be (W_k, W_ℓ) iff $i = [k, \ell]$ and $W_k \cap W_\ell = \{\vec{0}\}$ or letting (A_i, B_i) be $(space(W_k^n), space(W_\ell^n))$ where n is the least m such that $space(W_k^{m+1}) \cap space(W_\ell^{m+1}) \neq \{\vec{0}\}$ if $W_k \cap W_\ell \neq \{\vec{0}\}$.

Given the list $(A_0, B_0), (A_1, B_1), \ldots$, we shall construct W and V so that $W + V \neq A_i$ for any i such that $A_i + B_i = V_\infty$. Thus $W + V$ will not be recursive. In the construction that follows we will in fact construct two p–time height increasing disjoint independent sets K and L so that $W = space(K)$ and $V = space(L)$ will be our desired polynomial-time subspaces. Let $r_0, r_1, \ldots$ be a list of all prime numbers in increasing order. Our idea is to use the vectors $e_{r_i} + e_{r_i \cdot 2^n}$, $e_{r_i \cdot 2^n}$ where $n \geqslant 1$ to help us ensure that $A_i \neq W + V$ if $A_i + B_i = V_\infty$. The only vectors which will be placed into K will be of the form $e_{r_i} + e_{r_i \cdot 2^n}$ for some $i \geqslant 0$ and $n \geqslant 1$, and the only vectors which will be placed into L will be of the form $e_{r_i \cdot 2^n}$ for some $i \geqslant 0$ and $n \geqslant 1$. In fact, for any fixed i either

$$K \cap \{e_{r_i} + e_{r_i \cdot 2^n} : n \geqslant 1\} = \emptyset$$

and

$$L \cap \{e_{r_i \cdot 2^n} : n \geqslant 1\} = \emptyset,$$

or there will be an m such that

$$K \cap \{e_{r_i} + e_{r_i \cdot 2^n} : n \geqslant 1\} = \{e_{r_i} + e_{r_i \cdot 2^m}\}$$

and

$$L \cap \{e_{r_i \cdot 2^n} : n \geqslant 1\} = \{e_{r_i \cdot 2^m}\}.$$

Note that in the standard representation of V_∞, L will be a polynomial-time subset in the strongly p–time height increasing basis $\{st(e_n) : n > 0\}$ and K will be a polynomial-time subset of the strongly p–time height increasing independent set $\{e_k + e_{k \cdot 2^n} : k \text{ is odd and } n \geqslant 1\}$ so that L and K themselves will be strongly p–time independent sets. Thus by Theorem 7.26, W and V will be polynomial-time subspaces of $st(V_\infty)$. In the tally representation of V_∞, K and L will be polynomial-time height increasing independent sets so that by Theorem 7.6, W and V will be polynomial-time subspaces of $tal(V_\infty)$.

Now to decide if $e_{r_i} + e_{r_i \cdot 2^m} \in K$ and $e_{r_i \cdot 2^m} \in L$, we run the enumerations of A_i and B_i for m steps. Let A_i^m and B_i^m denote those elements enumerated into A_i and B_i respectively after m steps. If $m > |e_{r_i}|$ and $[space(A_i^m) + space(B_i^m)] \setminus [space(A_i^{m-1}) + space(B_i^{m-1})] \neq \emptyset$, then we place $e_{r_i} + e_{r_i \cdot 2^m}$ into K and $e_{r_i \cdot 2^m}$ into L iff $e_{r_i} \in [space(A_i^m) + space(B_i^m)] \setminus space(A_i^m)$. Otherwise we place neither $e_{r_i} + e_{r_i \cdot 2^m}$ into K nor $e_{r_i \cdot 2^m}$ into L. Using the fact that in m steps, we can at most enumerate m vectors which are of length at most m, and the fact that Gaussian elimination is polynomial-time in the dimensions of the matrix, it is easy to see that both K and L are p–time height increasing independent sets. Now suppose $e_{r_i} + e_{r_i \cdot 2^m} \in K$ and $e_{r_i \cdot 2^m} \in L$. Since $A_i \cap B_i = \{\vec{0}\}$, we know that each element

$v \in space(A_i) + space(B_i)$ has a unique expression in the form $v = a + b$ with $a \in space(A_i)$ and $b \in space(B_i)$. By our construction, it follows that $e_{r_i} \in [space(A_i^m) + space(B_i^m)] \setminus [space(A_i^m)]$ so that $e_{r_i} \notin space(A_i)$. But clearly $e_{r_i} \in space(K) + space(L)$, so that $A_i \neq space(K) + space(L)$.

Suppose there is no m such that $e_{r_i} + e_{r_i \cdot 2^m} \in K$ and $e_{r_i \cdot 2^m} \in L$. Then either there is no m such that $e_{r_i} \in space(A_i^m) + space(B_i^m)$ in which case $space(A_i) + space(B_i) \neq V_\infty$ so that we don't have to worry about A_i and B_i, or $e_{r_i} \in space(A_i^m)$ for some m (in which case $e_{r_i} \notin space(K) + space(L)$ so again $space(K) + space(L) \neq A_i$). □

Next we make some observations about the existence of subspaces V of $tal\,(V_\infty)$ which are in NP $\setminus$ P. We note that even with the assumption P $\neq$ NP, the existence of (NP $\setminus$ P)–subspaces requires further complexity theoretic assumptions. That is, in [34] Hartmanis proved that the existence of sparse sets in NP $\setminus$ P is equivalent to the separation of deterministic and nondeterministic exponential-time DEXT $\neq$ NEXT. Thus if DEXT = NEXT, then no (NP $\setminus$ P)–subspaces of $tal\,(V_\infty)$ can exist even if NP $\neq$ P. Since the existence of an oracle such that $\text{NP}^A \neq \text{P}^A$ and $\text{DEXT}^A = \text{NEXT}^A$ was proven by Wilson in [75], we have the following theorem.

Theorem 7.35 *There exists an oracle A such that $\text{NP}^A \neq \text{P}^A$ and no $(\text{NP}^A \setminus \text{P}^A)$–subspaces of $tal\,(V_\infty)$ exist.*

As a consequence of this theorem it follows that showing the existence of (NP $\setminus$ P)–subspaces is at least as hard as separating DEXT and NEXT. On the other hand it is sufficient to separate DOUBDEXT and DOUBNEXT to show the existence of (NP $\setminus$ P)–subspaces.

Theorem 7.36 ([5]) *If* DOUBDEXT $\neq$ DOUBNEXT, *then* (NP $\setminus$ P)*–subspaces of $tal\,(V_\infty)$ over finite fields exist.*

Sketch Proof. Let $A \in$ DOUBNEXT $\setminus$ DOUBDEXT and assume the underlying field F has k elements. Define $A_0 = \{0^{k^n} : \exists x \in A[n = 1\,x]\}$. Since $A \in$ DOUBNEXT $\setminus$ DOUBDEXT, it follows that $A_0 \in$ NP $\setminus$ P. But clearly $A_0 \subset \{tal\,(e_1),\, tal\,(e_2),\, \dots\}$ and and hence A_0 is a height increasing independent subset in NP $\setminus$ P. It thus follows from Theorems 7.6 and 7.9 that $space(A_0)$ is in NP $\setminus$ P. □

Corollary 7.37 *There exist recursive oracles A such that there are $(\text{NP}^A \setminus \text{P}^A)$–subspaces of $tal\,(V_\infty)$.*

Furthermore Mahaney [45] has shown that the existence of a sparse NP–complete set with respect to $\leqslant_m^{\mathrm{P}}$ implies NP = P. Thus, if P $\neq$ NP, then there cannot be a subspace V of $tal(V_\infty)$ which is NP–complete.

Next we turn our attention to the question of whether NP–maximal or NP–simple subspaces exist. We note that Breitbart [7] proved that if R is any infinite recursive set in $\{0,1\}^*$, then there exists a set S in P such that both $S \cap R$ and $R \setminus S$ are infinite. This results shows that there can be no NP–maximal sets since if $M \in$ NP and $R = \{0,1\}^* \setminus M$ is infinite, then certainly R is an infinite recursive set. Thus there is a set $S \in P$ such that both $S \cap R$ and $R \setminus S$ are infinite. But then $W = S \cup M$ is a set in NP such that both $W \setminus M$ and $\{0,1\}^* \setminus M$ are infinite so that M is not NP–maximal. Nerode and Remmel [55] proved that the analogue of Breidbart's splitting theorem holds for recursive subspaces of $tal(V_\infty)$ and $st(V_\infty)$.

Theorem 7.38 *Let V be an infinite dimensional recursive subspace of $tal(V_\infty)$ (respectively $st(V_\infty)$). Then there exist subspaces B_0 and B_1 in P such that $B_0 \cap B_1 = \{\vec{0}\}$, $B_0 + B_1 = tal(V_\infty)$ (resp. $B_0 + B_1 = st(V_\infty)$), and both $B_0 \cap V$ and $B_1 \cap V$ are infinite dimensional.*

We note that unlike the set case, Theorem 7.38 does not exclude the possibility of the existence of NP–maximal sets. That is, suppose V is an infinite and co-infinite dimensional subspace of $tal(V_\infty)$. Then the complementary subspace of V, $space(B_{\overline{V}})$, is certainly recursive, so that there exists a pair of polynomial-time complementary subspaces, U and W, so that $U \cap space(B_{\overline{V}})$ and $W \cap space(B_{\overline{V}})$ are infinite dimensional. However in this case, we cannot make the conclusion that $V + U$ is an NP–subspace, which witnesses that V is not NP–maximal for two reasons. First there is no guarantee that $V + U$ is co-infinite dimensional and second, in light of Theorem 7.34, there is no guarantee that $U + V$ is in NP. Indeed our next results will show that there are oracles A for which no NP^A–maximal sets exist. Similar remarks holds for $st(V_\infty)$.

First we show that the assumption that NP^X = co–NP^X also eliminates the possibility of the existence of NP^X–simple and NP^X–maximal subspaces of $tal(V_\infty)$.

Theorem 7.39 *Suppose that NP^X = co–NP^X and V is an NP^X subspace of $tal(V_\infty)$ such that $tal(V_\infty)/V$ is infinite dimensional. Then V is not NP^X–simple and V is not NP^X–maximal.*

Proof. By Theorem 7.10, it follows that $space(B_{\overline{V}}) \in \mathrm{NP}^X$ so that V is not NP^X–simple. To see that V is not NP^X–maximal, note that by our argument in Theorem 7.10, it follows that for any given $x \in \mathrm{NP}^X$, we can nondeterministically from an X oracle find a list of all elements $u_1 < \cdots < u_s$ of height $\leqslant ht(x)$ which are in B_V and a list of all elements $v_1 < \cdots < v_t$ of height $\leqslant ht(x)$ which are in $B_{\overline{V}}$. Thus we can form a new NP^X height increasing independent set C where $x \in C$ iff $x = u_i$ for some $i \leqslant s$ or $x = v_{2k}$ for some $2k \leqslant t$. It is then easy to see that both $tal\,(V_\infty)/space(C)$ and $space(C)/V$ are infinite dimensional.

It also follows from Theorem 7.10 that $space(C) \in \mathrm{NP}^X$ so that C witnesses that V is not NP^X–maximal. □

Since Baker, Gill and Solovay [3] produced recursive oracles X such that $\mathrm{NP}^X \neq \mathrm{P}^X$ but $\mathrm{NP}^X = \text{co–}\mathrm{NP}^X$, we have the following.

Theorem 7.40 *There exists a recursive oracle A such that $\mathrm{NP}^A \neq \mathrm{P}^A$ and there are no NP^A–simple or NP^A–maximal subspaces of $tal\,(V_\infty)$.*

We note that the construction of Theorem 7.39 does not construct a P^X–subspace W such that $W \cap V = \{\vec{0}\}$, since it is *a priori* possible that $space(B_{\overline{V}})$ does not contain an infinite dimensional subspace in P^X. Thus we do not automatically rule out the possibility of the existence of P^X–simple subspaces of $tal\,(V_\infty)$ with the assumption that $\mathrm{NP}^X = \text{co–}\mathrm{NP}^X$. We shall see a bit later that there exist oracles A such that no NP^A–simple, P^A–simple, or NP^A–maximal subspaces exists in $tal(V_\infty)$.

It is also the case that if a subspace V of $tal\,(V_\infty)$ has an infinite height increasing independent subset in P, then V is not P–simple or NP–simple.

Corollary 7.41 ([58]) *Let $V \in \mathrm{NP}$ be a subspace of $tal\,(V_\infty)$ such that V contains an infinite height increasing independent set C in P. Then V is not* NP*–simple or* P*–simple.*

Proof. We may assume that V is co-infinite dimensional since otherwise V cannot be NP^A–simple or P^A–simple. We can thus use the proof of Theorem 7.12 to construct a p–time infinite height increasing independent set D such that D is independent over V. It follows by Theorem 7.7, that $space(D)$ is a p–time subspace of $tal\,(V_\infty)$. Since D is independent over V, $space(D) \cap V = \{\vec{0}\}$ so that V is not NP–simple or P–simple. □

To prove that there exists a recursive oracle B such that $\text{NP}^B \neq \text{P}^B$ and yet no NP^B–maximal, NP^B–simple, or P^B–simple subspaces exist, we can again use the oracle from Theorem 7.16.

Theorem 7.42 *There is a recursive oracle B such that $\text{P}^B \neq \text{NP}^B$ and no NP^B–maximal, NP^B–simple, or P^B–simple subspaces of $tal\,(V_\infty)$ exist.*

Proof. Let B be the recursive oracle of Theorem 7.16. Let V be a NP^B subspace of $tal\,(V_\infty)$ such that the dimension of $tal\,(V_\infty)/V$ is infinite. By Theorem 7.6, $B_{\overline{V}}$ is p–time Turing reducible to V, so that $B_{\overline{V}}$ contains an infinite subset E in P^B. Thus E is an infinite height increasing independent set in P^B, so that by Theorem 7.6, $space(E)$ is an infinite dimensional subspace in P^B. Clearly, $space(E) \cap V = \{\vec{0}\}$ so that $space(E)$ witnesses that V is not P^B–simple or NP^B–simple. Moreover, since we can test whether $tal\,(e_1), \ldots, tal(e_n)$ are in E in polynomial time in $|tal\,(e_n)|$, the set

$$E_2 = \{tal\,(e_n) \in E : \text{card}\,(E \cap \{tal\,(e_1), \ldots, tal\,(e_n)\}) \text{ is even}\}$$

is also a p–time height increasing independent set. We claim that $W = space(V \cup E_2)$ is a subspace of $tal\,(V_\infty)$ which witnesses that V is not NP^B–maximal. Note that $B_V \cup E_2$ is a height increasing basis for W and that $E \setminus E_2 \subseteq B_{\overline{W}}$. Thus $W \supseteq V$ and the dimensions of both $tal\,(V_\infty)/W$ and W/V are infinite. Because $B_V \cup E_2$ is a height increasing basis for W, it follows that $x \in W$ iff there exists a $b \in V$ and an $e \in space(E_2)$ such that $x = b +_{tal} e$ and $ht(b), ht(e) \leqslant ht(x)$. Thus given a B–oracle, we can nondeterministically guess b and e of length $\leqslant k\,|x|$ and the computation which shows that $b \in V$, and then verify in polynomial time that $x = b +_{tal} e$ and $e \in space(E_2)$. Thus $W \in \text{NP}^B$ and hence V is not NP^B–maximal. □

Nerode and Remmel [58] showed that the assumption that $\text{NP}^X = \text{co–NP}^X$ also eliminates the possibility of the existence of NP^X–simple and NP^X–maximal sets in $st(V_\infty)/V$.

Theorem 7.43 *Suppose that $\text{NP}^X = \text{co–NP}^X$ and V is an NP^X subspace of $st(V_\infty)$ such that $st(V_\infty)/V$ is infinite dimensional. Then V is not NP^X–simple and V is not NP^X–maximal.*

As was the case for $tal(V_\infty)$, we can use the Baker-Gill-Solovay results to prove the following.

Theorem 7.44 *There exists a recursive oracle A such that $\mathrm{NP}^A \neq \mathrm{P}^A$ and there are no NP^A–simple or NP^A–maximal subspaces of $st(V_\infty)$.*

The analogue of Theorem 7.41 for $st(V_\infty)$ is the following.

Theorem 7.45 ([58]) *Let V be an* NP *co-infinite dimensional subspace of $st(V_\infty)$ such that V contains an infinite strongly p–time height increasing independent set C. Then V is not* NP*–simple or* P*–simple.*

Proof. Use the proof of Theorem 7.31 to construct a strongly p–time infinite height increasing independent set D such that D is independent over V. It follows by Theorem 7.26, that $space(D)$ is a p–time subspace of (V_∞). Since D is independent over V, $space(D) \cap V = \{\vec{0}\}$ so that V is not NP–simple or P–simple. □

One can again use the oracle of Theorem 7.16 to prove that there is an oracle B where no NP^B–maximal, NP^B–simple, nor P^B–simple subspaces of $st(V_\infty)$ exist.

Theorem 7.46 *There is a recursive oracle B such that $\mathrm{P}^B \neq \mathrm{NP}^B$ and no NP^B–maximal, NP^B–simple, or P^B–simple subspaces of $st(V_\infty)$ exist.*

In contrast to the set case, there are oracles X for which NP^X–maximal subspaces of $tal\,(V_\infty)$ and $st(V_\infty)$ exist. The proof requires a priority argument for the construction of the oracle. Such arguments are easier in $tal\,(V_\infty)$ than in $st(V_\infty)$. In $tal\,(V_\infty)$, one can naturally follow the usual practice in oracle constructions and make the desired NP^X–maximal subspace V be given by

$$V = \{1^n : (\exists \sigma \in X)\, |\sigma| = n\}.$$

This is not possible in $st(V_\infty)$. In $st(V_\infty)$, one constructs X so that there is an NP^X–independent set which generates the desired NP^X–maximal subspace. To see the difference between these two types of construction, we will give the full argument for $tal\,(V_\infty)$ and give just the construction for $st(V_\infty)$. We note that similar techniques are used to prove results in the standard and tally representations of the free Boolean algebra which are given in the next section.

Theorem 7.47 *There exists an r.e. oracle Y and a subspace V of $tal\,(V_\infty)$ which is both P^Y–simple and NP^Y–maximal.*

Proof. We shall construct Y so that

$$M = \{0\} \cup \{1^n : n > 0 \ \& \ (\exists \alpha \in \{0,1\}^*)(|\alpha| = n \text{ and } \alpha \in Y)\}$$

is our desired subspace. Clearly $M \in \text{NP}^Y$.

To ensure that M is co-infinite dimensional we must meet the following set of requirements.

$$T_j : \ \text{card}(\{n \ : \ Y \text{ contains no strings } \alpha \text{ with } k^n \leqslant |\alpha| < k^{n+1} - 1\}) \geqslant j$$

Thus T_j says there are at least j heights n so that M contains no strings of height n. So meeting requirement T_j ensures $\dim(V_\infty/M) \geqslant j$. To ensure that M is P^Y–simple, we shall meet the following set of requirements. Given any subset $V \subseteq tal(V_\infty)$, let $ht(V) = \{n : (\exists x \in V)\, ht(x) = n\}$, then

S_j : If N_j^Y is an infinite dimensional subspace of $tal(V_\infty)$ such that $ht(N_j^Y) \smallsetminus ht(M)$ is infinite, then $M \cap N_j^Y \neq \{\vec{0}\}$.

Now suppose that P_i^Y is an infinite dimensional subspace of $tal(V_\infty)$. Note that meeting all the requirements S_j will ensure that either $P_i^Y \cap M \neq \{\vec{0}\}$ or $ht(P_i^Y) \subseteq^* ht(M)$ where, for any two sets A and B, we write $A \subseteq^* B$ iff there is a finite set F such that $A \subseteq (B \cup F)$. Now suppose that $ht(P_i^Y) \subseteq^* ht(M)$, and let B_i be the standard height increasing basis for P_i^Y. By Lemma 7.6, B_i is in P^Y. Then clearly we can modify B_i by possibly deleting a finite set of elements to form a new height increasing basis C_i such that $ht(M) \supseteq \{n : (\exists x \in C_i)\, ht(x) = n\}$. Thus C_i will also be in P^Y and by Lemma 7.6, $space(C_i)$ will also be in P^Y. Hence if $ht(P_i^Y) \subseteq^* ht(M)$, then there exists some j such that P_j^Y is an infinite dimensional subspace of $tal(V_\infty)$ and $ht(P_j^Y) \subseteq ht(M)$. Thus to ensure that M is P^Y–simple, it will be enough to ensure that we meet the following set of requirements.

R_i : If P_i^Y is an infinite dimensional subspace of $tal(V_\infty)$, then $ht(P_i^Y) \not\subseteq ht(M)$.

Finally, to ensure M is NP–maximal, we shall meet the following set of requirements.

$Q_{[i,n]}$: If N_i^Y/M is infinite dimensional and $N_i^Y \supseteq M$, then there is an $x \in N_i^Y$ such that $x + tal(e_n) \in M$.

Note that if $N_i^Y \supseteq M$ and $\dim(N_i^Y/M)$ is infinite, then meeting all the requirements $Q_{[i,n]}$ will ensure that $tal(e_n) \in N_i^Y$ for all n so that $N_i^Y = tal(V_\infty)$. Thus in fact, M will be NP^Y–supermaximal.

We shall rank our requirements with those of highest priority coming first as $T_0, S_0, R_0, Q_0, T_1, S_1, R_1, Q_1, \ldots$.

In the construction that follows, we shall let Y_s denote the set of elements enumerated into Y by the end of stage s and

$$M_s = \{0\} \cup \{1^\ell : l > 0 \ \& \ (\exists \alpha \in \{0,1\}^*)(|\alpha| = \ell \ \& \ \alpha \in Y_s)\}.$$

We shall ensure that for each s, M_s is a finite dimensional subspace of $tal(V_\infty)$ and that $ht(M_s)$ is contained in $\{1, \ldots, s\}$. For any stage s, we let $CH_s = \{n_1^s < n_2^s < \cdots\}$ be the set of complementary heights for M_s, i.e., the set of all heights n so that there are no elements of $tal(V_\infty)$ of height n in M_s.

At any given stage s, we shall pick out at most one requirement A_j where A_j will be one of the requirements S_j, R_j, or Q_j and take an action to meet that requirement. The fact that the requirements T_j will be satisfied follows from the construction described below. For the otherrequirements, we shall then say that A_j *received attention* at stage s.

The action that we take to meet the requirement A_j of the form S_j or Q_j will always be of the same form. That is, we shall put some elements into Y at stage s and possibly restrain some elements from entering Y for the sake of the requirement. We shall let $res(A_j, s)$ denote the set of elements that are restrained from entering Y at stage s for the sake of requirement A_j. We say that requirement A_j of the form S_j or Q_j is *satisfied* at stage s, if there is a stage $s' < s$ such that A_j has received attention at stage s' and $res(A_j, s') \cap Y_s = \emptyset$.

The actions that we take to meet the requirements R_j will be slightly different. First, we shall declare that all R_j are in a *passive state* at the start of our construction. We would like to find an element $x \in P_j^{Y_s}$ of height n such that $n \notin ht(M_s)$. If we can find such an x, then we will restrain all y such that $k^{n-1} \leqslant |y| \leqslant k^n - 1$ plus all elements not in Y_s which are queried of the oracle Y_s during the computation of $P_j^{Y_s}(x)$ from entering Y for the sake of requirement R_j. Thus if we ensure that $res(R_j, s) \cap Y = \emptyset$, then M will have no elements of height n and $x \in P_j^Y$ so that $ht(P_j^Y) \nsubseteq ht(M)$. If we take such an action for R_j at stage s, then we will say that R_j has received attention at stage s and declare the state of R_j to be *active*. Then for all $t > s$, we will say that an active R_j is *satisfied* at stage t, if $res(R_j, s) \cap Y_t = \emptyset$. However if R_j is injured at some stage $t > s$ in the sense that $res(R_j, s) \cap Y_t \neq \emptyset$, then R_j will return to a passive state. If we cannot find such an x, we will attempt to force $ht(P_j^Y)$ to be finite. That is, since we will ensure that $ht(M_{s-1}) \subseteq \{0, \ldots, s-1\}$ for all s, M_{s-1} will have no elements of height s. Recall that we are assuming that for $n \geqslant 0$, the run time of computations of $P_j^X(y)$ for any oracle X is bounded $\max(2, n)^j$ for any string of length n. Then for $n \geqslant 2$, we let b_n be the largest i such that for all $k^{n-1} \leqslant r \leqslant k^n - 1$,

$$(k^n)^{(i+2)} < 2^{k^{n-2}}.$$

Note that it is easy to see that $\lim_{s\to\infty} b_s = \infty$. Our idea is that elements of height n in $tal\,(V_\infty)$ are of the form 1^r where $k^{n-1} \leqslant r \leqslant k^n - 1$. Our strategy at the end of stage $s-1$ for $s \geqslant 2$ will be to ensure that for all R_j with $j \leqslant b_s$ which are in a passive state and have the property that $P_j^{Y_{s-1}}(1^r) = 0$ for all $k^{s-1} \leqslant r \leqslant k^s - 1$, we restrain all elements which are not in Y_{s-1} and which are queried of the oracle Y_{s-1} in such computations from entering Y for the sake of R_j. This action will force $ht(P_j^Y)$ to be finite if R_j is in a passive state at stage s for all but finitely many s. For any fixed $j \leqslant b_s$, the maximum restraint imposed for R_j is if we restrained all elements not in Y_{s-1} which are queried of the oracle Y_{s-1} in some computation $P_j^{Y_{s-1}}(1^r) = 0$ with $1 \leqslant r \leqslant k^n - 1$. Since the total number of steps used in all these computations is at most

$$2^j + \sum_{i=2}^{k^s} i^j \leqslant k^s \cdot (k^s)^j = (k^s)^{(j+1)},$$

then clearly we could have restrained at most $(k^s)^{(j+1)}$ elements from entering Y for the sake of R_j. Thus at stage s, we will have restrained at most

$$\sum_{i=0}^{b_s} (k^n)^{(i+1)} < (k^n)^{(b_s+2)} < 2^{k^{(n-2)}}$$

elements for entering Y for the sake of some passive requirement R_j with $j \leqslant b_s$ at stage $s-1$. Hence for any given r with $k^{n-1} \leqslant r \leqslant k^n - 1$, we will have restrained at most 2^{r-1} elements of length r from entering Y for such R_j's.

Construction

STAGES 0, 1:

Let $Y_0 = Y_1 = \emptyset$ so that $M_0 = M_1 = \{\vec{0}\}$. Let $res(A_j, 0) = res(A_j, 1) = \emptyset$ for all requirements A_j of the form S_j, R_j, or Q_j.

STAGE $s \geqslant 2$:

Let A_j be the highest priority requirement among S_0, R_0, Q_0, ..., S_s, R_s, Q_s such that

CASE 1: $A_j = S_j$ and S_j is not satisfied at stage $s - 1$ and there exists an ℓ with $0 < ht(1^\ell) \leqslant s$ such that

(a) $1^\ell \in N_j^{Y_{s-1}}$,

(b) $ht(1^\ell) \in CH_{s-1}$ and $ht(1^\ell) > n_j^{s-1}$, and

(c) for each $1^n \in space(\{1^\ell\} \cup M_{s-1}) \setminus M_{s-1}$, there is a string $\alpha_n \in \{0,1\}^*$ such that $|\alpha_n| = |1^n| = n$ and α_n is not restrained from Y by any requirement of higher priority than S_j at stage $s - 1$ nor is α_n queried of the oracle in some fixed computation of $N_j^{Y_{s-1}}$ which accepts 1^ℓ.

CASE 2: $A_j = R_j$ and R_j is not satisfied at stage $s - 1$ and there exists an ℓ with $0 < ht(1^\ell) \leqslant s$ such that

(i) $1^\ell \in P_j^{Y_{s-1}}$, and

(ii) $ht(1^\ell) \in CH_{s-1}$ and $ht(1^\ell) > n_j^{s-1}$.

CASE 3: $A_j = Q_j$ and Q_j is not satisfied at stage $s - 1$, and if $j = [e, n]$, there exists an ℓ with $0 \leqslant ht(1^\ell) < s$ such that

(I) $1^\ell \in N_e^{Y_{s-1}}$,

(II) $ht(1^\ell) \in CH_{s-1}$ and $ht(1^\ell) > \max(n, n_j^{s-1})$, and

(III) for each $1^m \notin space(\{1^\ell +_{tal} tal\,(e_n)\} \cup M_{s-1}) \setminus M_{s-1}$, $ht(1^m) > n_j^{s-1}$ and there is a string α_m of length m in $\{0,1\}^*$ which is not restrained from Y by any requirement of higher priority than Q_j at stage $s - 1$ nor is α_m queried in some fixed computation of $N_e^{Y_{s-1}}$ which accepts 1^ℓ.

If there is no such requirement A_j, let $Y_s = Y_{s-1}$. Also for all requirements A_j of the form S_j or Q_j and for all requirements A_j of the form R_j where either R_j is satisfied at stage $s-1$ or $j > b_{s+1}$, let $res(A_j, s) = res(A_j, s-1)$. Declare that a requirement R_j is active at stage s iff R_j is active at stage $s-1$. For any R_j with $j \leqslant b_{s+1}$ which is currently passive and has the property that $P_j^{Y_s}(1^r) = 0$ for all $k^s \leqslant r \leqslant k^{s+1} - 1$, let $res(R_j, s)$ equal $res(R_j, s-1)$ union the set of all $y \notin Y_s$ such that y is queried of the oracle in one of the computations $P_j^{Y_s}(1^r)$ where $k^s \leqslant r \leqslant k^{s+1} - 1$.

If there is such a requirement A_j, we have three cases.

Case 1: $A_j = S_{j_s}$.

Let ℓ_s denote the least ℓ corresponding to S_{j_s}. Then for each $1^n \in space(\{1^{\ell_s}\} \cup M_{s-1}) \setminus M_{s-1}$, pick the least string α_n such that $|\alpha_n| = n$, α_n is not restrained from Y by any requirement of higher priority than S_{j_s} at stage $s-1$, nor is α_n queried of the oracle Y_{s-1} in the computation of $N_j^{Y_{s-1}}$ which accepts 1^{ℓ_s}, and put α_n into Y. This will ensure that if M_{s-1} is a finite dimensional subspace of V_∞, then M_s will also be a finite dimensional subspace of V_∞. Note that the assumption that $ht(1^{\ell_s}) \in CH_{s-1}$ ensures that all $1^n \in space(\{1^{\ell_s}\} \cup M_{s-1}) \setminus M_{s-1}$ have the property that $ht(1^n) \geqslant ht(1^{\ell_s})$. That is, such a 1^n must be of the form $1^n = \lambda \cdot_{tal} 1^{\ell_s} +_{tal} m$ where $m \in M_{s-1}$ and $\lambda \in F$. Then since $ht(m) \neq ht(1^{\ell_s})$, it must be the case that $ht(1^n) \geqslant ht(1^{\ell_s})$. Thus $ht(M_s) \cap \{n_1^{s-1}, \ldots, n_{j_s}^{s-1}\} = \emptyset$, and hence for all $i \leqslant j_s$, $n_i^{s-1} = n_i^s$. Let $res(S_{j_s}, s)$ equal the set of all strings not in Y_{s-1} which are queried of the oracle Y_{s-1} in the computation of $N_{j_s}^{Y_{s-1}}$ which accepts 1^{ℓ_s}, and say S_{j_s} receives attention at stage s. Also for all requirements A_j of the form S_j or Q_j and for all requirements A_j of the form R_j where either R_j is satisfied at stage $s-1$ or $j > b_{s+1}$, let $res(A_j, s) = res(A_j, s-1)$ if $Y_s \cap res(A_j, s-1) = \emptyset$ and let $res(A_j, s) = \emptyset$ if $Y_s \cap res(A_j, s-1) \neq \emptyset$. Declare that a requirement R_j is active at stage s iff R_j is active at stage $s-1$ and $Y_s \cap res(R_j, s-1) = \emptyset$. For any R_j with $j \leqslant b_{s+1}$ which is currently passive and has the property that $P_j^{Y_s}(1^r) = 0$ for all $k^s \leqslant r \leqslant k^{s+1} - 1$, let $res(R_j, s)$ equal $res(R_j, s-1)$ union the set of all $y \notin Y_s$ such that y is queried of the oracle Y_s in one of the computations $P_j^{Y_s}(1^r)$, where $k^s \leqslant r \leqslant k^{s+1} - 1$.

Case 2: $A_j = R_{j_s}$.

Let ℓ_s denote the least ℓ corresponding to j_s and $n_s = ht(1^{\ell_s})$. We then say that R_{j_s} is active and receives attention at stage s. We let $Y^s = Y^{s-1}$ and $res(R_{j_s}, s)$ consist of all elements y with $k^{n_s - 1} \leqslant |y| \leqslant k^{n_s} - 1$ and all elements which are not in Y_{s-1} and which are queried of the oracle Y_{s-1} in the computation $P_{j_{s-1}}^{Y_{s-1}}(1^{\ell_s})$. Note that if $res(R_{j_s}, s) \cap Y = \emptyset$, then M will have no elements of height $n_s = ht(1^{\ell_s})$ but $1^{\ell_s} \in P_{j_s}^{Y}$. Also for all requirements A_j of the form S_j or Q_j and for all requirements A_j of the form R_j where $j \neq j_s$ and where either R_j is satisfied at stage $s-1$ or $j > b_{s+1}$, let $res(A_j, s) = res(A_j, s-1)$. For $j \neq j_s$, declare that a requirement R_j is active at stage s iff R_j is active at stage $s-1$. For any R_j with $j \leqslant b_{s+1}$ which is currently passive and has the property that $P_j^{Y_s}(1^r) = 0$ for all $k^s \leqslant r \leqslant k^{s+1}-1$, let $res(R_j, s)$ equal $res(R_j, s-1)$ union the set of all $y \notin Y_s$ such that y is queried of the oracle Y_s in one of the computations $P_j^{Y_s}(1^r)$ where $k^s \leqslant r \leqslant k^{s+1}-1$.

Case 3: $A_j = Q_{j_s}$.

Let $j_s = [e_s, n_s]$ and ℓ_s denote the least ℓ corresponding to j_s. Then for each $1^m \in space(\{1^{\ell_s} +_{tal} tal\,(e_{n_s})\} \cup M_{s-1}) \setminus M_{s-1}$, pick the least string α_m such that $|\alpha_m| = m$, and α_m is not restrained from Y by any requirement of higher priority than Q_{j_s} at stage $s-1$ nor is α_m queried in the computation of $N_{e_s}^{Y_{s-1}}$ which accepts 1^{ℓ_s} and put α_m into Y. Once again this will ensure that M_s is a finite dimensional subspace of V_∞. Note that since $ht(1^{\ell_s}) > n_s = ht(tal\,(e_{n_s}))$, it follows that $ht(1^{\ell_s} +_{tal} tal\,(e_{n_s})) = ht(1^{\ell_s})$. Thus as in case 1, the assumption that $ht(1^{\ell_s}) \in CH_{s-1}$ ensures that all $1^n \in space(\{1^{\ell_s} +_{tal} tal\,(e_{n_s})\} \cup M_{s-1}) \setminus M_{s-1}$ have the property that $ht(1^n) \geqslant ht(1^{\ell_s})$. Let $res(Q_{j_s}, s)$ equal the set of all strings which are not in Y_{s-1} which are queried of the oracle in the computation of $N_{e_s}^{Y_{s-1}}$ which accepts 1^{ℓ_s} and say Q_{j_s} receives attention at stage s. Also for all requirements A_j of the form S_j or Q_j and for all requirements A_j of the form R_j where either R_j is satisfied at stage $s-1$ or $j > b_{s+1}$, let $res(A_j, s) = res(A_j, s-1)$ if $Y_s \cap res(A_j, s-1) = \emptyset$ and let $res(A_j, s) = \emptyset$ if $Y_s \cap res(A_j, s-1) \neq \emptyset$. Declare that a requirement R_j is active at stage s iff R_j is active at stage $s-1$ and $Y_s \cap res(R_j, s-1) = \emptyset$. For any R_j with $j \leqslant b_{s+1}$ which is currently passive and has the property that $P_j^{Y_s}(1^r) = 0$ for all $k^s \leqslant r \leqslant k^{s+1}-1$, let $res(R_j, s)$ equal $res(R_j, s-1)$ union the set of all $y \notin Y_s$ such that y is queried of the oracle Y_s in one of the computations $P_j^{Y_s}(1^r)$, where $k^s \leqslant r \leqslant k^{s+1}-1$.

This completes the construction of Y.

Lemma 7.48 *Each requirement of the form S_j, R_j, or Q_j receives attention at most finitely often.*

Proof. We proceed by induction on j. Suppose that s_0 is such that there is no stage $s \geqslant s_0$ such that one of $S_0, R_0, Q_0, \ldots, S_j, R_j, Q_j$ receives attention at stage s. Then if there is a $t > s_0$ such that S_{j+1} receives attention at stage t, then by construction S_{j+1} is satisfied at stage t and $res(S_{j+1}, t) \cap Y_t = \emptyset$. However, it is easy to see from our construction that for $s > t$, $res(S_{j+1}, s) = res(S_{j+1}, t)$ and $res(S_{j+1}, s) \cap Y_s = \emptyset$ unless some requirement of higher priority than S_{j+1} eceives attention at stage s. Since this never happens by our choice of s_0, S_{j+1} will be satisfied for $s > t$. Thus S_{j+1} can receive attention at most once after stage s_0. Thus there must be a stage s_1 such that there is no stage $s \geqslant s_1$ such that one of $S_0, R_0, Q_0, \ldots, S_j, R_j, Q_j, S_{j+1}$ receives attention at stage s. A similar argument will show that R_{j+1} can receive attention at most once after stage s_1. Thus there must be a stage s_2 such that there is no stage $s \geqslant s_2$ such that one of $S_0, R_0, Q_0, \ldots, S_j, R_j, Q_j, S_{j+1}, R_{j+1}$ receives attention at stage s. Finally a similar argument will show that Q_{j+1} can receive attention at most once after stage s_2. Thus each of the requirements S_j, R_j, or Q_j can receive attention only finitely often. □

Lemma 7.49 $\dim(tal\,(V_\infty)/M)$ *is infinite.*

Proof. We prove by induction that $\dim(tal\,(V_\infty)/M) \geqslant k$ for all k. That is, let t_0 be a stage such that no requirement $S_0, R_0, Q_0, \ldots, S_k, R_k, Q_k$ receives attention at any stage $s \geqslant t_0$. Since M_{t_0} is finite dimensional, $n_i^{t_0}$ is defined for all i. Hence M_t contains no strings of height n for $n = n_1^{t_0}, \ldots, n_k^{t_0}$. But no requirement S_j, R_j, or Q_j with $j > k$ can force elements of height $n \leqslant n_k^s$ into M at any stage s. Hence by our choice of t_0, there can be no strings of heights n for $n = n_1^{t_0}, \ldots, n_k^{t_0}$ in M. Thus $\dim(tal\,(V_\infty)/M) \geqslant k$. □

Lemma 7.50 *M is P^Y-simple.*

Proof. First we show that if N_j^Y is a subspace of $tal\,(V_\infty)$ such that $ht(N_j^Y) \setminus ht(M)$ is infinite, then $N_j^Y \cap M \neq \{\vec{0}\}$. For a contradiction, assume N_j^Y is such that $ht(N_j^Y) \setminus ht(M)$ is infinite and $N_j^Y \cap M = \{\vec{0}\}$. Note that since M is co-infinite dimensional by Lemma 7.49, it follows that $n_i = lim_{s\to\infty} n_i^s$ exists for all i. Let s_0 be a stage large enough so that $n_i^s = n_i$ for $i \leqslant j$ and none of the requirements $S_0, R_0, Q_0, \ldots, S_{j-1}, R_{j-1}, Q_{j-1}$ receives attention after stage s_0. Let U_{s_0} denote the set of all 1^n such that there exists a requirement A_i among $S_0, R_0, Q_0, \ldots, S_{j-1}, R_{j-1}, Q_{j-1}$ which is satisfied at stage s_0, such that there exists an $\alpha \in res(A_i, s_0)$ with $|\alpha| = n$. Our choice of s_0 ensures that if $n \notin U_{s_0}$, then no string α of length n is ever restrained from Y by a requirement of higher priority than S_j which is satisfied at some stage $t > s_0$. Also our choice of s_0 ensures that $n_i = n_i^t$ for all $i \leqslant j$ and $t > s_0$.

Next let $t_0 > s_0$ be such that

(1) $t_0 > \max(\{ht(y) : y \in U_{s_0}\} \cup \{2, s_0, n_j\})$,

(2) $b_{t_0} > j$, and

(3) $2^{r-1} > r^j$ for all $r > t_0$.

Note that for any $t > t_0$, our construction ensures that the number of strings of length r where $k^{t-1} \leqslant r \leqslant k^t - 1$ which are restrained by some requirement R_i with $i < j$ which is passive at stage t is less than 2^{r-1}. Moreover we are assuming that any successful computation of the oracle machine N_j^X for any oracle X on a string of length $r > 2$ takes at most r^j steps. Thus our choice of t_0 ensures that if $t > t_0$ and $1^x \in N_j^{Y_t}$ is string of height $> t_0$, then there is at least one string $\alpha_x \in \{0,1\}^*$ of length x which is not restrained from Y by any requirement of higher priority than S_j at stage t, nor is queried of the oracle Y_t in some fixed computation which shows that $1^x \in N_j^{Y_t}$. Since $ht(N_k^Y) \setminus ht(M)$ is infinite, there must exist a $1^n \in N_k^Y$ such that $ht(1^n) > t_0$ and $ht(1^n) \notin ht(M)$. Then there must be some stage $s > t_0$ such that $1^n \in N_j^{Y_{s-1}}$. Note that at stage s, each $1^m \in space(\{1^n\} \cup M_{s-1}) \setminus M_{s-1}$ has the property that $ht(1^m) \geqslant ht(1^n) > t_0$ and thus there is at least one string α_m of length m which is not restrained from Y by any requirement of higher priority than S_j at stage $s-1$, nor is queried of the oracle Y_{s-1} in some fixed computation which shows that $1^n \in N_j^{Y_{s-1}}$. Thus 1^n witnesses that S_j is a candidate to receive attention at stage s. Thus either S_j is satisfied at stage $s-1$ or S_j is highest priority requirement among S_0, R_0, Q_0, $\ldots$, S_s, R_s, Q_s which can receive attention at stage s. In either case, it follows that S_j will be satisfied at stage s. Thus there will be some $1^n \in (N_j^{Y_s} \cap M_s) \setminus \{\vec{0}\}$ such that all elements which are queried of the oracle Y_s in some computation which shows that $1^n \in N_j^{Y_s}$, and which are not in Y_s, are in $res(S_j, s)$. However our choice of t_0 ensures that we can never put any element of $res(S_j, s)$ into Y after stage s so that 1^n will witness that $N_j^Y \cap M \neq \{\vec{0}\}$.

Remark. We note that the assumption that $ht(N_j^Y) \setminus ht(M)$ is infinite seems to be crucial in this argument. That is, if we merely assume that $\dim(N_j^Y/M)$ is infinite, then it may be the case that whenever there exists a $1^n \in N_k^Y$ such that $ht(1^n) > t_0$ and $1^n \notin M$, then at a stage $s > t_0$ where $1^n \in N_k^{Y_{s-1}}$, there is some $1^m \in M_{s-1}$ such that $ht(1^m) = ht(1^n)$. In such a situation it is possible that $ht(1^n +_{tal} 1^m)$ is much less than $ht(1^n)$. That is, it may be possible that some element in $1^x \in space(\{1^n\} \cup M_{s-1}) \setminus M_{s-1}$ has height so small that all strings of length x are queried of the oracle during any computation which shows that $1^n \in N_k^{Y_{s-1}}$. Then it will be impossible to put a string of length x into Y_s so as to ensure that $1^x \in M_s$ while maintaining the computation to ensure that $1^n \in N_k^Y$.

To continue our proof of the lemma, we can now assume that if P_r^Y is an infinite dimensional subspace of $tal\,(V_\infty)$ such that $P_r^Y \cap M = \{\vec{0}\}$, then $ht(P_r^Y) \setminus ht(M)$ is finite. By our argument preceding the construction, it would then follow that there is some j such that P_j^Y is an infinite dimensional subspace of $tal\,(V_\infty)$ and $ht(P_j^Y) \subseteq ht(M)$. We shall now show that there can be no such j. For a contradiction, assume that P_j^Y is an infinite dimensional subspace of $tal\,(V_\infty)$ and $ht(P_j^Y) \subseteq ht(M)$. Let s_1 be a stage large enough so that $n_i^{s_1} = n_i$ for $i \leqslant j$ and none of the requirements S_0, R_0, Q_0, $\ldots$, S_{j-1}, R_{j-1}, Q_{j-1}, S_j receives attention after stage s_1. Let U_{s_1} denote the set of all 1^n such that there exists a requirement A_i among S_0, R_0, Q_0, $\ldots$, S_{j-1}, R_{j-1}, Q_{j-1}, S_j which is satisfied at stage s_1 and there exists an $\alpha \in res(A_i, s_1)$ with $|\alpha| = n$. Our choice of s_1 ensures that if $n \notin U_{s_1}$, then no string α of length n is ever restrained from Y by a requirement of higher priority than R_j which is satisfied at some stage $t > s_0$. Also our choice of s_1 ensures that $n_i = n_i^t$ for all $i \leqslant j$ and $t > s_0$.

Next let t_1 be such that

(1) $t_1 > \max(\{ht(y) : y \in U_{s_1}\} \cup \{2, s_1, n_{i-1}\})$,

(2) $b_{t_1} > j$, and

(3) $2^{r-1} > r^j$ for all $r > t_1$.

Now we claim that there can be no stage $t > t_1$ at which R_j is satisfied at stage t. That is, if R_j is satisfied at stage t, there must be some $s \leqslant t$ such that R_j receives attention at stage s, and there is a $1^x \in P_j^{Y_{s-1}}$ such that $q = ht(1^x) \in CH_{s-1}$, and $res(R_j, s) = res(R_j, t)$ contains all strings of length r where $k^{q-1} \leqslant r \leqslant k^q - 1$, and contains all strings which are not in Y_{s-1} which are queried of the oracle Y_{s-1} in the computation $P_j^{Y_{s-1}}(1^x) = 1$, and $res(R_j, s) \cap Y_t = \emptyset$. But then our choice of $t > t_1$ ensures that $res(R_j, s) \cap Y = \emptyset$, which means that M can have no strings of height q while $1^x \in P_j^Y$. But then 1^x witnesses that $ht(P_j^Y) \not\subseteq ht(M)$ which contradicts our assumption that $ht(P_j^Y) \subseteq ht(M)$. Thus it must be the case that for all stages $t > t_1$, R_j is in a passive state. It follows that for all $t > t_1$, there can be no r with $k^t \leqslant r \leqslant k^{t+1} - 1$ such that $P_j^{Y_t}(1^r) = 1$ since otherwise at stage $t+1$, there is some r with $k^t \leqslant r \leqslant k^{t+1} - 1$ such that $P_j^{Y_t}(1^r) = 1$. But then at stage $t+1$, 1^r witnesses that R_j is a candidate to receive attention at stage $t+1$. By our choice of $t > t_1$, it would follow that R_j is the highest priority requirement among $S_0, R_0, Q_0, \ldots, S_{t+1}, R_{t+1}, Q_{t+1}$ which could receive attention at stage $t+1$ so that R_j would receive attention at stage $t+1$ which we have already ruled out. Thus it must be the case that for all r with $k^t \leqslant r \leqslant k^{t+1} - 1$, $P_j^{Y_t}(1^r) = 0$. But then our choice of $t > t_1$ ensures that $j \leqslant b_{t+1}$ and hence all elements which are not in Y_t which are queried of the oracle Y_t during one of the computations $P_j^{Y_t}(1^r) = 0$ where $k^t \leqslant r \leqslant k^{t+1} - 1$ are put into $res(R_j, t)$. Again the fact that $t > t_1$ ensures that $res(R_j, t) \cap Y = \emptyset$ so that for all r with $k^t \leqslant r \leqslant k^{t+1} - 1$, $P_j^Y(1^r) = 0$. That is, P_j^Y has no strings of length $t+1$ for any $t > t_1$ and hence $ht(P_j^Y)$ is finite. Thus there can be no such P_j^Y such that P_j^Y is an infinite dimensional subspace of $tal(V_\infty)$ and $ht(P_j^Y) \subseteq ht(M)$. But this means that there can be no r such that P_r^Y is an infinite dimensional subspace of $tal(V_\infty)$ and $P_r^Y \cap M = \{\vec{0}\}$. Thus M is P^Y-simple as claimed. □

Lemma 7.51 *M is NP^Y-maximal.*

Proof. By our remarks preceding the construction, we need only show that we meet all the requirements $Q_{[e,n]}$. So assume N_e^Y is a subspace of $tal(V_\infty)$ such that (N_e^Y/M) is infinite dimensional and $N_e^Y \supseteq M$. Let $j = [e, n]$ and let s_2 be a stage such that $n_i = n_i^{s_2}$ for $i \leqslant j$ and none of the requirements $S_0, R_0, Q_0, \ldots, S_{j-1}, R_{j-1}, Q_{j-1}, S_j, R_j$ receive attention after stage s_2. Let U_{s_2} denote the set of all 1^n such that there exists a requirement A_i among $S_0, R_0, Q_0, \ldots, S_{j-1}, R_{j-1}, Q_{j-1}, S_j, R_j$ which is satisfied at stage s_2 and there exists an $\alpha \in res(A_i, s_2)$ with $|\alpha| = n$. Our choice of s_2 ensures that if $n \notin U_{s_2}$, then no string α of length n is ever restrained from Y by a requirement of higher priority than Q_j which is satisfied at some stage $t > s_2$. Also our choice of s_2 ensures that $n_i = n_i^t$ for all $i \leqslant j$ and $t > s_2$.

Next let t_2 be such that

(1) $t_2 > \max(\{ht(y) : y \in U_{s_2}\} \cup \{2, s_2, n_{i-1}\})$,

(2) $b_{t_2} > j$, and

(3) $2^{r-1} > r^j$ for all $r > t_2$.

Note that for any $t > t_2$, our construction ensures that the number of strings of length r, where $k^{t-1} \leqslant r \leqslant k^t - 1$, which are restrained by some requirement R_i with $i < j$ which is passive at stage t, is less than 2^{r-1}. Moreover we are assuming that any successful computation of the oracle machine N_j^X for any oracle X on a string of length $r \geqslant 2$ takes at most r^j steps. Thus our choice of t_2 ensures that if $t > t_2$ and $1^x \in N_j^{Y_t}$ is string of height $> t_2$, then there is at least one string $\alpha_x \in \{0,1\}^*$ of length x which is not restrained from Y by any requirement of higher priority than Q_j at stage t, nor is queried of the oracle Y_t in some fixed computation which shows that $1^x \in N_j^{Y_t}$.

Next observe that since $\dim(N_e^Y/M)$ is infinite and $N_e^Y \supseteq M$, it must be the case that $ht(N_e^Y) \setminus ht(M)$ is infinite. That is, let $A = \{a_0, a_1, \dots\}$ be an infinite set of elements of N_e^Y which is independent over M. Then consider some fixed $a_i \in A$ and suppose $a_i = \sum_{i=1}^{q} \lambda_i \cdot tal(e_{j_i})$ where $\lambda_i \in F$ for $i = 1, \dots, q$, $\lambda_q \neq 0$, and $j_1 < \cdots < j_q$. Thus $ht(a_i) = j_q$. Now if there exists an $m_1 \in M$ such that $ht(m) = ht(a_i)$, then

$$m_1 = \sum_{\ell \leqslant j_q} \beta_\ell \cdot tal\,(e_\ell)$$

where $\beta_\ell \in F$ for all ℓ and $\beta_{j_q} \neq 0$. But then

$$a_i^1 = a_i -_{tal} \frac{\lambda_{j_q}}{\beta_{j_q}} m_1$$

is an element of $N_e^Y \setminus M$ such $ht(a_i^1) < ht(a_i)$. Now if there exists an $m_2 \in M$ such that $ht(a_i^1) = ht(m_2)$, then once again there is some $\gamma \in F$ such that $a_i^2 = a_i^1 -_{tal} \gamma \cdot m_2$ is an element of $N_e^Y \setminus M$ with $ht(a_i^2) < ht(a_i^1) < ht(a_i)$. If we continue in this fashion, we must eventually find some $a_i^k = a_i +_{tal} v_k$ where $v_k \in M$ such that $ht(a_i^k) \notin ht(M)$. That is, we can replace our original independent set A over M by a set $A' = \{a_0', a_1', \dots\}$ where for all i, $a_i -_{tal} a'_i \in M$ and $ht(a'_i) \notin ht(M)$. But then A' is an infinite subset of N_e^Y which is independent over M. Thus there is no finite set F such that $space(M \cup F) \supseteq A'$. This implies that $ht(A') = \{ht(a'_i) : i \geqslant 0\}$ must be infinite, since otherwise there clearly would be a finite set F such that $space(M \cup F) \supseteq A'$. But by construction $ht(A') \subseteq ht(N_e^Y) \setminus ht(M)$ so that $ht(N_e^Y) \setminus ht(M)$ must be infinite.

Since $ht(N_e^Y) \setminus ht(M)$ is infinite, there must exist a $1^q \in N_e^Y$ such that $ht(1^q) > t_2$, $ht(1^q) > n$, and $ht(1^q) \notin ht(M)$. Then there must be some stage $s > t_2$ such that $1^q \in N_e^{Y_{s-1}}$. Note that at stage s, each

$$1^m \in space(\{1^q +_{tal} tal\,(e_n)\} \cup M_{s-1}) \setminus M_{s-1}$$

has the property that $ht(1^m) \geqslant ht(1^q +_{tal} tal\,(e_n)) = ht(1^q) > t_2$, and thus there is at least one string α_m of length m which is not restrained from Y by any requirement of higher priority than Q_j at stage $s-1$, nor is queried of the oracle Y_{s-1} in some fixed computation which shows that $1^q \in N_j^{Y_{s-1}}$. Thus 1^q witnesses that Q_j is a candidate to receive attention at stage s. Hence either Q_j is satisfied at stage $s-1$ or Q_j is highest priority requirement among S_0, R_0, Q_0, ..., S_s, R_s, Q_s which can receive attention at stage s. In either case, it follows that Q_j will be satisfied at stage s. Thus there will be some $1^q \in N_j^{Y_s}$ such that $1^q +_{tal} tal\,(e_n) \in M_s$ and all elements which are queried of the oracle in some computation which shows that $1^q \in N_j^{Y_s}$ and which are not Y_s, are in $res(Q_j, s)$. However our choice of t_2 ensures we can never put any element of $res(Q_j, s)$ into Y after stage s so that $1^q \in N_j^Y$ and hence requirement Q_j is met. Thus M will be NP^Y-supermaximal and hence will be NP^Y–maximal. This completes the proof of Lemma 7.51 and of Theorem 7.47. □ □

We note that M constructed in Theorem 7.47 has a number of interesting properties besides being NP^Y–maximal and P^Y–simple. First of all, it is easy to check that in meeting the requirements S_j, we made no use of the fact that N_j^Y was a subspace of $tal\,(V_\infty)$, but only that N_j^Y was a subset of $tal\,(V_\infty)$. Similarly, it is easy to check that in meeting the requirements R_j we made no use of the fact that P_j^Y was a subspace of $tal\,(V_\infty)$, but only that P_j^Y was a subset of V_∞. Thus meeting all the requirements R_j ensures that there is no infinite subset W of $tal\,(V_\infty)$ in P^Y such that $ht(W) \subseteq ht(M)$. Thus M does not contain any infinite P^Y set and hence M does not have a basis in P^Y. We also claim that $tal\,(V_\infty) \setminus M$ does not have any infinite subsets in P^Y. That is, suppose that $P_j^Y \subseteq tal(V_\infty) \smallsetminus M$. Now it cannot be that $ht(P_j^Y) \smallsetminus ht(M)$ is infinite, since otherwise there is an i such that $P_j^Y = N_i^Y$ and the fact that we met requirement S_i would mean that $P_j^Y \cap M \neq \{\vec{0}\}$. Thus $ht(P_j^Y) \subseteq^* ht(M)$. Let $Q = ht(space(A)) \smallsetminus ht(P_j^Y)$. Then clearly

$$S = \{x \in P_j^Y : ht(x) \notin Q\}$$

is an infinite set in P^Y such that $ht(S) \subseteq ht(space(A))$. Since meeting all the requirements R_j rules out the existence of such an S, $tal(V_\infty) \smallsetminus M$ does not contain an infinite set in P^Y. Recall that a set of strings S is called P^Y-*immune* if S has no infinite subset in P^Y. Thus both M and $tal(V_\infty) \smallsetminus M$ are P^Y-immune

Note also that by Theorem 7.40, the fact that M is NP^Y-maximal implies that $\mathrm{NP}^Y \neq \text{co-}\mathrm{NP}^Y$, and hence that $\mathrm{P}^Y \neq \mathrm{NP}^Y$. Thus we have proved the following.

Corollary 7.52 *There exists an r.e. oracle Y and a subspace M of $tal(V_\infty)$ such that*

(1) $\mathrm{P}^Y \neq \mathrm{NP}^Y$ *and* $\mathrm{NP}^Y \neq$ *co*-NP^Y,

(2) $M \in \mathrm{NP}^Y$,

(3) M *is* P^Y-*immune and hence has no basis in* P^Y,

(4) $tal(V_\infty) \smallsetminus M$ *is* P^Y-*immune, and*

(5) M *is both* P^Y-*simple and* NP^Y-*supermaximal.*

We next give an analogue of Theorem 7.47 for $st(V_\infty)$. Once again we shall think of $st(V_\infty)$ as the k-ary representation $B_k(V_\infty)$ so that for all $x \in st(V_\infty)$, $|x| = ht(x)$.

Theorem 7.53 *There exists an r.e. oracle D such that there exists an* NP^D-*supermaximal* P^D-*simple subspace in* $st(V_\infty)$.

Proof. The construction again proceeds in stages. We let D_s be the set of elements enumerated into D by the end of stage s. For any given $x \in \{0, \ldots, k-1\}^*$ with $|x| \geqslant 1$, we let C_x denote the set of all strings of length $8|x| + 2$ of $\{0, \ldots, k-1\}^*$ of the form $x\,1\,0^{4|x|}\,1\,\sigma$ where σ is any string of length $3|x|$ in $\{0, \ldots, k-1\}^*$. Note that there are $k^{3|x|}$ strings in C_x for any $x \in st(V_\infty)$. Let $C_\emptyset = \{\emptyset\}$. It is then easy to see that if $x \neq y$, then $C_x \cap C_y = \emptyset$.

We then define $A = \{x : C_x \cap D \neq \emptyset\}$. Thus A will be in NP^D. Our idea is to define D so that A is a height increasing independent subset of $st(V_\infty)$. Then by the relativized version of Theorem 7.28, $space(A) \in \mathrm{NP}^D$. Our construction of D will ensure that $space(A)$ is our desired P^D–simple NP^D–supermaximal space. Let $A_s = \{x : C_x \cap D_s \neq \emptyset\}$. At each stage s, we shall let $B_s = \{st(e_n) : A_s$ has no element of height $n\}$. Our construction will ensure that at each stage s, $A_s \cup B_s$ is a height increasing basis of $st(V_\infty)$. We define b_i^s for all i and s so that $B_s = \{b_0^s, b_1^s, \ldots\}$ where $ht(b_0^s) < ht(b_1^s) < \cdots$.

To ensure that $space(A)$ is co-infinite dimensional we must meet the following set of requirements.

$$T_j : \text{ card}\left(\{n : D \text{ contains no strings } \alpha \text{ with } |\alpha| = 8n+2\}\right) \geqslant j.$$

Thus T_j says there are at least j heights n so that A contains no strings of height n. So meeting requirement T_j ensures $\dim(V_\infty/space(A)) \geqslant j$. To ensure that $space(A)$ is P^D–simple, we shall meet the following set of requirements.

S_j : If N_i^D is an infinite dimensional subspace of $st(V_\infty)$ such that $ht(N_i^D) \setminus ht(space(A))$ is infinite, then $space(A) \cap N_i^D \neq \{\vec{0}\}$.

Now suppose that P_j^D generates an infinite dimensional subspace of $st(V_\infty)$ which is in NP^D. Note that meeting all the requirements S_j will ensure that either $space(P_i^D) \cap space(A) \neq \{\vec{0}\}$ or $ht(space(P_i^D)) \subseteq^* ht(space(A))$. Now suppose that $ht(space(P_i^D)) \subseteq^* ht(space(A))$, and let $U = ht(P_i^D) \setminus ht(space(A))$. If $U = \emptyset$, then $ht(P_i^D) \subseteq ht(space(A))$. Otherwise, U is a finite set, so let $U = \{n_0, \ldots, n_q\}$ and let $x_0, \ldots, x_q$ be elements of $space(P_i^D)$ such that $ht(x_i) = n_i$. Note that any $x \in st(V_\infty)$ is a string of the form $x = a_1 \cdots a_{|x|}$ where $a_j \in \{0, \ldots, k-1\}$. Then we define the full height of x, $fh(x) = \{n : 1 \leqslant n \leqslant |x| \text{ and } a_n \neq 0\}$. Then it is easy to see that given any $x \in space(P_i^D)$, there exists some $\lambda_1, \ldots, \lambda_q$ in F such that

$$fh\Big(x -_{st} \sum_{i=1}^{q} \lambda_i x_i\Big) \cap U = \emptyset.$$

That is, if $x = a_1 \cdots a_{|x|}$ where $|x| \geqslant n_q$ and $a_{n_q} \neq 0$ and $x_q = a_{1,q} \cdots a_{n_q,q}$ where $a_{n_q,q} \neq 0$, then

$$x' = x -_{st} \frac{a_q}{a_{n_q,q}} x_q = b_0 \cdots b_{|x|}$$

where $b_{n_q} = 0$ so that $n_q \notin fh(x')$.

Now if $b_{n_{q-1}} \neq 0$ and $x_{q-1} = a_{1,q-1} \cdots a_{n_{q-1},q-1}$ where $a_{n_{q-1},q-1} \neq 0$, then

$$x'' = x' -_{st} \frac{b_q}{a_{n_{q-1},q-1}} x_{q-1} = c_0 \cdots c_{|x|}$$

where $c_{n_q} = b_{n_q} = 0$ and $c_{n_{q-1}} = 0$ so that neither n_q nor n_{q-1} is in $fh(x'')$.

Continuing on in this way, we can construct our desired linear combination $\sum_{i=1}^{q} \lambda_i x_i$ such that $fh(x -_{st} \sum_{i=1}^{q} \lambda_i x_i) \cap U = \emptyset$.

Now let $Q = \{x \in space(P_i^D) : fh(x) \cap U = \emptyset\}$. It is easy to see that Q is a subspace of P_i^D and our argument above shows that $space(P_i^D) = space(\{x_1, \ldots, x_q\}) \oplus Q$. Thus Q is an infinite dimensional subspace of $st(V_\infty)$ such that $ht(Q) \subseteq ht(space(A))$. Let T be the set of all y such that $fh(y) \cap U = \emptyset$, $|y| > k^{|x_q|}$, and there exists an $x \in P_i^D$ and $z \in space(\{x_1, \ldots, x_q\})$ such that $x +_{st} z = y$. Note that $space(\{x_1, \ldots, x_q\})$ has exactly k^q elements since $\{x_1, \ldots, x_q\}$ is a height increasing basis for $space(\{x_1, \ldots, x_q\})$. Thus given any y with $|y| > k^{|x_q|}$, in polynomial time in $|y|$, we can find all $y +_{st} w$ such that $w \in space(\{x_1, \ldots, x_q\})$. Now for any $w \in space(\{x_1, \ldots, x_q\})$, $ht(w) \leqslant ht(x_q) = |x_q| < k^{|x_q|}$ so that $ht(y +_{st} w) = ht(y)$. Thus it takes at most $k^q(|y|^j)$ steps to test all such $y +_{st} w$ for membership in P_j^D given an oracle D. But then

$$y \in T \quad \text{iff} \quad \{y +_{st} w : w \in space(\{x_1, \ldots, x_q\})\} \cap P_j^D \neq \emptyset.$$

Thus it follows that T is in P^D and clearly T generates an infinite dimensional subspace of Q. Thus there must be some j such that P_j^D generates an infinite dimensional subspace of $st(V_\infty)$ and $ht(space(P_j^D)) \subseteq ht(space(A))$. Thus to ensure that $space(A)$ is P^D–simple, it will be enough to ensure that we meet the following set of requirements.

R_i : If P_i^D generates an infinite dimensional subspace of $st(V_\infty)$, then $ht(P_i^D) \nsubseteq ht(space(A))$.

Finally, to ensure that *space*(A) is NP–supermaximal, we shall meet the following set of requirements.

$Q_{[i,n]}$: If $N_i^D/space(A)$ is an infinite dimensional and $N_i^D \supseteq space(A)$, then there is an $x \in N_i^D$ such that $x + st(e_n) \in space(A)$.

Note that if $N_i^D \supseteq space(A)$ and $\dim(N_i^D/space(A))$ is infinite, then meeting all the requirements $Q_{[i,n]}$ will ensure that $st(e_n) \in N_i^D$ for all n, so that $N_i^D = st(V_\infty)$.

We shall rank our requirements with those of highest priority coming first as T_0, S_0, R_0, Q_0, T_1, S_1, R_1, Q_1, ...

As in the construction of Theorem 7.47, at any given stage s, we shall pick out at most one requirement E_j where E_j will be one of the requirements S_j, R_j, or Q_j and take an action to meet that requirement. We shall then say that E_j *received attention* at stage s. The action that we take to meet the requirement E_j of the form S_j or Q_j will always be of the same form. That is, we shall put some elements into D at stage s and possibly restrain some elements from entering D for the sake of the requirement. We shall let $res(E_j, s)$ denote the set of elements that are restrained from entering D at stage s for the sake of requirement E_j. We say that requirement E_j of the form S_j or Q_j is *satisfied* at stage s, if there is a stage $s' < s$ such that E_j has received attention at stage s' and $res(E_j, s') \cap D_s = \emptyset$.

The actions that we take to meet the requirements R_j will be essentially the same as in the construction of Theorem 7.47. First, we shall declare that all R_j are in a *passive* state at the start of our construction. We would like to find an element $x \in P_j^{D_s}$ of height n such that $n \notin ht(space(A_s))$. If we can find such an x, then we will restrain all y such that $|y| = 8n + 2$ and $y \in C_x$ for some $x \in st(V_\infty)$ of height n plus all elements not in D_s which are queried of the oracle during the computation of $P_j^{D_s}(x)$ from entering D for the sake of requirement R_j. Then if we ensure that $res(R_j, s) \cap D = \emptyset$, A will have no elements of height n and $x \in P_j^D$ so that $ht(P_j^D) \not\subseteq ht(space(A))$. If we take such an action for R_j at stage s, then we will say that R_j has received attention at stage s and declare the state of R_j to be *active*. Then for all $t > s$, we will say that an active R_j is *satisfied* at stage t, if $res(R_j, s) \cap D_t = \emptyset$. However, if R_j is injured at some stage $t > s$ in the sense that $res(R_j, s) \cap D_t \neq \emptyset$, then R_j will return to a passive state. If we cannot find such an x, we will attempt to force $ht(P_j^D)$ to be finite. That is, since we will ensure that

$ht(space(A_{s-1})) \subseteq \{0, \ldots, s-1\}$ for all s, A_{s-1} will have no elements of height s. Recall that we are assuming that for $n \geqslant 2$, the run time of computations of $P_j^X(y)$ for any oracle X is bounded $\max(2,n)^j$ for any string of length n. Then for $n \geqslant 2$, we let d_n be the largest i such that for all r,

$$n^{(i+2)} < k^n .$$

Note that it is easy to see that $\lim_{s\to\infty} d_s = \infty$. Our idea is that elements of height n in $st(V_\infty)$ are just the elements of length n. Our strategy at the end of stage $s-1$ for $s \geqslant 2$ is that for all R_j with $j \leqslant d_s$ which are in a passive state and have the property that $P_j^{D_{s-1}}(x) = 0$ for all $x \in st(V_\infty)$ of length s, we will restrain all elements which are not in D_{s-1} and which are queried in such computations from entering D for the sake of R_j. This action will force $ht(P_j^D)$ to be finite if R_j is in a passive state at stage s for all but finitely many s. For any fixed $j \leqslant d_s$, the maximum restraint imposed for R_j occurs if we restrained all elements not in D_{s-1} which are queried of the oracle D_{s-1} in some computation $P_j^{D_{s-1}}(x) = 0$ with $1 \leqslant |x| \leqslant n$ and $x \in st(V_\infty)$. Since the total number of steps used in all these computations is at most

$$2^j + \sum_{i=2}^{s} k^i i^j \leqslant s\,k^s \cdot s^j = k^s s^{(j+1)},$$

then clearly we could have restrained at most $k^s s^{(j+1)}$ elements from entering D for the sake of R_j. Thus at stage s, we will have restrained at most

$$\sum_{i=0}^{d_s} k^s s^{(i+1)} < k^s s^{(d_s+2)} < k^s k^s = k^{2s}$$

elements from entering D for the sake of some passive requirement R_j with $j \leqslant b_s$ at stage $s-1$. Hence for any given x with $|x| = n$, we will have restrained less than k^{2s} elements of C_x from entering D for such R_j's.

Construction

STAGES 0, 1:

Let $D_0 = D_1 = \emptyset$ so that $A_0 = A_1 = \emptyset$. Let $res(E_j, 0) = res(E_j, 1) = \emptyset$ for all requirements E_j of the form S_j, R_j, or Q_j.

STAGE $s \geqslant 2$:

Let E_j be the highest priority requirement among S_0, R_0, Q_0, ..., S_s, R_s, Q_s such that

CASE 1: $E_j = S_j$ and S_j is not satisfied at stage $s-1$ and there exists an $x \in st(V_\infty)$ with $0 < |x| \leqslant s$ such that

(a) $x \in N_j^{D_{s-1}}$,

(b) $|x| \notin ht(space(A_{s-1}))$ and $|x| > |b_j^{s-1}|$, and

(c) there exists a $y \in C_x$ such that y is not restrained from D by any requirement of higher priority than S_j at stage $s-1$ and y is not queried of the oracle D_{s-1} in some fixed computation which shows that $x \in N_j^{D_{s-1}}$.

CASE 2: $E_j = R_j$ and R_j is not satisfied at stage $s-1$ and there exists an $x \in st(V_\infty)$ with $0 \leqslant |x| \leqslant s$ such that

(i) $|x| \notin ht(space(A_{s-1})$, and

(ii) $x \in P_j^{D_{s-1}}$.

CASE 3: $A_j = Q_j$ and Q_j is not satisfied at stage $s-1$, and if $j = [e, n]$, there exists an x with $0 \leqslant |x| \leqslant s$ such that

(I) $x \in N_e^{D_{s-1}}$,

(II) $|x| \notin ht(space(A_{s-1})$, $|x| > |b_j^{s-1}|$, and $|x| > n$, and

(III) there exists a $y \in C_{x +_{st} st(e_n)}$ such that y is not restrained from D by any requirement of higher priority than S_j at stage $s-1$ and y is not queried of the oracle D_{s-1} in some fixed computation which shows that $N_e^{D_{s-1}}(x)$.

If there is no such requirement E_j, let $D_s = D_{s-1}$. Also for all requirements E_j of the form S_j or Q_j, and for all requirements E_j of the form R_j where either R_j is satisfied at stage $s-1$ or $j > d_{s+1}$, let $res(E_j, s) = res(E_j, s-1)$. Declare that a requirement R_j is active at stage s iff R_j is active at stage $s-1$. For any R_j with $j \leqslant d_{s+1}$ which is currently passive and has the property that $P_j^{D_s}(x) = 0$ for all $x \in st(V_\infty)$ of length $s+1$, let $res(R_j, s)$ equal $res(R_j, s-1)$ union the set of all $y \notin D_s$ such that y is queried of the oracle D_s in one of the computations $P_j^{D_s}(x)$ where $x \in st(V_\infty)$ of length $s+1$.

If there is such a requirement E_j, we have three cases.

Case 1: $E_j = S_{j_s}$.

Let x_s denote the least x corresponding to S_{j_s}. Then pick the least string $\alpha_{x_s} \in C_{x_s}$ such that α_{x_s} is not restrained from D by any requirement of higher priority than S_{j_s} at stage $s-1$, nor is α_{x_s} queried of the oracle D_{s-1} in the computation of $N_j^{D_{s-1}}$ which accepts x_s, and put α_{x_s} into D. Let $res(S_{j_s}, s)$ equal the set of all strings not in D_{s-1} which are queried of the oracle D_{s-1} in the computation of $N_{j_s}^{D_{s-1}}$ which accepts x_s, and say S_{j_s} receives attention at stage s. Also for all requirements E_j of the form S_j or Q_j, and for all requirements E_j of the form R_j where either R_j is satisfied at stage $s-1$ or $j > d_{s+1}$, let $res(E_j, s) = res(E_j, s-1)$ if $D_s \cap res(E_j, s-1) = \emptyset$ and let $res(E_j, s) = \emptyset$ if $D_s \cap res(E_j, s-1) \neq \emptyset$. Declare that a requirement R_j is active at stage s iff R_j is active at stage $s-1$ and $D_s \cap res(R_j, s-1) = \emptyset$. For any R_j with $j \leqslant d_{s+1}$ which is currently passive and has the property that $P_j^{D_s}(z) = 0$ for all $z \in st(V_\infty)$ of length $s+1$, let $res(R_j, s)$ equal $res(R_j, s-1)$ union the set of all $y \notin D_s$ such that y is queried of the oracle D_s in one of the computations $P_j^{D_s}(z)$, where $z \in st(V_\infty)$ and $|z| = s+1$.

Case 2: $E_j = R_{j_s}$.

Let x_s denote the least x corresponding to j_s. We then say that R_{j_s} is active and receives attention at stage s. We let $D_s = D_{s-1}$ and $res(R_{j_s}, s)$ consist of all elements y of length $8|x_s| + 2$ which are in some C_z such that $z \in st(V_\infty)$ and $|z| = |x_s|$, and all elements which are not in D_{s-1} and which are queried of the oracle D_{s-1} in the computation $P_{j_{s-1}}^{D_{s-1}}(x) = 1$. Note that if $res(R_{j_s}, s) \cap D = \emptyset$, then A will have no elements of height $|x_s|$ but $x_s \in P_{j_s}^D$. Also for all requirements E_j of the form S_j or Q_j, and for all requirements E_j of the form R_j where $j \neq j_s$ and where either R_j is satisfied at stage $s-1$ or $j > d_{s+1}$, let $res(E_j, s) = res(E_j, s-1)$. For $j \neq j_s$, declare that a requirement R_j is active at stage s iff R_j is active at stage $s-1$. For any R_j with $j \leqslant d_{s+1}$ which is currently passive and has the property that $P_j^{D_s}(x) = 0$ for all $x \in st(V_\infty)$ of length $s+1$, let $res(R_j, s)$ equal $res(R_j, s-1)$ union the set of all $y \notin D_s$ such that y is queried of the oracle D_s in one of the computations $P_j^{D_s}(x)$, where $x \in st(V_\infty)$ and $|x| = s+1$.

Case 3: $E_j = Q_{j_s}$.

Let $j_s = [e_s, n_s]$ and x_s denote the least x corresponding to j_s. Then pick the least string $\alpha_{x_s} \in C_{x_s}$ such that α_{x_s} is not restrained from D by any requirement of higher priority than Q_{j_s} at stage $s-1$, nor is α_{x_s} queried of the oracle D_{s-1} in the computation of $P_{e_s}^{D_{s-1}}(x_s)$, and put α_{x_s} into D. Let $res(Q_{j_s}, s)$ consist of all strings which are not in D_{s-1} which are queried of the oracle D_{s-1} in the computation of $P_{e_s}^{D_{s-1}}(x_s)$, and say Q_{j_s} receives attention at stage s. Also for all requirements E_j of the form S_j or Q_j and for all requirements E_j of the form R_j, where either R_j is satisfied at stage $s-1$ or $j > d_{s+1}$, let $res(E_j, s) = res(E_j, s-1)$ if $D_s \cap res(E_j, s-1) = \emptyset$ and let $res(E_j, s) = \emptyset$ if $D_s \cap res(E_j, s-1) \neq \emptyset$. Declare that a requirement R_j is active at stage s iff R_j is active at stage $s-1$ and $D_s \cap res(R_j, s-1) = \emptyset$. For any R_j with $j \leqslant d_{s+1}$ which is currently passive and has the property that $P_j^{D_s}(x) = 0$ for all $x \in st(V_\infty)$ of length $s+1$, let $res(R_j, s)$ equal $res(R_j, s-1)$ union the set of all $y \notin D_s$ such that y is queried of the oracle D_s in one of the computations $P_j^{D_s}(x)$, where $x \in st(V_\infty)$ and $|x| = s+1$.

This completes the construction of D. We note that A is a height increasing independent set in NP^Y, since our construction ensures that we can never put two elements of the same height in A. Thus by Theorem 7.28, $space(A) \in \mathrm{NP}^Y$. We then have to prove the same sequence of lemmas as in Theorem 7.47 to complete the proof the theorem. The details may be found in [58]. □

Again the $space(A)$ constructed in Theorem 7.53 has a number of interesting properties besides being NP^D–supermaximal and P^D–simple. First of all, meeting all the requirements R_j ensures that $space(A)$ is P^D–immune. That is, if P_i^D is an infinite subset of $space(A)$, then certainly P_i^D generates an infinite dimensional subspace of $st(V_\infty)$ and $ht(P_i^D) \subseteq ht(space(A))$, which would violate requirement R_i. Also, as in the construction of Theorem 7.47, it is easy to check that in meeting the requirements S_j we made no use of the fact that N_j^D was a subspace of $st(V_\infty)$, but only that N_j^D was a subset of $st(V_\infty)$. We claim that $st(V_\infty) \setminus space(A)$ does not have any infinite subsets in P^D. That is, suppose that $P_j^D \subseteq st(V_\infty) \setminus space(A)$. Now it cannot be that $ht(P_j^D) \setminus ht(space(A))$ is infinite since otherwise there is an i such that $P_j^D = N_i^D$ and the fact that we met requirement S_i would mean that $P_j^D \cap space(A) \neq \{\vec{0}\}$. Thus $ht(P_j^D) \subseteq^* ht(space(A))$. Let $Q = ht(space(A)) \setminus ht(P_j^D)$. Then clearly

$$S = \{x \in P_j^D : ht(x) \notin Q\}$$

is an infinite set in P^D which generates an infinite dimensional subspace of $st(V_\infty)$ and $ht(S) \subseteq ht(space(A))$. Since meeting all the requirements R_j rules out the existence of such an S, $st(V_\infty) \setminus space(A)$ does not contain an infinite set in P^D. Thus $space(A)$ and $st(V_\infty) \setminus space(A)$ are P^D–immune.

Note also that by Theorem 7.40, the fact that $space(A)$ is NP^D–maximal implies that $\mathrm{NP}^D \neq$ co–NP^D, and hence that $\mathrm{P}^D \neq \mathrm{NP}^D$. Thus we have proved the following.

Corollary 7.54 *There exists an r.e. oracle D and a subspace V of $st(V_\infty)$ such that*

(i) $\mathrm{P}^D \neq \mathrm{NP}^D$ *and* $\mathrm{NP}^D \neq$ *co–*NP^D,

(ii) $V \in \mathrm{NP}^D$,

(iii) V *is* P^D*–immune. and hence has no basis in* P^D,

(iv) $st(V_\infty) - V$ *is* P^D*–immune, and*

(v) V *is both* P^D*–simple and* NP^D*–supermaximal.*

Finally we observe that results about NP and P subspaces of $tal\,(V_\infty)$ naturally extend to results about NEXT and DEXT subspaces of $st(V_\infty)$ by Lemma 3.3. For a typical example, say that a subspace M of $st(V_\infty)$ is NEXT^A–maximal if $M \in \mathrm{NEXT}^A$, $\dim(st(V_\infty)/M)$ is infinite, and for any subspace W of $st(V_\infty)$ in NEXT^A containing M, either $\dim(st(V_\infty)/W)$ is finite or $\dim(W/M)$ is finite. Then Theorem 7.46 and Theorem 7.53 show that the question of the existence of NEXT–maximal subspaces is oracle dependent.

Theorem 7.55 *There is a recursive oracle A and an r.e. oracle B such that the following hold.*

(i) $\mathrm{NEXT}^A \neq \mathrm{DEXT}^A$ *and* $\mathrm{NEXT}^B \neq \mathrm{DEXT}^B$.

(ii) *There are no* NEXT^A*–maximal subspaces of* $st(V_\infty)$.

(iii) *There is a* NEXT^B*–maximal subspace W of* $st(V_\infty)$.

In the same way, all the results in this paper about P^X and NP^X subspaces of $tal\,(V_\infty)$ can be transfered to results about DEXT^X and NEXT^X subspaces of $st(V_\infty)$.

Next we consider some results on splitting theorems for $tal\,(V_\infty)$ due to Bäuerle. We note a result of Ash and Downey [1] that every r.e. subspace of V_∞ is the direct sum of two decidable spaces. In $tal\,(V_\infty)$ the property of being a direct sum of two p–time subspaces is equivalent to having a p–time basis.

Theorem 7.56 ([5]) *A subspace of V of $tal\,(V_\infty)$ can be split into two polynomial-time subspaces if and only if V has a basis in P.*

Note that by the results on bases and subspaces of $tal(V_\infty)$, we immediately get the following corollaries.

Corollary 7.57 *There is an exponential-time subspace W of $tal\,(V_\infty)$ that cannot be split into two polynomial time subspaces.*

Corollary 7.58 *For all r.e. degrees δ there is an r.e. subspace V of $tal\,(V_\infty)$ such that* $\deg(V) = \delta$, *and V can be split into two polynomial-time subspaces.*

Corollary 7.59 *There exists a recursive oracle A such that every $\mathrm{NP}^A \setminus \mathrm{P}^A$ subspace V of $tal\,(V_\infty)$ can be split into two P^A vector spaces.*

Corollary 7.60 *Let $\mathcal{F}$ be finite. There exists a recursive oracle B such that there is an $\mathrm{NP}^B \setminus \mathrm{P}^B$ subspace V of $tal\,(V_\infty)$ that cannot be split into two P^B vector spaces.*

Corollary 7.61 *Arguments valid under relativization are not sufficient to show $\mathrm{NP} \neq \mathrm{P} \longrightarrow$ every NP–subspace of $tal\,(V_\infty)$ can be split into two p–time subspaces. $\mathrm{NP} \neq \mathrm{P} \longrightarrow$ there exists an $(\mathrm{NP} \setminus \mathrm{P})$–subspace of $tal\,(V_\infty)$ which cannot be split into two p–time subspaces.*

In fact, Báuerle identifies three types of splittings by polynomial-time subspaces of $tal\,(V_\infty)$.

Definition 7.62 Let V be an r.e. vector space.

(1) V allows a P–*splitting* if there exist p–time spaces W_0 and W_1 such that $W_0 \cap W_1 = \{\vec{0}\}$ and $W_0 + W_1 = V$. We say that W_0 and W_1 P–split V.

(2) V allows an *induced* P–*splitting* if there exist p–time spaces W_0 and W_1 such that $W_0 \cap W_1 = \{\vec{0}\}$, $W_0 + W_1 = V_\infty$, $\dim(V \cap W_1) = \dim(V \cap W_0) = \infty$, and $(W_0 \cap V) + (W_1 \cap V) = V$. We say that W_0 and W_1 induce a P–splitting of V.

(3) V allows an *induced weak* P*–splitting* if there exist r.e. vector spaces W_0 and W_1 such that W_0 and W_1 have bases in P, $W_0 \cap W_1 = \{\vec{0}\}$, $W_0 + W_1 = V_\infty$, $\dim(V \cap W_1) = \dim(V \cap W_0) = \infty$, and $(W_0 \cap V) + (W_1 \cap V) = V$. We say that W_0 and W_1 induce a weak P–splitting of V.

Theorem 7.63 ([5]) *Let V be a subspace of tal (V_∞).*

(i) *If V allows a* P*–splitting, then V allows an induced* P*–splitting.*

(ii) *If V allows an induced* P*–splitting, then V allows an induced weak* P*–splitting.*

Theorem 7.64 ([5])

(i) *There exists an exponential-time subspace V of tal (V_∞) that allows an induced* P*–splitting but no* P*–splitting.*

(ii) *There exists an exponential-time subspace V of tal (V_∞) that does not allow an induced weak* P*–splitting*

This shows that the three notions of Definition 7.62 are increasingly weaker.

We end this section with some results of Bäuerle [5] on subspaces and superspaces of $(\mathrm{NP} \setminus \mathrm{P})$–subspaces of *tal* (V_∞).

Theorem 7.65 ([5]) *Every $(\mathrm{NP} \setminus \mathrm{P})$–subspace V of tal (V_∞) has a non-trivial $(\mathrm{NP} \setminus \mathrm{P})$–subspace W.*

Theorem 7.66 ([5]) *Let V be a subspace of tal(V_∞) such that $V \in \mathrm{NP} \setminus \mathrm{P}$. If V has a non-trivial superspace in* P*, then V has a non-trivial superspace in* $\mathrm{NP} \setminus \mathrm{P}$.

Theorem 7.67 ([5]) *There exists a recursive oracle C such that there exists a vector space $V \subset$ tal (V_∞) that satisfies the following properties:*

(1) $V \in \mathrm{NP}^C \setminus \mathrm{P}^C$,

(2) *V has a non-trivial superspace in* $\mathrm{NP}^C \setminus \mathrm{P}^C$,

(3) *V has no non-trivial superspaces in* P^C

Theorem 7.68 ([5]) *There is a recursive oracle C such that all non-trivial $(\mathrm{NP}^C \setminus \mathrm{P}^C)$-subspaces of $tal(V_\infty)$ have non-trivial superspaces in P^C and in $\mathrm{NP}^C \setminus \mathrm{P}^C$.*

Finally, under the assumption that $\mathrm{NP}^A = \text{co-}\mathrm{NP}^A$,

Theorem 7.69 ([5]) *Let A be an oracle such that $(\mathrm{NP}^A \setminus \mathrm{P}^A)$-subspaces of $tal(V_\infty)$ exist, and such that $\mathrm{NP}^A = co\text{-}\mathrm{NP}^A$. Then the following is true.*

(1) *For all $(\mathrm{NP}^A \setminus \mathrm{P}^A)$-subspaces V of $tal(V_\infty)$, their standard height increasing basis B_V is in $\mathrm{NP}^A \setminus \mathrm{P}^A$.*

(2) *For all $(\mathrm{NP}^A \setminus \mathrm{P}^A)$-subspaces V of $tal(V_\infty)$, their standard height increasing complementary basis $B_{\overline{V}}$ and their standard complement $(B_{\overline{V}})^*$ are in $\mathrm{NP}^A \setminus \mathrm{P}^A$.*

(3) *Every $(\mathrm{NP}^A \setminus \mathrm{P}^A)$-subspace V of $tal(V_\infty)$ can be split into two disjoint $\leqslant_T^{\mathrm{P}}$-incomparable $(\mathrm{NP}^A \setminus \mathrm{P}^A)$-subspaces.*

(4) *The set of $\leqslant_T^{\mathrm{P}}$ degrees with NP^A-subspaces of $tal(V_\infty)$ is dense.*

(5) *There exists a pair V, W of $(\mathrm{NP}^A \setminus \mathrm{P}^A)$-subspaces of $tal(V_\infty)$ such that, if $U \leqslant_T^{\mathrm{P}} V$ and $U \leqslant_T^{\mathrm{P}} W$, then $U \in \mathrm{P}^A$.*

(6) *The set of rationals $\mathcal{Q}$ can be embedded in the structures of $\leqslant_m^{\mathrm{P}}$- and $\leqslant_T^{\mathrm{P}}$-degrees of NP^A-subspaces of $tal(V_\infty)$.*

8 Polynomial-Time Boolean Algebras

In this section, we shall survey the results of Nerode and Remmel on the lower semilattice of NP–ideals of a polynomial-time presentation of the free Boolean algebra. Again we consider two natural representations of the free Boolean algebra called the tally and standard representation. We start by describing these two representations.

Let $\mathcal{P}([0,1))$ denote the Boolean algebra of all subsets of the rational left-closed right-open interval $[0,1)$ in the rational numbers $\mathbb{Q}$. The Boolean operations of meet, join, and complementation on $\mathcal{P}([0,1))$ are respectively intersection, union, and relative complement in $[0, 1)$. Let $\mathcal{B}([0,1))$ be the subalgebra of $\mathcal{P}([0,1))$ generated by the left-closed right-open intervals of the form

$$\left[\frac{i}{2^n}, \frac{j}{2^n}\right)$$

with $n \geqslant 0$ and $0 \leqslant i < j \leqslant 2^n$. For any subset $S \subseteq \mathcal{B}([0,1))$, $(S)^*$ denotes the subalgebra of $\mathcal{B}([0,1))$ generated by S and $I(S)$ denotes the ideal generated by S. Given a subalgebra $D \subseteq \mathcal{B}([0,1))$, we let $At(D)$ denote the set of atoms of D.

Next we define a natural generating sequence $a_1, a_2, \ldots$ for $\mathcal{B}([0,1))$ by induction:

$$\begin{aligned} a_1 &= [0, 1) \\ a_{2^{n-1}+m+1} &= \left[\frac{2m}{2^n}, \frac{2m+1}{2^n}\right), \quad \text{if } n \geqslant 1 \text{ and } 0 \leqslant m < 2^{n-1}. \end{aligned}$$

Thus,

$$a_2 = \left[0, \frac{1}{2}\right), \quad a_3 = \left[0, \frac{1}{4}\right), \quad a_4 = \left[\frac{1}{2}, \frac{3}{4}\right), \quad a_5 = \left[0, \frac{1}{8}\right),$$

$$a_6 = \left[\frac{1}{4}, \frac{3}{8}\right), \quad a_7 = \left[\frac{1}{2}, \frac{5}{8}\right), \quad a_8 = \left[\frac{3}{4}, \frac{7}{8}\right).$$

Let $A_n = \{a_1, \ldots, a_n\}^*$. Then it is not difficult to see that $A_1, A_2, A_3, \ldots$ is a strictly increasing sequence of subalgebras such that for each $n \geqslant 1$, there is a unique atom $x_n \in At(A_n)$ such that a_{n+1} splits x_n, i.e., $\emptyset \subset a_{n+1} \subset x_n$. In fact, one can easily show by induction that if k is of the form $2^{n-1} + m$ with $0 \leqslant m < 2^{n-1}$, then

$$At(A_k) = \left\{\left[\frac{i}{2^n}, \frac{i+1}{2^n}\right) : 0 \leqslant i < 2m\right\} \bigcup \left\{\left[\frac{j}{2^{n-1}}, \frac{j+1}{2^{n-1}}\right) : m \leqslant j < 2^{n-1}\right\}.$$

Hence

$$a_{k+1} = \left[\frac{2m}{2^n}, \frac{2m+1}{2^n}\right)$$

splits the atom

$$x_k = \left[\frac{m}{2^{n-1}}, \frac{m+1}{2^{n-1}}\right)$$

of A_k. It follows that A_n has exactly n atoms for each $n \geqslant 1$, so that A_n has exactly 2^n elements.

We use this generating sequence and its corresponding sequence of subalgebras

$$A_1 \subset A_2 \subset A_3 \subset \cdots$$

to define the standard and tally representations of $\mathcal{B}([0,1))$.

The Standard Representation of $\mathcal{B}([0,1))$.

First we describe a coding of the elements of $\mathcal{B}([0,1))$ which we call the *standard representation* of $\mathcal{B}([0,1))$. Our idea is to use binary numbers of length n to code the elements of $A_n \setminus A_{n-1}$ for $n > 1$. Formally, we define a one-to-one correspondence $\delta \to s_\delta$ between $Bin(\omega)$ and $\mathcal{B}([0,1))$ by induction.

For the base step, set

$$s_0 = \emptyset, \qquad s_1 = [0, 1).$$

For the induction step, assume that the correspondence $bin(k) \to s_k$ has been defined between $\{bin(k) : 0 < k < 2^n\}$ and A_n. We then extend our correspondence to A_{n+1} as follows. Given a binary number m of length $n+1$, let $m = k \cdot 2^i$ where k is odd, so that $bin(m) = 0^{i\frown} bin(k)$, and let

$$s_m = \begin{cases} s_k \cup a_{n+1} & \text{if } s_k \cap a_{n+1} = \emptyset \\ s_k \setminus a_{n+1} & \text{if } s_k \supseteq a_{n+1}. \end{cases} \tag{8.1}$$

Now let $At(A_n) = \{x_1, \ldots, x_{n-1}, x_n\}$ where x_n is the atom of A_n which is split by a_{n+1}. Then it is easy to see that every element of $A_{n+1} \setminus A_n$ is either of the form

$$a_{n+1} \cup \bigcup_{i \in S} x_i$$

or

$$(x_n \setminus a_{n+1}) \cup \bigcup_{i \in S} x_i$$

for some set $S \subseteq \{1, \ldots, n-1\}$. Thus (8.1) defines a one-to-one correspondence between $\{k : 2^n \leqslant k < 2^{n+1}\}$ and $A_{n+1} \setminus A_n$.

Indeed it is quite easy to use (8.1) to recursively construct s_n. We write $s_{bin(n)}$ for s_n in the following.

Example 8.1 Suppose $bin(n) = 0101101$. Then s_n can be constructed as follows.

$$s_0 = \emptyset.$$

$$s_{01} = a_2 = \left[0\,,\,\tfrac{1}{2}\right) \quad \text{since } a_2 \cap s_0 = \emptyset.$$

$$s_{0101} = s_{01} \cup a_4 = \left[0\,,\,\tfrac{1}{2}\right) \cup \left[\tfrac{1}{2}\,,\,\tfrac{3}{4}\right) = \left[0\,,\,\tfrac{3}{4}\right) \quad \text{since } a_4 \cap s_{01} = \emptyset.$$

$$s_{01011} = s_{0101} \setminus a_5 = \left[0\,,\,\tfrac{3}{4}\right) \setminus \left[0\,,\,\tfrac{1}{8}\right) = \left[\tfrac{1}{8}\,,\,\tfrac{3}{4}\right) \quad \text{since } a_4 \subseteq s_{0101}.$$

$$s_{0101101} = s_{01011} \setminus a_7 = \left[0\,,\,\tfrac{3}{4}\right) \setminus \left[\tfrac{1}{2}\,,\,\tfrac{5}{8}\right) = \left[\tfrac{1}{8}\,,\,\tfrac{1}{2}\right) \cup \left[\tfrac{5}{8}\,,\,\tfrac{3}{4}\right)$$
since $a_7 \subseteq s_{01011}$.

It is not difficult to show that given two σ and τ in $Bin(\omega)$ with $|\sigma| \leqslant |\tau|$, we can find α, β and γ in $Bin(\omega)$ such that

$$s_\alpha = s_\sigma \cup s_\tau, \qquad s_\beta = s_\sigma \cap s_\tau, \qquad s_\gamma = \left[0\,,\,1\right) \setminus s_\tau$$

in polynomial time in $|\tau|$. Furthermore, note that each of α, β and γ has length $\leqslant 2|\tau|$, since each of s_α, s_β and s_γ belongs to A_n if $s_\tau \in A_n$. See [54] for details. It follows that if we then define

$$\begin{aligned} \sigma \wedge_s \tau &= \alpha \\ \sigma \vee_s \tau &= \beta \\ \neg_s \tau &= \gamma \end{aligned}$$

then $st(\mathcal{B}) = (Bin(\omega)\,,\,\wedge_s\,,\,\vee_s\,,\,\neg_s)$ is a polynomial-time representation of the countable atomless Boolean algebra $\mathcal{B}([0,1))$ which we call the standard representation of $\mathcal{B}([0,1))$.

The Tally Representation of $\mathcal{B}([0,1))$.

The tally representation $tal(\mathcal{B})$ of $\mathcal{B}([0,1))$ can easily be defined from the binary representation $st(\mathcal{B}([0,1))$ to be the isomorphic image under the map taking $bin(n)$ to $tal(n)$ and is therefore a p–time structure by Lemma 3.4 in light of the note above.

Nerode and Remmel [54] studied three basic properties of ideals in recursive Boolean algebras. Here, given a Boolean algebra $\mathcal{B} = (B, \wedge_B, \vee_B, \neg_B)$, we say $I \subseteq B$ is an *ideal* if 0_B (the zero of B) is in I, if for all $x, y \in I$ then $x \vee_B y \in B$, and if for $x \in I$ and $z \in B$ then $x \wedge_B z \in I$. Then I is a *maximal ideal* if for all $z \in B$, either $z \in I$ or $\neg_B z \in I$.

Nerode and Remmel studied polynomial-time analogues of the following well known results on r.e. ideals in a recursive presentation of $\mathcal{B}([0,1))$.

(A) In a recursive Boolean algebra, every r.e. maximal ideal is recursive.

(B) Every proper recursive ideal is contained in a recursive maximal ideal.

(C) There exists an r.e. ideal of $\mathcal{B}([0,1))$ which is not extendible to a recursive ideal.

We note that (C) is equivalent to the proposition that there is an r.e. axiomatizable theory which is not contained in any decidable theory.

First consider (A). The fact that every r.e. maximal ideal of a recursive Boolean algebra is recursive is based on Kleene's lemma that a set which is r.e. and co–r.e. is automatically recursive. The obvious p–time analogue of (A) is that every NP maximal ideal of a p–time Boolean algebra is polynomial-time. However in this case, it is a long standing open problem whether $\mathrm{NP} \cap \mathrm{co\text{-}NP} = \mathrm{P}$. Moreover, by well known results of Baker-Gill–Solovay [3], there exist recursive oracles X and Y such that $\mathrm{NP}^X \cap \mathrm{co\text{-}NP}^X \neq \mathrm{P}^X$ and $\mathrm{NP}^Y \cap \mathrm{co\text{-}NP}^Y = \mathrm{P}^Y$, but $\mathrm{P}^Y \neq \mathrm{NP}^Y$. Thus it should come as no surprise that the analogue of (A) is oracle dependent. That is, Nerode and Remmel were able to modify the Baker-Gill-Solovay constructions to prove the following.

Theorem 8.1 *There exists a recursive oracle X such that there exists a maximal ideal I of $tal(\mathcal{B})$ such that $I \in \mathrm{NP}^X \setminus \mathrm{P}^X$.*

Our next corollary immediately follows from Theorem 8.1 and Lemma 3.3.

Corollary 8.2 *There exists a recursive oracle X such that there exists a maximal ideal I of $st(\mathcal{B})$ such that $I \in \mathrm{NEXT}^X \setminus \mathrm{DEXT}^X$.*

Theorem 8.3 *There exists a recursive oracle Y such that there exists a maximal ideal J of $st(\mathcal{B})$ such that $J \in \mathrm{NP}^X \setminus \mathrm{P}^X$.*

Theorem 8.4 *There exists a recursive oracle E such that $\mathrm{P}^E \neq \mathrm{NP}^E$ and the following hold.*

(i) *Every maximal ideal I of $tal\,(\mathcal{B})$ which is in NP^E is in P^E.*

(ii) *Every maximal ideal J of $st(\mathcal{B})$ which is in NEXT^E is in DEXT^E.*

(iii) *Every maximal ideal K of $st(\mathcal{B})$ which is in NP^E is in P^E.*

Theorems 8.1–8.4 then yield the following results.

Theorem 8.5 *Arguments which remain valid under relativization to oracles do not suffice to prove any of the following.*

(i) $\mathrm{P} \neq \mathrm{NP}$ *implies that every* NP *maximal ideal of $tal\,(\mathcal{B})$ is in* P.

(ii) $\mathrm{P} \neq \mathrm{NP}$ *implies that there is an* NP *maximal ideal of $tal\,(\mathcal{B})$ which is not in* P.

(iii) $\mathrm{P} \neq \mathrm{NP}$ *implies that every* NP *maximal ideal of $st(\mathcal{B})$ is in* P.

(iv) $\mathrm{P} \neq \mathrm{NP}$ *implies that there is an* NP *maximal ideal of $st(\mathcal{B})$ which is not in* P.

Next we turn to the analogues of (B). In this case, Nerode and Remmel proved that the obvious analogues of (B) are true for both $tal\,(\mathcal{B})$ and $st(\mathcal{B})$, although the argument requires a great deal more care. That is, Nerode and Remmel [54] proved the following.

Theorem 8.6 *Every proper ideal I of $st(\mathcal{B})$ which is in* P *can be extended to a maximal ideal J of $st(\mathcal{B})$ which is in* P.

Theorem 8.7 *Every proper ideal I of $st(\mathcal{B})$ which is in* $\mathrm{NP} \cap$ *co*–NP *can be extended to a maximal ideal J of $st(\mathcal{B})$ which is in* $\mathrm{NP} \cap$ *co*–NP.

Theorem 8.8 *Every proper ideal I of $st(\mathcal{B})$ which is in* DEXT *can be extended to a maximal ideal J of $st(\mathcal{B})$ which is in* DEXT.

Theorem 8.9 *Every proper ideal I of $st(\mathcal{B})$ which is in* $\mathrm{NEXT} \cap$ *co*–NEXT *can be extended to a maximal ideal J of $st(\mathcal{B})$ which is in* $\mathrm{NEXT} \cap$ *co*–NEXT.

Theorem 8.10 *Every proper ideal I of tal $(\mathcal{B})$ which is in* P *can be extended to a maximal ideal J of tal $(\mathcal{B})$ which is in* P.

Theorem 8.11 *Every proper ideal I of tal $(\mathcal{B})$ which is in* NP $\cap$ *co*–NP *can be extended to a maximal ideal J of tal $(\mathcal{B})$ which is in* NP $\cap$ *co*–NP.

Finally we turn to the analogues of (C). In this case the analogues are oracle dependent, despite Theorems 8.6–8.11.

Theorem 8.12 *There exists a recursive oracle A such that* $\mathrm{P}^A \neq \mathrm{NP}^A$ *and*

(i) *every proper ideal I_1 of $st(\mathcal{B})$ which is in* NP^A *is extendible to a maximal ideal J_1 of $st(\mathcal{B})$ which is in* NP^A,

(ii) *every proper ideal I_2 of tal $(\mathcal{B})$ which is in* NP^A *is extendible to a maximal ideal J_2 of tal $(\mathcal{B})$ which is in* NP^A.

(iii) *every proper ideal I_3 of $st(\mathcal{B})$ which is in* NEXT^A *is extendible to a maximal ideal J_3 of $st(\mathcal{B})$ which is in* NEXT^A.

Proof. Homer and Maass [36] constructed a recursive oracle A such that $\mathrm{P}^A \neq \mathrm{NP}^A$ but $\mathrm{NP}^A = \text{co–NP}^A$. Thus we can use the relativized version of Theorem 8.7 to prove part (i) and we can use the relativized version of Theorem 8.11 to prove part (ii). Finally part (iii) follows from part (i) and Lemma 3.3. □

The proof of the other direction of the oracle dependence requires a new construction which is much more subtle than any of the previous theorems on ideals in our p–time representation of $\mathcal{B}([0,1))$. The actual proofs can be found in [54].

Theorem 8.13

(i) *There exists a recursive oracle C and an ideal J_1 of tal $(\mathcal{B})$ which is in* NP^C *and which is not contained in any maximal ideal of tal $(\mathcal{B})$ which is in* NP^C.

(ii) *There exists a recursive oracle B and an ideal J_2 of $st(\mathcal{B})$ which is in* NP^B *and which is not contained in any maximal ideal of $st(\mathcal{B})$ which is in* NP^B.

Of course, we can combine Lemma 3.3 and Theorem 8.13 (i) to prove the following.

Theorem 8.14 *There exists a recursive oracle C and an ideal J_3 of $st(\mathcal{B})$ which is in* NEXT^C *and which is not contained in any maximal ideal of tal $(\mathcal{B})$ which is in* NEXT^C.

9 Conclusions and Future Directions

In this survey, we have presented the basic definitions of complexity theoretic algebra and model theory and have attempted to outline the current state of knowledge in the field. There is a great deal more which remains to be done. We will just mention four possible themes for future research.

First, we observe that the results on complexity theoretic algebra were limited to the study of ideals in the free Boolean algebras and subspaces of infinite dimensional vector spaces. There are many other algebraic structures that have been studied in recursive algebra, including fields, modules, subalgebras of Boolean algebras, subgroups of groups, etc.. Cenzer, Downey and Remmel [9] have recently investigated torsion-free Abelian groups. We have also given complexity theoretic results in combinatorics. Other related areas of mathematics such as geometry and number theory should also provide fruitful bases for investigation.

Second, most of our results concerned the notions of polynomial-time complexity, with some results given on linear time and on exponential-time complexity. There are many other interesting notions of complexity, including for example, PSPACE and LOGSPACE, which should provide both comparable and contrasting results.

Third, we gave only a few results involving the important complexity hypotheses of theoretical computer science, such as whether P = NP or NP = PSPACE. In the complexity theory of real functions, Ko [42] has provided many such results. For example, he gives a condition (not involving complexity) on a real function f showing that if P = NP and if f is a p–time computable function on the unit interval which satisfies this condition, then all roots of f are p–time computable. There should be similar results in complexity theoretic algebra.

Fourth, we have only begun the study of complexity theoretic model theory with a few results on relational structures and with the general notion of Scott sentences and categoricity. If one studies the recursive model theory survey by Harizanov [33], many problems suggest themselves. For example, the authors have recently investigated complexity theoretic versions of the effective completeness theorem in [17]. Decidability is also of interest in the study of prime and saturated models and in stability theory.

References

[1] C. J. Ash and R. G. Downey, Decidable subspaces and recursively enumerable subspaces, J. Symbolic Logic, **49** (1984) 1137–1145.

[2] C. J. Ash and A. Nerode, Intrinsically recursive relations, in: Aspects of Effective Algebra, (Proc. Conf. Monash Univ., Clayton, Australia, Aug. 1–4, 1979), J. N. Crossley (ed.), (Upside Down A Book Co., Yarra Glen, Victoria, Australia, 1981), 26–41.

[3] T. Baker, J. Gill and R. Solovay, Relativizations of the P = ? NP question, SIAM J. Comput., **4** (1975) 431–442.

[4] J. L. Balcázar, J. Díaz and J. Gabarró, Structural Complexity, Vol. 1, 2, (Springer-Verlag, Berlin, 1988, 1990), (2nd. rev. edn. 1995).

[5] F. A. Bäuerle, Complexity Theoretic Algebra, Ph.D. Thesis, Univ. California, San Diego, La Jolla, CA, (1994).

[6] C. H. Bennett and J. Gill, Relative to a random oracle A, $\mathrm{P}^A \neq \mathrm{NP}^A \neq$ co–NP^A with probability 1, SIAM J. Comput., **10** (1981) 96–113.

[7] S. Breidbart, On splitting recursive sets, J. Comput. System Sci., **17** (1978) 56–64.

[8] J. Cai, On Some Most Probable Separations of Complexity Classes, Ph.D. Thesis, Dept. Comput. Sci., Cornell Univ., Ithaca, NY, (1986).

[9] D. Cenzer, R. G. Downey and J. B. Remmel, Feasible torsion-free Abelian groups, (to appear).

[10] D. Cenzer and J. B. Remmel, Polynomial-time versus recursive models, Ann. Pure Appl. Logic, **54** (1991) 17–58.

[11] D. Cenzer and J. B. Remmel, Feasibly categorical abelian groups, in: Feasible Mathematics II, (Papers from 2nd. Workshop, May 28–30, 1992, Cornell Univ. Ithaca, NY), P. Clote and J. B. Remmel, (eds.), Progr. Comput. Sci. Appl. Logic, **13** (1992) 91–153.

[12] D. Cenzer and J. B. Remmel, Feasible graphs and colorings, Math. Logic Quart., **41** (1992) 327–352.

[13] D. Cenzer and J. B. Remmel, Polynomial-time abelian groups, Ann. Pure Appl. Logic, **56** (1992) 313–363.

[14] D. Cenzer and J. B. Remmel, Recursively presented games and strategies, Math. Social Sciences, **24** (1992) 117–139.

[15] D. Cenzer and J. B. Remmel, Feasibly categorical models, in: Logic and Computational Complexity, (Papers from Interntl. Workshop, LCC '94, Oct. 13–16, 1994, Indianapolis, IN), D. Leivant (ed.), Lecture Notes in Comput. Sci., **960** (1995) 300–312.

[16] D. Cenzer and J. B. Remmel, Complexity theoretic model theory and algebra, in: Handbook of Recursive Mathematics, Yu. L. Ershov, S. S. Goncharov, A. Nerode and J. B. Remmel, (eds.), (Elsevier, Amsterdam, 1998), 381–513.

[17] D. Cenzer and J. B. Remmel, Decidability and completeness and complexity, (to appear).

[18] D. Cenzer and J. B. Remmel, Feasible graphs with standard universe, (Special issue: Computability Theory, Proc. Conf., Oberwolfach 1996), Ann. Pure Appl. Logic, (to appear).

[19] A. Cobham, The intrinsic computational difficulty of functions, in: Proc. Inter. Cong. for Logic, Methodology, and Philos. Sci., (Jerusalem, 1984), North Holland, (1965) 24–30.

[20] S. A. Cook, The complexity of theorem proving procedures, ACM Sympos. Theory of Computation, (1971) 151–158.

[21] J. N. Crossley, (ed.), Aspects of Effective Algebra, (Proc. Conf. Monash Univ., Clayton, Australia, Aug. 1–4, 1979), (Upside Down A Book Co., Yarra Glen, Victoria, Australia, 1981).

[22] J. N. Crossley and J. B. Remmel, Polynomial-time combinatorial operators are polynomials. in: Feasible Mathematics, (Proc. Workshop, June 16–18, 1989, Cornell Univ. Ithaca, NY), S. Buss and P. J. Scott (eds.), Progr. Comput. Sci. Appl. Logic, **9** (1990) 99–130.

[23] J. N. Crossley and J. B. Remmel, Cancellation laws for polynomial-time p–isolated sets, Ann. Pure Appl. Logic, **56** (1992) 147–172.

[24] J. C. E. Dekker, Two notes on vector spaces with recursive operations, Notre Dame J. Formal Logic, **12** (1971) 329–334.

[25] R. G. Downey and M. R. Fellows, Fixed parameter tractability and completeness III: some structural aspects of the W–Hierarchy, in: Complexity Theory, (Papers arising from the Structure and Complexity Theory Workshop, Dagstuhl Internatl. Conf. Res. Ctr., Wadern, Germany, Feb. 2–8, 1992), K. Ambos-Spies, S. Homer and U. Schöning, (eds.), (Cambridge Univ. Press, 1993), 191–225.

[26] J. Edmonds, Paths, trees, and flowers, Canad. J. Math., **17** (1965) 449–467.

[27] K. Ko and H. M. Friedman, Computational complexity of real functions, Theoret. Comput. Sci., **20** (1982) 323–352.

[28] M. R. Garey and D. S. Johnson, Computers and Intractability, a guide to the theory of NP–completeness, (W. H. Freeman, San Francisco, 1979).

[29] S. S. Goncharov, Autostability and computable families of constructivizations (Russian), Algebra i Logika, **14** (1975) 647–680; [translated in: Algebra and Logic, **14** (1975) 392–409].

[30] S. S. Goncharov, Certain properties of the constructivization of Boolean algebras (Russian), Sibirsk. Math. Zh., **16** (1975) 264–278; [translated in: Siberian Math. J., **16** (1975) 203–214].

[31] S. S. Goncharov, Autostability of models and abelian groups (Russian), Algebra i Logika, **19** (1980) 23–44; [translated in: Algebra and Logic, **19** (1980) 13–27].

[32] S. Grigorieff, Every recursive linear ordering has a copy in DTIME–SPACE$(n, \log(n))$, J. Symbolic Logic, **55** (1990) 260–276.

[33] V. Harizanov, Pure recursive model theory, in: Handbook of Recursive Mathematics, Yu. L. Ershov, S. S. Goncharov, A. Nerode and J. B. Remmel, (eds.), (Elsevier, Amsterdam, 1998), 3–114.

[34] J. Hartmanis, On sparse sets in NP − P, Inform. Process. Lett., **16** (1983) 55–60.

[35] J. Hartmanis and R. E. Stearns, On the computational complexity of algorithms, Trans. Amer. Math. Soc., **117** (1965) 285–306.

[36] S. Homer and W. Maass, Oracle-dependent properties of the lattice of NP sets, Theoret. Comput. Sci., **24** (1983) 279–289.

[37] J. E. Hopcroft and J. D. Ullman, Formal Languages and their Relations to Automata, (Addison-Wesley, Reading, Mass, 1969).

[38] I. Kalantari and A. Retzlaff, Maximal vector spaces under automorphisms of the lattice of recursively enumerable vector spaces, J. Symbolic Logic, **42** (1977) 481–491.

[39] R. M. Karp, Reducibilities among combinatorial problems, in: Complexity of Computer Computations, (Proc. Sympos. IBM Watson Res. Ctr., Yorktown Heights, NY, Mar. 20–22, 1972), R. E. Miller, J. W. Thatcher and J. D. Bohlinger, (eds.), (Plenum Press, New York, 1972) 85–103.

[40] N. G. Khisamiev, Constructive abelian groups, in: Handbook of Recursive Mathematics, Yu. L. Ershov, S. S. Goncharov, A. Nerode and J. B. Remmel, (eds.), (Elsevier, Amsterdam, 1998), 1177–1231.

[41] B. M. Khoussainov and A. Nerode, Automatic presentations of structures, in: Logic and Computational Complexity, (Papers from Interntl. Workshop, LCC '94, Oct. 13–16, 1994, Indianapolis, IN), D. Leivant (ed.), Lecture Notes in Comput. Sci., **960** (1995) 367–392.

[42] K. Ko, Complexity theory of real functions, Progr. Theoret. Comput. Sci., (1991).

[43] K. Ko, Computational complexity of fixed points and intersection points, J. Complexity, **11** (1995) 265–292.

[44] P. La Roche, Recursively presented Boolean algebras, Notices Amer. Math. Soc., **24**:6 (1977) A 552–553 (abstract).

[45] S. R. Mahaney, Sparse complete sets for NP: solution of a conjecture of Berman and Hartmanis, J. Comput. System Sci., **25** (1982) 130–143.

[46] A. I. Mal'tsev, On recursive abelian groups (Russian), Dokl. Akad. Nauk SSSR, **146** (1962) 1009–1012; [translated in: Soviet Math. – Dokl., **3** (1962) 1431–1434; also in: The Metamathematics of Algebraic Systems, Collected Papers: 1936–1967, translated and edited by B. F. Wells III, Stud. Logic Found. Math., **66** (1971), Ch. 24, 282–286].

[47] D. A. Martin, Classes of recursively enumerable sets and degrees of unsolvability, Z. Math. Logik Grundlag. Math., **12** (1966) 295–310.

[48] G. Metakides and A. Nerode, Recursion theory and algebra, in: Algebra and Logic, (Proc. 14th. Summer Res. Inst. Austral. Math. Soc., Monash Univ., Clayton, Vic., Australia, Jan. 6 – Feb. 16, 1974), J. N. Crossley, (ed.), Lecture Notes in Math., **450** (1975) 209–219.

[49] G. Metakides and A. Nerode, Recursively enumerable vector spaces, Ann. Math. Logic, **11** (1977) 147–171.

[50] A. Nerode and J. B. Remmel, A survey of lattices of r.e. substructures, in: Recursion Theory, (Proc. AMS-ASL Summer Inst., Ithaca, NY, June 28 – July 16, 1982), A. Nerode and R. A. Shore, (eds.), Proc. Sympos. Pure Math., **42** (1985) 323–375.

[51] A. Nerode and J. B. Remmel, Recursion theory on matroids II, in: (Proc. Southeast Asian Conf. Logic, Singapore, 1981), C. T. Chong and M. J. Wicks, (eds.), Stud. Logic Found. Math., **111** (1983) 133–184.

[52] A. Nerode and J. B. Remmel, Complexity theoretic algebra I, Vector spaces over finite fields, in: Proc. Conf. Logic and Computer Science: New Trends and Applications, (Turin, 1986), Rend. Sem. Mat. Univ. Politec. Torino 1987, Special Issue, (1988) 1–10.

[53] A. Nerode and J. B. Remmel, Complexity theoretic algebra I: Vector spaces over finite fields, in: Structure in Complexity Theory, (Proc. 2nd. Annual IEEE Conf. Struct. Complexity Theory, Cornell Univ., Ithaca NY, June 16–19, 1987), (IEEE Comput. Soc. Press, Washington, DC, 1987), 218–239.

[54] A. Nerode and J. B. Remmel, Complexity-theoretic algebra II: Boolean algebras, (Special issue: 3rd. Asian Conf. Math. Logic, Beijing, Oct. 26–30, 1987), D. P. Yang, (ed.), Ann. Pure Appl. Logic, **44** (1989) 71–99.

[55] A. Nerode and J. B. Remmel, Complexity-theoretic algebra: vector space bases, in: Feasible Mathematics, (Proc. Workshop, June 16–18, 1989, Cornell Univ. Ithaca, NY), S. Buss and P. J. Scott (eds.), Progr. Comput. Sci. Appl. Logic, **9** (1990) 293–319.

[56] A. Nerode and J. B. Remmel, Polynomial time equivalence types, in: Logic and Computation, (Proc. Workshop, CMU, Pittsburg, June 30 – July 2, 1987), W. Sieg, (ed.), Contemp. Math., **106** (1990) 221–249.

[57] A. Nerode and J. B. Remmel, Polynomially isolated sets, in: Recursion Theory Week, (Proc. Conf., Oberwolfach, Mar. 19–25, 1989), K. Ambos-Spies, G. H. Müller and G. E. Sachs, (eds.), Lecture Notes in Math., **1432** (1990) 323–362.

[58] A. Nerode and J. B. Remmel, On the lattices of NP–subspaces of a polynomial time vector space over a finite field, Ann. Pure Appl. Logic, **81** (1996) 125–170.

[59] A. Nerode, J. B. Remmel and A. Ščedrov, Polynomially graded logic I a graded version of system T (preliminary report), in: Logic in Computer Science, (Proc. 4th. Annual IEEE Sympos. Logic in Comput. Sci.), LICS–89, (IEEE Comput. Soc. Press, Washington, DC, 1989) 375–385.

[60] A. T. Nurtazin Computable classes and algebraic criteria of autostability, summary of scientific thesis (Russian), (Math. Inst. Siberian Branch of SSSR Acad. Sci., Novosibirsk, 1974).

[61] P. Odifreddi, Classical Recursion Theory. The Theory of Functions and Sets of Natural Numbers, Stud. Logic Found. Math., **125** (1989).

[62] J. B. Remmel, Maximal and cohesive vector spaces, J. Symbolic Logic, **42** (1977) 400–418.

[63] J. B. Remmel, Recursively enumerable Boolean algebras, Ann. Math. Logic, **15** (1978) 75–107.

[64] J. B. Remmel, On r.e. and co-r.e. vector spaces with nonextendible bases, J. Symbolic Logic, **45** (1980) 20–34.

[65] J. B. Remmel, Recursion theory on algebraic structures with independent sets, Ann. Math. Logic, **18** (1980) 153–191.

[66] J. B. Remmel, Recursively categorical linear orderings, Proc. Amer. Math. Soc., **83** (1981) 387–391.

[67] J. B. Remmel, Recursive Boolean algebras, in: Handbook of Boolean Algebras, Vol. 3, J. D. Monk, R. Bonnet and S. Koppelberg, (eds.), (North-Holland, Amsterdam, NY, 1989), Ch. 25, 1097–1165.

[68] J. B. Remmel, When is every recursive linear ordering of type μ recursively isomorphic to a polynomial time linear ordering over the natural numbers in binary form?, in: Feasible Mathematics, (Proc. Workshop, June 16–18, 1989, Cornell Univ. Ithaca, NY), S. Buss and P. J. Scott (eds.), Progr. Comput. Sci. Appl. Logic, **9** (1990) 321–350.

[69] J. B. Remmel, Polynomial time categoricity and linear orderings, in: Logical Methods, (Papers from Conf. in honor of Anil Nerode's Sixtieth Birthday, June 1–3, 1992, Cornell Univ., Ithaca, NY), J. N. Crossley, J. B. Remmel, R. A. Shore and M. E. Sweedler, (eds.), Progr. Comput. Sci. Appl. Logic, **12** (1993) 713–746.

[70] J. H. Schmerl, Recursive colorings of graphs, Canad. J. Math., **32** (1980) 821–830.

[71] U. Schöning, Complexity and structure, Lecture Notes in Comput. Sci., **211** (1986).

[72] J. Shinoda and T. Slaman, On the theory of PTIME degrees of the recursive sets, J. Comput. System Sci., **41** (1990) 321–366.

[73] R. A. Shore and T. A. Slaman, The p–T–degrees of the recursive sets: lattice embeddings, extensions of embeddings and the two-quantifier theory, Theoret. Comput. Sci., **97** (1992) 263–284.

[74] R. L. Smith, Two theorems on autostability in p–groups, in: Logic Year 1979-1980, (Proc. Seminars and Conf. Math. Logic, Univ. Connecticut, Storrs, CT, 1979/80), M. Lerman, J. H. Schmerl and R. I. Soare, (eds.), Lecture Notes in Math., **859** (1981) 302–311.

[75] C. Wilson, Relativization, Reducibilities, and the Exponential-Time Hierarchy, Ph.D. Thesis, Univ. Toronto, Toronto, Canada, (1980).

[76] A. C. Yao, Separating the polynomial-time hierarchy by oracles, in: Foundations of Computer Science, (Proc. 26th. IEEE Sympos. Foundations Comput. Sci., Portland, OR, Oct. 21–23, 1985), FOCS-85, (IEEE Comput. Soc. Press, Washington, DC, 1985), 1–10.

Douglas Cenzer
Department of Mathematics
University of Florida
Gainesville, Fl 32611, USA
e-mail: cenzermath.ufl.edu

Jeffrey B. Remmel
Department of Mathematics
University of California at San Diego
La Jolla, CA 92093, USA
e-mail: jremmelucsd.edu

satisfied at stage [illegible] that there [illegible] with [illegible]
Our choice of [illegible] ensures that [illegible] then no string [illegible] is
ever restrained from [illegible] by a requirement of higher priority than [illegible] which is
satisfied at some stage [illegible]. Also our choice of [illegible] ensures that [illegible] for
all [illegible] and [illegible]

[illegible]

(1) [illegible]

(2) [illegible]

(3) [illegible]

Note that [illegible]
[illegible]

Chapter 11

A Bibliography of Recursive Algebra and Recursive Model Theory

I. Kalantari

Mathematics is replete with 'constructions'. It is natural to interpret 'constructions' in the mathematical world by 'constructions' in the physical world. For Euclid, a construction in geometry was permissible only if it was with ruler and compass. The development of arithmetic, analysis and algebra established construction of mathematical objects through 'computation'. When advanced and diverse branches of mathematics in the twentieth century found foundation and unification in the Cantorian school, much success was achieved, and so Hilbert proposed methodical lines of inquiry. However, this new school's 'nonconstructive' methods created obstacles for full acceptance. The theorems became general and the proofs elegant but often the mathematical objects could not be constructed in any computational sense.

When recursive functions found its origins in the works of Gödel, Church, Kleene, Turing and others, and when some of Hilbert's questions found resolution in 'incompleteness' and 'undecidability', it was natural to investigate the *recursiveness* or *non-recursiveness* of mathematical constructions.

It is this spirit that gave rise to the subjects of *recursive algebra* and *recursive model theory.*

Recursive algebra and recursive model theory have emerged over the last thirty years as lively and successful subjects pursued world-wide, largely due to the influence of Nerode and his school of thought and Ershov and his school of thought. In particular, the voluminous and novel works of Remmel in the West, and those of Goncharov in the East have helped establish great inroads in the field. A more detailed history can be found in the introductory section of this Handbook. Here, we leave the remaining permitted room for the long and self-explanatory bibliography.

HANDBOOK OF RECURSIVE MATHEMATICS
Edited by Yu. L. Ershov, S. S. Goncharov,
A. Nerode, and J. B. Remmel

This bibliography is a list of papers which are mostly either in recursive algebra or recursive model theory. Some other papers (and a few books) closely related (in the view of this collector) to these two subjects, such as those in recursive combinatorics, constructive algebra, reverse mathematics, recursion theory or logic in general, are also included.

A Bibliographer's Apology

This list was put together with velocity, and is certain to be imperfect. To the reader who finds any of the blemishes, I offer apologies. It is hoped that an electronic and expanded version of this list will be kept and updated with the help of the publishing company of this Handbook: Elsevier. To repair any oversight committed here, corrections, suggestions and remarks should be e-mailed to: i-kalantari@wiu.edu.

Acknowledgments

I thank Jeff Remmel for his encouragement for formation of this list. I am also grateful to Victor Marek, the managing editor of this volume, for his effective management and his reliable help in preparation of this bibliography. A first bibliography of recursive algebra was put together by Crossley and Miranda in 1981 (see [120]) and I used it as a starting point for the present list. Finally, I thank Bethany Webb, Jeanette Mumford and Sara Kalantari for their excellent help in various stages of preparation of this list.

Further Acknowledgments

An early draft of this list was circulated for reaction. Victor Marek made several valuable suggestions which lead to substantial improvements and I am indebted to him. I am also grateful to Doug Cenzer, John Chisholm, Rod Downey, Bill Gasarch, Sergey Goncharov, Valentina Harizanov, Tami Hummel, Bakhadyr Khoussainov, Julia Knight, Michael Moses, Anil Nerode, Jeff Remmel, Fred Richman, Nader Vakil, and Galen Weitkamp for their valuable remarks and additions leading to significant improvement of this list. Fred Richman's compiled bibliography of 61 sources in constructive algebra helped immensely in strengthening the list of the related work in that area; I am indebted to him too. Jeff Remmel deserves especial gratitude for his continued support and encouragement. His enthusiasm was contagious and fruitful. All remaining flaws are mine alone.

References

[1] P. H. G. Aczel and J. N. Crossley, Constructive order types. III, Arch. Math. Logik Grundlag., **9** (1966) 112–116.

[2] R. Aharoni, M. Magidor and R. A. Shore, On the strength of König's duality theorem for infinite bipartite graphs, J. Combin. Theory, Ser. B, **54** (1992) 257–290.

[3] D. A. Alton and E. W. Madison, Computability of Boolean algebras and their extensions, Ann. Math. Logic, **6** (1973/74) 95–128.

[4] K. R. Apt, Recursive embeddings of partial orderings, Canad. J. Math., **29** (1977) 349–359.

[5] E. Artin and O. Schreier, Algebraische Konstruktion reeller Körper, Abh. Math. Sem. Univ. Hamburg, **5** (1927) 85–99.

[6] C. J. Ash, Recursive labelling systems and stability of recursive structures in hyperarithmetical degrees, Trans. Amer. Math. Soc., **298** (1986) 497–514; errata ibid., **310** (1988) 851.

[7] C. J. Ash, Stability of recursive structures in the arithmetical degrees, Ann. Pure Appl. Logic, **32** (1986) 113–135.

[8] C. J. Ash, Categoricity in hyperarithmetical degrees, Ann. Pure Appl. Logic, **34** (1987) 1–14.

[9] C. J. Ash, Labelling systems and r.e. structures, Ann. Pure Appl. Logic, **47** (1990) 99–119.

[10] C. J. Ash, A construction for recursive linear orderings, J. Symbolic Logic, **56** (1991) 673–683.

[11] C. J. Ash and R. G. Downey, Decidable subspaces and recursively enumerable subspaces, J. Symbolic Logic, **49** (1984) 1137–1145.

[12] C. J. Ash and S. S. Goncharov, Strong Δ_2^0-categoricity (English), Algebra i Logika, **24** (1985) 718–727; [also in: Algebra and Logic, **24** (1985) 471–476].

[13] C. J. Ash, C. G. Jockusch Jr. and J. F. Knight, Jumps of orderings, Trans. Amer. Math. Soc., **319** (1990) 573–599.

[14] C. J. Ash and J. F. Knight, Pairs of recursive structures, Ann. Pure Appl. Logic, **46** (1990) 211–234.

[15] C. J. Ash and J. F. Knight, Relatively recursive expansions, Fund. Math., **140** (1992) 137–155.

[16] C. J. Ash and J. F. Knight, A completeness theorem for certain classes of recursive infinitary formulas, Math. Logic Quart., **40** (1994) 173–181.

[17] C. J. Ash and J. F. Knight, Mixed systems, J. Symbolic Logic, **59** (1994) 1383–1399.

[18] C. J. Ash and J. F. Knight, Recursive expansions, Fund. Math., **145** (1994) 153–169.

[19] C. J. Ash and J. F. Knight, Ramified systems, Ann. Pure Appl. Logic, **70** (1994) 205–221.

[20] C. J. Ash and J. F. Knight, Possible degrees in recursive copies, Ann. Pure Appl. Logic, **75** (1995) 215–221.

[21] C. J. Ash and J. F. Knight, Recursive structures and Ershov's hierarchy, Math. Logic Quart., **42** (1996) 461–468.

[22] C. J. Ash, J. F. Knight, M. Manasse and T. A. Slaman, Generic copies of countable structures, Ann. Pure Appl. Logic, **42** (1989) 195–205.

[23] C. J. Ash, J. F. Knight and J. B. Remmel, Quasi-simple relations in copies of a given recursive structure, Ann. Pure Appl. Logic, **86** (1997) 203–218.

[24] C. J. Ash, J. F. Knight and T. A. Slaman, Relatively recursive expansions II, Fund. Math., **142** (1993) 147–161.

[25] C. J. Ash and T. S. Millar, Persistently finite, persistently arithmetic theories, Proc. Amer. Math. Soc., **89** (1983) 487–492.

[26] C. J. Ash and A. Nerode, Functorial properties of algebraic closure and Skolemization, J. Austral. Math. Soc., **31** (1981) 136–141.

[27] C. J. Ash and A. Nerode, Intrinsically recursive relations, in: Aspects of Effective Algebra, (Proc. Conf. Monash Univ., Clayton, Australia, Aug. 1–4, 1979), J. N. Crossley (ed.), (Upside Down A Book Co., Yarra Glen, Victoria, Australia, 1981), 26–41.

[28] C. W. Ayoub, On constructing bases for ideals in polynomial rings over the integers, J. Number Theory, **17** (1983) 204–225.

[29] S. A. Badaev, On minimal enumerations, Siberian Adv. Math., **2** (1992) 1–30.

[30] S. A. Badaev and S. S. Goncharov, On computable minimal enumerations, in: Algeba, (Proc. 3rd. Internatl. Conf. Algebra, Krasnoyarsk, Aug. 23–28, 1993), (de Gruyter, Berlin and NY, 1996), 21–33.

[31] E. J. Barker, Intrinsically Σ^0_α relations, Ann. Pure Appl. Logic, **39** (1988) 105–130.

[32] E. J. Barker, Back and forth relations for reduced abelian p–groups, Ann. Pure Appl. Logic, **75** (1995) 223–249.

[33] J. Barwise, Admissible Sets and Structures, Perspect. in Math. Logic, (1975).

[34] J. Barwise and J. Schlipf, An introduction to recursively saturated and resplendent models, J. Symbolic Logic, **41** (1976) 531–536.

[35] J. Barwise and J. Schlipf, On recursively saturated models of arithmetic, in: Model Theory and Algebra, D. H. Saracino and V. B. Weispfenning, ed., Lecture Notes in Math., **498** (1981) 42–55.

[36] F. A. Bäuerle, Complexity Theoretic Algebra, Ph.D. Thesis, Univ. California, San Diego, La Jolla, CA, (1994).

[37] F. A. Bäuerle and J. B. Remmel, On speedable and levelable vector spaces, Ann. Pure Appl. Logic, **67** (1994) 61–112.

[38] G. Baumslag, F. B. Cannonito and C. F. Miller, III, Computable algebra and group embeddings, J. Algebra, **69** (1981) 186–212.

[39] G. Baumslag, F. B. Cannonito and D. J. S. Robinson, The algorithmic theory of finitely generated metabelian groups, Trans. Amer. Math. Soc., **344** (1994) 629–648.

[40] G. Baumslag, F. B. Cannonito, D. J. S. Robinson and D. Segal, The algorithmic theory of polycyclic-by-finite groups, J. Algebra, **142** (1991) 118–149.

[41] W. Baur, Rekursive Algebren mit Kettenbedingungen, Z. Math. Logik Grundlag. Math., **20** (1974) 37–46.

[42] W. Baur, Decidability and undecidability of theories of abelian groups with predicates for subgroups, Compositio Math., **31** (1975) 23–30.

[43] W. Baur, Undecidability of the theory of abelian groups with a subgroup, Proc. Amer. Math. Soc., **55** (1976) 125–128.

[44] W. Baur, On the elementary theory of quadruples of vector spaces, Ann. Math. Logic, **19** (1980) 243–262.

[45] D. R. Bean, Effective coloration, J. Symbolic Logic, **41** (1976) 469–480.

[46] D. R. Bean, Recursive Euler and Hamilton paths, Proc. Amer. Math. Soc., **55** (1976) 385–394.

[47] T. Becker and V. Weispfenning, Gröbner bases: a computational approach to commutative algebra, (in cooperation with H. Kredel), Grad. Texts in Math., **141** (1993).

[48] M. J. Beeson, Foundations of Constructive Mathematics: metamathematical studies, Ergeb. Math. Grenzgeb. Ser. 3, **6** (1985).

[49] P. Bernays and D. Hilbert, Grundlagen der Mathematik, Vols. I, II, (1934, 1939), Grundlehren Math. Wiss.; [2nd. edn. (1968, 1970)].

[50] R. Beigel and W. I. Gasarch, On the complexity of finding the chromatic number of a recursive graph I: The bounded case, Ann. Pure Appl. Logic, **45** (1989) 1–38.

[51] R. Beigel and W. I. Gasarch, On the complexity of finding the chromatic number of a recursive graph II: The unbounded case, Ann. Pure Appl. Logic, **45** (1989) 227–246.

[52] R. Beigel and W. I. Gasarch, Applications of binary search to recursive graph theory, Univ. Maryland Tech. Rep. TR 1804.

[53] E. Bishop, Foundations of Constructive Analysis, (McGraw-Hill, New York, Toronto, London, 1967).

[54] E. Bishop and D. S. Bridges, Constructive Analysis, Grundlehren Math. Wiss., **279** (1985).

[55] A. R. Blass, J. L. Hirst and S. G. Simpson, Logical analysis of some theorems of combinatorics and topological dynamics, in: Logic and Combinatorics, (Proc. AMS-IMS-SIAM Joint Summer Research Conf. Applications Math. Logic to Finite Combinatorics, Humboldt State Univ., Arcata, CA, Aug. 4–10, 1985), S. G. Simpson, (ed.), Contemp. Math., **65** (1987) 125–156.

[56] W. W. Boone, Certain simple unsolvable problems in group theory, I–VI, Nederl. Akad. Wentensch. Proc., Ser. A, **57** (1954) 231–237, 492–497; **58** (1955) 252–256, 571–577; **60** (1957) 22–27, 227–232.

[57] E. Borel, Leçons sur la théorie des fonctions, Gauthier-Villars et fils, Paris, (1898).

[58] S. Brackin, A summary of Ramsey-type theorems and their provability in weak formal systems, in: Logic and Combinatorics, (Proc. AMS-IMS-SIAM Joint Summer Research Conf. Applications Math. Logic to Finite Combinatorics, Humboldt State Univ., Arcata, CA, Aug. 4–10, 1985), S. G. Simpson, (ed.), Contemp. Math., **65** (1987) 169–178.

[59] D. S. Bridges and R. Mines, What is constructive mathematics?, Math. Intelligencer, **6** (1984) 32–38.

[60] D. S. Bridges and F. Richman, Varieties of Constructive Mathematics, London Math. Soc. Lecture Note Ser., **97** (1987).

[61] R. L. Brooks, On coloring the nodes of a network, Proc. Cambridge Philos. Soc., **37** (1941) 194–197, [also in: Classic Papers in Combinatorics, I. Gessel and G.-C. Rota (eds.), (Birkhäuser, Boston, 1987), 118–121].

[62] L. E. J. Brouwer, Collected Works, Vol. 1, Philosophy and Foundations of Mathematics, A. Heyting, (ed.), (North-Holland, Amsterdam, Oxford, 1975).

[63] L. E. J. Brouwer, Collected Works, Vol. 2, Geometry, Analysis, Topology and Mechanics, H. Freudenthal, (ed.), (North-Holland, Amsterdam, Oxford, 1976).

[64] L. E. J. Brouwer and B. de Loor, Intuitionistischer Beweis des Fundamentalsatzes der Algebra, Koninklijke Nederl. Akad. Wentensch. Proc., **27** (1924) 186–188.

[65] B. Buchberger, A theoretical basis for the reduction of polynomials to canonical forms, ACM SIGSAM Bull., **10** (1976) 19–29.

[66] B. Buchberger, A criterion for detecting unnecessary reductions in the construction of Gröbner-bases, in: Symbolic and algebraic computation, (EUROSAM '79, Internatl. Sympos., Marseille, June 1979), E. W. Ng, (ed.), Lecture Notes in Comput. Sci., **72** (1979) 3–21.

[67] B. Buchberger, A note on the complexity of constructing Gröbner-bases, in: Computer Algebra, (Proc. European Computer Algebra Conf., EUROCAL '83, March 28–30, 1983, London), J. A. van Halzen, (ed.), Lecture Notes in Comput. Sci., **162** (1983) 137–145.

[68] B. Buchberger, Gröbner bases: an introduction, in: Automata, Languages and Programming, (Proc. 19th. Internatl, Colloq. Technische Universität Wien, July 13–17, 1992), W. Kuich, (ed.), Lecture Notes in Comput. Sci., **623** (1992) 378–379.

[69] B. Buchberger, Symbolic computation: computer algebra and logic, in: Frontiers of Combining Systems, (Proc. Conf. March, 1996, Munich), F. Baader and K. U. Schulz, (eds.), Appl. Log. Ser., **3** (1996) 193–219.

[70] B. Buchberger and B. Roider, Input/output codings and transition functions in effective systems, Internat. J. Gen. Systems, **4** (1977/78) 201–209.

[71] S. A. Burr, Some undecidable problems involving the edge-coloring and vertex-coloring of graphs, Discrete Math., **50** (1984) 171–177.

[72] S. Burris, Decidable model companions, Z. Math. Logik Grundlag. Math., **35** (1989) 225–227.

[73] F. B. Cannonito and M. Finkelstein, On primitive recursive permutations and their universes, J. Symbolic Logic, **34** (1969) 634–638.

[74] J. S. Carroll, Some undecidability results for lattices in recursion theory, Pacific J. Math., **122** (1986) 319–331.

[75] H.-G. Carstens, The complexity of some combinatorial constructions, Z. Math. Logik Grundlag. Math., **23** (1977) 121–130.

[76] H.-G. Carstens and U. Golze, Recursive paths in cross-connected trees and an application to cell spaces, Math. Systems Theory, **15** (1982) 29–37.

[77] H.-G. Carstens and P. Päppinghaus, Recursive coloration of countable graphs, Ann. Pure Appl. Logic, **25** (1983) 19–45.

[78] H.-G. Carstens and P. Päppinghaus, Abstract construction of counterexamples in recursive graph theory, in: Computation and Proof Theory, (Proc. Logic Colloq., Aachen, July 18–23, 1983, Part II), M. M. Richter, E. Börger, W. Oberschelp, B. Schinzel and W. Thomas, (eds.), Lecture Notes in Math., **1104** (1984) 39–62.

[79] H.-G. Carstens and P. Päppinghaus, Extensible algorithms, in: Logic and Machines: Decision Problems and Complexity, (Proc. Sympos. Rekursive Kombinatorik, May 23–28, 1983, Univ. Münster, Westfalen), E. Börger, G. Hasenjaeger and D. Rödding, (eds.), Lecture Notes in Comput. Sci., **171** (1984) 162–182.

[80] J. Case, Sortability and extensibility of the graphs of recursively enumerable partial and total orders, Z. Math. Logik Grundlag. Math., **22** (1976) 1–18.

[81] J. Case, Effectivizing inseparability, Z. Math. Logik Grundlag. Math., **37** (1991) 97–111.

[82] C. Cellucci, Categorie ricorsive, Boll. Un. Mat. Ital., Ser. III, **19** (1964) 300–305.

[83] D. Cenzer, Effective real dynamics, in: Logical Methods, (Papers from Conf. in honor of Anil Nerode's Sixtieth Birthday, June 1–3, 1992, Cornell Univ., Ithaca, NY), J. N. Crossley, J. B. Remmel, R. A. Shore and M. E. Sweedler, (eds.), Progr. Comput. Sci. Appl. Logic, **12** (1993) 162–177.

[84] D. Cenzer, Π_1^0 classes, in: Handbook of Recursion Theory, E. R. Griffor, (ed.), North-Holland, (to appear).

[85] D. Cenzer, P. Clote, R. L. Smith, R. I. Soare and S. S. Wainer, Members of countable Π_1^0 classes, (Special issue: 2nd. Southeast Asian Logic Conf., Bangkok, 1984) Ann. Pure Appl. Logic, **31** (1986) 145–163.

[86] D. Cenzer, R. G. Downey, C. G. Jockusch and R. A. Shore, Countable thin Π_1^0 classes, Ann. Pure Appl. Logic, **59** (1993) 79–139.

[87] D. Cenzer and E. Howorka, On vertex k–partitions of certain infinite graphs, Discrete Math., **23** (1978) 105–113.

[88] D. Cenzer and J. B. Remmel, Polynomial-time versus recursive models, Ann. Pure Appl. Logic, **54** (1991) 17–58.

[89] D. Cenzer and J. B. Remmel, Feasible graphs and colorings, Math. Logic Quart., **41** (1992) 327–352.

[90] D. Cenzer and J. B. Remmel, Feasibly categorical abelian groups, in: Feasible Mathematics II, (Papers from 2nd. Workshop, May 28–30, 1992, Cornell Univ. Ithaca, NY), P. Clote and J. B. Remmel, (eds.), Progr. Comput. Sci. Appl. Logic, **13** (1992) 91–153.

[91] D. Cenzer and J. B. Remmel, Polynomial-time abelian groups, Ann. Pure Appl. Logic, **56** (1992) 313–363.

[92] D. Cenzer and J. B. Remmel, Recursively presented games and strategies, Math. Social Sciences, **24** (1992) 117–139.

[93] D. Cenzer and J. B. Remmel, Feasibly categorical models, in: Logic and Computational Complexity, (Papers from Interntl. Workshop, LCC '94, Oct. 13–16, 1994, Indianapolis, IN), D. Leivant (ed.), Lecture Notes in Comput. Sci., **960** (1995) 300–312.

[94] D. Cenzer and J. B. Remmel, Complexity and categoricity, Inform. and Comput., **140** (1998) 2–25.

[95] D. Cenzer and J. B. Remmel, Index sets for Π_1^0 classes, Ann. Pure Appl. Logic, (to appear).

[96] D. Cenzer and J. B. Remmel, Index sets in computable analysis, (to appear).

[97] D. Cenzer and R. L. Smith, On the ranked points of a Π_1^0 set, J. Symbolic Logic, **54** (1989) 975–991.

[98] C. C. Chang and H. J. Keisler, Model Theory, 3rd. edn., Stud. Logic Found. Math., **73** (1990); [1st. edn. 1973, 2nd. edn. 1977].

[99] K. H. Chen, Recursive well founded orderings, Ann. Math. Logic, **13** (1978) 117–147.

[100] J. Chisholm, Effective Model Theory vs. Recursive Model Theory, Ph.D. Thesis, Univ. Wisconsin, Madison, WI, (1988).

[101] J. Chisholm, The complexity of intrinsically r.e. subsets of existentially decidable models, J. Symbolic Logic, **55** (1990) 1213–1232.

[102] J. Chisholm, Effective model theory vs. recursive model theory, J. Symbolic Logic, **55** (1990) 1168–1191.

[103] J. Chisholm and M. Moses, Undecidable linear orderings that are n–recursive for each n, (in preparation).

[104] P. A. Cholak, S. S. Goncharov, B. M. Khoussainov and R. A. Shore, Computably categorical structures and expansions by constants, (to appear).

[105] P. A. Cholak, C. G. Jockusch Jr. and T. A. Slaman, Ramsey's theorem for pairs, Bulletin of Symbolic Logic, (1998) 63 (abstract).

[106] A. Church, Binary recursive arithmetic, J. Math. Pures Appl., Sér. 9, **36** (1957) 39–55.

[107] A. Church, An independence question in recursive arithmetic, in: Colloq. Found. Math., Math. Machines and Appl., (Akad. Kiadó, Budapest, Tihany, 1962), (1965) 21–26.

[108] A. Church and W. V. Quine, Some theorems on definability and decidability, J. Symbolic Logic, **17** (1952) 179–187.

[109] P. Clote, A note on decidable model theory, in: Model Theory and Arithmetic, (Paris, 1979/80), C. Berline, K. McAloon and J.-P. Ressyare, (eds.), Lecture Notes in Math., **890** (1981) 134–142.

[110] P. Clote, A recursion theoretic analysis of the clopen Ramsey theorem, J. Symbolic Logic, **49** (1984) 376–400.

[111] P. Clote, Weak partition relations, finite games, and independence results in Peano arithmetic, in: Model Theory of Algebra and Arithmetic, (Proc. Conf. Appl. Logic, Algebra and Arithmetic, Sept. 1–7, 1979, Karpacz, Poland), L. Packolski, J. Wierzejewski and A. J. Wilkie, (eds.), Lecture Notes in Math., **834** (1984) 92–107.

[112] P. Clote, On recursive trees with a unique infinite branch, Proc. Amer. Math. Soc., **93** (1985) 335–342.

[113] R. Coles, R. G. Downey and B. M. Khoussainov, Initial segments of computable linear orderings, (to appear).

[114] J. N. Crossley, Recursive equivalence: A survey, in: Proc. Summer School in Logic, (Leeds, 1967), Springer-Verlag, (1968) 241–251.

[115] J. N. Crossley, Constructive Order Types, (North-Holland, Amsterdam, 1969).

[116] J. N. Crossley, Recursive equivalence, Bull. London Math. Soc., **2** (1970) 129–151.

[117] J. N. Crossley, (ed.), Aspects of Effective Algebra, (Proc. Conf. Monash Univ., Clayton, Australia, Aug. 1–4, 1979), (Upside Down A Book Co., Yarra Glen, Victoria, Australia, 1981).

[118] J. N. Crossley, Fifty years of computability, Southeast Asian Bull. Math., **11** (1988) 81–99.

[119] J. N. Crossley, A. B. Manaster and M. F. Moses, Recursive categoricity and recursive stability, Logic Paper 49, (Monash Univ., Clayton, Victoria, Australia, 1983).

[120] J. N. Crossley and S. Miranda, A bibliography of effective algebra, in: Aspects of Effective Algebra, (Proc. Conf. Monash Univ., Clayton, Australia, Aug. 1–4, 1979), J. N. Crossley (ed.), (Upside Down A Book Co., Yarra Glen, Victoria, Australia, 1981)251–290.

[121] J. N. Crossley and A. Nerode, Combinatorial Functors, Ergeb. Math. Grenzgeb. Ser. 3, **81** (1974).

[122] J. N. Crossley and A. Nerode, Effective dimension, J. Algebra, **41** (1976) 398–412.

[123] J. N. Crossley and R. J. Parikh, On isomorphisms of recursive well-orderings, J. Symbolic Logic, **28** (1963) 308 (abstract).

[124] J. N. Crossley and J. B. Remmel, Undecidability and recursive equivalence I, in: (Proc. Southeast Asian Conf. Logic, Singapore, 1981), C. T. Chong and M. J. Wicks, (eds.), Stud. Logic Found. Math., **111** (1983) 37–53.

[125] J. N. Crossley and J. B. Remmel, Undecidability and recursive equivalence II, in: Computation and Proof Theory, (Proc. Logic Colloq., Aachen, July 18–23, 1983, Part II), M. M. Richter, E. Börger, W. Oberschelp, B. Schinzel and W. Thomas, (eds.), Lecture Notes in Math., **1104** (1984) 79–100.

[126] J. N. Crossley, J. B. Remmel, R. A. Shore and M. E. Sweedler, (eds.), Logical Methods, (Papers from Conf. in honor of Anil Nerode's Sixtieth Birthday, June 1–3, 1992, Cornell Univ., Ithaca, NY), Progr. Comput. Sci. Appl. Logic, **12** (1993).

[127] J. N. Crossley and J. Shepherdson, Extracting programs from proofs by an extension of the Curry-Howard process, in: Logical Methods, (Papers from Conf. in honor of Anil Nerode's Sixtieth Birthday, June 1–3, 1992, Cornell Univ., Ithaca, NY), J. N. Crossley, J. B. Remmel, R. A. Shore and M. E. Sweedler, (eds.), Progr. Comput. Sci. Appl. Logic, **12** (1993) 222–288.

[128] K. J. Davey, Inseparability in recursive copies, Ann. Pure Appl. Logic, **68** (1994) 1–52.

[129] M. D. Davis, Yu. Matijasevič and J. Robinson, Hilbert's tenth problem. Diophantine equations: positive aspects of a negative solution, in: Mathematical Developments Arising from Hilbert Problems, (N. Illinois Univ., De Kalb, Ill, 1974), Proc. Sympos. Pure Math., **28** (1976) 323–378.

[130] M. D. Davis, H. Putnam and J. Robinson, The decision problem for exponential diophantine equations, Ann. of Math., Ser. 2, **74** (1961) 425–436.

[131] M. D. Davis and E. J. Weyuker, Computability, Complexity and Languages, Fundamentals of Theoretical Computer Science, (Academic Press, New York, 1983), [2nd. edn. with R. Sigal, 1994].

[132] M. Dehn, Papers on group theory and topology, translated from the German and with introductions and an appendix by J. Stillwell, with an appendix by O. Schreier, (Springer-Verlag, 1987).

[133] J. C. E. Dekker, Countable vector spaces with recursive operations. Part I, J. Symbolic Logic, **34** (1969) 363–387.

[134] J. C. E. Dekker, Countable vector spaces with recursive operations. Part II, J. Symbolic Logic, **36** (1971) 477–493.

[135] J. C. E. Dekker, Two notes on vector spaces with recursive operations, Notre Dame J. Formal Logic, **12** (1971) 329–334.

[136] J. C. E. Dekker, Recursive equivalence types and cubes, in: Aspects of Effective Algebra, (Proc. Conf. Monash Univ., Clayton, Australia, Aug. 1–4, 1979), J. N. Crossley (ed.), (Upside Down A Book Co., Yarra Glen, Victoria, Australia, 1981) 87–121.

[137] J. C. E. Dekker, Twilight graphs, J. Symbolic Logic, **46** (1981) 539–571.

[138] J. C. E. Dekker and J. Myhill, Retraceable sets, Canad. J. Math., **10** (1958) 357–373.

[139] J. C. E. Dekker and J. Myhill, Recursive equivalence types, Publ. Math. Univ. California, **3** (1960) 67–214.

[140] J. Denef, Hilbert's tenth problem for quadratic rings, Proc. Amer. Math. Soc., **48** (1975) 214–220.

[141] J. Denef, Diophantine sets over algebraic integer rings, II, Trans. Amer. Math. Soc., **257** (1980) 227–236.

[142] J. Denef and L. Lipshitz, Diophantine sets over some rings of algebraic integers, J. London Math. Soc., Ser. 2, **18** (1978) 385–391.

[143] A. S. Denisov, Constructive homogeneous extensions, Sibirsk. Math. Zh., **25** (1984) 60–69 (Russian); [translated in: Siberian Math. J., **25** (1984) 879–888].

[144] A. S. Denisov, Every decidable theory has a $\mathbf{0}'$–strongly constructive homogeneous model (Russian), in: Computable Invariants in the Theory of

Algebraic Systems, V. N. Remeslennikov, (ed.), Akad. Nauk SSSR Sibirsk. Otdel., Vyčisl. Centr, Novosibirsk, (1987) 13–21.

[145] S. D. Denisov, Models of non-contradictory formulas and the Ershov hierarchy (Russian), Algebra i Logika, **11** (1972) 648–655, 736; [translated in: Algebra and Logic, **11** (1972) 359–362].

[146] R. P. Dilworth, A decomposition theorem for partially ordered sets, in: Classic Papers in Combinatorics, I. Gessel and G.-C. Rota (eds.), (Birkhäuser, Boston, 1987), 139–144.

[147] R. G. Downey, Abstract Dependence, Recursion Theory and the Lattice of Recursively Enumerable Filters, Ph.D. Thesis, Monash Univ., Clayton, Victoria, Australia, (1982).

[148] R. G. Downey, Abstract dependence, recursion theory and the lattice of recursively enumerable filters, Bull. Austral. Math. Soc., **27** (1983) 461–464.

[149] R. G. Downey, Nowhere simplicity in matroids, J. Austral. Math. Soc., Ser. A, **35** (1983) 28–45.

[150] R. G. Downey, On a question of A. Retzlaff, Z. Math. Logik Grundlag. Math., **29** (1983) 379–384.

[151] R. G. Downey, Bases of supermaximal subspaces and Steinitz systems. I, J. Symbolic Logic, **49** (1984) 1146–1159.

[152] R. G. Downey, Co-immune subspaces and complementation in V_∞, J. Symbolic Logic, **49** (1984) 528–538.

[153] R. G. Downey, A note on decompositions of recursively enumerable subspaces, Z. Math. Logik Grundlag. Math., **30** (1984) 465–470.

[154] R. G. Downey, Some remarks on a theorem of Iraj Kalantari concerning convexity and recursion theory, Z. Math. Logik Grundlag. Math., **30** (1984) 295–302.

[155] R. G. Downey, Bases of supermaximal subspaces and Steinitz systems. II, Z. Math. Logik Grundlag. Math., **32** (1986) 203–210.

[156] R. G. Downey, Undecidability of $L(F_\infty)$ and other lattices of r.e. substructures, Ann. Pure Appl. Logic, **32** (1986) 17–26; corr. ibid., **48** (1990) 299–301.

[157] R. G. Downey, Sound, totally sound, and unsound recursive equivalence types, Ann. Pure Appl. Logic, **31** (1986) 1–22.

[158] R. G. Downey, Maximal theories, Ann. Pure Appl. Logic, **33** (1987) 245–282.

[159] R. G. Downey, Orbits of creative subspaces, Proc. Amer. Math. Soc., **99** (1987) 163–170.

[160] R. G. Downey, On Π_1^0 classes and their ranked points, Notre Dame J. Formal Logic, **32** (1991) 499–512.

[161] R. G. Downey, An invitation to structural complexity, (New Zealand Math. Colloq., Dunedin, 1991), New Zealand J. Math., **21** (1992) 33–89.

[162] R. G. Downey, Every recursive Boolean algebra is isomorphic to one with incomplete atoms, Ann. Pure Appl. Logic, **60** (1993) 193–206.

[163] R. G. Downey, On presentations of algebraic structures, in: Complexity, Logic and Recursion Theory, A. Sorbi, (ed.), Lecture Notes in Pure and Appl. Math., **187** (1997) 157–205.

[164] R. G. Downey and M. R. Fellows, Parameterized computational feasibility, in: Feasible Mathematics II, (Papers from 2nd. Workshop, May 28–30, 1992, Cornell Univ. Ithaca, NY), P. Clote and J. B. Remmel, (eds.), Progr. Comput. Sci. Appl. Logic, **13** (1992) 219–244.

[165] R. G. Downey and M. R. Fellows, Parameterized Complexity, (Monograph, to appear), Springer-Verlag.

[166] R. G. Downey and G. Hird, Automorphisms of supermaximal subspaces, J. Symbolic Logic, **50** (1985) 1–9.

[167] R. G. Downey and C. G. Jockusch Jr., Every low Boolean algebra is isomorphic to a recursive one, Proc. Amer. Math. Soc., **122** (1994) 871–880.

[168] R. G. Downey and C. G. Jockusch Jr., Effective presentability of Boolean algebras of Cantor–Bendixson rank 1, J. Symbolic Logic, (to appear).

[169] R. G. Downey, C. G. Jockusch Jr. and M. Stob, Array nonrecursive sets and multiple permitting arguments, in: Recursion Theory Week, (Proc. Conf., Oberwolfach, Mar. 19–25, 1989), K. Ambos-Spies, G. H. Müller and G. E. Sachs, (eds.), Lecture Notes in Math., **1432** (1990) 141–173.

[170] R. G. Downey and I. Kalantari, Effective extensions of linear forms on a recursive vector space over a recursive field, Z. Math. Logik Grundlag. Math., **31** (1985) 193–200.

[171] R. G. Downey and J. F. Knight, Orderings with α–th jump degree $\mathbf{0}^{\alpha}$, Proc. Amer. Math. Soc., **114** (1992) 545–552.

[172] R. G. Downey and S. A. Kurtz, Recursion theory and ordered groups, Ann. Pure Appl. Logic, **32** (1986) 137–151.

[173] R. G. Downey and S. Lemmp, On the proof theoretical strength of the Dushnik-Miller theorem for countable linear orderings, (in preparation).

[174] R. G. Downey and M. F. Moses, On choice sets and strongly nontrivial self-embeddings of recursive linear orders, Z. Math. Logik Grundlag. Math., **35** (1989) 237–246.

[175] R. G. Downey and M. Moses, Recursive linear orders with incomplete successivities, Trans. Amer. Math. Soc., **326** (1991) 653–668.

[176] R. G. Downey and J. B. Remmel, The universal complementation property, J. Symbolic Logic, **49** (1984) 1125–1136.

[177] R. G. Downey and J. B. Remmel, Automorphisms and recursive structures, Z. Math. Logik Grundlag. Math., **33** (1987) 339–345.

[178] R. G. Downey and J. B. Remmel, Classification of degree classes associated with r.e. subspaces, Ann. Pure Appl. Logic, **42** (1989) 105–124.

[179] R. G. Downey and J. B. Remmel, Effectively and noneffectively nowhere simple subspaces, in: Logical Methods, (Papers from Conf. in honor of Anil Nerode's Sixtieth Birthday, June 1–3, 1992, Cornell Univ., Ithaca, NY), J. N. Crossley, J. B. Remmel, R. A. Shore and M. E. Sweedler, (eds.), Progr. Comput. Sci. Appl. Logic, **12** (1993) 314–351.

[180] R. G. Downey, J. B. Remmel and L. V. Welch, Degrees of splittings and bases of recursively enumerable subspaces, Trans. Amer. Math. Soc., **302** (1987) 683–714.

[181] L. van den Dries, New decidable fields of algebraic numbers, Proc. Amer. Math. Soc., **77** (1979) 251–256.

[182] B. N. Drobutun, Enumerations of simple models (Russian), Sibirsk. Math. Zh., **18** (1977) 1002–1014, 1205; [translated in: Siberian Math. J., **18** (1977) 707–716].

[183] Z. A. Dulatova, Extended theories of Boolean algebras (Russian), Sibirsk. Math. Zh., **25** (1984) 201–204.

[184] Z. A. Dulatova, On recursive automorphisms of atomless Boolean algebras (Russian), in: Proc. 9th. All-Union Conf. Math. Logic, Leningrad, (1988) 55.

[185] Z. A. Dulatova, Constructivity of Boolean algebras with a distinguished subalgebra (Russian), Mat. Zametki**46** (1989) 53–56, 127; [translated in: Math. Notes, **46** (1989) 924–926].

[186] V. D. Dzgoev, On the constructivization of certain structures (Abstract in Russian), (Dep. VINITI, No. 1606–79, Moscow, 1979), Sibirsk. Math. Zh., **21** (1980) 231.

[187] V. D. Dzgoev, Recursive automorphisms of constructive models (Russian), in: Proc. 15th. All-Union Algebraic Conf., (Novosibirsk, 1979), Part 2, (1979) 52.

[188] V. D. Dzgoev, Constructivizations of direct products of algebraic systems (Russian), Algebra i Logika, **21** (1982) 138–148; [translated in: Algebra and Logic, **21** (1982) 88–96].

[189] V. D. Dzgoev, Constructive Boolean algebras (Russian), Mat. Zametki, **44** (1988) 750–757, 861; [translated in: Math. Notes, **44** (1988) 896–901].

[190] V. D. Dzgoev, Constructization of Boolean lattices (Russian), Algebra i Logika, **27** (1988) 641–648, 736; [translated in: Algebra and Logic, **27** (1988) 395–400].

[191] V. D. Dzgoev and S. S. Goncharov, Autostability of models, Algebra i Logika, **19** (1980) 45–58, 132; [translated in: Algebra and Logic, **19** (1980) 28–37].

[192] H. M. Edwards, An appreciation of Kronecker, Math. Intelligencer, **9** (1987) 28–35.

[193] H. M. Edwards, Kronecker's views on the foundations of mathematics, in: The History of Modern Mathematics, Vol. I, Ideas and their Reception, (Proc. Sympos. Vassar College, Poughkeepsie, NY, June 20–24, 1989), D. E. Rowe and J. McClean, (eds.), (Academic Press, 1989) 67–77.

[194] H. M. Edwards, Kronecker's arithmetical theory of algebraic quantities, Jahresber. Deutsch. Math.-Verein., **94** (1992) 130–139.

[195] A. Ehrenfeucht, Separable theories, Bull. Acad. Polon. Sci. Sér. Sci. Math. Astronom. Phys., **9** (1961) 17–19.

[196] E. F. Eisenberg, Effective Isomorphisms of Algebraic Systems, Ph.D. Thesis, Cornell Univ., Ithaca, NY, (1974).

[197] E. F. Eisenberg and J. B. Remmel, Effective isomorphisms of algebraic structures, in: Patras Logic Symposion, (Proc. Logic Sympos. Patras, Greece, Aug. 18–22, 1980) G. Metakides (ed.), Stud. Logic Found. Math., **109** (1982) 95–122.

[198] D. Eisenbud, C. Huneke and W. Vasconcelos, Direct methods for primary decomposition, Invent. Math., **110** (1992) 207–235.

[199] E. Ellentuck, The first order properties of Dedekind finite integers, Fund. Math., **63** (1968) 7–25.

[200] E. Ellentuck, Nonrecursive combinatorial functions, J. Symbolic Logic, **37** (1972) 90–95.

[201] E. Ellentuck, A new proof that analytic sets are Ramsey, J. Symbolic Logic, **39** (1974) 163–165.

[202] Yu. L. Ershov, Decidability of the elementary theory of relatively complemented distributive lattices, and of the theory of filters (Russian), Algebra i Logika, **3** (1964) 17–38.

[203] Yu. L. Ershov, Numbered fields, in: Logic, Methodology and Philos. Sci. III, (Proc. 3rd. Internatl. Congr. Logic, Methodology and Philos. Sci., Amsterdam, 1967), B. van Rootselaar and J. F. Staal, (eds.), North-Holland, 1968) 31–34.

[204] Yu. L. Ershov, Constructive models (Russian), in: Izbr. Vopr. Alg. i Log., [Selected Questions in Algebra and Logic], Sbornik posvjascen. pamjati A. I. Mal'ceva, [A collection dedicated to the memory of A. I. Mal'tsev], Izdat. Nauk Sibirsk. Otdel., Novosibirsk, (1973) 111–130.

[205] Yu. L. Ershov, Skolem functions and constructive models (Russian), Algebra i Logika, **12** (1973) 644–654, 735; [translated in: Algebra and Logic, **12** (1973) 368–373].

[206] Yu. L. Ershov, Theory of Numerations III (Constructive Models) (Russian), Lib. Dept. Algebra and Math. Logic Novosibirsk Univ., **13**, (Novosibirsk Gosudarstr. Univ., Novosibirsk, 1974).

[207] Yu. L. Ershov, Hereditarily effective operations (Russian), Algebra i Logika, **15** (1976) 642–654. [translated in: Algebra and Logic, **15** (1976) 400–409.

[208] Yu. L. Ershov, Theory of Numerations (Russian), Monographs in Math. Logic and Foundations of Math., (Nauka, Moskva, 1977).

[209] Yu. L. Ershov, Theorie der Numerierungen, III, (translated from Russian and edited by G. Asser and H.-D. Hecker), Z. Math. Logik Grundlag. Math., **23** (1977) 289–371.

[210] Yu. L. Ershov, Decision Problems and Constructivizable Models (Russian), (Mathematical Logic and Foundations of Mathematics, Nauka, Moskva, 1980).

[211] Yu. L. Ershov, Regularly closed fields (Russian), Dokl. Akad. Nauk SSSR, **251** (1980) 783–785.

[212] Yu. L. Ershov, Undecidability of regularly closed fields (Russian), Algebra i Logika, **20** (1981) 389–394; [translated in: Algebra and Logic, **20** (1981) 257–260].

[213] Yu. L. Ershov, Eliminability of quantifiers in regularly closed fields (Russian), Dokl. Akad. Nauk SSSR, **258** (1981) 16–20.

[214] Yu. L. Ershov, On elementary theories of regularly closed fields (Russian), Dokl. Akad. Nauk SSSR, **257** (1981) 271–274; [translated in: Soviet Math. - Dokl., **23** (1981) 259–262].

[215] Yu. L. Ershov, Maximal RC_π–fields, Algebra i Logika, **32** (1993) 497–518, 586; [translated in: Algebra and Logic, **32** (1993) 267–278].

[216] Yu. L. Ershov, Decidability of the theory of the class of fields $\mathfrak{F}^f_*$, Dokl. Akad. Nauk, **336** (1994) 733–736; [translated in: Russian Acad. Sci. Dokl. Math., **49** (1994) 582–586].

[217] Yu. L. Ershov, Model-theoretic properties of RC^*–fields, Dokl. Akad. Nauk, **335** (1994) 138–141; [translated in: Russian Acad. Sci. Dokl. Math., **49** (1994) 255–259].

[218] Yu. L. Ershov, Definability and Computability (Russian), (Sibirsk. Shk. Alg. i Log., NII MIOONGU, Novosibirsk, 1996); [translated in: Siberian School of Algebra and Logic, (Consultants Bureau, Plenum, New York, 1996)].

[219] Yu. L. Ershov, I. A. Lavrov, A. D. Taĭmanov and M. A. Taĭtslin, Elementary theories (Russian), Uspekhi Mat. Nauk, **20** (1965) 37–108. [translated in: Russian Math. Surveys, **20** (1965) 35–105].

[220] Yu. L. Ershov and E. A. Palyutin, Mathematical Logic (Russian), 2nd. edn., (Nauka, Moskva, 1987). [English translation of 1st. edn., translated by V. Shokurov, (MIR Publishers, Moscow, 1984)].

[221] Yu. L. Ershov and M. A. Taitslin, Undecidability of certain theories, Algebra i Logika, **2** (1963) 37–41 (Russian).

[222] S. T. Fedoryaev, Structures of algebraic reducibility of positive numerations (Russian), Teor. Algoritm i ee Prilozhen, Vyčisl. Systemy, **129** (1989) 144–151, 197.

[223] S. T. Fedoryaev, Constructivizable models with linear structure of algebraic reducibility, Mat. Zametki, **48** (1990) 106–111; [translated in: Math. Notes, **48** (1990) 1245–1249].

[224] S. T. Fedoryaev, Some properties of algebraic reducibility of constructivizations (Russian), Algebra i Logika, **29** (1990) 597–612, 627; [translated in: Algebra and Logic, **29** (1990) 395–405].

[225] S. T. Fedoryaev, Countability of widths of algebraic reducibility structures for models in some classes, Siberian Adv. Math., **3** (1993) 81–102.

[226] S. T. Fedoryaev, Decidable algorithmic problems on relatively complemented distributive lattices which cannot be simultaneously decidable, Bulletin of Symbolic Logic, **1** (1995) 109 (abstract).

[227] S. T. Fedoryaev, Recursively incompatible algorithmic problems on 1-constructivizable distributive lattices with relative complements (Russian), Algebra i Logika, **34** (1995) 667–680, 729; [translated in: Algebra and Logic, **34** (1995) 371–378].

[228] L. J. Feiner, Orderings and Boolean Algebras not Isomorphic to Recursive Ones, Ph.D. Thesis, M.I.T., Cambridge, MA, (1967).

[229] L. J. Feiner, Hierarchies of Boolean algebras, J. Symbolic Logic, **35** (1970) 365–374.

[230] L. J. Feiner, Degrees of non-recursive presentability, Proc. Amer. Math. Soc., **38** (1973) 621–624.

[231] S. Fellner, Recursive and Finite Axiomatizability of Linear Orderings, Ph.D. Thesis, Rutgers Univ., New Brunswick, NJ, (1976).

[232] R. M. Friedberg, Two recursively enumerable sets of incomparable degrees of unsolvability, Proc. Nat. Acad. Sci. U.S.A., **43** (1957) 236–238.

[233] R. M. Friedberg, Three theorems on recursive enumeration: I decomposition, II maximal set, III enumeration without duplication, J. Symbolic Logic, **23** (1958) 309–316.

[234] H. M. Friedman, Higher set theory and mathematical practice, Ann. Math. Logic, **2** (1971) 325–357.

[235] H. M. Friedman, Some systems of second order arithmetic and their use, Proc. Internatl. Cong. Math. (Vancouver, BC, 1974) Canad. Math. Congress, Montréal, Que., **1** (1975) 235–242.

[236] H. M. Friedman and J. L. Hirst, Weak comparibility of well orderings and reverse mathematics, Ann. Pure Appl. Logic, **47** (1990) 11–29.

[237] H. M. Friedman, N. Robertson and P. Seymour, The metamathematics of the graph minor theorem, in: Logic and Combinatorics, (Proc. AMS-IMS-SIAM Joint Summer Research Conf. Applications Math. Logic to Finite Combinatorics, Humboldt State Univ., Arcata, CA, Aug. 4–10, 1985), S. G. Simpson, (ed.), Contemp. Math., **65** (1987) 229–261.

[238] H. M. Friedman, S. G. Simpson and R. L. Smith, Countable algebra and set existence axioms, Ann. Pure Appl. Logic, **25** (1983) 141–181; addendum, ibid. **27** (1985) 319–320.

[239] A. Fröhlich and J. C. Shepherdson, On the factorisation of polynomials in a finite number of steps, Math. Z., **62** (1955) 331–334.

[240] A. Fröhlich and J. C. Shepherdson, Effective procedures in field theory, Philos. Trans. Roy. Soc. London, Ser. A, **248** (1956) 407–432.

[241] F. Galvin and K. Prikry, Borel sets and Ramsey's theorem, J. Symbolic Logic, **38** (1973) 193–198.

[242] R. O. Gandy, Symposium on constructivity in mathematics: introductory remarks. in: Logic, Methodology and Philos. Sci. VI, (Proc. 6th. Internatl. Congr. Logic, Methodology, and Philos. Sci., Hannover, Aug. 22–29, 1979), L. J. Cohen, J. Łoś, H. Pfeiffer and K.-P. Podewski, (eds.), Stud. Logic Found. Math., **104** (1982) 141–143.

[243] R. O. Gandy, Limitations to mathematical knowledge, in: Logic Colloquium '80, (Papers intended for the European Summer Meeting of ASL to have been held in Prague, Aug. 24–30, 1980), D. van Dalen, D. Lascar and T. J. Smiley, (eds.), Stud. Logic Found. Math., 108 (1982) 129–146.

[244] R. O. Gandy, The confluence of ideas in 1936, in: The Universal Turing Machine: a Half-century Survey, R. Herken, (ed.), Oxford Univ. Press, (1988) 55–111; [2nd. edn., Computerkulture II, Springer-Verlag, Vienna, 1995].

[245] W. I. Gasarch, D. W. Keuker and D. M. Mount, Highly recursive rooted graphs, Congr. Numer., 69 (1989) 97–102; [also in: Proc. 19th. S. E. Conf. Combinatorics, Baton Rouge, LA, Feb. 15–19, 1988, F. Hoffman, R. C. Mullin, R. G. Stanton and K. B. Reid, (eds.),, Congr. Numer., **67** (1988)], also in: Proc. 18th. Manitoba Conf. Numer. Math. and Comput., Winnipeg, MB, Sept. 29 - Oct. 1, 1988), D. S. Meek, R. G. Stanton and G. H. J. van Rees, (eds.) Congr. Numer., **69** (1989)].

[246] W. I. Gasarch and A. Lee, On the finiteness of the recursive chromatic number, Ann. Pure Appl. Logic, (to appear).

[247] W. I. Gasarch and G. Martin, Index sets in recursive combinatorics, in: Logical Methods, (Papers from Conf. in honor of Anil Nerode's Sixtieth Birthday, June 1–3, 1992, Cornell Univ., Ithaca, NY), J. N. Crossley, J. B. Remmel, R. A. Shore and M. E. Sweedler, (eds.), Progr. Comput. Sci. Appl. Logic, **12** (1993) 352–385.

[248] K. Gödel, Über formal unentscheidbare Sätze der *Principia mathematica* und verwandter Systeme I, Monatsh. Math. Phys., **38** (1931) 173–198; [translated as: On formally undecidable propositions of *Principia Mathematica* and related systems I, (translated by B. Meltzer, with an introduction by R. B. Braithwaite), (Basic Books, NY, 1963, reprinted Dover, NY, 1992); also in: From Frege to Gödel: A Source Book in Logic, 1879–1931, J. van Heijenroot, (ed.), (Harvard Univ. Press, Cambridge, Mass. and Oxford Univ. Press, London, 1967) 592–617.]

[249] S. S. Goncharov, The constructivizability of superatomic Boolean algebras (Russian), Algebra i Logika, **12** (1973) 31–40, 120; [translated in: Algebra and Logic, **12** (1973) 17–22].

[250] S. S. Goncharov, Autostability and computable families of constructivizations (Russian), Algebra i Logika, **14** (1975) 647–680; [translated in: Algebra and Logic, **14** (1975) 392–409].

[251] S. S. Goncharov, Certain properties of the constructivization of Boolean algebras (Russian), Sibirsk. Math. Zh., **16** (1975) 264–278; [translated in: Siberian Math. J., **16** (1975) 203–214].

[252] S. S. Goncharov, Nonselfequivalent constructivizations of atomic Boolean algebras (Russian), Mat. Zametki, **19** (1976) 853–858; [translated in: Math. Notes, **19** (1976) 500–503].

[253] S. S. Goncharov, Restricted theories of constructive Boolean algebras (Russian), Sibirsk. Math. Zh., **17** (1976) 797–812; [translated in: Siberian Math. J., **17** (1976) 601–611].

[254] S. S. Goncharov, The quantity of nonautoequivalent constructivizations (Russian), Algebra i Logika, **16** (1977) 257–282, 377; [translated in: Algebra and Logic, **16** (1977) 169–185].

[255] S. S. Goncharov, Constructive models of $\aleph_1$–categorical theories (Russian), Mat. Zametki, **23** (1978) 885–888; [translated in: Math. Notes, **23** (1978) 486–487].

[256] S. S. Goncharov, Strong constructivizability of homogeneous models (Russian), Algebra i Logika, **17** (1978) 363–388, 490; [translated in: Algebra and Logic, **17** (1978) 247–263].

[257] S. S. Goncharov, Strongly constructive homogeneous models, in: Logic, Methodology and Philos. Sci. VI, (Proc. 6th. Internatl. Congr. Logic, Methodology, and Philos. Sci., Hannover, Aug. 22–29, 1979), L. J. Cohen, J. Łoś, H. Pfeiffer and K.-P. Podewski, (eds.), Stud. Logic Found. Math., **104** (1982) 108–112.

[258] S. S. Goncharov, Autostability of models and abelian groups (Russian), Algebra i Logika, **19** (1980) 23–44; [translated in: Algebra and Logic, **19** (1980) 13–27].

[259] S. S. Goncharov, Computable numerations (Russian), Algebra i Logika, **19** (1980) 507–551, 617; [translated in: Algebra and Logic, **19** (1980) 325–356].

[260] S. S. Goncharov, The problem of the number of nonautoequivalent constructivizations (Russian), Dokl. Akad. Nauk SSSR, **251** (1980) 271–274; [translated in: Soviet Math. - Dokl., **21** (1980) 411–414].

[261] S. S. Goncharov, On the problem of the number of nonautoequivalent constructivizations (Russian), Algebra i Logika, **19** (1980) 621–639, 745; [translated in: Algebra and Logic, **19** (1980) 401–414].

[262] S. S. Goncharov, A totally transcendental decidable theory without constructivizable homogeneous models (Russian), Algebra i Logika, **19** (1980) 137–149, 250; [translated in: Algebra and Logic, **19** (1980) 85–93].

[263] S. S. Goncharov, Totally transcendental theory with a nonconstructivizable prime model (Russian), Sibirsk. Math. Zh., **21** (1980) 44–51; [translated in: Siberian Math. J., **21** (1980) 32–37].

[264] S. S. Goncharov, Nonequivalent constructivizations (Russian), Proc. Math. Inst. Sib. Branch Acad. Sci., (Nauka, Novosibirsk, 1982).

[265] S. S. Goncharov, Universal recursively enumerable Boolean algebras (Russian), Sibirsk. Math. Zh., **24** (1983) 36–43; [translated in: Siberian Math. J., **24** (1983) 852–858].

[266] S. S. Goncharov, Constructive Models, Monash Univ., Clayton, Victoria, Australia, Monash Logic Papers, **56** (November 1984).

[267] S. S. Goncharov, Countable Boolean Algebras (Russian), (Nauka. Sibirsk. Otdel., Novosibirsk, 1988).

[268] S. S. Goncharov, Computable classes of constructivizations for models of finite constructivity type (Russian), Sibirsk. Math. Zh., **34** (1993) 23–37; [translated in: Siberian Math. J., **34** (1993) 812–824].

[269] S. S. Goncharov, Effectively infinite classes of weak constructivizations of models (Russian), Algebra i Logika, **32** (1993) 631–664, 712; [translated in: Algebra and Logic, **32** (1993) 342–360].

[270] S. S. Goncharov, A unique positive enumeration, Siberian Adv. Math., **4** (1994) 52–64.

[271] S. S. Goncharov, Countable Boolean algebras and decidability (Russian), Sibirsk. Skh. Alg. i Log. NII MIOONGU, Novosibirsk, (1996); [translated in: Siberian School of Algebra and Logic, (Consultants Bureau, Plenum, New York, 1997)].

[272] S. S. Goncharov, Morley's problem, Bulletin of Symbolic Logic, **3** (1997) 99 (abstract).

[273] S. S. Goncharov and B. N. Drobotun, Numerations of saturated and homogeneous models (Russian), Sibirsk. Math. Zh., **21** (1980) 25–41, 236; [translated in: Siberian Math. J., **21** (1980) 164–176].

[274] S. S. Goncharov and B. N. Drobotun, The algorithmic dimension of nilpotent groups (Russian), Sibirsk. Math. Zh., **30** (1989) 52–60, 225; [translated in: Siberian Math. J., **30** (1989) 210–217].

[275] S. S. Goncharov and V. D. Dzgoev, Autostability of models (Russian), Algebra i Logika, **19** (1980) 45–58, 132; [translated in: Algebra and Logic, **19** (1980) 28–37].

[276] S. S. Goncharov and B. M. Khoussainov, On the spectrum of degrees of decidable relations, Dokl. Akad. Nauk, **352** (1997) 301–303; [translated in: Soviet Math. - Dokl., **55** (1977) 55–57].

[277] S. S. Goncharov, A. V. Molokov and N. S. Romanovskii, Nilpotent groups of finite algorithmic dimension, Sibirsk. Math. Zh., **30** (1989) 82–88; [translated in: Siberian Math. J., **30** (1989) 63–68].

[278] S. S. Goncharov and A. A. Novikov, Examples of nonautostable systems (Russian), Sibirsk. Math. Zh., 25 (1984) 37–45; [translated in: Siberian Math. J., 25 (1984) 538–545].

[279] S. S. Goncharov and A. T. Nurtazin, Constructive models of complete decidable theories (Russian), Algebra i Logika, **12** (1973) 125–142, 243; [translated in: Algebra and Logic, **12** (1973) 67–77].

[280] H. Gonshor, Effective density types, Notre Dame J. Formal Logic, **17** (1976) 303–307.

[281] A. M. Grabfield, The question of the finitely many steps in the theory of polynomial ideals, M.A. Thesis, Wesleyan Univ., Middletown, CT, (1986).

[282] N. Greenleaf, Linear order in lattices: a constructive study, in: Studies in Foundations and Combinatorics, Adv. Math. Suppl. Stud., **1**, (Academic Press, New York and London, 1978), 11–30.

[283] R. Guhl, A theorem on recursively enumerable vector spaces, Notre Dame J. Formal Logic, **16** (1975) 357–362.

[284] R. Guhl, Two notes on recursively enumerable vector spaces, Notre Dame J. Formal Logic, **18** (1977) 295–298.

[285] D. Guichard, Automorphisms and Large Submodels in Effective Algebra, Ph.D. Thesis, Univ. Wisconsin, Madison, WI, (1982).

[286] D. Guichard, Automorphisms of substructure lattices in recursive algebra, Ann. Pure Appl. Logic, **25** (1983) 47–58.

[287] A. G. Hamilton, Bases and α–dimensions of countable vector spaces with recursive operations, J. Symbolic Logic, **35** (1970) 85–96.

[288] A. G. Hamilton, Recursive equivalence types of vector spaces, J. Austral. Math. Soc., **18** (1974) 376–384.

[289] D. Harel, Hamiltonian paths in infinite graphs, Israel J. Math., **76** (1991) 317–336.

[290] V. S. Harizanov, Degree Spectrum of a Recursive Relation on a Recursive Structure, Ph.D. Thesis, Univ. Wisconsin, Madison, WI, (1987).

[291] V. S. Harizanov, Some effects of Ash-Nerode and other decidability conditions on degree spectra, Ann. Pure Appl. Logic, **55** (1991) 51–65.

[292] V. S. Harizanov, Uncountable degree spectra, Ann. Pure Appl. Logic, **54** (1991) 255–263.

[293] V. S. Harizanov, The possible Turing degree of the nonzero member in a two element degree spectrum, Ann. Pure Appl. Logic, **60** (1993) 1–30.

[294] V. S. Harizanov, Turing degrees of certain isomorphic images of computable relations, Ann. Pure Appl. Logic, (to appear).

[295] L. Harrington, Recursively presentable prime models, J. Symbolic Logic, **39** (1974) 305–309.

[296] K. Hatzikiriakou, Algebraic disguises of Σ_1^0 induction, Arch. Math. Logic, **29** (1989) 47–51.

[297] K. Hatzikiriakou and S. G. Simpson, Countable valued fields in weak subsystems of second-order arithmetic, Ann. Pure Appl. Logic, **41** (1989) 27–32.

[298] L. S. Hay, A. B. Manaster and J. G. Rosenstein, Concerning partial recursive similarity transformations of linearly ordered sets, Pacific J. Math., **71** (1977) 57–70.

[299] G. T. Hermann, Die frage der endlich vielen Schritte in der Theorie der Polynomideale, Math. Ann., **95** (1926) 736–788.

[300] G. T. Hermann and S. D. Isard, Computability over arbitrary fields, J. London Math. Soc., Ser. 2, **2** (1970) 73–79.

[301] A. Heyting, Untersuchungen über intuitionistische Algebra, Verh. Nederl. Akad. Wetensch. Afd. Natuurk., Sect. 1, No. 2, Amsterdam, **18** (1941).

[302] A. Heyting, Intuitionism, An Introduction, (North-Holland, Amsterdam, 1956); [2nd. rev. edn., 1966].

[303] A. Heyting, Wat is berekenbaar? (Dutch), [What is computable?], Nieuw Arch. Wisk., Ser. 3, **17** (1969) 1–7.

[304] A. G. Higman, Subgroups of finitely presented groups, Proc. Roy. Soc. London Ser. A, **262** (1961) 455–475.+

[305] D. Hilbert, Mathematische Probleme, Archiv. f. Math. und Phys., Ser. 3, **1** (1901) 44–63, 213–237.

[306] P. Hingston, Effective Decomposition in Noetherian Rings, in: Aspects of Effective Algebra, (Proc. Conf. Monash Univ., Clayton, Australia, Aug. 1–4, 1979), J. N. Crossley (ed.), (Upside Down A Book Co., Yarra Glen, Victoria, Australia, 1981) 122–127.

[307] P. G. Hinman, Recursion-Theoretic Hierarchies, Perspect. in Math. Logic, (1978).

[308] G. Hird, Recursive Properties of Relations on Models, Ph.D. Thesis, Monash Univ., Clayton, Victoria, Australia, (1983).

[309] G. Hird, Recursive properties of intervals of recursive linear orders, in: Logical Methods, (Papers from Conf. in honor of Anil Nerode's Sixtieth Birthday, June 1–3, 1992, Cornell Univ., Ithaca, NY), J. N. Crossley, J. B. Remmel, R. A. Shore and M. E. Sweedler, (eds.), Progr. Comput. Sci. Appl. Logic, **12** (1993) 422–437.

[310] G. Hird, Recursive properties of relations on models, Ann. Pure Appl. Logic, **63** (1993) 241–269.

[311] J. L. Hirst, Marriage theorems and reverse mathematics, in: Logic and Computation, (Proc. Workshop, CMU, Pittsburg, June 30 – July 2, 1987), W. Sieg, (ed.), Contemp. Math., **106** (1990) 181–196.

[312] J. L. Hirst, Connected components of graphs and reverse mathematics, Arch. Math. Logic, 31 (1992) 183–192.

[313] T. Hirst and D. Harel, Taking it to the limit: on infinite variants of NP-complete problems, in: Proc. 8th. Annual Structure in Complexity Theory Conf., May 18–21, 1993, San Diego, CA, S. Homer, (ed.), J. Comput. System Sci., **53** (1996) 180–193; [another version in: Structure in Complexity Theory, (Proc. 8th. Annual IEEE Conf. Struct. Complexity Theory, June 19–23, 1993), (IEEE Comput. Sci. Press, Los Alamitos, CA, 1993), 292–304].

[314] W. Hodges, On the effectivity of some field constructions, Proc. London Math. Soc., Ser. 3, **32** (1976) 133–162.

[315] T. L. Hummel, Effective versions of Ramsey's theorem: avoiding the cone above $\mathbf{0}'$, J. Symbolic Logic, **59** (1994) 1301–1325.

[316] K. R. Hurlburt, Sufficiency conditions for theories with recursive models, Ann. Pure Appl. Logic, **55** (1992) 305–320.

[317] H. Ishihara, B. M. Khoussainov and A. Nerode, Decidable Kripke models of intuitionistic theories, Ann. Pure Appl. Logic, (to appear).

[318] H. Ishihara, B. M. Khoussainov and A. Nerode, Computable Kripke models and intermediate logics, (to appear).

[319] C. Jacobsson and C. Löfwall, Standard bases for general coefficient rings and a new constructive proof of Hilbert's basis theorem, J. Symbolic Comput., **12** (1991) 337–371.

[320] C. G. Jockusch Jr., Ramsey's theorem and recursion theory, J. Symbolic Logic, **37** (1972) 268–280.

[321] C. G. Jockusch Jr. and I. Kalantari, Recursively enumerable sets and van der Waerden's theorem on arithmetic progressions, Pacific J. Math., **115** (1984) 143–153.

[322] C. G. Jockusch Jr., A. A. Lewis and J. B. Remmel, Π^0_1-classes and Rado's selection principle, J. Symbolic Logic, **56** (1991) 684–693.

[323] C. G. Jockusch Jr. and T. G. McLaughlin, Countable retracing functions and Π^0_2 predicates, Pacific J. Math., **30** (1969) 67–93.

[324] C. G. Jockusch Jr. and A. Shlapentokh, Weak presentations of computable fields, J. Symbolic Logic, **60** (1995) 199–208.

[325] C. G. Jockusch Jr. and R. I. Soare, A minimal pair of Π_1^0 classes, J. Symbolic Logic, **36** (1971) 66–78.

[326] C. G. Jockusch Jr. and R. I. Soare, Degrees of members of Π_1^0 classes, Pacific J. Math., **40** (1972) 605–616.

[327] C. G. Jockusch Jr. and R. I. Soare, Π_1^0 classes and degrees of theories, Trans. Amer. Math. Soc., **173** (1972) 33–56.

[328] C. G. Jockusch Jr. and R. I. Soare, Degrees of orderings not isomorphic to recursive linear orderings, (Special issue: Internatl. Sympos. Math. Logic and its Applications, Nagoya, 1988), Ann. Pure Appl. Logic, **52** (1991) 39–64.

[329] C. G. Jockusch Jr. and R. I. Soare, Boolean algebras, Stone spaces, and the iterated Turing jump, J. Symbolic Logic, **59** (1994) 1121–1138.

[330] W. Julian, R. Mines and F. Richman, Algebraic numbers, a constructive development, Pacific J. Math., **74** (1978) 91–102.

[331] W. Julian, R. Mines and F. Richman, Alexander duality, Pacific J. Math., **106** (1983) 115–127.

[332] I. Kalantari, Structural Properties of the Lattice of Recursively Enumerable Vector Spaces, Ph.D. Thesis, Cornell Univ., Ithaca, NY, (1976).

[333] I. Kalantari, Major subspaces of recursively enumerable vector spaces, J. Symbolic Logic, **43** (1978) 293–303.

[334] I. Kalantari, Automorphisms of the lattice of recursively enumerable vector spaces, Z. Math. Logik Grundlag. Math., **25** (1979) 385–401.

[335] I. Kalantari, Effective content of the countable, geometric, form of the Hahn-Banach theorem, (unpublished manuscript, 1980).

[336] I. Kalantari, Effective content of a theorem of M. H. Stone, in: Aspects of Effective Algebra, (Proc. Conf. Monash Univ., Clayton, Australia, Aug. 1–4, 1979), J. N. Crossley (ed.), (Upside Down A Book Co., Yarra Glen, Victoria, Australia, 1981) 128–146.

[337] I. Kalantari and A. Retzlaff, Maximal vector spaces under automorphisms of the lattice of recursively enumerable vector spaces, J. Symbolic Logic, **42** (1977) 481–491.

[338] A. Kanamori and K. McAloon, On Gödel incompleteness and finite combinatorics, Ann. Pure Appl. Logic, **33** (1987) 23–41.

[339] A. L. Karp, Recursive vector spaces (Russian), in: Proc. 11th. Interrepublic Conf. Math. Logic, (Kazan', 1993), (1993) 68.

[340] L. C. Karpinski, Robert of Chester's Latin translation of the algebra of al-Khowarizmi, Univ. Michigan Studies, Humanistic Series, Vol. 11, Part 1, Contributions to the History of Science, pt. 1, (Macmillan, 1915).

[341] N. Kh. Kasymov, A dual problem in the theory of constructive models (Russian), Teor. Algoritm. i ee Prilozhen., Vyčisl. Sistemy, **129** (1989) 137–143, 197.

[342] N. Kh. Kasymov and B. M. Khoussainov, Finitely generated and absolutely locally finite algebras (Russian), Vičisl. Sistemy, Prikladn. Logika, Novosibirsk, **116** (1986) 3–15, 162.

[343] N. Kasymov and B. M. Khoussainov, Positive algebras and local positive equivalences, Dokl. Akad. Nauk UzSSR, **4** (1991).

N. Kasymov and B. M. Khoussainov, Positive equivalences with finite classes and related algebras (Russian), Sibirsk. Math. Zh., **33** (1992) 196–200; [translated in: Siberian Math. J., **33** (1992) 923–927].

[344] N. Kasymov and B. M. Khoussainov, Positive and negative enumerations of algebras (Russian), Vyčisl. Sistemy, Teor. Vyčhil. Yazyki Spetsif., **139** (1991) 103–110, 190; [translated in: Siberian Adv. Math., **3** (1993) 41–45].

[345] S. Kaufmann and M. Kummer, On a quantitative notion of uniformity, Fund. Inform., **25** (1996) 59–78.

[346] C. F. Kent, Constructive analogues of the group of permutations of the natural numbers, Trans. Amer. Math. Soc., **104** (1962) 347–362.

[347] N. G. Khisamiev, Strongly constructive models (Russian), Izv. Akad. Nauk Kaz. SSR, ser. Fiz.-Mat., **3** (1971) 59–63.

[348] N. G. Khisamiev, Strongly constructive models of a decidable theory (Russian), Izv. Akad. Nauk Kazakh. SSR, Ser. Fiz.-Mat., **1** (1974) 83–84.

[349] N. G. Khisamiev, Connection between constructivizability and strong constructivizability for different classes of abelian groups (Russian), Algebra i Logika, **23** (1984) 319–335; [translated in: Algebra and Logic, **23** (1984) 220–233.

[350] N. G. Khisamiev, Constructive abelian p–groups, Siberian Adv. Math., **2** (1992) 68–113.

[351] Z. G. Khisamiev and N. G. Khisamiev, Nonconstructivizability of the reduced part of a strongly constructive torsion-free abelian group (Russian), Algebra i Logika, **24** (1985) 108–118, 123; [translated in: Algebra and Logic, **24** (1985) 69–76].

[352] B. M. Khoussainov, Strongly effective unars and nonautoequivalent constructivizations, in: Some Problems in Differential Equations and Discrete Mathematics, Gos. Univ., Novosibirsk, (1986) 33–44.

[353] B. M. Khoussainov, Constants and classes of auto-equivalent constructivizations, in: Computable Invariants in the Theory of Algebraic Systems, V. N. Remeslennikov, (ed.), Akad. Nauk SSSR Sibirsk. Otdel., Vyčisl. Centr., Novosibirsk, (1987) 46–50.

[354] B. M. Khoussainov, Strongly ∀–finite theories of unars (Russian), Mat. Zametki, 41 (1987) 265–271, 288; [translated in: Math. Notes, **41** (1987) 151–154].

[355] B. M. Khoussainov, Algorithmic degree of unars (Russian), Algebra i Logika, **27** (1988) 479–494, 499; [translated in: Algebra and Logic, **27** (1988) 301–312].

[356] B. M. Khoussainov, Linear orbits and their constructivizations, Izv. Akad. Nauk UzSSR Ser., Fiz.-Mat. Nauk, (1990) 36–41, 97.

[357] B. M. Khoussainov, On some algorithmic properties of classical structures (Russian), Voprosy Vyčisl. i Prikl. Mat. (Tashkent), **90** (1990) 118–124, 157.

[358] B. M. Khoussainov, Finitely generated effectively representable models (Russian), Dokl. Akad. Nauk UzSSR, (1990) 8–9.

[359] B. M. Khoussainov, Positively presented groups and rings (Russian), Dokl. Akad. Nauk UzSSR, (1990) 6.

[360] B. M. Khoussainov, Invariant finitely generated submodels (Russian), Izv. Akad. Nauk UzSSR, Ser. Fiz.-Mat. Nauk, **6** (1990) 36–40, 80.

[361] B. M. Khoussainov, On positive enumerated models, Sibirsk. Math. Zh., **33** (1991) 184–191; [translated in: Siberian Math. J., **33** (1991) 507–513].

[362] B. M. Khoussainov, On the spectrum of autodimensions of homomorphic images of models, Dokl. Akad. Nauk UzSSR, (1991) 4–5.

[363] B. M. Khoussainov, Groups of automorphisms of models and their actions, Uzbek. Mat. Zh., **2** (1991) 48–52.

[364] B. M. Khoussainov, Solution of a problem of V. A. Uspenskiĭand A. L. Semenov, Dokl. Akad. Nauk UzSSR, (1991) 5–7.

[365] B. M. Khoussainov, Algorithmic dimensions of homomorphic images of models (Russian), Algebra i Logika, **31** (1992) 317–337; [translated in: Algebra and Logic, **31** (1992) 195–203].

[366] B. M. Khoussainov, Recursive unary algebras and trees, (Special issue: A selection of papers presented at the Sympos. Logic at Tvev '92, July 20–24, 1992), A. Nerode and M. A. Taĭtslin, (eds.), Ann. Pure Appl. Logic, 67 (1994) 213–268.

[367] B. M. Khoussainov, Randomness, computability, and algebraic specifications, Ann. Pure Appl. Logic, **91** (1998) 1–15.

[368] B. M. Khoussainov and R. Dadajanov, Algorithmic stability of models, in: Logical Methods, (Papers from Conf. in honor of Anil Nerode's Sixtieth Birthday, June 1–3, 1992, Cornell Univ., Ithaca, NY), J. N. Crossley, J. B. Remmel, R. A. Shore and M. E. Sweedler, (eds.), Progr. Comput. Sci. Appl. Logic, **12** (1993) 438–466.

[369] B. M. Khoussainov and A. Nerode, Automatic presentations of structures, in: Logic and Computational Complexity, (Papers from Interntl. Workshop, LCC '94, Oct. 13–16, 1994, Indianapolis, IN), D. Leivant (ed.), Lecture Notes in Comput. Sci., **960** (1995) 367–392.

[370] B. M. Khoussainov and A. Nerode, Effective properties of finitely generated r.e. algebras, in: Feasible Mathematics II, (Papers from 2nd. Workshop, May 28–30, 1992, Cornell Univ. Ithaca, NY), P. Clote and J. B. Remmel, (eds.), Progr. Comput. Sci. Appl. Logic, **13** (1992) 256–283.

[371] B. M. Khoussainov, A. Nies and R. A. Shore, COmputable models of theories with few models, Notre Dame J. Formal Logic, **38** (1997) 165–178.

[372] B. M. Khoussainov and R. A. Shore, Scott families and computable categoricity, in: Combinatorics, Complexity, and Logic (Singapore), (Proc. 1st. Internatl. Conf. Discrete Math. Theoret. Comput. Sci., DMTCS '96, Auckland, New Zealand, Dec. 1996), D. S. Bridges, C. S. Calude, J. Gibbons, S. Reeves and I. H. Witten, (eds.), Discrete Math. Theoret. Comput. Sci., (1997) 299–307.

[373] B. M. Khoussainov and R. A. Shore, Computable isomorphisms, degree spectra of relations, and Scott families, Ann. Pure Appl. Logic, (to appear).

[374] H. A. Kierstead, An effective version of Dilworth's theorem, Trans. Amer. Math. Soc., **268** (1981) 63–77.

[375] H. A. Kierstead, Recursive colorings of highly recursive graphs, Canad. J. Math., **33** (1981) 1279–1290.

[376] H. A. Kierstead, An effective version of Hall's theorem, Proc. Amer. Math. Soc., **88** (1983) 124–128.

[377] H. A. Kierstead, Recursive ordered sets, in: Combinatorics and Ordered Sets, (Proc. AMS-IMS-SIAM Joint Summer Research Conf., Humboldt State Univ., Arcata, CA, Aug. 11–17, 1985), I. Rival, (ed.), Contemp. Math., **57** (1986) 75–102.

[378] H. A. Kierstead, On Π_1–automorphisms of recursive linear orders, J. Symbolic Logic, **52** (1987) 681–688.

[379] H. A. Kierstead, Effective versions of the chain decomposition theorem, in: The Dilworth theorems, selected papers of Robert P. Dilworth, K. P. Bogart, R. Freese and J. P. S. Kung, (eds.), Contemp. Mathematicians, (1990) 36–38.

[380] H. A. Kierstead, G. McNulty and W. T. Trotter Jr., A theory of recursive dimension for ordered sets, Order, **1** (1984) 67–82.

[381] H. A. Kierstead, S. G. Penrice and W. T. Trotter Jr., On-line coloring and recursive graph theory, SIAM J. Discrete Math., **7** (1994) 72–89.

[382] H. A. Kierstead and J. B. Remmel, Indiscernibles and decidable models, J. Symbolic Logic, **48** (1983) 21–32.

[383] H. Kierstead and J. B. Remmel, Degrees of indiscernibles in decidable models, Trans. Amer. Math. Soc., **289** (1985) 41–57.

[384] H. A. Kierstead and W. T. Trotter Jr., An extremal problem in recursive combinatorics, in: Combinatorics, Graph Theory and Computing, Vol. 2, (Proc. 12th. S. E. Conf., Baton Rouge, LA, 1981, Congr. Numer., **33** (1981) 143–153.

[385] S. C. Kleene, Recursive predicates and quantifiers, Trans. Amer. Math. Soc., **53** (1943) 41–73.

[386] S. C. Kleene, Introduction to Metamathematics, (D. Van Nostrand, Princeton, 1952, and Elsevier, Amsterdam, 1964, 1971).

[387] S. C. Kleene, Hierarchies of number-theoretic predicates, Bull. Amer. Math. Soc., **61** (1955) 193–213.

[388] S. C. Kleene, Mathematical logic: constructive and non-constructive operations, in: 1960 Proc. Internat. Congr. Math., (Cambridge Univ. Press, 1960) 137–153.

[389] S. C. Kleene, Constructive functions in "*The foundations of intuitionistic mathematics*", in: Logic, Methodology and Philos. Sci. III, (Proc. 3rd. Internatl. Congr. Logic, Methodology and Philos. Sci., Amsterdam, 1967), B. van Rootselaar and J. F. Staal, (eds.), North-Holland, 1968) 137–144.

[390] S. C. Kleene, Origins of recursive function theory, in: Foundations of Computer Science, (Proc. 20th. Annual Sympos. Foundations Comput. Sci., San Juan, Puerto Rico, Oct. 21–31, 1979), FOCS-79, (IEEE Comput. Soc. Press, Washington, DC, 1979), 371–382.

[391] S. C. Kleene, Origins of recursive function theory, Ann. Hist. Comput., **3** (1981) 52–67.

[392] S. C. Kleene, The theory of recursive functions, approaching its centennial, Bull. Amer. Math. Soc. (N.S.), **5** (1981) 43–61.

[393] S. C. Kleene, Reflections on Church's thesis, Notre Dame J. Formal Logic, **28** (1987) 490–498.

[394] S. C. Kleene, Turing's analysis of computability, and major applications of it, in: The Universal Turing Machine: a half-century survey, R. Herken, (ed.), (Oxford Univ. Press, NY, 1988), 17–54.

[395] M. Kneser, Bemerkung über die Primpolynomzerlegung in endlich vielen Schritten, Math. Z., **57** (1953) 238–240.

[396] J. F. Knight, Theories whose resplendent models are homogeneous, Israel J. Math., **42** (1982) 151–161.

[397] J. F. Knight, Degrees of types and independent sequences, J. Symbolic Logic, **48** (1983) 1074–1081.

[398] J. F. Knight, Degrees coded into jumps of orderings, J. Symbolic Logic, **51** (1986) 1034–1042.

[399] J. F. Knight, Effective construction of models, in: Logic Colloquium '84, (Manchester, July 15–24, 1984), J. B. Paris, A. J. Wilkie and G. M. Wilmers, (eds.), Stud. Logic Found. Math., **120**, (1986) 105–119.

[400] J. F. Knight, Saturation of homogeneous resplendent models, J. Symbolic Logic, **51** (1986) 222–224.

[401] J. F. Knight, Degrees of models with prescribed Scott set, in: Classification Theory, (Proc. Joint U.S.–Israel Workshop on Model Theory in Math. Logic, Chicago, Dec. 15–19, 1985), J. T. Baldwin, (ed.), Lecture Notes in Math., **1292** (1987) 182–191.

[402] J. F. Knight, Constructions by transfinitely many workers, Ann. Pure Appl. Logic, **48** (1990) 237–259.

[403] J. F. Knight, A metatheorem for constructions by finitely many workers, J. Symbolic Logic, **55** (1990) 787–804.

[404] J. F. Knight, Nonarithmetical $\aleph_0$–categorical theories with recursive models, J. Symbolic Logic, **59** (1994) 106–112.

[405] J. F. Knight, Requirement systems, J. Symbolic Logic, **60** (1995) 222–245.

[406] J. F. Knight, A. H. Lachlan and R. I. Soare, Two theorems on degrees of models of true arithmetic, J. Symbolic Logic, **49** (1984) 425–436.

[407] J. F. Knight and M. Nadel, Expansions of models and Turing degrees, J. Symbolic Logic, **47** (1982) 587–604.

[408] J. König, Einleitung in die allgemeine Theorie der algebraischen Gröszen: aus dem Ungarischen ubertragen vom Verfasser, (Teubner, Leipzig, 1903).

[409] R. Kossak, H. Kotlarski and J. H. Schmerl, On maximal subgroups of the automorphism group of a countable recursively saturated model of PA, Ann. Pure Appl. Logic, **65** (1993) 125–148.

[410] G. Kreisel, Note on arithmetic models for consistent formulæ of the predicate calculus, Fund. Math., **37** (1950) 265–285.

[411] G. Kreisel, Note on arithmetic models for consistent formulae of the predicate calculus II, in: Proc. 11th. Internatl. Congr. Philos., Brussels, Aug. 20–26, 1953, Vol. 14, (North-Holland, Amsterdam and Éditions E. Nauwelaerts, Louvain, 1953), 39–49.

[412] G. Kreisel, A variant to Hilbert's theory of the foundations of arithmetic, British J. Philos. Sci., **4** (1953) 107–129.

[413] G. Kreisel, Analysis of the Cantor-Bendixson theorem by means of the analytic hierarchy, Bull. Acad. Polon. Sci. Sér. Sci. Math. Astronom. Phys., **7** (1959) 621–626.

[414] G. Kreisel, J. R. Shoenfield and H. Wang, Number theoretic concepts and recursive well-orderings, Arch. Math. Logik Grundlag., **5** (1960) 42–64.

[415] L. Kronecker, Grundzüge einer arithmetischen Theorie der algebraischen Grössen, J. Reine Angew. Math., **92** (1882) 1–122.

[416] W. Krull, Über Polynomzerlegung mit endlich vielen Schritten, Math. Z., **59** (1953) 57–60.

[417] W. Krull, Über Polynomzerlegung mit endlich vielen Schritten, II, Math. Z., **59** (1953) 295–298.

[418] W. Krull, Über Polynomzerlegung mit endlich vielen Schritten, III, Math. Z., **60** (1954) 109–111.

[419] J. Kruskal, The theory of well-quasi-ordering: a frequently discovered concept, J. Combin. Theory Ser. A, **13** (1972) 297–305.

[420] K. Zh. Kudaibergenov, A theory with two strongly constructivizable models (Russian), Algebra i Logika, **18** (1979) 176–185, 253; [translated in: Algebra and Logic, **18** (1979) 111–117].

[421] K. Zh. Kudaibergenov, Effectively homogenous models (Russian), Sibirsk. Math. Zh., 27 (1986) 180–182, 200.

[422] O. V. Kudinov, A criterion for the autostability of 1–decidable models (Russian), Algebra i Logika, **31** (1992) 479–492, 562; [translated in: Algebra and Logic, 31 (1992) 284–292].

[423] O. V. Kudinov, Algebraic dependencies and reducibilities of constructivizations in universal domains, Siberian Adv. Math., **3** (1993) 121–128.

[424] S. A. Kurtz, Notions of weak genericity, J. Symbolic Logic, **48** (1983) 764–770.

[425] V. A. Kuzicheva, Recursive endomorphisms of countable vector spaces with recursive operations (Russian), Dep. VINITI AN SSSR, (dep. No. 4175–84, 24 pages, June 1984).

[426] V. A. Kuzicheva, Minimal modules over the rings of recursive endomorphisms (Russian), in: Proc. 18th. Algebra Conf., (Kishinev, 1985), Part 1, (1985) 295.

[427] V. A. Kuzicheva, Inverse isomorphisms of rings of recursive endomorphisms (Russian), Vestnik Moskov. Univ. Ser. I Mat. Mekh., **41**:3 (1986) 91–93, 121; [translated in: Moscow Univ. Math. Bull., **41**:3 (1986) 82–84].

[428] A. H. Lachlan and E. W. Madison, Computable fields and arithmetically definable ordered fields, Proc. Amer. Math. Soc., **24** (1970) 803–807.

[429] W. M. Lambert Jr., A notion of effectiveness in arbitrary structures, J. Symbolic Logic, **33** (1968) 577–602.

[430] P. La Roche, Recursively presented Boolean algebras, Notices Amer. Math. Soc., **24**:6 (1977) A 552–553 (abstract).

[431] P. La Roche, Effective Galois theory, J. Symbolic Logic, **46** (1981) 385–392.

[432] S. Lempp and M. Lerman, Priority arguments using iterated trees of strategies, in: Recursion Theory Week, (Proc. Conf., Oberwolfach, Mar. 19–25, 1989), K. Ambos-Spies, G. H. Müller and G. E. Sachs, (eds.), Lecture Notes in Math., **1432** (1990) 277–296.

[433] M. Lerman, On recursive linear orderings, in: Logic Year 1979-1980, (Proc. Seminars and Conf. Math. Logic, Univ. Connecticut, Storrs, CT, 1979/80), M. Lerman, J. H. Schmerl and R. I. Soare, (eds.), Lecture Notes in Math., **859** (1981) 132–142.

[434] M. Lerman, Degrees of Unsolvability, Local and Global Theory, Perspect. in Math. Logic, (1983).

[435] M. Lerman and J. G. Rosenstein, Recursive linear orderings, in: Patras Logic Symposion, (Proc. Logic Sympos. Patras, Greece, Aug. 18–22, 1980) G. Metakides (ed.), Stud. Logic Found. Math., **109** (1982) 123–136.

[436] M. Lerman and J. H. Schmerl, Theories with recursive models, J. Symbolic Logic, **44** (1979) 59–76.

[437] M. Lerman, J. H. Schmerl and R. I. Soare (eds.), Logic Year 1979-1980, (Proc. Seminars and Conf. Math. Logic, Univ. Connecticut, Storrs, CT, 1979/80), Lecture Notes in Math., 859 (1981).

[438] A. A. Lewis, On effectively computable realizations of choice functions, Math. Social Sciences, **10** (1985) 43–80.

[439] A. A. Lewis, An infinite version of Arrow's theorem in the effective setting, Math. Social Sciences, **16** (1988) 41–48.

[440] C. R. Lin, Recursion Theory on Countable Abelian Groups, Ph.D. Thesis, Cornell Univ., Ithaca, NY, (1977).

[441] C. R. Lin, The effective content of Ulm's Theorem, in: Aspects of Effective Algebra, (Proc. Conf. Monash Univ., Clayton, Australia, Aug. 1–4, 1979), J. N. Crossley (ed.), (Upside Down A Book Co., Yarra Glen, Victoria, Australia, 1981) 147–160.

[442] C. R. Lin, Recursively presented abelian groups: effective p–group theory I, J. Symbolic Logic, **46** (1981) 617–624.

[443] M. Loebl and J. Matoušek, On undecidability of the weakened Kruskal Theorem, in: Logic and Combinatorics, (Proc. AMS-IMS-SIAM Joint Summer Research Conf. Applications Math. Logic to Finite Combinatorics, Humboldt State Univ., Arcata, CA, Aug. 4–10, 1985), S. G. Simpson, (ed.), Contemp. Math., **65** (1987) 275–280.

[444] J. Love, Stability among r.e. quotient algebras, Ann. Pure Appl. Logic, **59** (1993) 55–63.

[445] K. McAloon, Diagonal methods and strong cuts in models of arithmetic, in: Logic Colloq. '77, (Proc. Conf., Warsaw, Aug. 1–12, 1977, dedicated to the memory of A. Mostowski), A. Macintyre, L. Pacholski and J. Paris, (eds.), Stud. Logic Found. Math., **96** (1978) 171–181.

[446] C. McCarty, Realizability and recursive set theory, Ann. Pure Appl. Logic, **32** (1986) 153–183.

[447] A. Macintyre and D. Marker, Degrees of recursively saturated models, Trans. Amer. Math. Soc., **282** (1984) 539–554.

[448] T. G. McLaughlin, A note on effective ultrapowers: uniform failure of bounded collection, Math. Logic Quart., **39** (1993) 431–435.

[449] G. F. McNulty, Infinite ordered sets, a recursive perspective, in: Ordered Sets, (Proc. NATO Adv. Study Inst., Banff, Alta, Aug. 28 – Sept. 12, 1981), I. Rival (ed.), NATO Adv. Study Inst. Ser. C: Math. Phys. Sci., (Reidel, Dordrecht and Boston), **83** (1982) 299–330.

[450] E. W. Madison, Computable algebraic structures and nonstandard arithmetic, Trans. Amer. Math. Soc., **130** (1968) 38–54.

[451] E. W. Madison, A note on computable real fields, J. Symbolic Logic, **35** (1970) 239–241.

[452] E. W. Madison, Some remarks on computable (non-Archimedean) ordered fields, J. London Math. Soc., Ser. 2, **4** (1971) 304–308.

[453] E. W. Madison, The existence of countable totally nonconstructive extensions of the countable atomless Boolean algebra, J. Symbolic Logic, **48** (1983) 167–170.

[454] E. W. Madison, On Boolean algebras and their recursive completions, Z. Math. Logik Grundlag. Math., **31** (1985) 481–486.

[455] E. W. Madison and G. C. Nelson, Some examples of constructive and non-constructive extensions of the countable atomless Boolean algebra, J. London Math. Soc., Ser. 2, **11** (1975) 325–336.

[456] E. W. Madison and B. Zimmermann-Huisgen, Combinatorial and recursive aspects of the automorphism group of the countable atomless Boolean algebra, J. Symbolic Logic, **51** (1986) 292–301.

[457] A. I. Mal'tsev, Constructive algebras I (Russian), Uspekhi Mat. Nauk, 16 (1961) 3–60; [translated in: Constructive algebras I, Russian Math. Surveys, **16**:3 (1961) 77–129; also in: The Metamathematics of Algebraic Systems, Collected Papers: 1936–1967, translated and edited by B. F. Wells III, Stud. Logic Found. Math., **66** (1971), Ch. 18, 148–200].

[458] A. I. Mal'tsev, Algorithms and Recursive Functions (Russian), (Izdat. Nauka, Moscow, 1965); [translated: by L. F. Boron, L. E. Sanchis, J. Stilwell and K. Iseki, (Wolters-Noordhoff Publishing, Groningen, 1970)].

[459] A. I. Mal'tsev, Algebraic Systems (Russian), (posth. edn.), V. D. Smirnov and M. Taĭclin, (eds.), (Izdat. Nauka, Moscow, 1970); [translated: by B. D. Seckler and A. P. Doohovskoy, Grundlehren Math. Wiss., **192** (1973).]

[460] M. S. Manasse, Techniques and Counterexamples in Almost Categorical Recursive Model Theory, Ph.D. Thesis, Univ. Wisconsin, Madison, WI, (1982).

[461] A. B. Manaster and J. B. Remmel, Partial orderings of fixed finite dimension: model companions and density, J. Symbolic Logic, **46** (1981) 789–802.

[462] A. B. Manaster and J. B. Remmel, Some recursion theoretic aspects of dense two-dimensional partial orderings, in: Aspects of Effective Algebra, (Proc. Conf. Monash Univ., Clayton, Australia, Aug. 1–4, 1979), J. N. Crossley (ed.), (Upside Down A Book Co., Yarra Glen, Victoria, Australia, 1981) 161–188.

[463] A. B. Manaster, J. B. Remmel and J. H. Schmerl, Planarity and minimal path algorithms, J. Austral. Math. Soc., Ser. A, **40** (1986) 131–142.

[464] A. B. Manaster and J. G. Rosenstein, Effective matchmaking (recursion theoretical aspects of a theorem of Philip Hall, Proc. London Math. Soc., Ser. 3, **25** (1972) 615–654.

[465] A. B. Manaster and J. G. Rosenstein, Effective matchmaking and k–chromatic graphs, Proc. Amer. Math. Soc., **39** (1973) 371–378.

[466] A. B. Manaster and J. G. Rosenstein, Two-dimensional partial orderings: recursive model theory, J. Symbolic Logic, **45** (1980) 121–132.

A. B. Manaster and J. G. Rosenstein, Two-dimensional partial orderings: undecidability, J. Symbolic Logic, **45** (1980) 133–143.

[467] Yu. I. Manin, A course in mathematical logic, translated from the Russian by N. Koblitz. Grad. Texts in Math., **53** (1977).

[468] V. W. Marek, A. Nerode and J. B. Remmel, A theory of nonmonotonic rule systems I, Ann. Math. Artificial Intelligence, **1** (1990) 241–273; [also in: Logic in Computer Science, (Proc. 5th. Annual IEEE Sympos. Logic in Comput. Sci., Philadelphia, PA, June 4–7, 1990), LICS–90, (IEEE Comput. Sci. Press, Los Alamitos, CA, 1990), 79–94].

[469] V. W. Marek, A. Nerode and J. B. Remmel, A theory of nonmonotonic rule systems II, (Special issue: Artificial Intelligence and Mathematics, I, Proc. 1st. Internatl. Sympos., Fort Lauderdale, FL, Jan. 3–5, 1990), M. C. Golumbic and F. Hoffman, (eds.), Ann. Math. Artificial Intelligence, **5** (1992) 229–263.

[470] V. W. Marek, A. Nerode and J. B. Remmel, How complicated is the set of stable models of a recursive logic program?, Ann. Pure Appl. Logic, **56** (1992) 119–135.

[471] V. W. Marek, A. Nerode and J. B. Remmel, A context for belief revision: forward chaining-normal nonmonotonic rule systems, Ann. Pure Appl. Logic, **67** (1994) 269–323.

[472] V. W. Marek, A. Nerode and J. B. Remmel, The stable models of a predicate logic program, J. Logic Programming, **21** (1994) 129–154.

[473] V. W. Marek, A. Nerode and J. B. Remmel, Nonmonotonic rule systems with recursive sets of restraints, (Special issue: Proc. Sacks Sympos., Cambridge, Mass., 1993), Arch. Math. Logic, **36** (1997) 339–384.

[474] D. Marker, Degrees of models of true arithmetic, in: Logic Colloq. '81. (Proc. Herbrand Sympos., Marseilles, July, 16–24, 1981), J. Stern (ed.), Stud. Logic Found. Math., **107** (1982) 233–242.

[475] Yu. V. Matijasevič, Enumerable sets are diophantine (Russian), Dokl. Akad. Nauk SSSR, **191** (1970) 279–282; [translated in: Soviet Math. – Dokl., **11** (1970) 354–357].

[476] J. Mead, Recursive prime models for Boolean algebras, Colloq. Math., **41** (1979) 25–33.

[477] S. Merrin, Constructing bases for radicals and nilradicals of Lie algebras, Proc. Amer. Math. Soc., **119** (1993) 189–202.

[478] G. Metakides, (ed.), Patras Logic Symposion, (Proc. Logic Sympos. Patras, Greece, Aug. 18–22, 1980) Stud. Logic Found. Math., **109** (1982).

[479] G. Metakides and A. Nerode, Recursion theory and algebra, in: Algebra and Logic, (Proc. 14th. Summer Res. Inst. Austral. Math. Soc., Monash Univ., Clayton, Vic., Australia, Jan. 6 – Feb. 16, 1974), J. N. Crossley, (ed.), Lecture Notes in Math., **450** (1975) 209–219.

[480] G. Metakides and A. Nerode, Recursively enumerable vector spaces, Ann. Math. Logic, **11** (1977) 147–171.

[481] G. Metakides and A. Nerode, Effective content of field theory, Ann. Math. Logic, **17** (1979) 289–320.

[482] G. Metakides and A. Nerode, Recursion theory on fields and abstract dependence, J. Algebra, **65** (1980) 36–59.

[483] G. Metakides and A. Nerode, The introduction of nonrecursive methods into mathematics, in: The L. E. J. Brouwer Centenary Sympos., (Noordwijkerhout, June 8–13, 1981), A. S. Troelstra and D. van Dalen, (eds.), Stud. Logic Found. Math., **110** (1982) 319–335.

[484] G. Metakides, A. Nerode and R. A. Shore, Recursive limits on the Hahn-Banach Theorem, in: Errett Bishop: reflections on him and his research, (San Diego, Calif., 1983), M. Rosenblatt, (ed.), Contemp. Math., **39** (1985) 85–91.

[485] G. Metakides and J. B. Remmel, Recursion theory on orderings I, A model theoretic setting, J. Symbolic Logic, **44** (1979) 383–402.

[486] H. R. Mihara, Arrow's Theorem, Turing Computability, and Oracles, Ph.D. Thesis, Dept. Econ., Univ. Minnesota, MN, (1995).

[487] H. R. Mihara, Arrow's theorem and Turing computability, Econom. Theory, **10** (1997) 257–276.

[488] H. R. Mihara, Anonymity and neutrality in Arrow's theorem with restricted coalition algebras, Soc. Choice Welf., **14** (1997) 503–512.

[489] T. S. Millar, The Theory of Recursively Presented Models, Ph.D. Thesis, Cornell Univ., Ithaca, NY, (1976).

[490] T. S. Millar, Foundations of recursive model theory, Ann. Math. Logic, **13** (1978) 305–320.

[491] T. S. Millar, A complete, decidable theory with two decidable models, J. Symbolic Logic, **44** (1979) 307–312.

[492] T. S. Millar, Homogeneous models and decidability, Pacific J. Math., **91** (1980) 407–418.

[493] T. S. Millar, Counterexamples via model completions, in: Logic Year 1979-1980, (Proc. Seminars and Conf. Math. Logic, Univ. Connecticut, Storrs, CT, 1979/80), M. Lerman, J. H. Schmerl and R. I. Soare, (eds.), Lecture Notes in Math., **859** (1981) 215–229.

[494] T. S. Millar, Vaught's theorem recursively revisited, J. Symbolic Logic, **46** (1981) 397–411.

[495] T. S. Millar, Type structure complexity and decidability, Trans. Amer. Math. Soc., **271** (1982) 73–81.

[496] T. S. Millar, Omitting types, type spectrums, and decidability, J. Symbolic Logic, **48** (1983) 171–181.

[497] T. S. Millar, Persistently finite theories with hyperarithmetic models, Trans. Amer. Math. Soc., **278** (1983) 91–99.

[498] T. S. Millar, Decidability and the number of countable models, Ann. Pure Appl. Logic, **27** (1984) 137–153.

[499] T. S. Millar, Decidable Ehrenfeucht theories, in: Recursion Theory, (Proc. AMS-ASL Summer Inst., Ithaca, NY, June 28 – July 16, 1982), A. Nerode and R. A. Shore, (eds.), Proc. Sympos. Pure Math., **42** (1985) 311–321.

[500] T. S. Millar, Bad models in nice neighborhoods, J. Symbolic Logic, **51** (1986) 1043-1055.

[501] T. S. Millar, Prime models and almost decidability, J. Symbolic Logic, **51** (1986) 412–420.

[502] T. S. Millar, Recursive categoricity and persistence, J. Symbolic Logic, **51** (1986) 430–434.

[503] T. S. Millar, Homogeneous models and almost decidability, J. Austral. Math. Soc., Ser. A, **46** (1989) 343–355.

[504] T. S. Millar, Finite extensions and the number of countable models, J. Symbolic Logic, **54** (1989) 264–270.

[505] T. S. Millar, Tame theories with hyperarithmetic homogeneous models, Proc. Amer. Math. Soc., **105** (1989) 712–726.

[506] T. S. Millar, Abstract recursive model theory, in: Handbook of Recursion Theory, E. R. Griffor, (ed.), North-Holland, (to appear).

[507] T. S. Millar, Model completions and omitting types, J. Symbolic Logic, **60** (1995) 654–672.

[508] R. Mines, Completions of valuated abelian groups, in: Abelian Group Theory, (Proc. Conf. Univ. Hawaii, Honolulu, Hawaii, Dec. 28 – Jan. 4, 1982/83), R. Göbel, L. Lady and A. Marker, (eds.), Lecture Notes in Math., **1006** (1983) 569–581.

[509] R. Mines and F. Richman, Dedekind domains, in: Constructive Mathematics, (Proc. New Mexico State Univ. Conf., Aug. 11–15, 1980, Las Cruces, NM) F. Richman, (ed.), Lecture Notes in Math., **873** (1981) 16–30.

[510] R. Mines and F. Richman, Separability and factoring polynomials, Rocky Mountain J. Math., **12** (1982) 43–54.

[511] R. Mines and F. Richman, Valuation theory: a constructive view, J. Number Theory, **19** (1984) 40–62.

[512] R. Mines and F. Richman, Archimedean valuations, J. London Math. Soc., Ser. 2, **34** (1986) 403–410.

[513] R. Mines, F. Richman and W. Ruitenburg, A course in constructive algebra, Universitext, (Springer-Verlag, 1988).

[514] R. Mines and C. Vinsonhaler, Butler groups and Bext: a constructive view, in: Abelian Groups and Noncommutative Rings, A collection of papers in memory of R. B. Warfield Jr., L. Fuchs, K. R. Goodearl, J. T. Stafford and C. Vinsonhaler, (eds.), Contemp. Math., **130** (1992) 289–299.

[515] M. Minsky, Recursive unsolvability of Post's problem of "tag" and other topics in the theory of Turing machines, Ann. of Math., Ser. 2, **74** (1961) 437–455.

[516] D. Misercque, Problème des mariages et récursivité, Bull. Soc. Math. Belg. Sér. A, **30** (1978) 111–121.

[517] M. Morley, Decidable models, Israel J. Math., **25** (1976) 233–240.

[518] M. Morley and R. I. Soare, Boolean algebras, splitting theorems, and Δ_2^0 sets, Fund. Math., **90** (1975) 45–52.

[519] A. S. Morozov, Countable homogeneous Boolean algebras (Russian), Algebra i Logika, **21** (1982) 269–282; [translated in: Algebra and Logic, **21** (1982) 181–190.

[520] A. S. Morozov, Strong constructivizability of countable saturated Boolean algebras (Russian), Algebra i Logika, **21** (1982) 193–203; [translated in: Algebra and Logic, **21** (1982) 130–137.

[521] A. S. Morozov, Groups of recursive automorphisms of constructive Boolean algebras (Russian), Algebra i Logika, **22** (1983) 138–158; [translated in: Algebra and Logic, **22** (1983) 95–112].

[522] A. S. Morozov, Group $\mathrm{Aut}_r \langle Q, \leqslant \rangle$ is not constructivizable (Russian), Mat. Zametki, **36** (1984) 473–478; [translated in: Math. Notes, **36** (1984) 733–736].

[523] A. S. Morozov, Automorphisms of constructivizations of Boolean algebras (Russian), Sibirsk. Math. Zh., **26** (1985) 98–110; [translated in: Siberian Math. J., **26** (1985) 555–565].

[524] A. S. Morozov, Constructive Boolean algebras with almost-identical automorphisms (Russian), Mat. Zametki, **37** (1985) 478–482; [translated in: Math. Notes, **37** (1985) 266–268].

[525] A. S. Morozov, Computable groups of automorphisms of models (Russian), Algebra i Logika, **25** (1986) 415–424; [translated in: Algebra and Logic, **25** (1986) 261–266].

[526] A. S. Morozov, A countably categorical decidable model without nontrivial recursive automorphisms (Russian), **30** (1989) 221–224; [translated in: Siberian Math. J., 30 (1989) 346–348].

[527] A. S. Morozov, Elementary properties of groups of recursive permutations (Russian), Dokl. Akad. Nauk, **305** (1989) 274–276; [translated in: Soviet Math. - Dokl., **39** (1989) 282–284].

[528] A. S. Morozov, Recursive automorphisms of atomic Boolean algebras (Russian), Algebra i Logika, **29** (1990) 464–490; [translated in: Algebra and Logic, **29** (1990) 310–330].

[529] A. S. Morozov, Rigid constructive modules (Russian), Algebra i Logika, **28** (1989) 570–583; [translated in: Algebra and Logic, **28** (1989) 379–387].

[530] A. S. Morozov, On theories of classes of groups of recursive permutations (Russian), in: Proc. Inst. Math. Acad. Sibirsk. SSSR, Trudy Inst. Mat., Novosibirsk, **12** (1989) 91–104, 189; [translated in: Siberian Adv. Math., **1** (1991) 138–153].

[531] A. S. Morozov, Functional trees and automorphisms of models (Russian), Algebra i Logika, 32 (1993) 54–72; [translated in: Algebra and Logic, 32 (1993) 28–38].

[532] A. S. Morozov, Automorphism groups of decidable models (Russian), Algebra i Logika, **34** (1995) 437–447; [translated in: Algebra and Logic, **34** (1995) 242–248].

[533] Y. N. Moschovakis, Descriptive Set Theory, Stud. Logic Found. Math., **100** (1980).

[534] M. Moses, Recursive properties of isomorphism types, J. Austral. Math. Soc., Ser. A, **34** (1983) 269–286.

[535] M. Moses, Recursive linear orders with recursive successivities, Ann. Pure Appl. Logic, **27** (1984) 253–264.

[536] M. Moses, Relations intrinsically recursive in linear orders, Z. Math. Logik Grundlag. Math., **32** (1986) 467–472.

[537] M. Moses, Decidable discrete linear orders, J. Symbolic Logic, **53** (1988) 531–539.

[538] M. Moses, n–recursive linear orders without $(n+1)$–recursive copies, in: Logical Methods, (Papers from Conf. in honor of Anil Nerode's Sixtieth Birthday, June 1–3, 1992, Cornell Univ., Ithaca, NY), J. N. Crossley, J. B. Remmel, R. A. Shore and M. E. Sweedler, (eds.), Progr. Comput. Sci. Appl. Logic, **12** (1993) 572–592.

[539] A. Mostowski, On a system of axioms which has no recursively enumerable arithmetic model, Fund. Math., **40** (1953) 56–61.

[540] A. Mostowski, A formula with no recursively enumerable model, Fund. Math., **42** (1955) 125–140.

[541] A. A. Mučnik, On the unsolvability of the problem of reducibility in the theory of algorithms (Russian), Dokl. Akad. Nauk SSSR, **108** (1956) 194–197.

[542] J. Myhill and J. Shepherdson, Effective operations on partial recursive functions, Z. Math. Logik Grundlag. Math., **1** (1955) 310–317.

[543] C. St. J. A. Nash-Williams, On well-quasi-ordering finite trees, Proc. Cambridge Philos. Soc., **59** (1963) 833–835; [also in: Classic Papers in Combinatorics, I. Gessel and G.-C. Rota (eds.), (Birkhäuser, Boston, 1987), 329–331].

[544] A. Nerode, Composita, equations, and freely generated algebras, Trans. Amer. Math. Soc., **91** (1959) 139–151.

[545] A. Nerode, Some Stone spaces and recursion theory, Duke Math. J., **26** (1959) 397–406.

[546] A. Nerode, Extensions to isols, Ann. of Math., **73** (1961) 362–403.

[547] A. Nerode, Logic and foundations, (Special issue: Proc. 2nd. Simpos. Math., Univ. Malaya/Univ. Kebangsaan Malasia, Kuala Lumpur, 1974), Bull. Malaysian Math. Soc., (1975) 17–24.

[548] A. Nerode, Logic and foundations, in: Algebra and Logic, (Proc. 14th. Summer Res. Inst. Austral. Math. Soc., Monash Univ., Clayton, Vic., Australia, Jan. 6 – Feb. 16, 1974), J. N. Crossley, (ed.), Lecture Notes in Math., **450** (1975) 283–290.

[549] A. Nerode and J. B. Remmel, Recursion theory on matroids. in: Patras Logic Symposion, (Proc. Logic Sympos. Patras, Greece, Aug. 18–22, 1980) G. Metakides (ed.), Stud. Logic Found. Math., **109** (1982) 41–65.

[550] A. Nerode and J. B. Remmel, Recursion theory on matroids II, in: (Proc. Southeast Asian Conf. Logic, Singapore, 1981), C. T. Chong and M. J. Wicks, (eds.), Stud. Logic Found. Math., **111** (1983) 133–184.

[551] A. Nerode and J. B. Remmel, A survey of lattices of r.e. substructures, in: Recursion Theory, (Proc. AMS-ASL Summer Inst., Ithaca, NY, June 28 – July 16, 1982), A. Nerode and R. A. Shore, (eds.), Proc. Sympos. Pure Math., **42** (1985) 323–375.

[552] A. Nerode and J. B. Remmel, Generic objects in recursion theory, Lecture Notes in Math., **1141** (1985) 271–314.

[553] A. Nerode and J. B. Remmel, Generic objects in recursion theory II, (Special issue: 2nd. Southeast Asian Logic Conf., Bangkok, 1984), Ann. Pure Appl. Logic, **31** (1986) 257–288.

[554] A. Nerode and J. B. Remmel, Complexity theoretic algebra I, Vector spaces over finite fields, in: Proc. Conf. Logic and Computer Science: New Trends and Applications, (Turin, 1986), Rend. Sem. Mat. Univ. Politec. Torino 1987, Special Issue, (1988) 1–10.

[555] A. Nerode and J. B. Remmel, Complexity-theoretic algebra II: Boolean algebras, (Special issue: 3rd. Asian Conf. Math. Logic, Beijing, Oct. 26–30, 1987), D. P. Yang, (ed.), Ann. Pure Appl. Logic, **44** (1989) 71–99.

[556] A. Nerode and J. B. Remmel, Complexity-theoretic algebra: vector space bases, in: Feasible Mathematics, (Proc. Workshop, June 16–18, 1989, Cornell Univ. Ithaca, NY), S. Buss and P. J. Scott (eds.), Progr. Comput. Sci. Appl. Logic, **9** (1990) 293–319.

[557] A. Nerode and J. B. Remmel, Polynomial time equivalence types, in: Logic and Computation, (Proc. Workshop, CMU, Pittsburg, June 30 – July 2, 1987), W. Sieg, (ed.), Contemp. Math., **106** (1990) 221–249.

[558] A. Nerode and J. B. Remmel, Polynomially isolated sets, in: Recursion Theory Week, (Proc. Conf., Oberwolfach, Mar. 19–25, 1989), K. Ambos-Spies, G. H. Müller and G. E. Sachs, (eds.), Lecture Notes in Math., **1432** (1990) 323–362.

[559] A. Nerode and J. B. Remmel, On the lattices of NP–subspaces of a polynomial time vector space over a finite field, Ann. Pure Appl. Logic, **81** (1996) 125–170.

[560] A. Nerode and R. A. Shore, (eds.), Recursion theory, (Proc. AMS-ASL Summer Inst., Ithaca, NY, June 28 – July 16, 1982), Proc. Sympos. Pure Math., **42** (1985).

[561] A. Nerode and R. L. Smith, The undecidability of the lattice of recursively enumerable subspaces, in: Proc. 3rd. Brazilian Conf. on Math. Logic, (Inst. Math., Fed. Univ. Pernambuco, Recife, 1979), A. I. Arruda, N. C. A. da Costa and A. M. Sette, (eds.), Soc. Brasil Logica, Sao Paulo, (1980) 245–252.

[562] T. A. Nevins, Degrees of convex dependence in recursively enumerable vector spaces, Ann. Pure Appl. Logic, **60** (1993) 31–47.

[563] A. Nies and R. A. Shore, Interpreting true arithmetic in the theory of the r.e. truth table degrees, Ann. Pure Appl. Logic, **75** (1995) 269–311.

[564] E. Noether, Eliminationstheorie und allgemeine Idealtheorie, Math. Ann., **90** (1923) 229–262.

[565] P. S. Novikov, On the algorithmic unsolvability of the word problem in group theory (Russian), Trudy Mat. Inst. Steklov., **44** (1955).

[566] A. T. Nurtazin, Strong and weak constructivizations and computable families (Russian), Algebra i Logika, **13** (1974) 311–323, 364; [translated in: Algebra and Logic, **13** (1974) 177–184].

[567] A. T. Nurtazin, On constructive groups (Russian), in: Proc. 4th. All-Union Conf. Math. Logic, (Kishinev, 1976), (1976) 106.

[568] S. Oates, Jump Degrees of Groups, Ph.D. Thesis, Univ. Notre Dame, IN, (1989).

[569] P. Odifreddi, Classical Recursion Theory. The Theory of Functions and Sets of Natural Numbers, Stud. Logic Found. Math., **125** (1989).

[570] S. P. Odintsov, Atom-free ideals of constructive Boolean algebras (Russian), Algebra i Logika, **23** (1984) 278–295, 362; [translated in: Algebra and Logic, **23** (1984) 190–203].

[571] S. P. Odintsov, Lattice of a recursively enumerable subalgebras of recursive Boolean algebra (Russian), Algebra i Logika, **25** (1986) 631–642, 751; [translated in: Algebra and Logic, **25** (1986) 397–404].

[572] S. P. Odintsov, Restricted theories of constructive Boolean algebras of the lower stratum (Russian), Inst. of Math., Novosibirsk, (Preprint No. 21) (1986).

[573] S. P. Odintsov, Recursive Boolean algebras with a hyperhyperimmune set of atoms (Russian), Mat. Zametki, **44** (1988) 488–493, 557; [translated in: Math. Notes, **44** (1988) 747–749.]

[574] S. P. Odintsov, Hereditarily recursively enumerable subalgebras of recursive Boolean algebra (Russian), Algebra i Logika, **31** (1992) 38–46, 96; [translated in: Algebra and Logic, **31** (1992) 24–29].

[575] S. P. Odintsov and V. L. Selivanov, The arithmetical hierarchy and ideals of enumerated Boolean algebras (Russian), Sibirsk. Math. Zh., **30** (1989) 140–149; [translated in: Siberian Math. J., **30** (1989) 952–960].

[576] P. L. Olson, Difference Relations and Algebra: a Constructive Study, Ph.D. Thesis, Univ. Texas, Austin, TX, (1977).

[577] C. H. Papadimitriou and K. Steiglitz, Combinatorial Optimization: Algorithms and Complexity, (Prentice-Hall, Englewood Cliffs, New Jersey, 1982).

[578] C. H. Papadimitriou and M. Yannakakis, Optimization, approximation, and complexity classes, J. Comput. System Sci., **43** (1991) 425–440.

[579] J. Paris, Some Independence results for Peano arithmetic, J. Symbolic Logic, **43** (1978) 725–731.

[580] J. Paris and L. Harrington, A mathematical incompleteness in Peano arithmetic, in: J. Barwise, (ed.), Handbook of Mathematical Logic, Stud. Logic Found. Math., **90** (1977) 1133–1142.

[581] M. G. Peretyat'kin, Strongly constructive models and enumerations of the Boolean algebra of recursive sets (Russian), Algebra i Logika, **10** (1971) 535–557; [translated in: Algebra and Logic, **10** (1971) 332–345].

[582] M. G. Peretyat'kin, Complete theories with a finite number of countable models (Russian), Algebra i Logika, **12** (1973) 550–576, 618; [translated as: On complete theories with a finite number of denumerable models, Algebra and Logic, **12** (1973) 310–326].

[583] M. G. Peretyat'kin, Every recursively enumerable extension of a theory of linear order has a constructive model (Russian), Algebra i Logika, **12** (1973) 211–219, 244; [translated in: Algebra and Logic, **12** (1973) 120–124].

[584] M. G. Peretyat'kin, Strongly constructive model without elementary submodels and extensions (Russian), Algebra i Logika, **12** (1973) 312–322, 364; [translated in: Algebra and Logic, **12** (1973) 178–183].

[585] M. G. Peretyat'kin, A criterion for strong constructivizability of a homogeneous model (Russian), Algebra i Logika, **17** (1978) 436–454, 491; [translated in: Algebra and Logic, **17** (1978) 290–301.]

[586] M. G. Peretyat'kin, Semantically universal classes of models (Russian), Algebra i Logika, **30** (1991) 414–431, 507; [translated in: Algebra and Logic, **30** (1991) 271–282.]

[587] M. G. Peretyat'kin, Expressive power of finitely axiomatizable theories, I, II, III, Siberian Adv. Math., **3** (1993) no. 2, 153–197, no. 3, 123–145, no. 4, 131–201.

[588] T. Pheidas, Hilbert's Tenth Problem for a class of rings of algebraic integers, Proc. Amer. Math. Soc., **104** (1988) 611–620.

[589] A. G. Pinus, Efficient linear orders, Siberian Math. J., **16** (1975) 956–962.

[590] M. Pohst, P. Weiler and H. Zassenhaus, On effective computation of fundamental units. II, Math. Comp., **38** (1982) 293–329.

[591] M. Pohst and H. Zassenhaus, On effective computation of fundamental units. I, Math. Comp., **38** (1982) 275–291.

[592] E. L. Post, Recursively enumerable sets of positive integers and their decision problems, Bull. Amer. Math. Soc. (N.S.), **50** (1944) 284–316.

[593] M. B. Pour-El and J. I. Richards, Computability in Analysis and Physics, Perspect. in Math. Logic, (1989).

[594] M. Presburger, Über die Vollständigkeit eines gewissen Systems der Arithmetik ganzer Zahlen, in welchem die Addition als einzige Operation hervortritt, Comptes-Rendus du I Congres des Mathématiciens des Pays Slaves, Warszawa, (1929) 92–101, 393.

[595] M. O. Rabin, Effective computability of winning strategies, in: Contributions to the Theory of Games, Vol. 3, Annals of Math. Stud., **39** (1957) 147–157.

[596] M. O. Rabin, Recursive unsolvability of group theoretic problems, Ann. of Math., **67** (1958) 172–194.

[597] M. O. Rabin, Computable algebra, general theory and theory of computable fields, Trans. Amer. Math. Soc., **95** (1960) 341–360.

[598] M. O. Rabin, A simple method for undecidability proofs and some applications, in: Logic, Methodology and Philos. Sci. (Proc. 1964 Internat. Congr.), North-Holland, 1965) 58–68.

[599] M. O. Rabin, Decidability of second-order theories and automata on infinite trees, Bull. Amer. Math. Soc. (N.S.), **74** (1968) 1025–1029.

[600] M. O. Rabin, Decidability of second-order theories and automata on infinite trees, Trans. Amer. Math. Soc., **141** (1969) 1–35.

[601] R. C. Reed, A decidable Ehrenfeucht theory with exactly two hyperarithmetic models, Ann. Pure Appl. Logic, **53** (1991) 135–168.

[602] J. B. Remmel, Combinatorial Functors on Co-R.E. Structures, Ph.D. Thesis, Cornell Univ., Ithaca, NY, (1974).

[603] J. B. Remmel, Combinatorial functors on co-r.e. structures, Ann. Math. Logic, **10** (1976) 261–287.

[604] J. B. Remmel, Co-hypersimple structures, J. Symbolic Logic, **41** (1976) 611–625.

[605] J. B. Remmel, Maximal and cohesive vector spaces, J. Symbolic Logic, **42** (1977) 400–418.

[606] J. B. Remmel, An r–maximal vector space not contained in any maximal vector space, J. Symbolic Logic, **43** (1978) 430–441.

[607] J. B. Remmel, Realizing partial orderings by classes of co-simple sets, Pacific J. Math., **76** (1978) 169–184.

[608] J. B. Remmel, Recursively enumerable Boolean algebras, Ann. Math. Logic, **15** (1978) 75–107.

[609] J. B. Remmel, r–maximal Boolean algebras, J. Symbolic Logic, **44** (1979) 533–548.

[610] J. B. Remmel, Effective structures not contained in recursively enumerable structures, in: Aspects of Effective Algebra, (Proc. Conf. Monash Univ., Clayton, Australia, Aug. 1–4, 1979), J. N. Crossley (ed.), (Upside Down A Book Co., Yarra Glen, Victoria, Australia, 1981) 206–225.

[611] J. B. Remmel, On r.e. and co-r.e. vector spaces with nonextendible bases, J. Symbolic Logic, **45** (1980) 20–34.

[612] J. B. Remmel, Complementation in the lattice of subalgebras of a Boolean algebra, Algebra Universalis, **10** (1980) 48–64.

[613] J. B. Remmel, Recursion theory on algebraic structures with independent sets, Ann. Math. Logic, **18** (1980) 153–191.

[614] J. B. Remmel, Recursion theory on orderings II, J. Symbolic Logic, **45** (1980) 317–333.

[615] J. B. Remmel, On the effectiveness of the Schröder-Bernstein theorem, Proc. Amer. Math. Soc., **83** (1981) 379–386.

[616] J. B. Remmel, Recursive Boolean algebras with recursive atoms, J. Symbolic Logic, **46** (1981) 595–616.

[617] J. B. Remmel, Recursive isomorphism types of recursive Boolean algebras, J. Symbolic Logic, **46** (1981) 572–594.

[618] J. B. Remmel, Recursively categorical linear orderings, Proc. Amer. Math. Soc., **83** (1981) 387–391.

[619] J. B. Remmel, Graph colorings and recursively bounded Π_1^0-classes, Ann. Pure Appl. Logic, **32** (1986) 185–194.

[620] J. B. Remmel, Recursively rigid Boolean algebras, Ann. Pure Appl. Logic, **36** (1987) 39–52.

[621] J. B. Remmel, Recursive Boolean algebras, in: Handbook of Boolean Algebras, Vol. 3, J. D. Monk, R. Bonnet and S. Koppelberg, (eds.), (North-Holland, Amsterdam, NY, 1989), Ch. 25, 1097–1165.

[622] J. B. Remmel, When is every recursive linear ordering of type μ recursively isomorphic to a polynomial time linear ordering over the natural numbers in binary form?, in: Feasible Mathematics, (Proc. Workshop, June 16–18, 1989, Cornell Univ. Ithaca, NY), S. Buss and P. J. Scott (eds.), Progr. Comput. Sci. Appl. Logic, **9** (1990) 321–350.

[623] J. B. Remmel, Polynomial time categoricity and linear orderings, in: Logical Methods, (Papers from Conf. in honor of Anil Nerode's Sixtieth Birthday, June 1–3, 1992, Cornell Univ., Ithaca, NY), J. N. Crossley, J. B. Remmel, R. A. Shore and M. E. Sweedler, (eds.), Progr. Comput. Sci. Appl. Logic, **12** (1993) 713–746.

[624] J. B. Remmel and J. N. Crossley, The work of Anil Nerode: a retrospective, in: Logical Methods, (Papers from Conf. in honor of Anil Nerode's Sixtieth Birthday, June 1–3, 1992, Cornell Univ., Ithaca, NY), J. N. Crossley, J. B. Remmel, R. A. Shore and M. E. Sweedler, (eds.), Progr. Comput. Sci. Appl. Logic, **12** (1993) 1–85.

[625] J. B. Remmel and A. B. Manaster, Co-simple higher-order indecomposable isols, Z. Math. Logik Grundlag. Math., **26** (1980) 279–288.

[626] A. Retzlaff, Simple and hyperhypersimple vector spaces, J. Symbolic Logic, **43** (1978) 260–269.

[627] A. Retzlaff, Direct summands of recursively enumerable vector spaces, Z. Math. Logik Grundlag. Math., **25** (1979) 363–372.

[628] H. G. Rice, Recursive and recursively enumerable orders, Trans. Amer. Math. Soc., **83** (1956) 277–300.

[629] F. Richman, The constructive theory of countable abelian p-groups, Pacific J. Math., **45** (1973) 621–637.

[630] F. Richman, Constructive aspects of Noetherian rings, Proc. Amer. Math. Soc., **44** (1974) 436–441.

[631] F. Richman, The constructive theory of KT–modules, Pacific J. Math., **61** (1975) 263–274.

[632] F. Richman, A constructive modification of Vietoris homology, Fund. Math., **91** (1976) 231–240.

[633] F. Richman, Computing heights in Tor, Houston J. Math., **3** (1977) 267–270.

[634] F. Richman, Seidenberg's condition P, in: Constructive Mathematics, (Proc. New Mexico State Univ. Conf., Aug. 11–15, 1980, Las Cruces, NM) F. Richman, (ed.), Lecture Notes in Math., **873** (1981) 1–11.

[635] F. Richman, Finite-dimensional algebras over discrete fields, in: The L. E. J. Brouwer Centenary Sympos., (Noordwijkerhout, June 8–13, 1981), A. S. Troelstra and D. van Dalen, (eds.), Stud. Logic Found. Math., **110** (1982) 397–411.

[636] F. Richman, Nontrivial uses of trivial rings, Proc. Amer. Math. Soc., **103** (1988) 1012–1014.

[637] F. Richman, Intuitionism as generalization, Philos. Math., **5** (1990) 124–128.

[638] F. Richman, Polynomials and linear transformations, Linear Algebra Appl., **131** (1990) 131–137.

[639] F. Richman, The constructive theory of torsion-free abelian groups, Comm. Algebra, **18** (1990) 3913–3922.

[640] F. Richman, Separable extensions and diagonalizability, Amer. Math. Monthly, **97** (1990) 395–398.

[641] F. Richman, The constructive theory of countably generated Warfield modules, in: Abelian Groups and Noncommutative Rings, A collection of papers in memory of R. B. Warfield Jr., L. Fuchs, K. R. Goodearl, J. T. Stafford and C. Vinsonhaler, (eds.), Contemp. Math., **130** (1992) 371–383.

[642] F. Richman, Intuitionistic abelian group theory, in: Abelian Groups and Modules, (Proc. Internatl. Conf., Colorado College, Colorado Springs, CO, Aug. 7–12, 1995), D. M. Arnold and K. M. Rangaswamy, (eds.), Lecture Notes in Pure and Appl. Math., **182** (1996) 67–72.

[643] F. Richman, Interview with a constructive mathematician, Modern Logic, **6** (1996) 247–271.

[644] L. J. Richter, Degrees of structures, J. Symbolic Logic, **46** (1981) 723–731.

[645] L. Robbiano, (ed.), Computational aspects of commutative algebra, Reprint of J. Symbolic Comput., **6** (1982), no. 2-3, (Academic Press, London, 1989).

[646] J. Robinson, Definability and decision problems in arithmetic, J. Symbolic Logic, **14** (1949) 98–114.

[647] H. Rogers Jr., Theory of Recursive Functions and Effective Computability, (1st. edn., McGraw-Hill, New York-Toronto, Ont.-London, 1967; 2nd. edn., MIT Press, Cambridge, Mass., London, 1987).

[648] L. Rogers, Basic subgroups from a constructive viewpoint, in: Communications in algebra **8** (1980) 1903–1925.

[649] J. G. Rosenstein, $\aleph_0$ categoricity of groups, J. Algebra, **25** (1973) 435–467.

[650] J. G. Rosenstein, Linear orderings, Pure Appl. Math., **98** (1982).

[651] J. G. Rosenstein, Recursive linear orderings, in: Orders: Description and Roles, Ann. Discrete Math., **23** (1984) 465–475.

[652] J. B. Rosser, Extensions of some theorems of Gödel and Church, J. Symbolic Logic, **1** (1936) 87–91.

[653] D. K. Roy, R.e. presented linear orders, J. Symbolic Logic, **48** (1983) 369–376.

[654] D. K. Roy, Linear order types of nonrecursive presentability, Z. Math. Logik Grundlag. Math., **31** (1985) 495–501.

[655] D. K. Roy, Effective extensions of partial orders, Z. Math. Logik Grundlag. Math., **36** (1990) 233–236.

[656] D. K. Roy and R. Watnick, Finite condensations of recursive linear orders, Studia Logika, **47** (1988) 311–317.

[657] W. Ruitenburg, Constructions of finitely generated submodules of constructively Noetherian modules, Compositio Math., **62** (1987) 47–52.

[658] G. E. Sacks, The recursively enumerable degrees are dense, Ann. of Math., **80** (1964) 300–312.

[659] G. E. Sacks, Saturated Model Theory, (Mathematics Lecture Note Series, W. A. Benjamin, Inc., Reading, Mass., 1972).

[660] G. E. Sacks, Higher Recursion Theory, Perspect. in Math. Logic, (1990).

[661] A. Salomaa, Computation and Automata, (Cambridge Univ. Press, 1985).

[662] J. Schlipf, Toward model theory through recursive saturation, J. Symbolic Logic, **43** (1978) 183–206.

[663] J. H. Schmerl, Effectiveness and Vaught's gap ω two-cardinal theorem, Proc. Amer. Math. Soc., **58** (1976) 237–240.

[664] J. H. Schmerl, A decidable $\aleph_0$–categorical theory with a nonrecursive Ryll-Nardzewski function, Fund. Math., **98** (1978) 121–125.

[665] J. H. Schmerl, Decidability and $\aleph_0$–categoricity of theories of partially ordered sets, J. Symbolic Logic, **45** (1980) 585–611.

[666] J. H. Schmerl, Recursive colorings of graphs, Canad. J. Math., **32** (1980) 821–830.

[667] J. H. Schmerl, Arborescent structures, I: recursive models, in: Aspects of Effective Algebra, (Proc. Conf. Monash Univ., Clayton, Australia, Aug. 1–4, 1979), J. N. Crossley (ed.), (Upside Down A Book Co., Yarra Glen, Victoria, Australia, 1981) 226–231.

[668] J. H. Schmerl, The effective version of Brooks' theorem, Canad. J. Math., **34** (1982) 1036–1046.

[669] J. H. Schmerl, Recursion theoretic aspects of graphs and orders, in: Graphs and Order, The role of graphs in the theory of ordered sets and its applications, (Proc. NATO Adv. Study Inst., Banff, Alta., May 18–31, 1984), I. Rival, (ed.), NATO Adv. Study Inst. Ser. C: Math. Phys. Sci., (Reidel, Dordrecht and Boston), **147** (1985) 467–484.

[670] S. Schwarz, Recursive automorphisms of recursive linear orderings, Ann. Pure Appl. Logic, **26** (1984) 69–73.

[671] D. Scott, Algebra of sets binumerable in complete extensions of arithmetic, Proc. Sympos. Pure Math., **5** (1962) 117–121.

[672] D. Seetapun and T. A. Slaman, On the strength of Ramsey's Theorem, Notre Dame J. Formal Logic, **36** (1995) 570–582.

[673] A. Seidenberg, On k–constructible sets, k–elementary formulae, and elimination theory, J. Reine Angew. Math., **239/240** (1969) 256–267.

[674] A. Seidenberg, Construction of the integral closure of a finite integral domain, Rend. Sem. Mat. Fis. Milano, **40** (1970) 100–120.

[675] A. Seidenberg, On the length of a Hilbert ascending chain, Proc. Amer. Math. Soc., **29** (1971) 443–450.

[676] A. Seidenberg, Constructive proof of Hilbert's theorem on ascending chains, Trans. Amer. Math. Soc., **174** (1972) 305–312.

[677] A. Seidenberg, On the impossibility of some constructions in polynomial rings, Atti del Convegno Internazionale di Geometrica, Accademia Nazionale di Lincei, (1973) 77–85.

[678] A. Seidenberg, Constructions in algebra, Trans. Amer. Math. Soc., **197** (1974) 273–313.

[679] A. Seidenberg, What is Noetherian?, Rend. Sem. Mat. Fis. Milano, **44** (1974) 55–61.

[680] A. Seidenberg, Construction of the integral closure of a finite integral domain II, Proc. Amer. Math. Soc., **52** (1975) 368–372.

[681] A. Seidenberg, Constructions in a polynomial ring over the ring of integers, Amer. J. Math., **100** (1978) 685–703.

[682] A. Seidenberg, Survey of constructions in Noetherian rings, in: Recursion Theory, (Proc. AMS-ASL Summer Inst., Ithaca, NY, June 28 – July 16, 1982), A. Nerode and R. A. Shore, (eds.), Proc. Sympos. Pure Math., **42** (1985) 377–386.

[683] A. Seidenberg, On the Lasker-Noether decomposition theorem, Amer. J. Math., **106** (1984) 611–638.

[684] V. L. Selivanov, Two theorems on computable numerations (Russian), Algebra i Logika, **15** (1976) 470–484, 488.

[685] V. L. Selivanov, The Ershov hierarchy (Russian), Sibirsk. Math. Zh., **26** (1985) 134–149.

[686] V. L. Selivanov, Algorithmic complexity of algebraic systems (Russian), Mat. Zametki, **44** (1988) 823–832, 863; [translated in: Math. Notes, **44** (1988) 944–950].

[687] V. L. Selivanov, Precomplete numerations and functions without fixed points (Russian), Mat. Zametki, **51** (1992) 149–155.

[688] V. L. Selivanov, Recursiveness of ω–operations, Math. Logic Quart., **40** (1994) 204–206.

[689] V. L. Selivanov, On recursively enumerable structures, (Special issue: Papers in honor of the Sympos. Logical Foundations of Comput. Sci., Logic at St. Petersburg, St. Petersburg, 1994). Ann. Pure Appl. Logic, **78** (1996) 243–258.

[690] A. Shlapentokh, Hilbert's tenth problem for rings of algebraic functions in one variable over fields of constants of positive characteristic, Trans. Amer. Math. Soc., **333** (1992) 275–298.

[691] A. Shlapentokh, Diophantine equivalence and countable rings, J. Symbolic Logic, **59** (1994) 1068–1095.

[692] A. Shlapentokh, Diophantine undecidability in some rings of algebraic numbers of totally real infinite extensions of $\mathbb{Q}$, Ann. Pure Appl. Logic, **68** (1994) 299–325.

[693] A. Shlapentokh, Non-standard extensions of weak presentations, J. Algebra, **176** (1995) 735–749.

[694] A. Shlapentokh, Rational separability over a global field, Ann. Pure Appl. Logic, **79** (1996) 93–108.

[695] A. Shlapentokh, Algebraic and Turing separability of rings, J. Algebra, **185** (1996) 229–257.

[696] J. R. Shoenfield, Degrees of formal systems, J. Symbolic Logic, **23** (1958) 389–392.

[697] J. R. Shoenfield, Degrees of models, J. Symbolic Logic, **25** (1960) 233–237.

[698] J. R. Shoenfield, Priority constructions, Dedicated to the late Stephen Cole Kleene, Ann. Pure Appl. Logic, **81** (1996) 115–123.

[699] R. A. Shore, Controlling the dependence degree of a recursively enumerable vector space, J. Symbolic Logic, **43** (1978) 13–22.

[700] R. A. Shore, On the strength of Fraïssé's conjecture, in: Logical Methods, (Papers from Conf. in honor of Anil Nerode's Sixtieth Birthday, June 1–3, 1992, Cornell Univ., Ithaca, NY), J. N. Crossley, J. B. Remmel, R. A. Shore and M. E. Sweedler, (eds.), Progr. Comput. Sci. Appl. Logic, **12** (1993) 782–813.

[701] S. G. Simpson, Degrees of unsolvability: a survey of results, in: J. Barwise, (ed.), Handbook of Mathematical Logic, Stud. Logic Found. Math., **90** (1977) 631–652.

[702] S. G. Simpson, Nichtbeweisbarkeit von gewissen kobinatorischen Eigenschaften endlicher Baume, Arch. Math. Logik Grundlag., **25** (1985) 45–65; [translated as: Nonprovability of certain combinatorial properties of finite trees, in: Harvey Friedman's Research on the Found. of Math., L. A. Harrington, M. D. Morley, A. Ščedrov and S. C. Simpson, (eds.), Stud. Logic Found. Math., **117** (1985) 87–117].

[703] S. G. Simpson, Recursion theoretic aspects of the dual Ramsey theorem, in: Recursion Theory Week, (Proc. Conf., Oberwolfach, Apr. 15–21, 1984), H.-D. Ebbinghaus, G. H. Müller and G. E. Sachs, (eds.), Lecture Notes in Math., **1141** (1985) 357–371.

[704] S. G. Simpson, (ed.), Logic and Combinatorics, (Proc. AMS-IMS-SIAM Joint Summer Research Conf. Applications. Math. Logic to Finite Combinatorics, Humboldt State Univ., Arcata, CA, Aug. 4–10, 1985), Contemp. Math., **65** (1987).

[705] S. G. Simpson, Sets which do not have subsets of every higher degree, J. Symbolic Logic, **43** (1978) 135–138.

[706] S. G. Simpson, Subsystems of $\mathbb{Z}_2$ and reverse mathematics, appendix to Proof Theory, 2nd. edn., (by G. Takeuti) (North Holland, Amsterdam, 1987), 432–446.

[707] S. G. Simpson, Unprovable theorems and fast-growing functions, in: Logic and Combinatorics, (Proc. AMS-IMS-SIAM Joint Summer Research Conf. Applications Math. Logic to Finite Combinatorics, Humboldt State Univ., Arcata, CA, Aug. 4–10, 1985), S. G. Simpson, (ed.), Contemp. Math., **65** (1987) 359–394.

[708] S. G. Simpson, Subsystems of Second Order Arithmetic, (in preparation).

[709] S. G. Simpson and R. L. Smith, Factorization of polynomials and Σ_1^0 Induction, Ann. Pure Appl. Logic, **31** (1986) 289–306.

[710] R. L. Smith, Effective valuation theory, in: Aspects of Effective Algebra, (Proc. Conf. Monash Univ., Clayton, Australia, Aug. 1–4, 1979), J. N. Crossley (ed.), (Upside Down A Book Co., Yarra Glen, Victoria, Australia, 1981) 232–245.

[711] R. L. Smith, Effective aspects of profinite groups, J. Symbolic Logic, **46** (1981) 851–863.

[712] R. L. Smith, Two theorems on autostability in p-groups, in: Logic Year 1979-1980, (Proc. Seminars and Conf. Math. Logic, Univ. Connecticut, Storrs, CT, 1979/80), M. Lerman, J. H. Schmerl and R. I. Soare, (eds.), Lecture Notes in Math., **859** (1981) 302–311.

[713] R. L. Smith and L. van den Dries, Decidable regularly closed fields of algebraic numbers, J. Symbolic Logic, **50** (1985) 468–475.

[714] R. I. Soare, Cohesive sets and recursively enumerable Dedekind cuts, Pacific J. Math., **31** (1969) 215–231.

[715] R. I. Soare, Constructive order types on cuts, J. Symbolic Logic, **34** (1969) 285–289.

[716] R. I. Soare, Recursion theory and Dedekind cuts, Trans. Amer. Math. Soc., **140** (1969) 271–294.

[717] R. I. Soare, Automorphisms of the lattice of recursively enumerable sets, part I: maximal sets, Ann. of Math., **100** (1974) 80–120.

[718] R. I. Soare, Isomorphisms on countable vector spaces with recursive operations, J. Austral. Math. Soc., **18** (1974) 230–235.

[719] R. I. Soare, The infinite injury priority method, J. Symbolic Logic, **41** (1976) 513–530.

[720] R. I. Soare, Computational complexity, speedable and levelable sets, J. Symbolic Logic, **42** (1977) 545–563.

[721] R. I. Soare, Tree arguments in recursion theory and the $\mathbf{0}'''$–priority method, in: Recursion Theory, (Proc. AMS-ASL Summer Inst., Ithaca, NY, June 28 – July 16, 1982), A. Nerode and R. A. Shore, (eds.), Proc. Sympos. Pure Math., **42** (1985) 53–106.

[722] R. I. Soare, Recursively Enumerable Sets and Degrees. A Study of Computable Functions and Computably Generated Sets, Perspect. in Math. Logic, (1987).

[723] R. I. Soare, Computability and recursion, Bulletin of Symbolic Logic, **2** (1996) 284–321.

[724] D. R. Solomon, Reverse Mathematics and Ordered Groups, Ph.D. Thesis, Cornell Univ., Ithaca, NY, (1998).

[725] R. M. Solovay, Hyperarithmetically encodable sets, Trans. Amer. Math. Soc., **239** (1978) 99–122.

[726] V. D. Soloviev, Automorphisms of the structure of computational theories (Russian), in: Proc. 9th. All-Union Conf. Math. Logic, (Leningrad, 1988), (1988) 153.

[727] E. Specker, Ramsey's Theorem does not hold in recursive set theory, in: Logic Colloq. '69, (1971) 439–442.

[728] C. Spector, Recursive well-orderings, J. Symbolic Logic, **20** (1955) 151–163.

[729] J. Staples, On constructive fields, Proc. London Math. Soc., **23** (1971) 753–768.

[730] J. Staples, Axioms for constructive fields, Bull. Austral. Math. Soc., **8** (1973) 221–232.

[731] E. Steinitz, Algebraische theorie der Körper, (Chelsea Publishing Co., New York, N. Y., 1950).

[732] V. Stoltenberg-Hansen and J. V. Tucker, Effective algebras, in: Handbook of Logic in Computer Science, Vol. 4, Oxford Sci. Publ., (1995) 357–526.

[733] G. Stolzenberg, Constructive normalization of an algebraic variety, Bull. Amer. Math. Soc. (N.S.), **74** (1968) 595–599.

[734] W. Szmielew, Elementary properties of Abelian groups, Fund. Math., **41** (1955) 203–271.

[735] K. Tanaka, A game-theoretic proofs [proof] of analytic Ramsey theorem Z. Math. Logik Grundlag. Math., **38** (1992) 301–304.

[736] A. Tarski, Undecidability of group theory, J. Symbolic Logic, **14** (1949) 76–77.

[737] A. Tarski, A Decision method for elementary algebra and geometry, 2nd. edn., revised, (Univ. California Press, Berkeley and Los Angeles, CA, 1951).

[738] E. K. TeKolste, Some Results in R.E. Presented Vector Spaces, Ph.D. Thesis, Univ. Rochester, Rochester, NY, (1976).

[739] J. B. Tennenbaum, A Constructive Version of Hilbert's Basis Theorem, Ph.D. Thesis, Univ. California, San Diego, La Jolla, CA, (1973).

[740] S. Tennenbaum, Non-archimedean models for arithmetic, Notices Amer. Math. Soc., **6** (1959) 270.

[741] J. J. Thurber, Degrees of Boolean Algebras, Ph.D. Thesis, Univ. Notre Dame, Notre Dame, IN, (1994).

[742] J. J. Thurber, Recursive and r.e. quotient Boolean algebras, Arch. Math. Logic, **33** (1994) 121–129.

[743] B. A. Trakhtenbrot, Comparing the Church and Turing approaches: two prophetical messages, in: The universal Turing machine: a half-century survey, Oxford Sci. Publ., (Oxford Univ. Press, 1988) 603–630.

[744] A. S. Troelstra and D. van Dalen, Constructivism in Mathematics, Vol. 1, Stud. Logic Found. Math., **121** (1988).

[745] A. S. Troelstra and D. van Dalen, Constructivism in Mathematics, Vol. 2, Stud. Logic Found. Math., **123** (1988).

[746] J. V. Tucker, Computing in algebraic systems, in: Recursion theory: its generalisation and applications, (Proc. Logic Colloq., Univ. Leeds, Leeds, 1979), London Math. Soc. Lecture Note Ser., **45** (1980) 215–235.

[747] J. V. Tucker, Computability and the algebra of fields: some affine constructions, J. Symbolic Logic, **45** (1980) 103–120.

[748] A. M. Turing, On computable numbers, with an application to the Entscheidungsproblem, Proc. London Math. Soc., Ser. 2, **42** (1936) 230–265; corr. ibid., **43** (1937) 544–546.

[749] D. A. Tusupov, Numerations of homogeneous models of decidable complete theories with a computable family of types (Russian), Vyčisl. Systemy, **129**, (Akad. Nauk SSSR Sibirsk. Otdel., Inst. Mat., Novosibirsk) (1989) 152–171, 197–198.

[750] H. Tverberg, On Schmerl's effective version of Brooks's theorem, J. Combin. Theory, Ser. B, **37** (1984) 27–30.

[751] R. L. Vaught, Non-recursive-enumerability of the set of sentences true in all constructive models, Bull. Amer. Math. Soc. (N.S.), **63** (1957) 230.

[752] R. L. Vaught, Sentences true in all constructive models, J. Symbolic Logic, **25** (1960) 39–58.

[753] M. C. Venning, Type Structures of $\aleph_0$–Categorical Theories, Ph.D. Thesis, Cornell Univ., Ithaca, NY, (1977).

[754] Yu. G. Ventsov, Algorithmic properties of branching models (Russian), Algebra i Logika, **25** (1986) 369–383, 494; [translated in: Algebra and Logic, **25** (1986) 229–238].

[755] Yu. G. Ventsov, Nonuniform autostability of models (Russian), Algebra i Logika, **26** (1987) 684–714; [translated in: Algebra and Logic, **26** (1987) 422–440].

[756] Yu. G. Ventsov, Reducibility of bounded complexity between constructivizations of models (Russian), in: Computable Invariants in the Theory of Algebraic Systems, V. N. Remeslennikov, (ed.), (Akad. Nauk SSSR Sibirsk. Otdel., Vyčisl. Tsentr, Novosibirsk, 1987) 22–34.

[757] Yu. G. Ventsov, Algorithmic dimension of models (Russian), Dokl. Akad. Nauk, **305** (1989) 21–24; [translated in: Soviet Math. – Dokl., **39** (1989) 237–239].

[758] Yu. G. Ventsov, A problem on the effective choice of constructivizations, and recursive consistency of problems on constructive models (Russian), Algebra i Logika, **31** (1992) 3–20, 96; [translated in: Algebra and Logic, **31** (1992) 1–11].

[759] Yu. G. Ventsov, The effective choice problem for relations and reducibilities in classes of constructive and positive models (Russian), Algebra i Logika, **31** (1992) 101–118, 220; [translated in: Algebra and Logic, **31** (1992) 63–73].

[760] Yu. G. Ventsov, On effective choice of constructivizations, Siberian Adv. Math., **2** (1992) 198–203.

[761] V. N. Vlasov and S. S. Goncharov, Strong constructibility of Boolean algebras with elementary characteristic (1,1,0) (Russian), Algebra i Logika, **32** (1993) 618–630; [translated in: Algebra and Logic, **32** (1993) 334–341].

[762] B. L. van der Waerden, Eine Bemerkung über die Unzerlegbarkeit von Polynomen, Math. Ann., **102** (1930) 738–739.

[763] B. L. van der Waerden, Moderne Algebra, Parts I and II, (G. E. Stechert and Co., New York, 1943); [translated as: Modern Algebra, Vols. I and II, (translated from the 2nd. rev. German edition by F. Blum, with revisions and additions by the author), Frederick Ungar Publishing Co., NY, (1949)].

[764] B. L. van der Waerden, A history of algebra. From al-Khwârîzmî to Emmy Noether, Springer-Verlag, (1985).

[765] R. Watnik, Recursive and Constructive Linear Orderings, Ph.D. Thesis, Rutgers Univ., New Brunswick, NJ, (1980).

[766] R. Watnick, Constructive and recursive scattered order types, in: Logic Year 1979-1980, (Proc. Seminars and Conf. Math. Logic, Univ. Connecticut, Storrs, CT, 1979/80), M. Lerman, J. H. Schmerl and R. I. Soare, (eds.), Lecture Notes in Math., **859** (1981) 312–326.

[767] R. Watnick, A generalization of Tennenbaum's Theorem on effectively finite recursive linear orderings, J. Symbolic Logic, **49** (1984) 563–569.

[768] P. R. Young, Linear orderings under one–one reducibility, J. Symbolic Logic, **31** (1966) 70–85.

[769] H. Zassenhaus, On a theorem of Kronecker, Delta (Waukesha), **1** (1968/69) 1–14.

Iraj Kalantari
Department of Mathematics
Western Illinois University
One University Circle
Macomb, IL 61455, USA
i-kalantari@wiu.edu

[illegible]

Chapter 12

A Bibliography of Recursive Analysis and Recursive Topology

V. Brattka and I. Kalantari

Introduction

This Handbook's focus is on recursive algebra and recursive model theory with a list of references of work in those areas. In this note, we present a brief look into *recursive analysis* and *recursive topology*, and gather references for the work on these subjects since Turing's [344] original paper on computability of reals.

Recursive mathematics investigates the 'constructive' nature of mathematical results when 'constructive' is interpreted via recursive function theory or Turing computability, and while the classical laws of logic are adopted intact. *Recursive analysis* focuses this approach of study to classical analysis.

Recursive analysis begins by noting that a real x is *computable* (or *recursive*) if there is a computable (or recursive) Cauchy sequence $(r_n)_{n\in\omega}$ of rational numbers which converges *rapidly* to x, that is $|x-r_n| < 1/2^n$ for all n. Clearly, there are only countably many computable reals and thus uncountably many *noncomputable* ones. All algebraic real numbers and well-known constants such as π and e are computable in this sense. Turing [344] gave an equivalent and mathematically precise definition for 'computable reals' in his famous paper "*On computable numbers, ...* ". Indeed, the idea of a computable real was one of the basic motivations for the invention of his machine model of computability. Besides Cauchy sequence representation, there are several other representations of the real numbers (p–adic representations, continued fraction representation, representation by nested intervals) which, when effectivized, all induce the same notion of computability of real

HANDBOOK OF RECURSIVE MATHEMATICS
Edited by Yu. L. Ershov, S. S. Goncharov,
A. Nerode, and J. B. Remmel

numbers. In his original paper, Turing used the decimal expansion representation, but in a correction to his paper [344], he observed that the decimal representation is not appropriate to introduce computability of real-valued functions. For example, even an elementary function such as $\lambda x f(x) = 3x$ is not computable with respect to the decimal representation. Later, Mostowski [268] found that the notion of a *computable sequence* of real numbers, also, depends on the chosen representation for the reals. Computable reals have many interesting properties which warrant pursuit of the subject; for example, Rice [303] proved that computable real numbers form a real algebraically closed field.

In the next stage of development of recursive analysis, Grzegorczyk [119, 121] and Lacombe [240, 241, 242] defined *computability of real functions.* Their notion can be characterized as follows: a real-valued function f is *computable* if there is a Turing machine M which transforms each Cauchy sequence of rational numbers rapidly converging to x into a Cauchy sequence of rational numbers rapidly converging to $f(x)$. All polynomials with computable real coefficients and other well-known continuous functions, such as the trigonometric functions and the exponential function, are computable in this sense. It turns out that continuity is a *necessary* condition for computability of real functions and thus some classically familiar phenomena hold for this class of functions. However, further interest develops as some pathologies arise. For instance, a counterexample of Myhill [275] shows that the derivative of a computable function need not be computable, and one of Specker's [324] (also see Lacombe [244]) shows that a computable real function need not attain its maximum (a computable value) at a computable real. It is important to note that in the Grzegorczyk/Lacombe definition, computable real functions are functions with domains equal to all of the reals including the non-computable ones. In contrast to that, some authors in the 'Russian school' have studied computable functions whose domains are restricted to the computable reals. Their reports are, however, disparate and not over an apparent uniform platform. Recently, Kalantari and Welch [161, 162] have developed a setting that unifies the study and reveals many properties of such functions.

With attention paid to *sets* of real numbers, in a series of papers, Lacombe [244] and Kreisel and Lacombe [199] defined what is now called a *recursively enumerable* open subset of the real numbers. An open subset A of the real numbers is *recursively enumerable* if there is a computable sequence of pairs of rational numbers $(r_n, s_n)_{n \in \omega}$ such that the union of the open intervals (r_n, s_n)

is equal to A. This notion fuses nicely with recursion theory and yields interesting results in topology and analysis. (See Kalantari [153], Kalantari and Remmel [156] and Spreen [327] for a sample.)

There has been much further work on recursive analysis; see Aberth [5], Brattka [24, 26, 25], Caldwell and Pour-El [54], Ceĭtin [65, 66], Ceĭtin and Zaslavskiĭ [69], Ge and Nerode [101, 102], Goodstein [112], Grzegorczyk [120, 121], Hauck [123] to [137], Hertling [142, 141], Kalantari [153], Kalantari and Leggett [154, 155], Kalantari and Remmel [156], Kalantari and Retzlaff [157], Kalantari and Weitkamp [158, 159, 160], Kalantari and Welch [161] to [164], Ko [181, 184], Kreisel, Lacombe and Shoenfield [200], Kreitz and Weihrauch [205] to [208], and [361], Lachlan [238], Lacombe [240] to [247], Mazur [256], Metakides and Nerode [259], Metakides, Nerode and Shore [260], J. Rand Moschovakis [264, 265], Y. Moschovakis [266, 267], Myhill [275], Nerode and Huang [278], Orevkov [279] to [289], Pour-El and Richards [293] to [301], Rice [303], Šanin [308] to [312], Shepherdson [318], Soare [320, 321], Specker [323, 324, 325], Spreen [326] to [331], Troelstra [334] to [341], Weihrauch [350] to [359], Weihrauch and Zheng [365, 366], Zaslavskiĭ [372], Zhong [375], and Zhou [378, 379].

We sketch some of the more recent approaches:

The Sequence Approach

Pour-El and Richards [301] introduced an axiomatic approach to computability in Banach spaces. The axiomatized notion is that of a *computable sequence*. A subset of sequences of a Banach space is called a *computability structure* on that space if it fulfills three axioms. Roughly speaking, these axioms express that computable sequences are closed under linear combinations, under norm, and under limit (in a specific sense). This 'sequence approach' to computability has been inspired by the observation that a real function (defined on the unit interval, say) is computable in the sense of Grzegorczyk/Lacombe, if and only if it maps computable sequences to computable sequences and it admits a computable modulus of continuity. Pour-El and Richards applied their theory to L^p and l^p spaces, investigated computability of eigenvalues and formulated a criterion, for closed linear operators $f : X \to Y$ and Banach spaces X, Y, which characterizes the pathology that 'there is a computable point x such that $f(x)$ is non-computable'. When X is separable and there exists a computable sequence $(e_n)_{n\in\omega}$ whose linear span is dense in X, and further $(f(e_n))_{n\in\omega}$ is a computable sequence, their finding is: $f(x)$ is computable for each computable point x if and only if f is bounded.

There are some further results which are related to recursive functional analysis: Metakides, Nerode and Shore [260] have investigated the Hahn-Banach Theorem and Ge and Nerode [101] the Krein-Milman Theorem. In these papers *located sets*, subsets with computable distance functions, play a key role. Closed sets with this property have been investigated by Ge [100], Ge and Nerode [101] and Zhou [378]. An interesting example of use of located sets is in Metakides, Nerode and Shore [260]. They establish a recursive counterexample to the Hahn-Banach Theorem by showing that there is a recursively presentable Banach space B and a recursive linear functional λ defined on a recursively presented linear subspace V of B where λ has a recursive norm (through a recursively located kernel) but there is no recursively continuous linear functional on all of B which extends λ and has the same norm as λ.

The sequence approach of Pour-El and Richards has been generalized to Fréchet spaces by Washihara and Yasugi [349] and recently to metric spaces by Mori, Tsujii and Yasugi [263].

The Filter Approach

Kalantari and Welch [161, 162] have organized a systematic approach to recursive topology which uses *filters* as a device to define points, recursive points, functions and recursive functions for (well-behaved) topological spaces. In the filter approach, the objects of study are not points, they are 'neighborhoods'. The objects are 'pieces' of the space which approximate a location of the space as accurately as desired but not infinitely accurately. Their approach exposes explanation for existing phenomena on pathology of domains of recursive real functions while presenting new results and machinery to apply to recursive analysis.

After finding that for interesting spaces (such as $\mathbb{R}^n$) the filter approach produces a space homeomorphic to the original space, they show that the computable points and the computable functions obtained are precisely the traditional recursive points and recursive real functions. However, they also find a class of functions which, while perfectly computable and therefore computably continuous, have controllably small domains by 'avoiding' a certain class of points.

This approach leads to application of the theory to recursive analysis. As a consequence, they find that the collection of classical reals divide into the *three* disjoint sets of *recursive reals*, *avoidable reals* and *shadow reals*. The recursive reals are the traditional recursive points while avoidable points are those which, through recursive operations can be *excised* from the domain of a recursive function without affecting the effective continuity of the function. The shadow points are those which do not permit recursive excision; they act as if they are 'attached' to recursive points and therefore remain in the domain of every recursive function. Kalantari and Welch [161] name recursive functions with missing points from their domains *recursive quantum functions* and find a mathematically rich structure for them. For example in [162], they build a computable function on $[0, 1]$, generated by a recursive quantum *correspondence* (a partial function from basic open sets to basic open sets with controlled properties), that is not just partial, but *nonextendible* to a computable continuous function of larger domain. This function is of unbounded variation and its domain, which is an open set, can be made as small, in Lebesgue measure, as desired.

The Representation Approach

Another interesting approach to recursive analysis, the so called 'Type 2 Theory of Effectivity', has been developed by Kreitz and Weihrauch [206, 360]. The fundamental notion of this theory is the notion of 'representation'. Formally, a *representation* of a set X is a mapping from Cantor space (the space of all infinite sequences over a finite set) onto X, and intuitively, a representation is a description of how to code arbitrary points of X by infinite sequences of symbols. Examples of representations are the Cauchy representation and the decimal representation of the real numbers. In this approach, it is easy to define computability of points, sequences, and functions for a large class of spaces (T_0 spaces with countable bases). The best investigated spaces are the real numbers, the space of continuous functions, the hyperspace of closed subsets, L^p spaces, and measure spaces [29, 357, 358, 359]. The 'representation approach' offers tools to single out well-behaved (admissible) representations and is a natural way to find appropriate effectivity notions for many spaces. Recently, Hertling [143] has shown that the real number structure is *effectively categorical*: up to equivalence, there is only *one* representation which makes the basic operations of the structure computable.

In the 'representation approach', classical theorems can be expressed in highly effective versions. For example, Kreitz and Weihrauch [208] show that an effective version of the Heine-Borel Theorem holds and Hertling [141] shows that an effective version of the Riemann Mapping Theorem holds.

One further advantage of this approach is that it provides a direct way to investigate complexity in analysis. This is because representations allow a direct measurement of the amount of time and space that Turing machines use to carry out the desired computations; see Weihrauch [353], Müller [270, 271] , Schröder [316]. The results found there are polynomially equivalent to the results of Ko [181] who uses Turing machines with oracles to measure complexity. Besides studying other topics on complexity in analysis, Ko has also investigated representations of real numbers from a complexity point of view (e.g., [179]).

There is also a series of papers of Hauck ([123] to [133]), who independently developed a theory of representations for recursive analysis.

Representation-based computability for the space of reals, and more generally for complete separable metric spaces, can also be characterized through representation-free means: Brattka [24, 26] shows that, for such spaces, from some basic operations, and by applying certain closure schemes, (essentially) exactly the computable operations of Type 2 Theory can be generated.

Acknowledgments

We are grateful to Victor Marek and Jeff Remmel for bringing us together and for encouraging the completion of this list. We further wish to thank Xiaolin Ge for providing items for this bibliography.

References

[1] O. Aberth, Analysis in the computable number field, J. Assoc. Comput. Mach., **15** (1968) 275–299.

[2] O. Aberth, A chain of inclusion relations in computable analysis, Proc. Amer. Math. Soc., **22** (1969) 539–548.

[3] O. Aberth, Computable analysis and differential equations, in: Intuitionism and Proof Theory, (Proc. Conf., Buffalo, N.Y., 1968). A. Kino, J. Myhill and R. E. Vesley (eds.), Stud. Logic Found. Math., (1970) 47–52.

[4] O. Aberth, The failure in computable analysis of a classical existence theorem for differential equations, Proc. Amer. Math. Soc., **30** (1971) 151–156.

[5] O. Aberth, Computable analysis, (McGraw-Hill, New York, 1980).

[6] J. C. Baez, Recursivity in quantum mechanics, Trans. Amer. Math. Soc., **280** (1983) 339–350.

[7] G. Baigger, Die Nichtkonstruktivität des Brouwerschen Fixpunktsatzes, Arch. Math. Logik Grundlag., **25** (1985) 183–188. (German, The nonconstructiveness of Brouwer's fixed point theorem).

[8] R. Baire, Sur les fonctions des variables reeles, Annali di matematica pura ed applicata, Ser. 3, **3** (1899) 1–123.

[9] M. J. Beeson, Foundations of Constructive Mathematics: metamathematical studies, Ergeb. Math. Grenzgeb. Ser. 3, **6** (1985).

[10] E. Bishop, Constructive Methods in the Theory of Banach Algebras, in: Function Algebras, (Proc. Internat. Sympos. Function Algebras, Tulane Univ., New Orleans, LA, 1965). F. T. Birtel, (ed.), (Scott-Foresman, Chicago, 1966), 343–345.

[11] E. Bishop, Foundations of Constructive Analysis, (McGraw-Hill, New York, Toronto, London, 1967).

[12] E. Bishop, A constructive ergodic theorem, J. Math. Mech., **17** (1967/1968), 631–639.

[13] E. Bishop, The constructivization of abstract mathematical analysis, in: Proc. Internat. Congr. Math., Moscow, 1966, Izdat. Mir, Moscow, 1968, 308–313.

[14] E. Bishop and D. S. Bridges, Constructive Analysis, Grundlehren Math. Wiss., **279** (1985).

[15] E. Bishop and H. Cheng, Constructive Measure Theory, Mem. Amer. Math. Soc., (1972).

[16] M. Bläser, Uniform computational complexity of the derivatives of C^∞-functions, in: Computability and Complexity in Analysis (K. Ko and K. Weihrauch, eds.), Informatik-Berichte, FernUniversität Hagen, **190** (1995) 99–104.

[17] L. Blum, M. Shub and S. Smale, On a theory of computation and complexity over the real numbers: NP–completeness, recursive functions and universal machines, Bull. Amer. Math. Soc. (N.S.), **21** (1989) 1–46.

[18] B. Bolzano, Rein analytischer Beweis des Lehrsatzes, daß zwischen je zwei Werthen, die ein entgegengesetztes Resultat gewähren, wenigstens eine reelle Wurzel der Gleichung liege, Prague, 1817, in: Bernard Bolzano (1781–1848) Bicentenary, Early Mathematical Works, L. Nový, (ed.), Acta Historicæ Rerum Naturalium Nec Non Technicarum, special issue **12**, Inst. Czech. and General History (CSAS), Prague, 1981, 417–476.

[19] E. Borel, Leçons sur la théorie des fonctions, Gauthier-Villars et fils, Paris, (1898).

[20] E. Borel, Œuvres de Émile Borel, Vols. 1–4, (Éditions du CNRS, Paris, 1972).

[21] A. Borodin and I. Munro, The computational complexity of algebraic and numeric problems, (Elsevier, New York, 1975).

[22] S. S. Brady and J. B. Remmel, The undecidability of the lattice of r.e. closed subsets of an effective topological space, Ann. Pure Appl. Logic, **35** (1987) 193–203.

[23] V. Brattka, Computable selection in analysis, in: Computability and Complexity in Analysis, K. Ko and K. Weihrauch, (eds.), Informatik-Berichte, FernUniversität Hagen, **190** (1995) 125–138.

[24] V. Brattka, Recursive characterization of computable real-valued functions and relations, Theoret. Comput. Sci., **162** (1996) 45–77.

[25] V. Brattka, Computable invariance, in: Computing and Combinatorics, T. Jiang and D. T. Lee, (eds.), Lecture Notes in Comput. Sci., **1276** (1997) 146–155.

[26] V. Brattka, Order-free recursion on the real numbers, Math. Logic Quart., **43** (1997) 216–234.

[27] V. Brattka and P. Hertling, Continuity and computability of relations, Informatik Berichte, FernUniversität Hagen, Hagen, **164** (September 1994).

[28] V. Brattka and P. Hertling, Feasible real random access machines, in: Theory and Practice of Informatics, SOFSEM '96, K. G. Jeffrey, J. Král and M. Bartošek, (eds.), Lecture Notes in Comput. Sci., **1175** (1996) 335–342.

[29] V. Brattka and K. Weihrauch, Computability on subsets of Euclidean space I: closed and compact subsets, (submitted, 1997).

[30] D. S. Bridges, Some notes on continuity in constructive analysis, Bull. London Math. Soc., **8** (1976), 179–182.

[31] D. S. Bridges, Constructive Functional Analysis, Research Notes in Mathematics, **28**, Pitman (Advanced Publishing Program), Boston, Mass.-London, 1979.

[32] D. S. Bridges, Towards a constructive foundation for quantum mechanics, in: Constructive Mathematics, (Proc. New Mexico State Univ. Conf., Aug. 11–15, 1980, Las Cruces, NM) F. Richman, (ed.), Lecture Notes in Math., **873** (1981) 260–273.

[33] D. S. Bridges, Locatedness, convexity, and Lebesgue measurability, Quart. J. Math. Oxford Ser. 2, **39** (1988) 411–421.

[34] D. S. Bridges, A general constructive intermediate value theorem, Z. Math. Logik Grundlag. Math., **35** (1989) 433–435.

[35] D. S. Bridges, Sequential, pointwise, and uniform continuity: a constructive note, Math. Logic Quart., **39** (1993) 55–61.

[36] D. S. Bridges, Computability, a Mathematical Sketchbook, Grad. Texts in Math., **146** (1994).

[37] D. S. Bridges, A constructive look at the real number line. Real numbers, generalizations of the reals, and theories of continua, Synthese Lib., **242** (1994) 29–92.

[38] D. S. Bridges, Constructive mathematics and unbounded operators, J. Philos. Logic, **24** (1995) 549–561.

[39] D. S. Bridges, A. Calder, W. Julian, R. Mines, and F. Richman, Locating metric complements in $\mathbb{R}^n$, in: Constructive Mathematics, (Proc. New Mexico State Univ. Conf., Aug. 11–15, 1980, Las Cruces, NM) F. Richman, (ed.), Lecture Notes in Math., **873** (1981) 241–249.

[40] D. S. Bridges, A. Calder, W. Julian, R. Mines, and F. Richman, Picard's theorem, Trans. Amer. Math. Soc., **269** (1982) 513–520.

[41] D. S. Bridges and C. S. Calude, On recursive bounds for the exceptional values in speed-up, Theoret. Comput. Sci., **132** (1994) 387–394.

[42] D. S. Bridges, W. H. Julian, F. Richman, and R. Mines, Extensions and fixed points of contractive maps in $\mathbb{R}^n$, J. Math. Anal. Appl., **165** (1992) 438–456.

[43] D. S. Bridges and R. Mines, What is constructive mathematics?, Math. Intelligencer, **6** (1984) 32–38.

[44] D. S. Bridges and F. Richman, Varieties of Constructive Mathematics, London Math. Soc. Lecture Note Ser., **97** (1987).

[45] D. S. Bridges, F. Richman, and W. Yuchuan, Sets, complements and boundaries, Indag. Math. (N.S.), **7** (1996) 425–445.

[46] L. E. J. Brouwer, De ondetrouwbaarkeid der logische principes (Dutch), [The unreliability of logical principles], Tijdsch. Wijsbegeerte, **2** (1908) 152–158; [translated in: Collected Works, Vol. 1, Philosophy and Foundations of Mathematics, A. Heyting, (ed.), (North-Holland, Amsterdam, Oxford, 1975), 107–111].

[47] L. E. J. Brouwer, An intuitionist correction of the fixed-point theorem on the sphere, Proc. Roy. Soc. London, Ser. A, **213** (1952) 1–2; [also in: Collected Works, Vol. 1, Philosophy and Foundations of Mathematics, A. Heyting, (ed.), (North-Holland, Amsterdam, Oxford, 1975), 506–507].

[48] L. E. J. Brouwer, Door klassieke theorema's gesignaleerde pinkernen die onvindbaar zijn (Dutch), [Fixed cores which cannot be found, though they are claimed to exist by classical theorems], Koninklijke Nederl. Akad. Wentensch. Proc. Ser. A, **55** (1952) 443–445; [translated in: Collected Works, Vol. 1, Philosophy and Foundations of Mathematics, A. Heyting, (ed.), (North-Holland, Amsterdam, Oxford, 1975), 516–518].

[49] L. E. J. Brouwer, Points and spaces, Canad. J. Math., **6** (1954) 1–17; [also in: Collected Works, Vol. 1, Philosophy and Foundations of Mathematics, A. Heyting, (ed.), (North-Holland, Amsterdam, Oxford, 1975), 522–538].

[50] L. E. J. Brouwer, Intuitionistischer Beweis des Jordanshen Kurvensatzes, Koninklijke Nederl. Akad. Wentensch. Proc., **28** (1925) 503–508; [translated in: Collected Works, Vol. 1, Philosophy and Foundations of Mathematics, A. Heyting, (ed.), (North-Holland, Amsterdam, Oxford, 1975), 315–320; also translated to Russian in: Trudy Vyčisl. Centr. Akad. Nauk Armjan. SSR i Erevan. Gos. Univ., **5** (1968) 139–146].

[51] L. E. J. Brouwer, Collected Works, Vol. 1, Philosophy and Foundations of Mathematics, A. Heyting, (ed.), (North-Holland Publishing Co., Amsterdam, Oxford, 1975).

[52] L. E. J. Brouwer, Collected Works, Vol. 2, Geometry, Analysis, Topology and Mechanics, H. Freudenthal, (ed.), (North-Holland Publishing Co., Amsterdam, Oxford, 1976).

[53] L. E. J. Brouwer, Brouwer's Cambridge Lectures on Intuitionism, D. van Dalen, (ed.), (Cambridge Univ. Press, 1981).

[54] J. Caldwell and M. B. Pour-El, On a simple definition of computable function of a real variable — with applications to functions of a complex variable, Z. Math. Logik Grundlag. Math., **21** (1975) 1–19.

[55] C. S. Calude, Information and Randomness in Algorithmic Perspective, Monographs in Theoret. Comput. Sci., (Springer, Berlin, 1994).

[56] C. S. Calude, P. Hertling and B. M. Khoussainov, Do the zeroes of Riemann's Zeta-function form a random sequence?, Bull. European Assoc. Theoret. Comput. Sci., **62** (1997) 199–207.

[57] C. S. Calude, G. Istrate and M. Zimand, Recursive Baire classification and speedable functions, Z. Math. Logik Grundlag. Math., **38** (1992) 169–178.

[58] C. S. Calude and H. Jürgensen, Randomness is an invariant for number representations, in: Results and Trends in Theoretical Computer Science, (Proc. Colloq. in honor of Arto Aolomaa, Graz, Austria, June 10–11, 1994), H. Maurer, J. Karhumäki and G. Rosenberrg, (eds.), (Springer, Berlin, 1994), 44–66.

[59] C. S. Calude and M. Zimand, Effective category and measure in abstract complexity theory, Theoret. Comput. Sci., **154** (1996) 307–327.

[60] G. Cantor, Contribuzione al fondamenta della teoria degli insiemi transfiniti, Riv. Mat. Univ. Parma, **5** (1895) 129–162, [translated as: Contributions to the founding of the theory of transfinite numbers, Dover, New York, 1955].

[61] A. L. Cauchy, Œuvres completes d'Augustin Cauchy, (Gauthier-Villars, Paris, 1882–1974).

[62] A. L. Cauchy, Cours d'analyse de l'Ecole Royale Polytechnique, in: Œuvres completes d'Augustin Cauchy, Sér. 2, Gauthier-Villars, Paris, 1897, **3** (1921) 17–331.

[63] G. S. Ceĭtin, Algorithmic operators in constructive complete separable metric spaces (Russian), Dokl. Akad. Nauk SSSR, **128** (1959) 49–52.

[64] G. S. Ceĭtin, Algorithmic operators in constructive metric spaces (Russian), Trudy Mat. Inst. Steklov., **67** (1962) 295–361.

[65] G. S. Ceĭtin, Three theorems on constructive functions (Russian), Trudy Mat. Inst. Steklov., **72** (1964) 537–543.

[66] G. S. Ceĭtin, Mean-value theorems in constructive analysis (Russian), Trudy Mat. Inst. Steklov., **67** (1962) 362–384.

[67] G. S. Ceĭtin, On upper bounds of recursively enumerable sets of constructive real numbers, Trudy Mat. Inst. Steklov., **113** (1970) 102–172, [also in: Proc. Steklov Inst. Math., 113 (1970) 119–194].

[68] G. S. Ceĭtin, An algorithm for simplified syntactic analysis (Russian), Problemy Kibernet, **24** (1971) 227–242.

[69] G. S. Ceĭtin and I. D. Zaslavskiĭ, Singular coverings and properties of constructive functions connected with them (Russian), Trudy Mat. Inst. Steklov., **67** (1962) 458–502.

[70] G. S. Ceĭtin and I. D. Zaslavskiĭ, A criterion of the rectifiability of constructive plane curves (Russian), Izv. Akad. Nauk Armjan. SSR Ser. Mat., **5** (1970) 434–440.

[71] G. S. Ceĭtin and I. D. Zaslavskiĭ, Yet another constructive variant of the Cauchy theorem (Russian), Zap. Naučn. Sem. Leningrad. Otdel. Mat. Inst. Steklov (LOMI), **20** (1971) 36–39, 282–283.

[72] G. S. Ceĭtin, I. D. Zaslavskiĭand N. A. Šanin, Peculiarities of constructive mathematical analysis (Russian), in: Proc. Internat. Congr. Math. (Moscow, 1966), Izdat. Mir, Moscow, (1968) 253–261.

[73] D. Cenzer, Effective real dynamics, in: Logical Methods, (Papers from Conf. in honor of Anil Nerode's Sixtieth Birthday, June 1–3, 1992, Cornell Univ., Ithaca, NY), J. N. Crossley, J. B. Remmel, R. A. Shore and M. E. Sweedler, (eds.), Progr. Comput. Sci. Appl. Logic, **12** (1993) 162–177.

[74] D. Cenzer, P. Clote, R. L. Smith, R. I. Soare and S. S. Wainer, Members of countable Π_1^0 classes, (Special issue: 2nd. Southeast Asian Logic Conf., Bangkok, 1984) Ann. Pure Appl. Logic, **31** (1986) 145–163.

[75] G. J. Chaitin, Algorithmic Information Theory, with a foreword by J. T. Schwartz, Cambridge Tracts Theoret. Comput. Sci., **1** (1987).

[76] G. J. Chaitin, Incompleteness theorems for random reals, Adv. in Appl. Math., **8** (1987) 119–146.

[77] A. W. Chou, Some complexity issues in complex analysis, in: Computability and Complexity in Analysis (K. Ko and K. Weihrauch, eds.), Informatik-Berichte, FernUniversität Hagen, **190** (1995) 91–98.

[78] A. W. Chou and K. Ko, Computational complexity of two-dimensional regions, SIAM J. Comput., **24** (1995) 923–947.

[79] J. P. Cleave, The primitive recursive analysis of ordinary differential equations and the complexity of their solutions, J. Comput. System Sci., **3** (1969) 447–455.

[80] W. Collins, Provably recursive real numbers, Notre Dame J. Formal Logic, **19** (1978) 513–522.

[81] W. Collins and P. Young, Discontinuities of provably correct operators on the provably recursive real numbers, J. Symbolic Logic, **48** (1983) 913–920.

[82] D. van Dalen, Lectures on Intuitionism, in: (Proc. Cambridge Summer School Math. Logic, Aug. 1–21, 1971), A. R. D. Mathais and H. Rogers, (eds.), Lecture Notes in Math., **337** (1973) 1–94.

[83] D. van Dalen, An interpretation of instuitionistic analysis, with an appendix by A. S. Troelstra, Ann. Math. Logic, **13** (1978), 1–43.

[84] T. Deil, Darstellungen und Berechenbarkeit reeller Zahlen, Informatik Berichte 51, FernUniversität Hagen, Hagen, Dissertation, December 1984.

[85] H. J. Dettki, H. Schuster and K. Weihrauch, Type 2 recursion theory, in: Algebra, Combinatorics and Logic in Computer Science, Vol. 1, (Papers from Colloq., Sept. 12–16, 1983, Györ), J. Demetrovics, G. Katona and A. Salomaa, (eds.), Coll. Math. Soc. János Bolyai, **42** (1986) 355–361.

[86] D. Z. Du and K. Ko, Computational complexity of integration and differentiation of convex functions, Systems Sci. Math. Sci., **2** (1989) 70–79.

[87] D. Z. Du and K. Ko, A note on best fractions of a computable real number, J. Complexity, **8** (1992) 216–229.

[88] M. Dummett, The Philosophical Basis of Intuitionistic Logic, in: Logic Colloquium '73, (Bristol 1973), H. E. Rose and J. C. Shepherdson, (eds.), Stud. Logic Found. Math., **80** (1975) 5–40.

[89] M. Dummett, Elements of Intuitionism, (Oxford Univ. Press, 1977).

[90] S. Feferman, Systems of predicative analysis, J. Symbolic Logic, **29** (1964) 1–30.

[91] S. Feferman, Theories of finite type related to mathematical practice, in: J. Barwise, (ed.), Handbook of Mathematical Logic, Stud. Logic Found. Math., **90** (1977) 913–971.

[92] S. Feferman, Constructive Theories of Functions and Classes, in: Proc. Logic Colloq. '78, Mons, Aug. 24 – Sept. 1, 1978, M. Boffa, D. van Dalen and K. McAloon, (eds.), Stud. Logic Found. Math., **97** (1979) 159–224.

[93] S. Feferman, Between constructive and classical mathematics, in: Computation and Proof Theory, (Proc. Logic Colloq., Aachen, July 18–23, 1983, Part II), M. M. Richter, E. Börger, W. Oberschelp, B. Schinzel and W. Thomas, (eds.), Lecture Notes in Math., **1104** (1984) 143–162.

[94] G. Frege, Philosophical and Mathematical Correspondence, G. Gabriel, (ed.), (abridged from the German by B. McGuinness, translated by H. Kaal), (Chicago Univ. Press and Blackwell, Oxford, 1980).

[95] R. Freund, Real functions and numbers defined by Turing machines, Theoret. Comput. Sci., **23** (1983) 287–304.

[96] R. Freund and L. Staiger, Numbers defined by Turing machines, in: Computability and Complexity in Analysis K. Ko and K. Weihrauch, (eds.), Informatik-Berichte, FernUniversität Hagen, **190** (1995) 119–124; [also in: Coll. Logicum Ann. Kurt-Gödel-Soc, (Springer, Vienna), **2** (1996) 118–137.

[97] H. M. Friedman, Set theoretic foundations for constructive analysis, Ann. of Math., Ser. 2, **105** (1977) 1–28.

[98] H. M. Friedman, The computational complexity of maximization and integration, Adv. Math., **53** (1984) 80–98.

[99] D. Gaier, Konstruktive Methoden der konformen Abbildung, Tracts in Natural Philosophy, **3**, (Springer-Verlag, Berlin, 1964).

[100] X. Ge, Some Algorithms in Euclidean Space and Group Representations, Ph.D. Thesis, Univ. Minnesota, Minneapolis, MN, (1993).

[101] X. Ge and A. Nerode, On extreme points of convex compact Turing located sets, in: Logical Foundations of Computer Science, (Proc. 3rd. Internatl. Sympos., LFCS '94, St. Petersburg, July 11–14, 1994), (A. Nerode and Yu. V. Matijasevič, eds.), Lecture Notes in Comput. Sci., **813** (1994) 114–128.

[102] X. Ge and A. Nerode, Effective content of the calculus of variations I: semicontinuity and the chattering lemma, Ann. Pure Appl. Logic, **78** (1996) 127–146.

[103] X. Ge and J. I. Richards, Computability in unitary representations of compact groups, in: Logical Methods, (Papers from Conf. in honor of Anil Nerode's Sixtieth Birthday, June 1–3, 1992, Cornell Univ., Ithaca, NY), J. N. Crossley, J. B. Remmel, R. A. Shore and M. E. Sweedler, (eds.), Progr. Comput. Sci. Appl. Logic, **12** (1993) 386–421.

[104] V. I. Glivenko, Sur la logique de M. Brouwer, Roy. Belg. Bull. Cl. Sci., Sér. 5, **14** (1928) 225–228.

[105] V. I. Glivenko, Sur quelques points de la logique de M. Brouwer, Roy. Belg. Bull. Cl. Sci., Sér. 5, **15** (1929) 183–188.

[106] K. Gödel, Über eine bisher noch nicht benützte Erweiterung des finiten Standpunktes, Dialectica, **12** (1958) 280–287.

[107] N. D. Goodman and J. Myhill, The formalization of Bishop's constructive mathematics, in: Toposes, Algebraic Geometry and Logic, (Partial Report Conf. Connections between Category Theory, Algebraic Geometry and Intuitionistic Logic, Dalhousie Univ., Halifax, Nova Scotia, Jan. 16–19, 1971), F. W. Lawvere, (ed.), Lecture Notes in Math., **274** (1972) 83–96.

[108] R. L. Goodstein, Missing value theorems, Math. Gaz., **33** (1949) 19–25.

[109] R. L. Goodstein, Mean value theorems in recursive function theory I. Differential mean value theorems, Proc. London Math. Soc., **52** (1950) 81–106.

[110] R. L. Goodstein, The recursive irrationality of π, J. Symbolic Logic, **19** (1954) 267–274.

[111] R. L. Goodstein, On non-constructive theorems of analysis and the decision problem, Math. Scand., **3** (1956) 261–263.

[112] R. L. Goodstein, Recursive analysis, in: Constructivity in Mathematics, (Proc. Colloq., Amsterdam, Aug. 26–31, 1957), A. Heyting, (ed.), Stud. Logic Found. Math., (1959) 37–42.

[113] R. L. Goodstein, Recursive Analysis, Stud. Logic Found. Math., (1961).

[114] R. L. Goodstein, Existence in mathematics, Compositio Math., **20** (1968) 70–82.

[115] R. L. Goodstein, A constructive form for the second Gauss proof of the fundamental theorem of algebra, in: Constructive Aspects of the Fundamental Theorem of Algebra, (Proc. Sympos. Zürich–Rüschlikon, June 5–7, 1967), (Wiley-Interscience, NY, 1969), 69–76.

[116] R. L. Goodstein, Polynomials with computable coefficients, Notre Dame J. Formal Logic, **11** (1970) 447–448.

[117] R. L. Goodstein and J. Hooley, On recursive transcendence, Notre Dame J. Formal Logic, **1** (1960) 127–137.

[118] A. Grzegorczyk, Undecidability of some topological theories, Fund. Math., **38** (1951) 137–152.

[119] A. Grzegorczyk, Computable functionals, Fund. Math., **42** (1955) 168–202.

[120] A. Grzegorczyk, On the definition of computable functionals, Fund. Math., **42** (1955) 232–239.

[121] A. Grzegorczyk, On the definitions of computable real continuous functions, Fund. Math., **44** (1957) 61–71.

[122] A. Grzegorczyk, Some approaches to constructive analysis, in: Constructivity in Mathematics, (Proc. Colloq., Amsterdam, Aug. 26–31, 1957), A. Heyting, (ed.), Stud. Logic Found. Math., (1959) 43–61.

[123] J. Hauck, Ein Kriterium für die Annahme des Maximums in der berechenbaren Analysis, Z. Math. Logik Grundlag. Math., **17** (1971) 193–196.

[124] J. Hauck, Zur Präzisierung des Begriffes berechenbare reelle Funktion, Z. Math. Logik Grundlag. Math., **17** (1971) 295–300.

[125] J. Hauck, Funktional-Rekursion, Z. Math. Logik Grundlag. Math., **18** (1972) 31–36.

[126] J. Hauck, Berechenbare reelle Funktionen, Z. Math. Logik Grundlag. Math., **19** (1973) 121–140.

[127] J. Hauck, Berechenbare reelle Funktionenfolgen, Z. Math. Logik Grundlag. Math., **22** (1976) 265–282.

[128] J. Hauck, Konstruktive Darstellungen reeller Zahlen und Folgen, Z. Math. Logik Grundlag. Math., **24** (1978) 365–374.

[129] J. Hauck, Konstruktive Darstellungen in topologischen Räumen mit rekursiver Basis, Z. Math. Logik Grundlag. Math., **26** (1980) 565–576.

[130] J. Hauck, Stetigkeitseigenschaften berechenbarer reeller Funktionen, Z. Math. Logik Grundlag. Math., **26** (1980) 69–76.

[131] J. Hauck, Berechenbarkeit in topologischen Räumen mit rekursiver Basis, Z. Math. Logik Grundlag. Math., **27** (1981) 473–480.

[132] J. Hauck, Stetigkeitseigenschaften berechenbarer Funktionale, Z. Math. Logik Grundlag. Math., **28** (1982) 377–383.

[133] J. Hauck, Konstruktive reelle Funktionale und Operatoren, Z. Math. Logik Grundlag. Math., **29** (1983) 213–218.

[134] J. Hauck, Eine neue Definition berechenbarer reeller Funktionen, Z. Math. Logik Grundlag. Math., **30** (1984) 259–268.

[135] J. Hauck, Zur Wellengleichung mit konstruktiven Randbedingungen, Z. Math. Logik Grundlag. Math., **30** (1984) 561–566.

[136] J. Hauck, Ein Kriterium für die konstruktive Lösbarkeit der Differentialgleichung $y' = f(x, y)$, Z. Math. Logik Grundlag. Math., **31** (1985) 357–362.

[137] J. Hauck, Eine berechenbare Funktion mit rationalen Werten, die nicht rekursiv ist, Z. Math. Logik Grundlag. Math., **33** (1987) 255–256.

[138] J. van Heijenroot, From Frege to Gödel: A Source Book in Logic, 1879–1931, (Harvard Univ. Press, Cambridge, Mass. and Oxford Univ. Press, London, 1967).

[139] P. Hertling, Topological complexity with continuous operations, (Special issue: Foundations of Computational Math. Conf., Rio de Janeiro, Brazil, Jan. 5–12, 1997), F. Cucker and M. Shub, (eds.), J. Complexity, **12** (1996) 315–338.

[140] P. Hertling, Unstetigkeitsgrade von Funktionen in der effektiven Analysis, Informatik Berichte, FernUniversität Hagen, Hagen, Diss., **208** (November 1996).

[141] P. Hertling, The effective Riemann mapping theorem, in: Computability and Complexity in Analysis (K. Ko, A. Nerode, and K. Weihrauch, eds.), Dagstuhl-Seminar-Report, **176** (1997) 7 (abstract).

[142] P. Hertling, Effectivity and effective continuity of functions between computable metric spaces, in: Combinatorics, Complexity, and Logic (Singapore), (Proc. 1st. Internatl. Conf. Discrete Math. Theoret. Comput. Sci., DMTCS '96, Auckland, New Zealand, Dec. 1996), D. S. Bridges, C. S. Calude, J. Gibbons, S. Reeves and I. H. Witten, (eds.), Discrete Math. Theoret. Comput. Sci., (1997) 264–275.

[143] P. Hertling, The real number structure is effectively categorical, CDMTCS Research Report Series 057, (Univ. Auckland, Auckland, New Zealand, 1997).

[144] P. Hertling and K. Weihrauch, Levels of degeneracy and exact lower complexity bounds for geometric algorithms, in: Proc. 6th. Canad. Conf. Computational Geometry, Univ. Saskatchewan, (1994) 237–242.

[145] A. Heyting, Intuitionism, An Introduction, (North-Holland, Amsterdam, 1956); [2nd. rev. edn., 1966].

[146] P. G. Hinman and Y. N. Moschovakis, Computability over the continuum, in: Logic Colloq. '69, (Proc. Summer School and Colloq., Manchester, 1969), R. O. Gandy and C. M. E. Yates, (eds.), North-Holland, Amsterdam, (1971) 77–105.

[147] C.-K. Ho, Relatively recursive reals and real functions, Technical Report 1994-02, Dept. Comput. Sci., Univ. Chicago, 1994.

[148] C.-K. Ho, Beyond recursive real functions, Inform. and Comput., **124** (1996) 113–126.

[149] H. J. Hoover, Feasible Constructive Analysis, Ph.D. Thesis, Dept. Comput. Sci., Univ. Toronto, Toronto, Canada, (1987).

[150] J. M. E. Hyland, Applications of constructivity, in: Logic, Methodology and Philos. Sci. VI, (Proc. 6th. Internatl. Congr. Logic, Methodology, and Philos. Sci., Hannover, Aug. 22–29, 1979), L. J. Cohen, J. Łoś, H. Pfeiffer and K.-P. Podewski, (eds.), Stud. Logic Found. Math., **104** (1982) 145–152.

[151] D. Ilse, Zur Stetigkeit berechenbarer reeller Funktionen, Z. Math. Logik Grundlag. Math., **11** (1965) 297–342.

[152] P. T. Johnstone, Topos Theory, London Math. Soc. Monographs, **10**, (Academic Press, London, 1977).

[153] I. Kalantari, Major subsets in effective topology, in: Patras Logic Symposion, (Proc. Logic Sympos. Patras, Greece, Aug. 18–22, 1980) G. Metakides (ed.), Stud. Logic Found. Math., **109** (1982) 77–94.

[154] I. Kalantari and A. Leggett, Simplicity in effective topology, J. Symbolic Logic, **47** (1982) 169–183.

[155] I. Kalantari and A. Leggett, Maximality in effective topology, J. Symbolic Logic, **48** (1983) 100–112.

[156] I. Kalantari and J. B. Remmel, Degrees of recursively enumerable topological spaces, J. Symbolic Logic, **48** (1983) 610–622.

[157] I. Kalantari and A. Retzlaff, Recursive constructions in topological spaces, J. Symbolic Logic, **44** (1979) 609–625.

[158] I. Kalantari and G. Weitkamp, Effective topological spaces, I. A definability theory, Ann. Pure Appl. Logic, **29** (1985) 1–27.

[159] I. Kalantari and G. Weitkamp, Effective topological spaces, II. A hierarchy, Ann. Pure Appl. Logic, **29** (1985) 207–224.

[160] I. Kalantari and G. Weitkamp, Effective topological spaces, III. Forcing and definability, Ann. Pure Appl. Logic, **36** (1987) 17–27.

[161] I. Kalantari and L. Welch, Point-free topological spaces, functions, and recursive points. Filter foundation for recursive analysis, I, Ann. Pure Appl. Logic, (to appear).

[162] I. Kalantari and L. Welch, Recursive and nonextendible functions over the reals. Filter foundation for recursive analysis, II, Ann. Pure Appl. Logic, (submitted July 1996).

[163] I. Kalantari and L. Welch, Shadow points in recursive analysis. Filter foundation for recursive analysis, III, (in preparation, 1997).

[164] I. Kalantari and L. Welch, Effective content of some theorems of topology, (in preparation, 1998).

[165] M. I. Kanovič and B. A. Kušner, Complexity of algorithms, and Specker sequences (Russian), in: Studies in the Theory of Algorithms and Mathematical Logic, Vol. 2, Vyčisl. Centr. Akad. Nauk SSSR, Moscow, (1976) 77–83, 160.

[166] D. Klaua, Berechenbare Analysis, Z. Math. Logik Grundlag. Math., **2** (1956) 265–303.

[167] D. Klaua, Die Präzisierung des Berechenbarkeitsbegriffes in der Analysis mit Hilfe rationaler Funktionale, Z. Math. Logik Grundlag. Math., **5** (1959) 33–96.

[168] D. Klaua, Konstruktive Analysis, Mathematische Forschungsberichte, XI, (VEB Deutscher Verlag der Wissenschaften, Berlin, 1961).

[169] D. Klaua, Rational and real ordinal numbers, in: Real Numbers, Generalizations of the Reals, and Theories of Continua, Synthese Lib., **242** (1994) 259–276.

[170] S. C. Kleene and R. Vesli, The foundations of intuitionistic mathematics from the viewpoint of the theory of recursive functions (Russian), (translated from English by F. A. Kabakov and B. A. Kušner), Monographs in Math. Logic and Foundations of Math., (Izdat. Nauk, Moscow, 1978).

[171] K. Ko, The maximum value problem and NP real numbers, J. Comput. System Sci., **24** (1982) 15–35.

[172] K. Ko, Some negative results on the computational complexity of total variation and differentiation, Inform. Contr., **53** (1982) 21–31.

[173] K. Ko, On the computational complexity of ordinary differential equations, Inform. Contr., **58** (1983) 157–194.

[174] K. Ko, On the definitions of some complexity classes of real numbers, Math. Systems Theory, **16** (1983) 95–109.

[175] K. Ko, Reducibilities on real numbers, Theoret. Comput. Sci., **31** (1984) 101–123.

[176] K. Ko, Continuous optimization problems and a polynomial hierarchy of real functions, (Special issue: Complexity of Approximating Solved Problems, Papers from Sympos. Columbia Univ., Morningside Heights, NY, Apr. 17–19, 1985), J. F. Traub, (ed.), J. Complexity, **1** (1985) 210–231.

[177] K. Ko, Approximation to measurable functions and its relation to probabilistic computation, Ann. Pure Appl. Logic, **30** (1986) 173–200.

[178] K. Ko, On the computational complexity of best Chebyshev approximations, J. Complexity, **2** (1986) 95–120.

[179] K. Ko, On the continued fraction representation of computable real numbers, Theoret. Comput. Sci., **47** (1986) 299–313, corr. ibid., **54** (1987) 341–343.

[180] K. Ko, Inverting a one-to-one real function is inherently sequential, in: Feasible Mathematics, (Proc. Workshop, June 16–18, 1989, Cornell Univ. Ithaca, NY), S. Buss and P. J. Scott (eds.), Progr. Comput. Sci. Appl. Logic, **9** (1990) 239–257.

[181] K. Ko, Complexity theory of real functions, Progr. Theoret. Comput. Sci., (1991).

[182] K. Ko, On the computational complexity of integral equations, Ann. Pure Appl. Logic, **58** (1992) 201–228.

[183] K. Ko, Computational complexity of fixed points and intersection points, J. Complexity, **11** (1995) 265–292.

[184] K. Ko, A polynomial-time computable curve whose interior has a nonrecursive measure, Theoret. Comput. Sci., **145** (1995) 241–270.

[185] K. Ko, Recent progress on complexity theory of real functions, in: Computability and Complexity in Analysis, (CCA Workshop, Hagen, Aug. 19–20, 1995), K. Ko and K. Weihrauch, (eds.), Informatik-Berichte, FernUniversität Hagen, **190** (1995) 83–90.

[186] K. Ko, Fractals and complexity, in: Computability and Complexity in Analysis, K. Ko, N. Müller and K. Weihrauch, (eds.), Universität Trier, (1996) 43–48.

[187] K. Ko, On the computability of fractal dimensions and julia sets, Ann. Pure Appl. Logic, (to appear).

[188] K. Ko and H. Friedman, Computational complexity of real functions, Theoret. Comput. Sci., **20** (1982) 323–352.

[189] K. Ko and H. Friedman, Computing power series in polynomial time, Adv. in Appl. Math., **9** (1988) 40–50.

[190] K. Ko and K. Weihrauch, (eds.), Computability and Complexity in Analysis, Informatik Berichte, FernUniversität Hagen, Hagen, **190** (1995).

[191] K. Ko and K. Weihrauch, On the measure of two-dimensional regions with polynomial-time computable boundaries, in: Computational Complexity, (Proc. 11th. Annual IEEE Conf. Computational Complexity), S. Homer and J.-Y. Cai, (eds.), (IEEE Comput. Soc. Press, Los Alamitos, CA, 1996), 150–159.

[192] A. N. Kolmogorov, On the principle of *tertium non datur* (Russian), Mat. Sb., **32** (1925), 646–667; [translated in: From Frege to Gödel: A Source Book in Logic, 1879–1931, J. van Heijenroot, (ed.), (Harvard Univ. Press, Cambridge, Mass. and Oxford Univ. Press, London, 1967) 414–437].

[193] G. Kreisel, Interpretation of analysis by means of constructive functionals of finite types, in: Constructivity in Mathematics, (Proc. Colloq., Amsterdam, Aug. 26–31, 1957), A. Heyting, (ed.), Stud. Logic Found. Math., (1959) 101–128.

[194] G. Kreisel, Foundations of Intuitionistic Logic, in: Logic, Methodology, and Philosophy of Science (Proc. Internatl. Congr., Stanford, CA, 1960), (Stanford Univ. Press, 1962), 198–210.

[195] G. Kreisel, Church's thesis: A kind of reducibility axiom for constructive mathematics, in: Intuitionism and Proof Theory, (Proc. Conf., Buffalo, N.Y., 1968). A. Kino, J. Myhill and R. E. Vesley (eds.), Stud. Logic Found. Math., (1970) 121–150.

[196] G. Kreisel, Perspectives in the philosophy of pure mathematics, in: Logic, Methodology and Philos. Sci. IV, (Proc. 4th. Internatl. Congr. Logic, Methodology and Philos. Sci., Bucharest, Aug. 29 – Sept. 4, 1971), P. Suppes, L. Henkin, A. Joja and Gr. C. Moisil, (eds.), Stud. Logic Found. Math., **74** (1973) 255–277.

[197] G. Kreisel, Reviews: Pour-El and Richards 'A computable ordinary differential equation which possesses no computable solution' (1979), and 'The wave equation, with computable initial data such that its unique solution is not computable'(1981), J. Symbolic Logic, **47** (1982) 900–902.

[198] G. Kreisel, Mathematical logic: tool and object lesson for science. The present state of the problem of the foundations of mathematics, Florence, 1981, Synthese Lib., **62** (1985) 139–151.

[199] G. Kreisel and D. Lacombe, Ensembles récursivement mesurables et ensembles récursivement ouverts et fermés, C. R. Acad. Sci. Paris, **245** (1957) 1106–1109.

[200] G. Kreisel, D. Lacombe and J. R. Shoenfield, Fonctionnelles récursivement définissables et fonctionnelles récursives, C. R. Acad. Sci. Paris, **245** (1957) 399–402.

[201] G. Kreisel, D. Lacombe and J. R. Shoenfield, Partial recursive functionals and effective operations, in: Constructivity in Mathematics, (Proc. Colloq., Amsterdam, Aug. 26–31, 1957), A. Heyting, (ed.), Stud. Logic Found. Math., (1959) 290–297.

[202] G. Kreisel and A. Macintyre, Constructive logic versus algebraization, I, in: The L. E. J. Brouwer Centenary Sympos., (Noordwijkerhout, June 8–13, 1981), A. S. Troelstra and D. van Dalen, (eds.), Stud. Logic Found. Math., **110** (1982) 217–260.

[203] G. Kreisel and H. Putnam, Eine Unableitbarkeitsbeweismethode für den intuitionistischen Aussagenkalküil, Arch. Math. Logik Grundlag., **3** (1957) 74–78.

[204] G. Kreisel and A. S. Troelstra, Formal systems for some branches of intuitionistic analysis, Ann. Math. Logic, **1** (1970) 229–387.

[205] C. Kreitz and K. Weihrauch, Complexity theory on real numbers and functions, in: Theoretical Computer Science, (Papers intended for 6th. GI Conf., Dortmund, Jan. 5–7, 1983), A. B. Cremers and H.-P. Kriegel, (eds.), Lecture Notes in Comput. Sci., **145** (1982) 165–174.

[206] C. Kreitz and K. Weihrauch, A unified approach to constructive and recursive analysis, in: Computation and Proof Theory, (Proc. Logic Colloq., Aachen, July 18–23, 1983, Part II), M. M. Richter, E. Börger, W. Oberschelp, B. Schinzel and W. Thomas, (eds.), Lecture Notes in Math., **1104** (1984) 259–278.

[207] C. Kreitz and K. Weihrauch, Theory of representations, Theoret. Comput. Sci., **38** (1985) 35–53.

[208] C. Kreitz and K. Weihrauch, Compactness in constructive analysis revisited, Ann. Pure Appl. Logic, **36** (1987) 29–38.

[209] A. Kučera and B. Kušner, A type of recursive isomorphism of certain concepts of constructive analysis (Russian), Comment. Math. Univ. Carolinæ, **19** (1978) 97–105.

[210] M. Kummer and M. Schäfer, Computability of convex sets, in: STACS '95, (Proc. 12th. Annual Sympos. Theoret. Aspects of Comput. Sci., Munich, Mar. 2–4, 1995), (E. W. Mayr and C. Puech, eds.), Lecture Notes in Comput. Sci., **900** (1995) 550–561.

[211] B. Kušner, Riemann integration in constructive analysis (Russian), Dokl. Akad. Nauk SSSR, **156** (1964) 255–257; [translated in: Soviet Math. – Dokl., **5** (1964) 628–630].

[212] B. Kušner, Constructive theory of the Riemann integral (Russian), Dokl. Akad. Nauk SSSR, **165** (1965) 1238–1240; [translated in: Soviet Math. – Dokl., **6** (1965) 1584–1587].

[213] B. Kušner, The existence of unbounded constructive analytic functions (Russian), Dokl. Akad. Nauk SSSR, **160** (1965) 29–31; [translated in: Soviet Math. – Dokl., **6** (1965) 26–28].

[214] B. Kušner, On constructive antiderivatives (Russian), Mat. Zametki, **2** (1967) 157–166; [translated as: Primitive constructive functions, Math. Notes, **2** (1967) 577–582].

[215] B. Kušner, A remark on the domains of definition of constructive functions (Russian), Zap. Naučn. Sem. Leningrad. Otdel. Mat. Inst. Steklov., **8** (1968) 103–106.

[216] B. Kušner, Some mass problems connected with the integration of constructive functions (Russian), Trudy Mat. Inst. Steklov., **113** (1970) 39–72; [translated in: Proc. Steklov Inst. Math., **113** (1970) 42–83].

[217] B. Kušner, Computationally complex real numbers (Russian), Z. Math. Logik Grundlag. Math., **19** (1973) 447–452.

[218] B. Kušner, Continuity theorems for some types of computable operators (Russian), Dokl. Akad. Nauk SSSR, **208** (1973) 1031–1034; [translated in: Soviet Math. – Dokl., **14** (1973) 221–225].

[219] B. Kušner, Lectures on constructive mathematical analysis (Russian), Monographs in Math. Logic and Foundations of Math., (Izdat. Nauka, Moscow, 1973); [translated: by E. Mendelson, and edited by L. J. Leifman, Transl. Math. Monographs, **60** (1984)].

[220] B. Kušner, On a type of computable real function (Russian), Dokl. Akad. Nauk, **215** (1974) 259–262; [translated in: Soviet Math. – Dokl., **15** (1974) 466–470].

[221] B. Kušner, A constructive version of König's theorem; functions that are computable in the sense of Markov, Grzegorczyk and Lacombe (Russian), in: Theory of Algorithms and Mathematical Logic, (dedicated to A. A. Markov on the occasion of his seventieth birthday), (Vyčisl. Centr. Akad. Nauk SSSR, Moscow, 1974), 87–111, 215.

[222] B. Kušner, The computation of isolated roots of constructive functions (Russian), Z. Math. Logik Grundlag. Math., **22** (1976) 311–332.

[223] B. Kušner, On Grzegorczyk's theorem on the computability of an isolated extremum (Russian), in: Studies in the Theory of Algorithms and Mathematical Logic, Vol. 2, A. A. Markov and B. A. Kušner, (eds.), (Vyčisl. Centr. Akad. Nauk SSSR, Moscow, 1976), 112–121, 158.

[224] B. Kušner, Segmental coverings and uniform continuity of constructive functions (Russian), in: Studies in the Theory of Algorithms and Mathematical Logic, A. A. Markov and V. I. Homič, (eds.), (Nauka, Moscow, 1979), 62–69, 133.

[225] B. Kušner, Behavior of the general term of a Specker series (Russian), in: Mathematical Logic and Mathematical Linguistics, M. Gladkiĭ, (ed.), (Kalinin. Gos. Univ., Kalinin, 1981), 112–116.

[226] B. Kušner, On some topological properties of constructive plane curves (Russian), Dokl. Akad. Nauk SSSR, **260** (1981) 281–283; [translated in: Soviet Math. – Dokl., **24** (1981) 258–261].

[227] B. Kušner, Some extensions of Markov's constructive continuum and their applications to the theory of constructive functions, in: The L. E. J. Brouwer Centenary Sympos., (Noordwijkerhout, June 8–13, 1981), A. S. Troelstra and D. van Dalen, (eds.), Stud. Logic Found. Math., **110** (1982) 261–273.

[228] B. Kušner, A class of Specker sequences (Russian), in: Mathematical Logic, Mathematical Linguistics and Theory of Algorithms, A. V. Gladliĭ, (ed.), (Kalinin. Gos. Univ., Kalinin, 1983), 62–65.

[229] B. Kušner, Differentiability and uniform continuity of constructive functions (Russian), Dokl. Akad. Nauk, **281** (1985) 1314–1316; [translated in: Soviet Math. – Dokl., **31** (1985) 433–435].

[230] B. Kušner, The principle of bar-induction in Brouwer's theory of the continuum (Russian), in: Patterns in the Development of Modern Mathematics, Methodological Aspects, M. I. Panov, (ed.), (Nauka, Moscow, 1987), 230–250.

[231] B. Kušner, Effective meaning of a theorem on uniform limit transition (Russian), in: Problems in Mathematical Logic and in the Theory of Algorithms, N. M. Nagornyĭ, (ed.), (Akad. Nauk SSSR, Vyčisl. Centr., Moscow, 1988), 3–10.

[232] B. Kušner, Markov's constructive mathematical analysis: the expectations and the results, in: Mathematical Logic, (Proc. Heyting '88 Summer School, Chaika, Sept. 13–23, 1988), P. P. Petkov, (ed.), (Plenum, New York, 1990), 53–58.

[233] B. Kušner, The constructive mathematics of A. A. Markov: some reflections (Russian), Modern Logic, **3** (1993) 119–144.

[234] B. Kušner, Markov and Bishop: an essay in memory of A. A. Markov (1903–1979) and E. Bishop (1928–1983), in: Golden Years of Moscow Mathematics, S. Zdrovkovska and P. L. Duren, (eds.), Hist. Math., **6** (1993) 179–197.

[235] B. Kušner, Kurt Gödel and the constructive mathematics of A. A. Markov, in: Gödel '96, (Proc. Conf, Brno, August 1996), P. Hájek, (ed.), Lecture Notes Logic, **6** (1996) 50–63.

[236] S. Labhalla and H. Lombardi, Real numbers, continued fractions, and complexity classes, Ann. Pure Appl. Logic, **50** (1990) 1–28.

[237] S. Labhalla and H. Lombardi, Analyse de complexité pour un théorème de Hall sur les fractions continues, Math. Logic Quart., **42** (1996) 134–144.

[238] A. H. Lachlan, Recursive real numbers, J. Symbolic Logic, **28** (1963) 1–16.

[239] D. Lacombe, Classes récursivement fermées et fonctions majorantes, C. R. Acad. Sci. Paris, **240** (1955) 716–718.

[240] D. Lacombe, Extension de la notion de fonction récursive aux fonctions d'une ou plusieurs variables réelles I, C. R. Acad. Sci. Paris, **240** (1955) 2478–2480.

[241] D. Lacombe, Extension de la notion de fonction récursive aux fonctions d'une ou plusieurs variables réelles II, C. R. Acad. Sci. Paris, **241** (1955), 13–14.

[242] D. Lacombe, Extension de la notion de fonction récursive aux fonctions d'une ou plusieurs variables réelles III, C. R. Acad. Sci. Paris, **241** (1955) 151–153.

[243] D. Lacombe, Remarques sur les opérateurs récursifs et sur les fonctions récursives d'une variable réelle, C. R. Acad. Sci. Paris, **241** (1955) 1250–1252.

[244] D. Lacombe, Les ensembles récursivement ouverts ou fermés, et leurs applications á l'analyse récursive, C. R. Acad. Sci. Paris, **245** (1957) 1040–1043; **246** (1958) 28–31.

[245] D. Lacombe, Quelques propriétés d'analyse récursive, C. R. Acad. Sci. Paris, **244** (1957) 838–840, 996–997.

[246] D. Lacombe, Sur les possibilités d'extension de la notion de fonction récursive aux fonctions d'une ou plusieurs variables réelles, in: Le raissonement en mathématiques et en science expérimentales, (Colloq. Internat. CNRS), Éditions du CNRS, Paris, **70** (1958) 67–75.

[247] D. Lacombe, Quelques procédés de définition en topologie récursive, in: Constructivity in Mathematics, (Proc. Colloq., Amsterdam, Aug. 26–31, 1957), A. Heyting, (ed.), Stud. Logic Found. Math., (1959) 129–158.

[248] I. Lakatos, (ed.), Problems in the Philosophy of Mathematics, (Proc. Internatl. Colloq. Philos. Sci., Bedford College, London, July 11–17, 1965), Vol. 1, Stud. Logic Found. Math., (1967).

[249] A. A. Lewis, Some aspects of effectively constructive mathematics that are relevant to the foundations of neoclassical mathematical economics and the theory of games, Math. Social Sciences, **24** (1992) 209–235.

[250] M. Mandelkern, Connectivity of an interval, Proc. Amer. Math. Soc., **54** (1976) 170–172.

[251] M. Mandelkern, Continuity of monotone functions, Pacific J. Math., **99** (1982) 413–418.

[252] I. D. Zaslavskiĭand S. N. Manukjan, Partitionings of the plane by constructive curves (Russian), Trudy Vyčisl. Centr. Akad. Nauk Armjan. SSR i Erevan. Gos. Univ., **5** (1968) 26–138.

[253] A. A. Markov, On the continuity of constructive functions (Russian), Uspekhi Mat. Nauk (N.S.), **9** (1954) 226–230.

[254] P. Martin-Löf, Notes on constructive mathematics, (Almqvist and Wiksell, Stockholm, 1970).

[255] Yu. V. Matijasevič, G. E. Mints, V. P. Orevkov and A. O. Slisenko, "Nikolaĭ Aleksandrovich Shanin (on his seventieth birthday)" (Russian), Uspekhi Mat. Nauk, **45** (1990) 205–206; [translated in: Russian Math. Surveys, **45**:1 (1990) 239–240].

[256] S. Mazur, Computable Analysis, (based on lecture notes academic year 1949/50, Inst. Math., Polish Acad. Sci., Warsaw), A. Grzegorczyk and H. Rasiowa, (eds.), Rozprawy Matematyczne, Vol. 33, (Państwowe Wydawn. Naukowe, Warszawa, 1963).

[257] H. Meschkowski, Rekursive reelle Zahlen, Math. Z., **66** (1956) 189–202.

[258] H. Meschkowski, Zur rekursiven Funktionentheorie, Acta Math., **95** (1956) 9–23.

[259] G. Metakides and A. Nerode, The introduction of nonrecursive methods into mathematics, in: The L. E. J. Brouwer Centenary Sympos., (Noordwijkerhout, June 8–13, 1981), A. S. Troelstra and D. van Dalen, (eds.), Stud. Logic Found. Math., **110** (1982) 319–335.

[260] G. Metakides, A. Nerode and R. A. Shore, Recursive limits on the Hahn-Banach Theorem, in: Errett Bishop: reflections on him and his research, (San Diego, Calif., 1983), M. Rosenblatt, (ed.), Contemp. Math., **39** (1985) 85–91.

[261] W. Miller, Recursive function theory and numerical analysis, J. Comput. System Sci., **4** (1970) 465–472.

[262] A. F. Monna, The concept of function in the 19th and 20th centuries, in particular with regard to the discussions between Baire, Borel and Lebesgue, Arch. Hist. Exact Sci., **9** (1972) 57–84.

[263] T. Mori, Y. Tsujii and M. Yasugi, Computability structures on metric spaces, in: Combinatorics, Complexity, and Logic (Singapore), (Proc. 1st. Internatl. Conf. Discrete Math. Theoret. Comput. Sci., DMTCS '96, Auckland, New Zealand, Dec. 1996), D. S. Bridges, C. S. Calude, J. Gibbons, S. Reeves and I. H. Witten, (eds.), Discrete Math. Theoret. Comput. Sci., (1997) 351–362.

[264] J. R. Moschovakis, Can there be no nonrecursive functions?, J. Symbolic Logic, **36** (1971) 309–315.

[265] J. R. Moschovakis, A classical view of the intuitionistic continuum, Ann. Pure Appl. Logic, **81** (1996) 9–24.

[266] Y. N. Moschovakis, Recursive metric spaces, Fund. Math., **55** (1964) 215–238.

[267] Y. N. Moschovakis, Notation systems and recursive ordered fields, Compositio Math., **17** (1965) 40–71.

[268] A. Mostowski, On computable sequences, Fund. Math., 44 (1957) 37–51.

[269] A. Mostowski, On various degrees of constructivism, in: Constructivity in Mathematics, (Proc. Colloq., Amsterdam, Aug. 26–31, 1957), A. Heyting, (ed.), Stud. Logic Found. Math., (1959) 178–194.

[270] N. Th. Müller, Subpolynomial complexity classes of real functions and real numbers, in: Proc. 13th. Internatl. Colloq. Automata, Languages, and Programming, (Rennes, 1986) L. Kott, (ed.), Lecture Notes in Comput. Sci., **226** (1986) 284–293.

[271] N. Th. Müller, Uniform computational complexity of Taylor series, in: Proc. 14th. Internatl. Colloq. Automata, Languages, and Programming, (Karlsruhe, 1987) T. Ottmann, (ed.), Lecture Notes in Comput. Sci., **267** (1987) 435–444.

[272] N. Th. Müller, Polynomial-time computation of Taylor series, in: Proc. 22nd. JAIIO – Panel '93, Part 2, Buenos Aires, (1993) 259–281.

[273] N. Th. Müller, Constructive aspects of analytic functions, in: Computability and Complexity in Analysis, (CCA Workshop, Hagen, Aug. 19–20, 1995), K. Ko and K. Weihrauch, (eds.), Informatik-Berichte, FernUniversität Hagen, **190** (1995) 105–114.

[274] N. Th. Müller and B. Moiske, Solving initial value problems in polynomial time, in: Proc. 22nd. JAIIO – Panel '93, Part 2, Buenos Aires, (1993) 283–293.

[275] J. Myhill, A recursive function, defined on a compact interval and having a continuous derivative that is not recursive, Michigan Math. J., **18** (1971) 97–98.

[276] J. Myhill, What is a real number?, Amer. Math. Monthly, **79** (1972) 748–754.

[277] J. Myhill and J. Shepherdson, Effective operations on partial recursive functions, Z. Math. Logik Grundlag. Math., **1** (1955) 310–317.

[278] A. Nerode and W. Q. Huang, An application of pure recursion theory to recursive analysis (Chinese), Acta Math. Sinica, **28** (1985) 625–636.

[279] V. P. Orevkov, A constructive mapping of a square onto itself displacing every constructive point (Russian), Dokl. Akad. Nauk, **152** (1963) 55–58; [translated in: Soviet Math. – Dokl., **4** (1963) 1253-1256].

[280] V. P. Orevkov, Constructive mappings of polyhedra (Russian), Dokl. Akad. Nauk, **152** (1963) 278–281.

[281] V. P. Orevkov, Certain questions of the theory of polynomials with constructive real coefficients (Russian), Trudy Mat. Inst. Steklov., **72** (1964) 462–487.

[282] V. P. Orevkov, On constructive mappings of a circle into itself (Russian), Trudy Mat. Inst. Steklov., **72** (1964) 437–461.

[283] V. P. Orevkov, Certain types of continuity of constructive operators (Russian), Trudy Mat. Inst. Steklov., **93** (1967) 164–186.

[284] V. P. Orevkov, Constructive mappings of finite polyhedra (Russian), Trudy Mat. Inst. Steklov., **93** (1967), 142–163.

[285] V. P. Orevkov, Certain properties of homeomorphisms of constructive metric spaces (Russian), Zap. Naučn. Sem. Leningrad. Otdel. Mat. Inst. Steklov., **16** (1969) 157–164.

[286] V. P. Orevkov, The continuity of constructive functionals (Russian), Zap. Naučn. Sem. Leningrad. Otdel. Mat. Inst. Steklov., **20** (1971) 160–169, 287.

[287] V. P. Orevkov, The complexity of the expansion of algebraic irrationalities in continued fractions (Russian), Trudy Mat. Inst. Steklov., **129** (1973), 24–29, 267.

[288] V. P. Orevkov, A new proof of the uniqueness theorem for constructive differentiable functions of a complex variable (Russian), Zap. Naučn. Sem. Leningrad. Otdel. Mat. Inst. Steklov., **40** (1974) 119–126, 159.

[289] V. P. Orevkov, Problems in the constructive trend in mathematics, V. P. Orevkov and N. A. Šanin., (eds.), Proc. Steklov Inst. Math., **129** (1973), American Mathematical Society, Providence, R. I. (1975).

[290] H. Poincare, Les mathematiques et la logique, Revue de metaphysique et de morale, **14** (1906), 294–317.

[291] M. B. Pour-El, A comparison of five "computable" operators, Z. Math. Logik Grundlag. Math., **6** (1960) 325–340.

[292] M. B. Pour-El, Church's thesis and recursive analysis, in: Computability and Complexity in Analysis K. Ko and K. Weihrauch, (eds.), Informatik-Berichte, FernUniversität Hagen, **190** (1995) 115–118.

[293] M. B. Pour-El and J. I. Richards, Differentiability properties of computable functions — a summary, Acta Cybernet., **4** (1978/79) 123–125.

[294] M. B. Pour-El and J. I. Richards, A computable ordinary differential equation which possesses no computable solution, Ann. Math. Logic, **17** (1979) 61–90.

[295] M. B. Pour-El and J. I. Richards, The wave equation with computable initial data such that its unique solution is not computable, Adv. Math., **39** (1981) 215–239.

[296] M. B. Pour-El and J. I. Richards, Computability and noncomputability in classical analysis, Trans. Amer. Math. Soc., **275** (1983) 539–560.

[297] M. B. Pour-El and J. I. Richards, Noncomputability in analysis and physics: a complete determination of the class of noncomputable linear operators, Adv. Math., **48** (1983), 44–74.

[298] M. B. Pour-El and J. I. Richards, L^p-computability in recursive analysis, Proc. Amer. Math. Soc., **92** (1984) 93–97.

[299] M. B. Pour-El and J. I. Richards, Three theorems on the computability of linear operators, their eigenvalues and eigenvectors, Sūrikaisekikenkyūsho Kōkyūroku **588** (1986) 149–161, Logic and the foundations of mathematics, (Japanese).

[300] M. B. Pour-El and J. I. Richards, The eigenvalues of an effectively determined selfadjoint operator are computable, but the sequence of eigenvalues is not, Adv. Math., **63** (1987) 1–41.

[301] M. B. Pour-El and J. I. Richards, Computability in Analysis and Physics, Perspect. in Math. Logic, (1989).

[302] M. B. Pour-El and N. Zhong, The wave equation with computable initial data whose unique solution is nowhere computable, in: Computability and Complexity in Analysis K. Ko, N. Müller, and K. Weihrauch, (eds.), Universität Trier, (1996) 77–78; also in: Math. Logic Quart., **43** (1997) 499–509.

[303] H. G. Rice, Recursive real numbers, Proc. Amer. Math. Soc., **5** (1954) 784–791.

[304] D. B. Richardson, Some undecidable problems involving elementary functions of a real variable, J. Symbolic Logic, **33** (1968) 514–520.

[305] M. K. Richter and K.-C. Wong, Computable economic analysis, in: Computability and Complexity in Analysis, K. Ko, N. Müller and K. Weihrauch, (eds.), Universität Trier, (1996) 81–90.

[306] R. M. Robinson, Review of "Rekursive Funktionen" by R. Péter, J. Symbolic Logic, **16** (1951) 280–282.

[307] M. Rosenblatt, (ed.), Errett Bishop: Reflections on him and his research, Contemp. Math., 39 (1985).

[308] N. A. Šanin, Some problems of mathematical analysis in the light of constructive logic, Z. Math. Logik Grundlag. Math., **2** (1956) 27–36.

[309] N. A. Šanin, Constructive real numbers and constructive functional spaces (Russian), Trudy Mat. Inst. Steklov., **67** (1962) 15–294; [translated: by E. Mendelson, Transl. Math. Monographs, **21** (1968)].

[310] N. A. Šanin, A hierarchy of ways of understanding judgments in constructive mathematics (Russian), Trudy Mat. Inst. Steklov., **129** (1973), 203–266. Problems in the Constructive Trend in Mathematics, **6**.

[311] N. A. Šanin, The hierarchy of constructive Brouwer functionals (Russian), Zap. Naučn. Sem. Leningrad. Otdel. Mat. Inst. Steklov., **40** (1974) 142–147, Investigations in Constructive Mathematics and Mathematical Logic, VI, (dedicated to A. A. Markov on the occasion of his seventieth birthday).

[312] N. A. Šanin, Canonical recursive functions and operations (Russian), Zap. Naučn. Sem. Leningrad. Otdel. Mat. Inst. Steklov., **88** (1979) 218–235, 246–247. Studies in Constructive Mathematics and Mathematical Logic, VIII.

[313] N. A. Šanin, Georg Cantor as the author of constructions that play fundamental roles in constructive mathematics (Russian), Zap. Naučn. Sem. S.-Peterburg. Otdel. Mat. Inst. Steklov. POMI, **220** (1995) 5–22, 145, Issled. po Konstrukt. Mat. i Mat. Logike. IX.

[314] A. Ščedrov, Differential equations in constructive analysis and in the recursive realizability topos, J. Pure Appl. Algebra, **33** (1984), 69–80.

[315] A. Ščedrov, Diagonalization of continuous matrices as a representation of intuitionistic reals, Ann. Pure Appl. Logic, **30** (1986) 201–206.

[316] M. Schröder, Topological spaces allowing type 2 complexity theory, in: Computability and Complexity in Analysis, (CCA Workshop, Hagen, Aug. 19–20, 1995), K. Ko and K. Weihrauch, (eds.), Informatik-Berichte, FernUniversität Hagen, **190** (1995) 41–53.

[317] M. Schröder, Fast online multiplication of real numbers, in: STACS '97 (Lübeck) R. Reischuk and M. Morvan, (eds.), Lecture Notes in Comput. Sci., **1200** (1997) 81–92.

[318] J. C. Shepherdson, On the definition of computable functions of a real variable, Z. Math. Logik Grundlag. Math., **22** (1976) 391–402.

[319] S. G. Simpson, Which set existence axioms are needed to prove the Cauchy/Peano theorem for ordinary differential equations?, J. Symbolic Logic, **49** (1984) 783–802.

[320] R. I. Soare, Cohesive sets and recursively enumerable Dedekind cuts, Pacific J. Math., **31** (1969) 215–231.

[321] R. I. Soare, Recursion theory and Dedekind cuts, Trans. Amer. Math. Soc., **140** (1969) 271–294.

[322] R. I. Soare, Computability and recursion, Bulletin of Symbolic Logic, **2** (1996) 284–321.

[323] E. Specker, Nicht konstruktiv beweisbare Sätze der Analysis, J. Symbolic Logic, **14** (1949) 145–158.

[324] E. Specker, Der Satz vom Maximum in der rekursiven Analysis, in: Constructivity in Mathematics, (Proc. Colloq., Amsterdam, Aug. 26–31, 1957), A. Heyting, (ed.), Stud. Logic Found. Math., (1959) 254–265.

[325] E. Specker, The fundamental theorem of algebra in recursive analysis, in: Constructive Aspects of the Fundamental Theorem of Algebra, (Proc. Sympos. Zürich–Rüschlikon, June 5–7, 1967), (Wiley-Interscience, NY, 1969), 321–329.

[326] D. Spreen, Computable one-to-one enumerations of effective domains, Lecture Notes in Comput. Sci., **298** (1988) 372–384.

[327] D. Spreen, A characterization of effective topological spaces, Lecture Notes in Math., **1432** (1990) 363–387.

[328] D. Spreen, Computable one-to-one enumerations of effective domains, Inform. and Comput., **84** (1990) 26–46.

[329] D. Spreen, A characterization of effective topological spaces II, in: Topology and Category Theory in Computer Science (Oxford, 1989) (Oxford Sci. Publ., 1991) 231–255.

[330] D. Spreen, Effective operators and continuity revisited, Lecture Notes in Comput. Sci., **620** (1992) 459–469.

[331] D. Spreen, Effective inseparability in a topological setting, Ann. Pure Appl. Logic, **80** (1996) 257–275.

[332] D. Spreen and P. Young, Effective operators in a topological setting, Lecture Notes in Math., **1104** (1984) 437–451.

[333] J. Todd, Introduction to the Constructive Theory of Functions, (Academic Press, NY, 1963).

[334] A. S. Troelstra, Intuitionistic continuity, Nieuw Arch. Wisk., **15** (1967) 2–6.

[335] A. S. Troelstra, (ed.), Metamathematical Investigation of Intuitionistic Arithmetic and Analysis, Lecture Notes in Math., **344** (1973).

[336] A. S. Troelstra, Models and computability, in: Metamathematical Investigation of Intuitionistic Arithmetic and Analysis, A. S. Troelstra, (ed.), Lecture Notes in Math., **344** (1973) 97–174.

[337] A. S. Troelstra, A note on non-extensional operations in connection with continuity and recursivness, Nederl. Akad. Wentensch. Proc., Ser. A, **39** (1977) 455–462.

[338] A. S. Troelstra, Intuitionistic extensions of the reals, Nieuw Arch. Wisk., **28** (1980) 63–113.

[339] A. S. Troelstra, Intuitionistic extensions of the reals, II, Logic Colloq. '80 (Prague, 1980). Stud. Logic Found. Math., **108** (1982).

[340] A. S. Troelstra, Comparing the theory of representations and constructive mathematics, in: Computer Science Logic, E. Börger, G. Jäger, H. K. Büning, and M. M. Richter, (eds.), Lecture Notes in Comput. Sci., **626** (1992) 382–395.

[341] A. S. Troelstra, An intuitionistic look at the real numbers (Dutch), in: Summer course 1993: the real numbers, CWI Syllabi, **35** (Math. Centrum, 1993) 67–81.

[342] A. S. Troelstra and D. van Dalen, Constructivism in Mathematics, Vol. 1, Stud. Logic Found. Math., **121** (1988).

[343] A. S. Troelstra and D. van Dalen, Constructivism in Mathematics, Vol. 2, Stud. Logic Found. Math., **123** (1988).

[344] A. M. Turing, On computable numbers, with an application to the Entscheidungsproblem, Proc. London Math. Soc., Ser. 2, **42** (1936) 230–265; corr. ibid., **43** (1937) 544–546.

[345] A. M. Turing, Collected Works: Mechanical Intelligence, edited and with an introduction by D. C. Ince, with a preface by P. N. Furbank, (North-Holland Publishing Co., Amsterdam, 1992).

[346] W. Veldman, Investigations in Intuitionistic Hierarchy Theory, Ph.D. Thesis, Katholieke Universiteit, Nijmegen, The Netherlands, (1981).

[347] H. Wang, Specker's mathematical work from 1949 to 1979, Logic and Algorithmics (Zurich, 1980), Monograph. Enseign. Math., **30** (Geneva, 1982) 11–24.

[348] M. Washihara, Computability and Fréchet spaces, Math. Japon., **42** (1995) 1–13.

[349] M. Washihara and M. Yasugi, Computability and metrics in a Fréchet space, Math. Japon., **43** (1996) 431–443.

[350] K. Weihrauch, Computability on metric spaces, Informatik Berichte 21, FernUniversität Hagen, Hagen, (September 1981).

[351] K. Weihrauch, Type 2 recursion theory, Theoret. Comput. Sci., **38** (1985) 17–33.

[352] K. Weihrauch, Computability, EATCS Monographs on Theoretical Computer Science, **9**, Springer-Verlag, Berlin-New York, 1987.

[353] K. Weihrauch, On the complexity of online computations of real functions, J. Complexity, **7** (1991) 380–394.

[354] K. Weihrauch, The degrees of discontinuity of some translators between representations of the real numbers, Technical Report TR-92-050, International Computer Science Institute, Berkeley, (July 1992).

[355] K. Weihrauch, A simple and powerful approach for studying constructivity, computability, and complexity, in: Constructivity in Computer Science, (San Antonio, TX, June 19–22, 1991) J. P. Myers and M. J. O'Donnell, (eds.), Lecture Notes in Comput. Sci., **613** (1992) 228–246.

[356] K. Weihrauch, The TTE–interpretation of three hierarchies of omniscience principles, Informatik Berichte **130**, FernUniversität Hagen, Hagen, (September 1992).

[357] K. Weihrauch, Computability on computable metric spaces, Theoret. Comput. Sci., **113** (1993) 191–210. Fundamental Study.

[358] K. Weihrauch, A simple introduction to computable analysis, Informatik Berichte 171, FernUniversität Hagen, Hagen, July 1995, 2nd. edn..

[359] K. Weihrauch, Computability on the probability measures on the Borel sets of the unit interval, in: Automata, Languages and Programming, (Proc. 24th. Internatl. Colloq., ICALP '97, Bologna, July 7–11, 1997), P. Degano, R. Gorrieri and A. Marchetti-Spaccamela, (eds.), Lecture Notes in Comput. Sci., **1256** (1997) 166–176.

[360] K. Weihrauch, A foundation for computable analysis, in: Combinatorics, Complexity, and Logic (Singapore), (Proc. 1st. Internatl. Conf. Discrete Math. Theoret. Comput. Sci., DMTCS '96, Auckland, New Zealand, Dec. 1996), D. S. Bridges, C. S. Calude, J. Gibbons, S. Reeves and I. H. Witten, (eds.), Discrete Math. Theoret. Comput. Sci., (1997) 66–89.

[361] K. Weihrauch and C. Kreitz, Representations of the real numbers and of the open subsets of the set of real numbers, Ann. Pure Appl. Logic, **35** (1987) 247–260.

[362] K. Weihrauch and C. Kreitz, Type 2 computational complexity of functions on Cantor's space, Theoret. Comput. Sci., **82** (1991) 1–18. Fundamental Study.

[363] K. Weihrauch and G. Schäfer, Admissible representations of effective cpo's, Theoret. Comput. Sci., **26** (1983) 131–147.

[364] K. Weihrauch and U. Schreiber, Embedding metric spaces into cpo's, Theoret. Comput. Sci., **16** (1981) 5–24.

[365] K. Weihrauch and X. Zheng, Computability on continuous, lower semicontinuous and upper semi-continuous real functions, in: Computing and Combinatorics, T. Jiang and D. T. Lee, (eds.), Lecture Notes in Comput. Sci., **1276** (1997) 166–175.

[366] K. Weihrauch and X. Zheng, Effectiveness of the global modulus of continuity on metric spaces, in: Category Theory and Computer Science, E. Moggi and G. Rosolini, (eds.), Lecture Notes in Comput. Sci., **1290** (1997) 210–219.

[367] H. Weyl, Das Kontinuum, Veit, Leipzig, 1918.

[368] K.-C. Wong, Computability of minimizers and separating hyperplanes, Math. Logic Quart., **42** (1996) 564–568.

[369] T. Yamaguchi, On index sets of recursive reals, (draft, 1995).

[370] M. Yasugi, Computability in analysis, Workshop on Stochastic Numerics (Japanese), Sūrikaisekikenkyūsho Kōkyūroku, **850** (1993) 207–214.

[371] P. R. Young, An effective operator, continuous but not partial recursive, Proc. Amer. Math. Soc., **19** (1968) 103–108.

[372] I. D. Zaslavskiĭ, Some properties of constructive real numbers and constructive functions (Russian), Trudy Mat. Inst. Steklov., **67** (1962) 385–457.

[373] I. D. Zaslavskiĭ, On the differentiation and integration of constructive functions (Russian), Dokl. Akad. Nauk SSSR, **156** (1964) 25–27.

[374] I. D. Zaslavskiĭ, Rectifiability of constructive plane curves (Russian), Izv. Akad. Nauk Armjan. SSR Ser. Mat., **2** (1967) 69–82.

[375] N. Zhong, Effective separation axioms, Questions and Answers in General Topology, **14** (1996) 177–185.

[376] N. Zhong, Recursively enumerable subsets of $\mathbb{R}^q$ in two computing models: Blum-Shub-Smale machine and Turing machine, (preprint, 1996).

[377] N. Zhong and B.-Y. Zhang, Computable analysis of the Korteweg-de Vries equation, (preprint, 1996).

[378] Q. Zhou, Computable real-valued functions on recursive open and closed subsets of Euclidean space, Math. Logic Quart., **42** (1996) 379–409.

[379] Q. Zhou, Subclasses of computable real valued functions, in: Computing and Combinatorics, T. Jiang and D. T. Lee, (eds.), Lecture Notes in Comput. Sci., **1276** (1997) 156–165.

Vasco Brattka
Theoretische Informatik I
FernUniversität Hagen
D-58084 Hagen
Germany
Vasco.Brattka@FernUni-Hagen.de

Iraj Kalantari
Department of Mathematics
Western Illinois University
One University Circle
Macomb, IL 61455, USA
i-kalantari@wiu.edu